T0314259

Large Igneous Provinces

Geophysical Monograph Series

including

IUGG Volumes
Maurice Ewing Volumes
Mineral Physics Volumes

Geophysical Monograph Series

Geophysical Monograph 100

Large Igneous Provinces

Continental, Oceanic, and Planetary Flood Volcanism

John J. Mahoney
Millard F. Coffin

Editors

American Geophysical Union

Cover. Map showing global distribution of large igneous provinces (areas in bright red, from M. Coffin's global compilation). Image produced by Andrew Goodliffe and Fernando Martinez (SOEST, University of Hawai'i) using GMT (P. Wessel and W. H. F. Smith, Free software helps map and display data, *Eos Trans. AGU, 72,* 441, 1991). The data set contains combined global seafloor topography (W. H. F. Smith and D. T. Sandwell, Global seafloor topography from satellite altimetry and ship depth soundings, *Science, 277,* 1956-1962, 1997) and land topography derived from the USGS 30 arc-second digital elevation model.

Library of Congress Cataloging-in-Publication Data
Large igneous provinces : continental, oceanic, and planetary flood
    volcanism / John J. Mahoney, Millard F. Coffin, editors.
          p.   cm. -- (Geophysical Monograph ; 100)
      Includes bibliographical references.
      ISBN 0-87590-082-8
        1. Flood basalts. 2. Volcanism. I. Mahoney, John J., 1952- .
  II. Coffin, Millard F., 1955- . III. Series.
  QE462.B3L37 1997
  522' .26--dc21 97-43026

                                        CIP

ISBN 0-87590-082-8
ISSN 0065-8448

Copyright 1997 by the American Geophysical Union
2000 Florida Avenue, N.W.
Washington, DC 20009

CONTENTS

PREFACE

Continental flood basalts, volcanic passive margins, and oceanic plateaus represent the largest known volcanic episodes on our planet, yet they are not easily explained by plate tectonics. Indeed, some are likely to record periods when the outward transfer of material and energy from the Earth's interior operated in a significantly different mode than at present. In recent years, interest in large-scale mafic magmatism has surged as high-precision geochronological, detailed geochemical, and increasingly sophisticated geophysical data have become available for many provinces. However, the sheer amount of recent material, often in the form of detailed collaborative research projects, can overwhelm newcomers to the field and experts alike as the literature continues to grow dramatically. The need for an up-to-date review volume on a sizable subset of the major continental and oceanic flood basalt provinces, termed large igneous provinces, was recognized by the Commission on Large-Volume Basaltic Provinces (International Association of Volcanology and Chemistry of the Earth's Interior), and the co-editors were charged with organizing and implementing such a volume. We hope that this volume will be valuable to researchers and graduate students worldwide, particularly to petrologists, geochemists, geochronologists, geodynamicists, and plate-tectonics specialists; it may also interest planetologists, oceanographers, and atmospheric scientists.

Nearly a decade has passed since the publication of Continental Flood Basalts (edited by J.D. Macdougall; Kluwer, 1988), the volume that comes closest to being a predecessor of this one. Fundamental changes in understanding of large igneous provinces have occurred in the last ten years, including (1) the general acceptance of oceanic plateaus and volcanic passive margins as the submarine counterparts of continental flood basalts, (2) the realization that the major large igneous provinces are among the very largest igneous events on Earth in the last several hundred million years, (3) the wide application of the starting-plume hypothesis for the origin of many large igneous provinces, (4) a growing understanding of the role such provinces play in continental growth, and (5) the recognition that similar events have been important on the other terrestrial planets. More recent data on continental flood basalt provinces and the growing data base for volcanic passive margins and oceanic plateaus emphasize the need for a new book providing a broad perspective on the state of knowledge in this field.

Syntheses of recent and earlier work, combined with new results and interpretations, are presented here for many of the major provinces. Most of these chapters include up-to-date reviews of geologic setting, age and age distribution, petrology and geochemistry, petrogenesis, mantle sources, probable causes, and post-emplacement evolution. Several chapters include discussions of possibly related climatic, oceanographic, or biospheric effects, as well as effects on crustal evolution. Readers will quickly see that "one fits all" does not apply to large igneous provinces, at least at our current level of understanding. For example, although most authors now agree on the importance of plume involvement in the larger provinces, the role inferred for the plume can be quite different, even within different parts of a single province (e.g., from being the principal source of magma to a heat source driving lithospheric melting to the "carrier" of a halo of entrained mantle broadly like that feeding mid-ocean ridges).

Other papers focus on the extensive dike swarms that may be the only remaining record of older flood basalt events going back well into the Precambrian, on the geochemical differences (and their possible causes) between continental and oceanic flood basalts, on sulfide mineralization in flood basalt systems, on large igneous provinces elsewhere in the solar system, and on the physical mechanism of flood-basalt lava flow emplacement, which recent work suggests may be quite different than thought previously.

Numerous current and planned research projects promise that large igneous provinces will remain an active and fruitful area of scientific endeavor well into the next century. As the work presented in this volume demonstrates, studies of continental and oceanic large igneous provinces are complementary: on the continents, abundant samples are available for petrological, geochemical, geochronological, volcanological, and other studies, whereas in the oceans sampling is difficult but the lithospheric setting is relatively easy to characterize and geophysical imaging is quite effective. Nevertheless, study of flood basalt magmatism remains very much at a reconnaissance stage in comparison to such phenomena as mid-ocean ridge and arc magmatism. Given the potential that large igneous provinces have for illuminating mantle dynamics, as well as for effecting global environmental change, we hope that the papers presented here will stimulate new research endeavors.

We are grateful to the many reviewers (see acknowledgments in individual papers for names) who provided prompt, in-depth critical evaluations of the papers. The support of the International Association of Volcanology and Chemistry of the Earth's Interior (IAVCEI), in particular,

Wally Johnson, is appreciated in producing this volume. We also wish to thank all the lead authors for harnessing the intellect and energy of their co-authors, and for (mostly) meeting the deadlines imposed by stern editors. We are particularly indebted to Diane Henderson and the SOEST Publications staff for their invaluable assistance, including several rounds of copy editing, formatting, and advice with illustrations. Without their cheerful expertise, both of us would no doubt have long since wiggled out of editing this book.

John J. Mahoney
University of Hawaii

Millard F. Coffin
University of Texas at Austin

The Columbia River Flood Basalt Province: Current Status

Peter R. Hooper

Department of Geology, Washington State University, Pullman, Washington

The Columbia River flood basalt province is smaller by an order of magnitude than the Deccan, Karoo, Paraná, and Siberian continental flood basalt provinces. Its smaller size, relative youth (17–6 Ma), excellent exposure, and easy accessibility have allowed development of a flow-by-flow stratigraphy in which many flows can be traced across the Columbia Plateau, often linked directly to their strongly oriented feeder dikes in the southeast quadrant. The detailed stratigraphy provides a precise record of the changes in magma composition and volume with time and demonstrates more clearly here than in other provinces that single fissure eruptions had volumes in excess of 2,000 km^3 and flowed across the plateau for distances up to 600 km with negligible changes in chemical or mineralogical composition.

Current evidence suggests that the Columbia River flood basalts resulted from impingement of a small mantle plume, the Yellowstone hotspot, on the base of the lithosphere near the Nevada-Oregon-Idaho border at 16.5 Ma and that the main focus of eruption then moved rapidly north to the Washington-Oregon-Idaho border from where the main eruptions occurred. The rapid northerly translation of the main eruptive activity may have been controlled by weakened or thinned zones in the lithosphere. The few earliest flows have typical mantle plume compositions and the last, small-volume flows are contaminated by continental crust. In between, the great majority of flows carry a strong lithospheric signature, the source of which remains controversial—either an enriched continental lithospheric mantle or assimilated continental crust. The physical nature and rate of magma eruption are also controversial. Recent work suggests flows grew by internal injection rather than by turbulent surface flow and this has been used to imply significantly lower eruption rates than previously envisaged. However, the chemical and mineralogical homogeneity of single Columbia River basalt flows across many hundreds of kilometers implies that eruption and flow rates were still exceptionally high.

INTRODUCTION

The Columbia River Basalt Group (CRBG) forms a large intermontane plateau of 164,000 km^2 lying between the Cascade Range and the Rocky Mountains in southeast Washington, west-central Idaho, and northeast Oregon

Large Igneous Provinces: Continental, Oceanic, and Planetary Flood Volcanism
Geophysical Monograph 100

(Figure 1). Basalt flows dip gently west (<2°) in the northern part and are deformed in the south by mild east-west folds (including the Yakima fold belt on the west side of the Pasco Basin) which are associated with WNW-ESE and NNE-SSW strike-slip faults.

The plateau contains more than 300 individual basalt flows with an average volume of 500–600 km³ per flow and a total volume of about 175,000 km³ [*Tolan et al.*, 1989]. The earliest eruptions of CRBG-related tholeiites may be those of the Steens Basalt close to the Nevada/Oregon

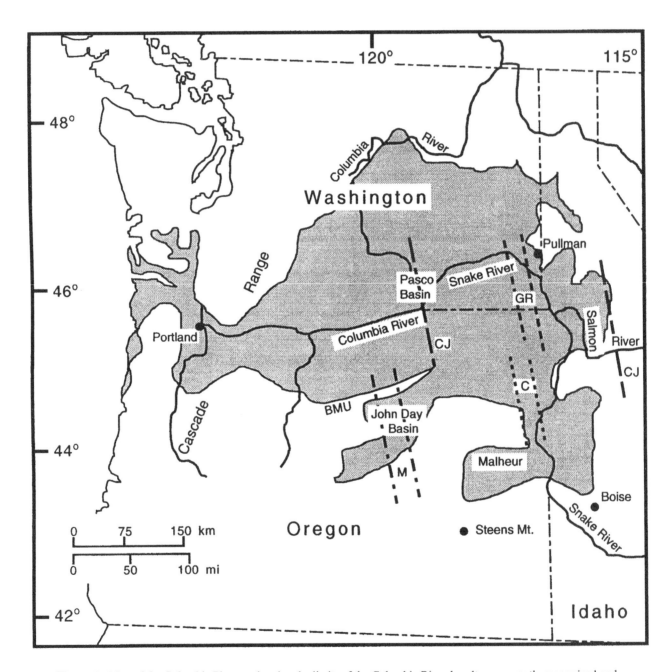

Figure 1. Map of the Columbia Plateau, showing the limits of the Columbia River basalts as currently recognized and the feeder dike systems. CJ marks the outer boundaries of the Chief Joseph dike swarm feeding the Clarkston (Imnaha, Grande Ronde and Wanapum) and Saddle Mountains Basalt. The Grande Ronde (GR) and Cornucopia (C) swarms are dike concentrations within the CJ. M is the Monument dike swarm which fed the Picture Gorge Basalt in the John Day basin. BMU is the Blue Mountains uplift.

border which erupted about 17.0–16.5 Ma [*Rytuba and McKee*, 1984; *Carlson and Hart*, 1987; *Swisher et al.*, 1990; *Lees*, 1994; *Zoback et al.*, 1994; *Camp*, 1995; *Hooper et al.*, 1995c]. If so, the eruptions then moved rapidly northward, the main volume of magma erupting between 16.5 and 15.5 Ma from strongly NNW-oriented fissures (Chief Joseph dike swarm) which crossed the eastern end of the Washington-Oregon border (Figure 1). Magmatic activity then decreased, finally ending around 6.0 Ma [*Tolan et al.*, 1989; *Baksi*, 1989] (Figure 2).

After pioneer mapping at the turn of the century [*Russell*, 1893, 1901; *Merriam*, 1901; *Smith*, 1901, 1903a,b; *Lindgren*, 1901], progress was hampered by the apparent similarity of the many flows exposed in the deep canyons of the Columbia, Snake, and Salmon Rivers. *Washington* [1922] analyzed major elements and compared the tholeiitic compositions of the flows to those of other classic flood basalt provinces such as the Indian Deccan. Later, *Fuller* [1931] described many of the physical aspects of the CRBs, including the remarkably intricate mixtures of basalt, altered glass, and sediment (palagonite/pillow complexes and "pepperites") formed when the CRB lavas ran into lakes created by the damming of the large rivers by earlier flows. Modern research on the CRBG was initiated by Waters and his students in the 1960s [*Waters*, 1961].

The problem of distinguishing individual flows has been largely overcome by application of magnetic polarity measurements and increasingly precise chemical analyses. *Campbell and Runcorn* [1951] demonstrated that magnetic polarity varied in some basalt sections, and more recent work has shown that individual flows possess unique magnetic properties which may be used as a reliable correlating tool [*Wells et al.*, 1989]. Use of major element analyses in conjunction with carefully measured sections permitted Waters and his students [*e.g. Schmincke*, 1967b] to develop the first outlines of a flow stratigraphy. Trace element analysis was first used to characterize flows by *Osawa and Goles* [1970] and has become the most reliable method of correlating individual flows across the province [*e.g., Beeson et al.*, 1985; *Reidel et al.*, 1989; *Hooper et al.*, 1995b].

Improved magnetic and analytical techniques have allowed flow-by-flow mapping, leading to relatively accurate estimates of the aerial extent and volume of individual eruptions and to the identification of many feeder dikes. This knowledge, and an increasingly mature understanding of the tectonic setting of this small but otherwise typical flood basalt province, makes the CRBG important in the development and evaluation of models which attempt to explain the association between large igneous provinces, mantle plumes, and lithospheric thinning. Current work on the earliest manifestations of

plume-related CRBG magmatism south of the Columbia Plateau suggests that the model of *White and McKenzie* [1989, 1995], in which both a mantle plume and active lithospheric extension are required to provide the exceptionally large melt volumes, needs some modification. Current evidence suggests that mantle plumes are the primary cause of continental flood basalt eruptions, abetted by the right tectonic environment in which weakened zones or thinspots in the lithosphere [*Thompson and Gibson*, 1991] are available to increase the melt volume. The form and location of the lithospheric thinspots may determine the exact location of the eruptions, sometimes funneling the magma away from the center of the underlying plume. The lithospheric extension typically associated with flood basalt eruptions, and which sometimes leads to plate separation, follows as a consequence of this plume and flood basalt activity.

PHYSICAL EVOLUTION

Eruptive Sequence

The Steens Basalt, a 1000-m-thick succession on Steens Mountain, Oregon (Figure 3), may be the earliest manifestation of the Columbia River flood basalt [*Carlson and Hart*, 1988]. Despite obvious lithological differences, such as thinner flows and more variable mineralogies, the sequence of flows on Steens Mountain is coarsely plagioclase-phyric and chemically similar to the Imnaha Basalt on the Columbia Plateau and to the lower basalt flows at Malheur Gorge (Figure 3). At the south end of the Pueblo Mountains, south of Steens Mountain, the base of the Steens Basalt sequence includes picritic basalt members (J. Evans, personal communication, 1995) of cumulate origin. These earliest CRBG-type eruptions are part of a long linear zone extending from possibly as far south as Roberts Mountain in Nevada [*Zoback et al.*, 1994] north through Steens Mountain to Pullman (Washington), a zone that may have been displaced by WNW trending strike-slip faults associated with later basin and range extension [*Lawrence*, 1976; *Hart and Carlson*, 1987].

The Steens Basalt is composed entirely of Imnaha-like basalt, the lower half magnetically reversed, the upper half normal [*Mankinen et al.*, 1987]. At Malheur Gorge, dominantly Imnaha Basalt of unknown magnetic polarity is overlain by a thin sequence of Grande Ronde Basalt. Farther north, along the southern edge of the previously recognized Columbia River basalt province (southern margin of the Wallowa Mountains and at Squaw Butte a few kilometers northwest of Boise, Idaho) only a few flows of magnetically reversed Imnaha Basalt underlie a thick sequence of magnetically normal Imnaha Basalt which is overlain by an

GROUP	SUB-GROUP	FORMATION	MEMBER	ISOTOPIC AGE (m.y.)	MAGNETIC POLARITY
COLUMBIA RIVER BASALT GROUP	CLARKSTON BASALT	SADDLE MOUNTAINS BASALT	LOWER MONUMENTAL MEMBER	6	N
			BASALT OF LOWER MONUMENTAL		
			BASALT OF TAMMANY C.		
			ICE HARBOR MEMBER	8.5	
			Basalt of Goose Island		N
			Basalt of Martindale		R
			Basalt of Basin City		N
			BUFORD MEMBER		R
			ELEPHANT MOUNTAIN MEMBER	10.5	R,T
			POMONA MEMBER	12	R
			EQUATZEL MEMER		N
			WEISSENFELS RIDGE MEMBER		
			Basalt of Slippery Creek		N
			Basalt of Tenmile Creek		N
			Basalt of Lewiston Orchards		N
			Basalt of Cloverland		N
			ASOTIN MEMBER	13	
			Basalt of Huntzinger		N
			WILBUR CREEK MEMBER		
			Basalt of Lapwai		N
			Basalt of Wahluke		N
			UMATILLA MEMBER		
			Basalt of Sillusi		N
			Basalt of Umatilla		N
		WANAPUM BASALT	PRIEST RAPIDS MEMBER	14.5	
			Basalt of Lolo		R
			Basalt of Rosalia		R
			ROZA MEMBER		T,R
			3 - 5 Flows		
			SHUMAKER CREEK MEMBER		N
			FRENCHMAN SPRINGS MEMBER		
			Basalt of Lyons Ferry		N
			Basalt of Sentinal Gap		N
			Basalt of Sand Hollow	15.3	N
			Basalt of Silver Falls		N,E
			Basalt of Ginkgo		E
			Basalt of Palouse Falls		E
			LOOKINGGLASS MEMBER		N
		ECKLER MOUNTAIN BASALT	DODGE MEMBER		
			Basalt of Dodge		N
			Basalt of Robinette Mountain		N
	PICTURE GORGE BASALT	GRANDE RONDE BASALT		15.6	N_2
					R_2
					N_1
					R_1
				16.5	T
	STEENS BASALT	IMNAHA BASALT			N_0
				17.5	R_0

Figure 2. Stratigraphic table of the Columbia River Basalt Group.

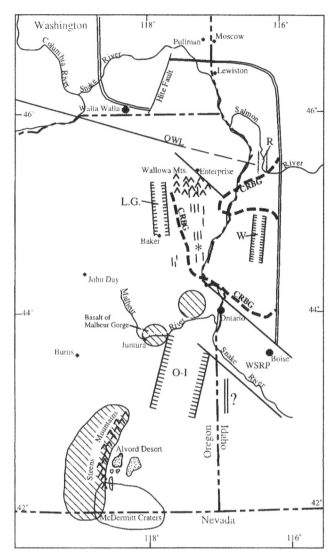

Figure 3. Aerial extent of the Imnaha Basalt, southeastern Columbia Plateau. LZ=suture zone between older cratonic lithosphere to north and east and accreted terranes to the south and west; SD=Sevin Devils Mts; LF=Limekiln Fault; FB=Farewell Bend; PM=Pedro Mt; LM = area of Imnaha vent scoria at Lookout Mountain; short solid lines represent the approximate positions of Imnaha and Grande Ronde dikes south of the Wallowa Mountains. OWL=Olympic-Wallowa Lineament; WSRP=western Snake River Plain; R=Riggins graben; LG = La Grande and Baker Grabens (14.–13 Ma). W = Weiser grabens. O-I = Oregon-Idaho graben. Note that the field of Steens Basalt shown is significantly smaller than indicated by *Hart and Carlson* [1987]; it probably extends much farther to the west, in particular, but its precise limits are poorly constrained.

increasingly thick (northwards) sequence of Grande Ronde Basalt. Still farther north, no Imnaha flows with reverse magnetic polarity are present. Assuming this to be the same reversed to normal upward polarity break, then the center of

magmatism moved northward with time [*Camp*, 1995] from Steens Mountain to the Washington-Oregon border. Present radiometric dates do not support this, but the northward migration may have been too rapid, in hundreds of thousands rather than in millions of years, to be detected by current absolute age dating techniques. NNW to N-S trending feeder dikes are present on Steens Mountain and what appears to be a large vent complex of Imnaha Basalt occurs on Lookout Mountain (Oregon) at the southern end of the Cornucopia dike swarm (LM, Figure 3 and Figure 4), which is composed primarily of Imnaha and Grande Ronde feeder dikes.

The phyric Imnaha flows were succeeded, conformably and without an apparent time gap, by the aphyric flows of Grande Ronde Basalt. The petrographic change was accompanied by a sharp increase in silica (Figure 5). Grande Ronde Basalt makes up 85% by volume of the CRBG and was erupted in less than one million years (16.5–15.6 Ma) [*Tolan et al.*, 1989]. These flows are basaltic andesites with a silica range between 53.0 and 57.5% (Figure 5). Chemical differences between the many Grande Ronde flows are small and they lack phenocrysts, making their individual identification particularly difficult. The flows do, however, exhibit three changes in magnetic polarity which can be recorded in the field by a simple, portable, fluxgate magnetometer and confirmed, where necessary, by more complete paleomagnetic characterization using drilled cores. The Grande Ronde Basalt is therefore divided into four magnetostratigraphic units from bottom to top, R_1, N_1, R_2 and N_2, which have been mapped across the province [Figure 2; *Choiniere and Swanson*, 1979; *Hooper et al.*, 1979; *Swanson et al.*, 1979, 1980, 1981; *Swanson and Wright*, 1983; *Reidel et al.*, 1989].

Most Grande Ronde flows were fed by members of the Grande Ronde and Cornucopia dike swarms (zones of concentrated dikes within the Chief Joseph dike swarm, Figure 1) in the most southeasterly corner of the province, but many Grand Ronde N_2 flows were derived from dikes farther west [*Tolan et al.*, 1989]. While Imnaha flows filled the deep canyons that predate the CRBG eruption and almost reached the peaks of the Seven Devils and Wallowa Mountains (Figure 3) [*Hooper and Camp*, 1981], the Grande Ronde lavas erupted over the newly developed flat plains of Imnaha Basalt. As a result, the Grande Ronde lavas flowed farther west down-slope and each flow is significantly thinner than most individual Imnaha flows. A few of the oldest Grande Ronde flows are present as far south as Farewell Bend and Malheur Gorge (Oregon) (Figure 3), but increasingly younger Grande Ronde units are found only farther and farther to the northwest, creating an "off-lap" effect [*Hooper and Camp*, 1981]. The uppermost

Figure 4. The Cornucopia dike swarm crossing pale granites and limestones of the Wallowa Mountains horst.

R_2 and all N_2 Grande Ronde flows erupted only on the northwestern (downthrow) side of the northeast trending Limekiln Fault (Figure 3), which formed towards the end of the R_2 magnetostratigraphic episode. It remains unclear to what extent this southeast to northwest offlap of flows was a consequence of the continuing northward migration of the feeder dike system [*Camp,* 1995] or of the continuing rise of the southeast corner of the Columbia Plateau. But it is evident that a continuously regenerating southeast-northwest slope was created across which new lava flowed towards the deepening Pasco Basin (Figure 1) [*Hooper and Camp,* 1981].

The apparently continuous rise of the southeast corner of the province is a facet of the rise of this part of Idaho documented throughout the Tertiary [*Axelrod,* 1968]. Correlation of the uplifted blocks with late Jurassic and Cretaceous granite intrusions suggests that uplift was due to the steady isostatic rise of these bodies [*Hooper and Camp,* 1981]. That this relative uplift continued into post-Miocene time is dramatically illustrated in the Wallowa and Seven Devils Mountains, where the basal flows of Imnaha basalt are now exposed on granite near the tops of those mountains, many hundreds of meters above the same flows in the surrounding low ground.

The Pasco Basin, at the low end of this continuously developing east-west slope, is filled with 3000 to 7000 m of sediment of Eocene age or older [*Campbell,* 1989]. The

sediment is overlain by an extraordinary 3500 m of CRBG [*Reidel et al.,* 1989], almost three times the thickness of basalt that accumulated around the feeder dikes supplying these flows to the southeast. The Pasco Basin has an abrupt N–S trending eastern edge against the gently westward dipping Palouse slope and is probably best regarded as a N–S oriented graben or rift. It was clearly a topographic low which acted as a catchment area throughout the Tertiary, first for sediment and then for successive eruptions of CRBG. Present topography in both the southeast part of the Columbia Plateau and the Pasco Basin suggests that this differential vertical movement across the CRB province, including the steady rise of the batholithic rocks in the southeast and the depression of the Pasco Basin, has continued to the present day.

At the height of the CRBG eruptions, R_2 and N_2 flows of Grande Ronde Basalt not only filled the Pasco Basin but overflowed into the lower Columbia River channel and continued through the Columbia Gorge across the rising Cascade arc to the Pacific Ocean [*Tolan et al.,* 1989]. These are remarkable distances (typically 300 to 600 km) for individual eruptions to cover and most workers have accepted in the past that the formation of these flows required very high eruption rates [*Shaw and Swanson,* 1970].

During the N_1 and R_2 eruption of Grande Ronde flows in the southeast of the Columbia Plateau, separate eruptions of

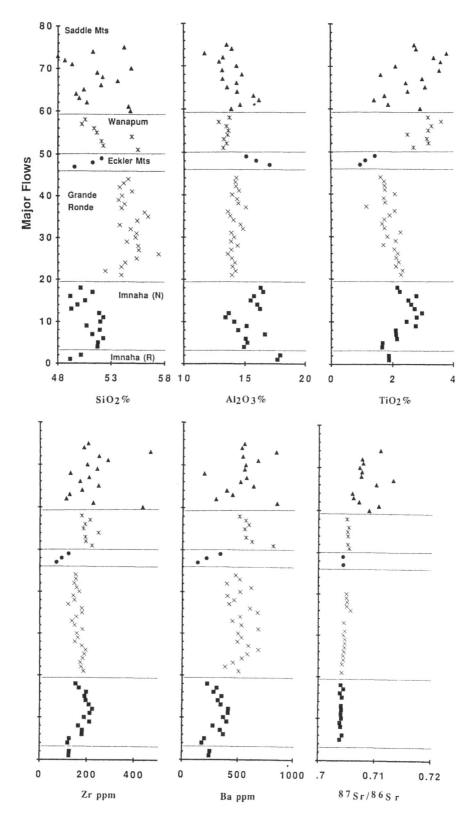

Figure 5. Columbia River Basalt Group. Chemical and isotopic parameters plotted against stratigraphic position for major flows, illustrating major chemical differences between subgroups and formations [*Hooper*, 1988a; *Hooper and Hawkesworth*, 1993].

Picture Gorge Basalt (PGB) [*Bailey*, 1989a,b] from the Monument dike swarm occurred in the John Day Basin (Figure 1) [*Fruchter and Baldwin*, 1975]. This basin was structurally isolated from the rest of the CRBG province by the NE–SW trending Blue Mountains anticlinal uplift (Figure 1).

The PGB subgroup differs from the Imnaha and Grande Ronde Basalts in being more chemically primitive (Figure 6a). The PGB also has distinctively lower concentrations of Sr (Figure 6b) and different incompatible trace element ratios which link it more obviously to the contemporaneous basin and range eruptions to the south rather than to the main sequence of the CRBG on the Columbia Plateau (Figure 6c) [*Hooper and Hawkesworth*, 1993]. The PGB subgroup is divided into three formations [*Bentley and Cockerham*, 1973; *Swanson et al.*, 1979; *Bailey*, 1989a]: from base to top, the coarsely plagioclase-phyric Twickenham Basalt (normal magnetic polarity); the aphyric to sparsely phyric Monument Mountain Basalt (normal polarity); and the predominantly phyric Dayville Basalt (changing from normal to reversed magnetic polarity)

[*Watkins and Baksi*, 1974]. The three formations can be clearly distinguished on the basis of phenocryst assemblage and major element composition (e.g., TiO2/MgO plots [*Bailey*, 1989a]). Across a low saddle in the Blue Mountains uplift, normal polarity PGB flows interfinger with N_1 Grande Ronde basalts which are capped by R_2 Grande Ronde basalts. Radiometric ages (16.5–15.6 Ma) [*Baksi*, 1989] confirm the correlation of the PGB with the N_1–R_2 magnetostratigraphic zones of the Grande Ronde Basalt (Figure 2).

The Picture Gorge Basalt subgroup is less than 1.5% by volume of the whole CRBG, but it mimics the sequence of the physical evolution of the main CRBG eruption [*Bailey*, 1989a]. The early plagioclase-phyric Twickenham Basalt filled in deep canyons of the earlier topography, as did the similarly phyric Imnaha Basalt to the east. The Twickenham Basalt was followed by much thinner and more widespread aphyric to sparsely phyric flows of Monument Mountain Basalt, equivalent to the Grande Ronde Basalt. The PGB ended in the smaller, predominantly phyric and chemically more variable flows of Dayville Basalt in much the same

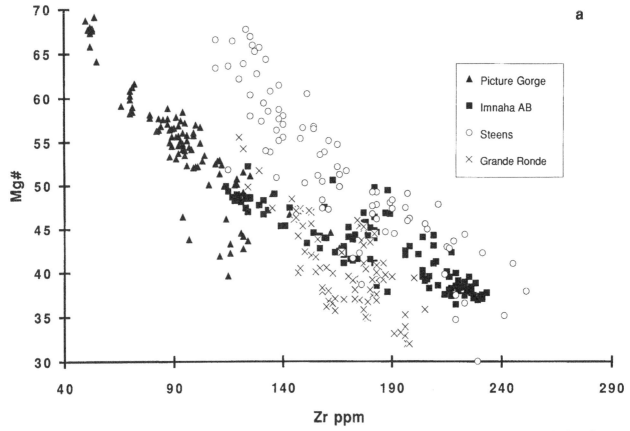

Figure 6. Chemical differences between Steens, Imnaha (Rock Creek (RC) and American Bar (AB) types), Grande Ronde, and Picture Gorge Basalts. (a) Mg# (Mg/(Mg+Fe2) vs Zr; (b) Sr vs SiO2; (c) Nb/Zr vs Zr/Y. [*Hooper*, 1988a; *Hooper and Hawkesworth*, 1993].

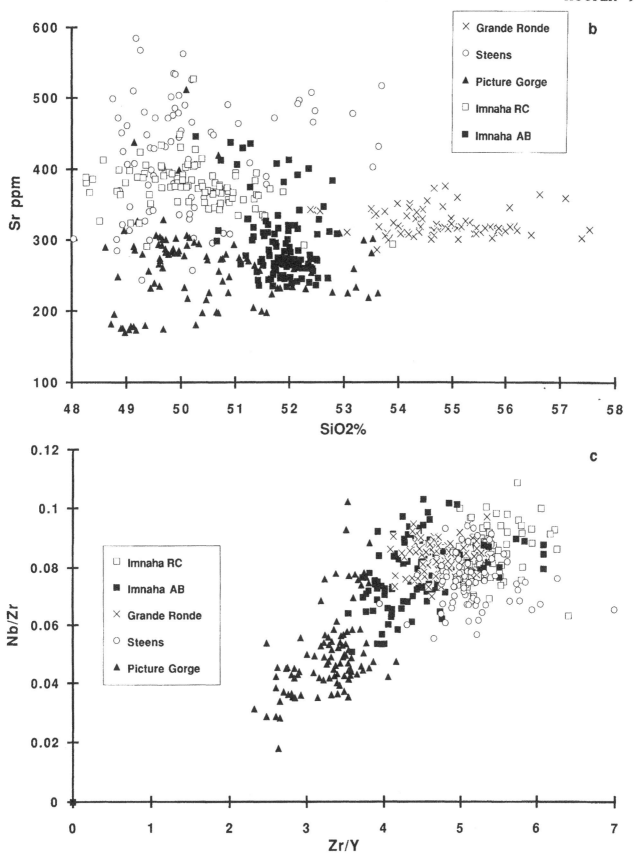

way as the main eruption to the east ended in the small, chemically variable Saddle Mountains Basalt. Although clearly from different sources, the similarities in these two sequences suggest that many aspects of the flow petrography, including the presence or absence of phenocrysts, were controlled by topographic features rather than by more fundamental differences in magma source or evolutionary paths. Large plagioclase phenocrysts appear to correlate with locally thickened canyon-filling flows, suggesting that the phenocrysts grew rapidly after eruption, rather than by the more normally assumed intratelluric crystallization in deep magma chambers before eruption [*Swanson and Wright*, 1981; *Wright et al.*, 1989].

In the main CRBG eruptions farther east, the large volumes of Grande Ronde Basalt were followed by a lull in magmatism. Only small eruptions of relatively primitive Eckler Mountain Basalt occurred along the eastern end of the Oregon-Washington border. These localized flows, the single flow of Robinette Mountain and the many Dodge flows, are typically underlain, and sometimes overlain, by saprolite horizons that correlate with saprolite and/or sedimentary units in other parts of the Columbia Plateau, all implying a significant decrease in volcanic activity.

A return to intense magmatic activity in the form of very large eruptions of Wanapum Basalt resumed thereafter (15–14.5 Ma [*Tolan et al.*, 1989]), albeit with magma significantly more enriched in iron, titanium, and phosphorus and, relative to Grande Ronde Basalt, depleted in silica (Figures 5 and 7a). The Wanapum began with the small, relatively evolved Lookingglass flow, followed by the many large flows of the Frenchman Springs Member, all erupted from the western margin of the Chief Joseph dike swarm [*Taubeneck*, 1970]. The subsequent Shumaker Creek, Powatka, and Roza Members [Figure 2; *Hooper and*

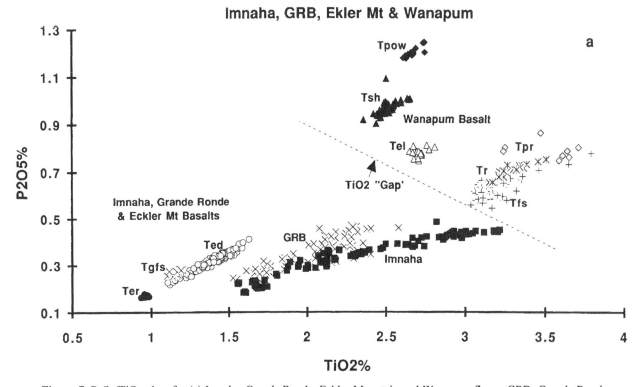

Figure 7. P$_2$O$_5$/TiO$_2$ plots for (a) Imnaha, Grande Ronde, Eckler Mountain and Wanapum flows. GRB=Grande Rond Basalt; Tgfs=Field Springs flow of the GRB; Ter=Robinette Mt flow (Eckler Mt); Ted=Dodge flows (Eckler Mt) Tel=Lookingglass flow; Tfs=Frenchman Springs Member; Tsh=Shumaker Creek Member; Tpow=Powatka flow; Tr=Roza Member and Tpr=Priest Rapids Member (see Figure 2). Note similar ranges for Grande Ronde (crosses) and Imnaha (squares) Basalts. (b) Saddle Mountain flows. Tu=Umatilla; Ta=Asotin; Tp=Pomona; Tem=Elephant Mountain; Tesq=Esquatzel; Tw=Wilbur Creek; Tws, Twc, Twt, and Twl=the Slippery Creek, Cloverland, Tenmile Creek, and Lewiston Orchards flows respectively of the Weissenfels Ridge Member; Tig, Tim, and Tib = the Goose Island, Martindale, and Basin City flows, respectively, of the Ice Harbor Member; Tn=Eden; and Tlm=Lower Monumental Member. The Saddle Mountains flows cover the same range as the lower flows of the CRBG but extend to more extreme values of P$_2$O$_5$ [*Hooper*, 1988a; *Hooper and Hawkesworth*, 1993]..

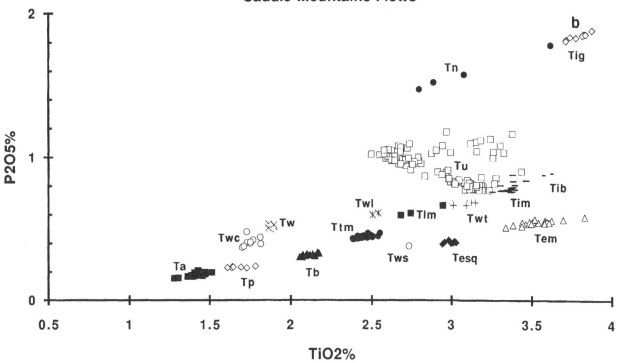

Saddle Mountains Flows

Figure 7. (continued)

Swanson, 1990] were erupted from the center of the Chief Joseph dike swarm, to be followed by flows of the Priest Rapids Member erupted from the eastern edge of that swarm in west-central Idaho. The three largest members of Wanapum Basalt (Frenchman Springs, Roza, and Priest Rapids Members) are chemically very similar, but are interposed with the smaller and much more siliceous Lookingglass, Shumaker Creek, and Powatka flows (Figures 2 and 5).

The CRBG eruption rate declined between 14.5 and 6.0 Ma during the eruption of the Saddle Mountains Basalt (Figure 7b). These mainly small flows (excepting the Pomona) filled valleys and canyons created in the older sheet flows of the CRBG by tectonic deformation and river erosion. Saddle Mountains Basalt flows are characterized by diverse compositions and by consistently more radiogenic Sr-isotope signatures (Figure 5).

Flow Volume and Homogeneity

Unusually large individual sheet-flows are the hallmark of a continental flood basalt province and all continental flood basalt provinces consist of exceptionally large volumes of tholeiitic basalt extruded over a short period, a process which requires higher than normal mantle temperatures. These features are well documented on the Columbia Plateau. Although the CRBG is substantially smaller than other classic continental flood basalts (an order of magnitude less than the estimated original volumes of the Deccan, Paraná, Karoo, and Siberian Trap provinces), 149,000 km³ of Grande Ronde Basalt (85% of the total CRBG) was erupted in less than one million years [Tolan et al., 1989]. Individual eruptions are remarkably homogeneous and may contain more than 700 km³ of magma [Shaw and Swanson, 1970; Tolan et al., 1989]; volumes greater than 2000 km³ have been suggested for some individual Grande Ronde flows [Reidel et al., 1989].

The Pomona flow, one of the largest in the Saddle Mountains Basalt, has an estimated volume of more than 700 km³ and can be traced for almost 600 km from its source in west-central Idaho. It flowed down the Snake River canyon and, like some of the earlier Grande Ronde flows, flooded the Pasco Basin, overflowed into the lower Columbia River channel, and plunged deep into wet sediments off the Pacific coast [Beeson et al., 1979; Pfaff and Beeson, 1989; Wells et al., 1989; Neim et al., 1994]. This may be the longest terrestrial lava flow documented and it shows no discernible variation in its major or trace element composition throughout its length [*Hooper*, 1982, 1988a]. Such large volumes of homogeneous magma available at one time suggests unusually rapid eruption rates from very large magma reservoirs. As there is no evidence of collapse structures accompanying the huge eruptions, the reservoirs probably formed near the base of the crust [*Swanson and Wright*, 1981; *Wright et al.*, 1989].

The only rationale for the formation of sheet-flows rather than shield volcanoes in continental flood basalt provinces is the eruption of an unusually large volume of low viscosity magma in a short time. Just how short a time has become a matter for debate. Big eruptions like those of the Roza Member were still largely liquid when they ran into water and chilled as glass some 300 km from their feeding fissures [*Swanson et al.*, 1975]. Most of their large plagioclase phenocrysts grew after eruption, implying little loss of heat over large distances. Recently, *Ho and Cashman* [1995] have shown that the Ginkgo flow (Frenchman Springs Member, Figure 2) lost only 20°C while travelling across 530 km of plateau (0.036°/km) and that the proportion of phenocrysts (15%) remained unchanged over this distance. *Shaw and Swanson* [1970] envisaged lava fronts 50 m high, perhaps 100 km long, moving across the very gently sloping Columbia Plateau at 3 to 5 km per hour [*Hooper*, 1982]. However, in analogy to the observed growth of flows on Hawaii, *Hon et al.* [1994] and *Self et al.* [this volume] argued that the CRBG flows were fed internally in a succession of large lobes, each new magma pulse raising the older flow crust, and that flow in such a process may have been significantly less rapid than the turbulent flow originally envisaged by *Shaw and Swanson* [1970]. Flow growth by inflation seems inevitable. This must have been the case for the Pomona flow, which moved down a narrow steep-walled canyon and which would surely have formed a solidified roof to become a well-insulated lava tube. A similar mechanism is probable for sheet-flows crossing the plateau surface, but it remains unclear just how fast flows that were fed in this manner could grow. It seems improbable that the detailed, small-scale models of Hawaii can be expanded by several orders of magnitude to fit the much higher eruption rates apparent on the Columbia Plateau [*Hon et al.*, 1994]. By assuming a lower viscosity and smaller initial flow thickness, later inflated, and the rate of crustal growth derived from the Hawaii study, *Self et al.* (this volume) conclude that CRBG flows covering areas of 700 to 2000 km² were formed over many years rather than the many days or weeks implied in the Shaw and Swanson model. The remarkable homogeneity of individual flows is more easily reconciled with the more rapid eruption models because it is difficult to understand how some evidence of crystal fractionation would not be detected in a magma body erupting over a period of many years. However, the eruption rates of the CRBG magmas remain unresolved.

Flow Morphology

A striking feature of the sheet-flows of the Columbia Plateau is the frequent development of well-defined colonnades and entablatures with vesicular tops [Figure 8]. Such simple flow structures are less common or even rare in other provinces (Deccan, Karoo, Paraná) where most of the flows are compound. (The term "compound flows" is used here to imply massive units built from many small lava flows or toes, each with pipe vesicles developed along their base and oxidized ropy upper surfaces; Figure 9; [*Walker*, 1968]). The entablature of a typical simple flow on the Columbia Plateau is finer grained and more glassy than the

Figure 8. Photo of a typical "simple" flow [*Walker*, 1968] on the Columbia Plateau, showing the colonnade at the bottom, entablature above, and a vesicular and scoracious top, Grand Coulee, Washington.

Figure 9. Lava lobes in a typical compound flow [*Walker*, 1968], Karoo basalts, Lesotho, southern Africa. Each lobe tends to have pipe vesicles at the base and an oxidized ropy top. Successive lobes are chemically identical, many making up a single compound flow. These structures are rare on the Columbia Plateau.

colonnade.   *Long and Wood* [1986] suggested that entablatures formed when fresh lava was covered by water which penetrated cracks in the congealing lava, causing rapid cooling from the surface downward. In this model the usually sharp break between entablature and colonnade is the line where the cooling surface rising from the base meets that descending from the top. Certainly the formation of lakes on top of these flows must have been common, because flow after flow obstructed and dammed the large rivers crossing the plateau. See *Self et al.* (this volume) for a more detailed discussion of the morphology of CRBG flows and alternative models for their origin.

Pillow-Palagonite Complexes and Invasive Flows

The presence of many lakes and rivers on the Columbia Plateau during the CRBG eruptions is evident from numerous examples of pillow-palagonite complexes formed when basalt lava invaded water and wet sediment (Figure 10) [*Fuller*, 1931; *Schmincke*, 1967a; *Swanson and Wright*, 1981].   Almost any road cut across the plateau in southeastern Washington exposes these complexes.   In many examples the elongation and slope of the pillows create the equivalent of foreset beds, demonstrating the direction of magma flow. Reactions between hot lava and

wet sediment may be complex, leading to intricate mixtures of small black glass fragments evenly mixed through the pale sediment (pepperite).   Often, solid masses of the younger basalt burrowed beneath the less dense and unconsolidated older sediment to form invasive (as opposed to intrusive) flows [*Byerly and Swanson*, 1978].   Recent work along the Pacific coast [*Beeson et al.*, 1979; *Neim and Neim*, 1985; *Pfaff and Beeson*, 1989; *Neim et al.*, 1994] has detailed extreme examples of this invasive mechanism. Many large CRBG flows (including the Pomona) swept down the Columbia River valley and invaded wet sediment as dikes and sills for hundreds of meters offshore. Careful mapping has demonstrated the continuity of CRBG flows on the plateau and flows through the Columbia Gorge, so establishing beyond reasonable doubt that the coastal dikes and sills are invasive into the wet sediments and are not local feeder dikes as originally interpreted [*Snavely et al.*, 1973].

Basalt Textures

The great majority of flows on the Columbia Plateau belong to the dominant Grande Ronde Basalt, which is aphyric, despite its evolved chemical signature.   Other formations are largely phyric, with plagioclase always the

Figure 10. Pillow-palagonite complex, Asotin Grade, southeast Washington. Such complexes are common across the Columbia Plateau, formed when the lavas flowed into river beds and lakes created by the damming of rivers by previous flows.

dominant phenocryst phase, followed by olivine, then augite. There is no simple correlation between the proportion and abundance of phenocrysts and such indices of fractionation as Mg# (Mg/{Mg+Fe^2}) or Zr abundance.

Porphyritic textures are traditionally regarded as the slow intratelluric growth of phenocrysts in reservoirs below the surface, surrounded by the finer grained, more rapidly cooled, matrix phases that formed as the residual liquid chilled on extrusion. However, an alternative scenario has been demonstrated experimentally [*Lofgren*, 1980] in which phenocrysts and groundmass grow at differing rates in a single cooling environment. Two lines of evidence suggest that this alternative mechanism has been effective on the Columbia Plateau: the correlation of porphyritic textures with unusually thick canyon-filling flows in the Imnaha and Twickenham Basalts, and the fewer, smaller phenocrysts created in the Roza flow when it plunged into water hundreds of kilometers from its eruption center [*Wright et al.*, 1989]. It is, therefore, possible that the typical phenocryst assemblage (plagioclase, followed by olivine,

then augite) observed is not the assemblage responsible for fractional crystallization in magma reservoirs at the base of the crust and should not be used uncritically as the assemblage to test fractionation models in the CRBG.

Linear Vent Systems

The earliest work on the Columbia Plateau suggested that the CRBG magma erupted from fissures. A detailed account of the highly oriented NNW-SSE feeder dikes and small vents for many flows was provided by *Swanson et al.* [1975] and this work has been augmented by numerous more detailed studies in the southeastern part of the plateau [e.g., *Hooper and Swanson*, 1990; *Reidel and Tolan*, 1992; *Hooper et al.*, 1995b]. Linear vent systems feeding individual eruptions or sequential eruptions of nearly identical composition, such as the Roza and Ice Harbor systems, are from 50 km to over 200 km long and only a few kilometers wide with a minimum length-to-width ratio of 15:1. Vents and cones of various sizes, complete with scoria, ash layers, cinders, and spatter can be studied between Clarkston and Pomeroy (Washington) [*Swanson and Wright*, 1978; *Swanson et al.*, 1980; *Hooper et al.*, 1995b].

CHEMICAL VARIATION AND PETROGENETIC MODELS

General

Most continental flood basalts tend toward higher silica and potassium concentrations than typical oceanic basalts. However, the high silica content of the Grande Ronde basaltic andesites (53.0 to 57.5 wt% SiO_2, normalized without volatiles) is unusual and, in the context of continental flood basalts, is rivaled only by the Tasmanian dolerites [e.g., *Hergt et al.*, 1989]. The scarcity of picritic compositions and the lack of rhyolitic units in the CRBG also contrast with most other flood basalt provinces.

Major units within the CRBG (subgroups and formations; Figure 2) are defined by differences in field appearance, petrography, and major element compositions (Figure 5) [*Waters*, 1961; *Wright et al.*, 1973]. These units also differ in their relative trace element abundances and isotopic ratios in a manner indicative of their derivation from a variety of source components in either the mantle or the crust [*Carlson et al.*, 1981; *Carlson*, 1984; *Hooper and Hawkesworth*, 1993; *Brandon et al.*, 1993]. The consistently basaltic character of all CRBG flows implies that they were ultimately derived from the partial melting of a mantle source. Most Columbia River basalts are relatively evolved and differ from oceanic basalts in that their large ion lithophile (LIL) and high field strength (HFS) incompatible trace elements are decoupled (as shown by negative Ta-Nb anomalies on a mid-ocean ridge basalt normalized element diagram). This trace element pattern is typical of subduction-related magmas and many continental crustal rocks and indicates a "lithospheric" (mantle or crustal) source component similar to, but generally less extreme than, that characteristic of subduction-related calc-alkaline sequences [e.g., *Prestvik and Goles*, 1985]. LIL/HFS element decoupling is generally attributed to the LIL enrichment of a previously depleted mantle wedge by the products of hydration from the subducting slab [e.g., *Pearce*, 1983]. Such a signature could have been acquired by the CRBG magmas either through crustal contamination or by partial melting of an older subcontinental lithospheric mantle (SCLM) enriched in LILs in an earlier subduction process. In all probability, both these source components were involved, but a clear distinction between them has not been found and their relative importance remains one of the most controversial aspects of the origin and evolution of the CRBG. The age of enrichment in crust or mantle will be reflected in the isotopic signature.

Sources

Before the effects of partial melting and crystal fractionation processes can be adequately assessed it is necessary to characterize the source components and to identify groups of flows that could be derived from single sources by partial melting and/or crystal fractionation. *Waters* [1961], using the few major element analyses then available, pointed out that partial melting and/or fractional crystallization processes alone were unable to create the chemical differences observed in the CRBG and that different sources were required. Thirty-five years later, with the benefit of thousands of major and trace element analyses and a comprehensive coverage of isotopic ratios [e.g., *Carlson*, 1984; *Church*, 1985; *Hooper*, 1988a; *Smith*, 1992; *Wright et al.*, 1989; *Hooper and Hawkesworth*, 1993; *Brandon et al.*, 1993; *Chamberlain and Lambert*, 1994; *Lambert et al.*, 1995], the precise nature of those sources remains only partially resolved. Most workers would agree that three or four different mantle sources, and a variable crustal component, can be identified with varying degrees of clarity.

Carlson [1984] and *Carlson and Hart* [1988] identified three mantle source components on the basis of Sr, Nd, and Pb isotopic arrays, including an enriched subcontinental lithospheric mantle source for the more primitive flows of the Saddle Mountains Basalt, similar to the source invoked for the Snake River tholeiites by *Leeman* [1975] and others. In addition, *Carlson and Hart* [1988] recognized a lithospheric component in the majority of the flows and, on the basis of high K/P ratios and variable $\delta^{18}O$ content with little variation in $^{87}Sr/^{86}Sr$, argued that this lithospheric component was probably derived from crustal contamination accompanied by crystal fractionation, most obviously in the Grande Ronde and Saddle Mountains flows.

On the basis of an expanded isotopic data base for the Imnaha, Grande Ronde, and Wanapum Basalts, *Hooper and Hawkesworth* [1993] identified a similar isotopic distribution, but interpreted the source components differently. *Carlson's* (1984) first source component (C-1; $^{87}Sr/^{86}Sr$ <0.7035; ε_{Nd}>+6.5; $^{206}Pb/^{204}Pb$=18.8; $^{207}Pb/^{204}Pb$=15.51; $^{208}Pb/^{204}Pb$=38.3; and $\delta^{18}O$ =5.6) is found primarily in the Picture Gorge Basalt and in other basalts erupted over a large part of the Oregon Plateau, including the ubiquitous high alumina olivine tholeiites (HAOTs) and the Steens Basalt. *Carlson and Hart* [1988] derived HAOTs from a depleted mantle which they associated with ocean island basalts and explained the lithospheric signature by a small addition of sediment to the source. In contrast, *Hooper and Hawkesworth* [1993] emphasized the strong decoupling of the LIL from the HFS incompatible trace elements in these rocks. They ascribed the decoupling to a depleted mantle source subsequently enriched by subduction-related fluids to create an enriched subcontinental lithospheric mantle component associated with the Blue Mountains accreted terranes, a source too young to display much variation in their depleted-mantle-type isotopic signature. This source was activated by decompressional melting during basin and range extension and it has been recognized over most of the Basin and Range Province [e.g., *Fitton et al.,* 1988; *Hooper et al.,* 1995a]. Both *Carlson and Hart* [1988] and *Hooper and Hawkesworth* [1993] emphasized the distinctness of this source from that responsible for the main Columbia River basalts.

Carlson's [1984] second (C-2) mantle source is dominated by the relatively restricted array of Imnaha Basalt isotopic ratios ($^{87}Sr/^{86}Sr$ =0.704; ε_{Nd} =+4.5; $^{206}Pb/^{204}Pb$ =19.09; $^{207}Pb/^{204}Pb$ =15.65; $^{208}Pb/^{204}Pb$ =38.7; and $\delta^{18}O$ >6.0). This is interpreted by Carlson as a mantle source enriched by a small percentage of sediment. However, on the basis of some of the earlier Imnaha flows lacking the LIL/HFS decoupling and showing overall similarities to Hawaiian tholeiites, *Hooper and Hawkesworth* [1993] related this component to an ocean island or mantle plume-related mantle source (Figure 11) and used it specifically as evidence that the CRBG was a consequence of mantle plume activity [see also *Brandon and Goles,* 1988].

The isotopic array formed by the Grande Ronde and Wanapum Basalts was interpreted by both *Carlson* [1984] and *Hooper and Hawkesworth* [1993] as a mixing array of which one end member is the Imnaha Basalt (Carlson's C-2). The other end member envisaged by *Carlson and Hart* [1988] is the same as for the more primitive Saddle Mountains Basalt (C-3; $^{87}Sr/^{86}Sr$ =0.7075; ε_{Nd} =-5; $^{206}Pb/^{204}Pb$ =18.09–19.11; $\delta^{18}O$ <+6), a subcontinental lithospheric mantle enriched in late Archaen time. *Hooper and Hawkesworth* [1993] also interpreted this end member as SCLM, but one enriched in the mid-Proterozoic and distinct from the SCLM responsible for the significantly more radiogenic Saddle Mountains Basalt. Unlike *Carlson and Hart* [1988], *Hooper and Hawkesworth* [1993] explained the lithospheric signature of the Grande Ronde and Wanapum Basalts as due primarily to this enriched SCLM component without resort to crustal contamination, for which field (crustal xenoliths) and petrographic (lack of

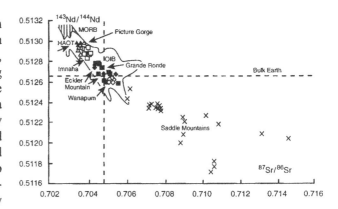

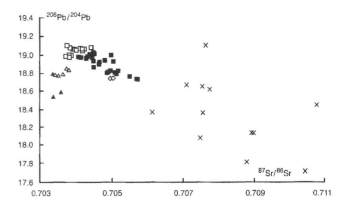

Figure 11. Radiogenic isotopic composition of the CRBG [after *Hooper and Hawkesworth,* 1993]. Triangles = Picture Gorge and high alumina olivine tholeiites (HAOT) from NE Oregon; open squares = Imnaha; filled squares = Grande Ronde; diamonds = Wanapum; crosses = Saddle Mountains Basalts.

equilibrium amongst phenocryst phases, for example, and lack of correlation between degree of evolution and the abundance and composition of the phenocryst assemblage) evidence is lacking. Both *Carlson and Hart* [1988] and *Hooper and Hawkesworth* [1993] agreed that the Saddle Mountains Basalt is probably derived from SCLM enriched in the late Archaen with some degree of crustal contamination, the latter most evident in flows like the Wilbur Creek with unusually high $^{87}Sr/^{86}Sr$ (0.713) values and erupted from fissures on or to the east of the boundary separating the accreted terranes and the older North American plate.

In a detailed isotopic study of the Picture Gorge Basalt, *Brandon et al.* [1993] found some variation in the Sr and Nd isotope ratios, the latter (but not the former) of which correlated with silica content. Assuming silica content was an index of crystal fractionation, they interpreted this correlation as evidence of crystal fractionation accompanied by crustal assimilation, so agreeing with earlier studies by *Bailey* [1989b] and *Carlson* [1984]. In the *Hooper and Hawkesworth* [1993] scenario, however, the use of silica as an index of fractionation was not convincing; the silica content would vary with both crustal assimilation and increasing additions of an enriched SCLM component.

Other isotopic studies include those of *Church* [1985; Pb isotopes], *Smith's* [1992] discussion of the origin of the CRBG flows along the Pacific coast of Oregon, and two studies [*Chamberlain and Lambert*, 1994; *Lambert et al.*, 1995] which discussed the isotopic variations in the Grande Ronde Basalt. Church suggested a depleted oceanic mantle source contaminated by subduction-derived fluids to form a hydrated peridotite, a concept echoed by *Hooper and Hawkesworth* [1993] using geochemical evidence. Smith emphasized the need for iron-rich (pyroxene-rich) subcontinental lithosphere as sources for most CRBG flows [see also *Wright et al.*, 1989], including the Grande Ronde Basalt, but denied any association between Imnaha Basalt (Carlson's C-2) and a plume-related source because, he claimed, such a component was available over a large part of the Pacific Northwest. *Chamberlain and Lambert* [1994] and *Lambert et al.* [1995] argued for at least six separate source components: a plume component, two depleted MORB-like source components slightly contaminated with sediment, two local crustal components, and a sixth, more complex crustal component.

In summary, the scatter of isotopic values requires that the CRBG is the product of many different source components. The evolved nature of even the most primitive flows requires either that the mantle-derived magmas underwent substantial fractionation in the crust before any flows erupted or that the mantle source was unusually enriched in iron [*Wright et al.*, 1989; *Smith*, 1992]. Beyond these basic points there appears to be little consensus as to the nature of the mantle sources and the degree to which the nearly ubiquitous lithospheric signature, as displayed by the decoupling of LIL from HFS incompatible trace elements, is the result of partial melting of an enriched MORB-type source or of crustal contamination. No clear way to distinguish between these two possibilities has yet been identified. Theoretically, the relative values of $\delta^{18}O$ and $^{87}Sr/^{86}Sr$ should provide such a distinction [*James*, 1981]. When applied to the Grande Ronde Basalt, for example, the data support the crustal contamination model [*Carlson and Hart*, 1988]. Unfortunately, the $\delta^{18}O$ values are only marginally above acceptable mantle values and all the data are from whole rocks, notoriously prone to increases in these values by small degrees of alteration, and so this evidence is also suspect.

Partial Melting, Crystal Fractionation, and Magma Mixing

The chemical variation caused by mixing of different source components, the dominant process in the formation of the Grande Ronde Basalt, makes it difficult to evaluate the effects of partial melting and fractional crystallization in those rocks. However, in the Imnaha Basalt the more uniform LIL/HFS element and isotopic ratios permit a single source model as a first approximation for their origin. The chemical variation in the Imnaha Basalt can be explained by a combination of partial melting, crystal fractionation, and magma mixing. The difference between the Imnaha and Grande Ronde Basalts is illustrated in Figure 12; the tight positive correlation between P and Zr in the Imnaha Basalt, as required in a system controlled primarily by partial melting and/or crystal fractionation, contrasts with the wider scatter of the Grande Ronde data, which reflect mixing.

Chemical variation within the Imnaha Basalt is considerable [*Hooper et al.*, 1984]. Two or three discrete sequences of Imnaha Basalt (Figure 13a,b), each with a significant range in HFS element concentrations, differ in their Sr content and Mg# at the same Zr abundance, and in such HFS/HFS ratios as Zr/Y. The variations suggest that each sequence was formed from a different parental magma and that each parental magma could represent different degrees of partial melting from a common source. Variation within each sequence is consistent with fractional crystallization of a gabbroic assemblage (plagioclase: augite: orthopyroxene 16:10:4 for the AB flows), augmented by recharge, in magma chambers in the lower crust [*Hooper*, 1988b; *Hooper and Hawkesworth*, 1993]. Although the process involved many cycles of 30–50% fractionation with 40–50% recharge, there is little increase in silica and little evidence of a negative europium anomaly.

A similar distinction between partial melting and subsequent gabbro fractionation can be made for the Eckler Mountain Basalt [*Hooper et al.*, 1995b]. In the Saddle Mountains Basalt, the detailed study of the Ice Harbor flows by *Helz* [1973, 1976] and *Helz and Wright* [1982] demonstrated that crystal fractionation played little part in the formation of those flows whereas the large variations in isotopic ratios in all Saddle Mountain Basalt flows implied that the chemical variation is principally a function of multiple (including crustal) sources.

In addition to the role of the mixing of partial melts from different sources (Grande Ronde Basalt) and mixing in the form of recharge of magma reservoirs which seems to have been a normal part of the gabbro fractionation process (Imnaha and Eckler Mountain Basalts), there are also clear

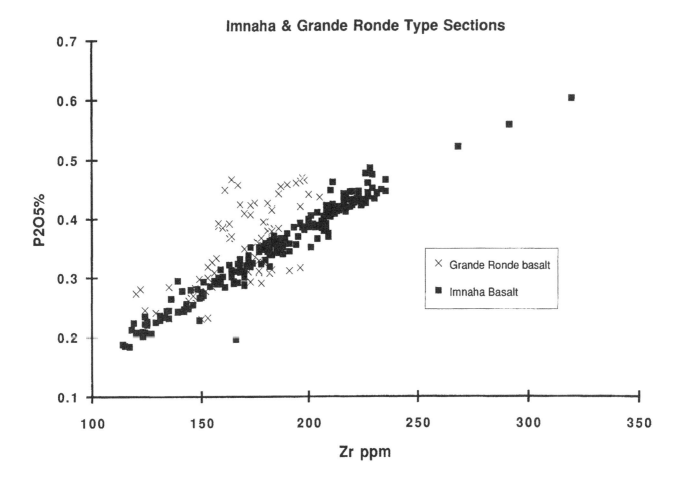

Figure 12. P_2O_5 versus Zr for Imnaha and Grande Ronde Basalt, illustrating the tight positive correlation in the Imnaha (filled squares), as required by simple partial melting + fractional crystallization and the greater scatter in the Grande Ronde Basalt (crosses) in which the chemical variation is ascribed in large part to mixing of two source components.

examples of magma mixing at later stages in the evolution of CRBG magmas. In a province where large volumes of magma erupted over a short time, magma mixing during eruption seems probable. Late magma mixing has been identified where the chemical differences between flows are clear, as between the Wilbur Creek and Asotin flows which mix to form the Lapwai or Huntzinger flow [*Hooper*, 1985; *Reidel and Fecht*, 1987], or between the two flows of the Umatilla Member (Saddle Mountains Basalt). However, in the more common scenario in which two magmas are of similar composition, mixing is difficult to demonstrate. Thus, mixing between the many rapidly erupted Grande Ronde flows seems probable, but has not yet been demonstrated because of their similar compositions.

Carlson and Hart [1988)] explained the chemical and isotopic variations within the Grande Ronde Basalt by a

combination of crystal fractionation and assimilation of a potassium-rich upper crustal component (AFC). In contrast *Helz and Wright* [1982] and *Wright et al.* [1989] argued against significant plagioclase-dominated (gabbroic) fractionation for the Grande Ronde. *Hooper and Hawkesworth* [1993] demonstrated a close positive correlation between Rb/Zr and Rb/Sr in the Grande Ronde Basalt which could not be explained by gabbro fractionation. They calculated that the maximum possible gabbro fractionation within the Grande Ronde was limited to 10%. This makes it difficult to accept the *Carlson and Hart* [1988] AFC model, in which assimilation of crustal rocks must be offset by substantial fractionation. Mixing of mantle-derived basaltic magma with a crustal melt would still be a feasible explanation, but one might then expect to see some indication of this silicic crustal melt on the

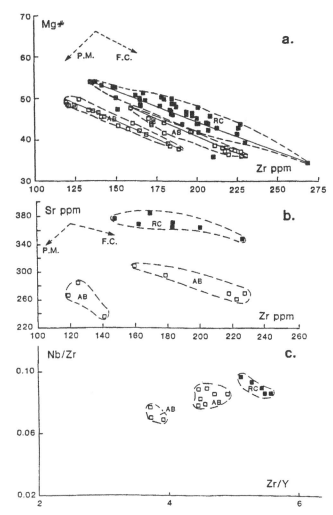

Figure 13. Imnaha Basalt. Plots to illustrate both partial melting (P.M.) and gabbro fractionation (F.C.). (a) Mg# vs Zr (data from *Hooper et al.* [1984]). (b) Sr vs Zr (average flow analyses from *Hooper* [1988b]); Nb/Zr vs Zr/Y (data from *Hooper and Hawkesworth* [1993]). Envelopes enclose data points believed to have formed by gabbro fractionation from a single parent magma. Data points outside these envelopes are interpreted as resulting from mixing between magmas derived from different parents.

surface. Such silicic magmas are specifically absent from the CRBG.

In summary, the most fundamental chemical and isotopic divisions within the CRBG result from different mantle and crustal source components. Within these divisions, however, smaller sequences with relatively coherent isotopic and trace element ratios can be isolated. Each of these sequences may be derived from a common source and their chemical variation may result from a combination of different degrees of partial melting and crustal (gabbroic) fractional crystallization (accompanied by recharge) in reservoirs at or near the base of the crust. Clear evidence of variable partial melting and gabbro fractionation can be demonstrated in the Imnaha and Eckler Mountain Basalts but appears to be largely absent in the Grande Ronde Basalt. Partial melting, crystal fractionation, and source mixing processes were probably augmented by the late mixing of magmas just prior to and during eruption.

TECTONIC SETTING

The main volume of the CRBG erupted through NNW-SSE oriented fissures that now form the Chief Joseph dike swarm (Figures 1 and 3). The dikes are confined to the southeast quadrant of the Columbia Plateau and extend almost as far south as Farewell Bend on the Snake River (Figure 3; P.R. Hooper, unpublished mapping, 1981; Howard Brookes, pers. comm. 1995). Circumstantial evidence [*Camp*, 1995; *Hooper et al.*, 1995c] suggests that the Steens Basalt may represent the earliest CRBG eruptions. If so, this earliest outburst (16.5 Ma) was of smaller but more frequent eruptions than the main fissure-fed CRBG eruptions to the north, and so formed the Steens Mountain shield volcanic edifice, rather than a true flood basalt plateau.

The Chief Joseph dike swarm is specifically confined to the relatively thin lithospheric wedge of the accreted Blue Mountains oceanic terrane that forms a deep embayment into the older and thicker North American plate (Figure 3) [e.g., *Vallier*, 1995]. The suture that forms the boundary between these two contrasting lithospheres (SZ in Figure 3) is marked by a strongly foliated and sheared zone separating strongly contrasting rock types [*Snee et al.*, 1995] with different $^{87}Sr/^{86}Sr$ signatures and intruded by ultramafic lenses [*Armstrong et al.*, 1977; *Fleck and Criss*, 1985; *Mohl and Thiessen*, 1995]. The thinner oceanic lithosphere west of this suture has been particularly prone to further thinning during subsequent east-west extension from the Eocene to the present [e.g., *Hooper et al.*, 1995a].

Deformation during and after the CRBG eruption is largely confined to that part of the Columbia Plateau overlying the accreted terranes. The basalts are undeformed north of the suture zone in eastern Washington (Figure 3) where they lie on the old cratonic crust. To the south, where they overlie the thinner, less competent, accreted terranes, they are folded and faulted (Figure 3) [*Reidel et al.*, 1994; *Hooper et al.*, 1995b]. Thickness and competency of the lithosphere thus appear to have played major roles in determining both where the basalts erupted and how they were subsequently deformed.

The east-west trending Yakima fold belt west of the Pasco Basin (Figure 1) and similar folds to the east, grew during and after the eruptions [*Hooper and Camp*, 1981; *Reidel*, 1984; *Price and Watkinson*, 1989; *Hooper et al.*, 1995b]. The highly oriented NNW-SSE feeder dikes of the Chief Joseph dike swarm testify to the consistent ENE-WSW extension throughout the eruptive period (17–6 Ma). Strike-slip strain with minor translation took place along steeply inclined NNE-SSW (left lateral) and WNW-ESE (right lateral) faults, including the Hite and associated faults [*Hooper et al.*, 1995b] and faults parallel with the Olympic-Wallowa Lineament (OWL) [*Hooper and Conrey*, 1989; *Mann and Meyer*, 1993]. The same strain pattern represented by these structures from 17 to 6 Ma can be measured directly on the Columbia Plateau today [*Kim et al.*, 1986].

The amount and timing of displacement on the OWL remains controversial [see *Mann and Meyer*, 1993; *Reidel and Tolan*, 1994]. Much of the translation may have taken place prior to the CRBG eruptions in a manner similar to the 50 km or more left-lateral displacement on the Hite Fault prior to Wanapum Basalt eruption (Figure 3) [*Reidel et al.*, 1994; *Sobczyk*, 1994; *Hooper et al.*, 1995b]. What is clear from these and other detailed analyses [*Reidel*, 1984; *Anderson*, 1987] is that the strain pattern observed across the whole Columbia Plateau is consistent with the pattern of east-west extension that created the Basin and Range Province and that the orientation of this pattern has not changed over the last 17 million years. If the 25° clockwise rotation of northeastern Washington [*Fox and Beck*, 1985] is restored, the type and orientation of strain appears to have changed little since the Eocene over the whole Pacific Northwest. For a contrary view, see *Barrash et al.* [1983].

North of the OWL, where the amount of extension and lithospheric thinning was minimal [*Taubeneck*, 1970], the NNW-SSE extensional fissures (dikes) and east-west folds dominate, with WNW-ESE right-lateral and NNE-SSW left-lateral strike-slip faulting forming a conjugate pair on a regional scale. On a strain ellipsoid, σ-1 is NNW-SSE and horizontal. Near the surface, with minimal crustal thickness, σ-3 is vertical, so σ-2 is WSW-ENE horizontal, the configuration required to create the east-west folds. Deeper into the crust, the vertical component increases, to become σ-2 and the WSW-ENE horizontal axis of the strain ellipsoid becomes σ-3, the configuration required to create conjugate WNW-ESE (right lateral) and NNW-SSE (left lateral) conjugate strike slip faults. Toward the base of the crust, in the vicinity of prospective magma reservoirs, the vertical pressure increased to equal the NNW-SSE horizontal pressure (=σ-1), with the WSW-ENE horizontal axis as σ-3, the geometry that would create vertical NNW-SSE extensional fissures, a plane of weakness which was used by the magma in forcing its way to the surface from reservoirs at the base of the crust.

South of the OWL significantly greater east-west extension and lithospheric thinning are demonstrated by the development of graben, horst, and complex pull-apart structures [*Gehrels*, 1981; *Hooper and Conrey*, 1989; *Mann*, 1989; *Mann and Meyer*, 1993]. Extension appears to have occurred sporadically; the La Grande, Baker and Weiser grabens formed between 13 and 14.5 Ma [*Bailey*, 1990] and farther south, current evidence suggests that extension increased at about 15 Ma [*Hooper et al.*, 1995c]. East-west folds of this age are not obvious. Here the strain ellipsoid has σ-1 vertical and σ-3 WSW-ENE horizontal, with the NNW-SSE horizontal axis representing σ-2. Prior to 14.5 Ma, when the east-west Aldridge Mountains anticline and parallel folds in northeast Oregon may have formed, the strain pattern was probably more similar to that north of the OWL .

The simplicity of this structural model and its application over 164,000 km^2 for ten million years are unusual and may reflect the relatively simple geology of the Columbia Plateau where, in the deformed southern part, oceanic crust is topped by a thin veneer of uniform basalt flows. Superimposed on this long-lived regional strain pattern is the isostatic rise of the Idaho batholith and other granitic bodies throughout the Tertiary [e.g., *Axelrod*, 1968]. The rise of the granitic rocks in the east and the continued deepening of the Pasco Basin to the west resulted in a constantly evolving east-to-west slope down which the CRBG lavas flowed.

Given this regional setting of thinned and attenuated lithosphere through which the CRBG was erupted, there has been discussion as to the genesis of the CRBG eruption. Was the eruption a consequence of extension, in the form of back-arc spreading 300 km to 400 km behind the active Cascade volcanic arc [e.g., *Carlson*, 1984; *Carlson and Hart*, 1988; *Smith*, 1992]? Or was the eruption due to the Yellowstone hotspot, whose trail across the eastern Snake River Plain places it beneath the McDermitt Craters on the south end of the Steens-Pueblo Mountain ridge at 16.5 Ma at the beginning of the CRBG eruption (Figure 3) [e.g., *Hooper*, 1984; *Brandon and Goles*, 1988; *White and McKenzie*, 1989, 1995; *Richards et al.*, 1989; *Carlson*, 1991; *Draper*, 1991; *Geist and Richards*, 1993; *Hooper and Hawkesworth*, 1993; *Camp*, 1995]?

There are obvious problems in attributing the CRBG eruption to a mantle plume in the form of the Yellowstone hotspot, in addition to our embarrassing ignorance as to what a hotspot actually is. First, the main eruption of the CRBG occurred 300 km to 400 km north of the supposed hotspot track and, second, the vast majority of the CRBG magma has a lithospheric, rather than an asthenospheric or potentially hotspot-related, geochemical and isotopic signature. Nevertheless, the association of hotspots with continental flood basalts in general is well established [e.g., *Duncan*, 1978; *Morgan*, 1981; *Macdougall*, 1988; *White and McKenzie*, 1989; *Campbell and Griffiths*, 1990; *Duncan and Richards*, 1991; *Basu et al.*, 1993] and the geographic displacement of the plume from the center of eruption may simply reflect the ability of the rising magma to seek out the weakest zones in the continental lithosphere [*Thompson and Gibson*, 1991].

Recent work along the southern margin of the Columbia River province [*Pierce and Morgan*, 1992; *Zobak et al.*, 1994; *Camp*, 1995] suggested that the first magmatic manifestation of the hotspot erupted as a south-to-north linear belt of Imnaha-like basalt adjacent to the western side of the lithospheric boundary (SZ in Figure 3), possibly following a series of older (Eocene?) rifts formed by the thinning of the extending oceanic lithosphere away from the older craton. In *Camp's* [1995] model, the plume head was distorted parallel to the overriding, more competent, craton [see also *Geist and Richards*, 1993] and the rapid ascent of the magma was facilitated along the weak boundaries in the accreted collage, the "thinspots" of *Thompson and Gibson* [1991]. Supporting evidence for this model comes from recent recognition that the "Unnamed Igneous Complex" of *Kittleman* [1973] in the Malheur River Gorge of eastern Oregon (Figure 3) is a southern extension of Imnaha and Grande Ronde Basalts and that the tholeiites of Steens Mountain are chemically similar to the earliest Imnaha flows [*Evans*, 1990; *Hooper et al.*, 1995c].

Using similar arguments, the subsequent overriding of the hotspot by the westward motion of the thicker and more competent North American plate, would have funneled the magma along the ancient and weaker tectonic boundary that underlies the eastern Snake River Plain and which conveniently parallels the approximate trace of the hotspot beneath the plate. For an alternative explanation of the sudden shift in the center of the hotspot magmatism see *Geist and Richards* (1993).

White and McKenzie [1989] argued persuasively that the large volume and rapid eruption rates of the CRBG and other continental flood basalts require a mantle hotspot in the literal sense. The chemical and isotopic similarity of early (RC type Imnaha) flows to Hawaiian tholeiites [*Hooper and Hawkesworth*, 1993] supports the presence of a mantle plume. However, the CRBG eruption clearly took place in an environment where east-west extension had been well established for 20 million years or more [e.g., *White*, 1992]. Extension immediately adjacent to the center of the CRBG eruption was minimal (perhaps less than 1%), but when the amount of extension on the south side of the OWL increased about 15 Ma to create typical basin and range faults and grabens, the physical and chemical nature of the

magmatism changed significantly [*Hooper*, 1990; *Hooper and Hawkesworth*, 1993].

Given that the model of *White and McKenzie* [1989, 1995] requires a thinner than normal lithosphere, in addition to the high temperatures of the mantle plume, to account for the exceptionally large volumes of melt involved in flood basalt provinces, those authors emphasized contemporaneous volcanism and extension. However, the field evidence on the Columbia Plateau, the Deccan, and the Siberian Traps indicates that extensional strain was absent immediately before and during the main flood basalt eruptions [*Hooper*, 1990; *Renne and Basu*, 1991]. The same is true for the Karoo province [*Marsh and Eales*, 1984; *Marsh et al.*, this volume] and, apparently, the Paraná [*Turner et al*, 1996]. Flood basalt eruptions seem to act as a catalyst for extension, leading to modest graben structures in some provinces (CRBG, Siberian Traps) and facilitating plate separation in others (Karoo, Deccan, Paraná, and NE Atlantic) 3–5 m.y. after the main magmatism.

We may conclude (1) that the large melt volumes of continental flood basalt provinces require both a mantle plume *and* weakened or thinned lithosphere, (2) that the eruptions through the lithosphere directly or indirectly trigger further extension which may lead to plate separation, and (3) that this subsequent extension is accompanied by magmatism which is of small volume and of more varied composition (calc-alkalic to bimodal and alkalic) than flood basalt volcanism. Both a weakened continental lithosphere *and* a mantle plume are essential to the formation of a continental flood basalt province, but the extension and possible plate separation, which are so often associated with flood basalt volcanism, follow and appear to be the consequence rather than the cause of mantle plume activity, in the manner originally advocated by *Morgan* [1981].

CONCLUSIONS

The three major tectonic factors most pertinent to the origin of the CRBG are (1) the Yellowstone hotspot; (2) back-arc spreading and consequent thinning of the continental lithosphere; and (3) physical proximity of the flood basalt eruption to the tectonic boundary between the thinner, denser, oceanic lithosphere of the accreted terranes and the thicker, more competent lithosphere of the old North American plate, recently welded and thickened further by intrusion of the Idaho batholith.

Almost all CRBG flows were erupted from fissures in the thinner accreted terranes and the first eruptions appear to have followed a narrow north-south zone, possibly a pre-Miocene graben, adjacent to the old crustal suture. Feeder dike orientation is NNW in all cases (Nevada Rift, Steens

Mountain, and the Chief Joseph dike swarm). This suggests that the present north-south to locally NNE alignment of these zones resulted from subsequent increases in basin and range east-west extension to the south, controlled by right-lateral displacement along WNW megashears, as argued by *Lawrence* [1976] and supported by more detailed work [*Hooper and Conrey*, 1989; *Mann*, 1989; *Mann and Meyer*, 1993].

The large volumes of tholeiite magma require the high mantle temperatures of a hotspot and the similarities between the earliest Imnaha Basalt and Hawaiian tholeiites support the derivation of these early magmas from a mantle plume.

In contrast, the continental lithospheric signature apparent in all but these few early CRBG flows suggests either significant crustal contamination and/or the entrainment of enriched subcontinental lithospheric mantle (SCLM) into the plume head. The lithospheric contribution varied with position and time and divides the CRBG into three fundamental subgroups, facts which appear to this author more readily explained by the incorporation of three distinct types of SCLM than by contamination by three different compositions of the same crust. First, the Imnaha and the Grande Ronde Basalts appear to have been derived directly from the mantle plume entraining and mixing with a lithospheric component which was enriched by subduction processes in the mid-Proterozoic. Second, the Picture Gorge Basalt, which erupted further west with a volume almost two orders of magnitude less than the combined Imnaha and Grande Ronde Basalts, may have been derived by decompressional melting from the more recently enriched SCLM below the Mesozoic accreted terranes during basin and range extension. And third, the later Saddle Mountains Basalt erupted as the North American craton overrode the hotspot, so melting an SCLM at the base of the North American plate, enriched in the late Archaen.

These mantle-generated magmas appear to have been stored in large reservoirs near the base of the crust from where they were periodically erupted rapidly to the surface through fissures whose strong NNW-SSE orientation reflects the prevailing regional ENE-WSW extensional strain. The exceptionally large individual eruptions may have been triggered by infusion of a new batch of melt entering the deep magma reservoirs from the mantle. The compositional variations in the dominant Grande Ronde Basalt have been explained either by a combination of crystal fractionation and crustal assimilation [*Carlson and Hart*, 1988] or by mixing of plume-derived magma and variable proportions of an SCLM, with only minor fractionation and crustal assimilation [*Hooper and*

Hawkesworth, 1993]. There is evidence for and against both these models. In contrast, the Imnaha and the Eckler Mountain Basalts can both be interpreted as evolving from their own single source by a combination of partial melting and gabbro fractionation processes. Magma mixing during or just prior to eruption can be demonstrated in the Saddle Mountains Basalt in at least two instances and such magma mixing may have been a common occurrence during the large eruptions of Grande Ronde Basalt.

There appears little immediate prospect of resolving the relative significance of the roles of the crust and an enriched subcontinental lithospheric mantle in the generation of the CRBG. Oxygen isotope data on the individual phases of the phyric Picture Gorge Basalt might help to confirm the data currently available on whole rock samples. In contrast, there is little doubt that a detailed study of the relations between structures and magmatism in the little-known area of eastern Oregon at the southern limits of the Columbia River basalt province would enhance our understanding of the relationships between continental flood basalts, mantle plumes, and lithospheric extension. Such an investigation, employing chemical-stratigraphic mapping and paleo-magnetic techniques, has the potential to resolve some of these fundamental problems at relatively low cost. A fuller understanding of the relative timing of extensional structures and volcanism in eastern Oregon could also help solve the contentious problems associated with dis-placement along such regional features as the Olympic Wallowa Lineament and the Brothers fault zone.

The apparent conflict between *White and McKenzie's* [1995] model and field evidence for the relative timing of the flood basalt eruptions and extension can be resolved by employing the *Thompson and Gibson* [1991] concept of thinspots in the continental lithosphere. It is the presence of a weak or thin lithosphere, not contemporaneous extension, which appears critical to the White and McKenzie thesis. In the case of the Columbia River basalts, such thinspots were almost certainly available because of earlier extension. Thin lithosphere, combined with the high temperatures of the plume, could have created the unusually large melt volumes. The geologic evidence indicates that extension increased following the huge tholeiitic eruptions, and as extension increased the nature of the magmatism also changed to that more typical of extensional continental terranes. The relative timing of extension and volcanism derived from the field evidence on the Columbia Plateau and several other flood basalt provinces points to the mantle plume and consequent flood basalt eruption acting as a stimulant to increased extension, which in some provinces has led to the separation of continents. The detailed knowledge available for the Columbia Plateau suggests that continental breakup was a potential consequence and not a precursor of flood basalt magmatism.

Acknowledgments. I am particularly grateful for the careful and thoughtful reviews of an earlier manuscript by Anita Grunder, Alan Brandon, and John Mahoney which contributed to significant improvements. I would also like to acknowledge the many years of encouragement and help from colleagues in the Geology Department and the Geoanalytical Laboratory at Washington State University.

REFERENCES

Anderson, J. L., The structural geology and ages of deformation of a portion of the southwest Columbia Plateau, Washington and Oregon, Ph.D. thesis, 283 pp., Univ. Southern California, Los Angeles, 1987.

Armstrong, R. L., W. H. Taubeneck, and P. O. Hales, Rb/Sr and K/Ar geochronometry of the Mesozoic granite rocks and their Sr isotope composition, Oregon, Washington and Idaho, *Geol. Soc. Am. Bull., 88,* 397-411, 1977.

Axelrod, D. I., Tertiary floras and topographic history of the Snake River basin, Idaho, *Geol. Soc. Am. Bull., 79,* 713-733, 1968.

Bailey, D. G., Geochemistry and petrogenesis of Miocene volcanic rocks in the Powder River Volcanic Field, northeastern Oregon, Ph.D. thesis, 341 pp., Washington State Univ., Pullman, 1990.

Bailey, M. M., Revisions to stratigraphic nomenclature of the Picture Gorge Basalt Subgroup, Columbia River Basalt group, in *Volcanism and Tectonism in the Columbia River Flood-Basalt Province, Spec. Pap. 239,* edited by S. P. Reidel and P. R. Hooper, pp. 67-84, Geological Society of America, Boulder, CO, 1989a.

Bailey, M. M., Evidence for magma recharge and assimilation in the Picture Gorge Basalt Subgroup, Columbia River Basalt Group, in *Volcanism and Tectonism in the Columbia River Flood-Basalt Province, Spec. Pap. 239,* edited by S. P. Reidel and P. R. Hooper, pp. 343-356, Geological Society of America, Boulder, CO, 1989b.

Baksi, A. K., Reevaluation of the timing and duration of extrusion of the Imnaha, Picture Gorge and Grande Ronde Basalts, Columbia River Basalt Group, in *Volcanism and Tectonism in the Columbia River Flood-Basalt Province, Spec. Pap. 239,* edited by S. P. Reidel and P. R. Hooper, pp. 105-112, Geological Society of America, Boulder, CO, 1989.

Barrash, W., J. G. Bond, and R. Venkatakrishnan, Structural evolution of the Columbia Plateau in Washington and Oregon, *Am. J. Sci., 283,* 897-935, 1983.

Basu, A. R., P. R. Renne, D. K. DasGupta, F. Teichmann, R. J. Poreda, Early and late alkali igneous pulses and a high ^{3}He Plume origin for the Deccan flood basalts, *Science, 261,* 902-905, 1993.

Beeson, M. H., K. R. Fecht, S. P. Reidel, and T. L. Tolan, Regional correlations within the Frenchman Springs member of the Columbia River Basalt Group: New insights into the Middle Miocene tectonics of northwestern Oregon, *Oreg. Geol., 47,* 87-96, 1985.

Beeson, M. H., R. Perttu, and J. Perttu, The origin of the Miocene basalts of coastal Oregon and Washington: An alternative hypothesis, *Oreg. Geol., 41,* 159-166, 1979.

Bentley, R. D., and R. S. Cockerham, The stratigraphy of the Picture Gorge Basalt (PGB), north central Oregon, *Geol. Soc. Am. Abstr. with Programs, 5,* 9, 1973.

Brandon, A. D., and G. G. Goles, A Miocene subcontinental plume in the Pacific Northwest: Geochemical evidence, *Earth Planet. Sci. Lett., 88,* 273-283, 1988.

Brandon, A. D., P. R. Hooper, G. G. Goles, and R. Lambert, Evaluating crustal contamination in continental basalts: the isotopic composition of the Picture Gorge Basalt of the Columbia River Basalt Group, *Contrib. Mineral. Petrol., 114,* 452-464, 1993.

Byerly, G., and D. A. Swanson, Invasive Columbia River basalt flows along the northwest margin of the Columbia Plateau, north-central Washington, *Geol. Soc. Am. Abstr. Programs, 10,* 98, 1978.

Camp, V. E., Mid-Miocene propagation of the Yellowstone mantle plume head beneath the Columbia River basalt source region, *Geology, 23,* 435-438, 1995.

Campbell, C. D., and S. I. Runcorn, Magnetization of the Columbia River basalt in Washington and north Oregon, *J. Geophys. Res., 61,* 449-458, 1951.

Campbell, I. H., and R. W. Griffiths, Implications of mantle plume structure for the evolution of flood basalts, *Earth Planet. Sci. Lett., 99,* 79-93, 1990.

Campbell, N. P., Structural and stratigraphic interpretation of rocks under the Yakima fold belt, Columbia Basin, based on recent surface mapping and well data, in *Volcanism and Tectonism in the Columbia River Flood-Basalt Province, Spec. Pap. 239,* edited by S. P. Reidel and P. R. Hooper, pp. 209-222, Geological Society of America, Boulder, CO, 1989.

Carlson, R. W., Isotopic constraints on Columbia River flood basalt genesis and the nature of the subcontinental mantle. *Geochim. Cosmochim. Acta, 48,* 2357-2372, 1984.

Carlson, R. W., Physical and chemical evidence on the cause and source characteristics of flood basalt volcanism, *Aust. J. Earth Sci., 38,* 525-544, 1991.

Carlson, R. W., and W. K. Hart, Crustal genesis on the Oregon Plateau, *J. Geophys. Res., 92,* 6191-6206, 1987.

Carlson, R. W., and W. K. Hart, Flood basalt volcanism in the Pacific Northwestern United States, in *Continental Flood Basalts,* edited by J. D. Macdougall, pp. 35-61, Kluwer Academic Publishers, Dordrecht, 1988.

Carlson, R. W., G. W. Lugmair, and J. D. Macdougall, Columbia River volcanism: the question of mantle heterogeneity or crustal contamination, *Geochim. Cosmochim. Acta, 45,* 2483-2499, 1981.

Chamberlain, V. E., and R. St. J. Lambert, Lead isotopes and the sources of the Columbia River Basalt Group, *J. Geophys. Res., 99,* 11805-11818, 1994.

Choiniere, S. R., and D. A. Swanson, Magnetostratigraphy and correlation of Miocene basalts of the northwestern American coast and Columbia Plateau, *Am. J. Sci., 279,* 755-777, 1979.

Church, S. E., Genetic interpretation of lead-isotope data from the Columbia River Basalt group, Oregon, Washington and Idaho, *Geol. Soc. Bull., 96,* 676-690, 1985.

Draper, D. S., Late Cenozoic bimodal magmatism in the northern

Basin and Range Province of southeastern Oregon, *J. Volcanol. Geotherm. Res., 47,* 299-328, 1991.

Duncan, R. A., Geochronology of basalts from the Nintyeast Ridge and continental dispersion in the eastern Indian Ocean, *J. Volcanol. Geotherm. Res., 4,* 283-305, 1978.

Duncan, R. A., and M. A. Richards, Hotspots, mantle plumes, flood deposits and true polar wander, *Rev. Geophys., 29,* 31, 1991.

Evans, J. G., Geology and mineral resources of the Jonesboro quadrangle, Malheur County, Oregon, *Oreg. Dept. Geol. Min. Ind., GMS-66,* scale 1:24,000, 1990.

Fitton, J. G., D. James, P. D. Kempton, D. S. Omerod, and W. P. Leeman, The role of lithospheric mantle in the generation of late Cenozoic basic magmas in the western United States, in *Oceanic and Continental Lithosphere: Similarities and Differences, Spec. Vol.,* edited by I. A. Menzies and K. G. Cox, pp. 331-350, Journal of Petrology, Oxford, 1988.

Fleck, R. J., and R. E. Criss, Strontium and oxygen isotope variations in Mesozoic and tertiary plutons of central Idaho, *Contrib. Mineral. Petrol., 90,* 291-308, 1985.

Fox, K. F., and M. E. Beck, Paleomagnetic results from Eocene volcanic rocks from northeastern Washington and the Tertiary tectonics of the Pacific Northwest. *Tectonics, 4,* 323-341, 1985.

Fruchter, J. S., and S. F. Baldwin, Correlations between dikes of the Monument swarm, central Oregon and Picture Gorge Basalt flows, *Geol. Soc. Am. Bull., 86,* 514-516, 1975.

Fuller, R. E., The aqueous chilling of basaltic lava on the Columbia River Plateau, *Am. J. Sci., 21,* 281-300, 1931.

Gehrels, G. E., The geology of the western half of the La Grande basin, northeastern Oregon, M.S. thesis, 97 pp., Univ. of Southern California, Los Angeles, 1981.

Geist, D., and M.A. Richards, Origin of the Columbia Plateau and Snake River Plain: Deflection of the Yellowstone plume, *Geology, 21,* 789-792, 1993.

Hart, W. K., and R. W. Carlson, Tectonic controls on magma genesis and evolution in the northwestern United States, *J. Volcanol. Geotherm. Res., 32,* 119-135, 1987.

Helz, R. T., Phase relations of basalts in their melting range at P_{H2O} =5 kb as a function of oxygen fugacity, *J. Petrol, 14,* 249-302, 1973.

Helz, R. T., Phase relations of basalts in their melting ranges at P_{H2O} =5 kb. Part II. Melt Compositions, *J. Petrol., 17,* 139-193, 1976.

Helz, R. T., and T. L. Wright, Inferred petrology of the source of the Yakima Basalt subgroup, (abstract), IAVCEI Meeting, Iceland, 1982.

Hergt, J. M., B. W. Chappell, M. T. McCulloch, I. McDougall, and A. R. Chivas, *J. Petrol., 30,* 841-883, 1989.

Ho, A., and K. V. Cashman, Geothermometry of the Ginkgo flow, Columbia River Basalt group, *Eos Trans. AGU, 76,* 46, Fall Meeting Suppl., p. F679, 1995.

Hon, K., J. Kauahikaua, R. Denlinger, and K. Mackay, Emplacement and inflation of Pahoehoe sheet flows - Observations and measurements of active lava flows on Kilauea Volcano, Hawaii, *Geol. Soc. Am. Bull., 106,* 351-370, 1994.

Hooper, P. R., The Columbia River basalts, *Science, 215,* 1463-1468, 1982.

Hooper, P. R., Physical and chemical constraints on the evolution of the Columbia River basalt, *Geology, 12,* 495-499, 1984.

Hooper, P. R., A case of simple magma mixing in the Columbia River Basalt Group: the Wilbur Creek, Lapwai and Asotin flows, Saddle Mountains Formation, *Contrib. Mineral. Petrol., 91,* 66-73, 1985.

Hooper, P. R., The Columbia River Basalt, in *Continental Flood Basalts,* edited by J. D. Macdougall, pp. 1-33, Kluwer Academic Publishers, Dordrecht, 1988a.

Hooper, P. R., Crystal fractionation and recharge (RFC) in the American Bar flows of the Imnaha Basalt, Columbia River basalt group, *J. Petrol., 29,* 1097-1118, 1988b.

Hooper, P. R., The timing of crustal extension and the eruption of continental flood basalts, *Nature, 345,* 246-249, 1990.

Hooper, P. R., and V. E. Camp, Deformation of the southeast part of the Columbia Plateau, *Geology, 9,* 323-328, 1981.

Hooper, P. R., and R. M. Conrey, A model for the tectonic setting of the Columbia River Basalt eruptions, in *Volcanism and Tectonism in the Columbia River Flood-Basalt Province, Spec. Pap. 239,* edited by S. P. Reidel and P. R. Hooper, pp. 293-306, Geological Society of America, Boulder, CO, 1989.

Hooper, P. R., and C. J. Hawkesworth, Isotopic and geochemical constraints on the origin and evolution of the Columbia River basalt, *J. Petrol., 34,* 1203-1246, 1993.

Hooper, P. R., and D. A. Swanson, The Columbia River Basalt group and associated volcanic rocks of the Blue Mountains Province, *U.S. Geol. Surv. Prof. Pap. 1437,* 63-99, 1990.

Hooper, P. R., C. R. Knowles, and N. D. Watkins, Magnetostratigraphy of the Imnaha and Grande Ronde Basalts in the southeast part of the Columbia Plateau, *Am. J. Sci, 279,* 737-745, 1979.

Hooper, P. R., W. D. Kleck, C. R. Knowles, S. P. Reidel, and R. L. Thiessen, Imnaha Basalt, Columbia River Basalt Group, *J. Petrol., 25,* 473-500, 1984.

Hooper, P. R., D. G. Bailey, and G. A. McCarley Holder, Tertiary calc-alkaline magmatism associated with lithospheric extension in the Pacific Northwest, *J. Geophys. Res., 100,* 10, 303-10, 319, 1995a.

Hooper, P. R., B. A. Gillespie, and M. E. Ross, The Eckler Mountain Basalts and associated flows, Columbia River Basalt group, *Can. J. Earth Sci., 32,* 410-423,1995b.

Hooper, P. R., C. J. Hawkesworth, K. Lees, M. Francis, J. Johnston, and B. Binger, The southern extension of the Columbia River basalts: Tectonic implications (abstract), *Eos Trans. AGU, 76,* 46, Fall Meeting Suppl., F698, 1995c.

James, D. E., The combined use of oxygen and radiogenic isotopes as indicators of crustal contamination, *Ann. Rev. Earth. Planet. Sci. 9,* 311-344, 1981.

Kim, K., S. A. Dischler, J. R. Aggson, and M. P. Hardy, The state of in situ stresses determined by hydraulic fracturing at the Hanford site, Richland, Washington, *Rockwell Hanford Operations Report RHO-BW-ST-73-P,* 1986.

Kittleman, L. R., Guide to the geology of the Owyhee region of Oregon, *Univ. Oreg. Mus. Nat. Hist., Bull., 21,* 45 pp., 1973.

Lambert, R. St J., V. E. Chamberlain, and J.G. Holland, Ferro-andesites in the Grande Ronde Basalt: their composition and significance in studies of the origin of the Columbia River Basalt group, *Can. J. Earth Sci., 32*, 424-436, 1995.

Lawrence, R. D., Strike-slip faulting terminates Basin and Range province in Oregon, *Geol. Soc. Am. Bull., 87*, 846-850, 1976.

Leeman, W. P. Radiogenic tracers applied to basalt genesis in the Snake River Plain-Yellowstone National Park region - evidence for a 2.7 B.Y. old upper mantle keel. *Geol. Soc. Am. Abstr. Prog., 7*, 1165, 1975.

Lees, K. R., Magmatic and tectonic changes through time in the Neogenic volcanic rocks of the Vale area, Oregon, North Western USA. Ph.D. thesis, 284 pp., Open University, Milton Keynes, U.K., 1994.

Lindgren, W., The gold belt of the Blue Mountains of Oregon, *U.S. Geol. Surv. 22nd Ann. Rept.*, 551-776, 1901.

Lofgren, G. Experimental studies on the dynamic crystallization of silicate melts, in *Physics of Magmatic Processes*, edited by R. B. Hargraves, pp. 487-551, Princeton University Press, Princeton, N. J., 1980.

Long, P. E., and B. J. Wood, Structures, textures and cooling histories of Columbia River basalt flows, *Geol. Soc. Am. Bull., 97*, 1144-1155, 1986.

Macdougall, J. D., Continental flood basalts and MORB: A brief discussion of similarities and differences in their petrogenesis, in *Continental Flood Basalts*, edited by J. D. Macdougall, pp. 331-341, Kluwer Academic Publishers, Dordrecht, 1988.

Mankinen, E. A., E. E. Larson, C. H. Gromme, M. Prevot, and R. S. Coe, The Steens Mountain (Oregon) geomagnetic polarity transition 3. Its regional significance. *J. Geophys. Res., 92*, 8057-8076, 1987.

Mann, G. M., Seismicity and late Cenozoic faulting in the Brownlee Dam area—Oregon-Idaho: A preliminary Report, *U.S. Geol. Surv. Open-File Report, 89-429*, 46 pp., 1989.

Mann, G. M., and C. E. Meyer, Late Cenozoic structure and correlation to seismicity along the Olympic-Wallowa Lineament, northwestern United States, *Geol. Soc. Am. Bull., 105*, 853-871, 1993.

Marsh, J. S., and H. V. Eales, The chemistry and petrogenesis of igneous rocks of the Karoo Central Area, southern Africa, *Geol. Soc. S. Afr. Spec. Publ. 13*, 27-68, 1984.

Merriam, J. C., A contribution to the geology of the John Day basin, *California Univ. Dept. Geol. Sciences Bull., 2*, 269-314, 1901.

Mohl, G. B., and R. L. Thiessen, Gravity studies of an island arc - continent suture in west-central Idaho and adjacent Washington, in *Geology of the Blue Mountains Region of Oregon, Idaho, and Washington: Petrology and Tectonic Evolution of Pre-Tertiary Rocks of the Blue Mountain Region, Prof. Pap. 1438*, edited by T. L. Vallier and H. C. Brooks, pp. 497-516, U.S. Geological Survey, Washington, D.C., 1995.

Morgan, W. J., Hotspot tracks and the opening of the Atlantic and Indian Oceans, in *The Sea, Vol. 7*, edited by C. Emiliani, pp. 443-487, Wiley-Interscience, New York, 1981.

Neim, A. R., and W. A. Neim, Oil and investigation of the Astoria basin, Clastop and northernmost Tillamook counties, northwest Oregon, Oil and Gas Investigations, 14, *Oreg. Dept. Geol. Min. Ind., Oil Gas. Invest. 14*, 1985.

Neim A. R., B. K. McKnight, and H. J. Meyer, Sedimentary,

volcanic, and tectonic framework of forearc basins and the Mist gas field, northwest Oregon, in *Geologic Field Trips in the Pacific Northwest*, edited by D. A. Swanson and R. A. Haugerud, Annual Meeting, IF 1-42, Geological Society of America, Boulder, CO, 1994.

Neim, A. R., and W. A. Neim, Oil and investigation of the Astoria basin, Clastop and northernmost Tillamook counties, northwest Oregon, Oil and Gas Investigations, 14, *Oreg. Dept. Geol. Min. Ind., Oil Gas. Invest. 14*, 1985.

Osawa, M., and G. G. Goles, Trace element abundances in Columbia River basalts, in *Proceedings, 2nd Columbia River Basalt Symposium*, edited by E. H. Gilmour and D. Stradling, pp. 55-71, Eastern Washington State College Press, Cheney, 1970.

Pearce, J. A., The role of subcontinental lithosphere in magma genesis at active continental margins. In *Continental Basalts and Mantle Xenoliths*, edited by C. J. Hawkesworth, and M. J. Norry, pp. 230-249, Shiva, Norwich, 1983.

Pfaff, V. J., and M. H. Beeson, Miocene basalt near Astoria, Oregon; Geophysical evidence for Columbia Plateau origin, *Geol. Soc. Am. Spec. Paper* in *Volcanism and Tectonism in the Columbia River Flood-Basalt Province, Spec. Pap. 239*, edited by S. P. Reidel and P. R. Hooper, pp. 143-156, Geological Society of America, Boulder, CO, 1989.

Pierce, K. L., and L. A. Morgan, The track of the Yellowstone hot spot: Volcanism, faulting and uplift, *Geol. Soc. Am. Mem. 179*, 1-53, 1992.

Prestvik, T., and G. Goles, Comments on petrogenesis and the tectonic setting of the Columbia River basalts, *Earth Planet. Sci Lett., 72*, 65-73, 1985.

Price, E. H., and A. J. Watkinson, Structural geometry and strain distribution within eastern Umtanum fold ridge, south-central Washington, in *Volcanism and Tectonism in the Columbia River Flood-Basalt Province, Spec. Pap. 239*, edited by S. P. Reidel and P. R. Hooper, pp. 265-282, Geological Society of America, Boulder, CO, 1989.

Reidel, S. P., The Saddle Mountains - The evolution of an anticline in the Yakima fold belt, *Am. J. Sci., 284*, 942-978, 1984.

Reidel, S. P., and K. R. Fecht, The Huntzinger flow; evidence of surface mixing of the Columbia River basalt and its petrogenetic implications, *Geol. Soc. Am. Bull., 98*, 664-677, 1987.

Reidel, S. P., and T. L. Tolan, Eruption and emplacement of flood basalt: An example from the large-volume Teepee Butte Member, Columbia River Basalt Group, *Geol. Soc. Am. Bull., 104*, 1650-1671, 1992.

Reidel, S. P., and T. L. Tolan, Late Cenozoic structure and correlation to seismicity along the Olympic-Wallowa Lineament, northwestern United States: Discussion and reply, *Geol. Soc. Am. Bull., 106*, 1634-1638, 1994.

Reidel, S. P., T. L. Tolan, P. R. Hooper, M. H. Beeson, K. R. Fecht, R. B. Bentley, and R. L. Anderson, The Grande Ronde Basalt, Columbia River basalt group; stratigraphic descriptions and correlations in Washington, Oregon and Idaho, in *Volcanism and Tectonism in the Columbia River Flood-Basalt Province, Spec. Pap. 239*, edited by S. P. Reidel and P. R.

Hooper, pp. 21-54, Geological Society of America, Boulder, CO, 1989.

Reidel, S. P., N. P. Campbell, K. R. Fecht, and K. A. Lindsey, Late Cenozoic structure and stratigraphy of south-central Washington, in *Regional Geology of Washington State, Bull. 80*, pp. 159-180, Washington Department of Natural Resources, Division of Geology and Earth Resources, Olympia, 1994.

Renne, P. R., and A. R. Basu, Rapid eruption of Siberian Traps flood basalts at the Permo-Triassic boundary, *Science, 253,* 176-179, 1991.

Richards, M. A., R. A. Duncan, and V. E. Courtillot, Flood basalts and hotspot tracks: Plume heads and tales, *Science, 246,* 103-107, 1989.

Russell, I. C., A geological reconnaissance in central Washington, *U.S. Geol. Surv. Bull., 108,* 108 pp., 1893.

Russell, I. C., Geology and water resources of Nez Perce County, Idaho, *U.S. Geol. Surv. Water-Supply Paper, 53,* 54 pp., 1901.

Rytuba, J. J., and E. H. McKee, Peralkaline ash-flow tuffs and calderas of the McDermitt volcanic field, southeast Oregon and north central Nevada, *J. Geophys. Res., 89,* 8616-8628, 1984.

Schmincke, H. -U. Fused tuff and pepperites in south-central Washington, *Geol. Soc. Am. Bull., 78,* 319-330, 1967a.

Schmincke, H. -U, Stratigraphy and petrography of four upper Yakima Basalt flows in south-central Washington, *Geol. Soc. Am. Bull., 78,* 1385-1422, 1967b.

Shaw, H. R., and D. A. Swanson, Eruption and flow rates of flood basalts, in *Proceedings, 2nd Columbia River Basalt Symposium*, edited by E. H. Gilmour and D. Stradling, pp. 271-299, Eastern Washington State College Press, Cheney, 1970.

Smith, A. D., Back-arc convection model for Columbia River basalt genesis, *Tectonophysics, 207,* 269-285, 1992.

Smith, G. O., Geology and water resources of a portion of Yakima County, Washington, *U.S. Geol. Surv. Water Supply-Paper, 55,* 68 pp., 1901.

Smith, G. O., Description of the Ellensberg quadrangle, Washington, *U.S. Geol. Surv. Geol. Atlas, Folio 86,* 7 pp., 1903a.

Smith, G. O., Geology and physiography of central Washington, *U.S. Geol. Surv. Prof. Pap. 19,* 1-39, 1903b.

Snavely, P. D., N. S. MacLeod, and H. C. Wagner, Miocene tholeiitic basalts of coastal Oregon and Washington and their relations to coeval basalts of the Columbia Plateau, *Geol. Soc. Am. Bull., 84,* 387-424, 1973.

Snee, L. W., K. Lund, J. F. Sutter, D. E. Balcer, and K. V. Evans, An ⁴⁰Ar/³⁹Ar chronicle of the tectonic development of the Salmon River suture zone, western Idaho, in *Geology of the Blue Mountains Region of Oregon, Idaho, and Washington: Petrology and Tectonic Evolution of Pre-Tertiary Rocks of the Blue Mountain Region, Prof. Pap. 1438,* edited by T. L. Vallier and H. C. Brooks, pp. 359-414, U.S. Geological Survey, Washington, D.C., 1995.

Sobczyk, S. M., Crustal thickness and structures of the Columbia Plateau using geophysical methods, unpubl. Ph.D. thesis, 208 pp., Washington State University, Pullman, 1994.

Swanson, D. A., and T. L. Wright, Field guide to field trip between Pasco and Pullman, Washington, emphasizing stratigraphy, vent areas and intra-canyon flows of the Yakima

Basalts, in *Proceedings, Geological Society of America, Cordilleran Section Meeting*, Pullman, 1978 (guide).

Swanson, D. A., and T. L. Wright, Guide to geologic field trip between Lewiston, Idaho, and Kimberly, Oregon, emphasizing the Columbia River Basalt group, in *Guides to Some Volcanic Terranes in Washington, Idaho, Oregon, and Northern California, Circ. 838,* edited by D. A. Johnston and J. Nolan, pp. 1-16, U.S. Geological Survey, Washington, D.C., 1981.

Swanson, D. A., and T. L. Wright, Geologic map of the Wenaha Tucannon Wilderness, Washington and Oregon, scale 1:48,000, *U.S. Geol. Surv. Misc. Field Studies Map MF-1536,* 1983.

Swanson, D. A., T. L. Wright, and R. T. Helz, Linear vent systems and estimated rates of magma production and eruption for the Yakima Basalt on the Columbia Plateau, *Am. J. Sci., 275,* 877-905, 1975.

Swanson, D. A., T. L. Wright, P. R. Hooper, and R. D. Bentley, Revisions in stratigraphic nomenclature of the Columbia River Basalt Group, *U.S. Geol. Survey Bull. 1457-G,* 59 pp., 1979.

Swanson, D. A., T. L. Wright, V. E. Camp, J. N. Gardner, R. T. Helz, S. M. Price, S. P. Reidel, and M. E. Ross, Reconnaissance geological map of the Columbia River Basalt Group, Pullman and Walla Walla quadrangles, southeast Washington and adjacent Idaho, scale 1:250,000, *U.S. Geol. Surv. Misc. Inves. Ser. Map I-1139,* 1980.

Swanson, D. A., J. L. Anderson, V. E. Camp, P. R. Hooper, W. H. Taubeneck, and T. L. Wright, Reconnaissance geological map of the Columbia River Basalt group, northern Oregon and western Idaho, scale 1:250,000, *U.S. Geol. Surv. Open-File Report 81-797,* 1981.

Swisher, C. C., J. A. Ach, and W. K. Hart, Laser fusion ⁴⁰Ar/³⁹Ar dating of the type Steens Mountain Basalt, southeastern Oregon and the age of the Steens geomagnetic polarity transition (abstract), *Eos Trans. AGU, 71,* 1296, 1990.

Taubeneck, W. H., Dikes of the Columbia River basalt in northeastern Oregon, western Idaho and southeastern Washington, in *Proceedings, 2nd Columbia River Basalt Symposium*, edited by E. H. Gilmour and D. Stradling, pp. 73-96, Eastern Washington State College Press, Cheney, 1970.

Thompson, R. N., and S. A. Gibson, Subcontinental mantle plumes, hotspots and pre-existing thinspots, *J. Geophys. Res., 148,* 973-977, 1991.

Tolan, T. L., S. P. Reidel, M. H. Beeson, J. L. Anderson, K. R. Fecht, and D. A. Swanson, Revisions to the estimates of the aerial extent and volume of the Columbia River Basalt Group, in *Volcanism and Tectonism in the Columbia River Flood-Basalt Province, Spec. Pap. 239,* edited by S. P. Reidel and P. R. Hooper, pp. 1-20, Geological Society of America, Boulder, CO, 1989.

Turner, S., C. Hawkesworth, K. Gallagher, K. Stewart, D.Peate, and M. Mantovani, Mantle plumes, flood basalts, and thermal models for melt generation beneath continents: Assessment of conductive heating model and application to the Paraná, *J. Geophys. Res., 101,* 11,503-11,518, 1996

Vallier, T. L., Petrology of the pre-Tertiary igneous rocks in the Blue mountains province of Oregon, Idaho and Washington: Implications for the geologic evolution of a complex island arc, in *Geology of the Blue Mountains Region of Oregon, Idaho, and Washington: Petrology and Tectonic Evolution of Pre-Tertiary Rocks of the Blue Mountain Region, Prof. Pap. 1438,* edited by T. L. Vallier and H. C. Brooks, pp. 125-510, U.S. Geological Survey, Washington, D.C., 1995.

Walker, G. P. L., Compound and simple lava flows and flood basalts, *Bull. Volcanol., 35,* 579-590, 1968.

Washington, H. S., Deccan traps and other plateau basalts, *Geol. Soc. Am. Bull., 33,* 765-804, 1922.

Waters, A. C., Stratigraphic and lithologic variations in the Columbia River basalt, *Am. J. Sci., 259,* 583-611, 1961.

Watkins, N. D., and A. K. Baksi, Magnetostratigraphy and oroclinal folding of the Columbia River, Steens and Owyhee basalts in Oregon, Washington and Idaho, *Am. J. Sci., 274,* 148-189, 1974.

Wells, R. E., R. W. Simpson, R. D. Bentley, M. H. Beeson, M. T. Mangan, and T. L. Wright, Correlation of Miocene flows of the Columbia River Basalt group from the central Columbia River Plateau to the coast of Oregon and Washington, in *Volcanism and Tectonism in the Columbia River Flood-Basalt Province, Spec. Pap. 239,* edited by S. P. Reidel and P. R. Hooper, pp. 113-130, Geological Society of America, Boulder, CO, 1989.

White, R. S., Magmatism during and after continental break-up, *Geol. Soc. London Spec. Publ., 68,* 1-16, 1992.

White, R. S., and D. P. McKenzie, Magmatism at rift zones: The generation of volcanic continental margins and flood basalts. *J. Geophys. Res., 94,* 7685-7729, 1989.

White, R. S., and D. P. McKenzie, Mantle plumes and flood basalts. *J. Geophys. Res., 100,* 17543-17586, 1995.

Wright, T. L., M. J. Grolier, and D. A. Swanson, Chemical variation related to the stratigraphy of the Columbia River basalt, *Geol. Soc. Am. Bull., 84,* 371-386, 1973.

Wright, T. L., M. Mangan, and D. A. Swanson, Chemical data for flows and feeder dikes of the Yakima Basalt subgroup, Columbia River Basalt Group, Washington, Oregon and Idaho, and their bearing on a petrogenetic model. *U.S. Geol. Survey Bull. 1821,* 71 pp., 1989.

Zoback, M. L., E. H. McKee, R. J. Blakely, and G. A. Thompson, The northern Nevada rift: Regional tectono-magmatic relations and middle Miocene stress direction, *Geol. Soc. Am. Bull., 106,* 371-382, 1994.

Evolution of the Red Sea Volcanic Margin, Western Yemen

Martin Menzies, Joel Baker, Gilles Chazot,[1] and Mohamed Al'Kadasi[2]

Department of Geology, Royal Holloway University of London, Egham, Surrey, United Kingdom

The temporal relationships between rifting processes associated with the breakup of the Afro-Arabian Plate and the opening of the Red Sea have been studied on the uplifted volcanic margin in western Yemen. Excellent exposure allows for the application of absolute and relative dating techniques. Magmatism: $^{40}Ar/^{39}Ar$ dating of feldspars in the volcanic rocks indicates that (a) volcanism began at 29-31 Ma, (b) a change from basic to silicic volcanism occurred at ca. 29 Ma, (c) large volume volcanism lasted until 26.5 Ma, and (d) eruption rates decreased with time. Exhumation: fission track (FT) analyses of apatites from basement metamorphic and sedimentary rocks reveal that the volcanic margin was rapidly and deeply exhumed, at the earliest, in the Oligo-Miocene. This result is consistent with a major erosional break in the volcanic stratigraphy, bracketed by $^{40}Ar/^{39}Ar$ dating as having formed between 26 and 19 Ma. Extension: field evidence indicates that extension was largely post-volcanic because the volcanic stratigraphy is devoid of major faults. In highly extended terranes $^{40}Ar/^{39}Ar$ ages of "hanging wall" volcanic rocks and apatite FT ages of "footwall" basement rocks demonstrate that extension occurred in the late Oligocene–early Miocene. Surface uplift: field evidence for surface uplift may exist in the change from marine to continental sedimentation found in the pre-volcanic sedimentary rocks (>31 Ma). Exhumation, the presumed response to surface uplift, occurred some 6 m.y. later, allowing that amount of time for the volcanic margin to be uplifted by >2–3 km. Integration of field and laboratory data reveals that the Red Sea volcanic margin evolved in response to contemporaneous surface uplift and volcanism that predated extension and exhumation by as much as 5 m.y.

---

[1]Also at the Ecole Normale Superieure, Lyon, France, and at BRGM, Orleans, France

[2]Also at the University of Sana'a, Sana'a, Yemen.

Large Igneous Provinces: Continental, Oceanic, and Planetary Flood Volcanism
Geophysical Monograph 100
Copyright 1997 by the American Geophysical Union

INTRODUCTION

In the last twenty years many different rifting models have been formulated to explain the opening of the Red Sea [e.g., *McGuire and Bohannon,* 1989; *Menzies et al.,* 1992; *Davison et al.,* 1994] (Figure 1). The young rift mountains of Yemen provide a rare opportunity to study extensive onshore exposures which encapsulate the rifting process and, ultimately, ocean basin formation. Many

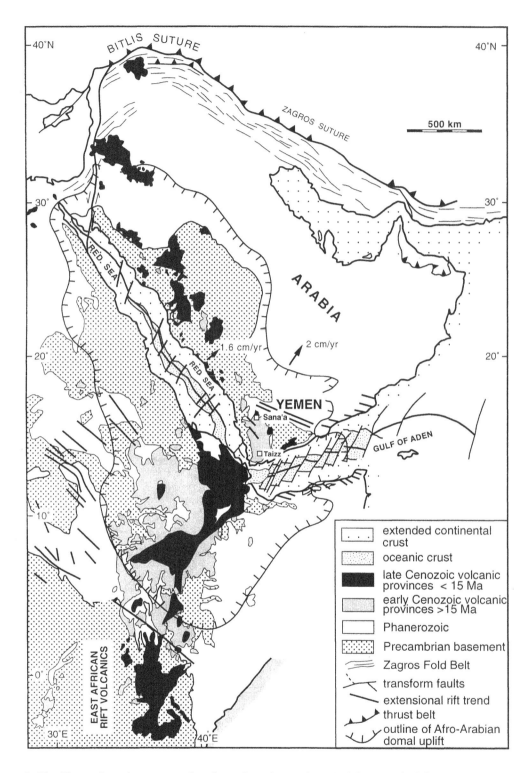

Figure 1. The Yemen large igneous province located on the southern Red Sea margin [after *Baker et al.,* 1996a]. Rates of plate movement and/or seafloor spreading are shown.

papers have focused on the flood volcanism of the Yemen Highlands [e.g., *Chiesa et al.,* 1983a,b; *Manetti et al.,* 1991; *Chazot and Bertrand,* 1993; *Baker et al.,* 1996a,b] which is believed to have contributions from deep-seated mantle plumes or shallower asthenospheric sources [e.g., *Almond,* 1986; *Camp and Roobol,* 1992]. Another aspect of the evolution of the rifted margin that has resulted in considerable controversy is the origin of the Afro-Arabian dome [*Cloos,* 1939; *Bohannon et al.,* 1989; *Dixon et al.,* 1989] and its relationship to upwelling mantle, volcanism, surface uplift, and mountain building. The Yemen margin has clearly undergone surface uplift and exhumation in an arid climate but how these relate, in time and space, to the volcanic activity and crustal extension on the margins is unknown. Whether or not rifted continental margins are plate driven (passive rift) or plume driven (active rift) is one aspect of rift margin development that continues to intrigue earth scientists.

In this paper, the methodology used to unravel the timing of surface uplift, magmatism, exhumation, and extension is summarised. On the southern Red Sea margin in the Yemen Republic, Precambrian basement is overlain by a cover of Mesozoic-Cenozoic limestones and sandstones (~1000–2000 m) and Tertiary basalts and rhyolites (~2500 m). Limited fossil evidence indicates that the pre-volcanic sandstones are Palaeocene-Eocene in age

[*Al'Subbary,* 1995]. Fieldwork provides an important basis for establishing the relative timing of rift processes, but a more precise chronology is dependent upon absolute dating of both the volcanic rocks (^{40}Ar/^{39}Ar) and the main period of crustal cooling (apatite fission track [FT] analyses).

MAGMATISM

The exposed volcanic stratigraphy is dominated, at the base, by a thick sequence of basalts overlain by an equally thick sequence of rhyolites-ignimbrites (Figure 2). The basalts thicken to the west and south and the uppermost section is dominated by more silicic compositions of rhyolitic airfall, ignimbritic pyroclastic deposits, rhyolitic lava flows, and volcaniclastic units.

Although basaltic magmatism was widespread on the southern Red Sea margin (Figure 1), it was not contemporaneous with extension, as there are no syn-volcanic extensional faults within the volcanic stratigraphy. Minor normal faults have been observed. Furthermore, there is little or no evidence for erosional breaks that would mark syn-volcanic exhumation with the exception of one near the top of the volcanic stratigraphy [*Baker et al.,* 1996a]. The presence of granitic intrusions

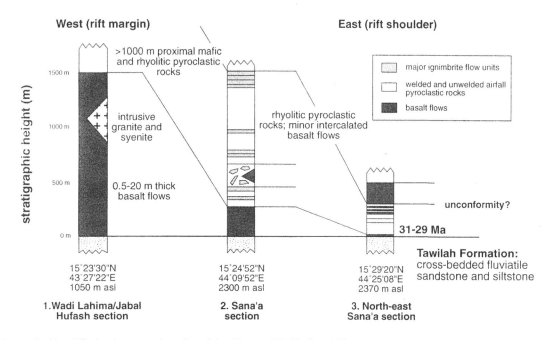

Figure 2. Simplified volcanostratigraphy of the Yemen LIP [*Baker,* 1995; *Baker et al.,* 1996b] along a west to east traverse from the Red Sea coast to Sana'a (Fig. 1). Note the lateral thinning of the LIP from the rift margin (west) to the rift shoulder (east). In addition, note the change from basaltic volcanism at the base to bimodal volcanism and eventually silicic volcanism at the top.

at the surface within the upper part of the volcanic stratigraphy (Figure 2) warrants attention because granitic intrusions normally crystallize at depths of at least 1000 m [*Blakey et al.,* 1994].

$^{40}Ar/^{39}Ar$ Dating

The $^{40}Ar/^{39}Ar$ dating of silicate phases separated from volcanic rocks defines a tightly constrained suite of late Oligocene ages [*Baker et al.,* 1994b, 1996a]. The silicic units are ideal for $^{40}Ar/^{39}Ar$ dating in that they contain potassium-rich phases (sanidine, anorthoclase, amphibole, and biotite). Whereas all the mineral samples analysed yielded plateau ages, whole-rock basaltic samples yielded no plateau ages and are therefore of limited value in Yemen geochronological studies. Mineral plateau and isochron ages, in the lowermost basaltic section, range from 29.2 to 30.9 Ma [*Baker et al.,* 1996a] and, although older basal flow ages are found in the southern part of the volcanic province, the youngest basal flow ages come from the northeast (Figure 3). A progressive decrease in age is observed upwards through the volcanic stratigraphy [*Baker et al.,* 1996a] (Figure 3), but the magnitude of this decrease differs significantly from that previously reported from K-Ar data [*Civetta et al.,* 1978]. In the northeast, the oldest rhyolites are ca. 29 Ma in age and the uppermost units have ages > 26 Ma. *Baker et al.* [1996a] evaluated whole-rock $^{40}Ar/^{39}Ar$ step-heating spectra and discussed the problems with such data. Major problems relate to post-crystallisation loss of radiogenic Ar from the groundmass of fine-grained and glassy rocks, and excess ^{40}Ar and ^{39}Ar recoil in some samples. The $^{40}Ar/^{39}Ar$ data constrain the period of volcanism to 30.9–26.5 Ma. In the southern part of the Yemen large igneous province (LIP), *Zumbo et al.* [1995] reported ages of 28.9 and 26.5 Ma for two basalts at the bottom and the top of a volcanic section, respectively. These data reiterate the conclusions of *Baker et al.* [1996a] that the uppermost volcanic units within the volcanostratigraphy have ages of 26.5 Ma. However, the lowermost flow in the section studied by *Zumbo et al.* [1995] is by no means basal as it coincides in age with the basaltic-silicic period in northern Yemen (i.e., 29–26 Ma). *Zumbo et al.* [1995] also obtained $^{40}Ar/^{39}Ar$ plateau ages on dikes cutting the LIP, and these range from 25.4 to 16.1 Ma, suggesting that LIP formation may have been active for a longer period than is indicated by the volcano-stratigraphy currently preserved in this LIP. In summary:

(a) Basaltic magmatism began throughout the area 31–29 m.y. ago and lasted for a maximum of 2 m.y. with a cumulative thickness of 1000-1500 metres,

(b) A switch from basaltic to bimodal magmatism occurred at ca. 29 Ma,

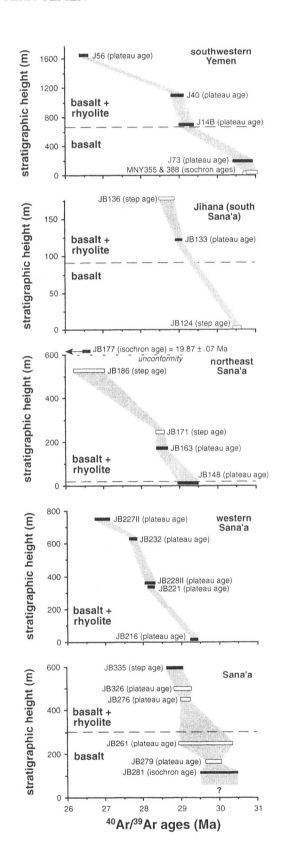

(c) Basaltic and silicic magmatism lasted from ca. 29–26 m.y. ago with a cumulative thickness of 1000 metres,

(d) The youngest exposed flows in the LIP have ages of 26.9–26.5 Ma,

(e) Plutonic and hypabyssal activity continued from ca. 25–16 Ma and one could argue that LIP formation must have accompanied such activity, albeit at a reduced eruption rate,

(f) A hiatus in volcanism occurred between 26 and 19 Ma and is evident as an erosional unconformity within the volcanic sequence.

K-Ar Dating

Volcanic rocks (e.g., basalts, basanites, and basaltic trachyandesites) at the base of the flood basalt-rhyolite sequence in western Yemen were dated using conventional K-Ar techniques. Sixty-three whole-rock analyses and one hornblende analysis gave a range in K-Ar dates of 14.4–63.6 Ma [*Al'Kadasi*, 1995], similar to that obtained from several published studies of the Yemen-Ethiopian LIP [e.g., *Civetta et al.*, 1978; *Manetti et al.*, 1991]. However, the applicability of these ages to primary petrogenetic processes must be carefully considered because of (1) the large range in radiogenic ^{40}Ar, (2) the highly variable weight loss on ignition (0–6 %), and (3) the unequivocal evidence for high-level contamination processes [*Baker et al.*, 1994a; 1996b].

The ^{40}Ar/^{39}Ar data allow the K-Ar analyses to be screened, a process that reveals that the majority of the K-Ar dates are of little value (J. Baker et al., Evaluating the usefulness of whole-rock K-Ar dating in a young large igneous province, ms. in progress). Whereas the most useful K-Ar dates are forthcoming from samples with >80% radiogenic ^{40}Ar, the screened K-Ar age range is consistent with, but lacks the resolution of, ^{40}Ar/^{39}Ar analyses from Yemen [*Baker et al.*, 1996a]. From the ^{40}Ar/^{39}Ar and K-Ar database it is apparent that many of the published K-Ar analyses for the Yemen flood basalts (and probably other volcanic fields marginal to the Red Sea) are erroneous and that on the basis of ^{40}Ar/^{39}Ar plateau ages flood volcanism lasted a mere 5 m.y. (31–26 Ma), at best half the time scale quoted in the literature on the basis of K-Ar dates.

Figure 3. The ^{40}Ar/^{39}Ar chronostratigraphy of the Yemen LIP [*Baker et al.*, 1996a]. The sections of the Yemen LIP that were studied occur along a west to east traverse from the Red Sea rift margin to Sana'a (rift shoulder). Along these sections volcanism began at 31–29 Ma, the switch from basaltic to silicic volcanism occurred at ca. 29 Ma and the youngest preserved volcanic units are 26.5 Ma in age.

EXTENSION

Detailed mapping [e.g., *Davison et al.*, 1994; *Al'Kadasi*, 1995; *Al'Subbary*, 1990, 1995; *Baker*, 1995] along traverses perpendicular to the present volcanic margins helped constrain the timing of extension. In the Tihama Plain east of Al Hudaydah (Figure 4), domino-fault blocks contain basement metamorphic rocks (Precambrian) overlain by a limestone-sandstone sequence (Jurassic-Tertiary) which in turn is overlain by basalts and rhyolites (Oligocene) containing contemporaneous granites (Oligo-Miocene?). Fieldwork reveals that extension is largely post-volcanic [*Menzies et al.*, 1992; *Davison et al.*, 1994].

Pre-LIP Sedimentsary Rocks and Hanging Wall Basalts

The ^{40}Ar/^{39}Ar dates [*Baker et al.*, 1996a] for the volcanic rocks allow evaluation of the timing of extension. Within the pre-volcanic sedimentary rocks (>31 Ma) there are no angular unconformities to indicate extensional episodes before the initiation of the LIP. Furthermore, within the volcanic stratigraphy (31–26 Ma) there is little or no fault control on the distribution of the volcanic rocks, and no angular unconformities, or faults, have been recognised within the volcanic stratigraphy, providing the most important conclusion that widespread extension must have largely postdated the Yemen LIP (i.e., <26 Ma). The hanging wall of rotated fault blocks found west of Sana'a contains late Oligocene lava flows overlying Cenozoic-Mesozoic sediments and in some cases Precambrian basement. This provides a maximum age for the onset of extension, placing the brittle extension of the pre-existent geology at <26 Ma.

Plutonic rocks

Plutonic rocks that intruded the western and southern coast of Yemen are mainly alkaline to peralkaline A-type granites of possible mantle origin [*Chazot and Bertrand*, 1995; *Blakey et al.*, 1994]. Various types of granites [*Blakey et al.*, 1994] within the basalt-rhyolite sequence were unroofed as a consequence of exhumation. More importantly, intrusive granites are an integral part of domino-fault blocks on the volcanic margin, a factor that is important as the age of the granites is considered to be different from that of the volcanic rocks used to constrain the "maximum" age of the onset of extension.

Zumbo et al. [1995] reported two ^{40}Ar/^{39}Ar plateau ages of 21.4 and 22.3 Ma for a plutonic body in the southern part of Yemen, and *Blakey et al.* [1994] reported a Rb-Sr age of ca. 25 Ma. In many cases the granites are somewhat

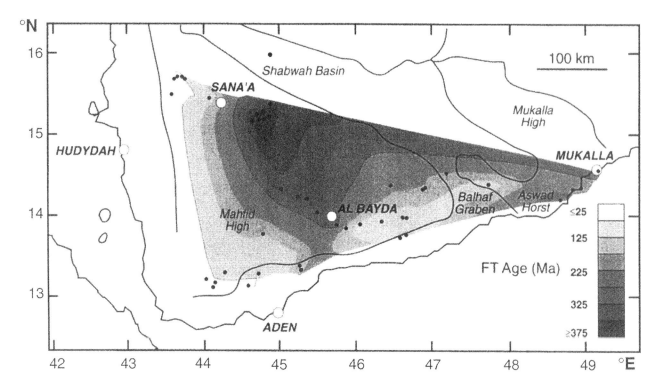

Figure 4. Contoured apatite fission track ages on the Red Sea and Gulf of Aden [*Menzies et al.,* 1997] for igneous, metamorphic and sedimentary basement rocks that underlie the Yemen LIP. The contours for 25 Ma and 175 Ma are picked out as thin black lines. Note that apatite fission track ages ≤25 Ma (with long mean track lengths, i.e., >14 μm) occur only on the Red Sea margin where the cooling event was rapid [*Menzies et al.,* 1997].

younger than the volcanic rocks [*Blakey et al.,* 1994] and their intrusion into the Yemen LIP occurred during a major magmatic event that occurred along the coast of Saudi Arabia between 21 and 24 Ma [*Sebai* 1989; *Féraud et al.,* 1991; *Sebai et al.,* 1991] and eruption of the Harrat As Sirat and Harrat Hadan in Saudi Arabia around 31–22 Ma and 28–27 Ma, respectively [*DuBray* et al., 1991; *Sebai et al.,* 1991]. These data point to a main, widespread period of extension between 24 and 21 Ma [*Féraud et al.,* 1991; *Sebai et al.,* 1991].

Footwall Basement Rocks

The zone of Oligo-Miocene exhumation defined on the basis of apatite FT analyses [*Menzies et al.,* 1997] is restricted in distribution. It is proximal to both the Red Sea volcanic margin and the Gulf of Aden non-volcanic margin, and it extends into the Ramlat As Sabatayn Graben (Figure 4), a Jurassic rift filled with >5000 m of sediment [*Davison et al.,* 1994]. Exhumation appears to be inextricably linked to areas of crustal extension and/or

reactivation of pre-existent lineaments of Gondwana or Jurassic age.

Apatite FT analyses [*Menzies et al.,* 1997] can provide additional information on the timing of extension if one accepts that, in part of the Red Sea and Gulf of Aden margins, tectonic denudation (driven by extension) may have been one of the most effective means of cooling the crust. While erosional denudation plays an important role, rapid cooling (as is evident in the track length data) would be best achieved by tectonic processes. Detailed mapping of the Red Sea [*Davison et al.,* 1994] and Gulf of Aden margins in Yemen demonstrates a coincidence between the presence of a belt of domino-fault blocks proximal to the present-day margins and the thermal domain characterised by rapid exhumation [*Yelland et al.,* 1994]. Aspects of the apatite FT analyses reflect tectonic, as well as erosional denudation, particularly where the samples were taken from the footwall exposures of fault blocks. As with the ^{40}Ar/^{39}Ar data, the apatite FT analyses constrain the timing of extension as having begun <26 m.y. ago. This is compatible with what has been deduced from other geological considerations [*Davison et al.,* 1994].

EXHUMATION

Fieldwork [*Menzies et al.*, 1992, 1994a,b; *Davison et al.*, 1994; *Baker*, 1995] reveals exhumational/erosional breaks near the top of the Yemen LIP, before the eruption of more recent volcanic rocks. Apatite FT analyses on samples collected from the Red Sea volcanic margin [*Menzies et al.*, 1997] highlight an area immediately adjacent to the Red Sea margin with apatite FT ages < 25 Ma (Figure 4) and track length distributions consistent with very rapid exhumation. Elevated Precambrian basement "highs" on the rift shoulder have apatite FT ages >>100 Ma (Figure 4) and track length distributions indicative of a complex pre-rift history. An intervening area along the Red Sea and Gulf of Aden margins, and inland along the Balhaf graben, has apatite FT ages of 25–100 Ma and track length distributions indicative of rapid exhumation (Figure 4). Modeling of the sample population with the longest track lengths (>14.0 μm) provides the most robust way of estimating the timing of the most recent cooling event [*Menzies et al.*, 1997]. While there is a significant spread in the timing of maximum temperatures prior to cooling, age and length best-fit paths for five samples that were modelled cluster at and around 31 Ma. Modeling of two older samples reveals that the timing of the latest cooling from temperatures in excess of 100°C for both samples was around 28–30 Ma. Consideration of all the apatite FT data [*Menzies et al.*, 1997] indicates that 25 Ma is an upper limit on the timing of the most recent significant cooling episode. Taken together, ^{40}Ar/^{39}Ar and apatite FT analyses indicate that crustal cooling exhumation was synchronous with LIP formation. If one assumes that surface uplift, in part, generated relief and drove exhumation, then surface uplift must have begun >30 m.y. ago, before eruption of the Yemen LIP and before major extension.

One could argue that the samples utilised for apatite FT analyses proximal to the Red Sea margin were buried under 2–3 km of volcanic rock and consequently any pre-existent record of pre-volcanic exhumation (if it existed) was effectively obliterated. The basement and sedimentary samples were annealed and subsequent crustal cooling triggered by crustal extension and tectonic denudation in the late Oligocene and Miocene "set" the fission track ages in this time period. In other words, it would have been highly unlikely that "older" FT, and thus exhumational ages, could have survived. Therefore the apatite FT analyses are biased by the very processes with which they are associated (i.e., volcanism).

While we acknowledge that this is a possibility, it fails to explain why an identical thermal history exists on the Gulf of Aden non-volcanic margin and within the Jurassic Ramlat As Sabatayn Graben (Figure 4) [*Menzies et al.*, 1997]. We prefer to interpret the apatite FT analyses to mean that regions proximal to the Gulf of Aden and Red Sea margins were rapidly exhumed in Oligocene and Miocene times, respectively, and that the apatite FT analyses record that history. The youngest apatite FT ages of 16–18 Ma imply that this cooling episode spanned several million years. Furthermore, the lack of interflow sediments indicates that exhumation must have been minimal during the main episode of magmatism (31–26 Ma) as surface uplift proceeded. However, as magmatism waned and extension became widespread exhumation became more pronounced. This is supported by the presence of minor sedimentary horizons toward the top of the volcanostratigraphy and an erosional unconformity at the top of the volcanic stratigraphy northeast of Sana'a [*Baker et al.*, 1996b].

MISSING VOLCANOSTRATIGRAPHY?

Although ~2500 m of volcanic rocks exist on the uplifted rift shoulders in western Yemen (Figure 2), this may be only part of the original stratigraphy [*Menzies et al.*, 1992]. Many aspects of the field and laboratory data lead us to this conclusion. Volatile-rich, amphibole-bearing granites intruded into the Oligocene volcanic rocks are now exposed at the surface, requiring, almost certainly, removal of 1000–1500 m of cover rocks by denudational processes. ^{40}Ar/^{39}Ar and Rb-Sr dating of these plutonic rocks points to intrusion between 25 and 21 Ma, in marked contrast to the age of the volcanic stratigraphy (31–26 Ma). In addition, dike swarms cutting the LIP indicate that magma transport by crack propagation though the LIP continued for 10 m.y. from 26 to 16 Ma [*Zumbo et al.*, 1995], again in contrast to the age of the preserved volcanic stratigraphy. Although unconformities are absent throughout most of the middle and lower volcanic section, there is a conspicuous unconformity in the uppermost volcanic stratigraphy. ^{40}Ar/^{39}Ar chronostratigraphy [*Baker et al.*, 1996a] of the flows above and below this unconformity indicates a time gap of 7 m.y. between 26 and 19 Ma. Ironically, this time gap corresponds with the main period of injection of plutonic rocks and dikes along the Red Sea margin indicating that the exhumation that removed part of the volcanostratigraphy may have been exacerbated by an important tectonic change around this time.

Apatite FT analyses indicated that the main period of exhumation was <30 Ma (Figure 4) and that exhumation became more active and widespread at <26 Ma. This happens to be the period of extension that relates to the opening of the proto-Gulf of Aden [*Courtillot et al.,* 1987] and formation of the earliest oceanic crust at ca. 20 Ma [*Sahota et al.,* 1995]. Such a major tectonic occurrence may relate to plate movement, so that the Yemen LIP moved away from the plume (i.e., Afar plume) but magma production from shallow magma chambers continued for 5 m.y. or more. Alternatively, rifting in the Gulf of Aden may have channeled plume-derived material eastward away from the Yemen LIP. However, once the eruption rates decreased [*Baker et al.,* 1996a] and the constructional aspects of LIP formation abated, extension led to collapse of the margin and contemporaneous exhumation unroofed the margin. One could speculate that about 1500 m of the volcanic section is missing, and was eroded during a period of exhumation that was triggered by the opening of the Gulf of Aden. Partial support for this idea is forthcoming from sediment budget analysis (I. Davison et al., pers. comm.) to explain the sedimentary rocks that underlie the Tihama Plain.

CRUST-MANTLE PROCESSES

Geochemistry can be used to understand crust-mantle processes and the provenance of individual batches of magma that constitute the Yemen LIP. We are uniquely placed in the southern Red Sea to assign a provenance to the LIP because of information available from several studies:

(a) The asthenosphere or depleted upper mantle from studies of MORB erupted along the Gulf of Aden ridge;

(b) A mantle plume from studies of continental volcanic rocks erupted in an area that, on the basis of heat flow and seismic tomography, may be a site of a deep mantle structure (i.e., the Afar plume);

(c) The lithospheric mantle from the study of ultramafic xenoliths derived from the shallow Arabian mantle; and

(d) The continental crust from studies of igneous and metamorphic rocks exposed at the surface and entrained by volcanic rocks erupted through crust of Archaean, Proterozoic, and Phanerozoic age.

It is widely accepted that Cenozoic volcanism associated with LIP formation in Ethiopia and Yemen [*Chiesa et al.,* 1989] was related to the impingement of a large mantle plume head beneath the Afro-Arabian lithosphere [e.g., *White et al.,* 1987; *White and McKenzie,* 1989]. Volcanism in Djibouti [*Vidal et al.,* 1991; *Deniel et al.,* 1994] is characterised by a "HIMU" isotopic and elemental signature that may be that of the Afar mantle plume. Moreover, geochemical studies of basalts erupted along the Red Sea [*Eissen et al.,* 1989; *Volker et al.,* 1993] and the Gulf of Aden [*Schilling et al.,* 1992] spreading axes have helped to define the isotopic composition of the local

asthenosphere. From these observations, the chemical composition of the magmas is related to mixing of different mantle sources. However, a very important non-oceanic isotopic component also exists, which is tentatively related to melting of the lithospheric mantle [*Chazot,* 1993; *Chazot and Bertrand,* 1993, 1995] or to contamination of the magmas by the continental crust [*Baker et al.,* 1996b]. *Baker et al.* [1996b] demonstrated that, in most cases, throughout 2500 m of volcanic stratigraphy, the volcanic rocks are contaminated with crust (Figure 5a,b). In some cases this includes highly magnesian magmas which contain zoned clinopyroxene phenocrysts that record the contamination process [*Baker et al.,* 1994a]. The isotopic data (Figure 5a,b) on the volcanic rocks from the Yemen LIP define a range in Sr and Nd isotope ratios from compositions similar to oceanic rocks (low $^{87}Sr/^{86}Sr$ and high $^{143}Nd/^{144}Nd$ ratios) to compositions similar to upper and lower crustal rocks (high $^{87}Sr/^{86}Sr$ and low $^{143}Nd/^{144}Nd$ ratios). Correlations between $^{87}Sr/^{86}Sr$ ratios and indices of fractionation (e.g., SiO_2 and Fe/Fe+Mg ratio) support a model of crustal contamination during fractionation in crustal magma chambers [*Baker et al.,* 1994, 1996b] (Figure 5c).

In the case of the Yemen LIP, elemental and isotopic data [*Baker et al.,* 1994, 1996a,b; *Chazot and Bertrand* 1993, 1995] define a temporal change in magma source/storage that, with time, changes from initial eruption of deep mantle-derived basic melts that are variably contaminated with lower crust, to eruption of silicic melts contaminated with upper crust through storage in shallow crustal reservoirs (Figure 5). This change is consistent with the involvement of a mantle plume (equiv. Afar) and asthenosphere (equiv. Gulf of Aden) early in the history of the LIP (31–29 Ma), ponding and underplating of basic magmas at the Moho and formation of shallow magma chambers after 2 m.y. of LIP formation, and caldera-forming eruptions late in the evolution of the LIP (<< 29 Ma).

SURFACE UPLIFT

With the use of $^{40}Ar/^{39}Ar$ and apatite FT analyses we can quantify the onset and termination of magmatism, the period of peak exhumation on the rifted margin, and the relative timing of extension. $^{40}Ar/^{39}Ar$ data (Figure 3) and apatite FT analyses (Figure 4) reveal a brief period of Oligocene magmatism and a protracted period of Oligo-Miocene exhumation/extension. While we are left with little or no information about the timing of surface uplift, onset of surface uplift in the Oligocene would be expected if it contributed to Oligo-Miocene exhumation.

Evidence for surface uplift, or base-level changes, may be quite subtle, especially if it was a transient feature immediately prior to the onset of volcanism. Much of this evidence is to be found in the pre-LIP sediments. The Yemen LIP is underlain by fluvial and marine sediments (Figure 6) overlain by thick paleosols [*Al'Kadasi*, 1995; *Al'Subbary*, 1995; *Baker et al.*, 1996a,b]. Palaeo-environmental interpretations indicate a nearshore depositional environment of very low relief (Figure 7), in contrast to the present topography of the rift-margin

mountains reaching 3600 m high. Vital information about surface uplift may be found in (1) shallow marine mudstones within the volcano-sedimentary stratigraphy, (2) palaeocurrent directions within the pre-volcanic sandstones, (3) any temporal palaeoenvironmental changes, and (4) palaeoaltitude data locked in plant remains within the sedimentary and volcanic stratigraphies.

With regard to the latter, *England and Molnar* [1990] stressed that one of the least controversial measures of surface uplift may be contained within the palaeoaltitude information that can be inferred from seasonality measurements in plant species existing over a wide range of altitudes. Although in western Yemen plant remains occur within the pre-volcanic sediments and the volcanic stratigraphy, our present knowledge of the extent of plant speciation indicates that their use is not conducive to the determination of palaeoaltitudes.

Shallow marine mudstones ?

In addition to marine sediments occurring immediately beneath the LIP, fossiliferous mudstones exist over a range of altitudes at the sediment-volcanic rock contact (Figure 6) and within the volcanic stratigraphy. They are useful indicators of surface uplift, provided they are marine in origin. The fossils in many of these sediments do not allow one to distinguish between lagoonal (near sea level) or lacustrine (any altitude) environments. However, a precise indication of the provenance of these sediments comes from a comparison of the $^{87}Sr/^{86}Sr$ ratio of the sediments with that of contemporaneous sea water. Fossils in sediments associated with the Yemen LIP have a range of $^{87}Sr/^{86}Sr$ isotopes different from that of Tertiary sea water, but similar to the Yemen volcanic rocks [*Baker*, 1995]. These fossils/sediments are therefore more likely to be freshwater in origin, derived from surface runoff. Their origin in a freshwater lacustrine environment is also

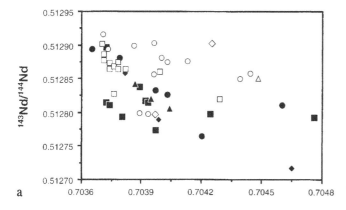

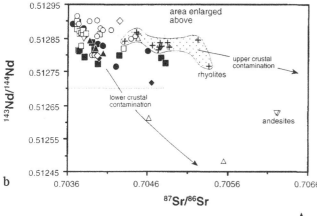

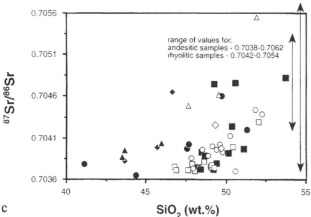

Figure 5. (a) Isotopic variation within the Yemen LIP [*Baker et al.*, 1996b] converges on a "mantle" composition with low $^{87}Sr/^{86}Sr$ ratio and high $^{143}Nd/^{143}Nd$ ratios [*Baker*, 1995; *Baker et al.*, 1996b] similar to that observed in oceanic rocks. (b) The spread in $^{87}Sr/^{86}Sr$ and $^{143}Nd/^{143}Nd$ isotopic ratios in the volcanic rocks of the Yemen LIP is believed to result from contamination with upper and lower crustal rocks. (c) Correlations between isotope ratios and indices of igneous fractionation (e.g., SiO_2) indicate that assimilation of crustal rocks occurred at the same time as fractionation in magma chambers at shallow levels in the crust. The rhyolites have a higher input of upper crustal material, perhaps due to residence and formation in crustal magma chambers. The least fractionated volcanic rock (lowest SiO_2) is the least contaminated.

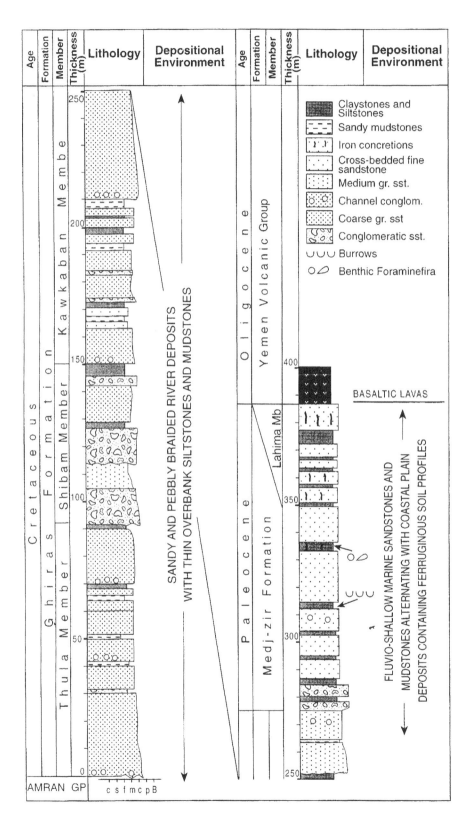

Figure 6. Nature of the Cretaceous-Tertiary sediments that immediately underlie the Yemen LIP [*Al'Kadasi*, 1995]. The development of iron concretions and thick paleosols at the contact and the overall change from marine-fluvial-subaerial (i.e., paleosols and lavas) palaeoenvironments [*Al'Subbar*, 1995] have implications for the timing of surface uplift. Essentially, this palaeoenvironmental change indicates a base-level change on the order of tens of metres that could well be the beginning of surface uplift.

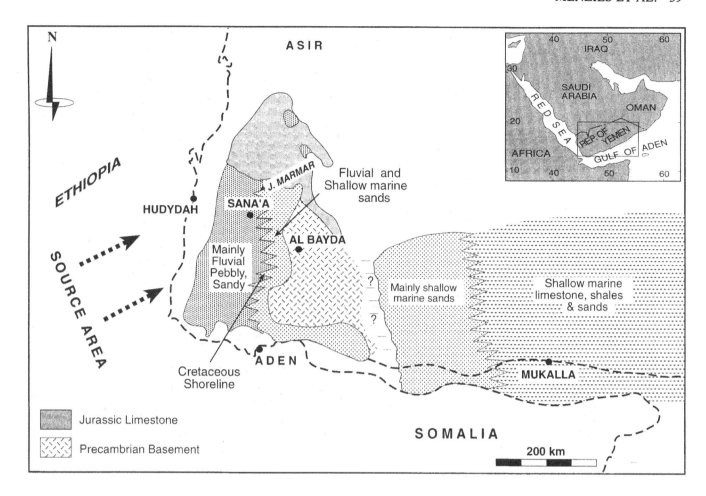

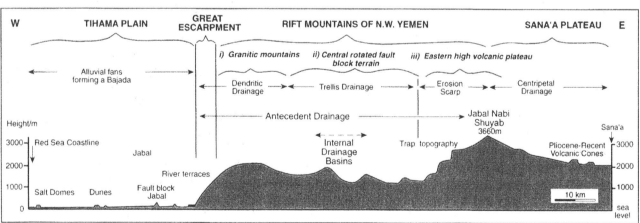

Figure 7. Palaeoenvironmental interpretations of the Tawilah sandstones (which underlie the Yemen LIP) [*Al'Subbary*, 1990; 1995] indicate a west to east paleoslope and a Cretaceous shoreline (top figure). The overall topography is a gentle west to east dipping palaeoslope which should be contrasted with the post-LIP topography (bottom figure). The latter reaches a maximum height of 3660 m with considerable relief westward to the Tihama Plain. Surface uplift of several thousand metres has occurred since deposition of the sediments but before extension and collapse of the margin in the late Oligocene to early Miocene [*Davison et al.*, 1994; *Menzies et al.*, 1994a,b].

consistent with the subaerial aspect of the Yemen LIP and their very localised occurrence as lenticular bodies.

Sandstones

Although palaeocurrent directions have been used to define domal surface uplift effectively in other LIPs [e.g., *Cox*, 1989], it should be noted that erosional processes on a broad low-relief dome (~1–1.5 km maximum elevation over areas of ~2 × 10⁶ km² [*White and McKenzie*, 1989]) would not be highly energetic. Therefore the erosional response to doming may be very subtle and largely lost, particularly in an extended volcanic margin that is subsequently broken up into rotated, domino-fault blocks. A detailed study of the pre-volcanic Cretaceous-Tertiary sandstones [*Al'Subbary* 1990, 1995] demonstrated a southwest provenance for the sandstones in present-day Sudan and/or Ethiopia, as well as the presence of a west-east palaeoslope (Figure 7) which is based on a continental aspect to the sediments in western Yemen and a marine aspect to the sediments in eastern Yemen. The palaeocurrent data [*Al'Subbary*, 1995] do not reveal a pattern consistent with domal uplift. However, comparison of the pre-volcanic topography with the present-day topography emphasises (Figure 7) the topographic change and the need for surface uplift.

Paleosols

The nature of the sedimentary rocks underlying the Oligocene volcanic rocks is pivotal to establishing subtle changes in the palaeoenvironment. Detailed work on these sediments by *Al'Kadasi* [1995] and *Al'Subbary* [1995] indicates (1) widespread development of paleosols or ferricreted clastic sediments immediately beneath the basal volcanic units (Figure 6), (2) a considerable variation in the thickness of the paleosols from 5–70 m, and (3) that ferruginous paleosol development is more marked toward the top of the pre-LIP fluvial to shallow marine succession [*Al'Subbary*, 1990, 1995].

Summerfield [1991] stated that low local relief was an essential factor in duricrust (paleosol/ferricrete) formation and thus a lack of intense denudation and sediment starvation were vital prerequisites. We concur with *Summerfield* [1991] and argue that the widespread occurrence of paleosols indicates that sediment starvation affected the whole area of study after a period (Cretaceous-Tertiary) of sandstone deposition [*Al'Subbary*, 1990, 1995]. Moreover, sediment starvation continued for the next 6 m.y. when about 3000 m of volcanic rocks were erupted with few intervening sedimentary units.

One of the major problems is to establish how much time is locked up in the formation of the pre-LIP sediments and the paleosols. We have precise age information on the overlying volcanic rocks (onset at 29–31 Ma) but the age information from the sediments in which the paleosols formed is less precise. The Medj-Zir Formation is believed to be Paleocene in age [*Al'Subbary*, 1995] and, if correct, this could mean that 30 m.y. may have separated formation of the uppermost pre-volcanic sediments and the first basaltic eruptions.

Palaeoenvironmental Change

An overall palaeoenvironmental change is evident in the pre-LIP (Palaeocene-Eocene) sediments and the Oligocene volcanic rocks. A temporal change can be summarised as follows, although steps (a) through (d) have no accurate age control: (a) Cretaceous-Tertiary - deposition of fluvial sandstones on a peneplain (low-relief continent) (Figure 6); (b) Cretaceous-Tertiary - development of marine sediments near the top of the fluvial sandstone sequence (shallow marine, continental shelf); (c) Paleocene - deposition of fluvial sandstones and development of paleosols (low relief continent) (Figure 6); (d) Paleocene - localised ponding of lacustrine deposits (low-relief continent); (e) Oligocene - eruption of thousands of metres of subaerial volcanic rocks (continent) (Figures 2 and 3).

Given our concerns about the time gap in the pre-LIP sediments two interpretations are possible.

(1) The lack of any disturbance between the marine, fluvial, paleosol, and lacustrine facies strongly indicates that changes in base level were slight and on the order of tens of metres. In fact, the development of thick paleosols reflects tectonic stability, sediment starvation and passage of time. All of these could be used to argue against significant surface uplift. The base level change observed at the top of the pre-LIP sediments could have happened any time between ca. 60 Ma and 31 Ma. There could be a major time break between the sediments and the LIP, on the order of tens of millions of years, in which case the change in palaeoenvironment is not relevant to the volcanic margin history involving plume impingement.

(2) Alternatively, these facies changes could be interpreted as indicative of a temporal change from an aqueous (i.e., shallow marine/continental margin) to a sub-aerial environment (i.e., emergent continent). On a developing volcanic margin this could represent the geological expression of the early stages of broad surface uplift. Initially this would have been on the order of tens of metres but it eventually accounted for >2000 m of surface uplift.

We propose that initiation of surface uplift occurred prior to (>31 Ma) the onset of LIP formation and must have continued throughout the period of peak LIP magmatism until 26 Ma. The protracted period of exhumation was the erosional response to surface uplift exacerbated by tectonic denudation, once extension led to collapse of the margin some 26 m.y. ago. This is consistent with the apatite FT analyses on the basement rocks.

SUMMARY

Application of absolute dating techniques (i.e., $^{40}Ar/^{39}Ar$ and apatite FT analyses) to the Yemen LIP indicates

(a) initiation of magmatism at 31 Ma,

(b) rapid exhumation of the LIP in the Oligo-Miocene,

(c) extension in the Miocene (<26 Ma),

(d) contemporaneous extension and plutonic/hypabyssal activity (<26Ma).

Unfortunately, no unequivocal evidence can be found for the effects of surface uplift in the sedimentary or the volcanic record other than a very general palaeo-environmental change. While erosional denudation requires surface uplift to enhance topographic relief, in many instances tectonic denudation driven by extension can more effectively cool the crust. Extension and exhumation may relate to widespread extension (<26 Ma) due to the opening of the Gulf of Aden. Exhumation was minimal for 5 m.y. during eruption of the Yemen LIP (31–26 Ma), but accelerated around 26 Ma, the time of widespread extension and plutonic activity.

We conclude from an integration of field observations and absolute dating that continental rifting and LIP formation in the southern Red Sea occurred in response to surface uplift and flood volcanism with minimal extension, followed by a protracted period of exhumation and extension. This model requires that previous models based on preliminary apatite FT analyses and K-Ar data be revised [*Menzies et al.*, 1992, *Davison et al.*, 1994] and is consistent with the ideas of *Richards et al.* [1989].

Acknowledgments. Geological research in Yemen would not have been possible without the enthusiasm of Jim McGrath (British Council) and Dr. Al'Kirby (University of Sana'a). We acknowledge financial support from the British Council (Al'Kadasi & Baker), the European Union (Chazot), CNRS-INSU (Chazot), the NERC (Menzies and Hurford), the Department of Geology, Royal Holloway University of London (Baker & Al'Kadasi) and the Arab League (Al'Kadasi). British Petroleum and the Royal Society are thanked for supporting a major expedition to Yemen in 1991. This work, undertaken over the last five years, would have been impossible without the logistical support provided by the University of Sana'a. Thanks finally to Dr. Abdulkarim Al'Subbary (University of Sana'a) for continual assistance with field research in Yemen. Victor Camp, John Mahoney, and John Pallister are thanked for their constructive comments which helped improve presentation of our ideas. Gary Nichols' input regarding fluvial sediments is appreciated.

REFERENCES

Al'Kadasi, M., Temporal and spatial evolution of the basal flows of the Yemen Volcanic Group. Unpublished doctoral thesis, Royal Holloway University of London, 285 pp., 1995.

Almond, D. C., Geological evolution of the Afro-Arabian dome, *Tectonophysics, 131*, 301-332, 1986.

Al Subbary, A. A., Stratigraphic and sedimentological studies of the Tawilah Group, Al-Ghiras area, northeast Sana'a, Yemen Arab Republic, Unpublished MSc thesis, University of Sana'a, 184 pp., 1990.

Al'Subbary, A. A., The sedimentology and stratigraphy of the Cretaceous-early Tertiary Tawilah group, western Yemen. Unpublished doctoral thesis, Royal Holloway University of London, 153 pp., 1995.

Baker, J., Stratigraphy, $^{40}Ar/^{39}Ar$ geochronology and geochemistry of Cenozoic volcanism in Yemen. Unpublished doctoral thesis, Royal Holloway University of London, 386 pp., 1995.

Baker, J., C. Macpherson, D. Mattey, M. Menzies, and M. Thirlwall, Laser fluorination oxygen isotope analysis of phenocryst phases from Oligo-Miocene flood basalts and Quaternary intraplate basalts in Yemen. *U. S. Geol. Surv. Circ. 107,* 18, 1994a.

Baker, J., M. Menzies, L. Snee, and M. Thirlwall, Stratigraphy, $^{40}Ar/^{39}Ar$ geochronology and geochemistry of flood volcanism in Yemen. *Mineral. Mag., 58A,* 42-43, 1994b.

Baker, J., L. Snee, and M. A. Menzies, A brief Oligocene period of flood volcanism in Yemen; implications for the duration and rate of continental flood volcanism at the Afro-Arabian triple junction. *Earth Planet. Sci. Lett., 138*, 39-55, 1996a.

Baker, J. A., M. F. Thirlwall, and M. A. Menzies, Sr-Nd-Pb and trace element evidence for crustal contamination of a mantle plume: Oligocene flood volcanism in western Yemen, *Geochim. Cosmochim. Acta, 60*, 2559-2581, 1996b.

Blakey, S., M. A. Menzies, and M. F. Thirlwall, Geochemistry of Tertiary within-plate (A-type) granitoids at a passive margin, southern Red Sea (abstract), International Association of Volcanology and Chemistry of the Earth's Interior, Ankara, 1994.

Bohannon, R. G., C. W. Naeser, D. L. Schmidt, and R. A. Zimmerman, The timing of uplift, volcanism and rifting peripheral to the Red Sea: A case for passive rifting, *J. Geophys. Res., 94*, 1683-1701, 1989.

Camp, V. E., and M. J. Roobol, Upwelling asthenosphere beneath western Arabia and its regional implications, *J. Geophys. Res., 97*, 15,255-15,271, 1992.

Chazot, G., Evolution geochimique du magmatisme Cenozoique au Yemen: interactions entre le rift Mer Rouge - Aden et le point chaud Afar, Unpublished doctoral thesis, Univ. Claude Bernard, Lyon I., 159 pp., 1993.

Chazot, G., and H. Bertrand, Mantle sources and magma-continental crust interactions during early Red Sea-Aden rifting in southern Yemen: elemental and Sr, Nd, Pb isotopic evidence, *J. Geophys. Res., 98*, 1819-1835, 1993.

Chazot, G., and H. Bertrand, Genesis of silicic magmas during Tertiary continental rifting in Yemen, *Lithos, 36*, 69-84, 1995.

Chiesa, S., L. La Volpe, L. Lirer, and G. Orsi, Geological and structural outline of Yemen plateau: Yemen Arab Republic, *N. Jb. Geol. Paleont. Mh., 11*, 641-656, 1983a.

Chiesa, S., L. La Volpe, L. Lirer, and G. Orsi, Geology of the Dhamar-Rada' volcanic field, Yemen Arab Republic, *N. Jb. Geol. Paleont. Mh., 8*, 481-494, 1983b.

Chiesa, S., L. Civetta, M. De Fino, L. La Volpe, and G. Orsi, The Yemen Trap Series: genesis and evolution of a continental flood basalt province, *J. Volcanol. Geotherm. Res., 36*, 337-350, 1989.

Civetta, L., L. La Volpe, and L. Lirer, K-Ar ages of the Yemen plateau, *J. Volcanol. Geotherm. Res., 4*, 307-314, 1978.

Cloos, H., Hebung-Spaltung-Vulkanismus, *Geol. Rundsch., 30*, 405-527, 1939.

Courtillot, V., R. Armijo, and P. Tapponnier, Kinematics of the Sinai triple junction and a two-phase model of Arabia-Africa rifting, in *Continental Extensional Tectonics, Spec. Publ., 28*, edited by M. P. Coward, J. F. Dewey, and P. L. Hancock, pp. 559-573, The Geological Society, London, 1987.

Cox, K. G., The role of mantle plumes in the development of continental drainage patterns, *Nature, 342*, 873-877, 1989.

Davison, I., M. Al-Kadasi, S. Al-Khirbash, A. Al-Subbary, J. Baker, S. Blakey, D. Bosence, C. Dart, L. Owen, M. Menzies, K. McClay, G. Nichols, and A. Yelland, Geological Evolution of the southern Red Sea rift margin - Republic of Yemen. *Geol. Soc. Am. Bull., 106*, 1474-1493, 1994.

Deniel, C., P. Vidal, C. Coulon, P. J. Vellutini, and P. Piguet, Temporal evolution of mantle sources during continental rifting: the volcanism of Djibouti (Afar), *J. Geophys. Res., 99*, 2853-2869, 1994.

Dixon, T. H., E. R. Ivins, and B. J. Franklin, Topographic and volcanic asymmetry around the Red Sea: constraints on rift models. *Tectonics, 8*, 1193-1216, 1989.

DuBray, E. A., D. B. Stoeser, and E. H. McKee, Age and petrology of the Tertiary As Sarat volcanic field, southwestern Saudi Arabia, *Tectonophysics, 198*, 155-180, 1991.

Eissen, J. P., T. Juteau, J. L. Joron , B. Dupre, E. Humler, and A. Al'Mukhamedov, Petrology and geochemistry of basalts from Red Sea axial rift at 18° North, *J. Petrol., 30*, 791-839, 1989.

England, P., and P. Molnar, Surface uplift, uplift of rocks, and exhumation of rocks, *Geology, 18*, 1173-1177, 1990.

Féraud, G., V. Zumbo, A. Sebai, and H. Bertrand, [40]Ar/[39]Ar age and duration of tholeiitic magmatism related to the early opening of the Red Sea rift, *Geophys. Res. Lett., 18*, 195-198, 1991.

Manetti, P., G. Capaldi, S. Chiesa, L. Civetta, S. Conticelli, M. Gasparon, L. La Volpe, and G. Orsi, Magmatism of the eastern Red Sea margin in the northern part of Yemen from Oligocene to present, *Tectonophysics, 198*, 181-202, 1991.

McGuire, A. V., and R. G. Bohannon, Timing of mantle upwelling: evidence for a passive origin for the Red Sea rift, *J. Geophys. Res., 94*, 1677-1682, 1989.

Menzies, M. A., J. Baker, D. Bosence, C. Dart, I. Davison, A. Hurford, M. Al'Kadasi, K. McClay, G. Nichols, A. Al'Subbary, and A. Yelland, The timing of magmatism, uplift and crustal extension: preliminary observations from Yemen, in *Magmatism and the Causes of Continental Break-up, Spec. Publ. 68*, edited by B. C. Storey, T. Alabaster, and R. J. Pankhurst, pp. 293-304, The Geological Society, London, 1992.

Menzies, M. A., A. Yelland, J. Baker, S. Blakey, G. Chazot, M. Al'Kadasi, and C. Rundle, Evolution of the Red Sea volcanic margin - a multi-isotopic approach, *U. S. Geol. Surv. Circ., 1107*, 216, 1994a.

Menzies, M.A., M. Al-Kadasi, S. Al-Khirbash, A. Al-Subbary, J. Baker, S. Blakey, D. Bosence, I. Davison, C. Dart, L. Owen, K. McClay, G. Nichols, A. Yelland, and F. Watchorn, Geology of the Republic of Yemen, in *The Geology and Mineral Resources of Yemen*, edited by D. A. McCombe, G. L. Fernette, and A. J. Alawi, Tech. Rep., pp. 21-48, Ministry of Oil and Mineral Resources, Geological Survey and Minerals Exploration Board, Yemen Mineral Sector Project, 1994b.

Menzies, M. A., K. Gallagher, A. Yelland, and A. J. Hurford, Volcanic and non-volcanic rifted margins of the Red Sea and Gulf of Aden: Crustal cooling and margin evolution in Yemen, *Geochim. Cosmochim. Acta*, in press, 1997.

Richards, M. A., R. A. Duncan, and V. E. Courtillot, Flood basalts and hot-spot tracks - plume heads and tails, *Science*, 246, 105-107, 1989.

Sahota, G., P. Stules, and K. Gerdes, Evolution of the Gulf of Aden and implications for the development of the Red Sea. *"Rift Sedimentation and Tectonics in the Red Sea-Gulf of Aden Region", Sana'a University, Yemen, 56*, 1995.

Schilling, J. G., R. H. Kingsley, B. B. Hanan, and B. L. McCully, Nd-Sr-Pb isotopic variations along the Gulf of Aden: evidence for Afar mantle plume-continental lithosphere interaction, *J. Geophys. Res., 97*, 10,927-10,966, 1992.

Sebai, A., V. Zumbo, G. Feraud, H. Bertrand, A. G. Hussain, G. Giannerini, and R. Campredon, [40]Ar/[39]Ar dating of alkaline and tholeiitic magmatism of Saudi Arabia related to the early Red Sea rifting, *Earth Planet. Sci. Lett., 104*, 473-487, 1991.

Sebai, A., Datation [40]Ar/[39]Ar du magmatisme lie aux stades precoces de l'ouverture des rifts continentaux: exemples de l'Atlantique Central et de la Mer Rouge, Unpublished doctoral thesis, Univ. Nice-Sophia Antipolis, 1989.

Summerfield, M. A., Sub-aerial denudation of passive margins: regional elevation versus local relief models, *Earth Planet. Sci. Lett., 102*, 460-469, 1991.

Vidal, P., C. Deniel, P. J. Vellutini, P. Piguet, C. Coulon, J. Vincent, and J. Audin, Changes of mantle sources in the course of a rift evolution: the Afar case, *Geophys. Res. Lett., 18*, 1913-1916, 1991.

Volker, F., M. T. McCulloch, and R. Altherr, Submarine basalts from the Red Sea: new Pb, Sr and Nd isotopic data, *Geophys. Res. Lett., 20*, 927-930, 1993.

White, R., and D. McKenzie, Magmatism at rift zones: the

generation of volcanic continental margins and flood basalts, *J. Geophys. Res.*, *94*, 7685-7729, 1989.

White, R. S., G. D. Spence, S. R. Fowler, D. P. McKenzie, G. K. Westbrook, and A. N. Bowen, Magmatism at rifted continental margins, *Nature*, *330*, 439-444, 1987.

Yelland, A., M. A. Menzies, and A. Hurford, Exhumation of the Red Sea and Gulf of Aden rift-flanks: fission track contrasts within the volcanic and non-volcanic margins of Yemen, *U. S. Geol. Surv. Circular, 1107*, 360, 1994.

Zumbo, V., G. Feraud, H. Bertrand, and G. Chazot, $^{40}Ar/^{39}Ar$ chronology of the tertiary magmatic activity in Southern Yemen during the early Red Sea - Aden rifting, *J. Volcanol. Geotherm. Res.*, *65*, 265-279, 1995.

Mohamed Al'Kadasi, Joel Baker, Gilles Chazot, and Martin Menzies, Department of Geology, Royal Holloway University of London, Egham, Surrey, TW20 OEX, England

The North Atlantic Igneous Province

A. D. Saunders

Department of Geology, University of Leicester, Leicester, United Kingdom

J. G. Fitton

Department of Geology and Geophysics, University of Edinburgh, West Mains Road, Edinburgh, United Kingdom

A. C. Kerr, M. J. Norry, and R. W. Kent

Department of Geology, University of Leicester, Leicester, United Kingdom

The North Atlantic Igneous Province extends from eastern Canada to the British Isles, a pre-drift distance of almost 2000 km. The igneous rocks are predominantly basaltic, but differentiates and anatectic melts are also represented. Two major phases of igneous activity can be discerned. Phase 1 began about 62 m.y. ago with continent-based magmatism in Baffin Island, W and SE Greenland, the British Isles, and possibly central E Greenland (the Lower Basalts around Kangerlussuaq). Phase 2 began about 56 m.y. ago and is represented by seaward-dipping reflector sequences (SDRS) along the continental margins, the Main Series basalts in central E Greenland, the Greenland-Faeroes Ridge, and Iceland. Contamination by continental crust was prevalent during Phase 1 but also occurred during Phase 2, especially during the formation of the early SDRS. Although it is unnecessary to involve the continental lithosphere mantle in the formation of Phase 1 or Phase 2 magmas, it is not possible to completely exclude it. We argue that the Iceland plume played a pivotal role in the formation of the North Atlantic Igneous Province because (1) the simultaneous and widespread initiation of activity requires a major thermal event in the mantle; (2) some of the magmas associated with Phase 1 were highly magnesian, indicating that the liquids and, by implication, the mantle source regions were unusually hot; (3) the SDRS were emplaced subaerially or into shallow water, indicating buoyant support by the mantle during rifting and breakup; and (4) the isotopic and compositional diversity recorded in present-day Icelandic basalts is observed in many of the Palaeocene sequences, after crustal contamination and pressure of melt segregation are taken into account. The widespread and simultaneous activity of Phase 1 activity requires an abnormally high mantle flux rate. This may be associated with the arrival of a start-up plume, but alternatively it represents the arrival of a pulse of hot mantle, following a period of weak plume activity during the Cretaceous. In either scenario, the igneous activity appears to have been focused along lines of weakness in the lithosphere. Phase 2 activity is closely linked to continental breakup. Forced mantle convection, caused by hot mantle flowing into the developing rift zones during continent breakup, may have led to the very high magma production rates (two to three times those observed in present-day Iceland) that formed the SDRS and associated deep crustal intrusions.

1. INTRODUCTION

Large Igneous Provinces: Continental, Oceanic, and Planetary Flood Volcanism
Geophysical Monograph 100

The widespread magmatism that occurred throughout the North Atlantic region during Palaeogene times constitutes one of the first large igneous provinces (LIPs) to be recognised. *Giekie* [1880] noted the similarity of the basaltic successions in the British Isles, the Faeroes and Iceland, and the idea of a broadly related, originally contiguous province was further developed by *Holmes* [1918]. His Brito-Arctic Province encompassed the now dispersed basaltic lava piles and associated differentiates and plutonic rocks found in NW Europe, the Faeroes, Iceland, and East and West Greenland. In 1935, *Richey* introduced the term Thulean Province, and recently the province has incorporated the igneous rocks found in eastern Canada and along the margins of the North Atlantic basin (Figure 1). In this paper, because we are focusing on the development and influence of the Iceland thermal anomaly throughout Tertiary to Recent times, we use the general term North Atlantic Igneous Province.

The North Atlantic Igneous Province includes the basaltic and picritic lavas of Baffin Island and West Greenland; the >5-km-thick sequences of continental flood basalts in the Scoresby Sund-Blosseville Kyst region of East Greenland; the seaward-dipping reflectors associated with the Greenland and NW European volcanic rifted margins; the aseismic ridges of the Greenland-Iceland-Faeroes Ridges; and, of course, Iceland itself. Indeed, Iceland provides a possible model for the formation of oceanic plateaus, which are the most enigmatic of LIPs [e.g., *Saunders et al., 1996*]. In addition to the large volume of extrusive rocks, the North Atlantic Igneous Province also boasts an impressive diversity of intrusive bodies.

The region has a long history of exploration and study, both academic and commercial, ensuring a large database of geological information. Recently, interest in the region has grown, with a succession of Ocean Drilling Program (ODP) legs (104, 152, and 163) on the Vøring Plateau and the SE Greenland margin; seismic profiling associated with the Faeroes-Iceland Ridge Experiment (FIRE) [*Staples et al., 1996*]; seismic profiling and on-land investigations along the Greenland margins by the Danish Lithosphere Centre [*Larsen et al., 1995*]; and detailed petrological and geophysical investigations of the Iceland neovolcanic zones and the Reykjanes Ridge. Because of the diversity of its magmatic products and the extensive information presently available, the North Atlantic Igneous Province provides an excellent opportunity to study several facets of LIP development, including the duration of magmatism, the interaction between lithosphere and sub-lithospheric mantle, and the role of a mantle plume in LIP formation.

Considered individually, the onshore basalt sequences of the North Atlantic Igneous Province cover smaller areas than many continental flood basalts. However, when we

include the offshore sequences, such as the basalts along the continent-ocean transition and on the major offshore plateaus (e.g., the Vøring and Rockall Plateaus), the total area increases to about 1.3×10^6 km^2 [*Eldholm and Grue, 1994*]. This is comparable with the estimated original area of the Deccan Province in India [e.g., *Mahoney, 1988*]. Inclusion of the thickened crust of the Greenland-Faeroes Ridge and Iceland makes the North Atlantic Igneous Province one of the largest in the world [*Macdougall, 1988*].

An outstanding problem of LIP formation is the extent to which mantle plumes are involved in their formation. Many workers now accept that plumes have a pivotal role in the formation of continental flood basalts [e.g., *Morgan, 1971, 1972; White and McKenzie, 1989*], oceanic plateaus and aseismic ridges [e.g., *Vink, 1984; Coffin and Eldholm, 1994*], and volcanic rifted margins [e.g., *White et al., 1987; Coffin and Eldholm, 1994*]. The short duration of the bulk of the magmatism associated with some individual LIPs (e.g., Deccan [*Courtillot et al., 1988*]; North Atlantic [*White, 1989*]) implies an event of almost cataclysmic proportions. No consensus exists, however, whether LIPs represent the arrival of a plume head at the base of the lithosphere, with consequent outburst of magma [*Richards et al., 1989; Griffiths and Campbell, 1990; Campbell and Griffiths, 1990*], or whether a plume head grows more slowly ('incubates'), and the magmatic release is due to extension of the overlying lithosphere [e.g., *White and McKenzie, 1989; Kent et al., 1992*] (see discussion by *Saunders et al.* [1992]).

Other workers have relegated the role of mantle plumes to insignificance or dismissed them altogether. *Anderson et al.* [1992], for example, invoked 'hotcells'—broad regions of the mantle which have low seismic velocity and which by implication are hotter than adjacent regions. In this model, the buildup of heat is accomplished by thermal blanketing by the overlying lithosphere, and adjacent regions are cooled by subduction. The geochemical characteristics of hotspot-related magmas may be provided by delamination of the continental lithosphere [e.g., *McKenzie and O'Nions, 1983; Allègre and Turcotte, 1985; Smith, 1993*], thus eliminating the requirement for plumes connected to an exotic, deep-seated mantle source. *Mutter et al.* [1988] and *Zehnder et al.* [1990] dispensed altogether with the requirement for thermal anomalies, suggesting that the excess magmatism associated with volcanic rifted margins was a result of enhanced mantle convection, triggered by the thermal contrast between old, cold lithosphere and the warmer asthenosphere. Such a process is necessary, they argued, where there is no clear evidence for a mantle plume. Which (if any) of these

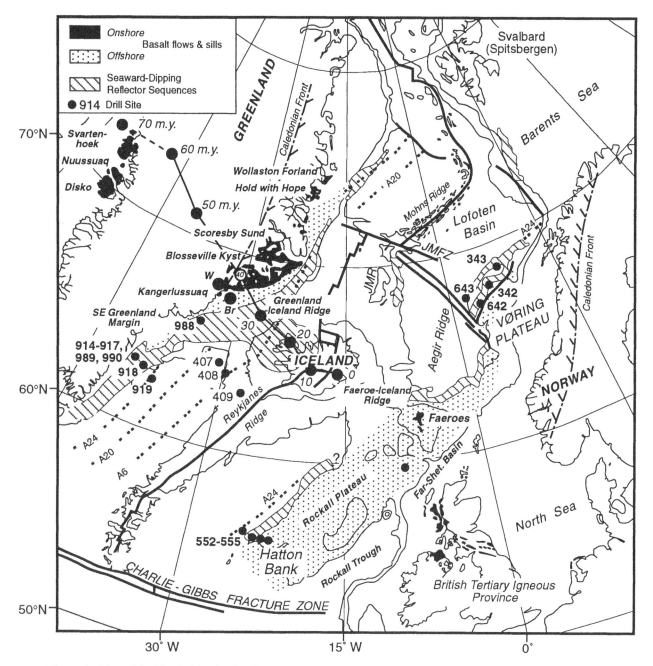

Figure 1. Map of the North Atlantic showing the extent of the North Atlantic Igneous Province, and the locations of the main areas discussed in the text (Baffin Island not shown) [modified after *Larsen et al., 1994*]. Estimated positions of the ancestral Iceland plume are taken from *Lawver and Müller* [1994] (dashed line between 60 and 70 Ma indicates uncertainty about existence of the plume prior to 62 m.y. ago). Alternative locations of plume axis are shown by W (*White and McKenzie* [1989], at anomaly 24 time) and Br (*Brooks* [1973b]) ca. 60 m.y. ago). Small filled circles indicate location of DSDP and ODP basement sites. JMR: Jan Mayen Ridge; JMFZ: Jan Mayen Fracture Zone; Far-Shet Basin: Faeroes-Shetland Basin.

models is correct has serious implications for understanding the thermal structure of the mantle and its evolution through time. We shall return to these models in Section 7.

An important aspect of the North Atlantic Igneous Province is the relative ease with which we may gain information about the present-day hotspot. Many LIPs have been related to an extant hotspot (e.g., Deccan-

Réunion; Kerguelen Plateau-Isles Kerguelen; Paraná-Tristan da Cuhna; Madagascar-Marion/Prince Edward Islands), but in all cases the plume thought to be responsible for the formation of the LIP is now located beneath thick lithosphere. Consequently, the volume and composition of the magmas are restricted by the lithospheric lid to moderate- to high-pressure melting [e.g., *Ellam,* 1992]. In the case of Iceland, however, the proposed plume is ascending very close to the Mid-Atlantic Ridge (MAR), and the mantle is thus melting at least partly in the spinel field. It is therefore possible to sample low-pressure melts from the present day hotspot. Furthermore, sampling is not restricted to a focused point, as on most ocean islands, but is possible across virtually the entire width of the thermal anomaly. As we shall describe below, this has provided an opportunity to investigate not only the thermal structure of the plume, but its compositional structure as well.

As a working hypothesis, we follow several previous workers in accepting that a mantle plume is presently located beneath eastern Iceland, and that this is the gun, still firing, not merely smoking, responsible for much of the activity of the North Atlantic Igneous Province [e.g., *White and McKenzie,* 1989, 1995]. In Section 7 we discuss other possible hypotheses. These are the main questions, common to many LIPs, that we seek to address in this paper:

• What are the age and duration of the magmatism?

• How was the composition of magma controlled by the lithosphere and/or by more deep-seated processes (for example, a plume or a hotcell)?

• What does the magmatism tell us about the evolution and structure of the sublithospheric mantle beneath the North Atlantic?

• Did the arrival of a hot plume initiate magmatism and lead to continental breakup, or did lithospheric thinning release magma from a pre-existing thermal anomaly? Was this anomaly created by a slowly incubating plume or a hotcell?

2. TIME SCALE USED IN THIS CHAPTER

Much of the igneous activity reported in this chapter occurred in the Cenozoic, mostly during Palaeocene and Eocene times. Several time scales are used in the literature, namely those of *Harland et al.* [1982, 1990], *Cande and Kent* [1992, 1995], and *Berggren et al.* [1985, 1995]. There are considerable differences between some of these time scales, so it is important to standardise to one of them. *Cande and Kent* [1992] presented a detailed revised Cenozoic magnetochronology based on an evaluation of

seafloor magnetic anomalies in the South Atlantic and fast spreading regions in the Indian and Pacific Oceans. They used a Cretaceous-Palaeogene boundary calibration age of 66 m.y., based on high-precision laser fusion ^{40}Ar-^{39}Ar sanidine dates. Subsequent studies have shown, however, that this age may be too old, and a consensus age of 65 m.y. has emerged for the Cretaceous-Palaeogene boundary. *Gradstein et al.* [1994] also used the 65 m.y. age as an anchor point for their Mesozoic time scale. This date was subsequently used to revise the geomagnetic polarity time scale [*Cande and Kent,* 1995], which formed the basis of the geochronology and chronostratigraphy used by *Berggren et al.* [1995]. We use the *Berggren et al.* [1995] time scale throughout this chapter (Figure 2).

As a preface to Sections 3 and 4 we stress that there are few high-precision dates for many regions of the North Atlantic Igneous Province (although it may be one of the better-dated LIPs). We have tried to restrict our discussion to ^{40}Ar-^{39}Ar ages (step-degassing of whole-rock samples or laser fusion of feldspars from volcanic rocks). K-Ar ages are less reliable [e.g., *Fitch et al.,* 1988], although such dates are often the only ones available.

3. THE TECTONIC DEVELOPMENT OF THE NORTH ATLANTIC BASINS

For the purposes of this review we shall focus on the area of the North Atlantic to the north of latitude 50°N, including the continental borderlands of NW Europe, Greenland and NE Canada (approximately the region shown on Figure 1). *Srivastava and Tapscott* [1986] provided a chronology of the tectonic development of the North Atlantic region. Rifting successively propagated northward, initially into the Labrador Sea and Rockall Trough, and then into the NE Atlantic. In some instances the line of eventual separation was guided by pre-existing anisotropy in the lithosphere, but the thermal swell associated with the ancestral Iceland plume may have exerted an important control during the Palaeogene [*Hill,* 1991].

A series of basins characterises the pre-Cenozoic palaeogeography of the North Atlantic region. These basins were partly precursors to plate separation, but they were also remnants of basins that had been developing throughout the Mesozoic. In the British Isles, the Tertiary lavas erupted onto a varied substrate with a low topography, probably developed during the Upper Cretaceous [e.g., *Watson,* 1985]. Many of the eruptive centres are either in, or along the margins of, Mesozoic basins sitting on Proterozoic or Archaean crust [*Thompson and Gibson,* 1991]. Deep sedimentary basins occur along

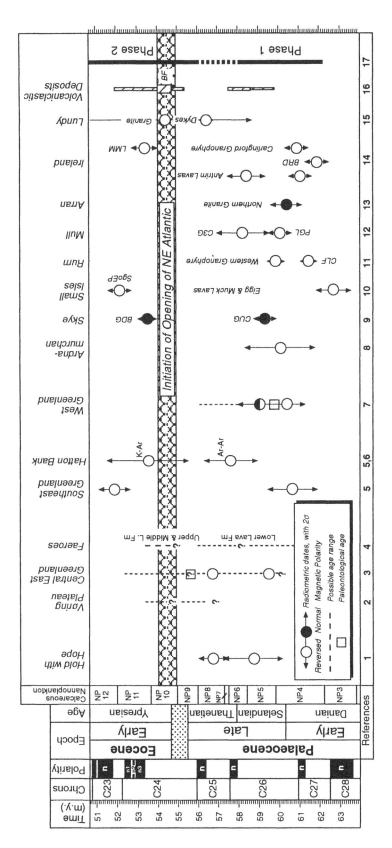

Figure 2. Time scale and event chart for the North Atlantic Igneous Province. Time scale from *Berggren et al.* [1995]. References and comments: 1 - *Upton et al.* [1995]; 2 - *Schönharting and Abrahamsen* [1989]; 3 - *Soper et al.* [1976a,b], *Noble et al.* [1988], *Hansen et al.* [1989], and *Storey et al.* [1996]; 4 - *Waagstein* [1988]; 5 - *Sinton and Duncan* [1996]; 6 - *Macintyre and Hamilton* [1984]; 7- *Parrott* [1976], *Clarke et al.* [1983], *Piasecki et al.* [1992], and *Storey et al.* [1996]; 8 - *Walsh et al.* [1979]; 9 - *Dickin* [1981] (CUG: Coire Uaigneich granite; BDG: Beinn an Dubhaich granite); 10 - *Dagley and Mussett* [1986], *Dickin and Jones* [1983], and *Pearson et al.* [1996] (SgoEP: Sgurr of Eigg pitchstone); 11 - *Mussett* [1984] (CLF: Canna Lava Formation); 12 - *Mussett* [1986] and *Walsh et al.* [1979] (C3G: Centre 3 granite); 13 - *Dickin at al.* [1981] and *Evans et al.* [1973]; 14 - *Thompson* [1986], *Wallace et al.* [1994], and *Thompson et al.* [1987] (BRD: Blind Rock Dyke; LMM: late Mourne Mountains granites); 15 - *Hampton and Taylor* [1983] and *Mussett et al.* [1976]; 16 - *Knox and Morton* [1983, 1988] (BF: Balder Formation); 17 - This paper (see text). Age of initial opening of NE Atlantic from *Vogt and Avery* [1974].

the East Greenland margin (for example, the Jameson Land Basin) although, like in the British Isles, lavas were erupted through crust of different ages and composition. In West Greenland, a complex graben system was already developed in the Mesozoic, between the future Baffin Island and Greenland [*Henderson et al.*, 1976]. These graben contain thick sequences of Cretaceous sediments and perhaps Jurassic sediments as well. Subsidence continued after the start of volcanism in the Palaeocene.

Seafloor spreading was underway in the region to the north of the Azores-Gibraltar Fracture Zone by 126 Ma, slowly propagating northwards so that by the Cenomanian (Anomaly M0-34, 95 Ma), seafloor spreading was active in all regions south of the Charlie Gibbs Fracture Zone (~53°N). Farther north, rifting had already begun in the region of the future Rockall Trough during the Aptian (ca. 118 Ma), before developing into seafloor spreading during the Albian or Cenomanian [*Roberts et al.*, 1981]. Seafloor spreading between Canada and Greenland was preceded by a long period of slow extension [*Chalmers*, 1991].

Seafloor spreading was underway in the Atlantic in all regions south of Greenland by Campanian times (ca. 84 Ma). Spreading in the Rockall Trough stopped completely during the Campanian or Maastrichtian [*Kristoffersen*, 1978; *Roberts et al.*, 1981]. There is some debate about when seafloor spreading began in the Labrador Sea. *Srivastava* [1978] introduced the commonly accepted plate kinematic model, suggesting that the earliest identifiable anomaly in the northern Labrador Sea is anomaly 31. In *Roest and Srivastava's* [1989] recent revision, however, the oldest preserved anomaly is 33 (ca. 80 Ma), but *Chalmers* [1991] suggested that in the northern Labrador Sea spreading probably began as late as the Palaeocene (anomaly 27). Seafloor spreading slowed down considerably by 50 Ma, and had stopped by 36 Ma. The main axis of spreading migrated into the NE Atlantic, between Europe and Greenland, during the early Eocene, Chron 24 (C24) time [*Vogt and Avery*, 1974].

Prior to about 62 Ma there is little evidence of large-volume magmatism in the North Atlantic region. There is no evidence that the Cretaceous ocean crust is thicker than normal (6–8 km). Anton Dohrn [*Jones et al.*, 1994] and Rosemary Bank [*Hitchen and Richie*, 1993; *Morton et al.*, 1995], Late Cretaceous seamounts at the northern end of the Rockall Trough and a sill of tentative Campanian age in the Faeroe-Shetland Basin [*Fitch et al.*, 1988] indicate sporadic outbursts of pre-Cenozoic magmatic activity in the region, over and above that associated with rifting and extension in the Rockall Trough. The precise extent and cause of this Late Cretaceous activity, and its relationship to the later Palaeogene outbursts, are presently unknown.

After the Cretaceous, however, the style of magmatism and sedimentation changed dramatically. Broadly speaking, two episodes or phases of magmatism occurred during the Palaeogene [e.g., *White and McKenzie*, 1989]. An episode of continent-based magmatism at 62–58 Ma (Phase 1) produced large volumes of basalt and associated intrusive rocks in the future eastern Canada (Baffin Island), West Greenland, SE Greenland, the British Isles, and possibly parts of the central East Greenland successions. This activity preceded the main breakup and separation of Greenland from NW Europe by about 4 m.y. Local uplift around several basins, such as the northern North Sea Basin and the Faeroe-Shetland Basin, is indicated by the input of clastic sediments into what previously had been low-energy depositional environments [*Anderton*, 1993]. This uplift began in Danian times, reached its peak in the Thanetian (58–55 Ma) and was followed in some areas by rapid subsidence [e.g., *England et al.*, 1993; *Turner and Scrutton*, 1993]

Plate separation and ocean crust formation, which began at approximately C24r time (56–53.5 Ma) [*Vogt and Avery*, 1974], was accompanied by the formation of thick, seaward-dipping reflector sequences (SDRS) along much of the E Greenland and NW European margins (Phase 2). *Larsen and Jakobsdóttir* [1988] have estimated, on the evidence provided by seismic profiles across the SDRS in East Greenland, that during breakup, volcanic productivity along the North Atlantic rift zone was almost three times higher than in present-day Iceland. Eruption rates during the initial stages of plate separation were higher than any subsequently recorded in the North Atlantic basin. Where recovered, most of the lavas of the SDRS were erupted in a subaerial or shallow aqueous environment, indicating dynamic support of the plate margins [*Hinz*, 1981; *Clift et al.*, 1995]. Anomalously thick basaltic crust (up to 35 km thick [*Bott*, 1983; *White et al.*, 1995; *Staples et al.*, 1996]) continued to be produced between Greenland and the Faeroes to form the Greenland-Iceland-Faeroes Ridges and Iceland itself.

Plate tectonic reconstructions place the centre of the Iceland hotspot beneath central Greenland during the period 60–50 Ma, although there is some uncertainty about its precise location. *Lawver and Müller* [1994], on the basis of relative plate motions, located the plume axis close to the centre of Greenland between 60 and 50 Ma (Figure 1), whereas *Brooks* [1973b], *Brooks and Nielsen* [1982b] and *White and McKenzie* [1989] placed it farther to the east, close to Kangerlussuaq, during the time of eruption of the East Greenland basalts. There is, however, little evidence of a hotspot 'track' until the formation of the Greenland-Faeroes Ridge, but this may simply reflect the lack of surface expression of magmatism due to the thick cratonic lithosphere beneath central Greenland. The Phase 1 magmatism occurred over a widespread area, some 2000 km across (outline by the bold dash-dot line in Figure 3), requiring a broad thermal anomaly [*White*, *1988*], and perhaps associated with a startup plume [*Griffiths and Campbell*, 1990; *Campbell and Griffiths*, 1990]. The Phase 2 magmatism was initially concentrated along the lines of plate separation, but still required a broad thermal anomaly to account for the dynamic support of the plate margins and the huge, widespread volumes of melt [*White et al.*, 1987]. We return to this point in Section 7. Magmatism continued longer in East Greenland than in the British Isles, reflecting the proximity of the ancestral Iceland plume after plate separation.

4. THE NORTH ATLANTIC IGNEOUS PROVINCE: A REVIEW OF AGE, DURATION, AND MAIN CHARACTERISTICS OF THE MAGMATISM

Reviews of the magmatism in the North Atlantic Igneous Province are provided by *Dickin* [1988], *Mussett et al.* [1988], *Upton* [1988], *White and McKenzie* [1989], *Larsen et al.* [1994], and *Ritchie and Hitchen* [1996]. Each of the following subsections will briefly consider the geographical extent of the sub-province, the age and duration of the magmatic activity, the magmatic productivity, and provide a summary of the geochemistry of the main igneous units. We concentrate on the basaltic lavas, because they are often the most visible expressions of magmatic activity and the most extensively sampled and studied.

Lavas were erupted in three different tectonic environments: (1) onto continental crust, forming continental flood basalts; (2) along the continental margins, where they form the off-lapping lava piles of the SDRS in a transitional or fully oceanic setting; and (3) in a truly oceanic setting along the Greenland-Iceland-Faeroes Ridges. This difference in eruptive setting is reflected in variations in composition, the continental flood basalts often showing evidence of modification by the continental lithosphere, and in timing, the continental sequences generally being older than the SDRS.

4.1. Iceland

For the purpose of this review, Iceland is important for three reasons. Firstly, the crustal structure and magmatic plumbing systems provide potential analogues for the formation of Palaeogene SDRS along the margins of the Atlantic Basin. Both environments are characterised by

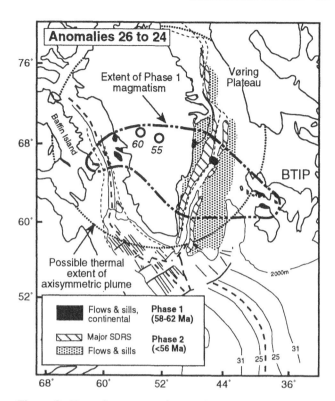

Figure 3. Tectonic reconstruction at Chron 24 time (ca. 55 Ma), but including the magmatic activity from C26 through C24 time. Location of continental blocks and plate boundaries from *Srivastava and Tapscott* [1986]; estimated positions of the ancestral Iceland plume axis at 60 and 55 Ma from *Lawver and Müller* [1994]. Note that *White and McKenzie* [1989] place the axis of the plume at 55 Ma closer to the central East Greenland volcanic successions. Ornamentation of the volcanic successions is based on whether the activity is Phase 1 or Phase 2, rather than continent- or ocean-based. The bold dash-dot line indicates the extent of the early (Phase 1) magmatism. The dotted circle is drawn assuming an axisymmetric plume head centred on the *Lawver and Müller* [1994] 55 Ma position of the plume axis.

excess magmatism, over and above that at normal mid-ocean ridges. Secondly, the basalts from Iceland and the adjacent ridge segments provide compositional data about the underlying plume. They thus provide reference points on Sr, Nd and Pb isotope, and chemical variation diagrams. Thirdly, gravity and topographic (and bathymetric) data across the island and adjacent seafloor help to constrain the thermal structure of the plume.

The geology of Iceland can be divided into axial rift and marginal zones [e.g., *Saemundsson*, 1986]. The axial rift zones or neovolcanic zones, defined on the basis that they are younger than the Brunhes Chron (0.7 Ma), are the main regions of crustal formation. *Jakobsson* [1972] has estimated that the volcanic discharge in the neovolcanic zones is in the order of 40 km^3/1000 y (approximately 1.0 × 10^{-4} km^3 per km of rift per year for a 400-km-long rift zone), with higher effusion rates near the centre of the island. Nearly 90% of the lavas are tholeiitic basalt, with small amounts of acid and intermediate rocks. Both central and fissure eruptions are found. Zones of limited extension occur as branches to the main rift zones in three regions. One of these is the Snaefellsnes Peninsula, where the average discharge rate of lavas is lower than in the main rift zones [*Jakobsson*, 1972], and where the lavas are mainly transitional to alkalic in composition, consistent with small-degree melts derived from greater depths than the tholeiites of the main rift zones.

The marginal zones, representing crust more than 0.7 m.y. old, make up more than two-thirds of the area of the island. The oldest crust in Iceland is found in the northwest part of the island, where basalts of up to 15 m.y. old have been recovered [*Moorbath et al.*, 1968; *McDougall et al.*, 1984]. In eastern Iceland the oldest crust is approximately 13–14 m.y. old [*Watkins and Walker*, 1977].

The Icelandic crust is approximately 25–30 km thick [*Bjarnason et al.*, 1993; *Bott*, 1983; *White et al.*, 1995, 1996], and at least the upper 10 km comprises lava flows. The flows dip gently towards extant and extinct rift zones, the dips increasing in the vicinity of the axial zones [*Walker*, 1960]. The metamorphic grade increases through the lava pile, such that in the Tertiary sequences it is not uncommon for the flows exposed on mountain tops to be unmetamorphosed, whereas flows from valley floors preserve zeolite or lower greenschist grades [*Walker*, 1960]. In the main sequences of Tertiary lavas, the geothermal gradient is estimated to be approximately 40–60°C/km; in the axial zones this locally rises to 100°C/km [*Pálmason and Saemundsson*, 1974; *Pálmason et al.*, 1978].

Pálmason's [1980, 1986] model probably best describes crustal formation in Iceland, albeit with modifications suggested by *Menke and Levin* [1994]. In this model, the excess magmatism along the axial zone causes overloading of the lithosphere, leading to subsidence of the crust. Lava flows erupted closest to the axis will subside the greatest amount and be buried beneath subsequent flows. These rift-proximal flows subside and form the deeper portions of the Icelandic crust and undergo progressive zeolite, amphibolite and granulite facies metamorphism. The distal fringes of the largest flows escaping the rift zone, however, avoid burial and remain close to the surface. Smaller flows unable to escape the axial rift will not remain at the surface.

The model predicts that the axial zone is underlain by a partial melt zone at about 12 km depth, consistent with low resistivity values at depths greater than about 12 km under the axial zone [*Beblo and Björnsson*, 1980]. Recent geophysical studies, however, suggest that the lower crust may be cooler, even in the axial zone [*Menke and Levin*, 1994]. It is possible that there is far greater heat loss via hydrothermal circulation from high-level magma bodies in the axial zone, than predicted by the Pálmason model.

As pointed out by *Imsland* [1983] and *Hardarson and Fitton* [1994], the basalts preserved in the surficial lava sequences of the marginal zones will provide a preferential sampling of the larger eruptive units. Material erupted solely in the rift axis will not be preserved for near-surface sampling. This is important, because it is evident that the spectrum of compositions seen in the neovolcanic zones is far wider than in the Tertiary lavas on Iceland. Basalts from the neovolcanic zones range from light-rare-earth-element (REE)-enriched to strongly light-REE-depleted (Figure 4), and Zr/Nb ranges from 5 to 35. In comparison, the Tertiary sequences have a much more restricted range of Zr/Nb (9±2 [*Hardarson and Fitton*, 1994]), and the REE show uniform, slight light-REE enrichment.

Schilling et al. [1982] suggested that this variation is due to changes in the composition of the mantle source (a greater involvement of light-REE-depleted mid-ocean ridge basalt (MORB)-like mantle in more recent Icelandic eruptions), but it is more likely that the earlier Tertiary sequences result from homogenising of small, compositionally varied, magma batches in large chambers. The two tectonic environments on Iceland, axial and marginal zones, thus provide different types of information about the Icelandic source and the magmatic plumbing systems.

The Pálmason model also adequately explains the morphology and formation of Palaeogene SDRS, the offlapping appearance of SDRS being similar to the rift-dipping structure of the Icelandic lava piles [e.g., *Gibson and Gibbs*, 1987; *Larsen and Jakobsdóttir*, 1988]. By analogy with the Icelandic setting, the accessible portions of the Palaeogene SDRS may not provide details of the full range of available magma types; lavas erupted close to the rift axis will have subsided to depths well beyond the reach of present-day drilling.

4.1.1. *Temperature of the Iceland plume.* Although there is general agreement that the thick crust associated with the Greenland-Iceland-Faeroes Ridge is the result of melting of hot mantle, there is no consensus about the precise temperature of the mantle beneath Iceland. On the one hand, *White et al.* [1995] suggested that the excess temperature in the core of the plume is of the order of 200°C, with a disk-shaped thermal anomaly extending some 1350 km from the plume axis. In this model, the

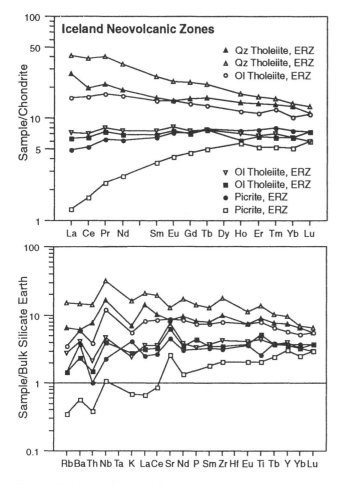

Figure 4. Chondrite-normalised REE data, and bulk-silicate Earth (BSE)-normalised trace element data, for lavas from the Iceland neovolcanic zones. Data source: *Hémond et al.* [1993]. BSE normalising values from *McDonough and Sun* [1995]. Chondrite values from *Nakamura* [1974].

plume has a narrow vertical conduit, some 50–100 km in diameter [e.g., *Vogt,* 1983; *White,* 1988], and the disk-like head cools progressively to ambient temperatures at its margins. *Ribe et al.* [1995], on the other hand, modelled the same system assuming a diffuse zone, at least 600 km in diameter, of warm ($\Delta T = 90°C$), upwelling mantle.

Three direct observations provide the key input parameters for these models. These are the thickness of the crust, the positive residual depth anomaly of zero-age crust along the MAR and on Iceland, and the composition of erupted basalts (see Figure 5). *White et al.* [1995] used a combination of these observations to derive their model. The volume of melt (or crustal thickness) is related to the temperature of the mantle source [*McKenzie and Bickle,* 1988]. However, the parameters used by *White et al.* [1995] assume that the mantle decompresses and melts

passively in response to plate separation. Although this is valid in the case of normal ocean crust, it is unlikely to be true for Iceland, where part of the uplift of the island is thought to be caused by dynamic uplift from the plume. Consequently, the mantle flow rates through the melting zone are likely to be enhanced convectively, producing greater volumes of magma for a given potential temperature and implying that the mantle is cooler than predicted by the passive upwelling model. Indeed, *White et al.* [1995] noted that the predicted volume of melt from REE inversion modelling of data from Krafla, close to the axis of the plume, is only 16 km. This is significantly less than the observed 25 km crustal thickness, and implies forced convection.

Ribe et al. [1995] applied numerical models for plume buoyancy, flow, and melting to derive their temperature contrasts. When they used *Schilling's* [1991] estimates of the plume flux (1390 kg s^{-1}) and excess temperature

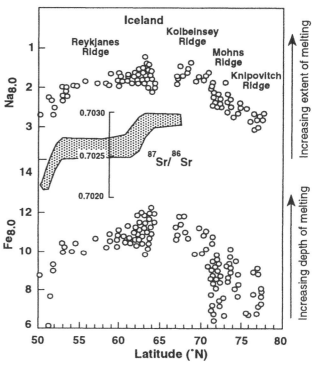

Figure 5. Variation of $^{87}Sr/^{86}Sr$ and fractionation-corrected Na and Fe along the Mid-Atlantic Ridge, in the vicinity of Iceland. Note the decrease in $^{87}Sr/^{86}Sr$ to MORB-like values at about 60°N, whereas Fe$_{8.0}$ remains elevated until about 52°N. If high Fe$_{8.0}$ (and low Na$_{8.0}$) reflects deeper, more extensive melting as suggested by *Klein and Langmuir* [1987], this diagram implies that the thermal signature of the plume is more extensive than the isotopic signature. La/Yb shows a similar distribution to that of $^{87}Sr/^{86}Sr$. Data sources: Fe and Na from *Klein and Langmuir* [1987]; $^{87}Sr/^{86}Sr$ from *Hart et al.* [1973].

(ΔT=263°C), the predicted topography above the plume axis was 10 km, corresponding to a crustal thickness of 68 km. As they said, "this far exceeds the crustal thickness of 20-34 km predicted for the same ΔT by purely passive upwelling [*McKenzie and Bickle*, 1988; *Klein and Langmuir*, 1987]." Their preferred model required a plume with a ΔT of 93°C, a buoyancy flux of 1370 kg s^{-1}, a volume flux of 193 m^{3}s^{-1}, and a mean upwelling velocity of 2 cm y^{-1}. The predicted topographic anomaly more closely resembled the observed one, although a cooler plume would probably have given an even better fit.

It is difficult to assess the *Ribe et al.* [1995] and *White et al.* [1995] models using independent chemical criteria. Recently, however, *Wolfe et al.* [1997] presented the results of a regional broadband seismic experiment which indicated a cylinder of low velocity material beneath Iceland. Their data suggest that the cylinder has a diameter of approximately 300 km at between 100 and 400 km depth, which is wider than the 'narrow stem' model of *White et al.* [1995]. Excess temperature may be as much as 200-300°C in the model of *Wolfe et al.* [1997], but prediction of temperature from P and S waves is fraught with uncertainty.

The models mentioned here provide possible end-member conditions; in reality, intermediate temperatures and fluxes probably pertain. The likelihood of forced convection means that the *White et al.* [1995] and *Wolfe et al.* [1997] estimates of mantle temperature are too high. It also means that the existing REE inversion models are inaccurate for the plume-ridge environment. Melt volumes will be increased over those in purely passive upwelling models by a factor approximately equivalent to (rate of mantle upwelling)/(rate of plate separation). In Iceland this factor will be >1 and, during the initial stages of plate breakup above a mantle plume, may be >>1. The excessive melt thicknesses predicted by numerical studies may be alleviated if some of the plume material flows along the adjacent mid-ocean ridge, but it is difficult to see how effective this mechanism would be in removing excess thermal energy. It is also likely that the temperature and plume flux have varied slightly with time. V-shaped ridges in the ocean crust to the south of Iceland imply short-lived increases in magma production rates, which equate to transient temperature increases of about 30°C if the decompression is entirely passive [*White et al.*, 1995], less if there is a component of forced convection.

4.1.2. *The composition of the Iceland plume.* An extensive literature exists on the composition of Icelandic basalts and differentiates [e.g., *Jakobsson*, 1972; *Jakobsson et al.*, 1978; *Imsland*, 1983; *Steinthorsson et al.*, 1985; *Oskarsson et al.*, 1985; *Hémond et al.*, 1993]. There

is a wide range of compositions, from strongly light-REE-depleted picrites to moderately light-REE-enriched tholeiites, alkali basalts, trachytes and rhyolites (Figure 4). Multielement plots show positive Nb anomalies, a feature of most ocean island basalts [e.g., *Tarney et al.*, 1980; *Hofmann et al.*, 1986; *Saunders et al.*, 1988]. The more primitive basalts and picrites show positive Ba and Sr anomalies that may be due to crustal assimilation [e.g., *Hémond et al.*, 1993] but which alternatively may be a source characteristic.

It has been long recognised that the Icelandic basalts are sampling a 'depleted' mantle source; see, for example, the εNd - ^{87}Sr/^{86}Sr diagram (Figure 6A) [e.g., *O'Nions et al.*, 1977; *Zindler et al.*, 1979; *Hémond et al.*, 1993]. Data for many picrites and tholeiites from the neovolcanic zones overlap with North Atlantic mid-ocean ridge basalt (MORB) compositions, but other samples, including alkali basalts from the Snaefellsnes Peninsula and elsewhere, extend the Iceland data field to higher ^{87}Sr/^{86}Sr and lower εNd values. Like many other ocean islands, therefore, the Iceland basalts do not define a single isotopic composition but form an array in isotope space [e.g., *Sun and Jahn*, 1975; *Zindler and Hart*, 1986; *Furman et al.*, 1995]. To produce the low εNd values by modification of a melt from a MORB source (for example, by assimilation of Icelandic crust) requires a component much older than Iceland, so the mantle source beneath Iceland must be isotopically heterogeneous [*Hémond et al.*, 1993].

Trace element abundances are controlled in part by fractionation processes (partial melting and fractional crystallisation) but incompatible element ratios probably reflect more closely the composition of their source. Nonetheless, if the amount of partial melting is sufficiently small, even those trace elements with very similar and low bulk distribution coefficients in mantle lherzolite will be fractionated from one another. Similarly, melt extraction will leave a residue with drastically changed element ratios, so remelting this residue can, in theory, produce liquids substantially depleted in the more incompatible elements. This was the principle behind the dynamic partial melting models suggested by *Wood* [1979b] and *Elliott et al.* [1991] to account for incompatible-element depleted basalts and picrites in Iceland. Despite these caveats, however, there is a clear correlation between trace element abundances and isotope ratios in Icelandic basalts (for example, Zr/Nb and εNd, Figure 7A), implying that the range of incompatible element patterns cannot be produced by partial melting processes alone.

The implication is that the Iceland plume comprises a 'depleted' end-member and a 'less-depleted' end-member, relative to estimates of primordial mantle. These end-

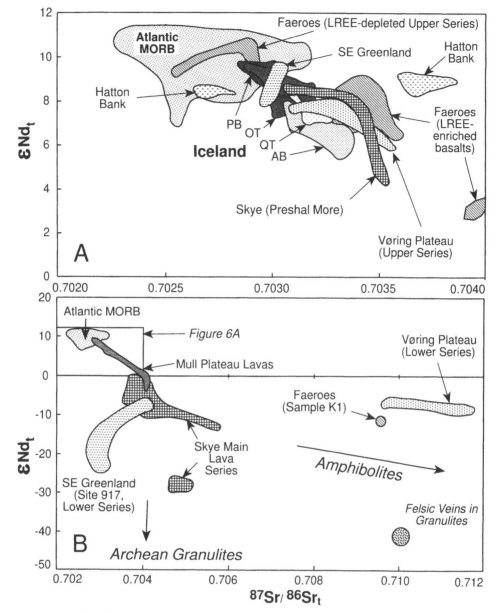

Figure 6. εNd$_t$ versus ^{87}Sr/^{86}Sr$_t$ for basalts and related rocks from the North Atlantic Igneous Province. A: expanded part of the diagram showing the variation in Icelandic basalts and picrites (PB - picritic basalts; OT - olivine tholeiites; QT- quartz tholeiites; AB - alkali basalts), various SDRS basalts (SE Greenland, Hatton Bank and the Upper Series at Site 642, Vøring Plateau), basalts from the Faeroe Islands ('Faeroes light-REE-enriched basalts' are from the Lower, Middle and Upper Series of the Faeroe Plateau Lava Group), and Skye (Preshal More basalts). 'SE Greenland' basalts are from Site 918, and selected samples from the Upper Series at Site 917. B. Main diagram illustrating the wide variation of εNd and ^{87}Sr/^{86}Sr in the North Atlantic Igneous Province, and mostly reflecting the effects of crustal contamination. Faeroes sample K1 is a highly contaminated basalt from the Upper Series of the FPLG. Vøring Plateau (Lower Series) rocks are predominantly dacites. Qualitative contamination vectors for amphibolite and granulite crustal contamination are shown. Data sources: Atlantic MORB - *Ito et al.* [1987]; Faeroes - *Gariépy et al.* [1983]; Iceland - *Hémond et al.* [1993] and references therein; Vøring Plateau -*Taylor and Morton* [1989]; Hatton Bank - *Macintyre and Hamilton* [1984]; Mull Plateau lavas and Archaean felsic veins from granulite - *Kerr* [1995a]; Skye lavas - *Dickin* [1981], *Thompson et al.* [1972, 1980, 1982, 1984]; SE Greenland - *Fitton et al.* [1996a]. Samples age-corrected to approximate age of emplacement where data allow; otherwise present-day values are plotted. The shift due to age correction is small compared to the overall range of values on these diagrams.

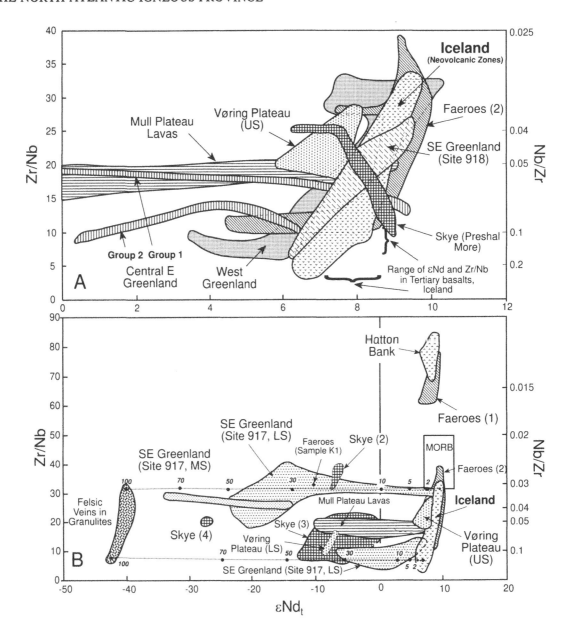

Figure 7. Zr/Nb verses εNd in lavas from the North Atlantic Igneous Province. (A) Detail showing the variation for samples with εNd > 0, and the clear overlap between Iceland and some basalt suites from the North Atlantic Igneous Province. Note the positive correlation of between Zr/Nb and εNd in Icelandic basalts. Groups 1 and 2 refer to basalts and related rocks from the Lower Basalts, Kangerlussuaq, central East Greenland [*Holm*, 1988]. (B) Plot showing the full scale of variation. Note the subhorizontal dispersion indicative of mixing between a high εNd sublithospheric source and a low-εNd contaminant. Two bulk assimilation lines are shown, between Archaean felsic veins and two end-members of the Iceland array (figures give percentage of contaminant). Faeroes (1) refers to light-REE-depleted basalts the Upper Series of the Faeroe Plateau Lava Group; Faeroes (2) represents the remainder of the Faeroes basalts, except for sample K1, a highly contaminated basalt, which is shown separately. 'Skye 1': Preshal More basalts; 'Skye 2, 3 and 4': Skye Main Lava Series. Data sources: as for Figure 6, plus West Greenland - *Holm et al.* [1992; 1993]; SE Greenland (Sites 917 and 918) from *Fitton et al.* [1996a,b]; Hatton Bank Zr/Nb values - Brodie and Fitton [1996]. Range of Tertiary Iceland basalt compositions from *Hardarson and Fitton* [1994]. Field for MORB assembled from data of *Ito et al.* [1987] and unpublished data of the authors.

members are approximately delineated by the picrites and alkali basalts, respectively, at least in isotopic terms. The depleted end-member superficially resembles MORB but *Thirlwall et al.* [1994] have shown that Icelandic basalts and picrites have higher $^{208}Pb/^{204}Pb$ at a given $^{207}Pb/^{204}Pb$ than North Atlantic MORB. *Fitton et al.* [1996b, 1997] have also shown from Zr-Nb-Y correlations that the depleted Iceland end-member is different from MORB (see Section 7). It would appear, therefore, that the depleted end-member is an intrinsic part of the Iceland plume, as suggested by *Hémond et al.* [1993], *Thirlwall et al.* [1994], and *Kerr et al.* [1995b], and not entrained or advected MORB mantle.

There are several implications of this two-component mantle model that have a bearing on studies of other parts of the North Atlantic Igneous Province, if the ancestral Iceland plume was indeed responsible for much of the magmatism.

• It is necessary to consider the Iceland plume not as a single point in isotope or element space but as a spectrum of compositions (e.g., Figure 7). Both 'depleted' and 'less-depleted' basalts may be derived from the plume.

• The 'less-depleted' end-member may be associated with a mantle lithology that has a lower melting point than the depleted end-member; for example as veins [*Wood*, 1979a; *Tarney et al.*, 1980] or streaks [*Fitton and Dunlop*, 1985] in more depleted peridotite. The proportion of 'less-depleted' to 'depleted' components contributing to the melt will be a function of the conditions of melting as well as the composition of the source. Deeper, small-degree melting will enhance the effects of the 'less-depleted' end-member, as suggested by *Hémond et al.* [1993] and *Hards et al.* [1995] for the Icelandic alkali basalts. More extensive melting will homogenise the system.

• The extent of melting will tend, on average, to be greatest in the axial rift zones near the axis of the plume, where the hottest mantle is allowed to decompress the most. Melt segregation will preferentially extract the more fusible material and progressively change the bulk composition of the source. This process may produce a radial chemical and isotopic gradient in the plume source.

We shall return to these points in Section 7, but they need to be borne in mind for the remainder of Section 4.

4.2. British Tertiary Igneous Province

A record of Palaeogene igneous activity is preserved in NW Scotland (including the classic areas of Ardnamurchan, Skye, Mull, Rum, Eigg, Muck, Canna, and Arran), Ireland (Antrim and the Mourne Mountains), Lundy, and offshore regions such as the Rockall Plateau,

Irish Sea and the Faeroe-Shetland Basin. Tertiary volcanism in western Britain tends to be focused around central intrusive complexes, which comprise a wide variety of igneous rocks ranging from peridotite to granite. These complexes will not be discussed in detail here; excellent reviews can be found in *Emeleus* [1982, 1991] and *Thompson* [1982]. Basalt dyke swarms, often focused towards the complexes, indicate local crustal dilation of up to 25% [*Speight et al.*, 1982] and have cross-cutting relationships that show emplacement at different times throughout the province's history.

The igneous rocks of the British Tertiary Igneous Province, or Hebridean Province, have a long history of investigation. In the latter half of the nineteenth century, discussion of the British Tertiary Igneous Province was dominated by Archibald Giekie and John Judd. *Giekie* [1867, 1888] believed that the lavas erupted through fissures, which are now represented by the great dyke swarms. Alternatively, *Judd* [1874, 1889] proposed that the igneous complexes of Western Scotland were the 'eroded basal wrecks' of large central Tertiary volcanoes. During this period *Giekie* [1880] was the first to recognise that the British Tertiary province was only a part of a much more extensive region of Tertiary volcanic activity encompassing the Faeroes and Iceland.

The publication of the Mull memoir of the Geological Survey of Scotland [*Bailey et al.*, 1924] was a milestone in the history of igneous petrology. It established for the first time the concept of 'magma types' and 'magma series', based on the igneous rocks of Mull. Two main magma types were identified by *Bailey et al.* [1924]: the Non-Porphyritic Central Type and the Plateau Type. These two types later became known as the tholeiitic [*Kennedy*, 1933] and alkali olivine basalt [*Tilley*, 1950; *Tilley and Muir*, 1962] types, respectively.

The relationship between these two magma types was debated by authors such as *Bowen* [1928] and *Kennedy* [1933], and the British Tertiary Igneous Province featured prominently in these discussions. In his study of Hawaiian and Hebridean lavas, *Wager* [1956] proposed that tholeiitic lavas were the result of partial melting of a layer of peridotite at a high structural level, and that alkali basalts were derived by partial melting at a much deeper level. Wager, therefore, proposed depth of melting as a critical factor in determining magma type, a concept which is now central to models of magma genesis and composition.

Like the lavas in East Greenland, those of the British Tertiary Volcanic Province were erupted onto a varied surface of Precambrian to Cretaceous age. Most of the landscape was eroded to low relief by Late Cretaceous times. Thin sedimentary deposits of that age are not uncommon, although there is good evidence that in some areas even this cover had been removed or thinned before the lavas were erupted. The extensive development of clay-with-flints deposits (for example, in Antrim, Northern Ireland) implies a period of subaerial weathering and erosion during Late Cretaceous/Early Palaeocene times [e.g., *Wilson and Manning*, 1978]. Pre-basalt Tertiary sediments are localised and scarce, and the environment of deposition varies considerably. Sediments associated with the Eigg Lava Formation [*Emeleus*, 1997], for example, are generally low-energy deposits associated with sluggish streams and shallow freshwater lakes. Terrestrial sediments containing lignite, leaf-beds and sandstones are occasionally interbedded with lavas on Skye, Mull and elsewhere [*Richey*, 1935]. The slightly younger sediments associated with the Canna Lava Formation and the Skye Main Lava Series are high energy conglomerates and sedimentary breccias, possibly related to the development of the Rum Central Complex, a substantial volcanic edifice at that time. On Mull there is a mudstone at the base of the lava pile, which may represent a lateritised tuff [*Bailey et al.*, 1924], and the basal Tertiary sediments contain a few metres of sandstone with grains of aeolian origin [*Bailey*, 1924].

The diverse substrate is reflected by the style of eruption of the earliest magmas. Most of the British Tertiary Igneous Province lavas were erupted subaerially on dry land, but local occurrences of pillow lavas, vitric tuffs and hyaloclastites (for example, at the base of the Skye Main Lava Series [*Anderson and Dunham*, 1966]) are consistent with emplacement into shallow water. There is no clear indication from the terrestrial deposits whether strong uplift began significantly prior to basalt eruption. Palaeogene uplift of the order of 300 to 1000 m is recorded by apatite fission tracks in sediments from the East Irish Sea Basin [*Hardman et al.*, 1993]. This uplift appears to have been accompanied by a heating event and emplacement of the Fleetwood Dyke Group. The latter is a suite of dolerites that form part of the Irish Sea dyke swarm [*Kirton and Donato*, 1985], and which have yielded K-Ar dates of 65.5±1.0 and 61.5±0.8 Ma [*Arter and Fagin*, 1993]. Post-Cretaceous uplift and tilting of western Scotland is recorded in the stratigraphy of the Inner Moray Firth Basin in eastern Scotland [*Underhill*, 1991] and this could be linked with the development of the proto-Iceland thermal anomaly and associated rifting in the west [*Thomson and Underhill*, 1993]. Thermal uplift may also be responsible for other uplift events (for example, the uplift and denudation of up to 3 km of sediments from parts of northern Britain [*Lewis et al.*, 1992]).

4.2.1. *Skye.* The Skye lava succession covers an area of some 1500 km² , has a cumulative thickness of ~1200 m [*England*, 1994], and has been divided into three magma types: the Skye Main Lava Series, the Fairy Bridge magma type, and the Preshal More magma type. The Skye Main Lava Series is composed predominantly of transitional alkali basalts, mostly with $La_n/Nd_n \leq 1$ and $Nd_n/Yb_n = 3–5$ [*Thompson et al.*, 1972, 1980] (Figure 8), although some 20% of the lava pile contains intercalated hawaiites-mugearites-benmoreites-trachytes [*Anderson and Dunham*, 1966; *England*, 1994]. (Note that La_n/Nd_n refers to chondrite-normalised La/Nd.) The Fairy Bridge magma type is found occasionally within the upper half of the Skye lava pile and comprises basalts with flat REE patterns but otherwise has major element chemistry similar to the Skye Main Lava Series [*Thompson et al.*, 1980; *Scarrow*, 1992]. This type is also well represented in the Skye dyke swarm [*Mattey et al.*, 1977]. The Preshal More magma type is represented by several basaltic flows near the top of the Skye lava succession, by occasional dykes, and by some intrusions in the Cuillin intrusive complex [*Thompson*, 1982]. The magma type is characterised by a tholeiitic major element chemistry with light-REE-depleted patterns [*Esson et al.*, 1975; *Mattey et al.*, 1977; *Thompson et al.*, 1980] (Figure 8).

4.2.2. *Mull.* The Mull lava succession covers an area of 840 km² on the Island of Mull and the adjoining mainland area of Morvern. *Bailey et al.* [1924] and *Emeleus* [1991] estimated that the lava succession has an approximate aggregate thickness of 1800 m. *Kerr* [1994, 1995a] identified three magma types within the Mull succession. These three magma types are broadly similar to those identified on Skye, the difference being that on Mull a successive relationship between the three types can be seen clearly. The earliest lavas, the Mull Plateau Group, are transitional tholeiitic-alkalic picritic-basalts (up to 15 wt.% MgO) to hawaiites with similar REE patterns to the Skye Main Lava Series (Figure 8). Trachytes and benmoreites near the top of the Mull Plateau Group account for less than 5% of the total preserved lava volume. The Coire Gorm magma type which overlies the trachytes of the Mull Plateau Group comprises transitional basalts with chondritic REE profiles, like the Skye Fairy Bridge magma type. The Central Mull tholeiites are the youngest lava type found on Mull; they possess flat to LREE-depleted REE patterns and are compositionally similar to the Preshal More lavas from Skye (Figure 8). Like the equivalent basalts from the Skye Main Lava Series, some of the Mull Plateau Basalts have characteristic trace element signatures, including high Ba/Rb and Ba/Nb, and low εNd values (Figures 5 and 8), consistent with assimi-

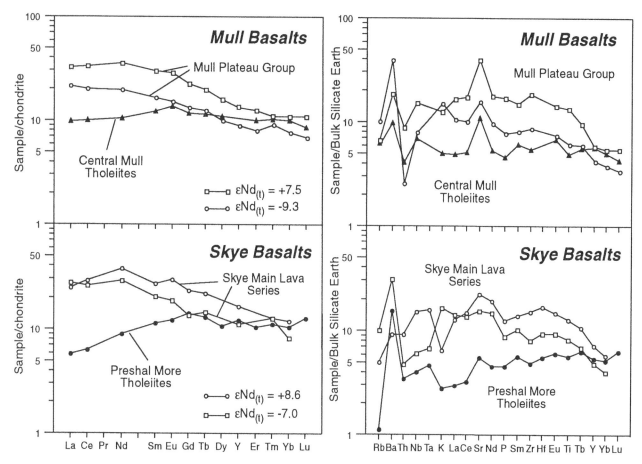

Figure 8. Chondrite-normalised REE and bulk silicate Earth-normalised trace element patterns for basalts from the two main lava successions on Mull and Skye, Scotland. Data sources: Mull - *Kerr* [1995a], *Kerr et al.* [1995a]; Skye - *Thompson et al.* [1972, 1980, 1982, 1984]. Central Mull tholeiite and Preshal More basalt patterns are average values; Mull Plateau Group basalts are represented by samples BR24 and BR8; and Skye Main Lava Series is represented by samples SK906 and SK976. εNd values calculated at t=60 Ma.

lation of Archaean Lewisian granulites (see Section 6).

4.2.3. *The Small Isles.* The thin lava sequences found on Eigg, Muck, Rum, and Canna are composed of two formations, the Eigg Lava and the Canna Lava Formations [*Emeleus, 1997*]. The Eigg Lava Formation, erupted before the emplacement of the Rum Central Complex, consists of transitional olivine basalts and, low in the succession, several flows of mugearite. Feldspar-phyric basaltic hawaiites occur near the top. The Canna Lava Formation was erupted after the Rum Central Complex and comprises four members with rock types ranging from olivine basalt to hawaiite and icelandite. Details of the chemistry of the lavas are provided by *Emeleus* [1985, 1997]. The most dramatic manifestation of the Tertiary activity in this area is, however, the Rum Central Complex, the remnants of a cyclically layered magma chamber originally derived from a picritic magma [e.g., *Emeleus*, 1987; *Young et al.*, 1988].

4.2.4. *Northern Ireland.* The Antrim Lavas represent the largest remnant of the British Tertiary Igneous Province, cover an area of 3500 km², and achieve a thickness of up to 800 m. The lava succession consists of three formations, the Lower, Middle and Upper Formations [*Old, 1975; Wilson and Manning, 1978*]. The Middle Formation includes the basalts of the Giant's Causeway (the Causeway Member). The basalts of the Upper and Lower Formations, although transitional basalts, are more tholeiitic than the basalts of the Mull Plateau Group and the Skye Main Lava Series [*Lyle, 1980; 1985*]. Basalts of the Lower Formation, and most of those of the Upper Formation, are light-REE-enriched with convex-upwards patterns [*Barrat and Nesbitt, 1996*]. The basalts of the Causeway Member show great variation in the degree of light-REE enrichment (La_n/Yb_n ranges from <0.6 to >3) and in isotope ratios (e.g., $\varepsilon Nd_{t=60}$ ranges from -11 to +8.5), probably caused by assimilation of variable amounts of Dalradian crust of Late Proterozoic-Lower Palaeozoic age [*Wallace et al.*, 1994; *Barrat and Nesbitt*, 1996]. The light-REE-depleted basalts from the Causeway Member resemble the Preshal More and Central Mull Tholeiites.

4.2.5. *Age of the British Tertiary Igneous Province.* For detailed accounts of the age of the British Tertiary Igneous Province, the reader is referred to *Mussett et al.* [1988], *Dickin* [1988] and *Ritchie and Hitchen* [1996]. The oldest ages of any reliability for the British province are 63.0±3.4 and 63.3±1.8 Ma (^{40}Ar-^{39}Ar ages) for reversely magnetised basaltic lavas from Eigg and Muck [*Dagley and Mussett,* 1986] (Figure 2). These ages have recently been confirmed and refined by step-heating ^{40}Ar-^{39}Ar ages (62.8±0.6 and 62.4±0.6 Ma) on sanidines from tuffs intercalated with the Eigg Lava Formation [*Pearson et al.,* 1996] These are among the oldest ages for the entire North Atlantic Igneous Province. A K-Ar age of 81± 4 Ma for a dolerite sill from the Faeroe-Shetland Intrusive Complex (Well 219/28-2 [*Fitch et al.*, 1988]) requires confirmation by ^{40}Ar-^{39}Ar methods; the bulk of the complex gives cooling ages of 55–53 Ma [*Hitchen and Richie,* 1993]. Lavas with reversed magnetic polarity from the older Plateau Group on Mull give an ^{40}Ar-^{39}Ar age of 60±0.5 Ma [*Mussett,* 1986]; the Centre 3 Granite on Mull gives an Rb-Sr age of 58.2±1.3 Ma [*Walsh et al.,* 1979]; and the centres on Ardnamurchan, which again have reversed polarity, give an Rb-Sr age of 60.0±1.7 Ma [*Walsh et al.,* 1979]. The Western Granophyre on Rum (reversed polarity) has an Rb-Sr age of 59.8±0.4 Ma [*Mussett,* 1984]. No reliable ages are available for the lavas from Skye or Arran, although it is possible that the Main Series lavas from Skye are coeval with the Canna Lava Formation on Rum (61.4±0.4 Ma [*Mussett,* 1984; *Mussett et al.,* 1988; *Bell and Williamson,* 1994]). Note, however, that pebbles of the Rum granophyre occur in conglomerates between the lavas on Canna, and the lavas sit unconformably on the granophyre [*Black,* 1952], implying that either the 61.4 Ma age for the basalt is too old, or the 59.8 Ma age for the granophyre is too young.

Most of the Palaeogene activity in Ireland occurred during a reversed polarity event. Dingle Dyke (ca. 59 Ma), Blind Rock Dyke (61.7±0.5 Ma), and the Carlingford Granophyre (60.9±0.5 Ma) are among the oldest dated events [*Thompson,* 1986]. Few reliable ages are available for the Antrim Lavas (58.3±1.1 to 61.0±0.6 Ma [*Thompson,* 1986; *Wallace et al.,* 1994]).

The majority of the dates are therefore consistent with emplacement of magmas during C26r (57.95–60.9 Ma, according to the time scale of *Berggren et al.* [1995]), although it is possible that some activity (e.g., on Muck

and Eigg) occurred during C27r. Not all of the dated material falls into this neat pattern, however. The Northern Granite of Arran, for example, has an Rb-Sr age of 60.3±0.8 Ma (and a virtually identical ^{40}Ar-^{39}Ar age) [*Evans et al.*, 1973; *Dickin et al.,* 1981] but shows a normal magnetic polarity. Coire Uaigneich Granite on Skye (59.3±0.4 Ma) is also magnetically normal. It is unclear if these bodies were emplaced during C27n or 26n.

In some areas, activity continued for several million years after the main event; for example, Beinn an Dubhaich Granite, Skye (53.5±0.4 Ma [*Dickin,* 1981]), the Loch Ba Felsite, Mull (56.5±1 to 58.2±1.3 Ma [*Mussett,* 1986; *Walsh et al.,* 1979]), the Sgurr of Eigg pitchstone obsidian (52.1±0.5 Ma [*Dickin and Jones,* 1983]), and the later granites in the Mourne Mountains (53.3±0.6 Ma [*Thompson et al.,* 1987]). The Lundy Granite, the most southerly known part of the North Atlantic Igneous Province, has an Rb-Sr age of 54±4 Ma [*Hampton and Taylor,* 1983]. It is intruded by dykes that are predominantly magnetically reversed, and that have been dated by ^{40}Ar-^{39}Ar at 56.4±0.3 Ma [*Mussett et al.,* 1976].

4.3. *Offshore UK*

An important record of widespread Palaeogene volcanic and intrusive activity is preserved on the continental shelf and basins around the British Isles. Indeed, the volume of material far exceeds that preserved on the mainland, and the extensive commercial exploration that has been underway for the last three decades has ensured reasonable sampling density and seismic correlations. A review of the igneous activity to the northwest of the UK is provided by *Ritchie and Hitchen* [1996], who allocated the activity to seven major categories, based on location, age, structure and genetic relationship: (i) the Faeroe Plateau Lava Group (FPLG, see Section 4.7), (ii) the North Rockall Trough - Hebrides Lavas Group; (iii) central igneous complexes, such as St Kilda, Rockall and Erlend; (iv) the Faeroe-Shetland Intrusive Complex; (v) volcaniclastic deposits; (vi) the Minch region; and (vii) the Wyville-Thomson Ridge.

The FPLG (see Section 4.7) and the North Rockall Trough - Hebrides Lavas Group are both part of a much more widespread, in parts discontinuous, subcrop of lavas and sills that extend from the southern end of the Rockall Plateau to the Vøring Plateau (Figure 1). As discussed in the previous section, the age of the FPLG is poorly constrained but it was probably emplaced during C26r-n and C24r times. On the basis of K-Ar and biostratigraphic ages, the North Rockall Trough - Hebrides Lavas Group activity spans 63 to 50 Ma [*Ritchie and Hitchen,* 1996], but this may narrow when ^{40}Ar-^{39}Ar measurements are made. Most of the activity associated with the Faeroe-Shetland Intrusive Complex, a belt of intrusive rocks that covers an area of approximately 40,000 km^2, occurred between 55 and 53 Ma [*Hitchen and Ritchie, 1987; Fitch et al., 1988; Ritchie and Hitchen, 1996*]. Earlier activity recorded in this complex, such as the 80-m.y.-old basalt sill recovered from Well 219/28-2 [*Fitch et al., 1988*] may be related to the Cretaceous magmatism in the Rockall Trough (e.g., Rosemary Bank [*Hitchen and Ritchie, 1993; Morton et al., 1995*] and Anton Dohrn Seamount [*Jones et al., 1994*]).

Volcaniclastic deposits are common in offshore boreholes, and provide important correlation horizons in the North Sea and to the northwest of Britain. Because they can often be dated biostratigraphically, they provide a useful means of dating major volcanic episodes. *Knox and Morton* [1983, 1988] showed that there were two distinct phases of volcaniclastic sedimentation in the North Sea Basin, the first during nannofossil zones NP5-6 (60–57.5 Ma on the *Berggren et al.* [1995] time scale), the second and most voluminous during zones NP9 to 13 (56–50 Ma) (Figure 2). Basic, acid, tholeiitic and alkaline varieties are found. The most extensive horizons occur in the Balder Formation (NP10), which are predominantly derived from Fe-Ti-rich basalt precursors [*Knox and Morton, 1988; Morton and Evans, 1987*]. Eruptive (source) centres were broadly to the west of the British Isles and in the Faeroe-Greenland region during the first phase of activity, but appear to have been restricted to the Faeroe-Greenland area during the second phase [*Knox and Morton, 1988*].

4.4. West Greenland and Baffin Island

Basalts and picrites of Palaeocene age crop out extensively in central West Greenland and in the region of Cape Dyer on Baffin Island [*Clarke and Pedersen, 1976; Clarke, 1977; Larsen et al., 1992*]. The two successions were probably contemporaneous and contiguous. The West Greenland lavas cover an area of approximately 55,000 km^2 [*Clarke and Pedersen, 1976*], and the lava pile may exceed 5 km in thickness on Ubekendt Ejland, although faulting makes accurate determination of the thickness difficult [*Larsen, 1977*]. There are insufficient thickness and age data to make an accurate assessment of their volume or their eruption rates.

The West Greenland lavas have been divided into three lithostratigraphical units by *Hald and Pedersen* [1975]. These are, from old to young: the Vaigat Formation, consisting of lavas and hyaloclastites, mostly of picritic composition; the Maligât Formation, which is dominated

by feldspar-phyric tholeiitic basalts; and the Hareøen Formation, consisting mostly of olivine-phyric basalts. Most of the lavas were erupted subaerially, but the presence of hyaloclastites, breccias and marine mudstones in the Vaigat Formation implies eruption into shallow seawater during the earliest stages of activity. The outcrops at Cape Dyer and along the Baffin coast to the north are generally much thinner than those in West Greenland.

Beckinsale et al. [1974] published a Rb-Sr isochron of 67±5 Ma for a small intrusion emplaced into picrites at Ubekendt Ejland. *Parrott* [1976, in *Clarke et al., 1983*] argued that the bulk of the activity on Ubekendt Ejland occurred between 60 and 56 Ma, on the basis of ^{40}Ar-^{39}Ar dates. The bulk of the West Greenland lavas are reversely magnetised [e.g., *Larsen et al., 1992*], apart from a sequence of normal polarity lavas in the lower part of the Vaigat Formation which *Athavale and Sharma* [1975] tentatively correlated with C25n. On the basis of palynological data [*Piasecki et al., 1992*], however, the bulk of the Vaigat and Maligât Formations appear to have been erupted during C26r, suggesting that the normal event in the Vaigat Formation is C27n. In earlier studies, dinoflagellates recovered from mudstone on Nuussuaq suggested that the earliest lavas of the Vaigat Formation correspond to nannoplankton zone NP3 [*Jurgensen and Mikkelsen, 1974; Larsen et al., 1992*], although it is now thought that this assignment may be too old (L.M. Larsen, pers. comm.). This has been confirmed by recent ^{40}Ar-^{39}Ar dates of 60–60.5 Ma for the oldest West Greenland basalts [*Storey et al., 1996*]. On the basis of the latest age determinations and palynological data, therefore, the main phase of magmatism in West Greenland was essentially contemporaneous with the earliest activity of the British Tertiary Igneous Province and SE Greenland (Figure 2).

Seafloor spreading at a transect at 57–62°N, 600 km to the southeast of Disko, probably began during anomaly 27 time [*Chalmers, 1991*], although *Roest and Srivastava* [1989] have argued for earlier initiation of seafloor spreading (anomaly 33: Late Cretaceous). Detailed information about the structure of the seafloor between Cape Dyer and Disko is not available to the authors. Nonetheless, it appears, on the basis of the interpretation of *Chalmers* [1991], that the main pulse of flood basalt magmatism in the Disko area and the initiation of seafloor spreading at 57–62°N were approximately contemporaneous. This is an important point, because if it could be shown that the onshore magmatism pre-dated the seafloor spreading, then it would demonstrate that the thermal anomaly was a pre-existing feature. Conversely, if the flood basalts and picrites substantially postdate the seafloor spreading, it would indicate the arrival of the thermal anomaly beneath this region.

Onset of seafloor spreading in the Palaeocene is also suggested by the sediments on the Labrador margin and in the Disko-Nuussuaq area [Chalmers, 1991]. No significant unconformity is seen in sediments of Campanian age (C33), as would be expected if rifting and seafloor spreading had occurred at this time. There is, however, a major hiatus in the Danian in the Labrador margin sequences [Balkwill, 1987] and at the top of the Cretaceous in the Disko-Nuussuaq area [Henderson et al., 1976], consistent with footwall uplift and possibly the effects of a mantle thermal anomaly. The main eruptive centres appear to have been seaward of both the Baffin Island and West Greenland successions [Upton, 1988], implying that what is now a graben structure in the Davis Strait was a structural high during the Danian. Southeast-directed syn- and post-volcanic tilting of Disko, and lava thicknesses that indicate flow from the west, also indicate syn-magmatic uplift in the region of Davis Strait [Larsen and Pedersen, 1990, 1992].

An unusual feature of the West Greenland and Baffin Island lavas is the high proportion of picrites, between 30 and 50% of the total lava pile [e.g., Clarke, 1970; Clarke and Upton, 1971; Clarke and Pedersen, 1976; Francis, 1985; Pedersen, 1985; Holm et al., 1993; Gill et al., 1992; Larsen et al., 1992], which is substantially greater than in either East Greenland (~15% of the Lower Basalts; see below) or the British Tertiary Igneous Province. The composition of the parental liquids responsible for the picrites has been the subject of considerable debate; were they primary high-MgO liquids [e.g., Clarke, 1970; Clarke and O'Hara, 1979], or did the liquids undergo olivine accumulation [Hart and Davis, 1978]? Analysed olivines in lavas from the Vaigat Formation have forsterite contents in excess of Fo_{92} [Pedersen, 1985] which implies equilibrium liquid MgO contents of about 19%. Gill et al. [1992] used this figure to estimate potential temperatures in the mantle source between 1540 and 1600°C, assuming a depth of melt segregation equivalent to 2.0 GPa, and anhydrous melting. This implies an excess mantle temperature of between 240 and 300°C, a surprisingly high figure given the distal nature of these lavas in relation to the proposed plume axis [Gill et al., 1992; Chalmers et al., 1995] (see Section 7).

The lavas on Baffin Island are almost exclusively picrites or olivine tholeiites [Francis, 1985; Robillard et al., 1992]. Robillard et al. [1992] identified two compositional types that are stratigraphically interbedded: (i) a light-REE-depleted suite, with La_n/Sm_n~0.6–0.7 and $^{87}Sr/^{86}Sr_{present\ day(pd)}$ (unleached) 0.7031–0.7032 and (ii) a suite that shows slight light-REE enrichment (La_n/Sm_n~1–1.2) and slightly more radiogenic Sr isotope ratios (0.7032–0.7039). They compared these two suites with depleted, normal (N) -MORB, and enriched (E) -MORB, respectively, and argued that they are derived from two distinct mantle sources on the periphery of the ancestral Iceland plume. Picrites from the lowermost parts of the successions on Disko overlap with present-day North Atlantic MORB ($^{87}Sr/^{86}Sr_{(pd)}$ 0.7030–0.7036 and $\varepsilon Nd_{t=60}$ +7.3 to +10.1), although most other picrites from Ubekendt Ejland and Svartenhuk Halvø overlap with MORB and with basalts from Iceland ($\varepsilon Nd_{t=60}$ >+3.4 and $^{87}Sr/^{86}Sr_{(pd)}$ >0.7031) [Holm et al., 1993].

The West Greenland and Baffin picrites and olivine basalts provide evidence of rapid ascent of magma, with minimal interaction with the crust or storage in long-lived magma bodies [e.g., Upton, 1988]. Nonetheless, contamination of picritic and basaltic magma by shale and sandstone in assimilation-fractional crystallisation (AFC)-type processes was demonstrated by Pedersen and Pedersen [1987], who analysed a range of lava types from the Vaigat and Maligât Formations. Rhyolites from the successions appear to represent anatectic crustal melts rather than the products of fractional crystallisation. Igneous activity continued in the West Greenland area, but at declining rates, through the Palaeogene. Alkaline lavas of the Erquâ Formation and a suite of lamprophyres of Oligocene age were emplaced on Ubekendt Ejland [Parrot and Reynolds, 1975; Larsen, 1977].

4.5. Central East Greenland

The on-land portion of the East Greenland magmatic province stretches from Kap Gustav Holm in the south to Shannon Island in the north, a distance of some 1200 km. Figure 1 shows the locations of the main outcrops. The most northerly outcrops, around Wollaston Forland, Hold with Hope, and Shannon Island, will be described in subsection 4.6. Wager [1934], Brooks [1973a], Deer [1976], Noe-Nygaard [1974, 1976], Upton [1988], and Larsen et al. [1989] provided key descriptions.

The province is dominated volumetrically by basalt lavas which comprise the spectacular landscape in central East Greenland, along the Blosseville Kyst between Kangerlussuaq and Scoresby Sund, where the lava sequences may be as much as 7 km thick and individual flows may have volumes of up to 300 km^3 [Nielsen and Brooks, 1981; Larsen et al., 1989]. The total volume of extrusive material preserved in central East Greenland between Kangerlussuaq and Scoresby Sund is approximately 160,000 km^3 [Nielsen and Brooks, 1981;

Larsen et al., 1989]. A further 10,000 km³ may have covered Jameson Land and areas to the north, and up to 60,000 km³ of basalt may have remained on the conjugate plate boundary following plate separation. This gives a total volume of approximately 230,000 km³. In addition, the region is characterised by later intrusive centres, the most famous of which is the gabbro-granophyre Skaergaard intrusion [*Wager and Deer,* 1939], and extensive dyke swarms along the coastal margin.

The basalts of central East Greenland have been divided into two series, the Lower Basalts and the Main Series, comprising a total of 10 formations, some of which are tentatively assigned as lateral equivalents [*Larsen et al.,* 1989]. The oldest preserved part of the sequence, the Lower Lavas or Lower Basalts, is exposed near Kangerlussuaq Fjord, at the southern end of the main outcrop. The Lower Basalts have an estimated thickness of 1.5 km and approximately 15% of the lavas are high-MgO basalts or picrites [*Nielsen et al.,* 1981; *Brooks and Nielsen,* 1982a,b; *Fram and Lesher,* 1996]. The basalts have undergone considerable secondary alteration, in places to greenschist grade, making direct age determination difficult. The Main Series (or Plateau) Basalts, exposed along the Blosseville Kyst and around Scoresby Sund, are predominantly tholeiites. Although Mg-rich varieties occur, none are as picritic as those found in the Lower Basalts. High-Si varieties of basalt occur near the bottom of the sequence and in at least one flow near the top of the Main Series in the Skraenterne Formation [e.g., *Larsen et al.,* 1989].

Where the base of the lava pile is exposed, the basalts sit either on Palaeogene sediments or lap onto the Precambrian basement in the west and north. Most of the lavas were erupted subaerially, through fissures, although some of the earliest flows, those in the Vandsfaldsdalen Formation, were erupted in a shallow marine environment [*Soper et al.,* 1976b], resulting in thick hyaloclastite deposits. *Larsen and Watt* [1985] believed that Mesozoic sediments underlie the entire eastern half of the lava pile. Facies changes indicate a shallowing of the marine basin in the Danian [*Soper et al.,* 1976b]. This is succeeded by an unconformity passing up into coarse sands and volcanogenic sediments of the basal part of the Vandsfaldsdalen Formation. There is no indication in the sediments of strong uplift in this region, or in adjacent source areas, prior to Danian times (pre-65 Ma).

There is uncertainty about the age of the basalts from central East Greenland. The main constraints are provided by microfossils and palaeomagnetism, neither of which provide an absolute age. All of the basalts, including the Lower Basalts, were erupted during a period of reversed polarity [e.g., *Tarling,* 1967; *Soper et al.,* 1976b]. A marine dinoflagellate (*Apectodinium homomorphum*) is found in shales interbedded with hyaloclastites at the base of the Lower Basalts in the Kangerlussuaq region [*Soper et al.,* 1976a,b]. Previous studies have suggested that the dinoflagellate has a range from mid-Thanetian to early Bartonian (ca. 56–40 Ma on the time scale of *Berggren et al.* [1995]) and that its base corresponds to the base of nannoplankton zone NP9 (the so-called 'Base Apectodinium Datum') [*Powell,* 1988], implying that the basalts were erupted during C24r [e.g., *Berggren et al.,* 1985]. However, this biostratigraphic control should be used with caution (D. Jolley, pers. comm.). Several occurrences of *Apectodinium* spp. have been reported at stratigraphic levels below the 'Base Apectodinium Datum', and *A. homomorphum* occurs in strata of Danian age in the Maureen Formation of the central North Sea [*Thomas,* 1996]. Given these uncertainties, the Lower Basalts may have been erupted during either C26r, C25r, or C24r. *Noble et al.* [1988] suggested that the Main Series basalts along the Blosseville Kyst were erupted between 53 and 57 Ma (K-Ar dates). *Hansen et al.* [1989] obtained an incremental heating ^{40}Ar-^{39}Ar age of 56.7±4.3 Ma for a basalt from the Main Series in the Scoresby Sund area, and *Storey et al.* [1996] suggested, on the basis of ^{40}Ar-^{39}Ar data, that the basalts of the Lower Series may be 59–60 m.y. old.

The basalts from the Scoresby Sund region are predominantly tholeiites that underwent extensive differentiation in mid- to upper-crustal magma chambers, resulting in the production of high-TiO_2 ferrobasalts or titano-tholeiites. The depth of fractionation is indicated by the displacement of the basalts to the low-clinopyroxene side of the 1-atmosphere cotectic on the normative *ol-di-hy* triangle of *Thompson* [1982], consistent with fractionation at about 0.35 GPa (11 km) [*Larsen et al.,* 1989]. That these magma chambers were open is indicated by the cyclical eruption of high Ti, Fe tholeiites and, indeed, by the large volumes of individual flows [*Hogg et al.,* 1989; *Brooks et al.,* 1991]. Evacuation of a realistic portion of a magma body (~1% [*O'Hara and Mathews,* 1981]) to produce flows with a typical volume of 20 to 60 km³ requires a chamber of up to 12,000 km³ [*Larsen et al.,* 1989]. As pointed out by *Larsen et al.* [1989], a sill-like elliptical chamber with dimensions 150 × 30 × 5 km would fit easily within the area of the postulated feeder dyke swarms (~200 × 30 km).

There are few published isotope or comprehensive trace element data for the basalts from central East Greenland. The bulk of the data are for the Lower Basalts at Kangerlussuaq [*Holm,* 1988; *Gill et al.,* 1988]. *Larsen et al.* [1989] published large amounts of major and some trace element data for the Main Series basalts, and *Holm* [1988] included isotopic and trace element analyses for eight Main Series basalts. The Fladø Dykes, emplaced immediately to the south of Kangerlussuaq, are considered to be the hypabyssal equivalents of the Lower Basalts, and at the time of writing they provide the most complete trace element data set for the central East Greenland basalts [*Gill et al.*, 1988].

Values of $^{87}Sr/^{86}Sr_{(pd)}$ and $\varepsilon Nd_{(pd)}$ range from 0.7032 to 0.7094 and +7.8 to -5.8, respectively, for the entire central East Greenland basalt province [*Carter et al.*, 1979; *Holm*, 1988; *Larsen et al.*, 1989]. The lowest $^{87}Sr/^{86}Sr$ and highest εNd values overlap those of present-day ratios erupted in the active rift zones of Iceland (e.g., Figure 7); most of the Scoresby Sund basalts fall within the range 0.7034⁻0.7038. Therefore, the bulk of the basalts have isotopic signatures suggesting derivation from a sublithospheric source, although the (limited) Pb isotope data suggest some contamination of the magmas with unradiogenic Pb [*Holm*, 1988]. The greatest isotopic range is in the Vandsfaldsdalen Formation in the Lower Basalts, but high $^{87}Sr/^{86}Sr$, high-SiO$_2$ basalts are found elsewhere in the succession; for example, in the Skraenterne Formation [*Larsen et al.*, 1989]. Similar, high-SiO$_2$ and high $^{87}Sr/^{86}Sr$ basalts are found in West Greenland [*Pedersen*, 1985; *Pedersen and Pedersen*, 1987] and the Faeroes [*Hald and Waagstein*, 1983; *Gariépy et al.*, 1983]. The Lower Basalts include picrites and ankaramites, equivalents of which are also found in the Fladø dykes [*Holm*, 1988; *Gill et al.*, 1988]. Isotope data are available for only one ankaramite ($^{87}Sr/^{86}Sr_{(pd)}$ = 0.7064) and three picrites (0.7031–0.7044), which have values overlapping those of lower-Mg lavas in the Lower Basalts.

Many of the picrites and ankaramites of the Lower Basalts and Fladø Dykes have high abundances of the highly incompatible elements (e.g., Rb to Ce on Figure 9), and steep REE profiles [*Gill et al.*, 1988]. Some tholeiites of the Lower Basalts share this characteristic, but tend to have lower La$_n$/Yb$_n$ ratios. The majority of the tholeiites of the Lower Basalts have flatter REE patterns and show relative depletion of the highly incompatible elements (e.g., samples GGU267909 and GM20332 on Figure 9). The limited trace element data for the Main Series Basalts [*Holm*, 1988; *Larsen et al.*, 1989] suggest that they share the characteristics of the 'depleted' tholeiites of the Lower Basalts (for example, similar Zr/Y ratios and similar trace element patterns from Rb to Ti), although the absolute abundances of incompatible elements may be high, especially in the evolved ferrobasalts and titano-tholeiites from Scoresby Sund [*Larsen et al.*, 1989]. On the basis of

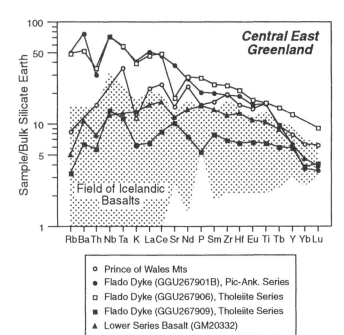

Figure 9. Chondrite-normalised REE and bulk silicate Earth-normalised trace element patterns for basalts from Kangerlussuaq (Flado Dykes and Lower Series Basalts) and Prince of Wales Mountains (average value). Data sources: Kangerlussuaq - *Gill et al.* [1988]; Prince of Wales Mountains - *Hogg et al.* [1989]. Field for Iceland taken from Figure 4.

parameters such as La/Yb and Zr/Nb, *Holm* [1988] also recognised two groups in the Lower Basalts, which broadly correspond to this bipartite division into 'enriched' and 'depleted' types. Both groups show a range of isotope values (e.g., εNd: Figure 7), which *Holm* [1988] attributed to mixing between an Icelandic-type mantle source and old continental lithospheric mantle. Although there can be little doubt that crustal contamination was involved in producing the high $^{87}Sr/^{86}Sr$, high-SiO$_2$ characteristics of some basalts from East Greenland, we believe that the evidence for the involvement of continental lithospheric mantle is far from clear-cut, and return to this point in Section 6.

4.6. *Northeast Greenland*

Scattered outcrops of Tertiary basalts crop out along the East Greenland margin between 72° and 76°N. In the Gauss-Halvø - Hold with Hope region, the lava succession is ~800 m thick [*Upton et al.*, 1980, 1995] and has been divided into Lower and Upper Series. A recent ^{40}Ar-^{39}Ar date on a basal nephelinite from Hold with Hope gives an eruption age of 58.7 ± 1.4 Ma [*Upton et al.*, 1995]. Two

dykes which postdate the Upper Series give ages of 56.7 ±
0.7 and 56.6 ± 1.9 Ma, respectively. *Upton et al.* [1995]
argued that these dates, combined with the available
palaeomagnetic evidence, indicate that the Lower Series
and the earliest part of the Upper Series belong to C24r,
with succeeding magnetically normal polarity lavas
belonging to Subchron 24n.3. However, it is more likely
that eruption occurred during C26r-26n or 25r-25n on the
basis of the revised time scale of *Berggren et al.* [1995]. A
late sheet from the Myggbukta Complex gives ^{40}Ar-^{39}Ar
age of 32.7 ± 2.9 Ma [*Upton et al.,* 1995]. The causes of
this late magmatic event are unclear, but *Upton et al.*
[1995] related it to minor tectonic adjustments along the
continental margin.

The Lower Series basalts have been correlated with
lavas from the upper part of the succession recovered from
Hole 642E on the Vøring Plateau (see below). They are
mildly light-REE-enriched quartz tholeiites (La_n/Yb_n=2
[*Thirlwall et al.,* 1994], Figure 10), with Nd, Sr and Pb
isotopic ratios similar to Icelandic tholeiites. Pb isotopes
suggest that the basalts of the Lower Series were not
contaminated by continental crust. The lavas may have
been erupted in an oceanic setting, a suggestion supported
by the moderate La_n/Yb_n ratios that imply shallow melting
and melt segregation [*Thirlwall et al.,* 1994]. The Upper
Series lavas, on the other hand, are regarded as higher-
pressure melts of a similar source and have variable
amounts of contamination. Like the lavas from the
Scoresby Sund-Kangerlussuaq region, the basal units were
erupted into shallow water, but the remainder of the
basalts were subaerial.

Two periods of Tertiary magmatism have also been
identified at Traill Ø, approximately 150 km south of Hold
with Hope [*Price et al.,* 1996]. The first comprises a series
of tholeiitic sills, which have an emplacement age of
approximately 54 Ma (^{40}Ar-^{39}Ar step heating). The second
comprises smaller volumes of alkalic dykes and two large
syenite complexes, emplaced at about 36 Ma and of a
similar age to the Myggbukta Dykes.

4.7. Faeroe Islands

The Faeroe Islands are the exposed part of the much
larger Faeroes Block, a fragment of continental crust
capped by a thick sequence of basalt lava flows, the Faeroe
Plateau Lava Group (FPLG) [*Casten,* 1973; *Casten and
Nielsen,* 1975; *Bott et al.,* 1974; *Hald and Waagstein,*
1984; *Ritchie and Hitchen,* 1996]. To the northwest, the
crust becomes entirely oceanic in structure, as it grades
into the Faeroes-Iceland Ridge. Southwestwards, the block
bounds the Rockall Plateau, a fragment of Precambrian

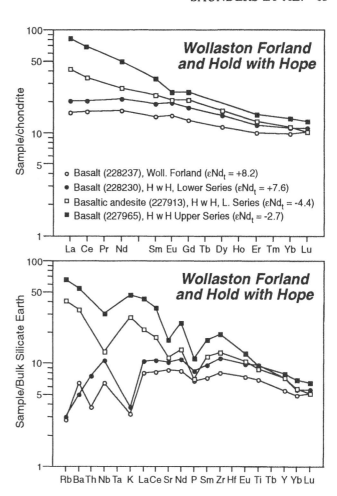

Figure 10. Chondrite-normalised REE and bulk silicate Earth-
normalised trace element patterns for basalts and basaltic andesite
from Wollaston Forland, and Lower and Upper Series, Hold with
Hope, NE Greenland. Data source: *Thirlwall et al.* [1994]. Note
the similarity of the patterns of the high εNd Lower Series lava
with the basalts from the Upper Series from the Vøring Plateau
(Figure 13). εNd calculated assuming t=50 Ma.

lithosphere that has a capping of Palaeocene basalts and
Cenozoic sediments [*Roberts,* 1975]. The Faeroe Island
lavas cover an area of approximately 1400 km^2 on land,
but extend considerably farther than this offshore. The
lavas have an exposed thickness of about 3 km, but the
base has not been recovered, even by drilling to a depth of
2 km in the Lopra-1 borehole and to 660 m in the
Vestamanna-1 borehole. Therefore, the entire sequence
exceeds 5 km in thickness [*Hald and Waagstein,* 1984;
Waagstein and Hald, 1984; *Waagstein,* 1988].

The FPLG has been divided into three series, Lower,
Middle, and Upper, on the basis of minor unconformities
[*Noe-Nygaard and Rasmussen,* 1968]. All of the lavas
were erupted subaerially, and progressive uplift westwards

of the islands resulted in gentle tilting towards the east. There was a hiatus between the Lower and Middle Series lavas, marked by a 10-metre-thick layer of clays and coal. The latter has been dated as late Palaeocene [*Lund*, 1983]. There are presently no reliable radiometric ages available for the FPLG. All of the Middle and Upper Series lavas are magnetically reversed, but the Lower Series contains at least two normal polarity events [*Nielsen*, 1983]. *Waagstein* [1988], on the basis of comparisons with marine magnetic anomaly data, proposed that the lavas of the Lower Series were emplaced during the period early 26r to early 24r, and that the Middle and Upper Series were emplaced during C24r.

All of the analysed FPLG are tholeiites. Those from the Lower Series are silica-oversaturated, whereas the Middle and Upper Series basalts are olivine tholeiites. A perhaps more fundamental change in composition occurs near the boundary between the Middle and Upper Series, where there is a change from entirely light-REE-enriched to a mixture of light-REE-depleted and light-REE-enriched tholeiites [*Noe-Nygaard and Rasmussen*, 1968; *Schilling and Noe-Nygaard*, 1974; *Bollingberg et al.*, 1975; *Gariépy et al.*, 1983] (Figure 11). Some of this variation in REE patterns may be due to contamination of magmas by continental crust (e.g., sample K-1), but it is likely that variations in the composition of the source or depth of melting also played an important role (see Sections 6 and 7) [*Schilling and Noe-Nygaard*, 1974; *Wood*, 1979a,b].

4.8. Seaward-Dipping Reflector Sequences

Seaward-dipping reflector sequences (SDRS) have a characteristic, off-lapping architecture on seismic profiles. Their volcanic nature is now firmly established by deep-sea drilling [*Roberts et al.*, 1984; *Eldholm et al.*, 1987; *Larsen et al.*, 1994]. The SDRS in any one region may be up to 6 km thick, and they are complemented by thick sequences of basic intrusive sequences at lower to middle crustal level [e.g., *White et al.*, 1987]. Furthermore, the majority of the lavas, where sampled, were erupted subaerially or in shallow water, testifying to substantial support of the margin during rifting and plate separation.

In addition to seismic reflection and refraction studies, the seaward-dipping reflector sequences in the North Atlantic have been the target of six Deep Sea Drilling Project (DSDP) and Ocean Drilling Program (ODP) cruises: Legs 48 and 81 on the Rockall Plateau; Legs 38 and 104 on the Vøring Plateau; and Legs 152 and 163 on the SE Greenland margin.

4.8.1. Rockall Plateau/Hatton Bank (DSDP Legs 48 and 81). Leg 48 failed to reach igneous basement due to

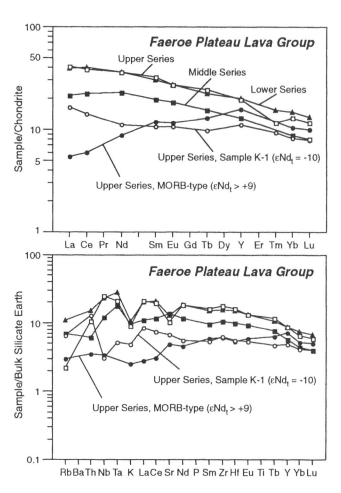

Figure 11. Chondrite-normalised REE and bulk silicate Earth-normalised trace element patterns for basalts from the Faeroe Islands. Average values for the Upper (excluding high-εNd, MORB-like depleted basalts), Middle and Lower Series are shown. High εNd, MORB-like basalts from the Upper Series are shown as a separate average value. Sample K-1 is a highly contaminated basalt from the Upper Series, and is plotted on Figures 6 and 7. Data source: *Gariépy et al.* [1983]. εNd calculated assuming t=60 Ma.

technical difficulties. During Leg 81, however, basaltic basement was reached at all of the drilled sites (552 through 555) [*Roberts et al.*, 1984]. Sites 552 and 553 were located on the main SDRS; Site 555 on the flanks of Hatton Bank on the most 'landward' or eastern part of the SDRS; and Site 554 was on the western edge of the SDRS. Penetration of the basaltic basement was in excess of 100 m at three of the four sites.

The basalts from the upper part of the sequence at Site 555 (the most landward of the four sites) are reversely magnetised and gave K-Ar ages of 52.3±1.7 and 54.5±2.0 Ma [*Macintyre and Hamilton*, 1984], possibly

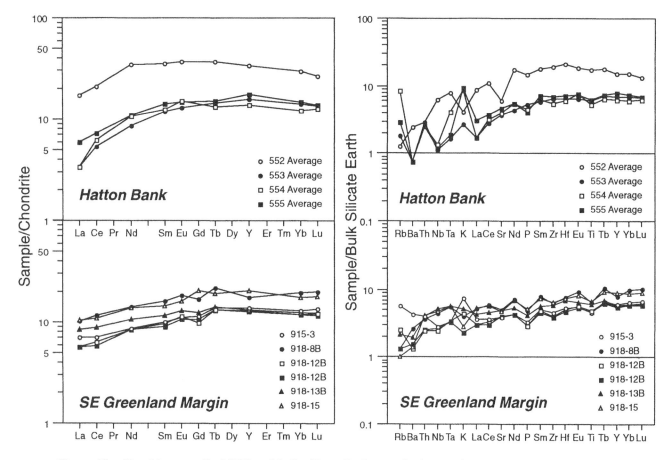

Figure 12. Chondrite-normalised REE and bulk silicate Earth-normalised trace element patterns for basalts from Hatton Bank on the SW Rockall Plateau, and SE Greenland. Data for Hatton Bank represent averages by ODP Site (552 through 555). Data for SE Greenland represent individual samples. Data sources: Hatton Bank - *Merriman et al.* [1988], with revised Nb values from *Brodie and Fitton* [1996]; SE Greenland - *Fitton et al.* [1996b].

corresponding to C24r. Basalt samples analysed ^{40}Ar-^{39}Ar by *Sinton and Duncan* [1996] all show disturbed age spectra, although results for two lavas from Site 555 which indicate an eruption age of 57.6±1.3 and 57.1±5.6 Ma, respectively, are considered to be reliable; these ages would correspond to C25r. These basalts lie above sediments belonging to nannofossil Zone NP9 (upper Palaeocene) [*Backman,* 1984], so although the K-Ar age is consistent with the biostratigraphic data, the ^{40}Ar-^{39}Ar ages are significantly older. The NP9 sediments are predominantly volcanogenic and were succeeded by sediments deposited in a brackish, intertidal lagoonal environment. The basalts at Sites 552 and 554 were erupted in a shallow marine environment, whereas those at Site 553 were erupted subaerially. The basalts at Site 552 are overlain by sediments of NP11 and possibly NP10 (early Eocene) age, and there are abundant tuffs interbedded with the overlying sediments. Basalts at Sites 553 and 554 are at least NP11 (early Eocene) age.

The basalts recovered from Hatton Bank are all tholeiites that show strong depletion of the light REE (Figure 12), similar to parts of the Upper Lavas of the Faeroes or the Preshal More basalts from Skye [*Joron et al.*, 1984; *Merriman et al.*, 1988]. Despite the strong light REE depletion, however, it is apparent from the Pb isotopes that the basalts from Site 553 have suffered minor contamination by material with low ^{206}Pb/^{204}Pb (<17), but MORB-like ^{207}Pb/^{204}Pb, possibly mid-Proterozoic, Laxfordian continental crust [*Morton and Taylor,* 1987; *Merriman et al.,* 1988]. Such material has been dredged from the Rockall Bank [*Miller et al.*, 1973; *Morton and Taylor,* 1991].

4.8.2. Vøring Margin: DSDP Leg 38 and ODP Leg 104. Three sites were drilled seaward of the Vøring Plateau escarpment during Leg 38 [*Talwani et al.,* 1976]. Short sections of altered basalt were recovered at all three sites, although the importance of these rocks was not fully realised until *Hinz* [1981] proposed that the SDRS

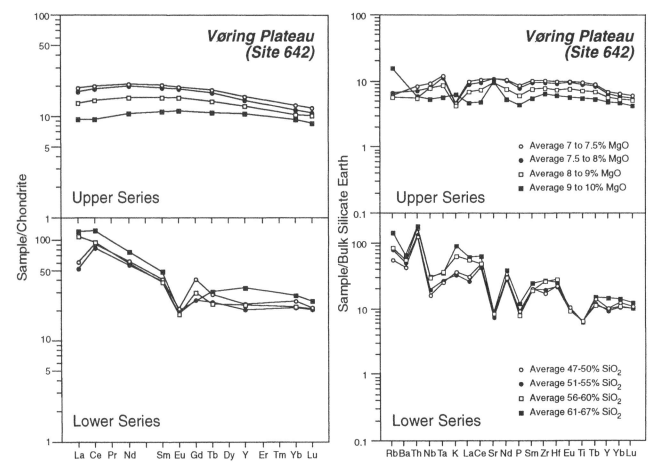

Figure 13. Chondrite-normalised REE and bulk silicate Earth-normalised trace element patterns for basalts and evolved lava compositions from the Vøring Plateau (ODP Leg 104, Site 642). Samples are averages on the basis of MgO content (for the Upper Series basalts) and SiO_2 content (for the Lower Series basalts and dacites). Data sources: *Viereck et al.* [1988, 1989].

represent accumulations of basaltic lavas. A 900-m-thick Eocene volcanic section was drilled on the Vøring Plateau at Hole 642E during Leg 104 [*Eldholm et al.,* 1987]. The volcanic succession was divided into Upper and Lower Series, separated by about 7 m of estuarine, volcaniclastic sediments. All of the lavas recovered from the Vøring Plateau were erupted in a subaerial environment, except for the some pillow basalts from Site 643 on the outer edge of the SDRS [*Eldholm et al.,* 1987].

The basalts from the Upper Series at Site 642 include tholeiites with flat to slightly light-REE-enriched profiles, and they are compositionally similar to basalts from the Reykjanes Ridge [*Viereck et al.,* 1988] (Figure 13). The $\varepsilon Nd_{t=60}$ values (+6 to +7.5) are slightly lower than in Icelandic basalts of equivalent Zr/Nb (Figure 7). The Lower Series drilled at Site 642 comprises basalts, intermediate rocks and 13 peraluminous, cordierite-bearing dacite flows. The dacites show strong light-REE

enrichment, low abundances of Nb, Sr, P, Eu and Ti (Figure 13), elevated $^{87}Sr/^{86}Sr$ (0.7088 to 0.7118) and low $\varepsilon Nd_{t=60}$ (-7 to -9) (Figure 6). These data are consistent with assimilation of substantial amounts of metaluminous continental basement or overlying sediments [*Viereck et al.,* 1988].

The lavas of the Lower Series are all normally magnetised, whereas those of the Upper Series are reversely magnetised. They have been tentatively assigned to Chrons 25n-24r [*Schönharting and Abrahamsen,* 1989] but without radiometric age data this assignment must be considered as speculative. Various workers have suggested, on the basis of geochemical criteria, that the basalts from the Vøring margin correlate with basalts from the conjugate margin at Hold with Hope and Wollaston Forland [*Schilling,* 1976; *Viereck et al.,* 1988; *Upton et al.,* 1995]. Because the age of the NE Greenland basalts is well constrained, an implication of this correlation, if correct, is that the Upper Series basalts from Site 642 are no younger than C25r.

4.8.3. *SE Greenland Margin: ODP Legs 152 and 163.* A transect of five basement sites was drilled at 63°N on the SE Greenland margin during Legs 152 and 163 [*Larsen et al.*, 1994; *Duncan et al.*, 1996]. This margin was chosen for study because of its relatively simple structure, detailed seismic coverage, and the presence of a 150-km-wide subcrop of SDRS. The occurrence in the distal portion of the SDRS of a magnetic anomaly corresponding to C24n indicated that the oldest basalts from the sequence were at least C24r.

The most successful penetration and recovery was at Hole 917A, close to the inner part of the SDRS on the continental shelf, where 91 extrusive units were identified from the 749-m-deep hole. The sequence was divided into three series [*Larsen et al.*, 1994]. Lavas from the Lower Series include basalts that show significant lithospheric contamination, but some have preserved a sublithospheric chemical signature and resemble tholeiites from Iceland [*Fitton et al.*, 1996b] (these can be seen on Figure 7B). The Middle Series lavas, which include dacites, are clearly contaminated by crust [*Fitton et al.*, 1996a,b; *L.M. Larsen et al.*, 1996]. The Upper Series, which exhibits much less contamination, contains magnesian basalts and picrites [*Fram et al.*, 1996]. The Upper Series is separated from the Middle Series by a thin sedimentary horizon, implying a marked hiatus in eruptive activity. All of the lavas recovered at Site 917 were erupted in a subaerial environment, and are reversely magnetised. On the basis of ^{40}Ar-^{39}Ar ages, the basalts of the Lower and Middle Series were erupted approximately 61–62 m.y. ago (C27r according to the time scale of *Berggren et al.* [1995]) [*Sinton and Duncan*, 1996; *Werner et al.*, 1996]. There are no radiometric ages for the lavas of the Upper Series, which means that the age of picritic magmatism at this site is not precisely constrained.

Basalts were also recovered at Site 989, inboard of Site 917; Sites 915 and 990, approximately 3 km to the SE of Site 917; and at Site 918, located close to the centre of the outcrop of the dipping reflectors. All of the basaltic flows recovered from Site 918 appear, on the basis of Sr, Nd and Pb isotopes, to be uncontaminated by continental crust and show strong light-REE depletion (Figure 12) [*Fitton et al.*, 1996a,b]. In many respects they resemble the basalts recovered from Hatton Bank, and the data for both areas are plotted on Figure 12 for direct comparison. Like those from Site 917, the basalts recovered at Site 918 were erupted in a subaerial setting, and most are reversely magnetised. The uppermost unit at Site 918 provides a reliable ^{40}Ar-^{39}Ar age of 52 Ma [*Sinton and Duncan*,

1996]. This unit may be a sill, so the underlying basaltic flows may be significantly older, but the seafloor magnetic lineations suggest that the bulk of the SDRS were erupted during C24r [e.g., *Larsen and Jakobsdóttir*, 1988].

4.8.4. *Deep seismic profiling of the continental margins.* An important feature of the volcanic rifted margins of the North Atlantic is the presence of thick prisms of material with high seismic velocities (7.3–7.4 km s^{-1}) at lower crustal depths. These prisms, which have been observed under the Rockall [*White et al.*, 1987; *Fowler et al.*, 1989; *Morgan et al.*, 1989; *Barton and White*, 1995], Vøring [*Mutter et al.*, 1982; *Skogseid and Eldholm*, 1988; *Mutter and Zehnder*, 1988], SE Greenland [*Larsen and Jakobsdóttir*, 1988], and Lofoten [*Goldschmidt-Rokita et al.*, 1994] margins, have been interpreted as underplated olivine gabbro or as continental crust with a high proportion of gabbroic intrusions; they can be considered as the plutonic equivalents, perhaps with a substantial cumulate component, of the SDRS. The thickness of the underplated prisms varies considerably, ranging from a few kilometres at the distal regions of the province (e.g., 5 km at the Edoras Bank, on the southeast flank of the Rockall Plateau [*Barton and White*, 1995]), to more than 10 km on the Hatton Bank and Vøring margins [e.g., *White et al.*, 1987; *Mutter and Zehnder*, 1988; *Mutter et al.*, 1988]. The large volumes of underplated material strongly influence calculations of total magma production. The increase in the thickness of the igneous rocks along the European margin as the Faeroes-Iceland Ridge is approached implies either an increase in potential temperature and/or an increase in mantle convection towards the ancestral plume axis [*Barton and White*, 1995].

4.9. *Summary Statement*

The activity associated with the North Atlantic Igneous Province appears to have begun in the Early Palaeocene, approximately 62 m.y. ago, during magnetic reversal 27r or 26r (Figure 2). Earlier, possibly Cretaceous, magmatism is recorded on the Anton Dohrn Seamount, Rosemary Bank, and the Faeroes-Shetland Basin, but their relationship with the later activity is unclear. The possibility that the North Atlantic Igneous Province had precursor magmatism substantially before 62 Ma cannot be precluded, although the volumes are likely to have been small.

The available age data indicate two main phases of igneous activity within the North Atlantic Igneous Province as a whole. Phase 1, recorded in the terrestrial, continent-based lava sequences in West Greenland, SE

Greenland, NE Greenland, and the British Isles, occurred during C26r (and possibly began during C27r) and lasted for about 4 m.y., from 62 to 58 Ma. The Lower Series lavas from central East Greenland may also belong to Phase 1. Many Phase 1 basalts show evidence of contamination as they ascended through the continental lithosphere (see Section 6), although within individual subprovinces it is not unusual to find lavas that have undergone minimal contamination (e.g., some of the basalts on Skye and Mull). This first phase of activity was also characterised by frequent eruption of picritic magmas, implying a hot mantle source, a point we return to below. Residual activity continued in several regions.

Phase 2 began at about 56 Ma, in the late Palaeocene, and is linked with the breakup of the NE Atlantic, formation of the bulk of the SDRS along the continental margins, and the eruption of at least the Main Series basalts in central East Greenland. The Phase 2 activity was localised along the lines of continental breakup and, in the case of the central East Greenland flood basalts, close to the ancestral plume axis. The timing of emplacement was therefore controlled in part by the timing of plate separation. The formation of the SDRS was essentially diachronous. Phase 2 continues at the present day in Iceland but with (1) more focused activity and (2) lower magma production rates than accompanied plate breakup during the Palaeocene/Eocene. Many of the magmas associated with Phase 2 are uncontaminated by continental lithosphere, presumably because they were erupted during or after plate rupture and separation.

We emphasise, however, that the age of the emplacement of the SDRS is poorly constrained. For example, the only SDRS that have been drilled to underlying basement (Site 917 on the SE Greenland margin) revealed an unexpected series of older basalts (26r instead of the anticipated 24r). This result does not negate the observations in the previous paragraph, because there is no reason why the lower SDRS in a region are not pre-breakup continental flood basalts, preserved by the subsequent rifting, subsidence and burial. It is likely, however, that as more data become available, Phases 1 and 2 will become contiguous, but the present data suggest that there was a short hiatus between the two. A further anomaly that concerns us is the age of the central East Greenland basalts. It is not clear why eruption did not begin in 26r or 27r times. The ancestral plume axis was closer to Kangerlussuaq than either SE Greenland or NW Britain at 62–58 Ma, whether either *Lawver and Müller's* [1994] or *White and McKenzie's* [1989] positioning of the hotspot is used. We suggest that the Lower Basalts of the Kangerlussuaq area may in fact be older than indicated by

the published biostratigraphic evidence. Confirming this by radiometric methods will be difficult, given the extent of metamorphism of these lavas, but we note that preliminary age determinations support an early period of eruption [*Storey et al.,* 1996].

5. RATES OF MAGMA PRODUCTION AND CRUSTAL GENERATION

A characteristic of many LIPs is the transient nature of the bulk of the magmatism. It has long been recognised that the rate of magma production in the North Atlantic increased during breakup and then sharply declined, but these rates are only order-of-magnitude estimates. This impression simply reflects the uncertainties in the volumes of magma involved, and in the timing and duration of emplacement. *Roberts et al.* [1984] estimated the total volume of Palaeocene to early Eocene basalt to be 2×10^6 km^3, whereas *White et al.* [1987] suggested a total igneous volume, including erupted material and additions to the deeper crust, of between 5×10^6 and 1×10^7 km^3. The most accurate estimates currently available are probably those of *Eldholm and Grue* [1994]. Their estimate of the total crustal volume is 6.6×10^6 km^3, close to the lower estimate of *White et al.* [1987] and, by their own admission, a conservative figure. This volume comprises the SDRS and deep crustal prisms associated with the Atlantic margins and the East Greenland basalts but excludes the magmatism associated with the Greenland-Iceland-Faeroes Ridge and the Phase 1 activity in West Greenland, Baffin Bay, and the British Isles, which clearly predates the breakup magmatism by several million years. It also excludes dykes and material that may have been eroded or buried beneath the Greenland ice cap.

Calculating eruption rates and duration of activity is fraught with uncertainty. The problem is compounded by a paucity of age data. On a broad scale, *Eldholm and Grue* [1994] estimated the mean extrusion rate to have been about 0.6 km^3/yr, or 2.3×10^{-4} km^3 per km of rift/yr, assuming steady magma production over a 3 m.y. period along a 2560 km rift zone. These figures refer only to the volcanic portions of the margins, and they increase dramatically if the bulk of the magma was produced during the earliest stages of breakup; if two-thirds of the basalts were emplaced within 1.0 or 0.5 m.y. after breakup, the initial eruption rates would be 2 to 4 times higher [*Eldholm et al.,* 1987; *Eldholm and Grue,* 1994]. *Larsen et al.* [1989] calculated the effusion rates in central East Greenland to be 2×10^{-4} km^3 per km of rift/yr, similar to the figures calculated by *Nielsen and Brooks* [1981]. Similarly, *Larsen and Jakobsdóttir* [1988] estimated that eruption rates of 4.5–4.6 × 10^{-4} km^3 per km of rift/yr were involved in the formation of the SDRS along the East Greenland margin. For comparison, *Pálmason* [1986] estimated that the eruption rates in Iceland are of the order of 1.33 × 10^{-4} km^3 per km of rift/yr, with a maximum near the centre of the island of 2.5 × 10^{-4} km^3 per km of rift/yr [*Jakobsson, 1972*].

An alternative way of evaluating the magmatic productivity is to consider the total output per kilometre of rift axis, including lavas and plutonic rocks. If the Icelandic crust has an average thickness of 25 km [*Bott*, 1983; *White et al.*, 1995], then the mean magmatic productivity over the last 15 m.y. has been 4 to 5 × 10^{-4} km^3 per km of rift/yr. By comparison, if the entire igneous crust of the North Atlantic margins (6.3 × 10^6 km^3) was emplaced in 3 m.y., during C24r, then the total magmatic productivity was on the order of 8.2 × 10^{-4} km^3 per km of rift/yr. If the bulk of the magmatism occurred during the initial 1 m.y. after breakup, then this figure would be higher still. Along-margin variations in crustal thickness also imply that total magmatic productivity varied as function of distance from the plume axis [e.g., *Barton and White*, 1995].

Clearly, far more data are required to verify and refine these estimates; in particular, deep crustal seismic data to ascertain the volumes of deep crustal intrusive bodies, and reliable ages for the SDRS are needed. At 63°N on the SE Greenland margin, recent drilling and age determinations [*Larsen et al.*, 1994; *Sinton and Duncan*, 1996] have revealed that a substantial portion of the lowermost SDRS were erupted earlier than C24r, implying that the effusion rates are lower than those originally estimated by *Larsen and Jakobsdóttir* [1988]. Nonetheless, there appears to be a consistent message that the magmatic productivity during breakup was 2 to 3 times greater than on present-day Iceland.

6. THE INFLUENCE OF THE CONTINENTAL LITHOSPHERE

The volumes and composition of magmas reflect a complex interplay between the lithosphere and the underlying asthenosphere. The lithosphere not only serves to truncate the low-pressure regions of any melting column (even to the extent of precluding melting altogether) but also may modify magmas by contamination and crustal-level fractionation. Questions that are central to flood basalt formation include the following. What is the role of the continental lithosphere in flood basalt genesis? Is it simply a physical control of the melt zones in the underlying asthenosphere, or does the lithosphere melt to

produce significant volumes of magma? To what extent does pre-existing topography at the base of the lithosphere influence the channelling of hot, buoyant plume material, or how do thinspots in the lithosphere influence the location and volume of magmatism? To what degree are magmas contaminated as they ascend through the lithosphere (especially by continental crust)? What is the extent of medium- to low-pressure fractionation in intracrustal magma chambers, especially in controlling eruption of picrites?

6.1. *Crustal Contamination*

Many lavas associated with North Atlantic activity were erupted through and onto continental crust, so crustal contamination of magmas is expected. Furthermore, a substantial portion of the province has been erupted through lithosphere of Archaean or Proterozoic age (for example, all of the Scottish province north of the Great Glen Fault), with distinctive trace element and isotopic characteristics. Here, we briefly review evidence for contamination, beginning with the British Tertiary Igneous Province, because it has received the greatest attention.

Crustal contamination of asthenosphere-derived Skye and Mull lavas has been postulated by many authors including *Moorbath and Bell* [1965], *Moorbath and Welke* [1969], *Carter et al.* [1978], *Moorbath and Thompson* [1980], *Dickin* [1981, 1988], *Thompson et al.* [1982], *Thirlwall and Jones* [1983], and *Kerr et al.* [1995a]. Of interest here is the identification of the contaminants and the mechanism of their incorporation in the magmas. *Carter et al.* [1978] identified two fundamentally different contaminants by using Sr and Nd isotope data. The premise of their argument was that the two contaminants, granulite- and amphibolite-facies rocks, have very different isotopic characteristics, the granulite gneiss in particular having low $^{87}Sr/^{86}Sr$ ratios reflecting low time-integrated Rb/Sr ratios. They concluded that in general the basic lavas from Skye, Mull and the Small Isles were contaminated by granulite-facies crust of Lewisian (Archaean) age, whereas the Skye granites contain a large component of amphibolite-facies Lewisian crust. The low Rb and Th and high Ba content of many basalts from the British Tertiary Igneous Province can be explained by the incorporation of small amounts of Lewisian granulite. This is illustrated by Figures 6, 7, and 8. Note in particular the displacement to low εNd values, but at constant Zr/Nb, of the Mull and Skye suites on Figure 7. Most putative crustal contaminants exert relatively minor influence on Zr/Nb, whereas Precambrian crustal rocks, in particular Archaean granulites and amphibolites, have dramatically lower εNd values than magmas derived from sublithospheric sources. Similar displacements are also shown by magmas from central East Greenland, the Faeroes, and the SE Greenland and Vøring Plateau SDRS Lower Series lavas.

Subsequent studies have refined this model using a variety of isotope and trace element tracers and discriminants. *Dickin* [1981] showed that the magmas of the Skye Main Lava Series were contaminated primarily by granulite-facies crust, but that the magmas that produced the succeeding Preshal More basalts and Cuillin gabbros were contaminated by amphibolite-facies crust. One interpretation is that the site of crustal assimilation migrated upwards through the crust with time. Other parts of the Hebridean province have also been 'fingerprinted' in this way. Minor pitchstone and felsite intrusions in the Eigg lava pile show contamination by granulite-type material [*Dickin and Jones,* 1983], and the layered gabbros of the Rum intrusion show evidence of upper crustal contamination [*Palacz,* 1984]. *Thompson et al.* [1986] identified possible contamination from Moine schists in several basal tholeiites from SW Mull, and *Wallace et al.* [1994] recently demonstrated that the crustal contaminant of the Tertiary basalts in Northern Ireland is probably of Proterozoic age, reflecting the younger age of the basement below that region.

Crustal assimilation models of different levels of sophistication have been proposed. These range from the combined assimilation-fractional crystallisation (AFC) model of *DePaolo* [1981] to those involving partial melting of gneisses of intermediate composition [e.g., *Thirlwall and Jones,* 1983] or the melting of more fusible components in the Lewisian basement [*Moorbath and Thompson,* 1980; *Dickin,* 1981; *Thompson et al.,* 1982]. Selective extraction of melts from the crust appears necessary because of the frequent absence of a correlation between the degree of silica saturation and $^{87}Sr/^{86}Sr$ or εNd. Using combined Nd and Ce isotope data for Skye basalts, *Dickin et al.* [1987] argued that the contaminant precursors were granitic sheets, selectively extracted during the assimilation process, rather than partially melted intermediate gneisses. Note, however, that *Reiners et al.* [1995] have suggested that under certain conditions AFC-type processes can lead to significant contamination of relatively unfractionated basaltic melts.

Thirlwall and Jones [1983] and *Kerr et al.* [1995a] noted that within the Mull Plateau Group and Skye Main Lava Series the most magnesian (i.e., hottest) lavas are the most contaminated. This observation led *Kerr et al.* [1995a] to suggest that the hottest magmas had undergone assimilation of continental crust during their turbulent ascent in thin sheet-like magma conduits, with the more

evolved magma being too cool to assimilate much crust.

The pattern of contamination observed in many of the Faeroese basalts is similar to that recorded by British Tertiary Igneous Province lavas. Sample K-1 of *Gariépy et al.* [1983], recovered from the Upper Series, has abnormally high $^{87}Sr/^{86}Sr$ and $^{208}Pb/^{204}Pb$ and low εNd, $^{206}Pb/^{204}Pb$, and $^{207}Pb/^{204}Pb$ ratios (Figures 6, 7, and 12), consistent with contamination of the parental magma by Precambrian amphibolite gneiss. Less contaminated lavas from the Faeroes fall within the 'depleted' quadrant on the Nd-Sr mantle array (Figure 6), but many of these basalts have anomalously high $^{208}Pb/^{204}Pb$ ratios, again suggesting assimilation of high Pb/Nd crustal material. Some samples from the Lower, Middle, and Upper Series fall on the so-called oceanic mantle arrays on the Nd-Sr and Pb-Pb diagrams suggesting that these samples suffered insignificant contamination.

Other suites that show evidence of contamination that has been attributed to assimilation of continental crust include the following:

• Contamination by Precambrian, possibly mid-Proterozoic Laxfordian crust, to produce low $^{206}Pb/^{204}Pb$ values in SDRS basalts from Site 553 on the Rockall Plateau [*Morton and Taylor,* 1987].

• Evidence of contamination by and bulk assimilation of granulite and amphibolite gneisses in the basalts and dacites recovered at Site 917, especially in the Lower and Middle Series, on the SE Greenland margin [*Larsen et al.,* 1994; *Fitton et al.,* 1995; 1996a,b].

• Bulk assimilation of crust to produce peraluminous, cordierite-bearing, high $^{87}Sr/^{86}Sr$ and low εNd dacites in the Lower Series at Site 642 on the Vøring margin [*Viereck et al.,* 1988] (Figures 6, 7, and 13) and to produce dacites recovered from Well 163/6-1A on the Rockall Plateau [*Morton et al.,* 1988].

• Upper Series lavas from Hold with Hope (some containing more than 10% MgO) which have high $^{207}Pb/^{204}Pb$ and $^{208}Pb/^{204}Pb$[*Thirlwall et al.,* 1994].

• High-SiO$_2$, high $^{87}Sr/^{86}Sr$ basalts in the plateau basalts from Scoresby Sund [*Larsen et al.,* 1989].

• Selective contamination of basaltic magma by Precambrian basement in some dykes along the East Greenland margin [*Blichert-Toft et al.,* 1992].

• Evidence of extensive contamination of picritic and basaltic magmas of the Vaigat and Maligât Formations, West Greenland, by shales and sandstones [*Pedersen and Pedersen,* 1987]. In the more extreme examples, this has produced rhyolites [*Pedersen and Pedersen,* 1987] and native magmatic iron [e.g., *Pedersen,* 1981].

The main conclusion to be drawn from these studies is that crustal contamination was virtually a ubiquitous process, at least during Phase 1 activity, but also during some of the Phase 2 activity. An implication is that the contamination needs to be identified and accounted for before any attempt is made to evaluate the composition of the sublithospheric source, or estimates the conditions of melting by, for example, REE inversion methods [e.g., *White and McKenzie, 1995*].

Although it may be argued that some of the contamination may have occurred as the magmas ascended through the continental lithospheric mantle, a potential reservoir for long-term development of radiogenic Sr and unradiogenic Nd (Pb may be either radiogenic or unradiogenic), there is little evidence for this being a major process in the North Atlantic Igneous Province. *Holm* [1988] argued that the low εNd, high $^{87}Sr/^{86}Sr$ lavas belonging to the Lower Basalts from Kangerlussuaq represent mixing between an incompatible-element-enriched component in ancient continental lithospheric mantle, and melts from an Icelandic-type source (hence producing the horizontal arrays on Figure 7). He rejected the suggestion of contamination of the magmas by granulite- or amphibolite-dominated crust partly on the basis that the Rb/Nb ratios of the lavas did not corroborate the isotope systematics. However, the lavas have undergone greenschist grade metamorphism, rendering the Rb abundances unreliable.

The involvement of hydrous continental lithospheric mantle in basalt production is nevertheless an attractive option. Several workers [e.g., *Gallagher and Hawkesworth*, 1992] have argued that the continental lithospheric mantle undergoes wholesale melting during extension-driven decompression. *Thompson and Morrison* [1988] and *Kerr* [1993] have proposed that the mantle lithosphere has also contaminated some of the Tertiary basalts of Skye and Mull. *Dickin* [1981] argued that the extremely low $^{206}Pb/^{204}Pb$ and $^{207}Pb/^{204}Pb$ end-member in some basalts could not be generated in the mantle; the most likely source of such material, he argued, would be Lewisian granulite (i.e., ancient, U-poor rock). However, small-degree melts percolating from the asthenosphere may lead to modification of the lithospheric mantle, for example, selective enrichment of incompatible elements and hydrous phases [e.g., *Foley et al.*, 1987; *McKenzie*, 1989; *Arndt and Christensen*, 1992]. Given that many of the Tertiary magmas were emplaced into lithosphere that had been essentially unaffected by tectonism since Proterozoic and in some areas Archaean times, there had been a long time for the lithospheric mantle to have become modified in this way. It is questionable, however, whether very low U/Pb ratios can be generated and maintained within the continental lithospheric mantle.

Weaver and Tarney [1983] argued that evidence for crustal contamination in many Phanerozoic flood basalts is ambiguous. The crust and the underlying continental lithosphere can develop together, and may inherit similar trace element and isotopic signatures. The clear advantage of invoking the continental lithosphere is that it explains the absence of a correlation between SiO_2 content and, for example, Pb or Sr isotopes. This "problem" can, however, be circumvented by using selective contamination models, or open-system magma chambers, as mentioned above.

6.2. The Lithosphere as a Mechanical Filter

Not only may the crust contaminate basaltic magmas, but it can cause ascending magmas to stall, either at the Moho or at shallower discontinuities, and form magma chambers. *Thompson et al.* [1972] demonstrated that the lavas from the Skye Main Lava Series could be either Hy- or Ne-normative, implying that the lavas had fractionated to their present compositions at pressures of ~1.0 GPa, where the low-pressure silica-saturation thermal barrier is not operational. *Thompson* [1982] later showed that the Skye Main Lava Series basalts plotted along the 0.9 GPa olivine-plagioclase-clinopyroxene cotectic, possibly reflecting processes operating in Moho magma chambers.

The majority of the basalts along the Blosseville Kyst and near Scoresby Sund were erupted to the east of the Caledonian front, whereas the Kangerlussuaq region lies to the west of this belt. In addition, whereas the Kangerlussuaq region has small, generally shallow, Mesozoic and early Tertiary basins overlying the stable Proterozoic and Archaean basement, the region to the north of Scoresby Sund is characterised by a deep crustal basin, and relatively thin continental crust—the Jameson Land Basin [*Larsen and Marcussen*, 1992]. This crustal anisotropy may have controlled the nature of the erupted magmas. Lava sequences from near Scoresby Sund show repeated development of FeTi basalts, consistent with fractionation and mixing in upper- and mid-crustal level magma chambers [*Larsen et al., 1989*; *Brooks et al., 1991*]. Seismic reflection data for the Jameson Land Basin indicate not only that the basin sits on top of anomalously thin basement (6-9 km), but also that there are extensive sill and dyke complexes within the basin [*Larsen and Marcussen*, 1992]. These may be the mid- to upper-crustal magma bodies that supplied the thick lava sequences to the Scoresby Sund region via lateral injection, although geochemical comparisons are needed to corroborate this suggestion.

Lavas from the Kangerlussuaq Lower Basalts, however, appear to have had a different eruption history. Not only are a high proportion of the lavas highly magnesian, implying more or less direct ascent through the crust, but also more evolved magmas show strong evidence of high pressure (~0.9 GPa) pyroxene fractionation, presumably at the base of the crust [*Fram and Lesher*, 1996], and similar to that reported in the Skye Main Lava Series of the British Tertiary Igneous Province. Crustal structure has clearly played an important role in the emplacement history of the Greenland basalts.

Once magma chambers have become established, either at the base of or within the crust or beneath oceanic spreading centres, they will act as effective density traps, hindering the eruption of high-density picritic magmas [e.g., *Huppert and Sparks,* 1980; *Stolper and Walker,* 1980]. This may explain why such magmas are frequently restricted to the initial episodes of continental breakup (e.g., in West Greenland, the early activity in SE Greenland, the Lower Basalts in Kangerlussuaq, and parts of the British Tertiary province such as Rum) before the establishment of steady-state magma bodies [*Fitton et al.,* 1995]. The alternative possibility is that the temperature of the mantle source has changed with time, a point returned to in Section 7.

The lithosphere also acts as a lid, capping and truncating any melting events in the underlying asthenosphere [e.g., *McKenzie and Bickle,* 1988; *Ellam,* 1992; *Saunders et al.,* 1992; *Fram and Lesher,* 1993]. At its most extreme, thick lithosphere will curtail all melting, but the introduction of hot mantle may raise the temperature of the mantle to the point where the solidus is reached and melting begins. The variation in mantle temperature and lithosphere thickness can produce four 'end-member' conditions. (i) Thick lithosphere (>150 km) and ambient temperature mantle, where melting will be restricted to small-degree, incompatible-element-enriched, hydrous melts. Most of these melts probably never reach the surface, being 'frozen' in the overlying lithospheric mantle. (ii) Thick lithosphere and hot mantle (excess potential temperature, ΔT_p > 100°C) cause melting to be restricted to high pressure. This will produce high-MgO, high-FeO and incompatible element-enriched liquids with strongly fractionated REE patterns because of the retention of the heavy REE in garnet. (iii) Zero-thickness lithosphere and normal mantle temperatures leading to a system dominated by shallow-level melting and producing normal ocean crust. (iv) Zero-thickness lithosphere and elevated mantle temperature; like condition (ii), the melting will begin at greater depth than (i) and (iii), thus producing some melts with high FeO contents and fractionated REE patterns. Melting continues to a much higher level, however, resulting in very large volumes of

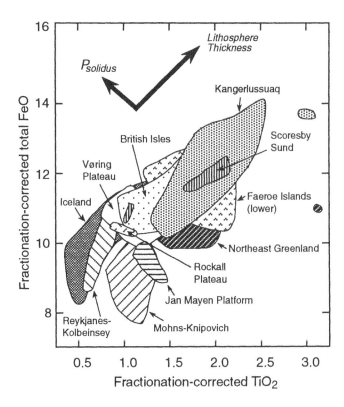

Figure 14. Fractionation-corrected total iron (FeO) versus fractionation-corrected TiO_2 in basalts from the North Atlantic Igneous Province. Arrows indicate qualitatively the effect of changing the thickness of the lithosphere (i.e. changing the pressure at the top of the melting zone) and changing the pressure at which melting begins (which is a function of temperature and volatile content of the source). From *Fram and Lesher* [1993].

liquid and dilution of the high-pressure signature. Thick oceanic crust results (Iceland).

The "lid effect" is evident in the composition of many North Atlantic basalts [e.g., *Fram and Lesher,* 1993]. The various REE and multielement plots presented in this paper reveal the fractionation of light from heavy REE, and Y and Ti from the more incompatible elements. This is likely to result from melting in the presence of garnet, which preferentially retains the heavy REE (e.g., Tm, Yb and Lu) and Y and Sc. *Fram and Lesher* [1993] attributed correlations between FeO and TiO_2 and between Dy/Yb and TiO_2 to lithosphere-induced effects (Figure 14). By using only primitive basalts to minimise the effects of low pressure fractionation, all of their data were then corrected for any slight fractionation by numerically adding equilibrium olivine to bring the Mg number (i.e., $Mg/Mg+Fe^{2+}$) of the rocks to 0.70. These fractionation-corrected TiO_2 and FeO values show a clear positive correlation.

This trend is the opposite to what is seen in regionally averaged MAR and Icelandic basalts, where there is an inverse correlation between fractionation-corrected Fe and moderately incompatible elements such as Na_2O [*Klein and Langmuir*, 1987] and TiO_2. The iron content of the primary melts increases with the mean pressure of melting [*Langmuir and Hanson*, 1980; *Klein and Langmuir*, 1987], but the content of Ti and Na decreases with increasing mean extent of melting [*Klein and Langmuir*, 1987]. Decreasing Na and increasing Fe are indeed observed in MAR basalts as Iceland is approached, presumably reflecting increasing depth and extent of melting associated with the Iceland thermal anomaly (Figure 5).

Why then do the data for basalts from the entire North Atlantic Igneous Province show positive correlations between FeO, TiO_2 and Dy/Yb? *Fram and Lesher* [1993] argued that the lithosphere restricts mantle upwelling, effectively increasing the mean pressure of melting. Samples with high fractionation-corrected TiO_2, FeO and Dy/Yb are thus derived from melts that retained the high pressure melting signature. In a general sense, the values of TiO_2, FeO and Dy/Yb decrease with time, such that the younger basalts have the lower values. This, they argued, is due to progressive thinning of the lithosphere associated with continental breakup and development of an oceanic system.

Similar arguments can be presented using Sc and Zr (Figure 15). Scandium is an element that is partitioned into clinopyroxene ($D_{Sc(2.8GPa)}$ = 0.51) and garnet ($D_{Sc(2.8GPa)}$ = 2.27) (where D=mineral-melt partition coefficient) [*Ulmer*, 1989]; it behaves in a similar fashion to the heavy REE. During low-pressure melting in the field of spinel stability, Sc/Zr of the melt increases with increasing extent of melting, reflecting the progressive removal of pyroxene (Figure 15, curves 1 and 2). During melting of garnet peridotite, however, Sc is retained by garnet in the source until all of the garnet is melted, either isobarically or, more likely, as the melting matrix ascends above the garnet-spinel transition. The basalts from the SDRS along the Hatton Bank and SE Greenland margins show the highest concentrations of Sc; along with basalts from the Upper Series in the Faeroes and Vøring Plateau, they have Sc concentrations as high as or higher than MORB. The Sc data thus corroborate the REE data, which imply that all of these basalts were generated by large-degree melts at shallow levels (curve 1 on Figure 15). Even these conditions are insufficient alone to produce the high Sc abundances found in some SDRS basalts; the melts must have undergone substantial fractionation of plagioclase and olivine (e.g., curve 1a). Note that Tertiary basalts from Mull, central East Greenland, and Iceland have lower Sc

abundances and lower Sc/Zr, implying that the 'average' depth of melting is greater than during the formation of the SDRS.

This change in magma composition, apparently reflecting the shift from high (garnet present) to moderate (garnet minor or absent) pressure of melting, is observed locally in at least three areas. *Ellam* [1992] and *Kerr* [1994] suggested that the heavy-REE-depleted patterns observed in Skye Main Lava Series and the Mull Plateau Group (Figure 8) were generated below a lithospheric 'lid' which was at least 60–80 km thick and restricted melting to a source region containing garnet. Later, more extensive melting of a similar source beneath thinner lithosphere (<60 km), entirely within the spinel lherzolite stability field, resulted in the relatively flat heavy REE patterns of the Fairy Bridge/Coire Gorm and Preshal More/Central Mull Tholeiite types. A similar shift in chemistry is observed in the Antrim Lavas (when passing from the Lower Formation to the Causeway Member of the Middle Formation [*Barrat and Nesbitt*, 1996]). In a third site-specific study, *Fram et al.* [1996] also calculated that the mean pressure of melting decreased, and the extent of melting increased, during the rifting and breakup of the SE Greenland margin.

An implication of *Kerr's* [1994] model is that the lithosphere below the Hebrides had to thin by at least 15 km in 2 to 3 m.y. It is difficult to see how this thinning could be achieved by lithosphere stretching alone, unlike the SE Greenland margin [*Fram et al.*, 1996] where the magmatism and breakup were more or less in the same place. Most of the crustal extension in the British Tertiary province is associated with the Central Complexes (where locally it may be up to 25%; for example, on Skye [*Speight et al.*, 1982]), but the regional extension is generally much less (~1% [*England*, 1988]). Consequently, *Kerr* [1994] and *Barrat and Nesbitt* [1996] have suggested that the lower lithosphere was eroded by a strongly advecting plume mantle. Removal of 15 km of lithosphere by hot plume mantle would, however, cause substantial surface uplift (R. England, pers. comm., 1996), of which there is no evidence. Long-distance, lateral transport of magmas to the Skye and Mull fields cannot be completely eliminated, and neither can changing source compositions.

7. THE ROLE OF THE ICELAND PLUME

7.1. *Thermal Aspects*

The voluminous magmatism that accompanied plate separation was mostly short-lived and often restricted to

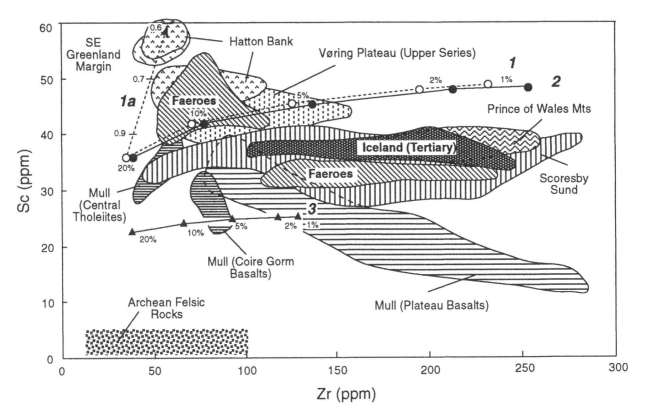

Figure 15. Sc versus Zr for North Atlantic Igneous Province basalts. The field of Archaean felsic rocks (mainly veins in granulites [*Kerr*, 1995a]) suggests that contamination has played an insignificant role in controlling the abundances of Sc and Zr in these samples. Data sources as for Figures 6 and 7, with additional data from *Fitton et al.* [1996b] for the SE Greenland margin basalts, and *Brodie and Fitton* [1996] for Hatton Bank basalts. Iceland basalt data from A. D. Saunders, J. G. Fitton and B. S. Hardarson (unpublished neutron activation (Sc) and X-ray fluorescence (Zr) analyses). Modelling curves assume fractional melting of spinel peridotite (curves 1 and 2) and garnet peridotite (curve 3). Open circles: 'depleted' mantle; filled circles and triangles: 50:50 mixture of 'depleted' and 'primitive' mantle (assumes an equal contribution from each mantle type during melting). Compositions and definitions of 'primitive' and 'depleted' mantle, mineral proportions and partition coefficients from *McKenzie and O'Nions* [1991]. Line 1a represents evolution by fractional crystallisation of a liquid derived by 20% melting of 'depleted' spinel peridotite. Fractionating phases assumed to be plagioclase plus olivine (Sc and Zr will each have partition coefficients approaching zero). Fractions of liquid remaining are shown. Inclusion of clinopyroxene in the crystallising assemblage will lead to decreasing Sc/Zr ratios in the liquid.

no more than one polarity event in any one locality. Shortly after plate separation the excessive magmatism was restricted to the Greenland-Faeroes Ridge, and latterly to Iceland, on the axis of the present-day plume. There is strong evidence from seafloor bathymetry [*Haigh*, 1973], from the major element composition of basalts erupted along the MAR south of Iceland [*Klein and Langmuir*, 1987], and from the thickness of the ocean crust around Iceland [*White et al.*, 1995] that the thermal effects of the present-day Iceland plume extend beyond the boundaries of Iceland itself [e.g., *Vogt*, 1983]. Furthermore, it can be shown that the thermal effects have a more widespread extent and influence than the compositional (trace element,

isotopic) effects (Figure 5). Interestingly, it would appear that the thermal influence of the Iceland plume may be as widespread, at least to the south of Iceland, as it was during early Tertiary times.

Recent activity on Iceland provides a useful model with which to explore further the Palaeocene magmatism along the margins of the North Atlantic. Certainly, the likely presence of a plume beneath Iceland (see Section 4.1) lends powerful support to those who advocate a plume-type origin for the bulk of the North Atlantic Igneous Province. However, there are several anomalies that have to be accounted for if we follow the plume paradigm.

Firstly, as discussed in Section 5, the rate of magma

generation during the formation of the SDRS was up to three times greater than in present-day Iceland. If the rate of magmatism was simply a function of magma temperature and extension (decompression) [e.g., *McKenzie and Bickle,* 1988], then this would imply that during the Eocene large volumes of the plume were substantially hotter than at the present day. Is there any independent evidence for this? Secondly, some basalts from the SDRS (e.g., from the SE Greenland margin and Hatton Bank), and the more distal on-shore portions of the North Atlantic Igneous Province (West and SE Greenland; the British Tertiary Igneous Province) are at least superficially similar to MORB and thus appear to have been derived from a depleted mantle source, [e.g., *Joron et al.,* 1984; *Robillard et al.,* 1992; *Holm et al.,* 1993; *Kerr et al.,* 1995b; *Fitton et al.,* 1996b, 1997]. How could this occur if a plume was responsible for transporting the thermal energy to the site of magma generation? Before we address these two points, it is worth considering alternatives to the plume model.

The excess magmatism that occurred during the initial stages of plate separation may be a result of enhanced convection in the underlying asthenosphere [*Mutter et al.,* 1988]. Fundamental to this model is the nature of the initial plate separation; one requirement is a sharp thermal boundary layer at the edges of the separating plates. The asthenosphere adjacent to this boundary or 'edge' cools, thus setting up a local convection cell beneath the developing rift axis. Increased flux of mantle through the melt zone results in increased magma generation and the formation of a volcanic rifted margin. Developing plate boundaries that do not have this marked 'edge' should not develop the necessary convection systems and remain essentially 'non-volcanic' (e.g., Biscay margin).

Although enhanced convection may be an important method of generating large volumes of magma from hot mantle (see below), it is difficult to see how it can work in ambient asthenosphere (potential temperature, T_p, ~1300°C). The convection system requires heat extraction from the downwelling limb of the cell into the adjacent, cool lithosphere. This effect, coupled with energy loss during melting, will tend to be self-limiting, and it is unlikely that any more melt will be produced than during normal seafloor spreading, once plate separation has occurred. Numerical models [e.g., *Boutilier and Keen,* 1995] tend to support this prediction. On the other hand, if the mantle beneath the lithosphere is hotter than ambient, then enhanced convection could occur as the hot mantle flows from beneath the lithosphere and is channelled into the developing rift axis.

Knowledge about the temperature and thermal evolution of the sublithospheric mantle during Palaeocene and Eocene times is crucial to these discussions. The volume of melt generated at a rift zone is dependent on the amount of decompression (controlled by lithosphere thickness and amount of extension) and the temperature of the underlying asthenosphere [*McKenzie and Bickle,* 1988]. Estimates about the thermal state of the mantle may therefore be obtained from the volume of lavas and intrusive rocks, providing we have an approximate understanding of the melting conditions (anhydrous or hydrous; passive or active?) and the composition of the source material. Alternatively, we may assess temperature directly from the lavas themselves.

The most cited evidence for the existence of hotter-than-ambient mantle is the occurrence of picritic magmas. If it can be shown that erupted liquids had a high MgO content (either by direct analysis of aphyric lavas or by extrapolation from equilibrium olivine compositions) and were anhydrous, then this implies elevated melt temperatures. Whether or not picrites can be used as temperature probes is, however, less clear. Although picrites are found associated with continental flood basalts [e.g., *Campbell and Griffiths,* 1990], and have thus been linked with anomalously hot mantle, their frequency of eruption in these and other settings (e.g., mid-ocean ridges) may be a function of 'eruptive potential' as much as actual frequency of production: picrites are dense melts which will tend to stall in basaltic magma chambers [e.g., *Stolper and Walker,* 1980].

In the North Atlantic Igneous Province, picritic parental magmas have been inferred for Baffin Island [e.g., *Clarke and Pedersen,* 1976; *Francis,* 1985], West Greenland [*Pedersen,* 1985; *Gill et al.,* 1992], central East Greenland [*Nielsen et al.,* 1981; *Gill et al.,* 1988; *Fram and Lesher,* 1996], SE Greenland [*Larsen et al.,* 1994; *Fram et al.,* 1996; *L.M. Larsen et al.,* 1996; *Thy et al.,* 1996], the British Tertiary Igneous Province (especially the Rum complex) [e.g., *Kent,* 1995], and Iceland [e.g., *Elliott et al.,* 1991]. Parental magmas of the Skye Main Lava Series and the Mull Plateau Group may have contained as much as 15% MgO [*Kerr,* 1995a; *Scarrow and Cox,* 1995]. The potential temperature of the upwelling mantle that would have produced high-MgO basalts in Mull has been estimated to have been 1420–1460°C [*Kerr,* 1995a], some 120–160°C hotter than ambient. The Rum liquids would have required an even hotter source. Olivines from a peridotitic dyke on Ard Nev, Rum [*McClurg,* 1982], may have been in equilibrium with a dry melt containing 18–20 % MgO, at a temperature of 1420–1460°C (corresponding to an approximate source temperature of 1540°C at 2 GPa) [*Kent,* 1995]. Equivalent potential temperatures for the source of the West Greenland picrites may have been as high as 1540–1600°C [*Gill et al.,* 1992].

A full evaluation of the Palaeocene picrites is long overdue, but two observations are pertinent. (i) The majority of the picrites were erupted during the early stages of the North Atlantic Igneous Province, during times of initial rifting and prior to breakup. With the possible exception of the picrites recovered at Site 917 on the SE Greenland margin (presently undated), no high-MgO liquids have been found in the SDRS associated with the Phase 2 magmatism. As pointed out above, this should be treated with caution, however, because steady-state magma bodies may have filtered the dense picrite liquids. (ii) The picrites are not restricted in space to any one locality that may, for example, have corresponded with a hot plume stem [*Campbell and Griffiths,* 1990] (see discussion by *Larsen et al.* [1992] and *Chalmers et al.* [1995]). The eruption sites are scattered from Baffin Island in the west to the British Tertiary Province in the east, a distance of some 2000 km (Figure 3).

Although appeal may be made to the development of a hotcell [*Anderson et al.,* 1992] beneath Baffin Island, Greenland, and NW Europe to explain the wide distribution of the Palaeocene magmatism, we find this explanation unsatisfactory for several reasons. Firstly, the hotcell hypothesis advocates thermal blanketing by the lithosphere, and cooling by adjacent subduction zones, to generate horizontal thermal gradients in the asthenosphere. The time span required to develop high potential temperatures suggested by picritic magmas is likely to be several millions if not tens of millions of years, if indeed such high temperatures could ever be achieved. Such slow build-up of heat is not consistent with the absence of widespread magmatism in developing Cretaceous rifts such as the Labrador Sea-Davis Strait or Rockall Trough. Secondly, the early phase of magmatism appears to have occurred simultaneously over a widespread area. This is again inconsistent with the long period of asthenosphere heating. Thirdly, the presence of a thermal, gravity, chemical, and topographic anomaly beneath present-day Iceland lends powerful circumstantial support that a similar anomaly existed in the past.

The uplift recorded in the Danian sediments in central East Greenland and in the Forties Field provides strong corroborative evidence for the development of a widespread thermal anomaly and resultant buoyant support of the lithosphere. The more distal effects of this anomaly are recorded in the subsidence history of the North Sea Basin, where there is evidence for a transient thermal pulse between 65 and 52 Ma [*Nadin et al.,* 1995]. The simultaneous occurrence of magmatism in West Greenland

and the British Isles implies phenomenally high mantle flux rates if the plume was axisymmetric, indicating the arrival of a large plume head or blob. These flux rates could be reduced somewhat if the plume was channelled by pre-existing lithospheric topography, or ascended in a sheet-like form rather than as a tube. Whatever model is invoked, for the thermal anomaly to be emplaced so quickly and over such a wide area implies that it had low viscosity and, by implication, was very hot. This could account for the restriction of high-MgO liquids to the Phase 1 activity; later magmatism was derived from a system that had already begun to cool (Figure 16).

What could be responsible for the burst of magmatism that produced the SDRS and the complementary plutonic, deep crustal prisms? Plate rupture and separation and decompression melting of the hot asthenosphere [*White and McKenzie,* 1989; *Barton and White,* 1995] appear to offer the best explanation. *Clift et al.* [1995], using subsidence curves constrained by sedimentological and biostratigraphic data, estimated that the excess mantle temperature at the Hatton Bank margin was of the order of 75 to 100°C at the time of rifting. The equivalent ΔT at Site 643 on the Vøring margin was between 30 and 75°C, and 110–190°C at Site 918 on the SE Greenland margin. Their subsidence model predicted a peak mantle temperature at 42 Ma, rather than at the time of breakup, at Site 918. The temperature excesses are lower than the estimates derived from high-MgO liquids and provide support for the idea that the system had begun to cool down by the time of plate breakup.

Enhancement of melt production by active upwelling of hot mantle from beneath the separating plates may also contribute to the very large volumes of melt. As the plates separated, hot mantle would tend to flow back towards the developing North Atlantic rift axis and generate, by forced convection, the large volumes of magma associated with the SDRS. (This mechanism would be in addition to any mantle channelling directly along the rift axis from the plume, but which would be unable to melt if it was travelling essentially horizontally and isobarically.) We have already mentioned, in Section 4.1, that forced convection is likely in Iceland and reduces the modelled temperature of the plume. Note that the hot mantle associated with the developing plume head would cool rapidly, not only by heat loss through conduction into the overlying lithosphere, but also through magma production. Supply of new, hot mantle could not have continued at the initial high flux rates.

It is interesting to speculate that mantle backflow to the developing rift axes also provides a mechanism whereby the North Sea Basin returned to its normal subsidence

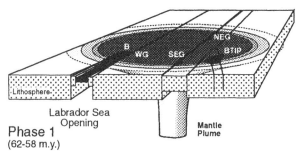

Phase 1
(62-58 m.y.)
• Hot mantle plume
• Mainly continent-based magmatism
  at widespread localities
• Picrites and high-Mg basalts common

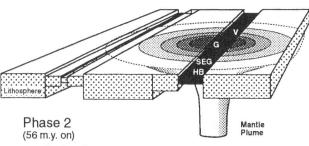

Phase 2
(56 m.y. on)
• Cooler plume
• Large volumes of melt
  produced along lines of plate separation
• Plume flows back to rift zones?

Figure 16. Schematic illustration of the development of the Iceland plume during Palaeocene and Eocene times. Phase 1. Rapid development of a widespread plume head and eruption of basalts and picrites over a large area. Sites of eruption may be controlled by pre-existing thinspots (BTIP) or a developing rift system (B, WG, SEG). Abbreviations: NEG - Northeast Greenland; BTIP - British Tertiary Igneous Province; SEG - Southeast Greenland; WG - West Greenland; B - Baffin Island. Phase 2. This phase is associated with the main rifting and breakup event, and subsequent activity forming the Faeroes-Greenland Ridge and Iceland. By the time of plate separation, plume flux rates had diminished, the plume head may have cooled, and activity was focused along the developing rifted margins. Abbreviations: V - Vøring Plateau; G - central East Greenland; SEG - Southeast Greenland margin; HB - Hatton Bank margin.

curve only 7 or 8 m.y. after the thermal pulse arrived; the hot mantle reached its maximum extent by about 55 Ma, then partially retreated as the main North Atlantic Basin opened.

The absence of any obvious plume activity in the North Atlantic region prior to 62 Ma suggests that the plume was not present in the region prior to that time, or the plume was very weak, or the region had not migrated into the plume's sphere of influence. The plume may have been

present to the northwest of Greenland in the Cretaceous. The Alpha Ridge, a region of anomalously thick crust in the Arctic Basin, and Cretaceous flood basalts in the high Arctic on Axel Heiberg Island indicate some form of plume activity [*Lawver and Müller,* 1994; *Tarduno,* 1996]. There are, however, at present no geochemical data or radiometric age dates with which to confirm or refute the involvement of the ancestral Iceland plume in the formation of these magmatic subprovinces.

If the mantle flux rates predicted for the Palaeogene sequences had been operating throughout the Cretaceous, we would expect to find voluminous magmatism in the Labrador Sea/Davis Strait regions where basin development had been operating on and off throughout the Mesozoic. The only magmatism in this area appears to be that associated with the Disko-Baffin Island sequences, all of which are of Palaeocene age, and which erupted suddenly at about 60–61 Ma. Phase 1 magmatism appears, therefore, to be associated with a sudden burst of plume activity. It is not possible to say, at present, whether this activity represents the arrival and impact of a new plume [e.g., *Richards et al.,* 1989; *Campbell and Griffiths,* 1990], or a sudden increase in the flux rate of what previously had been a relatively weak plume system. Although there is little doubt that both phases of activity exploited lines of weakness (e.g., thinspots) in the lithosphere, the location and nature of the Phase 2 magmatism was controlled to a far greater degree by the developing rift axes and lines of plate separation. During Phase 1, magmatism appears to have been an active response to the plume; during Phase 2, which continues to the present day, magmatism has been a subtle (and changing) interplay between a passive upwelling in response to plate separation and active convection.

7.2. Compositional Structure of the Iceland Plume

Plumes are thought to originate from a thermal boundary layer deep within the Earth. The precise location of this boundary layer is disputed and may not be the same for all plumes. Equally contentious is the role of the boundary layer in isolating material of different compositions. At the one extreme, it could be argued that plume material is derived from a boundary layer that is compositionally identical to the material into which it ascends. The plume would thus be distinguished solely on the basis of its thermal signature (T, rates of melt generation, buoyancy and gravity effects). However, we may expect any erupted magmas to be compositionally different (for example, in terms of major elements and many trace element ratios) from melts derived from cooler, ambient mantle, simply because the geometry of the melting column would be affected by an increase in potential temperature. If both the plume and the surrounding asthenospheric mantle are internally homogeneous and have the same bulk composition, we may reasonably expect, however, the mantle and the erupted products to have the same Sr, Nd, Pb, and He isotopic signatures.

At the other extreme, the plume may originate in a boundary layer that is compositionally different from the asthenosphere into which it rises, or the plume may entrain mantle material as it ascends (especially during its initial stages) [e.g., *Griffiths and Campbell*, 1990]. In either of these cases, there is no reason to assume that the surrounding ambient asthenosphere and the plume have the same isotopic compositions. Indeed, it is the long-recognised isotopic diversity of magmas from oceanic islands that has led to complex, often contradictory, chemodynamic models of Earth structure, involving plumes originating from isotopically distinct, long-lived reservoirs [e.g., *Zindler and Hart*, 1986].

The North Atlantic offers a unique opportunity to study the compositional structure of a large mantle plume. Not only is there the juxtaposition of the plume and a spreading ridge, thus providing material from the entire width of the plume head, but there is also excellent preservation of material from the Palaeogene to the present day. It is generally accepted that the axis of the plume is presently located beneath eastern Iceland and that the thermal effects of this plume extend to approximately 50°N on the MAR [*e.g., White et al.*, 1995, and references therein]. *Schilling et al.* [1973] and *Hart et al.* [1973] were the first to report the geochemical gradient along the Reykjanes Ridge - MAR. Various compositional parameters, such as $^{87}Sr/^{86}Sr$, $^{206}Pb/^{204}Pb$, and La/Sm, correlate inversely with water depth, and these workers interpreted the correlations as a result of mixing between an 'enriched' Iceland plume and a 'depleted' asthenosphere (MORB mantle). As noted above, the extent of the thermal effects of the plume is greater than that of the elemental and isotopic effects, implying some form of decoupling within the plume structure. The distal part of the plume, between 62° and 54°N on the MAR, appears to comprise hotter-than-ambient mantle (with high fractionation-corrected FeO values [*Klein and Langmuir*, 1987]), but which also possesses MORB-like isotope and trace element ratios.

As mentioned in Section 4.1, the Iceland plume is isotopically heterogeneous, producing a range of 'depleted' and 'less depleted' magma types. This 'depleted' end-member is not the same as the MORB source, and to minimise confusion we call it the depleted plume end-member. *Thirlwall et al.* [1994] and *Thirlwall* [1995] noted that all analysed basalts from Iceland have distinctly higher $^{208}Pb/^{204}Pb$ at a given $^{207}Pb/^{204}Pb$ than North Atlantic MORB. The least crustally contaminated Palaeocene basalts from Hold with Hope also show this characteristic, leading *Thirlwall et al.* [1994] to suggest that the plume can be 'fingerprinted'. On the basis of these data *Thirlwall* [1995], *Hards et al.* [1995] and *Kerr et al.* [1995b] also suggested that the depleted end-member found in Icelandic basalts originates from within the plume. Thus any model of the Iceland plume structure needs to consider three components: (i) the depleted and (ii) the less depleted plume end-members, and (iii) the heated, adjacent asthenosphere (usually the MORB source). Can we detect all three components in the Palaeogene basalts?

The incompatible-element-depleted nature of uncontaminated Hebridean, Hatton Bank, and Faeroes Tertiary lavas has long been recognised [*Carter et al.*, 1979; *Wood*, 1979b; *Morrison et al.*, 1980; *Ridley*, 1973; *Thompson et al.*, 1980; *Gariépy et al.*, 1983] (see, for example, Figures 8, 11, and 12). *Morrison et al.* [1980] suggested that a small-degree melt fraction had been extracted from the mantle source region of the Hebridean lavas during the Permo-Carboniferous, and that remelting of this Palaeozoic residuum produced the depleted Tertiary basalts. Subsequent analysis of the Nd isotope ratios of British Tertiary Igneous Province basalts [*Thirlwall and Jones*, 1983; *Kerr et al.*, 1995a] has shown, however, that some uncontaminated lavas possess $\varepsilon Nd_{t=60}$ as high as +9. Thus, the Hebridean basalts have a long history of depletion which is inconsistent with a single-stage depletion event as recent as the Permo-Carboniferous. The same problem exists for *Wood's* [1979b] dynamic melting model, which advocated that the melting and extraction episodes were close in time.

Although data for many basaltic suites plot to the low-εNd side of the Iceland array on Figure 7, indicating variable amounts of lithospheric contamination, many data plot on the array or project back to it. The implication of these data is that the majority of the primary melts were originally derived from sources with εNd and Zr/Nb lying on or close to the Iceland-MORB array. Basalts from the Faeroes, West Greenland, and Site 918 on the SE Greenland margin encompass much of the Iceland array. The Mull Plateau Lavas and the Group 1 basalts from central East Greenland may have tapped sublithospheric mantle with high Zr/Nb (~25) and εNd (~+9) whereas basalts belonging to Group 2 from central East Greenland tapped less depleted mantle with a lower Zr/Nb (10–12) and εNd (~+7). The data shown on Figure 7 are consistent with the ultimate derivation of the majority of the basalts from mantle with an Icelandic signature.

The two exceptions are the light-REE-depleted basalts from Hatton Bank and from the Faeroes Upper Series. Both sets of lavas have high Zr/Nb, far above the Iceland array. These basalts appear to have been derived from hot MORB mantle (the elevated temperatures being necessary for the formation of the SDRS and the dynamic support of the margin during rifting). The Hatton Bank basalts are among the farthest from any putative plume axis, suggesting that the plume had a MORB-like annulus in Eocene times, much as it appears to do today [*Fitton et al.,* 1996b, 1997]. This may be overly simplistic, however, because the Faeroes basalts were erupted closer to the ancestral plume axis, implying that MORB mantle was partially entrained within the body of the plume, and not just forming an annular rim.

These observations confirm the work by *Fitton et al.* [1996b, 1997], who have devised a discriminant between Icelandic basalts and MORB using Nb/Y versus Zr/Y (Figure 17). This diagram relies on the unusual distribution of Nb in MORB and ocean island basalts. In ocean island basalts, Nb is invariably enriched relative to Zr. When data for basalts from the Iceland neovolcanic zones are plotted on this diagram, they fall within a tightly defined array. Data for Icelandic Tertiary basalts form a tight cluster but are still within the array. All normal MORB for which we have reliable data fall below the array (enriched, or E-type, MORB project from normal MORB towards the array, but are not plotted here). The discriminant is less sensitive than Pb isotopes to the effects of crustal contamination, because crust and MORB plot on the same side of the array (i.e., towards low Nb/Y), and the Nb/Pb ratio of most continental crust is significantly lower than that of mantle-derived liquids.

The diagram demonstrates that basalts from Site 918 on the SE Greenland margin plot at the 'depleted' end of the Iceland array, despite being light-REE-depleted and otherwise resembling normal MORB. This is the same conclusion as that drawn from Figure 7. Light-REE-depleted basalts from the Upper Series on the Faeroes and from Hatton Bank fall within the MORB field, however, suggesting that the mantle source for these basalts is not Icelandic. Other basalts from the Faeroe Islands plot in or close to the Iceland field; the slight displacement below the lower part of the array shown by some of the Middle Series basalts is probably a result of crustal contamination. Light-REE-depleted samples from West Greenland, a region also on the periphery of the Palaeocene plume, also plot below the Iceland array, indicating derivation from a

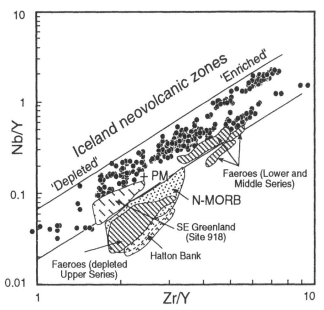

Figure 17. Nb/Y versus Zr/Y in basalts from the Iceland neovolcanic zones, normal (N-) MORB, SE Greenland margin, Hatton Bank, and the Faeroes Islands. Modified after, and data from, *Fitton et al.* [1996b, 1997] and *Gariépy et al.* [1983]. 'Depleted' and 'enriched' refer to basalts from the Iceland neovolcanic zones that have low Zr/Y and high Zr/Y, respectively. Note that the Iceland data define a tight array, whereas N-MORB fall below this array, despite having similar REE profiles to depleted Icelandic picrites (see, for example, Figure 4). PM - primordial mantle, from *Sun and McDonough* [1989].

MORB-like source [*Fitton et al.,* 1997b,c]

The data for the Icelandic and the Palaeogene sequences provide strong circumstantial support for the involvement of a heterogeneous mantle plume in the formation of the North Atlantic Igneous Province. Heating of MORB-like asthenosphere alone will not produce the diversity of isotopic compositions, when the effects of the lithosphere (pressure control and contamination) are accounted for. Although far more data are required to constrain speculations about the nature of the plume, it would appear, on the basis of the available data, that the plume is intrinsically heterogeneous, with both 'less depleted' and 'depleted' components. Either or both of these components may have originated from the lower mantle [e.g., *Hart et al.,* 1992]. Furthermore, the plume has an annulus of hot, depleted MORB mantle that appears to have been the source of magmas erupted on the periphery of the plume [*Fitton et al.,* 1997]. Whether such an annulus exists in other plumes is difficult to ascertain; normally, such material would be trapped beneath lithosphere and be unable to readily melt.

8. MAIN CONCLUSIONS AND SUGGESTIONS FOR FUTURE WORK

• The North Atlantic Igneous Province is one of the largest LIPs, occurring in places as far apart as Baffin Island and West Greenland in the west, the British Isles in the east, and Hold with Hope in NE Greenland in the north, a distance of more than 2000 km. The total volume of Palaeocene and Eocene magmas probably exceeded 6×10^6 km^3 and may be as much as 10×10^6 km^3.

• The bulk of the activity occurred in two main phases. Phase 1 began about 62 m.y. ago, and lasted for 2 to 4 m.y; most of the magmas associated with Phase 1 were erupted through and onto continental crust. Phase 2 began at about 56 Ma with the onset of plate breakup and separation and continues at the present time in Iceland. No rocks dated by ^{40}Ar-^{39}Ar methods and with reliable age spectra have ages greater than about 62 Ma.

• The main product of the activity is tholeiitic basalt, although alkali basalts are common. Differentiated rocks are frequently found on continental areas and along the rifted margins.

• The lavas erupted on Iceland, especially those in the neovolcanic zones, attest to the heterogeneous nature of the underlying mantle. This heterogeneity appears to be replicated throughout the province. Just as the Icelandic basalts do not originate from a simple, one-component source but originate from depleted and enriched mantle end-members, so the Palaeogene basalts are derived from a spectrum of sublithospheric mantle compositions. The composition of the sublithospheric melt depends on the conditions of melting and may have been influenced by the distance from the ancestral plume axis.

• There is good evidence that the lithosphere has strongly influenced the composition of the erupted rocks, both by truncating the melting column and by contaminating many of the magmas as they ascended to the surface. The continental crust has been the main contaminant, but the evidence for involvement of continental lithospheric mantle is equivocal. The fact that either could be involved as a contaminant, however, suggests that caution should be exercised when inversion modelling of REE data is undertaken.

• The thick oceanic crust associated with Iceland and the Greenland-Faeroes Ridge is convincing evidence that a thermal anomaly, presumably a mantle plume, resides beneath Iceland. The precise temperature of the plume is not known; estimates range from about 70° to 200°C above ambient temperature. Estimates based solely on passive upwelling of mantle in response to plate separation must be treated with caution, because forced convection will reduce the temperature required to produce the calculated melt volumes.

• We envisage that the activity of Phase 1 was associated with a transient, very hot, low-viscosity plume. This would account for the simultaneous eruption of basalts and picrites in Greenland, Canada and the British Isles (a distance of more than 2000 km). Very high mantle flux rates were probably associated with this transient event, which may represent the arrival of a new plume, or a sudden increase in flux rates of a pre-existing weaker plume. Channelling of the hot mantle into pre-existing thinspots may have restricted magmatism to localised areas.

• Phase 2 magmatism was linked with plate separation, and may well have tapped mantle material that had already lost substantial heat via conduction and via the melt generation associated with Phase 1. The bulk of the SDRS and mid- and lower-crustal level intrusive rocks of the volcanic rifted margins were produced during this time. Magma supply rates were possibly 2 to 3 times that in present day Iceland, suggesting that forced convection operated during the time of initial rifting. Flow of the plume mantle into the rift axis from beneath the adjacent subcontinental areas may have been an important process, because this would have enhanced the convection process.

• The plume may have been compositionally zoned. The recovery of light-REE-depleted basalts from Hatton Bank on the southeast tip of the Rockall Plateau, the furthest point from the Palaeocene/Eocene plume axis, suggests that the plume may have had an annulus of heated MORB-type mantle. This MORB-type mantle was hotter than ambient, given the large volumes of melt to form the SDRS, and the dynamic support of the margin (the lavas were erupted close to or above sea-level).

• Although there is a large amount of published data and text on the North Atlantic Igneous Province, there are still several important gaps in our knowledge that make definitive assessment difficult. Reliable ages for the basalt successions in central East Greenland, the Faeroes Islands, Vøring Plateau, Skye, Mull and other Scottish Islands are absent. Further recovery of samples from the SDRS in transects closer to the ancestral plume axis (aborted during ODP Leg 163) is required to determine the compositional gradients in the Palaeocene plume. Detailed studies of the Cretaceous-Tertiary sedimentation history of the basins offshore of Europe may help constrain the uplift history of the region and date more precisely the fluctuations in the flux of the plume. Despite these gaps, however, the North Atlantic Igneous Province offers a virtually unique opportunity to understand the interactions between a mantle plume and the lithosphere, and to understand the thermal and compositional history of the plume itself. It is among the best, if not the most accessible, of natural laboratories.

Acknowledgments. In such a paper, it is not possible to give full credit to all workers or to represent their ideas with complete accuracy. We hope that any omissions or inaccuracies are considered as no more than accidental. We have been helped by thoughtful and thorough (in some cases very detailed) reviews by Rick Carlson, Henry Emeleus, Richard England, Lotte Melchior Larsen, Chip Lesher, John Mahoney, Dan McKenzie, and Matthew Thirlwall. Research in the North Atlantic region has been supported in part by the NERC, UK (Grants GST/02/673 JGF and ADS, and GT5/F/92/GS/4 to RWK).

REFERENCES

Allègre, C. J., and D. L. Turcotte, Geodynamic mixing in the mesosphere boundary layer and the origin of oceanic islands, *Geophys. Res. Lett., 12*, 207-210, 1985.

Anderson, D. L., Y.-S. Zhang, and T. Tanimoto, Plume heads, continental lithosphere, flood basalts and tomography, in *Magmatism and the Causes of Continental Break-up, Spec. Publ., 68*, edited by B. C. Storey, T. Alabaster and R. J. Pankhurst, pp. 99-124, The Geological Society, London, 1992.

Anderson, F. W., and K. C. Dunham, *The Geology of Northern Skye*, HMSO, Edinburgh, 1966.

Anderton, R., Sedimentation and basin evolution in the Paleogene of the Northern North Sea and Faeroe-Shetland Basin, in *Petroleum Geology of Northwest Europe*, edited by J. R. Parker, p 31 The Geological Society, London, 1993.

Arndt, N. T., and U. Christensen, Role of lithospheric mantle in continental volcanism: thermal and geochemical constraints, *J. Geophys. Res., 97*, 10,967-10,981, 1992.

Arter, G., and S. W. Fagin, The Fleetwood Dyke and the Tynwald fault zone, Block 113.27, East Irish Sea Basin, in *Petroleum Geology of Northwest Europe*, edited by J. R. Parker, pp. 835-843, The Geological Society, London, 1993.

Athavale, R. N., and P. V. Sharma, Palaeomagnetic results on early Tertiary lava flows from West Greenland and their bearing on the evolution history of Baffin Bay-Labrador Sea region., *Can. J. Earth Sci., 12*, 1-18, 1975.

Backman, J., Cenozoic calcareous nannofossil biostratigraphy from the northeastern Atlantic Ocean - Deep Sea Drilling Project Leg 81, *Init. Repts. Deep Sea Drill. Proj., 81*, 403-428, 1984.

Bailey, E. B., The desert shores of the Chalk Seas, *Geol. Mag., 61*, 102-116, 1924.

Bailey, E. B., C. T. Clough, W. B. Wright, J. E. Richey, and G. V. Wilson, Tertiary and post-Tertiary geology of Mull, Loch Aline and Oban, *Geol. Surv. G. B. Mem. Geol. Surv. Scotl.,* 1924.

Balkwill, H. R., Labrador Basin: structural and stratigraphic style, in *Sedimentary Basins and Basin-Forming Mechanisms, Mem. 12*, edited by C. Beaumont and A. J. Tankard, pp. 17-43, Canadian Society of Petroleum Geologists, Calgary, 1987.

Barrat, J. A., and R. W. Nesbitt, Geochemistry of the Tertiary volcanism of Northern Ireland, *Chem. Geol., 129*, 15-38, 1996.

Barton, A. J., and R. S. White, The Edoras Bank margin; continental break-up in the presence of a mantle plume, *J. Geol. Soc. Lond., 152*, 971-974, 1995.

Beblo, M., and A. Björnsson, A model of electrical resistivity beneath NE-Iceland, correlation with temperature, *J. Geophys., 47*, 184-190, 1980.

Beckinsale, R. D., R. N. Thompson, and J. J. Durham, Petrogenetic significance of initial $^{87}Sr/^{86}Sr$ ratios in the North Atlantic Tertiary igneous province in the light of Rb-Sr, K-Ar and ^{18}O abundance studies of the Sarqâta qáqâ intrusive complex, Ubekendt Island, West Greenland, *J. Petrol., 15*, 525-538, 1974.

Bell, B. R., and I. T. Williamson, Picritic basalts from the Palaeocene lava field of west-central Skye, Scotland; evidence for parental magma compositions, *Min. Mag., 58*, 347-356, 1994.

Berggren, W. A., D. V. Kent, and J. J. Flynn, Jurassic to Paleogene: Part 2. Paleogene geochronology and chronostratigraphy, in *The Chronology of the Geological Record, Spec. Publ. 10*, edited by N. J. Snelling, pp. 141-195, The Geological Society, London, 1985.

Berggren, W. A., D. V. Kent, C. C. Swisher III, and M.-P. Aubry, A revised Cenozoic geochronology and chronostratigraphy, in *Geochronology, Time Scales and Global Stratigraphic Correlation, Spec. Publ. 54*, edited by W. A. Berggren, D. V. Kent, M.-P. Aubry and J. Hardenbol, pp. 129-212, (SEPM) Society for Sedimentary Geology, Tulsa, OK, 1995.

Bjarnasson, I. T., W. Menke, Ó. G. Flóvenz, and D. Caress, Tomographic image of the mid-Atlantic plate boundary in southwestern Iceland, *J. Geophys. Res., 98*, 6607-6622, 1993.

Black, G. P., The age relationship of the granophyre and basalt of Orval, Isle of Rhum, *Geol. Mag., 89*, 106-112, 1952.

Blichert-Toft, J., C. E. Lesher, and M. T. Rosing, Selectively contaminated magmas of the Tertiary East Greenland macrodike complex, *Contrib. Mineral. Petrol., 110*, 154-172, 1992.

Bollingberg, H., C. K. Brooks, and A. Noe-Nygaard, Trace element variations in Faeroese basalts and their possible relationships to ocean floor spreading history, *Bull. Geol. Soc. Denmark, 24*, 55-60, 1975.

Bott, M. P. H., The crust beneath the Iceland-Faeroe Ridge, in *Structure and Development of the Greenland-Scotland Ridge. New Methods and Concepts*, edited by M. H. P. Bott, S. Saxov, M. Talwani and J. Thiede, pp. 63-75, Plenum Press, New York and London, 1983.

Bott, M. H., J. Sunderland, U. Casten, and S. Saxov, Evidence for continental crust beneath the Faeroe Islands, *Nature, 248*, 202-204, 1974.

Boutilier, R. R., and C. E. Keen, Models of small scale convection and melting below divergent plate boundaries (abstract), *Eos Trans. AGU*, F546, 1995.

Bowen, N. L., *The Evolution of the Igneous Rocks.* Princeton University Press, 1928.

Brodie, J., and J. G. Fitton, Composition of basaltic lavas from the seaward-dipping reflector sequence recovered during DSDP Leg 81 (Hatton Bank), *Proc. Ocean Drill. Prog., Sci. Results, 152,* in press, 1996.

Brooks, C. K., Tertiary of Greenland - a volcanic and plutonic record of continental break-up, *Mem. Amer. Ass. Petrol. Geol., 19,* 150-160, 1973a.

Brooks, C. K., Rifting and doming in southern east Greenland, *Nature, 244,* 23-25, 1973b.

Brooks, C. K., and T. F. D. Nielsen, The Phanerozoic development of the Kangerdlugssuaq area, East Greenland, *Medd. Grønl., Geosci., 9,* 30 pp., 1982a.

Brooks, C. K., and T. D. F. Nielsen, The E Greenland continental margin: a transition between oceanic and continental magmatism, *J. Geol. Soc. Lond., 139,* 265-275, 1982b.

Brooks, C. K., L. M. Larsen, and T. F. D. Nielsen, Importance of iron-rich tholeiitic magmas at divergent plate margins: a reappraisal, *Geology, 19,* 269-272, 1991.

Campbell, I. H., and R. W. Griffiths, Implications of mantle plume structure for the evolution of flood basalts, *Earth Planet. Sci. Lett., 99,* 79-93, 1990.

Cande, S. C., and D. V. Kent, A new geomagnetic polarity time scale for the Late Cretaceous and Cenozoic, *J. Geophys. Res., 97,* 13,917-13,951, 1992.

Cande, S. C., and D. V. Kent, Revised calibration of the geomagnetic polarity timescale for the late Cretaceous and Cenozoic, *J. Geophys. Res., 100,* 6093-6095, 1995.

Carter, S. R., N. M. Evensen, P. J. Hamilton, and R. K. O'Nions, Neodymium and strontium isotope evidence for crustal contamination of continental volcanics, *Science, 202,* 743-747, 1978.

Carter, S. R., N. M. Evensen, P. J. Hamilton, and R. K. O'Nions, Basalt magma sources during the opening of the North Atlantic, *Nature, 281,* 28-31, 1979.

Casten, U., The crust beneath the Faeroe Islands, *Nature, 241,* 83-84, 1973.

Casten, U., and P. H. Nielsen, Faeroe Islands - a microcontinental fragment? *J. Geophys., 41,* 357-366, 1975.

Chalmers, J. A., New evidence on the structure of the Labrador Sea/Greenland continental margin, *J. Geol. Soc. Lond., 148,* 899-908, 1991.

Chalmers, J. A., L. M. Larsen, and A. K. Pedersen, Widespread Palaeocene volcanism around the northern North Atlantic and Labrador Sea: evidence for a large, hot, early plume head, *J. Geol. Soc. Lond., 152,* 965-969, 1995.

Clarke, D. B., Tertiary basalts of Baffin Bay: Possible primary magma from the mantle, *Contrib. Mineral. Petrol., 25,* 203-224, 1970.

Clarke, D. B., The Tertiary volcanic province of Baffin Bay, *Geol. Ass. Canada, Spec. Paper, 16,* 445-460, 1977.

Clarke, D. B., and M. J. O'Hara, Nickel, and the existence of high-MgO liquids in nature, *Earth Planet. Sci. Lett., 44,* 153-158, 1979.

Clarke, D. B., and A. K. Pedersen, Tertiary volcanic province of Baffin Bay, in *Geology of Greenland,* edited by A. Escher and W. S. Watts, pp. 365-385, Geological Survey of Greenland, Copenhagen, 1976.

Clarke, D. B. and B. G. J. Upton, Tertiary basalts of Baffin Island: field relations and tectonic setting, *Can. J. Earth Sci., 8,* 248-258, 1971.

Clarke, D. B., G. K. Muecke, and G. Pe-Piper, The lamprophyres of Ubekendt Island, West Greenland: products of renewed partial melting or extreme differentiation?, *Contrib. Mineral. Petrol., 83,* 117-127, 1983.

Clift, P. D., J. Turner, and Ocean Drilling Program Leg 152 Scientific Party, Dynamic support by the Icelandic plume and vertical tectonics of the northeast Atlantic continental margins, *J. Geophys. Res., 100,* 24,473-24, 846, 1995.

Coffin, M. F. and O. Eldholm, Large igneous provinces: crustal structure, dimensions, and external consequences, *Rev. Geophys., 32,* 1-36, 1994.

Courtillot, V., G. Féraud, H. Maluski, D. Vandamme, M. G. Moreau, and J. Besse, Deccan flood basalts and the Cretaceous/Tertiary boundary, *Nature, 333,* 843-846, 1988.

Dagley, P., and A. E. Mussett, Palaeomagnetism and radiometric dating of the British Tertiary Igneous Province: Muck and Eigg, *Geophys. J. R. Astron. Soc., 85,* 221-242, 1986.

Deer, W. A., Tertiary igneous rocks between Scoresby Sund and Kap Gustav Holm, East Greenland, in *Geology of Greenland,* edited by A. Escher and W. S. Watt, pp. 404-429, Geological Survey of Greenland, Copenhagen, 1976.

DePaolo, D. J., Trace element and isotopic effects of combined whole rock assimilation and fractional crystallization, *Earth Planet. Sci. Lett., 53,* 189-202, 1981.

Dickin, A. P., Isotope geochemistry of Tertiary igneous rocks from the Isle of Skye, N.W. Scotland, *J. Petrol., 22,* 155-189, 1981.

Dickin, A. P., The North Atlantic Tertiary Province, in *Continental Flood Basalts,* edited by J. D. Macdougall, pp. 111-149, Kluwer Academic Publishers, Dordrecht, Netherlands, 1988.

Dickin, A. P., and N. W. Jones, Isotopic evidence for the age and origin of pitchstones and felsites, Isle of Eigg, NW Scotland, *J. Geol. Soc. Lond., 140,* 691-700, 1983.

Dickin, A. P., S. Moorbath and W. H.J., Isotope, trace element and major element geochemistry of Tertiary igneous rocks, Isle of Arran, Scotland, *Trans. Roy. Soc. Edin., 72,* 159-170, 1981.

Dickin, A. P., N. W. Jones, M. F. Thirlwall, and R. N. Thompson, A Ce/Nd isotope study of crustal contamination processes affecting Palaeocene magmas in Skye, Northwest Scotland, *Contrib. Mineral. Petrol., 96,* 455-464, 1987.

Duncan, R. A., H. C. Larsen, J. Allen, et al., *Proc. Ocean Drill. Prog., Init. Repts., 163,* 296 pp., 1996.

Eldholm, O., J. Theide, and E. Taylor, *Proc. Ocean Drill. Prog., Init. Repts., 104,* 783 pp., 1987.

Eldholm, O., and K. Grue, North Atlantic volcanic margins: dimensions and production rates, *J. Geophys. Res., 99,* 2955-2968, 1994.

Ellam, R.M., Lithospheric thickness as a control on basalt geochemistry, *Geology, 20,* 153-156, 1992.

Elliott, T. R., C. J. Hawkesworth, and K. Grönvold, Dynamic melting of the Iceland plume, *Nature, 351,* 201-206, 1991.

Emeleus, C. H., The central complexes, in *Igneous Rocks of the British Isles,* edited by D.S. Sutherland, pp. 369-414, John

Wiley & Sons Ltd., 1982.

Emeleus, C. H., The Tertiary lavas and sediments of north-west Rhum, Inner Hebrides, *Geol. Mag., 122*, 419-437, 1985.

Emeleus, C. H., The Rhum layered complex, Inner Hebrides, Scotland, in *Origins of Igneous Layering*, edited by I. Parsons, pp. 263-286, Reidel, Dordrecht, 1987.

Emeleus, C. H., Tertiary igneous activity, in *Geology of Scotland*, edited by G. Y. Craig, pp. 455-502, The Geological Society, London, 1991.

Emeleus, C. H., Geology of Rum and the adjacent Islands, Memoir, British Geological Survey, in press, 1997.

England, R. W., The early Tertiary stress regime in NW Britain: evidence from the patterns of volcanic activity, in *Early Tertiary Volcanism and the Opening of the NE Atlantic, Spec. Publ. 39*, edited by A. C. Morton and L. M. Parson, pp. 381-389, The Geological Society, London, 1988.

England, R. W., The structure of the Skye lava field, *Scott. J. Geol., 30*, 33-37, 1994.

England, R. W., R. W. H. Butler, and D. H. W. Hutton, The role of Paleocene magmatism in the Tertiary evolution of basins on the NW seaboard, in *Petroleum Geology of Northwest Europe*, pp. 97-105, The Geological Society, London, 1993.

Esson, J., A. C. Dunham, and R. N. Thompson, Low alkali, high Ca, olivine tholeiite lavas from the Isle of Skye, Scotland, *J. Petrol., 16*, 488-497, 1975.

Evans, A. L., F. J. Fitch, and J. A. Millar, Potassium-argon age determinations on some British Tertiary igneous rocks, *J. Geol. Soc. Lond., 129*, 419-443, 1973.

Fitch, F. J., G. L. Heard, and J. A. Miller, Basaltic magmatism of late Cretaceous and Palaeogene age recorded in wells NNE of the Shetlands, in *Early Tertiary Volcanism and the Opening of the NE Atlantic, Spec. Publ. 39*, edited by A. C. Morton and L. M. Parson, pp. 253-262, The Geological Society, London, 1988.

Fitton, J. G., and H. M. Dunlop, The Cameroon line, West Africa, and its bearing on the origin of oceanic and continental alkali basalt, *Earth Planet. Sci. Lett., 72*, 23-38, 1985.

Fitton, J. G., A. D. Saunders, L. M. Larsen, M. S. Fram, A. Demant, C. Sinton, and Leg 152 Scientific Party, Magma sources and plumbing systems during break-up of the SE Greenland margin: results from ODP Leg 152, *J. Geol. Soc. Lond., 152*, 985-990, 1995.

Fitton, J. G., B. S. Hardarson, R. M. Ellam, and G. Rogers, Sr-, Nd-, and Pb-isotopic composition of volcanic rocks from the southeast Greenland margin at 63°N: temporal variation in crustal contamination during continental break-up, in *Proc. Ocean Drill. Prog., Sci. Results, 152*, in press, 1996a.

Fitton, J. G., A. D. Saunders, L. M. Larsen, B. S. Hardarson, and M. J. Norry, Volcanic rocks from the East Greenland margin at 63°N: composition, petrogenesis and mantle sources, in *Proc. Ocean Drill. Prog., Sci. Results, 152*, in press, 1996b.

Fitton, J. G., A. D. Saunders, M. J. Norry, and B. S. Hardarson, Thermal and chemical structure of the Iceland plume, *Earth Planet. Sci. Lett.*, in press, 1997.

Foley, S. F., G. Venturelli, D. H. Green, and L. Toscani, The ultrapotassic rocks: characteristics, classification, and constraints from petrogenetic models, *Earth-Sci. Rev., 24*, 84-

134, 1987.

Fowler, S. R., R. S. White, G. D. Spence, and G. D. Westbrook, The Hatton Bank continental margin II. Structure from two-ship expanding spread profiles, *Geophys. J., 96*, 295-309, 1989.

Fram, M., and C. E. Lesher, Geochemical constraints on mantle melting during creation of the North Atlantic basin, *Nature, 363*, 712-715, 1993.

Fram, M. S., and C. E. Lesher, Polybaric differentiation of magmas of the East Greenland early Tertiary flood basalt province, *J. Petrol.*, in press, 1996.

Fram, M. S., C. E. Lesher, and A. M. Volpe, Mantle melting systematics: the transition from continental to oceanic volcanism on the southeast Greenland margin, in *Proc. Ocean Drill. Prog., Sci. Results, 152*, in press, 1996.

Francis, D., The Baffin Bay lavas and the value of picrites as analogues of primary magmas, *Contrib. Mineral. Petrol., 89*, 144-155, 1985.

Furman, T., F. Frey, and K.-H. Park, The scale of source heterogeneity beneath the Eastern neovolcanic zone, Iceland, *J. Geol. Soc. Lond., 152*, 997-1002, 1995.

Gallagher, K., and C. J. Hawkesworth, Dehydration melting and the generation of continental flood basalts, *Nature, 358*, 57-59, 1992.

Gariépy, C., J. Ludden, and C. Brooks, Isotopic and trace element constraints on the genesis of the Faeroe lava pile, *Earth Planet. Sci. Lett., 63*, 257-272, 1983.

Gibson, I. L., and A. D. Gibbs, Accretionary volcanic processes and the crustal structure of Iceland, *Tectonophysics, 133*, 57-64, 1987.

Giekie, A., On the Tertiary volcanic rocks of the British Islands, *Geol. Mag., 4*, 316-319, 1867.

Giekie, A., The lava fields of North-Western Europe. *Nature, 23*, 3-5, 1880.

Giekie, A., The history of volcanic action during the Tertiary period in the British Isles, *Trans. Roy. Soc. Edin., 35*, 21-184, 1888.

Gill, R. C. O., T. F. D. Nielsen, C. K. Brooks, and G. A. Ingram, Tertiary volcanism in the Kangerdlugssuaq region, E Greenland: trace-element geochemistry of the Lower Basalts and tholeiitic dyke swarms, in *Early Tertiary Volcanism and the Opening of the NE Atlantic, Spec. Publ. 39*, edited by A. C. Morton and L. M. Parson, pp. 161-179, The Geological Society, London, 1988.

Gill, R. C. O., A. K. Pedersen, and J. G. Larsen, Tertiary picrites from West Greenland: melting at the periphery of a plume? in *Magmatism and the Causes of Continental Break-up, Spec. Publ. 68*, edited by B. C. Storey, T. Alabaster and R. J. Pankhurst, pp. 335-348, The Geological Society, London, 1992.

Goldschmidt-Rokita, A., K. F. J. Hansch, H. B. Hirschleber, T. Iwasaki, T. Kanazawa, H. Shimarura, and M. A. Sellevol, The ocean/continent transition along a profile through the Lofoten Basin, northern Norway, *Mar. Geophys. Res., 16*, 201-224, 1994.

Gradstein, F. M., F. P. Agterberg, J. G. Ogg, J. Hardelbol, P. van Veen, J. Thierry, and Z. Huang, A Mesozoic time scale, *J.*

Geophys. Res., 99, 24,051-24,074, 1994.

Griffiths, R. W., and I. H. Campbell, Stirring and structure in mantle starting plumes, *Earth Planet. Sci. Lett., 99*, 66-78, 1990.

Haigh, B. I. R., North Atlantic oceanic topography and lateral variations in the upper mantle, *Geophys. J. R. Astron. Soc., 33*, 405-420, 1973.

Hald, N., and A. K. Pedersen, Lithostratigraphy of the early Tertiary volcanic rocks of central West Greenland, *Rapp. Grønlands Geol. Unders., 69*, 17-24, 1975.

Hald, N., and R. Waagstein, Silicic basalts from the Faeroe Islands: Evidence of crustal contamination, in *Structure and Development of the Greenland-Scotland Ridge*, edited by M. H. P. Bott, S. Saxov, M. Talwani and J. Thiede, pp. 343-349, Plenum Press, New York, 1983.

Hald, N., and R. Waagstein, Lithology and chemistry of a 2-km sequence of Lower Tertiary tholeiitic lavas drilled on Suduroy, Faeroe Islands (Lopra-1), in *The Deep Drilling Project 1980-1981 in the Faeroe Islands*, edited by O. Berthelsen, A. Noe-Nygaard and J. Rasmussen, pp. 15-38, Føroya Fródskaparfelag, Tórshavn, 1984.

Hampton, C. M., and P. N. Taylor, The age and nature of the basement of southern Britain: evidence from Sr and Pb isotopes in some English granites, *J. Geol. Soc. Lond., 140*, 499-509, 1983.

Hansen, H., Rex, D.C., P.G. Guise, and C.K. Brooks, [40]Ar/[39]Ar ages on Tertiary East Greenland basalts from the Scoresby Sund area (abstract), *Eos Trans. AGU, 74*, 625, 1989.

Hardarson, B. S., and J. G. Fitton, Geochemical variation in the Tertiary basalts of Iceland, *Min. Mag., 58A*, 372-373, 1994.

Hardman, M., J. Buchanan, P. Herrington, and A. Carr, Geochemical modelling of the East Irish Sea Basin: its influence on predicting hydrocarbon type and quality, in *Petroleum Geology of Northwest Europe*, edited by J. R. Parker, pp. 809-821, The Geological Society, London, 1993.

Hards, V. L., P. D. Kempton, and R. N. Thompson, The heterogeneous Iceland plume: new insights from the alkaline basalts of the Snaefell volcanic centre, *J. Geol. Soc. Lond., 152*, 1003-1009, 1995.

Harland, W. B., A. V. Cox, P. G. Llewellyn, C. A. G. Pickton, A. G. Smith, and R. Walters, *A Geologic Time Scale*, 131 pp., Cambridge University Press, Cambridge, 1982.

Harland, W. B., R. L. Armstrong, A. V. Cox, L. E. Craig, A. G. Smith, and D. G. Smith, *A Geologic Time Scale 1989*, 263 pp., Cambridge University Press, Cambridge, 1990.

Hart, S. R., and K. E. Davis, Nickel partitioning between olivine and silicate melt, *Earth Planet. Sci. Lett., 40*, 203-219, 1978.

Hart, S. R., J.-G. Schilling, and J. L. Powell, Basalts from Iceland and along the Reykjanes Ridge: Sr isotope geochemistry, *Nature, 246*, 104-107, 1973.

Hart, S. R., E. H. Hauri, L. A. Oschmann, and J. A. Whitehead, Mantle plumes and entrainment: isotopic evidence, *Science, 256*, 517-520, 1992.

Hémond, C., N. T. Arndt, and A. W. Hofmann, The heterogeneous Iceland plume: Nd-Sr-O and trace element constraints, *J. Geophys. Res., 98*, 15,833-15850, 1993.

Henderson, G., A. Rosenkrantz, and E. J. Schiener, Cretaceous-Tertiary sedimentary rocks of West Greenland, in *Geology of Greenland*, edited by A. Escher and W. S. Watt, pp. 340-362, Geological Survey of Greenland, Copenhagen, 1976.

Hill, R. I., Starting plumes and continental break-up, *Earth Planet. Sci. Lett., 104*, 398-416, 1991.

Hinz, K., A hypothesis of terrestrial catastrophes. Wedges of very thick oceanward dipping layers beneath passive continental margins – their origin and palaeoenvironmental significance, *Geol. Jahrb., E22*, 3-28, 1981.

Hitchen, K., and J. D. Ritchie, Geological review of the West Shetland area, in *Petroleum Geology of North West Europe*, edited by J. Brooks and K. W. Glennie, pp. 737-749, Graham and Trotman, London, 1987.

Hitchen, K., and J. D. Ritchie, New K-Ar ages, and a provisional chronology, for the offshore part of the British Tertiary Igneous Province, *Scott. J. Geol., 29*, 73-85, 1993.

Hofmann, A. W., K. P. Jochum, M. Seufert, and W. M. White, Nb and Pb in oceanic basalts: new constraints on mantle evolution, *Earth Planet. Sci. Lett., 79*, 33-45, 1986.

Hogg, A. J., J. J. Fawcett, J. Gittins, and M. P. Gorton, Cyclical variation in composition in continental tholeiites of East Greenland, *Can. J. Earth Sci., 26*, 534-543, 1989.

Holm, P. M., Nd, Sr and Pb isotope geochemistry of the Lower Lavas, E Greenland Tertiary Igneous Province, in *Early Tertiary Volcanism and the Opening of the NE Atlantic, Spec. Publ. 39*, edited by A. C. Morton and L. M. Parson, pp. 181-195, The Geological Society, London, 1988.

Holm, P. M., N. Hald, and T. F. D. Nielsen, Contrasts in composition and evolution of Tertiary CFBs between West and East Greenland and their relations to the establishment of the Icelandic mantle plume, in *Magmatism and the Causes of Continental Break-up, Spec. Publ. 68*, edited by B. C. Storey, T. Alabaster and R. J. Pankhurst, pp. 349-364, The Geological Society, London, 1992.

Holm, P. M., R. C. O. Gill, A. K. Pedersen, J. G. Larsen, N. Hald, T. F. D. Nielsen, and M. F. Thirlwall, The Tertiary picrites of West Greenland: contributions from 'Icelandic' and other sources, *Earth Planet. Sci. Lett., 115*, 227-244, 1993.

Holmes, A., The basaltic rocks of the Arctic region, *Min. Mag., 18*, 180-222, 1918.

Huppert, H. E., and R. S. J. Sparks, The fluid dynamics of a basaltic magma chamber replenished by influx of hot, dense ultrabasic magma, *Contrib. Mineral. Petrol., 75*, 279-289, 1980.

Imsland, P., Iceland and the ocean floor. Comparison of chemical characteristics of the magmatic rocks and some volcanic features, *Contrib. Mineral. Petrol., 83*, 31-37, 1983.

Ito, E., W. M. White, and C. Göpel, The O, Sr, Nd and Pb isotope geochemistry of MORB, *Chem. Geol., 62*, 157-176, 1987.

Jakobsson, S. P., Chemistry and distribution patterns of Recent volcanic rocks in Iceland, *Lithos, 5*, 365-386, 1972.

Jakobsson, S. P., J. Jónsson, and F. Shido, Petrology of the western Reykjanes Peninsula, Iceland, *J. Petrol., 19*, 669-705, 1978.

Jones, E. J. W., R. Siddall, M. F. Thirlwall, P. N. Chroston, and A. J. Lloyd, Anton Dohrn seamount and the evolution of the Rockall Trough, *Ocean. Acta, 17*, 237-247, 1994.

Joron, J.-L., H. Bougault, R. C. Maury, M. Bohn, and A. Desprairies, Strongly depleted tholeiites from the Rockall Plateau margin, North Atlantic: geochemistry and mineralogy, *Init. Repts. Deep Sea Drill. Proj., 81,* 783-794, 1984.

Judd, J. W., On the ancient volcanoes of the Highlands and the relations of their products to the Mesozoic strata, *Q. J. Geol. Soc. Lond., 30,* 220-302, 1874.

Judd, J. W., The Tertiary volcanoes of the Western Isles of Scotland, *Q. J. Geol. Soc. Lond., 45,* 187-218, 1889.

Jürgensen, T., and N. Mikkelsen, Coccoliths from volcanic sediments (Danian) in Nugssuaq, West Greenland, *Bull. Geol. Soc. Denmark, 23,* 225-230, 1974.

Kennedy, W. Q., Trends of differentiation in basaltic magmas, *Am. J. Sci. 25,* 239-256, 1933.

Kent, R. W., Magnesian basalts from the Hebrides, Scotland: Chemical composition and relationship to the Iceland plume, *J. Geol. Soc. Lond., 152,* 979-983, 1995.

Kent, R. W., M. Storey, and A. D. Saunders, Large igneous provinces: sites of plume impact or plume incubation? *Geology, 20,* 891-894, 1992.

Kerr, A. C., Elemental evidence for an enriched small-fraction melt input into Tertiary Mull basalts, Western Scotland, *J. Geol. Soc. Lond., 150,* 763-769, 1993.

Kerr, A. C., Lithospheric thinning during the evolution of continental Large Igneous Provinces: a case study from the North Atlantic Tertiary Province, *Geology, 22,* 1027-1030, 1994.

Kerr, A. C., The geochemistry of the Mull-Morvern Tertiary lava succession, NW Scotland: an assessment of mantle sources during plume-related volcanism, *Chem. Geol., 122,* 43-58, 1995a.

Kerr, A. C., The melting processes and composition of the North Atlantic (Iceland) plume: geochemical evidence from the early Tertiary lavas, *J. Geol. Soc. Lond., 152,* 975-978, 1995b.

Kerr, A. C., P. D. Kempton, and R. N. Thompson, Crustal assimilation during turbulent magma ascent [ATA]; new isotopic evidence from the Mull Tertiary lava succession, N.W. Scotland, *Contrib. Mineral. Petrol., 119,* 142-154, 1995a.

Kerr, A. C., A. D. Saunders, V. L. Hards, J. Tarney, and N. H. Berry, Depleted mantle plume geochemical signatures: No paradox for plume theories, *Geology, 23,* 843-846, 1995b.

Kirton, S. R., and J. A. Donato, Some buried Tertiary dykes of Britain and surrounding waters deduced by magnetic modelling and seismic reflection methods, *J. Geol. Soc. Lond., 142,* 1047-1057, 1985.

Klein, E. M., and C. H. Langmuir, Global correlation of ocean ridge basalt chemistry with axial depth and crustal thickness, *J. Geophys. Res., 92,* 8089-8115, 1987.

Knox, R. W. O., and A. C. Morton, Stratigraphical distribution of Early Palaeogene pyroclastic deposits in the North Sea Basin, *Proc. Yorks. Geol. Soc., 44,* 355-363, 1983.

Knox, R. W., and A. C. Morton, The record of early Tertiary N Atlantic volcanism in sediments of the North Sea Basin, in *Early Tertiary Volcanism and the Opening of the NE Atlantic, Spec. Publ. 39,* edited by A. C. Morton and L. M. Parson, pp. 407-419, The Geological Society, London, 1988.

Kristoffersen, Y., Sea-floor spreading and the early opening of the North Atlantic, *Earth Planet. Sci. Lett., 38,* 273-290, 1978.

Langmuir, C. H., and G. N. Hanson, An evaluation of major element heterogeneity in the mantle sources of basalts, *Phil. Trans. R. Soc. Lond., A297,* 383-407, 1980.

Larsen, H. C., and C. Marcussen, Sill-intrusion, flood basalt emplacement and deep crustal structure of the Scoresby Sund region, East Greenland, in *Magmatism and the Causes of Continental Break-up, Spec. Publ. 68,* edited by B. C. Storey, T. Alabaster and R. J. Pankhurst, pp. 365-386, The Geological Society, London, 1992.

Larsen, H. C., and S. Jakobsdóttir, Distribution, crustal properties and significance of seawards-dipping sub-basement reflectors off E Greenland, in *Early Tertiary Volcanism and the Opening of the NE Atlantic, Spec. Publ. 39,* edited by A. C. Morton and L. M. Parson, pp. 95-114, The Geological Society, London, 1988.

Larsen, H. C., C. K. Brooks, J. R. Hopper, T. Dahl-Jensen, A. K. Pedersen, T. F. D. Nielsen, and field parties, The Tertiary opening of the North Atlantic: DLC investigations along the east coast of Greenland, *Rapp. Grønlands Geol. Unders., 165,* 106-115, 1995.

Larsen, H. C., A. D. Saunders, P. D. Clift, et al., *Proc. Ocean Drill. Prog., Init. Repts., 152,* 977 pp., 1994.

Larsen, J. G., Transition from low potassium olivine tholeiites to alkali basalts on Ubekendt Ejland, *Medd. Grøn., 200,* 1-42, 1977.

Larsen, L. M., and A. K. Pedersen, Volcanic marker horizons in the Maligât Formation on Disko and Nûgssuaq, and implications for the development of the southern part of the West Greenland basin in the early Tertiary, *Rapp. Grønlands Geol. Unders., 148,* 65-73, 1990.

Larsen, L. M., and A. K. Pedersen, Volcanic marker horizons in the upper part of the Maligât Formation on eastern Disko and Nuussuaq, Tertiary of West Greenland: syn- to post-volcanic basin movements, *Rapp. Grønlands Geol. Unders., 155,* 85-93, 1992.

Larsen, L. M., and S. W. Watt, Episodic volcanism during break-up of the North Atlantic: evidence from the East Greenland plateau basalts, *Earth Planet. Sci. Lett., 73,* 105-116, 1985.

Larsen, L. M., J. G. Fitton, and M. S. Fram, Volcanic rocks of the southeast Greenland margin in comparison with other parts of the North Atlantic Tertiary igneous province, in *Proc. Ocean Drill. Prog., Sci. Results, 152,* in press, 1996.

Larsen, L. M., W. S. Watt, and M. Watt, Geology and petrology of the Lower Tertiary plateau basalts of the Scoresby Sund region, East Greenland, *Bull. Geol. Surv. Green., 157,* 1-164, 1989.

Larsen, L. M., A. K. Pedersen, G. K. Pedersen, and S. Piasecki, Timing and duration of Early Tertiary volcanism in the North Atlantic: new evidence from West Greenland, in *Magmatism and the Causes of Continental Break-up, Spec. Publ. 68,* edited by B. C. Storey, T. Alabaster and R. J. Pankhurst, pp. 321-333, The Geological Society, London, 1992.

Lawver, L. A., and R. D. Müller, Iceland hotspot track, *Geology, 22,* 311-314, 1994.

Lewis, C. A., P. F. Green, A. Carter, and A. J. Hurford, Elevated

K/T palaeotemperatures throughout northwest England: three kilometres of Tertiary erosion? *Earth Planet. Sci. Lett., 112,* 132-145, 1992.

Lund, J., Biostratigraphy of interbasaltic coals, in *Structure and Development of the Greenland-Scotland Ridge. New Methods and Concepts,* edited by M. H. P. Bott, S. Saxov, M. Talwani, and J. Thiede, pp. 417-423, Plenum Press, New York and London, 1983.

Lyle, P., A petrological and geochemical study of the Tertiary basaltic rocks of NE Ireland, *J. Earth Sci. R. Dub. Soc, 2,* 137-152, 1980.

Lyle, P., The petrogenesis of Tertiary basaltic and intermediate lavas of NE Ireland, *Scott. J. Geol, 21,* 71-84, 1985.

Macdougall, J. D. (Ed.), *Continental Flood Basalts,* 341 pp., Kluwer Academic Publishers, Dordrecht, Netherlands, 1988.

Macintyre, R. M. and P. J. Hamilton, Isotopic geochemistry of lavas from Sites 553 and 555, *Init. Repts. Deep Sea Drill. Proj., 81,* 775-781, 1984.

Mahoney, J. J., Deccan Traps, in *Continental Flood Basalts,* edited by J. D. Macdougall, pp. 151-194, Kluwer Academic Publishers, Dordrecht, Netherlands, 1988.

Mattey, D. P., I. L. Gibson, G. F. Marriner, and R. N. Thompson, The diagnostic geochemistry, relative abundance and spatial distribution of high-Ca, low-alkali tholeiite dykes in the lower Tertiary regional swarm of the Isle of Skye, NW Scotland, *Min. Mag., 41,* 273-285, 1977.

McClurg, J. E., *Petrology and evolution of the northern part of the Rhum ultrabasic complex,* Unpublished PhD thesis, University of Edinburgh, 1982.

McDonough, W. F. and S.-s. Sun, The composition of the Earth, *Chem. Geol., 120,* 223-253, 1995.

McDougall, I., L. Kristjansson and K. Saemundsson, Magnetostratigraphy and geochronology of NW-Iceland, *J. Geophys. Res., 89,* 7029-7060, 1984.

McKenzie, D., Some remarks on the movement of small melt fractions in the mantle, *Earth Planet. Sci. Lett., 95,* 53-72, 1989.

McKenzie, D., and M. J. Bickle, The volume and composition of melt generated by extension of the lithosphere, *J. Petrol., 29,* 625-679, 1988.

McKenzie, D., and R. K. O'Nions, Partial melt distributions from inversion of rare earth element concentrations, *J. Petrol., 32,* 1021-1091, 1991.

McKenzie, D., and R. K. O'Nions, Mantle reservoirs and ocean island basalts, *Nature, 301,* 229-231, 1983.

Menke, W. and V. Levin, Cold crust in a hot spot, *G. Res. Lett., 21,* 1967-1970, 1994.

Merriman, R. J., P. N. Taylor and A. C. Morton, Petrochemistry and isotope geochemistry of early Palaeogene basalts forming the dipping reflector sequence SW of Rockall Plateau, NE Atlantic, in *Early Tertiary Volcanism and the Opening of the NE Atlantic, Spec. Publ. 39,* edited by A. C. Morton and L. M. Parson, pp. 123-134, The Geological Society, London, 1988.

Miller, J. A., D. H. Matthews, and D. G. Roberts, Rocks of Grenville age from Rockall Bank, *Nature, 246,* 61, 1973.

Moorbath, S., and J. D. Bell, Strontium isotope abundance studies and Rb-Sr age determinations on Tertiary Igneous rocks from the Isle of Skye, NW Scotland, *J. Petrol., 6,* 37-66, 1965.

Moorbath, S., and R. N. Thompson, Strontium isotope geochemistry and petrogenesis of the early Tertiary lava pile of the Isle of Skye, Scotland, and other basic rocks of the British Tertiary Province: an example of magma-crust interaction, *J. Petrol., 21,* 295-321, 1980.

Moorbath, S., and H. Welke, Lead isotope studies on igneous rocks from the Isle of Skye, NW Scotland, *Earth Planet. Sci. Lett., 5,* 217-230, 1969.

Moorbath, S., H. Sigurdsson and R. Goodwin, K-Ar ages of oldest exposed rocks in Iceland, *Earth Planet. Sci. Lett., 4,* 197-205, 1968.

Morgan, J. V., P. J. Barton, and R. S. White, The Hatton Bank continental margin III, Structure from wide-angle OBS and multichannel seismic refraction profiles, *Geophys. J. Int., 98,* 367-384, 1989.

Morgan, W. J., Convection plumes in the lower mantle, *Nature, 230,* 42-43, 1971.

Morgan, W. J., Plate motions and deep mantle convection, *Geol. Soc. Am. Mem., 132,* 7-22, 1972.

Morrison, M. A., R. N. Thompson, I. L. Gibson, and G. F. Marriner, Lateral chemical heterogeneity in the Palaeocene upper mantle beneath the Scottish Hebrides, *Phil. Trans. Royal Soc. Lond., A297,* 229-244, 1980.

Morton, A. C., and J. A. Evans, Geochemistry of basaltic ash beds from the Fur Formation, Island of Fur, Denmark, *Bull. Geol. Soc. Denmark, 37,* 1-9, 1987.

Morton, A. C., and P. N. Taylor, Lead isotope evidence for the structure of the Rockall dipping-reflector passive margin, *Nature, 326,* 381-383, 1987.

Morton, A. C., and P. N. Taylor, Geochemical and isotopic constraints on the nature and age of basement rocks from Rockall Bank, NE Atlantic, *J. Geol. Soc. Lond., 148,* 630-634, 1991.

Morton, A. C., J. E. Dixon, J. G. Fitton, R. M. Macintyre, D. K. Smythe and P. N. Taylor, Early Tertiary volcanic rocks in Well 163/6-1A, Rockall Trough, in *Early Tertiary Volcanism and the Opening of the NE Atlantic, Spec. Publ. 39,* edited by A. C. Morton and L. M. Parson, pp. 293-308, The Geological Society, London, 1988.

Morton, A. C., K. Hitchen, J. D. Ritchie, N. M. Hine, M. Whitehouse, and S. G. Carter, Late Cretaceous basalts from Rosemary Bank, northern Rockall Trough, *J. Geol. Soc. Lond., 152,* 947-952, 1995.

Mussett, A. E., Time and duration of Tertiary igneous activity of Rhum and adjacent areas, *Scott. J. Geol., 20,* 273-279, 1984.

Mussett, A. E., ^{40}Ar/^{39}Ar step-heating ages of the Tertiary igneous rocks of Mull, Scotland, *J. Geol. Soc. Lond., 143,* 887-896, 1986.

Mussett, A. E., P. Dagley, and M. Eckford, The British Tertiary Igneous Province: palaeomagnetism and ages of dykes, Lundy Island, Bristol Channel, *Geophys. J. R. Astron. Soc., 46,* 595-603, 1976.

Mussett, A. E., P. Dagley, and R. R. Skelhorn, Time and duration of activity in the British Tertiary igneous province, in *Early Tertiary Volcanism and the Opening of the NE Atlantic, Spec.*

Publ. 39, edited by A. C. Morton and L. M. Parson, pp. 337-348, The Geological Society, London, 1988.

Mutter, J. C., and C. M. Zehnder, Deep crustal structure and magmatic processes: the inception of seafloor spreading in the Norwegian-Greenland sea, in *Early Tertiary Volcanism and the Opening of the NE Atlantic, Spec. Publ. 39*, edited by A. C. Morton and L. M. Parson, pp. 34-38, The Geological Society, London, 1988.

Mutter, J. C., W. R. Buck, and C. M. Zehnder, Convective partial melting. 1. A model for the formation of thick basaltic sequences during the initiation of spreading, *J. Geophys. Res., B93*, 1031-1048, 1988.

Mutter, J. C., M. Talwani, and P. L. Stoffa, Origin of seaward-dipping reflectors in oceanic crust off the Norwegian margin by "subaerial seafloor spreading," *Geology, 10*, 353-357, 1982.

Nadin, P. A., N. J. Kusznir, and J. Toth, Transient regional uplift in the Early Tertiary of the northern North Sea and the development of the Iceland plume, *J. Geol. Soc. Lond., 152*, 953-958, 1995.

Nakamura, N., Determination of REE, Ba, Fe, Mg, Na and K in carbonaceous and ordinary chondrites, *Geochim. Cosmochim. Acta, 38*, 757-775, 1974.

Nielsen, P. H., Geology and crustal structure of the Faeroe Islands - a review, in *Structure and Development of the Greenland-Scotland Ridge. New Methods and Concepts*, edited by M. H. P. Bott, S. Saxov, M. Talwani, and J. Thiede, pp. 77-86, Plenum press, New York and London, 1983.

Nielsen, T. F. D., and C. K. Brooks, The E Greenland rifted continental margin: an examination of the coastal flexure, *J. Geol. Soc. Lond., 138*, 559-568, 1981.

Nielsen, T. F. D., N. J. Soper, C. K. Brooks, A. M. Faller, A. C. Higgins, and D. W. Matthews, The pre-basaltic sediments and the lower basalts at Kangerdlugssuaq, East Greenland: their stratigraphy, lithology, palaeomagnetism and petrology, *Medd. Grøn., Geosci., 6*, 25 pp., 1981.

Noble, R. H., R. M. Macintyre, and P. E. Brown, Age constraints on Atlantic evolution: timing of magmatic activity along the E Greenland continental margin, in *Early Tertiary Volcanism and the Opening of the NE Atlantic, Spec. Publ. 39*, edited by A. C. Morton and L. M. Parson, pp. 210-214, The Geological Society, London, 1988.

Noe-Nygaard, A., Cenozoic to Recent volcanism in and around the North Atlantic Basin, in *The Ocean Basins and Margins*, edited by A. Nairn and F. G. Stehli, pp. 391-443, Plenum Press, New York, 1974.

Noe-Nygaard, A., Tertiary igneous rocks between Shannon and Scoresby Sund, East Greenland, in *Geology of Greenland*, edited by A. Escher and W. S. Watt, pp. 386-402, Geological Survey of Greenland, Copenhagen, 1976.

Noe-Nygaard, A., and J. Rasmussen, Petrology of a 3000-metre sequence of basaltic lavas in the Faeroe Islands, *Lithos, 1*, 268-304, 1968.

Old, R. A., The age and field relationships of the Tardree Tertiary rhyolite complex, County Antrim, N, Ireland, *Bull. Geol. Surv. G. B., 51*, 21-40, 1975.

O'Hara, M. J., and R. E. Mathews, Geochemical evolution in an advancing, periodically replenished, periodically tapped, continuously fractionated magma chamber, *J. Geol. Soc. Lond., 138*, 237-277, 1981.

O'Nions, R. K., P. J. Hamilton and N. M. Evensen, Variations in $^{143}Nd/^{144}Nd$ and $^{87}Sr/^{86}Sr$ ratios in oceanic basalts, *Earth Planet. Sci. Lett., 34*, 13-22, 1977.

Oskarsson, N., S. Steinthorsson, and G. E. Sigvaldason, Iceland geochemical anomaly: origin, volcanotectonics, chemical fractionation and isotope evolution of the crust, *J. Geophys. Res., 90*, 10,011-10,025, 1985.

Palacz, Z. A., Isotopic and geochemical evidence for the evolution of a cyclical unit in the Rhum intrusion, north-west Scotland, *Nature, 307*, 618-620, 1984.

Pálmason, G., A continuum model of crustal generation in Iceland: kinematic aspects, *J. Geophys., 47*, 7-18, 1980.

Pálmason, G., Model of crustal formation in Iceland, and application to submarine mid-ocean ridges, in *The Western North Atlantic Region, GNAM Ser.*, edited by P. R. Vogt and B. E. Tucholke, pp. 87-97, Geological Society of America, Boulder, CO, 1986.

Pálmason, G., and K. Saemundsson, Iceland in relation to the Mid-Atlantic Ridge, *Ann. Rev. Earth Planet. Sci., 2*, 25-50, 1974.

Pálmason, G., S. Arnorsson, I. B. Fridleifsson, H. Kristmannsdóttir, K. Saemundsson, V. Stefansson, B. Steingrímsson, J. Tomasson, and L. Kristjansson, The Iceland crust: evidence from drillhole data on structure and processes, in *Deep Drilling Results in the Atlantic Ocean; Ocean Crust, Maurice Ewing Ser. 3*, edited by M. Talwin, C. G. Harrison, and D. E. Hayes, pp. 43-65, AGU, Washington, D. C., 1978.

Parrott, R. J. E., $^{40}Ar/^{39}Ar$ dating on Labrador Sea volcanics and their relation to sea-floor spreading, M.Sc., Dalhousie University, Halifax, 1976.

Parrott, R. J. E. and P. H. Reynolds, Argon 40/argon 39 geochronology: age determinations of basalts from the Labrador Sea area, *Geol. Soc. Amer. Bull., 7*, 835, 1975.

Pearson, D. G., C. H. Emeleus and S. P. Kelley, Precise $^{40}Ar/^{39}Ar$ ages for the initiation of igneous activity in the Small Isles, Inner Hebrides and implications for the timing of magmatism in the British Tertiary Volcanic Province, *J. Geol. Soc. Lond., 153*, 815-818, 1996.

Pedersen, A. K., Armacolite-bearing Fe-Ti oxide assemblages in graphite-equilibrated salic volcanic rocks with native iron from Disko, central West Greenland, *Contrib. Mineral. Petrol., 77*, 307-324, 1981.

Pedersen, A. K., Reaction between picritic magma and continental crust: early Tertiary silicic basalts and magnesian andesites from Disko, West Greenland, *Bull. Geol. Soc. Greenland, 152*, 126 pp., 1985.

Pedersen, A. K. and S. Pedersen, Sr isotope chemistry of contaminated Tertiary volcanic rocks from Disko, central West Greenland, *Bull. Geol. Soc. Denmark, 36*, 315-336, 1987.

Piasecki, S., L. M. Larsen, A. K. Pedersen and G. K. Pedersen, Palynostratigraphy of the lower Tertiary volcanics and marine clastic sediments in the southern part of the West Greenland Basin: implications for the timing and duration of the volcanism, *Rapp. Grønlands Geol. Unders., 154*, 13-31, 1992.

Powell, A. J., A modified dinoflagellate cyst biozonation for latest Palaeocene and earliest Eocene sediments from the central North Sea, *Rev. Palaeobot. Palaen., 56*, 327-344, 1988.

Price, S., J. Brodie, A. Witham, and R.W. Kent, Mid-Tertiary rifting and magmatism in the Traill Ø region, East Greenland. *J. Geol. Soc. Lond.*, in press, 1996.

Reiners, P. W., B. K. Nelson, and M. S. Ghiorso, Assimilation of felsic crust by basaltic magma: thermal limits and extents of crustal contamination of mantle-derived melts, *Geology, 23*, 563-566, 1995.

Ribe, N. M., U. R. Christensen, and J. Theissing, The dynamics of plume-ridge interactions, 1: ridge-centred plumes, *Earth Planet. Sci. Lett., 134*, 155-168, 1995.

Richards, M. A., R. A. Duncan, and V. E. Courtillot, Flood basalts and hot-spot tracks: plume heads and tails, *Science, 246*, 103-107, 1989.

Richey, J. E., *Scotland: the Tertiary Volcanic Districts*, His Majesty's Stationery Office, Edinburgh, 120 pp., 1935.

Ridley, W. I., The petrology of volcanic rocks from the Small Isles of Inverness-shire, *Report of the Institute of Geological Sciences*, 1973.

Ritchie, J. D. and K. Hitchen, Early Paleogene offshore igneous activity to the northwest of the UK and its relationship to the North Atlantic Igneous Province, in *Correlation of the Early Paleogene in Northwest Europe, Spec. Publ. 101*, edited by R. W. O. Knox, R. M. Corfield and R. E. Dunay, pp. 63-78, The Geological Society, London, 1996.

Roberts, D. G., Marine geology of the Rockall Plateau and Trough, *Phil. Trans. Roy. Soc. London., A278*, 447-509, 1975.

Roberts, D. G., D. G. Masson, and P. R. Miles, Age and structure of the southern Rockall Trough: new evidence, *Earth Planet. Sci. Lett., 52*, 115-128, 1981.

Roberts, D. G., and D. Schnitker, et al., *Init. Repts. Deep Sea Drill. Proj., 81*, 923 pp., 1984.

Robillard, I., D., Francis and J. N. Ludden, The relationship between E- and N-type magmas in the Baffin Bay lavas, *Contrib. Mineral. Petrol., 112*, 230-241, 1992.

Roest, W. R., and S. P. Srivastava, Sea-floor spreading in the Labrador Sea: a new reconstruction, *Geology, 17*, 1000-1003, 1989.

Saemundsson, K., Subaerial volcanism in the western North Atlantic, in *The Western North Atlantic Region*, edited by P. R. Vogt and B. E. Tucholke, pp. 69-86, Geol. Soc. Amer., 1986.

Saunders, A. D., M. J. Norry, and J. Tarney, Origin of MORB and chemically-depleted mantle reservoirs: trace element constraints, in *Oceanic and Continental Lithosphere: Similarities and Differences*, edited by M. A. Menzies and K. G. Cox, pp. 415-445, Journal of Petrology Special Issue, Oxford University Press, 1988.

Saunders, A. D., M. Storey, R. W. Kent, and M. J. Norry, Consequences of plume-lithosphere interactions, in *Magmatism and the Causes of Continental Break-up*, Spec. Publ. 68, edited by B. C. Storey, T. Alabaster and R. J. Pankhurst, pp. 41-60, The Geological Society, London, 1992.

Saunders, A. D., J. Tarney, A. C. Kerr, and R. W. Kent, The formation and fate of large oceanic igneous provinces, *Lithos, 37*, 81-95, 1996.

Scarrow, J. H., Petrogenesis of the Tertiary lavas, Isle of Skye, N.W. Scotland, Unpublished D. Phil. thesis, University of Oxford, 1992.

Scarrow, J. H., and K. G. Cox, Basalts generated by decompressive adiabatic melting of a mantle plume - a case study from the Isle of Skye, NW Scotland, *J. Petrol, 36*, 3-22, 1995.

Schilling, J.-G., Iceland mantle plume: geochemical study of Reykjanes Ridge, *242*, 565-571, 1973.

Schilling, J.-G., Rare-earth, Sc, Cr, Fe, Co, and Na abundances in DSDP Leg 38 basement basalts: some additional evidence on the evolution of the Thulean Volcanic Province, *Init. Repts. Deep Sea Drill. Proj., 38*, 741-750, 1976.

Schilling, J.-G., Fluxes and excess temperatures of mantle plumes inferred from their interaction with migrating mid-ocean ridges, *Nature, 352*, 397-403, 1991.

Schilling, J.-G., and A. Noe-Nygaard, Faeroe-Iceland plume: rare earth evidence, *Earth Planet. Sci. Lett., 24*, 1-14, 1974.

Schilling, J.-G., P. S. Meyer, and R. H. Kingsley, Evolution of the Iceland hotspot, Nature, *296*, 313-320, 1982.

Schönharting, G. and N. Abrahamsen, Paleomagnetism of the volcanic sequence in Hole 642E, ODP Leg 104, Vøring Plateau, and correlation with Early Tertiary basalts in the North Atlantic, in *Proc. Ocean Drill. Prog., Sci. Results, 104*, 911-920, 1989.

Sinton, C. W. and R. A. Duncan, ^{40}Ar-^{39}Ar ages of lavas from the southeast Greenland margin, ODP Leg 152 and the Rockall Plateau, DSDP Leg 81, in *Proc. Ocean Drill. Prog., Sci. Results, 152*, in press, 1996.

Skogseid, J. and O. Eldholm, Early Cainozoic evolution of the Norwegian volcanic passive margin and the formation of marginal highs, in *Early Tertiary Volcanism and the Opening of the NE Atlantic, Spec. Publ. 39*, edited by A. C. Morton and L. M. Parson, pp. 49-56, The Geological Society, London, 1988.

Smith, A. D., The continental mantle as a source for hotspot volcanism, *Terra Nova, 5*, 452-460, 1993.

Soper, N. J., C. Downie, A. C. Higgins, and L. I. Costa, Biostratigraphic ages of Tertiary basalts on the east Greenland continental margin and their relationship to plate separation in the northeast Atlantic, *Earth Planet. Sci. Lett., 32*, 149-157, 1976a.

Soper, N. J., A. C. Higgins, C. Downie, D. W. Matthews and P. E. Brown, Late Cretaceous - early Tertiary stratigraphy of the Kangerdlugssuaq area, east Greenland, and the age of opening of the north-east Atlantic, *J. Geol. Soc. Lond., 132*, 85-104, 1976b.

Speight, J. M., R. R. Skelhorn, T. Sloan, and R. J. Knaap, The dyke swarms of Scotland, in *Igneous Rocks of the British Isles*, edited by D. S. Sutherland, pp. 449-459, John Wiley and Sons Ltd., 1982.

Srivastava, S. P., Evolution of the Labrador Sea and its bearing on the early evolution of the North Atlantic, *J. R. Astron. Soc., 52*, 313-357, 1978.

Srivastava, S. P., and C. R. Tapscott, Plate kinematics of the North Atlantic, in *The Western North Atlantic Region,* GNAM Ser., edited by P. R. Vogt and B. E. Tucholke, pp. 379-404, Geological Society of America, Boulder, CO, 1986.

Staples, R. K., R. S. White, B. Brandsdóttir, W. H. Menke, P. K. H. Maguire, J. McBride, and J. Smallwood, Faero-Iceland Ridge Experiment - 1. The crustal structure of north-eastern Iceland, *J. Geophys. Res.,* in press, 1996.

Steinthorsson, S., N. Oskarsson, and G. E. Sigvaldason, Origin of alkali basalts in Iceland: a plate tectonic model, *J. Geophys. Res., 90,* 10,027-10,042, 1985.

Stolper, E., and D. Walker, Melt density and the average composition of basalt, *Contrib. Mineral. Petrol., 74,* 7-12, 1980.

Storey, M., R. A. Duncan, H. C. Larsen, A. K. Pedersen, R. Waagstein, L. M. Larsen, C. Tegner and C. A. Lesher, Impact and rapid flow of the Iceland plume beneath Greenland at 61 Ma (abstract), *Eos Trans. AGU, 77,* F839, 1996.

Sun, S.-S., and B.-m. Jahn, Lead and strontium isotopes in post-glacial basalts from Iceland, *Nature, 255,* 527-530, 1975.

Sun, S.-s., and W. F. McDonough, Chemical and isotopic systematics of oceanic basalts: implications for mantle composition and processes, *Magmatism in the Ocean Basins, Spec. Publ. 41,* edited by A. D. Saunders and M. J. Norry, pp. 313-345, The Geological Society, London, 1989.

Talwani, M., and G. Udintsev et al., *Init. Repts. Deep Sea Drill. Proj., 38,* 1256 pp., 1976.

Tarduno, J. A., Arctic flood basalt volcanism: examining the hypothesis of Cretaceous activity at the Iceland hotspot (abstract), *Eos Trans. AGU, 77,* F844, 1996.

Tarling, D. H., The palaeomagnetic properties of some Tertiary lavas from East Greenland, *Earth Planet. Sci. Lett., 3,* 81-88, 1967.

Tarney, J., D. A. Wood, A. D. Saunders, J. R. Cann, and J. Varet, Nature of mantle heterogeneity in the North Atlantic: evidence from deep sea drilling, *Phil. Trans. Roy. Soc. Lond., A297,* 179-202, 1980.

Taylor, P. N., and A. C. Morton, Sr, Nd, and Pb isotope geochemistry of the upper and lower volcanic series at Site 642, *Proc. Ocean Drill. Prog., Sci. Results, 104,* 429-435, 1989.

Thirlwall, M. F., Generation of Pb isotopic characteristics of the Iceland plume, *J. Geol. Soc. Lond., 152,* 991-996, 1995.

Thirlwall, M. F., and N. W. Jones, Isotope geochemistry and contamination mechanics of Tertiary lavas from Skye, in *Continental Basalts and Mantle Xenoliths,* edited by C. J. Hawkesworth and M. J. Norry, pp. 186-208, Shiva, Nantwich, 1983.

Thirlwall, M. F., B. G. J. Upton, and C. Jenkins, Interaction between continental lithosphere and the Iceland plume - Sr-Nd-Pb isotope geochemistry of Tertiary basalts, NE Greenland, *J. Petrol., 35,* 839-879, 1994.

Thomas, J., The occurrence of the dinoflagellate cyst *Apectodinium* (Costa and Downie 1976) Lentin and Williams 1977 in the Moray and Montrose Groups (Danian to Thanetian) of the UK central North Sea, in *Correlation of the Early Paleogene in Northwest Europe, Spec. Publ. 101,* edited by R. W. O. Knox, R. M. Corfield, and R. E. Dunay, pp. 115-120, The Geological Society, London, 1996.

Thompson, P., *Dating the British Tertiary Igneous Province in Ireland by the $^{40}Ar/^{39}Ar$ stepwise degassing method,* unpublished PhD thesis, University of Liverpool, 1986.

Thompson, P., A. E. Mussett and P. Dagley, Revised ^{40}Ar-^{39}Ar age for granites of the Mourne Mountains, Ireland, *Scott. J. Geol., 23,* 215-220, 1987.

Thompson, R. N., Magmatism of the British Tertiary Volcanic Province, *Scott. J. Geol., 18,* 49-107, 1982.

Thompson, R. N. and S. A. Gibson, Subcontinental mantle plumes, hotspots and pre-existing thinspots, *J. Geol. Soc. Lond., 148,* 973-977, 1991.

Thompson, R. N., and M. A. Morrison, Asthenospheric and lower-lithospheric mantle contributions to continental extensional magmatism: an example from the British Tertiary Province, *Chem. Geol., 68,* 1-15, 1988.

Thompson, R. N., J. Esson, and A. C. Dunham, Major element chemical variation in the Eocene lavas of the Isle of Skye, *J. Petrol., 13,* 219-253, 1972.

Thompson, R. N., A. P. Dickin, I. L. Gibson, and M. A. Morrison, Elemental fingerprints of isotopic contamination of Hebridean Palaeocene mantle-derived magmas by Archaean sial, *Contrib. Mineral. Petrol., 79,* 159-168, 1982.

Thompson, R. N., M. A. Morrison, G. L. Hendry, and S. J. Parry, An assessment of the relative roles of crust and mantle in magma genesis: an elemental approach, *Phil. Trans. R. Soc. Lond., A310,* 549-590, 1984.

Thompson, R. N., I. L. Gibson, G. F. Marriner, D. P. Mattey, and M. A. Morrison, Trace-element evidence of multistage mantle fusion and polybaric fractional crystallisation in the Palaeocene lavas of Skye, N W Scotland, *J. Petrol., 21,* 265-293, 1980.

Thompson, R. N., M. A. Morrison, A. P. Dickin, I. L. Gibson, and R. S. Harmon, Two contrasted styles of interaction between basic magmas continental crust in the British Tertiary Volcanic Province, *J. Geophys. Res., 91,* 5985-5997, 1986.

Thomson, K. and J. R. Underhill, Controls on the development and evolution of structural styles in the Inner Moray Firth Basin, in *Petroleum Geology of Northwest Europe,* edited by J. R. Parker, pp. 1167-1178, The Geological Society, London, 1993.

Thy, P., C. E. Lesher, and M. S. Fram, Low pressure experimental constraints on the evolution of basaltic lavas from Site 917, southeast Greenland continental margin, in *Proc. Ocean Drill. Prog., Sci. Results, 152,* in press, 1996.

Tilley, C. E., Some aspects of magmatic evolution, *Q. J. Geol. Soc. Lond., 106,* 37-61, 1950.

Tilley, C. E., and I. D. Muir, The Hebridean Plateau Magma Type, *Trans. Edin. Geol. Soc. 19,* 208-215, 1962.

Turner, J. D. and R. A. Scrutton, Subsidence patterns in the western margin basins: evidence from the Faeroe-Shetland Basin, in *Petroleum Geology of Northwest Europe,* edited by J. R. Parker, pp. 975-983, The Geological Society, London, 1993.

Ulmer, P., Partitioning of high-field strength elements among olivine, pyroxene, garnet and calc-alkaline picrobasalt:

experimental results and application, *Carnegie Inst. Wash. Yrbook, 1988-1989*, 42-47, 1989.

Underhill, J. R., Implications of Mesozoic-Recent basin development in the western Inner Moray Firth, UK, *Mar. Petrol. Geol., 8*, 359-369, 1991.

Upton, B. G. J., History of Tertiary igneous activity in the N Atlantic borderlands, in *Early Tertiary Volcanism and the Opening of the NE Atlantic, Spec. Publ. 39*, edited by A. C. Morton and L. M. Parson, pp.429-453, The Geological Society, London, 1988.

Upton, B. G. J., C. H. Emeleus, and N. Hald, Tertiary volcanism between 74° and 76°N, NE Greenland: Hold with Hope and Gauss Halvø, *J. Geol. Soc. Lond., 137*, 491-508, 1980.

Upton, B. G. J., C. H. Emeleus, D. C. Rex, and M. F. Thirlwall, Early Tertiary magmatism in NE Greenland, *J. Geol. Soc. Lond., 152*, 959-964, 1995.

Viereck, L. G., J. Hertogen, L. M. Parson, A. C. Morton, D. Love, and I. L. Gibson, Chemical stratigraphy and petrology of the Vøring plateau tholeiitic lava and interlayered volcaniclastic sediments at ODP Hole 624E, *Proc. Ocean Drill. Prog., Sci. Results, 104*, 367-396, 1989.

Viereck, L. G., P. N. Taylor, L. M. Parson, A. C. Morton, J. Hertogen, I. L. Gibson, and O. S. Party, Origin of the Palaeogene Vøring Plateau volcanic sequence, in *Early Tertiary Volcanism and the Opening of the NE Atlantic, Spec. Publ. 39*, edited by A. C. Morton and L. M. Parson, pp. 69-83, The Geological Society, London, 1988.

Vink, G. E., A hotspot model for Iceland and the Vøring Plateau, *J. Geophys. Res., 89*, 9949-9959, 1984.

Vogt, P. R., The Iceland mantle plume: status of the hypothesis after a decade of new work, in *Structure and Development of the Greenland-Scotland Ridge. New Methods and Concepts*, edited by M. H. P. Bott, S. Saxov, M. Talwani and J. Thiede, pp. 191-213, Plenum Press, New York and London, 1983.

Vogt, P. R., and O. E. Avery, Detailed magnetic surveys in the north-east Atlantic and Labrador Sea, *J. Geophys. Res., 79*, 363-389, 1974.

Waagstein, R., Structure, composition and age of the Faeroe basalt plateau, in *Early Tertiary Volcanism and the Opening of the NE Atlantic, Spec. Publ. 39*, edited by A. C. Morton and L. M. Parson, pp. 225-238, The Geological Society, London, 1988.

Waagstein, R., and N. Hald, Structure and petrography of the 660 m lava sequence in the Vestamanna-1 drill hole, lower and middle basalt series, Faeroe Islands, in *The Deep Drilling Project 1980-1981 in the Faeroe Islands*, edited by O. Berthelesen, A. Noe-Nygaard and J. Rasmussen, pp. 39-70, Føroya Fródskaparfelag, Tórshavn, 1984.

Wager, L. R., Geological investigations in East Greenland. Part 1. General geology from Angmagsalik to Kap Dalton, *Medd. Grøn., 105*, 46 pp., 1934.

Wager, L. R., A chemical definition of fractionation stages as a basis for comparison of Hawaiian, Hebridean and other basic lavas. *Geochim. Cosmochim. Acta, 9*, 217-248, 1956.

Wager, L. R., and W. A. Deer, Geological investigations in East Greenland. III. The petrology of the Skaergaard intrusion, Kangerdlugssuaq, *Medd. Grøn., 105*, 353 pp., 1939.

Walker, G. P. L., Zeolite zones and dyke distribution in relation to the structure of basalts in eastern Iceland, *J. Geol., 68*, 515-528, 1960.

Wallace, J. M., R. M. Ellam, I. G. Meighan, P. Lyle, and N. W. Rogers, Sr Isotope data for the Tertiary lavas of Northern Ireland: Evidence for open system petrogenesis, *J. Geol. Soc. Lond., 151*, 869-877, 1994.

Walsh, J. N., R. D. Beckinsale, R. R. Skelhorn, and R. S. Thorpe, Geochemistry and petrogenesis of Tertiary granitic rocks from the Island of Mull, northwest Scotland, *Contrib. Mineral. Petrol., 71*, 99-116, 1979.

Watkins, N. D., and G. P. L. Walker, Magnetostratigraphy of eastern Iceland, *Am. J. Sci., 277*, 513-584, 1977.

Watson, J., Northern Scotland as an Atlantic-North Sea divide, *J. Geol. Soc. Lond., 142*, 221-244, 1985.

Weaver, B. L., and J. Tarney, Chemistry of the subcontinental mantle: inferences from Archaean and Proterozoic dykes and continental flood basalts, in *Continental Basalts and Mantle Xenoliths*, edited by C. J. Hawkesworth and M. J. Norry, pp. 158-185, Shiva, Nantwich, Cheshire, U.K., 1983.

Werner, R., P. van den Bogaard, C. Lacasse, and H.-U. Schmincke, Chemical composition, age, and source of volcaniclastic sediments from Sites 917 and 918, in *Proc. Ocean Drill. Prog., Sci. Results, 152*, in press, 1996.

White, R. S., A hot-spot model for the early Tertiary volcanism in the N Atlantic, in *Early Tertiary Volcanism and the Opening of the NE Atlantic, Spec. Publ. 39*, edited by A. C. Morton and L. M. Parson, pp. 3-13, The Geological Society, London, 1988.

White, R. S., Igneous outbursts and mass extinctions, *Eos Trans. AGU*, 70(46), 1480-1483, 1989.

White, R. S., and D. McKenzie, Magmatism at rift zones: the generation of volcanic continental margins and flood basalts, *J. Geophys. Res., 94*, 7685-7729, 1989.

White, R. S., and D. McKenzie, Mantle plumes and flood basalts, *J. Geophys. Res., 100*, 17,543-17,585, 1995.

White, R. S., J. W. Bown, and J. R. Smallwood, The temperature of the Iceland plume and origin of outward-propagating V-shaped ridges, *J. Geol. Soc. Lond., 152*, 1039-1045, 1995.

White, R. S., G. D. Spence, S. R. Fowler, D. P. McKenzie, G. K. Westbrook, and A. N. Bowen, Magmatism at rifted continental margins, *Nature, 330*, 439-444, 1987.

White, R. S., J. H. McBride, P. K. H. Maguire, B. Brandsdóttir, W. Menke, T. A. Minshull, K. R. Richardson, J. R. Smallwood, R. K. Staples, and F. Group, FIRE: Faeroe-Iceland Ridge Experiment, *Eos Trans. AGU, 77*, 197, 200-201, 1996.

Wilson, H. E. and P. I. Manning, Geology of the Causeway Coast, *Memoir of the Geological Society of Northern Ireland*, Belfast (H.M.S.O.), 72 pp., 1978.

Wolfe, C. J., I. Th. Bjarnason, J. C. VanDecar, and S. C. Solomon, Seismic structure of the Iceland mantle plume, *Nature, 385*, 245-247, 1997.

Wood, D. A., A variably veined suboceanic upper mantle - genetic significance for mid-ocean ridge basalts from geochemical evidence, *Geology, 7*, 499-503, 1979a.

Wood, D. A., Dynamic partial melting: its applications to the petrogeneses of basalts erupted in Iceland, the Faeroe Islands, the Isle of Skye (Scotland) and the Troodos Massif (Cyprus), *Geochim. Cosmochim. Acta, 43*, 1031-1046, 1979b.

Young, I. M., R. C. Greenwood, and C. H. Donaldson, Formation of the Eastern Layered Series of the Rhum Complex, northwest Scotland, *Can. Mineral., 26*, 225-233, 1988.

Zehnder, C. M., J. C. Mutter, and P. H. Buhl, Deep seismic and geochemical constraints on the nature of rift-induced magmatism during break-up of the North Atlantic, *Tectonophysics, 173*, 545-565, 1990.

Zindler, A., and S. R. Hart, Chemical Geodynamics, *Ann. Rev. Earth Plan. Sci., 14*, 493-571, 1986.

Zindler, A., S. R. Hart, and F. A. Frey, Nd and Sr isotope ratios and rare earth element abundances in Reykjanes Peninsula basalts: evidence for mantle heterogeneity beneath Iceland, *Earth Planet. Sci. Lett., 45*, 249-262, 1979.

R. W. Kent, A. C. Kerr, M. J. Norry, and A. D. Saunders, Department of Geology, University of Leicester, Leicester, LE1 7RH, United Kingdom.

J. G. Fitton, Department of Geology and Geophysics, University of Edinburgh, West Mains Road, Edinburgh, EH9 3JW, United Kingdom.

Cretaceous Basalts in Madagascar and the Transition Between Plume and Continental Lithosphere Mantle Sources

Michael Storey

Danish Lithosphere Centre, Copenhagen, Denmark

John J. Mahoney

School of Ocean and Earth Science and Technology, University of Hawaii, Honolulu, Hawaii

Andrew D. Saunders

Department of Geology, University of Leicester, Leicester, United Kingdom

Isotopic data on Cretaceous basalts from three transects along the 1500-km length of the rifted eastern margin of Madagascar reveal systematic along-axis variations that suggest the Marion plume was an important source of melts at the southern end of the rifted margin. The relative melt contribution from the lithospheric mantle and a normal mid-ocean ridge basalt mantle source appears to have increased northward along the continental rift away from the estimated plume center. Data on basalt flows from the Volcan de l'Androy massif and on the associated Ejeda-Bekily dike swarm, situated inland from the rifted margin in the south of the island but above the postulated focal point of the plume, show signatures attributed to the continental mantle. These systematic spatial geochemical variations place bounds on the composition and thermal structure of the Marion mantle plume at the time of the formation of the Madagascar igneous province some 88 million years ago.

INTRODUCTION

The trace of mantle plume activity on the ocean crust can sometimes be backtracked to continental volcanic rifted margins composed of flood basalts and seaward-dipping reflectors [e.g., *Hinz,* 1981; *Morgan,* 1981; *White and McKenzie,* 1989]. Volcanic rocks erupted in such settings show a spectrum of elemental and isotopic compositions that range well beyond those observed in either mid-ocean ridge basalts (MORB) or ocean island basalts (OIB) [e.g., *Hawkesworth et al.,* 1990; *Saunders et al.,* 1992]. Some of these geochemical features have been interpreted to indicate that the continental mantle lithosphere is an important site of melt generation in the formation of flood basalt provinces [e.g., *Hawkesworth et al.,* 1990; *Hergt et al.,* 1991]. A different view is that the observed compositions arise through contamination of asthenosphere-derived melts by the continental mantle and crust [e.g., *Carlson et al.,* 1981; *McKenzie and Bickle,* 1988; *Ellam and Cox,* 1991; *Arndt and Christensen,* 1992]. If the continental mantle is essentially anhydrous then it is likely that only small amounts of melt can be generated within the lithosphere [*Arndt and Christensen,* 1992; *Farnetani*

Large Igneous Provinces: Continental, Oceanic, and Planetary Flood Volcanism
Geophysical Monograph 100

and Richards, 1994]. On the other hand, if portions of the continental mantle are hydrated then the lower solidus temperature, compared to dry peridotite, will increase the possibility of partial melting of the lithosphere during rifting [e.g., *Gallagher and Hawkesworth,* 1992; *Harry and Leeman,* 1995; *McKenzie and O'Nions,* 1995]. Geothermometry and geobarometry studies of continental mantle nodule suites indicate that the majority of nodules last equilibrated at pressures and temperatures close to the wet solidus [see *McKenzie and O'Nions,* 1995]. If parts of the continental lithospheric mantle have been metasomatised by fluids or small-degree melts originating from the MORB-source mantle or from subduction of oceanic lithosphere, then these regions are potential sites of melt generation (in addition to the asthenosphere) along rifted margins. Which type of source predominates depends on the interplay of factors such as the major element composition of the mantle lithosphere, the volume of hydrated mantle present, the thickness of the lithosphere, and the potential temperature of the underlying asthenosphere.

The island of Madagascar provides an opportunity to evaluate the relative roles of different mantle sources to flood basalt magmatism related to continental breakup. The island was the site of widespread voluminous magmatism during the rifting of Madagascar from India at ca. 88 Ma [*Storey et al.,* 1995]. Plate reconstructions place the Marion (Prince Edward) hotspot close to the southern tip of the continental rift of Madagascar at this time (Figure 1). The rock types mainly consist of basalt flows and dikes, but also include rhyolite flows and rarer microgranites and microgabbro intrusions. In this paper we describe the setting of Madagascar in the Late Cretaceous and the nature and timing of the igneous activity. In particular, we focus on published and new chemical and isotopic data on basalts from three transects along the 1500 km length of the rifted eastern margin.

REGIONAL SETTING OF MADAGASCAR

Madagascar (Figure 1) is the largest island in the Indian Ocean, separated from the African mainland by the Mozambique Channel and to the north by the Western Somali and Comores Basins. To the east of the island the water depth increases rapidly to ≈5 km in the Mascarene Basin. Whereas the Western Somali Basin is floored by oceanic crust [*Coffin et al.,* 1986], the crustal structure of the Mozambique Channel is less clear; it may be partly continental [e.g., *Mougenot et al.,* 1986]. A pronounced aseismic ridge, the Madagascar Plateau, runs from the southern tip of the island toward the Southwest Indian Ridge (SWIR) [e.g., *Sinha et al.,* 1981]. This plateau

represents the trace of the Marion hotspot, which is now situated beneath Marion and Prince Edward Islands on the Antarctic Plate south of the SWIR (Figure 1).

The island is the product of two major rifting events associated with the progressive breakup of the Gondwana supercontinent. The first was the separation of Madagascar/India from Africa, which began with seafloor spreading in the Western Somali and Comores Basins during Bajocian times (≈180 Ma) [*Coffin and Rabinowitz,* 1988], although earlier rifting may have provided important depocenters for Karoo, Permo-Triassic and Early Jurassic sediments in the Morondava, Majunga, and Diego Basins (Figure 2) [*Nichols and Daly,* 1989]. The pre-dispersal location of Madagascar against mainland Africa has been the subject of considerable debate. However, from seafloor magnetic anomalies and fracture zone patterns, it is now generally accepted that prior to Middle Jurassic times it lay adjacent to the coast of Somalia, Kenya, and Tanzania [*Smith and Hallam,* 1970; *McElhinny et al.,* 1976; *Scrutton et al.,* 1981; *Coffin et al.,* 1986; *Coffin and Rabinowitz* 1988; *de Wit et al.,* 1988] rather than against the coast of Mozambique [*Green,* 1972; *Flores,* 1984] or as a stationary island in a fixed location relative to Africa [*Dixey,* 1960]. Jurassic and Early Cretaceous southward motion of Madagascar relative to Africa was accommodated by a major transform fault, the Davie Ridge [*Coffin and Rabinowitz,* 1987; *Bassias and Leclaire,* 1990]. This motion resulted in the sedimentary basins of western Madagascar being faulted against slivers of older ocean crust to form the Mozambique Channel. Madagascar/India was stable relative to Africa by the time of the Albian/Aptian boundary (112 Ma).

The second rifting event occurred in the Late Cretaceous as seafloor spreading in the Mascarene Basin resulted in the separation of Greater India from Madagascar. Regarding the pre-dispersal location of Madagascar against India, *Crawford* [1978] speculated that the Proterozoic Narmada-Son lineament of central-west India can be traced into northern Madagascar. *Katz and Premoli* [1979] linked the Bhavani lineament in southern India with either the Itremo or Ranotsara lineaments in central eastern Madagascar, and *Agrawal et al.* [1992] suggested that Madagascar was part of the Dharwar craton of India prior to fragmentation. A detailed discussion of this subject has been given by *Windley et al.* [1994].

The oldest magnetic lineations recognized in the southern Mascarene Basin are anomalies 33, 33r, and 34 [*Dyment,* 1991]. They trend to the northwest and are displaced by a series of northeast-southwest-trending fracture zones (Figure 1). Anomaly 34 (≈84 Ma) varies from 50 km to 200 km from the continental edge of southeast Madagascar, as determined from satellite-derived gravity

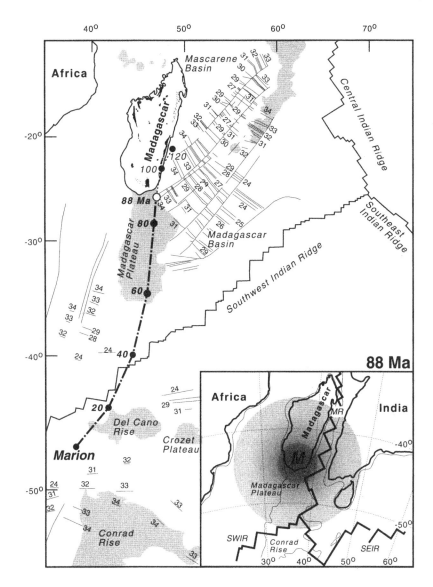

Figure 1. Madagascar and principal structures in the Southwest Indian Ocean. Magnetic lineations in the Mascarene and Madagascar basins are from *Dyment* [1991]. The Marion hotspot track was calculated using the model of *Müller et al.* [1993]. The hypothetical 88 to 120 Ma track is shown by the faint dashed line. The longitudinal error in the model for the post-84 Ma hotspot track is of the order of several hundred kilometers. The uncertainties increase for the older portion of the track. Inset is a plate reconstruction for ≈88 Ma showing relative positions of Africa, India, and Madagascar and the paleoridge system: Mascarene Ridge (MR), Southwest Indian Ridge (SWIR), Southeast Indian Ridge (SEIR). Shaded circle has a radius of approximately 1000 km and illustrates the possible center (M) and lateral extent of the Marion plume top. From *Storey et al.* [1995].

data. The linearity of the rifted east coast margin is suggestive of strike-slip faulting prior to the opening of the Mascarene Basin [*Dyment*, 1991]. Such faulting would have provided a zone of weakness during the breakup of Madagascar and India, as would the strong anisotropy in the basement fabric [*Boast and Nairn*, 1982].

Calculations based on the plate motion model of *Müller et al.* [1993] place the Marion hotspot about 100 km south of Madagascar around the time of continental breakup with India [*Storey et al.*, 1995]. It is also noteworthy that the postulated 88 to 120 Ma track of the Marion hotspot closely parallels the rifted eastern margin of Madagascar (Figure 1). Whether the Marion hotspot existed before 88 Ma is unknown, although it has been speculated that it was the source of the earlier Karoo flood basalts of southern Africa [e.g., *Morgan*, 1981]. If the plume did exist well

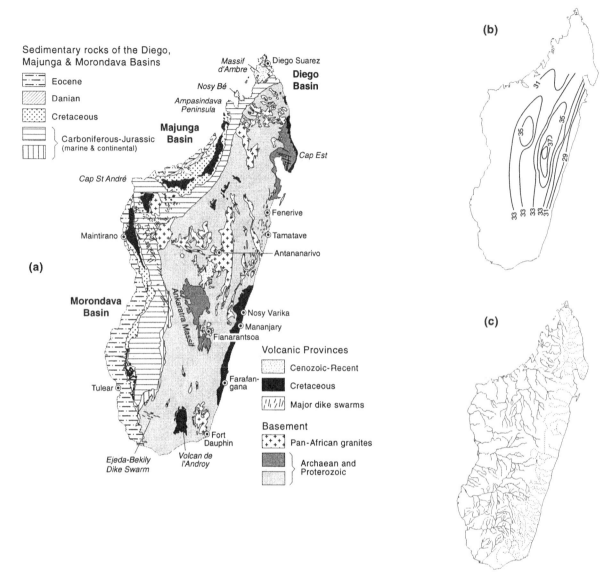

Figure 2. (a) Geological sketch map of Madagascar based on *Besairie* [1964]. Cretaceous volcanic rocks mainly crop out along the eastern rifted margin, at Volcan de l'Androy in the south, and in the Majunga and Morondava basins in the west. (b) Map of the Moho topography of Madagascar (contours are in km), showing the crustal thickness decreasing progressively from the central highlands to the eastern rifted margin [from *Fourno and Roussel,* 1994]. (c) The drainage pattern of Madagascar: the dotted lines are eastward-draining rivers and the solid lines are westward-draining rivers; note the easterly position of the main watershed divide.

before 88 Ma, then it could have played an active role in the breakup of Madagascar and India by providing a preferred path for propagation of the Mascarene Ridge. The trigger for breakup itself may have been the northward propagation of the Mascarene Ridge as a consequence of the ≈88 Ma capture of the spreading ridge system to the south by the Marion hotspot, as the trailing edge of Madagascan lithosphere passed north of the plume axis [*Storey et al.,* 1995].

Unlike the earlier breakup with Africa, prodigious amounts of magma were erupted over much of Madagascar in association with this rifting event. Between the opening of the Mascarene Basin and anomaly 34 time, the plume also seems to have interacted with the triple junction of the SWIR, Southeast Indian Ridge and nascent Mascarene Ridge, as indicated by the presence of large oceanic plateaus [*Goslin et al.,* 1980; *Sinha et al.,* 1981] flanked by anomaly 34 (Figure 1). The Conrad Rise and the western and southern part of the Madagascar Plateau are conjugate with respect to the SWIR; the northeastern part of the Madagascar Plateau and the bathymetric high on the west side of the southern tip of India [*Laughton*, 1975] are conjugate with respect to the Mascarene Ridge, as suggested by the structural trends of these features [*Dyment*, 1991]. Collectively, the Madagascar flood basalts and these oceanic plateaus constitute a single large igneous province [*Coffin and Eldholm*, 1994] with a probable original area of more than 1×10^6 km^2.

SUMMARY OF THE GEOLOGY OF MADAGASCAR

Main Physiographic Features

The physiography of the island reflects fairly accurately its underlying geology. The mountainous backbone of the island, rising to 2638 m in the Ankaratra Massif, is composed mainly of Precambrian basement with local Cenozoic volcanic provinces (Figure 2). To the east, a narrow coastal plain and a remarkably straight coastline are defined by Mesozoic and Cenozoic faulting, whereas to the west a series of embayments mark the onshore outcrops of the Morondava and Majunga sedimentary basins. These basins, which began developing in the Carboniferous and Permian, continued as important marine and continental depocenters throughout the Jurassic and Cretaceous and into the Tertiary [e.g., *Besairie and Collignon*, 1972]. They deepen toward the west and northwest. The smaller Diego Basin is in the far north of the island.

The island has a remarkable physiographic asymmetry, with the main watershed divide running close to the eastern margin (Figure 2c). Eastward from the divide, the land surface descends rapidly across a series of fault scarps to the narrow eastern coastal plain, a narrow continental shelf, and into the deep Mascarene Basin. Westward, the topography gradually drops toward the Morondava and Majunga basins, although there is a set of cuestas formed from basinward-dipping, resistant sedimentary and volcanic rocks. It is tempting to suggest that the physiographic asymmetry results from uplift associated with the Cretaceous magmatic underplating and rifting (cf. Paraná and Deccan [*Cox*, 1989]).

Precambrian Basement

The Precambrian basement of Madagascar makes up nearly two-thirds of the island [*Besairie*, 1964; *Besairie*, 1967]. The Precambrian terrain is divided into two major blocks by the NW-SE Bongolavo-Ranotsara shear zone [see *Windley et al.*, 1994]. North of this zone the basement is mainly granitic or gneissic, whereas south of it supracrustal rocks predominate. However, the Androyan

System to the south of the shear zone is characterized by granulites and migmatites and is considered to be Late Archean to Early Proterozoic in age [*Paquette et al.*, 1994; *Windley et al.*, 1994]. Large areas of the Madagascan Precambrian crust were reworked by the Pan-African event around 550 Ma [*Windley et al.*, 1994]. East-west profiles of Bouguer anomalies across the Precambrian basement of Madagascar reveal rapid shallowing of the Moho from a crustal domain of normal thickness (35–40 km) in the center of the island to a thickness of 25–27 km along the east coast (Figure 2b) [*Fourno and Roussel*, 1994].

Cretaceous Magmatism

The Cretaceous volcanic and intrusive rocks of Madagascar crop out semicontinuously along the 1500-km length of the east coast, which marks the rifted margin, and in the Majunga and Morondava basins in western Madagascar (Figure 2). The rocks include basalt flows, dikes, and some rhyolite flows. Along the rifted margin, the flows lie mainly upon the Precambrian basement, whereas most of the dikes are coast-parallel. The lavas generally form shallowly seaward-dipping piles. Whether any seaward-dipping basalt reflectors are present offshore is unknown; however, the narrowness and steepness of the continental margin appear to preclude the existence of a large volcanic wedge of the sort that typifies other volcanic rifted margins, such as East Greenland [e.g., *Larsen and Jakobsdóttir*, 1988]. Several portions of the rifted eastern margin of Madagascar are devoid of volcanic rocks but this may be a function of a lack of preservation rather than a reflection of the magmatic activity. For example, lavas are found infrequently in the region north of Tamatave, but dikes are very common. Elsewhere, the basalt is reduced to thick piles of saprolite and laterite covered by heavy vegetation.

The Volcan de l'Androy complex in southern Madagascar contains the thickest sequence of Cretaceous volcanic rocks exposed on the island; this massif forms an oval-shaped outcrop some 50 km wide by 90 km long and consists of interbedded flows of basalt and rhyolite, with microgranite intrusions exposed at the northern and western margins of the complex (Figure 3) [*Battistini*, 1959]. The massif sits within a much wider basin formed by headwater erosion by the Mandrare River system. This massif is located more than 100 km west of the continental rift in an area that has apparently undergone little crustal extension. The total thickness of the flows may be more than 2000 m [*Battistini*, 1959]. The lowest flows lie unconformably on Archean granitic gneisses of the Androyan System. Substantial erosion has left a series of rhyolite-capped table mountain remnants (e.g., Vohit-

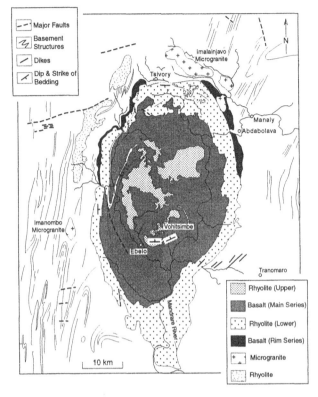

Figure 3. Left: Geological map of the Volcan de l'Androy basalt-rhyolite lava massif. Vohitsimbe is a prominent rhyolite-capped table mountain in the center of the massif. The dip arrows indicate that the complex has a saucer-shaped structure. Map modified after *Battistini* [1959] using the Landsat image shown at right.

simbe), and it is unclear how much material has been stripped from the top of these plateaus. In many areas, the lowest flows are rhyolite, forming an annular outcrop pattern, but there are local outcrops of underlying basalt that we have termed the "rim series" (Figure 3). Above the lower rhyolites is a layer of flat-lying or gently inwardly dipping tholeiitic basalt flows, interbedded with and overlain by the upper rhyolites. The proportion of rhyolite to basalt increases markedly up section, the topmost units forming columnar-jointed rhyolite flows up to 50 m thick. The flows appear to have been derived locally, as indicated by microgranites that partly encircle the massif and which are compositionally identical to the lower rhyolites; the overall form resembles a large caldera, although there are no obvious ring faults, which are normally associated with such structures. To the west of the Volcan de l'Androy massif is the Ejeda-Bekily dike swarm, which trends toward the massif (Figure 2). These dikes are unusual among the Cretaceous rocks in that they are predominantly nepheline-normative [*Dostal et al.,* 1992].

Basalt flows, interbedded with Upper Cretaceous sedimentary rocks in the Majunga and Morondava basins, crop out for 700 km in the northwest and 200 km in the southwest. Sedimentation began during the Carboniferous in the Morondava Basin, but was slightly later (late Permian) in the Majunga Basin [*Besairie,* 1966; *Boast and Nairn,* 1982]. Sedimentation alternated between marine and non-marine, reflecting the proximity of the basin margins. In the Morondava Basin, the largest of the basins, igneous rocks are most prominent in the north and south. Lava flows (predominantly basaltic) make up a pile as much as 100 m thick, but 30 m is more common. Rhyolite flows and microgabbro intrusions are present, and a dike swarm crops out adjacent to the basalt flows in the northern part of the basin (Figure 2). The Majunga Basin extends along the west coast of Madagascar from the southern side of the Ampasindava Peninsula to Cap St. André [*Boast and Nairn,* 1982] (Figure 2). Basaltic lavas were erupted over a wide area of the Majunga Basin during the Turonian and in the north of the basin are overlain by sediments with an Upper Turonian marine fauna. The lavas share many of the features of those found in the Morondava Basin, although they are compositionally different. They are widespread, average approximately 50 m in total thickness (up to 200 m in some areas), and are predominantly basaltic.

Small outliers of basalt, reported as Cretaceous by *Besairie* [1964], rest on Precambrian basement in north-central Madagascar (Figure 2). These outliers are important because, if they are Cretaceous, they indicate that the extent of the magmatism was far greater than that presently preserved along the coastal margins. They suggest that basalt flows travelled across the island, from the eastern rift zone into the western basins.

TIMING OF CRETACEOUS MAGMATISM IN MADAGASCAR AND THE BREAKUP FROM INDIA

Although the Cretaceous flood basalts of Madagascar can be related to the track of the Marion hotspot (Figure 1), the precise timing of volcanism was poorly known until recently. K-Ar dates range from 31 to 97 Ma [*Dostal et al.*, 1992; *Storetvedt et al.*, 1992], whereas paleontological evidence [*Besairie and Collignon*, 1972; *Boast and Nairn*, 1982] suggests a Maastrichtian (65.4 to 71.3 Ma [*Obradovich*, 1993]) to Turonian (88.7 to 93.3 Ma) age range for the igneous activity. An ^{40}Ar-^{39}Ar dating study [*Storey et al.*, 1995], however, showed that the Madagascar Cretaceous volcanic province formed over no more than 6 m.y., a much shorter interval than indicated by the published K-Ar dates. This study demonstrated that the volcanic rocks of the 1500-km-long rifted margin show virtually no statistically significant differences in age (Figure 4); the weighted mean of the isochron ages is 87.6 ± 0.6 Ma. Two tholeiitic basalt flows from the Majunga Basin gave ages of 87.6 ± 2.9 Ma and 88.5 ± 1.3 Ma, respectively, within error of the age determinations for the rifted margin rocks. A rhyolite dike intruded into the base of the Volcan de l'Androy lavas has an age of 86.3 ± 1.9 Ma. An age of 84.4 ± 0.4 Ma was given by sanidine phenocrysts from a rhyolite flow at the top of Volcan de l'Androy, which is slightly younger than the mean of the rifted margin and Majunga Basin dates. Similar young ages were also shown by a basalt sample of the Ejeda-Bekily dike swarm (84.8 ± 1.3 Ma), and by a basalt flow from the southwest part of Morondava Basin (84.5 ± 0.7 Ma). These data suggest that Madagascan volcanism ceased first in the north [*Storey et al.*, 1995].

Because breakup between Madagascar and Greater India occurred at the time of the Cretaceous Quiet Zone, the precise age of the rift-to-drift transition cannot be determined from seafloor magnetic lineations. However, seafloor-spreading was clearly well organized in the southern Mascarene Basin shortly before 84 Ma, the age of anomaly 34 [*Harland et al.*, 1989]. The establishment of regular seafloor spreading in the Mascarene Basin thus appears to have coincided with the eruption of the Madagascar flood basalt province.

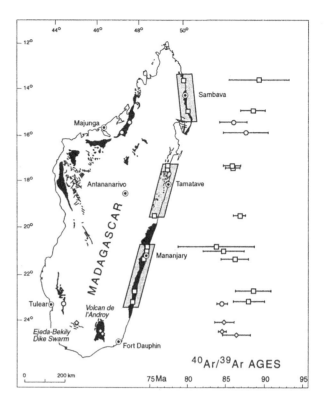

Figure 4. The outcrop pattern of Madagascar Cretaceous volcanic rocks (black) with shaded boxes showing our three sample transects along the eastern rifted margin, centered around Mananjary (south), Tamatave (central) and Sambava (north), respectively. Also shown are ^{40}Ar-^{39}Ar age determinations on the Cretaceous volcanic rocks. Sample symbols: squares = eastern rifted margin; circles = Majunga and Morondava basins; diamonds = Volcan de l'Androy and Ejeda-Bekily dike swarm. The error bars are ± 1s. From *Storey et al.* [1995].

GEOCHEMISTRY OF THE CRETACEOUS MAGMAS OF MADAGASCAR

Geochemical studies of the Cretaceous volcanic rocks of Madagascar are few. A reconnaissance isotopic study was carried out by *Mahoney et al.* [1991] on samples from the Ejeda-Bekily dike swarm and from the flows east of Tulear and along the southeast coast near Mananjary. Major and trace element data for the same sample set were given by *Dostal et al.* [1992]. Here, we review this work and present results of our new geochemical investigations.

Samples

Samples for our present study were collected from three transects encompassing most of the length of the volcanic rifted margin of eastern Madagascar (Figure 4). From south to north they are the Mananjary transect, the

Tamatave transect, and the Sambava transect. Petrographic descriptions of a subset of the samples are given in Appendix 1. In addition, we also sampled the Volcan de l'Androy massif and the Ejeda-Bekily dike swarm, essentially forming one transect perpendicular to the rifted margin (J. Mahoney et al., ms. in prep.).

Because of limited exposure it is not possible to state with certainty the environment of emplacement, although the presence of vesicular blocks suggests that most units were probably emplaced as thick flows rather than intrusive sheets. Around Tamatave, the sampled units were predominantly dikes. Textures range from fine-grained, almost aphanitic basalts to ophitic dolerites and, in some cases (MAN 90-17), accumulative gabbros.

The majority of the Mananjary samples are coarse-grained basalts or dolerites with significantly less than 5% phenocrysts and are thus likely to approximate liquid compositions. Some dolerites contain trace or small (<5%) amounts of partially resorbed plagioclase. However, many fine-grained basalts contain small amounts of clinopyroxene microphenocrysts in addition to plagioclase phenocrysts, implying that similar phases may be present in the dolerites but cannot be readily distinguished. A few samples contain olivine microphenocrysts, now completely replaced by clays. Clinopyroxene is often purplish to brown in thin section, reflecting the high Ti and Fe content of many samples (e.g., MAN 90-80). Some samples show no visible signs of alteration (e.g., MAN 90-20), which is also reflected in their low total weight loss on ignition. The majority of samples, however, show patchy replacement of the mesostasis by yellow-brown or green-brown clays and, rarely, chlorite. The majority of the primary silicate phases are unaffected by alteration. Vesicles, where present, are often filled with clays, chlorite or, in one sample (MAN 90-15), prehnite and pumpellyite.

All of the Tamatave samples, with the exception of TAM 92-43 which may be from a flow, were taken from dikes. The samples are aphyric, and either pale-green or brown clays replace the mesostasis. Calcite is a common secondary phase.

The samples from the Sambava transect were collected from dikes and flows. Like the samples from Mananjary, they exhibit a range of textures, but a common type is an "ophimottled" texture, where pyroxene, mostly clinopyroxene but also some orthopyroxene, forms glomero-ophitic patches. Samples may be aphyric (e.g., SAM 92-1) or plagioclase-phyric (SAM 92-3). Clinopyroxene is a microphenocryst phase in some fine-grained samples (e.g., SAM 92-20D). Alteration is usually restricted to replacement of mesostasis by brown clays. No calcite is observed, and some samples (e.g., SAM 92-3) show no visible signs of alteration.

Analytical Methods

X ray fluorescence (XRF) measurements and instrumental neutron activation analysis (INAA) of the Madagascar samples were carried out at the University of Leicester. For XRF analysis, samples were crushed in an agate shatterbox. For determination of the trace elements V, Cr, Ba, Nb, Zr, Y, Sr, Rb, and Ni the rock powder was made into 46-mm-diameter powder briquettes by adding several drops of a 7% solution of polyvinyl alcohol to 15 g of rock powder and subjecting the mixture to a pressure of 15 tons in a steel die. Analysis was carried out using a W-anode tube on a Philips PW 1400 X ray spectrometer. For major-element analysis, rock powders were dried overnight at 120°C before determining weight loss on ignition (LOI) at 800°C. Glass disks were made by fusing (at ≈1200°C) a mixture of 1 g of ignited rock powder and 5 g of lithium metaborate/lithium tetraborate flux (Johnson Matthey Spectroflux 100B). Analysis of the glass disks was carried out using a Rh-anode tube on an ARL 8420+ X ray spectrometer.

Rare-earth elements (REE) and Ta, Th, Hf, Sc, and Co concentrations were determined by instrumental neutron activation on 0.2-g splits of the Madagascar samples and also of five lavas from Marion and Prince Edward islands, and Funk Seamount (a volcano of the Marion group). Samples were irradiated for approximately 30 hours in a thermal neutron flux of 1×10^{12} n cm^{-2} in the Imperial College Reactor Centre and activities of the radioisotopes counted on EG & G ORTEC coaxial and loaxial high-purity germanium detectors. Reactor flux variation was monitored and corrected by interspersing wafer-thin iron foils between the samples. Elemental abundances were determined by reference to the standard AC(II) [Potts et al., 1981]. Because of the low abundances of Sc and Co in AC(II), the standard JB-1a [Ando et al., 1987] was included for Sc and Co determinations. The standards BOB-1 and JB-1a (excluding Sc and Co in the latter) were run as "unknowns" in each sample batch to monitor analytical uncertainty. A subset of our results appears in Tables 1 and 2.

Sr, Nd, and Pb isotopic determinations, as well as isotope-dilution abundance measurements, were carried out at the Open University, England, and at the University of Hawaii. At the Open University, rock chips (<1 mm), rather than powders, were selected for Pb-isotopic analysis to avoid contamination introduced through the crushing procedure. To remove possible surface contamination, the rock chips were cleaned for 10 minutes in an ultrasonic bath using ultrapure 6M HCl followed by ultrapure H$_2$O. This procedure was repeated at least twice. Dissolution was by a HF-HNO$_3$ mixture, the residue being converted

TABLE 1. Major Element and Trace Element Data for Rocks of the Rifted Eastern Margin of Madagascar

	Mananjary high-Mg-Ti rocks				Mananjary Fe-Ti series (TiO₂ > 3 wt. %)						
Sample	MAN90-45	MAN90-47	MAN90-85	MAN90-86	MAN90-1B	MAN90-6	MAN90-8	MAN90-31	MAN90-43	MAN90-49	MAN90-57
SiO_2	42.6	42.6	49.8	51.3	48.0	48.7	50.3	49.4	47.6	47.8	46.4
TiO_2	2.60	3.38	3.29	3.77	3.65	3.49	4.08	3.08	3.97	3.68	4.91
Al_2O_3	9.35	8.99	10.39	11.51	12.24	11.76	12.83	12.83	11.35	11.97	11.21
Fe_2O_3[a]	17.73	18.63	13.80	13.16	16.24	16.21	13.92	15.64	17.50	16.28	17.69
MnO	0.21	0.20	0.16	0.14	0.21	0.21	0.15	0.22	0.21	0.22	0.23
MgO	14.92	13.50	10.61	7.27	6.03	5.76	4.52	6.13	5.59	6.48	5.58
CaO	10.38	9.05	9.31	7.03	10.28	10.40	8.68	10.59	9.61	10.00	9.78
Na_2O	1.80	2.28	1.45	3.28	2.27	2.20	2.86	2.25	2.34	2.30	2.35
K_2O	0.72	1.03	0.13	1.53	0.30	0.25	0.31	0.20	0.43	0.46	0.54
P_2O_5	0.33	0.45	0.33	0.39	0.33	0.28	0.48	0.27	0.36	0.33	0.46
Total	100.64	100.12	99.24	99.42	99.55	99.31	98.15	100.65	99.02	99.51	99.15
LOI	0.42	-0.19	2.65	2.04	0.01	-0.07	0.84	0.22	0.07	-0.18	-0.05
XRF (ppm)											
V	423	420	302	335	377	413	364	374	449	394	421
Cr	764	558	656	372	140	121	37	110	87	179	139
Ba	289	413	61	434	100	86	159	84	117	109	165
Nb	20	26	19	23	15	13	19	12	16	14	19
Zr	169	247	254	315	233	220	343	192	264	257	319
Y	22	27	30	38	45	46	43	45	62	47	64
Sr	667	869	359	408	254	202	431	225	218	254	212
Rb	23	23	3	25	7	6	8	6	9	9	13
Ni	583	574	359	149	91	63	50	82	66	100	67
INAA (ppm)											
La	19.6	26.9	19.1	24.6	14.0	12.7	28.3	12.9	17.1	15.3	19.2
Ce	43.5	63.7	45.8	58.1	35.4	32.6	66.7	31.5	43.7	39.0	49.7
Nd	28.3	38.9	30.6	36.6	25.7	24.0	41.8	22.7	31.2	27.8	36.2
Sm	6.8	9.3	7.4	8.6	7.1	6.9	9.5	6.5	8.7	7.8	9.8
Eu	2.4	3.2	2.6	3.0	2.3	2.3	3.3	2.2	2.9	2.6	3.1
Gd	6.6	8.9	7.2	8.4	7.3	7.7	9.8	7.0	9.2	8.2	11.1
Tb	1.20	1.27	1.19	1.31	1.32	1.27	1.46	1.22	1.74	1.59	2.05
Yb	1.30	1.45	2.01	2.51	3.51	3.81	3.09	4.01	5.20	3.83	5.17
Lu	0.18	0.17	0.28	0.35	0.51	0.56	0.43	0.60	0.80	0.56	0.76
Ta	1.15	1.49	1.11	1.33	0.95	0.97	1.21	0.82	1.14	0.92	1.46
Th	1.70	2.21	1.45	2.02	1.37	1.24	2.98	1.11	1.53	1.24	1.81
Hf	4.40	6.17	5.91	7.21	5.52	5.49	8.18	4.82	6.35	6.47	8.12
Sc	24.3	21.2	26.0	31.0	38.5	44.3	23.2	40.6	42.1	37.5	41.3
Co	94	97	61	49	52	52	50	57	59	57	55
Isotope Dilution (ppm)											
U			0.288								
Th			0.804								
Pb	2.40	3.86	2.06		1.56	1.29		1.20	1.79	1.32	1.97

TABLE 1. (continued)

Sample	Mananjary Fe-Ti series (TiO$_2$>3 wt.%)					Mananjary Fe-Ti series (TiO$_2$<2 wt.%)						
	MAN90-63	MAN90-66	MAN90-71	MAN90-80	MAN90-81	MAN90-2	MAN90-15	MAN90-16	MAN90-17	MAN90-20	MAN90-22/2	MAN90-77
SiO$_2$	47.6	48.2	48.6	46.9	47.9	50.2	49.2	49.7	49.5	49.7	49.7	49.9
TiO$_2$	3.79	3.54	3.55	4.76	3.37	1.60	1.99	1.27	0.81	1.44	1.35	1.66
Al$_2$O$_3$	13.07	12.25	12.25	11.13	12.41	14.32	14.75	14.98	18.05	19.15	14.11	14.63
Fe$_2$O$_3$[a]	16.00	16.20	16.71	17.49	16.34	12.69	12.18	11.44	9.30	8.87	11.94	11.93
MnO	0.21	0.21	0.21	0.23	0.22	0.19	0.16	0.16	0.13	0.12	0.17	0.17
MgO	5.25	6.34	5.28	5.38	6.11	7.11	8.15	8.71	7.59	5.23	9.59	8.31
CaO	10.09	10.30	9.88	9.52	10.52	11.03	10.29	12.07	12.20	12.05	11.54	11.66
Na$_2$O	2.37	2.25	2.43	2.68	2.26	2.41	2.82	2.01	2.68	2.71	1.80	2.02
K$_2$O	0.38	0.25	0.35	0.55	0.34	0.25	0.61	0.10	0.19	0.11	0.20	0.09
P$_2$O$_5$	0.37	0.28	0.34	0.45	0.33	0.15	0.18	0.10	0.07	0.13	0.11	0.14
Total	99.13	99.82	99.64	99.04	99.76	99.95	100.35	100.57	100.51	99.48	100.54	100.54
LOI	-0.11	0.18	0.01	-0.80	0.45	1.81	3.09	1.56	-0.09	0.00	3.51	0.46
XRF (ppm)												
V	386	387	365	469	385	341	267	255	178	205	250	269
Cr	108	163	100	113	128	106	371	286	242	209	476	362
Ba	95	79	112	183	137	49	86	47	49	93	35	32
Nb	11	10	13	20	13	5	6	3	2	3	5	5
Zr	266	200	254	329	210	100	126	73	33	34	80	106
Y	62	49	53	59	48	27	36	21	10	14	18	29
Sr	212	229	276	248	214	242	334	209	353	407	265	215
Rb	7	6	10	12	7	7	14	3	3	2	5	2
Ni	92	89	70	67	77	60	186	139	107	97	181	175
INAA (ppm)												
La	14.0	11.3	15.3	19.7	14.3	5.9	8.2	4.7	3.6	4.8	4.5	6.1
Ce	36.8	29.2	39.8	49.2	34.7	14.7	20.0	11.9	8.8	11.7	11.4	15.4
Nd	28.8	21.7	30.1	34.8	25.2	11.2	15.1	8.8	5.9	8.4	9.1	11.9
Sm	8.4	6.4	8.0	9.7	6.8	3.4	4.4	2.6	1.7	2.4	2.8	3.7
Eu	2.6	2.2	2.7	3.0	2.3	1.4	1.7	1.1	1.0	1.3	1.1	1.4
Gd	9.6	7.1	9.5	9.7	8.2	3.8	4.9	3.0	2.0	2.8	3.1	4.3
Tb	1.64	1.28	1.57	1.70	1.35	0.74	0.92	0.65	0.36	0.46	0.57	0.79
Yb	5.23	3.41	3.86	4.37	4.04	2.44	2.24	1.64	0.89	1.20	1.61	2.41
Lu	0.78	0.51	0.57	0.65	0.59	0.37	0.34	0.25	0.13	0.18	0.24	0.36
Ta	0.83	0.80	0.92	1.28	0.92	0.31	0.44	0.25	0.17	0.20	0.26	0.34
Th	1.06	0.96	1.24	1.81	1.30	0.52	0.60	0.39	0.33	0.20	0.35	0.48
Hf	6.65	4.88	6.50	7.60	5.42	2.52	3.19	1.97	1.18	1.30	2.03	2.67
Sc	42.1	39.6	36.2	40.8	43.4	45.3	31.9	35.4	29.6	23.1	36.9	37.9
Co	52	55	56	50	53	46	54	54	49	39	55	57
Isotope Dilution (ppm)												
U								0.0576				
Th								0.211				
Pb	1.32	1.39	1.40	1.91		0.806	1.98	0.507	0.475		0.513	0.584

TABLE 1. (continued)

Sample	Tamatave				Sambava					
	TAM92-6	TAM92-27	TAM92-30	TAM92-43	SAM92-1	SAM92-3	SAM92-10	SAM92-16	SAM92-20D	SAM92-33B
SiO_2	44.5	49.2	47.9	47.7	49.9	49.1	49.5	49.8	49.6	45.1
TiO_2	5.39	2.85	3.72	3.98	1.89	2.20	2.05	2.11	2.47	0.49
Al_2O_3	12.58	12.93	12.66	12.25	14.17	14.71	13.79	13.78	14.19	19.82
Fe_2O_3[a]	17.35	16.22	18.45	18.05	13.21	13.08	13.44	13.86	13.81	10.60
MnO	0.21	0.23	0.28	0.23	0.19	0.19	0.21	0.21	0.19	0.12
MgO	6.62	5.71	4.17	5.17	7.09	6.23	6.89	7.04	6.30	10.48
CaO	10.04	9.47	8.44	8.66	11.37	11.17	11.03	11.01	11.16	10.92
Na_2O	2.88	2.93	3.31	3.12	2.60	2.82	2.51	2.51	2.76	2.29
K_2O	0.56	0.61	0.60	0.54	0.19	0.23	0.28	0.27	0.25	0.10
P_2O_5	0.36	0.29	0.60	0.43	0.19	0.28	0.20	0.20	0.28	0.03
Total	100.45	100.47	100.11	100.10	100.76	100.01	99.92	100.80	100.99	99.92
LOI	1.35	0.53	1.25	0.70	0.05	0.12	0.73	-0.04	0.41	0.16
XRF (ppm)										
V	520	405	374	449	312	307	293	304	340	89
Cr	42	70	34	39	197	158	169	205	176	221
Ba	260	143	230	377	119	130	81	93	98	87
Nb	13	13	19	15	8	10	7	8	10	1
Zr	145	209	308	266	124	163	135	144	160	6
Y	26	45	64	59	29	32	31	34	33	6
Sr	641	245	310	281	256	338	235	230	305	271
Rb	15	9	12	7	3	5	1	4	9	1
Ni	105	63	36	66	91	89	109	96	86	192
INAA (ppm)										
La	13.2	14.2	22.7	19.7	9.7	13.2	8.3	9.3	12.6	1.7
Ce	32.5	35.8	56.3	47.7	23.5	31.3	20.9	23.0	31.8	3.5
Nd	23.2	25.0	40.8	32.6	16.1	21.8	16.5	17.1	22.3	2.0
Sm	6.0	6.7	10.5	8.7	4.3	5.7	4.6	4.9	5.7	0.6
Eu	2.4	2.3	3.6	3.0	1.6	2.1	1.7	1.8	2.2	0.4
Gd	6.5	7.7	11.9	9.2	5.1	6.3	4.9	5.9	5.6	0.4
Tb	1.00	1.30	2.01	1.70	0.84	1.07	0.94	0.99	1.11	0.17
Yb	1.66	3.54	5.04	4.49	2.43	2.74	2.51	2.63	2.78	0.47
Lu	0.22	0.53	0.72	0.69	0.36	0.39	0.36	0.39	0.39	0.07
Ta	0.96	0.80	1.20	0.99	0.45	0.60	0.46	0.48	0.61	0.02
Th	1.04	1.21	1.67	1.36	1.02	1.10	0.68	0.74	1.06	0.05
Hf	4.32	5.40	7.73	6.66	3.20	4.17	3.51	3.79	4.23	0.16
Sc	30.1	39.9	35.2	38.3	39.4	36.2	36.6	38.9	35.6	18.4
Co	58	52	47	57	50	46	54	55	48	63
Isotope Dilution (ppm)										
U	0.150	0.255	0.535	0.304	0.199	0.254	0.157	0.201	0.258	0.0098
Th	0.688	0.839	1.93	1.11	0.505	0.999	0.539	0.684	0.981	0.0754
Pb	7.92	1.43	1.16	1.45	1.51	1.62	0.602	0.964	1.65	0.582

Major elements are all in weight percent. LOI = loss on ignition. Fe_2O_3[a] = all Fe as Fe_2O_3. Estimated relative errors on major and minor elements are 1%. For trace elements analysed by XRF and INAA the precision is estimated to be generally better than 5% [Fitton et al., 1997]. For the trace elements determined by isotope dilution the analytical uncertainty is $\approx 0.5\%$ for Pb, $\approx 1\%$ for U and < 2% for Th.

TABLE 2. Major and Trace Element Contents of Volcanic Rocks from Funk Seamount and Marion and Prince Edward Islands

Location	Marion			Funk Seamount		Prince Edward
Sample	WJM-50	WJM-49	WJM-21	D1-B1	D1-E	PREI-13
SiO_2	46.9	48.5	49.5	44.7	44.9	
TiO_2	3.44	3.36	2.74	4.05	4.01	
Al_2O_3	16.48	15.13	16.55	16.03	16.10	
FeO [a]	11.97	12.33	11.05	13.59	13.53	
MnO	0.17	0.18	0.19	0.16	0.15	
MgO	5.11	5.53	3.94	5.59	5.88	
CaO	10.29	9.90	7.22	8.09	8.05	
Na_2O	3.32	3.11	4.29	4.65	4.37	
K_2O	1.30	1.18	1.86	1.66	1.69	
P_2O_5	0.54	0.51	0.83	1.10	1.08	
LOI	0.81	0.65	0.61	0.79	0.77	
H_2O-	0.03	0.05	0.05	0.13	0.14	
Total	100.36	100.43	98.84	100.53	100.71	
XRF (ppm)						
V	288	264	104	187	175	
Cr	32	29	3	73	73	
Ba	278	256	448	391	374	
Nb	41	36	57	59		
Zr	251	250	356	329	390	
Y	30	32	42	38	27	
Sr	614	507	777	1099	1074	
Rb	22	20	33	26	29	
Ni	41	<2	108	49	41	
Sc	23	25	14	13	14	
Cu	58	11	445	18	17	
Zn	108	111	338	141	147	
Co	42	47	20	48	44	
INAA (ppm)						
La	32.0	27.4	48.1	43.9	45.2	16.4
Ce	65.4	59.2	95.8	95.0	98.3	37.1
Nd	35.4	34.0	48.7	56.2	58.8	24.6
Sm	7.8	7.5	10.9	13.0	13.0	6.1
Eu	2.6	2.7	3.8	4.4	4.5	2.7
Gd						
Tb	1.04	1.09	1.44	1.65	1.65	0.91
Yb	2.26	2.32	2.94	1.92	2.00	1.92
Lu	0.32	0.32	0.38	0.23	0.24	0.27
Ta	2.40	2.05	3.33	2.87	3.07	2.71
Th	3.27	2.83	4.48	3.29	3.59	1.56
Hf	6.17	6.23	8.63	8.73	9.29	8.46
Sc	24.6	27.1	15.6	14.1	13.6	21.2
Co	48	47	23	47	47	32

INAA analyses were made during this study. The XRF trace and major element data are from *Reid and le Roex* [1988] (Funk Seamount) and *Mahoney et al.* [1992] (Marion).
[a] All Fe as FeO.

to nitrates and then chlorides. Pb was separated using a microcolumn technique for which the total procedural blank was less than 0.5 ng g^{-1}. Sr and Nd were separated using conventional ion-exchange procedures. Isotopic compositions were measured on a MAT 261 multicollector mass-spectrometer. A VG Sector multicollector mass-spectrometer was used at the University of Hawaii following sample preparation and analytical procedures outlined by *Mahoney et al.* [1991]. For interlaboratory comparison, standard rock BCR-1 was run at both the Open University and the University of Hawaii. Isotopic compositions of a subset of samples are given in Table 3 and the U, Pb, and Th isotope-dilution data in Table 1. Also shown in Table 3 are isotopic values age-corrected to 88 Ma on the basis of measured parent-daughter ratios and published ^{40}Ar-^{39}Ar ages [*Storey et al.,* 1995]. For the Open University Pb isotopic data, the age correction was derived from the Th/Pb ratio and assuming Th/U = 4. The University of Hawaii results were age-corrected using U, Th, Pb, Sr, Rb, Sm, and Nd isotope-dilution data (not shown in Table 1 for Nd, Sm, Rb, or Sr) determined on the same sample splits analyzed for isotope ratios.

East Coast Basalts and Dikes

Chemical results. The great majority of data for rocks from the rifted eastern margin of Madagascar fall on a low-pressure fractionation trend of strong Fe- and Ti-enrichment with decreasing MgO (Fe-Ti series), which shows a maximum between 4 and 6 wt.% MgO (Figure 5). The Fe-Ti enrichment trend is most pronounced in samples from the Mananjary and Tamatave transects. It is accompanied by a slight decrease in SiO$_2$ and near-constant CaO/Al$_2$O$_3$, with plagioclase fractionation indicated by varying degrees of Sr and Eu depletion (Figure 6). The lavas and dikes of the Fe-Ti series show strong compositional similarities to evolved iron-rich, silica-poor liquids described for the East Greenland volcanic rifted margin [*Brooks et al.,* 1991] but contrast with recent alkalic lavas from Funk Seamount and Marion Island, where low CaO/Al$_2$O$_3$ suggests that fractionation occurred at higher pressures [see *Reid and le Roex,* 1988]. The rift-related Fe-Ti basalts of Madagascar are enriched in the light REE (Figure 6). In primitive-mantle-normalized trace element plots of samples from the Tamatave and Sambava transects, the presence of a pronounced positive Ba spike is the most notable difference from the Mananjary Fe-Ti series patterns.

A number of basalts, particularly from the Mananjary and Sambava transects, fall below the main Fe-Ti enrichment, low-pressure fractionation trend and also

typically have higher SiO$_2$, as well as a strong depletion in Nb and Ta relative to other highly incompatible elements (Figures 5 and 6). There is no obvious spatial or temporal distinction between the location of the Fe-Ti series basalts and these low Fe-Ti basalts along the east coast and the two types are occasionally interlayered. The low Fe-Ti basalts show elevated and widely ranging initial ^{87}Sr/^{86}Sr and low ε_{Nd} values (M. Storey et al., unpubl. data) similar to those of the Morondava Basin in southwestern Madagascar [*Mahoney et al.,* 1991]. The Morondava Basin tholeiites have been interpreted as being variably but highly contaminated by ancient crustal material, broadly like that affecting the Bushe and Poladpur Formations of the later (66 Ma) Deccan Traps in India [*Mahoney et al.,* 1991]. Likewise, we consider the compositional and isotopic features of most of the low Fe-Ti basalts to be an indication that they have experienced variable amounts of crustal contamination.

Lastly, we recognize a rare group of high Mg-Ti basalts from the Mananjary transect (Figure 5). These basalts are strongly enriched in the light REE, resembling Marion hotspot lavas. Three of the samples analysed have a pronounced Ba spike in their primitive-mantle-normalized patterns (Figure 6). Two features common to both the Marion lavas and the Mananjary high Mg-Ti basalts are a steep slope between the middle and heavy REE (Figure 6) and low Sc abundances (<25 ppm; Tables 1 and 2). These observations indicate residual garnet during melting.

Isotopes. New isotopic data for the east coast samples that lack chemical signs of probable significant crustal contamination (e.g., high SiO$_2$ and La/Nb) are given in Table 3. In Pb isotopic space (Figure 7) most of the data plot above the Northern Hemisphere Reference Line of *Hart* [1984], a general feature of the Indian Ocean mantle domain [e.g., *Dupré and Allègre,* 1983; *Hamelin et al.,* 1986; *Price et al.,* 1986; *Mahoney et al.,* 1989, 1992]. The basalts of the southernmost (Mananjary) transect on the east coast fall mainly into two isotopically distinct groups, which correlate broadly with major element compositions. Data for the more evolved members (TiO$_2$>3 wt.%) of the Mananjary Fe-Ti series lavas mostly plot to the right of the geochron and show a positive correlation between ^{206}Pb/^{204}Pb and both ^{207}Pb/^{204}Pb and ^{208}Pb/^{204}Pb (Figure 7). The highest ^{206}Pb/^{204}Pb ratios ($\approx$18.6) are comparable to those measured in recent Marion hotspot lavas, although the latter are characterized by slightly lower ^{207}Pb/^{204}Pb and ^{208}Pb/^{204}Pb ratios (Figure 7). Curiously, the less evolved members (TiO$_2$<2 wt.%) of the Mananjary Fe-Ti series basalts exhibit a much narrower range in ^{206}Pb/^{204}Pb ratios (17.7-17.9), substantially lower than

TABLE 3. Sr, Nd, and Pb Isotopic Compositions of Rocks from the Rifted Eastern Margin of Madagascar

Transect Sample	$^{87}Sr/^{86}Sr$ measured	$^{87}Sr/^{86}Sr$ t= 88 Ma	$^{143}Nd/^{144}Nd$ measured	εNd t=88 Ma	$^{206}Pb/^{204}Pb$ measured	$^{207}Pb/^{204}Pb$ measured	$^{208}Pb/^{204}Pb$ measured	$^{206}Pb/^{204}Pb$ t=88 Ma	$^{207}Pb/^{204}Pb$ t=88 Ma	$^{208}Pb/^{204}Pb$ t=88 Ma
Mananjary										
Mananjary high-Mg-Ti rocks										
MAN90-45	0.70379	0.70366	0.513023	+8.1	17.946	15.548	37.712	17.794	15.541	37.513
MAN90-47	0.70380	0.70370	0.512847	+4.7	17.822	15.580	37.889	17.699	15.574	37.727
MAN90-85 [a]	0.70338	0.70336	0.512754	+2.7	16.921	15.175	37.084	16.805	15.169	36.977
MAN90-86	0.70424	0.70422	0.512720	+2.2	16.918	15.153	37.677	16.765	15.146	37.478
Mananjary Fe-Ti series (TiO$_2$>3 wt.%)										
MAN90-1B	0.70357	0.70347	0.512925	+5.9	18.448	15.578	38.813	18.256	15.569	38.561
MAN90-6	0.70363	0.70351	0.512930	+6.0	18.762	15.607	38.928	18.549	15.596	38.649
MAN90-8	0.70603	0.70596			15.891	15.013	37.017	15.823	15.010	36.928
MAN90-31	0.70355	0.70345			18.820	15.605	38.856	18.615	15.595	38.587
MAN90-43	0.70356	0.70340	0.512929	+6.0	18.807	15.588	38.764	18.619	15.579	38.517
MAN90-49	0.70308	0.70295	0.513000	+7.4	18.348	15.506	38.141	18.144	15.496	37.873
MAN90-57	0.70377	0.70354	0.512894	+5.4	18.695	15.615	38.899	18.492	15.605	38.633
MAN90-63	0.70325	0.70313			18.260	15.544	38.086	18.086	15.536	37.858
MAN90-66	0.70348	0.70339	0.512913	+5.6	18.574	15.572	38.591	18.422	15.565	38.392
MAN90-71	0.70349	0.70336			17.967	15.483	38.153	17.776	15.474	37.902
MAN90-80	0.70347	0.70330	0.512957	+6.5	18.389	15.552	38.616	18.182	15.542	38.345
MAN90-81	0.70359	0.70347			18.421	15.582	38.636	18.218	15.572	38.370
Mananjary Fe-Ti series (TiO$_2$<2 wt.%)										
MAN90-2	0.70366	0.70356	0.512994	+7.1	18.012	15.479	37.971	17.874	15.473	37.788
MAN90-15	0.70350	0.70335			17.847	15.485	37.963	17.782	15.481	37.878
MAN90-16 [a]	0.70355	0.70354	0.512919	+5.5	17.892	15.517	37.957	17.794	15.512	37.840
MAN90-17	0.70333	0.70330			17.982	15.571	38.254	17.831	15.564	38.057
MAN90-20	0.70491	0.70489	0.512880	+5.0	16.876	15.558	37.614	16.834	15.556	37.559
MAN90-22/2	0.70438	0.70431	0.512903	+5.3	17.991	15.510	38.006	17.844	15.503	37.813
MAN90-77	0.70360	0.70357	0.512986	+6.9	18.041	15.487	38.051	17.864	15.478	37.818
Tamatave										
TAM92-6 [a]	0.70577	0.70569	0.512685	+1.3	16.718	15.160	37.980	16.702	15.159	37.956
TAM92-27 [a]	0.70431	0.70418	0.512900	+5.4	17.606	15.416	37.912	17.454	15.409	37.748
TAM92-30 [a]	0.70499	0.70478	0.512851	+4.5	17.767	15.426	38.123	17.372	15.407	37.656
TAM92-43 [a]	0.70446	0.70435	0.512847	+4.4	17.515	15.421	37.846	17.336	15.412	37.632
Sambava										
SAM92-1 [a]	0.70425	0.70420	0.512700	+1.5	17.305	15.496	38.078	17.192	15.491	37.984
SAM92-3 [a]	0.70386	0.70382	0.512752	+2.6	17.120	15.391	37.738	16.987	15.385	37.567
SAM92-10 [a]	0.70340	0.70338	0.512891	+5.1	17.645	15.476	37.884	17.421	15.465	37.633
SAM92-16 [a]	0.70346	0.70339	0.512885	+5.0	17.718	15.506	38.007	17.539	15.497	37.807
SAM92-20D [a]	0.70385	0.70374	0.512787	+3.3	17.328	15.433	37.877	17.196	15.427	37.711
SAM92-33B [a]	0.70682	0.70681	0.512353	-5.3	18.173	15.782	38.852	18.158	15.781	38.815
USGS standard										
BCR-1	0.70498		0.512627		18.816	15.640	38.741			
BCR-1 [a]	0.70502		0.512633		18.807	15.621	38.678			

[a] Analyzed at the University of Hawaii.

Pb isotopic ratios are corrected for fractionation using the NBS 981 standard values of *Todt et al.* [1996]. Errors on the Hawaii data are $\leq \pm 0.000012$ (0.2 ε units) for $^{143}Nd/^{144}Nd$, ± 0.00002 for $^{87}Sr/^{86}Sr$, ± 0.010 for $^{206}Pb/^{204}Pb$ and $^{207}Pb/^{204}Pb$, and ± 0.032 for $^{208}Pb/^{204}Pb$. Maximum errors on the Open University data are comparable, except for samples MAN90-22/2, -45, and -66, which have errors $\leq \pm 0.000021$ for $^{143}Nd/^{144}Nd$.

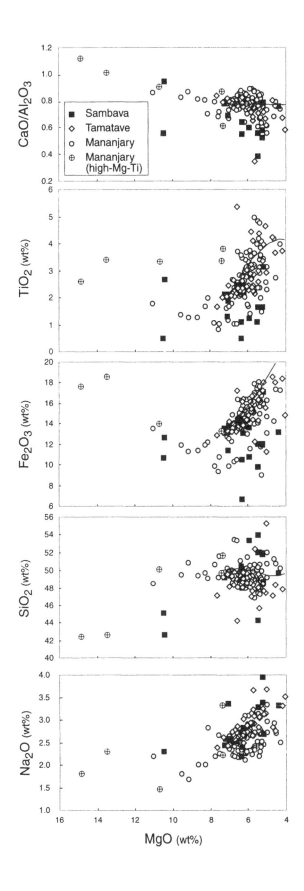

seen in Marion hotspot lavas. Basalts from the Tamatave and Sambava transects, as well as the Mananjary high Mg-Ti basalts, are characterized by low $^{206}Pb/^{204}Pb$ ratios, with the majority having values between 16.6 and 17.8. These low $^{206}Pb/^{204}Pb$ basalts form two populations on the basis of their $^{207}Pb/^{204}Pb$ ratios. Interestingly, the group of low $^{206}Pb/^{204}Pb$ basalts which have high $^{207}Pb/^{204}Pb$ ratios overlaps with data for MORB from the anomalous SWIR ridge segment between 39° and 41°E [*Mahoney et al.*, 1992].

In $^{206}Pb/^{204}Pb$-ε_{Nd} and $^{206}Pb/^{204}Pb$-$^{87}Sr/^{86}Sr$ isotope space the evolved members of the Mananjary Fe-Ti series basalts show an opposite trend from the other east coast basalts, with initial ε_{Nd} decreasing from +7.4 to +5.4 with increasing $^{206}Pb/^{204}Pb$. This array (Trend I) spans a range of values from those for normal MORB from the 32-39°E segment of the SWIR to Marion hotspot compositions (Figure 8). Initial $^{87}Sr/^{86}Sr$ increases from 0.7031 to 0.7036, overlapping with the high $^{87}Sr/^{86}Sr$ region of the Marion hotspot field. These $^{87}Sr/^{86}Sr$ ratios are less than the lowest values measured for basalts from Réunion and Mauritius for closely comparable $^{206}Pb/^{204}Pb$.

The Sambava and Tamatave basalts and the other Mananjary samples, including the high Mg-Ti basalts, form a loosely defined array (Trend II in Figure 8) in which ε_{Nd} decreases and $^{87}Sr/^{86}Sr$ increases with decreasing $^{206}Pb/^{204}Pb$. A similar trend is shown by isotopic data for the Ejeda-Bekily dike swarm [*Mahoney et al.*, 1991]. At the high $^{206}Pb/^{204}Pb$ end of the array, Trend II approaches, but does not quite reach, the values of normal MORB from the SWIR (32-39°E). At the low $^{206}Pb/^{204}Pb$ end, the array partly overlaps with the anomalous MORB compositions reported for the SWIR ridge segment between 39° and 41°E [*Mahoney et al.*, 1992].

Origin of Trend I

Trend I is consistent with derivation of the evolved Mananjary Fe-Ti series basalts from mixtures of Marion plume mantle and Indian MORB mantle, the latter isotopically similar to the source of MORB from the 32-39°E segment of the SWIR. The variation of $^{206}Pb/^{204}Pb$ with $^{87}Sr/^{86}Sr$ and ε_{Nd} is suggestive of two-component

Figure 5. Selected major element oxides and ratios versus MgO for basalts from the eastern rifted margin. The solid heavy line represents the low-pressure (0.5 kb) liquid line of descent calculated using the program MELTS [*Ghiorso and Sack*, 1995]. Note that for a given MgO content, the Mananjary basalts on average have lower Na_2O than those from Tamatave and Sambava.

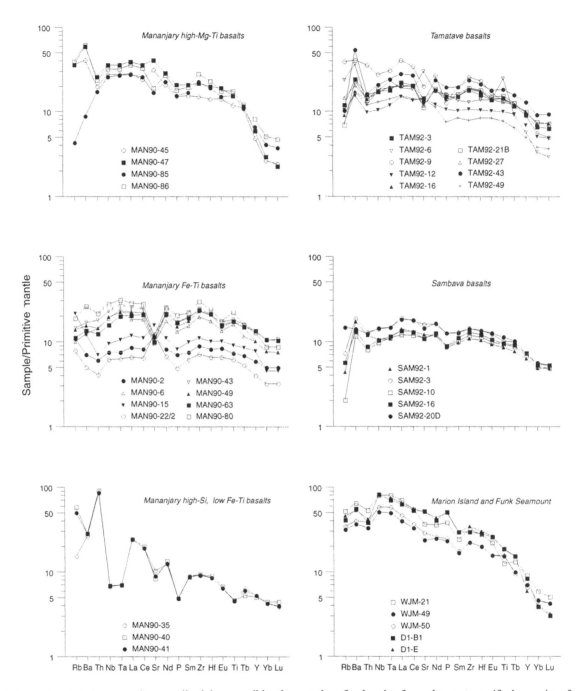

Figure 6. Primitive-mantle-normalized incompatible element data for basalts from the eastern rifted margin of Madagascar and lavas of Marion Island and Funk Seamount (normalizing values from *Sun and McDonough* [1989]). Note the pronounced Ba spike shown by the high-Mg-Ti basalts from Mananjary and also most basalts from the Tamatave and Sambava transects. Many of the high-Fe-Ti basalts from Mananjary show pronounced Sr and Eu anomalies, indicating that they have suffered significant amounts of plagioclase fractionation, whereas others are clearly accumulative in plagioclase. The high-Si, low-Fe-Ti basalts from Mananjary are strongly depleted in Nb and Ta relative to other highly incompatible elements, consistent with significant contamination by continental crust.

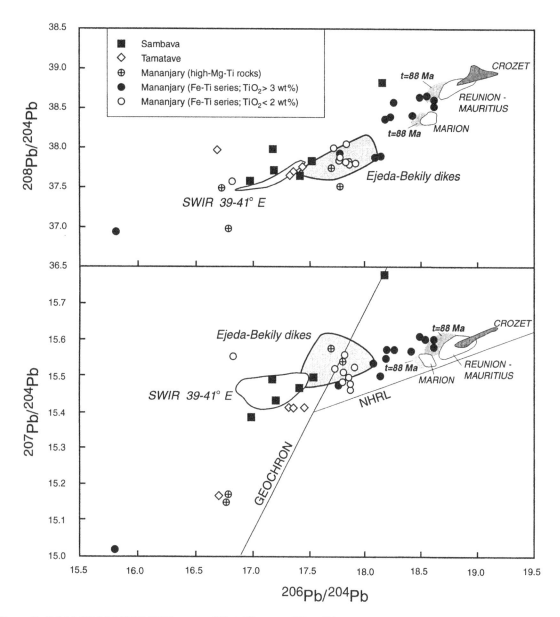

Figure 7. Initial (88 Ma) $^{206}Pb/^{204}Pb$ versus $^{207}Pb/^{204}Pb$ and $^{208}Pb/^{204}Pb$ for basalts from the eastern rifted margin of Madagascar showing data from Table 3 and *Mahoney et al.* [1991]. Note that although some of the high-Fe-Ti basalts from Mananjary have $^{206}Pb/^{204}Pb$ ratios which overlap with those of Marion, they are also characterized by higher $^{207}Pb/^{204}Pb$ and $^{208}Pb/^{204}Pb$ ratios. The field for the Ejeda-Bekily dike swarm is from *Mahoney et al.* [1991]; the fields for the SWIR (39-41°E) and the Marion Réunion-Mauritius and Crozet hotspots are from *Mahoney et al.* [1992] and references therein. Also shown is the the estimated range in Pb isotopic composition of the Marion and Réunion plumes at 88 Ma, assuming μ values for the respective sources of 10 and 15, and $\kappa = 3.3$ [see *Peng and Mahoney, 1995*]. Northern Hemisphere Reference Line (NHRL) is from *Hart* [1984].

mixing [e.g., *Langmuir et al.,* 1978; *Barling and Goldstein,* 1990] between a high-ε_{Nd} component with low Nd/Pb and Sr/Pb, and a low-ε_{Nd} component with high Nd/Pb and Sr/Pb. However, there are two discrepancies. The first is that the variation in Nd/Pb and Sr/Pb ratios of Trend I basalts is small and these elemental ratios do not correlate

well with isotopic composition. Secondly, these basalts have higher $^{207}Pb/^{204}Pb$ and higher $^{208}Pb/^{204}Pb$ than those from the Marion hotspot at comparable $^{206}Pb/^{204}Pb$. One explanation is that minor amounts of continental Pb, with high $^{207}Pb/^{204}Pb$ and $^{208}Pb/^{204}Pb$ ratios, were incorporated into these magmas during ascent through the

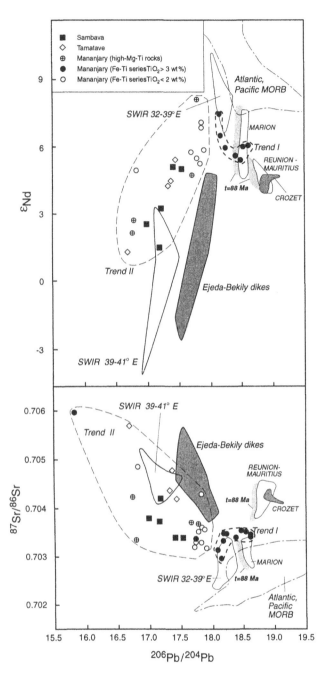

Figure 8. Initial ε_{Nd} and $^{87}Sr/^{86}Sr$ versus $^{206}Pb/^{204}Pb$ for basalts from the eastern rifted margin of Madagascar. The fields for the SWIR (32–39°E) and Atlantic and Pacific MORB are from *Mahoney et al.* [1992] and references therein. Also shown are the estimated range in isotopic composition of the Marion and Réunion plumes at 88 Ma. Other data sources are as Figure 7.

$^{207}Pb/^{204}Pb$ and $^{208}Pb/^{204}Pb$ without significantly changing the $^{206}Pb/^{204}Pb$ composition is demonstrated by the strongly contaminated lavas in southern Madagascar, which have high $^{207}Pb/^{204}Pb$ and $^{208}Pb/^{204}Pb$ for $^{206}Pb/^{204}Pb$ ratios ranging between ≈ 18 and 18.8 [*Mahoney et al.,* 1991; ms. in prep.]. An alternative explanation is that the early Marion plume activity involved the melting out of a component with slightly higher $^{207}Pb/^{204}Pb$ and $^{208}Pb/^{204}Pb$ than seen in the present-day hotspot. If this were the case it would make Marion more conformable with other Indian Ocean hotspot islands such as Réunion and Crozet.

Origin of the Low $^{206}Pb/^{204}Pb$ Basalts (Trend II)

Trend II shows decreasing $^{206}Pb/^{204}Pb$ with increasing $^{87}Sr/^{86}Sr$, contrasting with the general trend shown by MORB and OIB of increasing $^{206}Pb/^{204}Pb$ with increasing $^{87}Sr/^{86}Sr$ (Figure 8). Low $^{206}Pb/^{204}Pb$ is commonly observed in lamproites and continental flood basalts, and there is evidence that it can be a compositional feature of both continental lithospheric mantle [e.g., *Hawkesworth et al.,* 1990] and lower crust [e.g., *Moorbath and Welke,* 1969]. The low $^{206}Pb/^{204}Pb$ basalts from the east coast of Madagascar offer no unique explanation for their origin, although the general lack of strong elemental evidence for crustal contamination in the samples analysed suggests that the source of the unradiogenic Pb is more likely to be the mantle lithosphere. The presence of both low $^{207}Pb/^{204}Pb$ and high $^{207}Pb/^{204}Pb$ types implies continental components of different ages; a single-stage Pb growth model suggests the involvement of Early to Middle Proterozoic lithosphere in the origin of the high $^{207}Pb/^{204}Pb$ - low $^{206}Pb/^{204}Pb$ basalts from the Tamatave and Sambava transects, whereas the low $^{207}Pb/^{204}Pb$ - $^{206}Pb/^{204}Pb$ basalts indicate a role for Archean lithosphere. Trend II can thus be viewed as being broadly consistent with mixing of low $^{206}Pb/^{204}Pb$ Madagascan mantle-lithosphere-derived components of different ages with an Indian Ocean normal-MORB mantle component (and possibly minor Marion hotspot influence).

Mantle with high $^{207}Pb/^{204}Pb$ - low $^{206}Pb/^{204}Pb$ isotopic compositions is being tapped today along the SWIR between 39° and 41°E (Figures 7 and 8). A proposed origin for these low $^{206}Pb/^{204}Pb$ MORBs is that they represent the remnants of ancient continental mantle, thermally eroded by the Marion plume from Indo-Madagascar in the middle Cretaceous [*Mahoney et al.,* 1992]. The presence of isotopically similar basalts from the Tamatave and Sambava transects supports this idea.

lithosphere. The presence in Madagascar of continental lithosphere with the necessary isotopic composition to lift

Volcan de l'Androy

The Volcan de l'Androy massif is a bimodal basalt-rhyolite association with volumetrically minor hybrid magmas. The basalts are generally more evolved than those of the east coast, although some sparsely olivine-phyric high-MgO (11–12 wt.%) lavas are present. The Volcan de l'Androy basalts form two compositionally distinct groups (Groups I and II). Group I (the most abundant) has lower total iron, TiO_2, and Nb than the Group II basalts, at given MgO contents (Figure 9). In mantle-normalized diagrams the Group I basalts show pronounced negative Nb, Ta, and Sr anomalies and are light-REE enriched (Figure 10). The Sr anomalies may be a result of plagioclase fractionation, as Sr/Nd decreases with decreasing MgO. The Group I magmas, which have low initial ε_{Nd} and very high initial $^{87}Sr/^{86}Sr$ (>0.71; J. Mahoney et al., ms. in prep.) almost certainly assimilated continental crust.

The Group II basalts, which are mainly found in the "rim series" (Figure 3) on the eastern side of the massif, lack Nb, Ta, and Sr anomalies (Figure 10). Their mantle-normalized patterns show strong similarities with those of OIB, indicating that small-degree, alkalic mantle melts were available during the formation of the Volcan de l'Androy complex. The $^{87}Sr/^{86}Sr$ ratios of the Group II basalts range from 0.7059 to 0.7064, which is substantially less than those for Group I but significantly higher than for Marion hotspot basalts (Figure 8). The Pb isotopes are also distinct from those of the Marion hotspot, with much higher $^{208}Pb/^{204}Pb$ ratios (>39; J. Mahoney et al., ms. in prep.). This result leads us to propose that the Group II basalts represent small-degree alkalic melts from the continental lithospheric mantle, as suggested for the associated Ejeda-Bekily dike swarm [*Mahoney et al.,* 1991].

The absence of eutaxitic textures in the thick rhyolites that are especially prominent in the upper part of the sequence suggests that these are lava flows rather than pyroclastic deposits, despite having aspect ratios in excess of 50:1. Most of the rhyolites are phenocryst-poor. As with the basalts, two distinct sets of rhyolites are present. Group I rhyolites, like their basalt counterparts, have relatively low abundances of Nb, Ta, Zr, and Hf. (La/Yb)$_N$ (N = primitive-mantle-normalized) ratios are about 8, and these rhyolites have moderate negative Eu anomalies (Figure 10). SiO_2 ranges from 65 to 80 wt.%, the highest values probably reflecting alteration (Figure 11). The microgranites surrounding the massif are compositionally identical to the Group I rhyolites. The Group II rhyolites, on the other hand, have much higher contents of Nb, Ta,

Zr, and Hf, flatter REE profiles [(La/Yb)$_N$ ~ 3] and very pronounced, negative Ba, Eu, Sr, P, and Ti anomalies. SiO_2 ranges from 71 to 76 wt.%. Group II rhyolites are

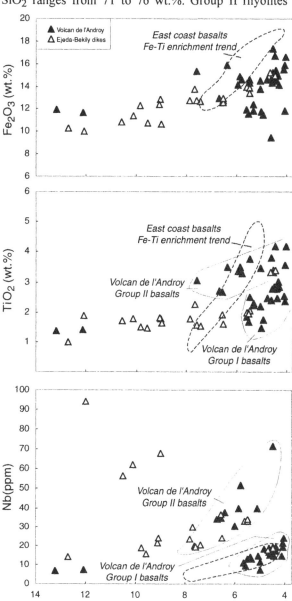

Figure 9. Fe_2O_3, TiO_2 and Nb versus MgO in basalts from the Volcan de l'Androy complex and the associated Ejeda-Bekily dike swarm in the south of Madagascar. Data for Volcan de l'Androy are from our work and will be presented in full elsewhere (J. Mahoney et al., ms. in prep.). The data for the Ejeda-Bekily dikes are from *Dostal et al.* [1992] and A. D. Saunders et al. (unpub. data). Note that the Volcan de l'Androy basalts fall into two compositionally distinct groups. Shown for comparison is the compositional range exhibited by basalts from the eastern rifted margin (dashed field).

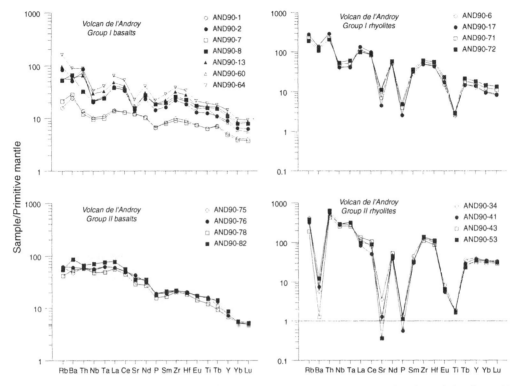

Figure 10. Primitive-mantle-normalized incompatible element data for Groups I and II basalts and rhyolites of Volcan de l'Androy. Note the general similarity between the Group II basalts and the Ejeda-Bekily dikes (shown by shaded field; data from *Dostal et al.* [1992]).

present on Vohitsimbe, where they are found interbedded with both Group I rhyolites and basalts. Group II basalts are found in nearby dikes. Several of the Group II rhyolites are mildly peralkaline. The relatively low concentrations of Ba, light- and middle-REE, Eu, Sr, P, and Ti are consistent with extraction of K-feldspar, plagioclase, apatite and titanomagnetite, ± amphibole. These rhyolites are almost certainly related to the Group II basalts by differentiation (Figure 11), although isotopic data (J. Mahoney et al., ms. in prep.) also indicate some crustal assimilation. Conversely, the Group I rhyolites may be differentiates of the Group I basalts or some mixture of basalt and crustal melts; the Nb and Zr abundances in the Group I rhyolites (27–67 ppm and >500 ppm, respectively) are too high for them to be pure crustal melts.

Ejeda-Bekily Dike Swarm

These dikes (Figure 2) have been discussed extensively by *Mahoney et al.* [1991] and *Dostal et al.* [1992]. They have low initial ε_{Nd} and $^{206}Pb/^{204}Pb$ ratios and are unusual among the Madagascar basalts in being nepheline-normative. In terms of their trace elements, they are similar to the Group II basalts of Volcan de l'Androy but have higher Ba/Ta and Ba/La ratios (Figure 10). *Mahoney et al.*

[1991] noted that some of the Ejeda-Bekily dikes have similar isotopic compositions to the low $^{206}Pb/^{204}Pb$ MORBs found on the SWIR to the south of Madagascar.

Majunga and Morondava Basins

Isotope and trace element data for a suite of tholeiites from the southern part of the Morondava Basin were reported by *Mahoney et al.* [1991] and *Dostal et al.* [1992]. The rocks have similar compositions to the Group I basalts of Volcan de l'Androy, with negative Nb, Ta, Sr, and P anomalies, high initial $^{87}Sr/^{86}Sr$ (>0.71), and low ε_{Nd} (<-6), rather similar to the Bushe Formation of the Deccan Traps in India [*Mahoney et al.,* 1991]. The isotopic data and variations in REE ratios are entirely consistent with the magmas having assimilated sialic crust.

The compositional similarity between the Volcan de l'Androy and southern Morondava Basin basalts (Figure 12, top) suggests that lavas may well have been able to travel across the island and into the western basins. The distance from Volcan de l'Androy to the region southeast of Tulear is about 200 km. The northernmost outcrops in the Tulear area would have necessitated flows some 300-350 km long. These are not excessive distances when compared with large flood basalt flows, such as those in

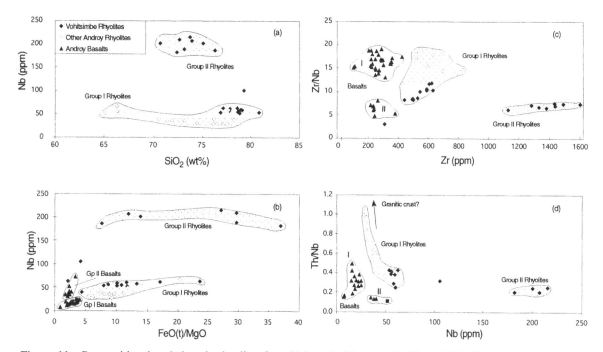

Figure 11. Compositional variations in rhyolites from Volcan de l'Androy. (a) Nb vs. SiO_2, illustrating the marked difference between the Group I and II rhyolites. Note that both types of rhyolite are interbedded with basalts in the Vohitsimbe lava succession. (b) FeO(t)/MgO versus Nb, showing the relationship between the Group I and II rhyolites, and the Group I and II basalts (FeO(t)=total Fe expressed as FeO). (c) Zr versus Zr/Nb. The very high Zr concentrations of the Group II rhyolites suggest that they are the products of large amounts of crystal fractionation of a basaltic parent. (d) Nb versus Th/Nb. Note the trends to high Th/Nb in both the basalts and rhyolites of Group I, indicating crustal contamination.

the Columbia River province [e.g., *Tolan et al.,* 1989; *Hooper*, this volume].

The basalts of the Majunga Basin, in the northwest of Madagascar, are compositionally different from the Tulear basalts. They show greater diversity although this may simply reflect the wider area sampled. As a whole, they straddle the normative quartz-, olivine-tholeiite divide, but in terms of trace elements can be divided into three groups: a group with large negative Nb and Ta anomalies, a group with convex-upward mantle-normalized patterns and low abundances of heavy REE, and a group with convex-upward mantle-normalized patterns and greater abundances of heavy REE (Figure 12) [cf. *Melluso et al.,* 1997]. The last two groups can be related by different pressures of melting; the first group may have undergone some crustal assimilation.

DISCUSSION

Marion Plume Mantle in Fe-Ti Basalt Petrogenesis

Despite the evidence that mantle plumes play an important role in flood basalt formation, few geochemical studies have provided unequivocal evidence that hotspots contribute materially to the magmatism. Exceptions are isotopic studies on the northwestern Deccan Traps [*Peng and Mahoney*, 1995] and East Greenland Tertiary tholeiites [*Thirlwall et al.,* 1994] which respectively revealed the presence of Réunion and Iceland mantle plume components in the erupted magmas. The Mananjary Fe-Ti basalts also provide good isotopic evidence for a plume-type component. Trend I (Figure 8) can largely be accounted for by mixing between a mantle end-member somewhat similar to that being tapped today at Marion, Prince Edward, and Funk volcanoes, and a high ε_{Nd}-low $^{87}Sr/^{86}Sr$ component similar to normal SWIR MORB mantle.

That mantle rather similar in isotopic composition to the present-day Marion hotspot magmas was important at 88 Ma in Madagascar is of relevance in understanding the Pb isotopic evolution of Indian Ocean mantle plumes. It has been proposed that large $^{206}Pb/^{204}Pb$ variations along the Ninetyeast Ridge, presumed to be the 82–38 Ma track of the Kerguelen plume, are essentially an aging effect resulting from a high μ ($^{238}U/^{204}Pb$) value in the plume source [*Class et al.,* 1993]. This idea was disputed for the Réunion hotspot case by *Peng and Mahoney* [1995] on the

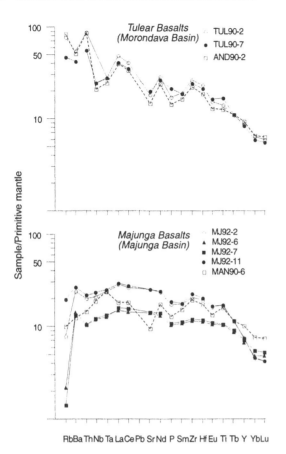

Figure 12. Primitive-mantle-normalized incompatible element data for some basalts of the Majunga and Morondava basins (A. D. Saunders et al., unpub. data). Note the similarity of the Tulear patterns and some Group I basalts from Volcan de l'Androy (e.g., AND90-2). Likewise, the patterns of the Majunga basalts show similarities with the east coast Fe-Ti basalts (e.g., MAN90-6).

basis of isotopic data for basalts and picrites of the northwest Deccan Traps. There, the Pb isotopic composition of the plume appears to have changed little since the Late Cretaceous. The Madagascar Fe-Ti basalts also do not require large changes in $^{206}Pb/^{204}Pb$ in the Marion plume in the last 88 m.y. The implication is that the Réunion and Marion plumes have much smaller μ values (<15) than required by the *Class et al.* [1993] model ($\mu > 30$). This is supported by measured μ values in Réunion subalkalic basalts, which are ≤ 20, in general [*Peng and Mahoney,* 1995].

Significance of Spatial Compositional Variations in the Basalts

A prediction from dry-melting models for a mantle plume that underlies a zone of continental extension is that the amount of melting should decrease systematically with increasing distance along the rift axis away from the plume center, because of the decreasing potential temperature of the sublithospheric mantle (Figure 13). If hydrated portions of the continental mantle are also undergoing melting in response to extension and conductive heating by the plume, then the amount of melt produced may be less sensitive to changes in potential temperature of the asthenosphere beneath a continental rift (Figure 13). In this scenario, the proportion of melt generated from the lithospheric mantle relative to melt produced from the plume should become higher with increasing distance along the rift axis away from the plume center.

Likewise, as the degree to which the lithosphere has been thinned by extension decreases inland from the rift, the extent of melting within the plume will become increasingly restricted by the "lid" effect and melts generated within the continental mantle may predominate. If the lithosphere is thicker than about 130 km, then most of the melt will be generated by "wet" continental mantle as the "dry" plume solidus is not crossed (Figure 13). Do the Madagascar Cretaceous volcanic rocks fit with these predictions? Here we consider the significance of the regional geochemical variations in the basalts along the rifted margin and inland at Volcan de l'Androy.

Using the plate motion model of *Müller et al.* [1993], *Storey et al.* [1995] placed the Marion hotspot within about 100 km south of Madagascar around the time of continental breakup with India. They suggested that, within the uncertainty of the plate reconstruction, Volcan de l'Androy marked the focal point of the Marion plume at 88 Ma. The isotopic evidence for a Marion plume component in the Mananjary Fe-Ti basalts is consistent with the plume being situated near the southern end of the continental rift. In contrast, all of the basalts from the two more northerly transects (Tamatave and Sambava) have low $^{206}Pb/^{204}Pb$ ratios, reflecting significant continental lithospheric influences. For comparable MgO contents, the Tamatave and Sambava basalts have higher Na_2O contents (Figure 5), which may indicate a decrease in the amount of melting northward along the rift axis, although source heterogeneity is another possible explanation for the Na variation.

The total absence of Marion-like isotopic compositions in the basalts of the central and northern part of the eastern rifted margin suggests that plume melts were only available in the south of the island, or that the plume signature was masked by lithospheric and/or MORB-type mantle components. The isotopic data indicate that the sources of the basalts in the north are mainly a mixture of continental mantle lithosphere and Indian MORB-source mantle. Support for a greater involvement of lithospheric mantle in the north is also provided by the positive Ba

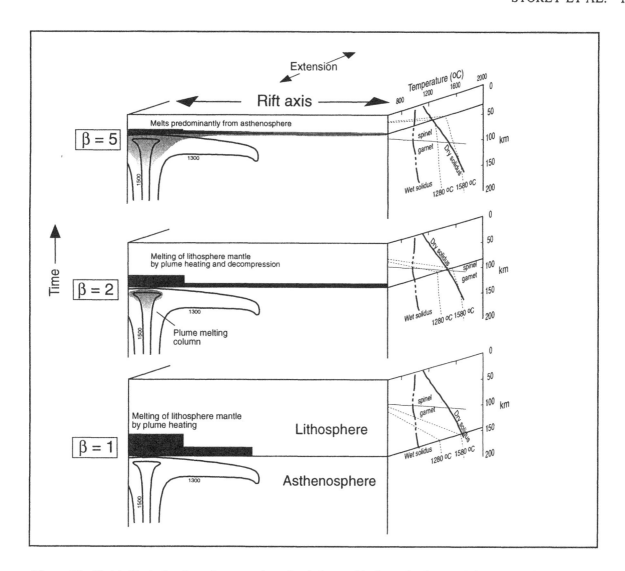

Figure 13. Sketch illustrating how the proportion of melt from a "dry" mantle plume, relative to melt derived from "wet" lithospheric mantle, may vary beneath a developing continental rift. If the lithosphere is thicker than about 130 km, then most melt will be generated within the lithosphere because the "dry" plume solidus (potential temperature of $\approx 1580°C$) is not crossed. However, with time continued extension and thinning of the lithosphere will result in enhanced decompressional melting of the plume, such that plume-derived melts will show an increasing dominance over those derived from the continental mantle. In addition to this temporal variation, there also may be a spatial variation along the rift axis with a decrease in the proportion of plume-derived melts with increasing distance from the plume center. β is the original lithosphere thickness divided by the extended thickness. Plume isotherms are from *Watson and McKenzie* [1991]. Solidus curves and the garnet-spinel transition depth are from *McKenzie and Bickle* [1988] and *McKenzie and O'Nions* [1995].

anomalies shown by the Sambava and Tamatave basalts in primitive-mantle-normalized diagrams (Figure 6). For example, high Ba/Nb is observed in late Cenozoic basalts erupted in an extensional environment in the western United States [e.g., *Fitton et al.*, 1988]. Some of these Cenozoic lavas exhibit a negative correlation between $^{143}Nd/^{144}Nd$ and La/Nb, a feature interpreted to indicate

mixing between asthenosphere-derived magmas and magmas (high Ba/Nb) derived from the continental lithospheric mantle [*Fitton et al.*, 1988]. The east-coast rifted margin basalts of Madagascar may record the transition between a plume-dominated source in the south and a mixture of continental lithospheric mantle and MORB mantle sources in the north.

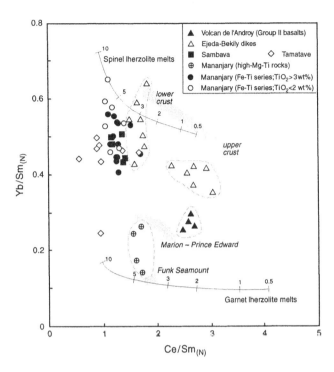

Figure 14. The variation of the light to middle REE (Ce/Sm) and heavy to middle REE (Yb/Sm) ratios for Madagascar Cretaceous volcanic rocks, normalized (N) to primitive-mantle values. The compositions of aggregated melts produced by different degrees of melting of a spinel lherzolite and garnet lherzolite source are shown. The source and melting modes for spinel lherzolite are from *Kelemen et al.* [1992], the source mode for garnet lherzolite is from S. Bernstein et al. (Plume-related, post-rift basaltic magmatism along the East Greenland margin, ms. submitted to *Earth Planet. Sci. Lett.*) and the garnet lherzolite melting mode is from *Walter and Presnall* [1994]. The melting model is based on the fractional melting equations of *Gast* [1968] and *Shaw* [1970], and partition coefficient data from *Kelemen et al.* [1993], *Shimizu and Kushiro* [1975], and *Hart and Dunn* [1993]. The composition of the model mantle is one of slight REE depletion (Ce/Sm$_{(N)}$ = 0.83). All of the Madagascar data shown are from our work, except for some additional analyses of the Ejeda-Bekily dikes from *Dostal et al.* [1992]. Continental crust compositions are from *Taylor and McLennan* [1985].

In addition to these along-rift-axis variations, isotopic data on the Group II basalt flows of Volcan de l'Androy (J. Mahoney et al., ms. in prep.) and the associated Ejeda-Bekily dike swarm (Figure 8), situated inland from the rifted margin in the south of the island but above the postulated focal point of the plume, are dominated by components from the continental mantle. The lack of Marion hotspot compositions in the Volcan de l'Androy basalts and the Ejeda-Bekily dikes may be caused by the thickness of the lithosphere restricting the amount of melting of the plume.

Figure 14 shows a primitive-mantle-normalized ratio of a light REE to a middle REE (Ce/Sm$_{(N)}$) plotted against a ratio of a heavy REE to a middle REE (Yb/Sm$_{(N)}$) for basalts of the eastern rifted margin, Volcan de l'Androy, and the Ejeda-Bekily dike swarm. Samples with obvious signs of crustal contamination (e.g., high SiO$_2$ and La/Nb) are not plotted. Also shown are the calculated compositions of aggregated melts produced by fractional melting of garnet lherzolite and spinel lherzolite. Because the ratio on each axis has a common denominator, trends caused by mixing of melts produced from garnet lherzolite with melts produced from spinel lherzolite will be linear [e.g., *Thirlwall et al.*, 1994], as suggested by the steep array defined by the Fe-Ti basalts and the high Mg-Ti basalts from the rifted margin. Interestingly, this array also overlaps compositions of Marion Island and Funk Seamount lavas, perhaps suggesting that the potential temperature of the Marion plume has not changed significantly with time. The lower melt productivity of the present-day Marion hotspot may result from a combination of its intraplate setting [cf. *Ellam*, 1992] and a lower flux of plume material. The intercepts on the melting curves could suggest that the Fe-Ti basalts formed by mixing of melts produced by ≈4% melting of garnet lherzolite and ≈10% melting of spinel lherzolite. The Volcan de l'Androy Group II basalts show evidence for residual garnet in their source, and they appear to have been produced by lower extents of melting than the east coast lavas. The Ejeda Bekily dikes define two compositional groups. One group appears to represent small amounts (≈3%) of melting of spinel lherzolite; the other group suggests contributions of melt produced by small-degree melting of both garnet lherzolite and spinel lherzolite sources. These calculations neglect the source heterogeneity indicated by the range in Sr, Nd, and Pb isotopes.

In summary, the REE variations are consistent with the largest extent of melting occurring along the rifted margin, in response to extension of the Madagascan lithosphere and mantle decompression. The lower Na$_2$O contents of the Mananjary Fe-Ti basalts may also suggest an increase in mantle temperatures and degree of melting to the south. Paradoxically, despite likely higher sublithospheric potential temperatures, smaller melt fractions were generated at Volcan de l'Androy, which is presumed to mark the center of the Marion plume at 88 Ma. Isotopic and elemental data on the Volcan de l'Androy Group II basalts and the Ejeda Bekily dikes suggest that they represent small-degree melts of spinel lherzolite and garnet lherzolite of the continental lithosphere.

The observed spatial variations in isotopic and REE compositions of the basalts that form the Madagascar Cretaceous volcanic province better fit with the notion they were produced from a smaller thermal and compositional mantle anomaly than the large starting-plume head envisaged for the origin of some flood basalt provinces.

APPENDIX 1 (PETROGRAPHIC DESCRIPTIONS)

MAN 90-1B. Granular to subophitic basalt. Trace quantities of brown smectite pseudomorphs, possibly after olivine. Abundant titanomagnetite. Clinopyroxene is a pink-brown color, reflecting the high Ti content of the sample.

MAN 90-2. Fine-grained granular-textured basalt. Contains approximately 3% partially resorbed and fractured plagioclase phenocrysts. Alteration is restricted to localized patches of green clays in the groundmass.

MAN 90-6. Fine-grained, sparsely plagioclase- and clinopyroxene-phyric basalt. The majority of the plagioclase phenocrysts are found in glomerocrysts with subordinate amounts of clinopyroxene. Isolated phenocrysts of plagioclase are partially resorbed. The groundmass comprises plagioclase and granular clinopyroxene and titanomagnetite. Alteration is slight, restricted to dark-brown clay pseudomorphs, possibly after olivine (<1% of rock).

MAN 90-8. A very fine grained, nearly aphyric, almost aphanitic rock. Contains plagioclase, equant opaque grains, and clinopyroxene.

MAN 90-15. Aphyric basalt, with possible infilling of vesicles with prehnite and pumpellyite.

MAN 90-16. Very fine-grained basalt, with approximately 5% clinopyroxene phenocrysts. The rock shows little evidence for secondary alteration.

MAN 90-17. Olivine gabbro. This is a plagioclase orthocumulate, with strongly colored (brown) intercumulus clinopyroxene. Abundant (≈5%) well-rounded olivine crystals. Alteration is slight and restricted to localised formation of clay.

MAN 90-20. Strongly plagioclase-phyric basalt with trace of clinopyroxene. Probably partially accumulative. No visible sign of alteration.

MAN 90-22/2. Vesicular, variolitic, aphyric basalt. Vesicles are filled with very pale-green chlorite/smectite.

MAN 90-31. Sparsely plagioclase-phyric dolerite, with brown clays replacing the mesostasis. Some clays may be pseudomorphing olivine.

MAN 90-43, -49. Aphyric dolerite. Yellow-brown clays form patches in mesostasis. Trace amounts of partially resorbed plagioclase phenocrysts.

MAN 90-45, -47. Olivine-clinopyroxene-phyric basalts.

MAN 90-57. Aphyric dolerite.

MAN 90-63, -71. Sparsely plagioclase-phyric dolerites. Green-brown smectite is present in mesostasis.

MAN 90-66. Medium- to coarse-grained plagioclase-rich dolerite. Subophitic texture between clinopyroxene and plagioclase; ≈3% opaque minerals, probably mostly titanomagnetite. Several patches of mesostasis interstitial between plagioclase and clinopyroxene are replaced by yellow-green clay (≈2% of rock).

MAN 90-77. Plagioclase-rich dolerite with an "ophimottled" appearance caused by abundant ophitic glomerocrysts of plagioclase and clinopyroxene. The areas around the glomerocrysts comprise plagioclase, clinopyroxene, titanomagnetite and an altered mesostasis (now yellow-green clay, ≈4% of rock).

MAN 90-80. Plagioclase-, titanaugite-phyric basalt. Groundmass is hyalopilitic, with abundant opaques. No visible signs of alteration.

MAN 90-81. Aphyric dolerite, with dark-green smectite in patches of mesostasis.

MAN 90-85. Sparsely clinopyroxene- and plagioclase-microphyric basalt. Unusual opaque-rimmed ocelli with cores filled with an unidentified, low-birefringence mineral.

MAN 90-86. Medium-grained clinopyroxene-microphyric basalt. Chlorite-filled vesicles.

TAM 92-6. Aphyric, plagioclase-rich dolerite, with abundant titanomagnetite. Mesostasis replaced by pale green smectite, and calcite.

TAM 92-27. Coarse-grained, aphyric basalt; 5-10% brown smectite clays replace mesostasis.

TAM 92-30A. Thin section shows segregation of mafic (clinopyroxene) and felsic (plagioclase) phases. Alteration is pervasive, with green-colored clays, calcite and occasionally quartz. The plagioclase is fractured and partially altered to clays.

TAM 92-43. Aphyric, coarse-grained basalt. Brown clays replace the mesostasis.

SAM 92-1. Aphyric, fine-grained, granular-textured basalt. Brown clay replaces the mesostasis.

SAM 92-3. Plagioclase-phyric basalt, with ≈10% plagioclase phenocrysts. The groundmass has an ophimottled texture, with clinopyroxene forming glomero-ophitic clusters. No alteration is visible.

SAM 92-10, -16. An aphyric basalt, although clinopyroxene forms glomero-ophitic clusters, giving an ophimottled appearance. Some of the pyroxene may be orthopyroxene or low-Ca pyroxene. Orange-brown clays have replaced part of the mesostasis.

SAM 92-20D. Fine-grained basalt. Abundant pseudomorphs of pale-green clay may be replacing microphenocrysts of olivine and/or clinopyroxene. The iron-rich mesostasis is almost opaque.

SAM 92-33B. Plagioclase-phyric basalt. The groundmass plagioclase is strongly flow-aligned. The groundmass also contains granules of clinopyroxene, olivine, and orthopyroxene.

Acknowledgments. Field work in Madagascar was made possible through A. Razafiniparany (University of Antananarivo). A. Randriamanantenasoa and M. Rajaoherinirina provided field support. Anton le Roex donated the Marion and Funk samples. Chris Hawkesworth provided generous access to the Open University isotope laboratory. Peter Van Calsteren, Janet Hergt and Mabs Klunker gave analytical advice and assistance. Nick Marsh is thanked for carrying out XRF analyses of the Madagascar rocks, and. Stefan Bernstein for discussion on modelling of REE data. Rob Ellam, Godfrey Fitton, and Chris

Harris critically reviewed the manuscript. This work was supported by the NERC, NSF, and Danish National Research Council.

REFERENCES

Agrawal, P. K., O. P. Pandey, and J. G. Negi, Madagascar: a continental fragment of the paleosuper Dharwar craton of India, *Geology, 20*, 543-546, 1992.

Ando, A., N. Mita, and S. Terashima, 1986 values for fifteen GSJ rock reference samples, "Igneous rock series", *Geostand. Newslett., 11*, 159-166, 1987.

Arndt, N. T., and U. Christensen, The role of lithospheric mantle in continental flood volcanism - thermal and geochemical constraints, *J. Geophys. Res., 97*, 10,967-10,981, 1992.

Barling, J., and S. L. Goldstein, Extreme isotopic variations in Heard Island lavas and the nature of mantle reservoirs, *Nature, 348*, 59-62, 1990.

Bassias, Y., and L. Leclaire, The Davie Ridge in the Mozambique Channel: Crystalline basement and intraplate magmatism, *N. jb. Geol. Palaont. Mh., 2*, 67-90, 1990.

Battistini, R., La structure du massif volcanique de l'Androy (Madagascar), *Bull. Geol. Soc. France, 7*, 187-191, 1959.

Besairie, H., Geological map of Madagascar (3 Sheets). *Serv. Géol. Madagascar*, 1964.

Besairie, H., Report 172, *Serv. Geol. Madagascar*, 1966.

Besairie, H., The Precambrian of Madagascar, in *The Precambrian*, edited by K. Rankama, pp. 133-142, Wiley, London, 1967.

Besairie, H., and M. Collignon, Géologie de Madagascar - I, les terrains sédimentaires, *Ann. Géol. Madagascar, 35*, 463 pp., 1972.

Boast, J., and A. E. M. Nairn, An outline of the geology of Madagascar, in *The Ocean Basins and Margins*, vol. 6, edited by A. E. M. Nairn and F. G. Stehli, pp. 649-696, Plenum, New York, 1982.

Brooks, C. K., L. M. Larsen, and T. F. D. Nielsen, Importance of iron-rich tholeiitic magmas at divergent plate margins: A reappraisal, *Geology, 19*, 269-272, 1991.

Carlson, R. W., G. W. Lugmair, and J. D. Macdougall, Columbia River volcanism: the question of mantle heterogeneity or crustal contamination, *Geochim. Cosmochim. Acta, 45*, 2483-2499, 1981.

Class, C., S. L. Goldstein, S. J. G. Galer, and D. Weis, Young formation age of a mantle plume source, *Nature, 362*, 715-721, 1993.

Coffin, M. F., and O. Eldholm, Large igneous provinces: Crustal structure, dimensions, and external consequences, *Rev. Geophys., 32*, 1-36, 1994.

Coffin, M. F., and P. D. Rabinowitz, Reconstruction of Madagascar and Africa - evidence from the Davie fracture-zone and Western Somali Basin, *J. Geophys. Res., 92*, 9385-9406, 1987.

Coffin, M. F., and P. D. Rabinowitz, Evolution of the conjugate East African-Madagascan margins and the western Somali Basin, *Geol. Soc. Am. Spec. Pap., 226*, 78 pp., 1988.

Coffin, M. F., P. D. Rabinowitz, and R. E. Houtz, Crustal structure in the Western Somali Basin, *Geophys. J. R. Astron. Soc., 86*, 331-369, 1986.

Cox, K. G., The role of mantle plumes in the development of continental drainage patterns, *Nature, 342*, 873-876, 1989.

Crawford, A. R., Narmada-Son lineament of India traced into Madagascar, *J. Geol. Soc. Ind., 19*, 144-153, 1978.

de Wit, M. J., M. Jeffrey, H. Bregh, and L. Nicolaysen, Geological Map of Sections of Gondwana Reconstructed to their Disposition ca 150 Ma, Univ. Witwatersrand, Johannesburg, 1988.

Dixey, F., The geology and geomorphology of Madagascar and a comparison with eastern Africa, *J. Geol. Soc. Lond., 116*, 255-268, 1960.

Dostal, J., C. Dupuy, C. Nicollet, and J. M. Cantagrel, Geochemistry and petrogenesis of Upper Cretaceous basaltic rocks from southern Malagasy, *Chem. Geol., 97*, 199-218, 1992.

Dupré, B., and C. J. Allègre, Pb-Sr isotope variation in Indian Ocean basalts and mixing phenomena, *Nature, 303*, 142-146, 1983.

Dyment, J., Structure et évolution de la lithospherère océanique dans l'océan Indien: Apport des anomalies Magnétiques, Thèse doct., 374 pp., Univ. Louis Pasteur, Strasbourg, France, 1991.

Ellam, R. M., Lithospheric thickness as a control on basalt geochemistry, *Geology, 20*, 153-156, 1992.

Ellam, R. M., and K. G. Cox, An interpretation of Karoo picrite basalts in terms of interaction between asthenospheric magmas and the mantle lithosphere, *Earth Planet. Sci. Lett., 105*, 330-342, 1991.

Farnetani, C. G., and M. A. Richards, Numerical investigations of the mantle plume initiation model for flood basalt events, *J. Geophys. Res., 99*, 13,813-13,833, 1994.

Fitton, J. G., G. James, P. D. Kempton, D. S. Ormerod, and W. P. Leeman, The role of lithospheric mantle in the generation of Late Cenozoic basic magmas in the western United States, *J. Petrol., Spec. Vol.*, 331-351, 1988.

Fitton, J. G., A. D. Saunders, L. M. Larsen, B. S. Hardarson, and M. J. Norry, Volcanic rocks from the southeast Greenland margin at 63°N: composition, petrogenesis and mantle sources, in *Proc. Ocean Drill Prog., Sci. Results, 152*, edited by H. C. Larsen, A. D. Saunders and S. Wise, Ocean Drilling Program, College Station, Tex., in press, 1997..

Flores, G., The SE Africa triple junction and the drift of Madagascar, *J. Petrol. Geol., 7*, 403-418, 1984.

Fourno, J. P., and J. Roussel, Imaging the Moho depth in Madagascar through the inversion of gravity data: geodynamic implications, *Terra Nova, 6*, 512-519, 1994.

Gallagher, K., and C. J. Hawkesworth, Dehydration melting and the generation of continental flood basalts, *Nature, 358*, 57-59, 1992.

Gast, P. W., Trace element fractionation and the origin of tholeiitic and alkaline magma types, *Geochim. Cosmochim. Acta., 32*, 1057-1068, 1968.

Ghiorso, M. S., and R. O. Sack, Chemical mass transfer in magmatic processes, IV. A revised and internally consistent thermodynamic model for the interpolation and extrapolation of liquid-solid equilibria in magmatic systems at elevated temperatures and pressures, *Contrib. Mineral. Petrol., 119,* 197-212, 1995.

Goslin, J. J., J. Ségoufin, R. Schlich, and R. L. Fisher, Submarine topography and shallow structure of the Madagascar Ridge, *Bull. Geol. Soc. Am., 91,* 741-753, 1980.

Green, A. G., Sea floor spreading in the Mozambique Channel, *Nature, 236,* 19-21, 1972.

Hamelin, B., B. Dupré, and C. J. Allègre, Pb-Sr-Nd isotopic data of Indian Ocean ridges: new evidence of large scale mapping of mantle heterogeneities, *Earth Planet. Sci. Lett., 76,* 288-298, 1986.

Harland, W. B., R. L. Armstrong, A. V. Cox, L. E. Craig, A. G. Smith, and D. G. Smith, *A Geologic Time Scale,* 263 pp., Cambridge University Press, Cambridge, 1989.

Harry, D. L., and W. P. Leeman, Partial melting of melt metasomatised subcontinental mantle and the magma source potential of the lower lithosphere, *J. Geophys. Res., 100,* 10,255-10,269, 1995.

Hart, S. R., A large scale isotopic mantle anomaly in the Southern Hemisphere, *Nature, 309,* 753-757, 1984.

Hart, S. R., and T. Dunn, Experimental cpx/melt partitioning of 24 trace elements, *Contrib. Mineral. Petrol., 113,* 1-8, 1993.

Hawkesworth, C. J., P. D. Kempton, N. W. Rogers, R. W. Ellam, and P. W. Van Calsteren, Continental mantle lithosphere, and shallow level enrichment processes in the Earth's mantle, *Earth Planet. Sci. Lett., 96,* 256-268, 1990.

Hergt, J. M., D. W. Peate, and C. J. Hawkesworth, The petrogenesis of Mesozoic Gondwana low-Ti flood basalts, *Earth Planet. Sci. Lett., 105,* 134-148, 1991.

Hinz, K., A hypothesis on terrestrial catastrophes. Wedges of very thick oceanward dipping layers beneath passive continental margins, *Geol. J., E22,* 3-28, 1981.

Katz, M. B., and C. Premoli, India and Madagascar in Gondwanaland based on matching Precambrian lineaments, *Nature, 279,* 312-315, 1979.

Kelemen, P. B., H. J. B. Dick, and J. E. Quick, Formation of harzburgite by pervasive melt/rock reaction in the upper mantle, *Nature, 358,* 635-641, 1992.

Kelemen, P. B., N. Shimizu, and T. Dunn, Relative depletion of niobium in some arc magmas and the continental crust: Partitioning of K, Nb, La and Ce during melt/rock reaction in the upper mantle, *Earth Planet. Sci. Lett., 120,* 111-134, 1993.

Larsen, H. C., and S. Jakobsdóttir, Distribution, crustal properties and significance of seawards-dipping sub-basement reflectors of E Greenland, in *Early Tertiary Volcanism and the Opening of the NE Atlantic, Spec. Publ. 39,* edited by A. C. Morton and L. M. Parson, pp. 95-114, The Geological Society, London, 1988.

Langmuir, C. H., R. D. Vocke Jr., G. N. Hanson, and S. R. Hart, A general mixing equation with applications to Icelandic basalts, *Earth Planet. Sci. Lett., 37,* 380-392, 1978.

Laughton, A. S., General Bathymetric Chart of the Oceans, *Canadian Hydrographic Service, sheet 5-05,* 1975.

Mahoney, J. J., J. H. Natland, W. M. White, R. Poreda, S. H. Bloomer, R. L. Fisher, and A. N. Baxter, Isotopic and geochemical provinces of the western Indian Ocean spreading centers, *J. Geophys. Res., 94,* 4033-4052, 1989.

Mahoney, J., C. Nicollet, and C. Dupuy, Madagascar basalts: tracking oceanic and continental sources, *Earth Planet. Sci. Lett., 104,* 350-363, 1991.

Mahoney, J. J., A. P. le Roex, Z. X. Peng, R. L. Fisher, and J. H. Natland, Southwestern limits of Indian ocean ridge mantle and the origin of low $^{206}Pb/^{204}Pb$ mid-ocean ridge basalt: isotope systematics of the central Southwest Indian Ridge (17-50°E), *J. Geophys. Res., 97,* 19,771-19,790, 1992.

McElhinny, M. W., B. J. J. Embleton, L. Daly, and J. P. Pozzi, Palaeomagnetic evidence for the location of Madagascar in Gondwanaland, *Geology, 4,* 455-458, 1976.

McKenzie, D., and M. J. Bickle, The volume and composition of melt generated by extension of the lithosphere, *J. Petrol., 29,* 625-679, 1988.

McKenzie, D., and R. K. O'Nions, The source regions of ocean island basalts, *J. Petrol., 36,* 133-159, 1995.

Melluso, L., V. Morra, P. Brotzu, A. Razafiniparany, V. Ratrimo, and D. Razafimahatratra, Geochemistry and Sr-isotopic composition of the Late Cretaceous flood basalt sequence of northern Madagascar: petrogenetic and geodynamic implications, *J. Afr. Earth Sci., 24,* in press, 1997.

Moorbath, S., and H. Welke, Lead isotope studies on igneous rocks from the Isle of Skye, northwestern Scotland, *Earth Planet Sci. Lett., 5,* 217-230, 1969.

Morgan, W. J., Hotspot tracks and the opening of the Atlantic and Indian Oceans, in *The Sea,* vol. 7, edited by C. Emiliani, pp. 443-487, Wiley, New York, 1981.

Mougenot, D., M. Recq, P. Virlogeux, C. Lepvrier, Seaward extension of the East-African rift, *Nature, 321,* 599-603, 1986.

Müller, R. D., J.-Y. Royer, and L. A. Lawver, Revised plate motions relative to the hotspots from combined Atlantic and Indian Ocean hotspot tracks, *Geology, 21,* 275-278, 1993.

Nichols, G. J., and M. C. Daly, Sedimentation in an intracratonic extensional basin: the Karoo of the central Morondava Basin, Madagascar, *Geol. Mag., 126,* 339-354, 1989.

Obradovich, J. D., A Cretaceous time scale, in *Evolution of the Western Interior Basin, Spec. Pap. 39,* edited by W. G. E. Caldwell, pp. 379-396, Geological Association of Canada, Waterloo, ON, 1993.

Paquette, J. L., A. Nédélec, B. Moine, and M. Makotondrazafy, U-Pb, single zircon Pb-evaporation, and Sm-Nd isotopic study of a granulite domain in SE Madagascar, *J. Geol., 102,* 523-538, 1994.

Peng, Z. X., and J. J. Mahoney, Drillhole lavas from the northwestern Deccan Traps, and the evolution of Réunion hotspot mantle, *Earth Planet Sci. Lett., 134,* 169-185, 1995.

Potts, P. J., O. Williams-Thorpe, and J. S. Watson, Determination of the rare-earth element abundances in 29 international rock standards by instrumental neutron activation analysis: Critical appraisal of calibration errors, *Chem. Geol., 34,* 331-352, 1981.

Price, R. C., A. K. Kennedy, M. Riggs-Sneeringer, and F. A. Frey, Geochemistry of basalts from the Indian Ocean triple junction: Implication for the generation and evolution of Indian Ocean ridge basalts, *Earth Planet. Sci. Lett., 78,* 279-296, 1986.

Reid, A. M., and A. P. le Roex, Kaersutite-bearing xenoliths and megacrysts in volcanic rocks from Funk Seamount in the southwest Indian Ocean, *Mineral. Mag.*, 52, 359-370, 1988.

Saunders, A. D., M. Storey, R. W. Kent, and M. J. Norry, Consequences of plume-lithosphere interactions, in *Magmatism and the Causes of Continental Break-up, Spec. Publ. 68*, edited by B. C. Storey, T. Alabaster, and R. J. Pankhurst, pp. 41-60, The Geological Society, London, 1992.

Scrutton, R. A., W. B. Heptonstall, and J. H. Peacock, Constraints on the motion of Madagascar with respect to Africa, *Mar. Geol.*, 43, 1-20, 1981.

Shaw, D. M., Trace element fractionation during anatexis, *Geochim. Cosmochim. Acta.*, 34, 237-243, 1970.

Shimizu, N., and Kushiro, I., The partitioning of rare earth elements between garnet and liquid at high pressures: preliminary experiments, *Geophys. Res. Lett.*, 2, 414-416, 1975.

Sinha, M. C., K. E. Louden, and B. Parsons, The crustal structure of the Madagascar Ridge, *Geophys. J. R. Astron. Soc.*, 66, 351-377, 1981.

Smith, A. G., and A. Hallam, The fit of the southern continents, *Nature*, 225, 139-144, 1970.

Storetvedt, K. M., J. G. Mitchell, M. C. Abranches, S. Maaloe, and G. Robin, The coast-parallel dolerite dykes of east Madagascar; age of intrusion, remagnetization and tectonic aspects, *J. Afr. Earth Sci.*, 15, 237-249, 1992.

Storey, M., J. J. Mahoney, A. D. Saunders, R. A. Duncan, S. P. Kelley, and M. F. Coffin, Timing of hot-spot related volcanism and the breakup of Madagascar and India, *Science*, 267, 852-855, 1995.

Sun, S.-s. and W. F. McDonough, Chemical and isotopic systematics of oceanic basalts: implications for mantle composition and processes, in *Magmatism in the Ocean Basins, Spec. Publ. 42*, edited by A. D. Saunders and M. J. Norry, pp. 313-345, The Geological Society, London, 1989.

Taylor, S. R., and S. M. McLennan, *The Continental Crust: Its Composition and Evolution*, 312 pp., Blackwell, Oxford, 1985.

Thirlwall, M. F., B. G. J. Upton, and C. Jenkins, Interaction between continental lithosphere and the Iceland plume; Sr-Nd-Pb isotope geochemistry of Tertiary basalts, NE Greenland, *J. Petrol.*, 35, 839-879, 1994.

Todt, W., R. A. Cliff, A. Hanser, and A. W. Hofmann, Evaluation of a ^{202}Pb-^{205}Pb double spike for high-precision lead isotope analysis, in *Earth Processes: Reading the Isotopic Code, Geophys. Monogr. Ser.*, vol. 95, edited by A. Basu and S. R. Hart, pp. 429-437, AGU, Washington, D.C., 1996.

Tolan, T. L., and S. P. Reidel, Structure map of a portion of the Columbia river flood-basalt province, in *Volcanism and Tectonism in the Columbia River Flood-Basalt Province, Spec. Pap. 239*, edited by S. P. Riedel and P. R. Hooper, enclosed map, Geological Society of America, Boulder, Colo., 1989.

Walter, M. J., and D. C. Presnall, Melting behaviour of simplified lherzolite in the system CaO-MgO-Al_2O_3-SiO_2-Na_2O from 7 to 35 kbar, *J. Petrol.*, 35, 329-359, 1994.

Watson, S., and D. P. McKenzie, Melt generation by plumes: a study of Hawaiian volcanism, *J. Petrol.*, 32, 1991.

White, R. S., and D. P. McKenzie, Magmatism at rift zones: the generation of volcanic continental margins and flood basalts, *J. Geophys. Res.*, 94, 7685-7730, 1989.

Windley, B. F., A. Razafiniparany, T. Razakamanana, and D. Ackermand, Tectonic framework of the Precambrian of Madagascar and its Gondwana connections: a review and reappraisal, *Geol. Rundsch.*, 83, 642-659, 1994.

John J. Mahoney, School of Ocean and Earth Science and Technology, University of Hawaii, 2525 Correa Road, Honolulu, Hawaii, 96822, USA.

Andrew D. Saunders, Department of Geology, University of Leicester, University Road, Leicester, LE1 7RH, United Kingdom.

Michael Storey, Danish Lithosphere Centre, Øster Voldgade 10, DK-1350 Copenhagen K, Denmark (e-mail: ms@dlc.ku.dk).

The Caribbean-Colombian Cretaceous Igneous Province: The Internal Anatomy of an Oceanic Plateau

Andrew C. Kerr[1], John Tarney[1], Giselle F. Marriner[2], Alvaro Nivia[3], Andrew D. Saunders[4]

The Late Cretaceous Caribbean–Colombian igneous province is one of the world's best-exposed examples of a plume-derived oceanic plateau. The buoyancy of the plateau (resulting from residual heat and thick crust) kept it from being totally subducted as it moved eastward with the Farallon Plate from its site of generation in the eastern Pacific and encountered a destructive plate margin. In effect, the plateau makes up much of the Caribbean Plate; it is well exposed around its margins, but more so in accreted terranes in western Colombia (including the well-known Gorgona komatiites and Bolívar mafic/ultramafic cumulates). Compositionally, the lavas of the plateau form three groups: (a) basalts, picrites, and komatiites with light-rare-earth-element (LREE)-depleted chondrite-normalised patterns; (b) basalts with LREE-enriched patterns; and (c) basalts with essentially flat REE patterns (the most dominant type) similar to many of the basalts from the Ontong Java Plateau. These three types demonstrate the heterogeneous nature of the mantle plume source region. The picrites and the komatiites seem to lie nearer the base of the plateau than the more homogeneous basalts; thus, the more MgO-rich melts may have been erupted before large magma chambers had a chance to develop. A reconstructed crustal cross section through the plateau consists of dunitic and pyroxenitic cumulates near the base which are overlain by layered olivine-rich gabbros and more isotropic gabbros. The lowermost eruptive sequence comprises compositionally heterogeneous picrites/komatiites overlain by more homogeneous pillow basalts. Spectacular hornblende-plagioclase veins cut the Bolívar assemblage and these may represent local partial melts of the plateau's base as it was thrusted onto the continent. Subduction-related batholiths and extrusive rocks found around the margin of the province are of two distinct ages; one suite represents pre-plateau collision-related volcanism whereas the other suite, slightly younger than the plateau, may be associated with obduction.

[1]Department of Geology, University of Leicester, University Road, Leicester LE1 7RH, UK

[2]Department of Geology, Royal Holloway University of London, Egham, Surrey, TW20 0EX, UK

[3]Ingeominas - Regional Pacifico, AA9724, Cali, Colombia

[4]Department of Geology, University of Leicester, University Road, Leicester LE1 7RH, UK

Large Igneous Provinces: Continental, Oceanic, and Planetary Flood Volcanism
Geophysical Monograph 100
Copyright 1997 by the American Geophysical Union

1. INTRODUCTION

Oceanic large igneous provinces such as the Ontong Java Plateau, the Nauru Basin, Manihiki Plateau, and the Kerguelen Plateau, with 10-40 km thick crustal sections [e.g., *Coffin and Eldholm*, 1994], are vast outpourings of basaltic magma (often with erupted volumes of basalt $>4 \times 10^6$ km^3). Despite their obvious volumetric importance we still know relatively little about their structure and composition. Several of these plateaus and basins have been sampled by the Deep Sea Drilling Project/Ocean Drilling Program (DSDP/ODP); however, the greatest depth to which any of them has been drilled is 700 m into the Nauru Basin by DSDP legs 61 and 89 [e.g., *Saunders*, 1985]. Because these over-thickened areas of oceanic crust tend to be more buoyant than normal ocean crust (due to both thermal and density differences) [*Burke et al.*, 1978; *Nur and Ben Avraham*, 1982; *Cloos*, 1993], they are not so easily subducted. These oceanic plateaus have, therefore, the potential to be obducted onto continental margins [*Ben Avraham et al.*, 1981], thus increasing both the possibility of their preservation and accessibility for detailed study [e.g., *Saunders et al.*, 1996].

One such area where it is believed that an oceanic plateau has been obducted onto the margin of a continent is northwestern South America [*Marriner and Millward*, 1984; *Millward et al.*, 1984; *Aspden et al.*, 1987; *Kerr et al.*, 1996a]. This review will show that the accreted volcanic successions in Colombia and Ecuador, along with the thickened oceanic crust of the Caribbean Plate, and other smaller obducted basaltic fragments around the margins of the Caribbean (Figure 1) are the remnants of a Late Cretaceous oceanic plateau. The processes of accretion, imbrication, and tectonic uplift of the Caribbean-Colombian oceanic plateau have locally exposed its lower reaches, providing information on the internal structure and composition of one of these vast oceanic plateaus.

2. STRUCTURE AND TECTONICS OF THE CARIBBEAN REGION

The preserved water-covered area of the Caribbean–Colombian Cretaceous Igneous Province (CCCIP) is about 6×10^5 km²; however, because a significant portion of the plateau appears to have accreted onto the western continental margin of Colombia and Ecuador (and some may have subcreted), the oceanic plateau may originally have been more than twice this size [*Burke*, 1988].

The seismic refraction studies initiated by *Edgar et al.* [1971], which were integrated with gravity data by *Case et al.* [1990] and summarised by *Donnelly* [1994], reveal the anomalously thickened nature of the oceanic crust in the submerged portion of the Caribbean Plate. The crust of the Caribbean varies from ~8 to >20 km thick [see Figure 3.5 of *Donnelly*, 1989], well in excess of the 6-7 km of oceanic crust produced at a normal mid-ocean ridge. This thickened oceanic crust was drilled in several places by DSDP Leg 15. From the results of this drilling, *Donnelly* [1973] and *Donnelly et al.* [1973] proposed that a large flood basalt event had occurred in the Caribbean region in

the Late Cretaceous. This 'flood basalt event' is now recognised as the remnant of a major oceanic plateau [*Burke et al.*, 1978; *Duncan and Hargraves*, 1984; *Donnelly et al.*, 1990; *Hill*, 1993; *Kerr et al.*, 1996a], which *Duncan and Hargraves* [1984] and *Klaver* [1987] proposed had formed as a consequence of uprise of a deep mantle plume.

Throughout the history of geological research in the Caribbean region there has been considerable controversy as to whether the Caribbean Plate formed in situ or whether it was transported to its present position from a westerly (Pacific) direction. The former view is an older one [e.g., *Schubert*, 1935; *Meyerhoff and Meyerhoff*, 1972] and has received relatively little attention in recent years, and most modern authors accept the premise that the components of the Caribbean Plate have had a highly mobile history [*Burke et al.*, 1984; *Pindell*, 1990]. Nevertheless, within this model, opinions differ as to the timing and extent of movements [e.g., *Burke*, 1988; *Donnelly*, 1989; *Pindell and Barrett*, 1990]. However, the origin of the Caribbean Plate in the eastern Pacific as an oceanic plateau appears to be generally accepted.

Using a fixed hotspot reference frame, *Duncan and Hargraves* [1984] and *Hill* [1993] suggested that the CCCIP was produced by melting during the initial 'plume head' phase of the Galápagos hotspot. Eastward movement of the Farallon Plate in the Late Cretaceous–Early Tertiary brought the northern part of the plateau into the continental gap which had opened up between North and South America since the Jurassic. The eastward moving plateau appears to have been too buoyant (due to remnant heat and crustal thickness) to be subducted [*Burke et al.*, 1978; *Hill*, 1993]. This 'clogging' of the subduction zone led to a 'flip' in the direction of subduction from east to west, and the Atlantic Plate began to be consumed by subduction as opposed to the Farallon/Caribbean Plate [*Lebrón and Perfit*, 1994]. The northern part of the Caribbean Plate is bounded by a series of approximately east-west trending strike-slip faults, whereas its southern margin is a broad complex zone of convergence and right lateral strike-slip faulting [*Pindell et al.*, 1988; *Ladd et al.*, 1990]. This tectonic activity has exhumed and exposed portions of the oceanic plateau crust at the margins of the Caribbean Plate.

In contrast, the southern part of the Caribbean–Colombian plateau began to interact with the northwestern continental margin of South America. In this region, docking of the plateau with the continent also resulted in the jamming of the subduction zone, but rather than subduction flip this led to progressive, westward back-stepping of the subduction zone, leading to the formation of accreted oceanic plateau terranes along the northwestern

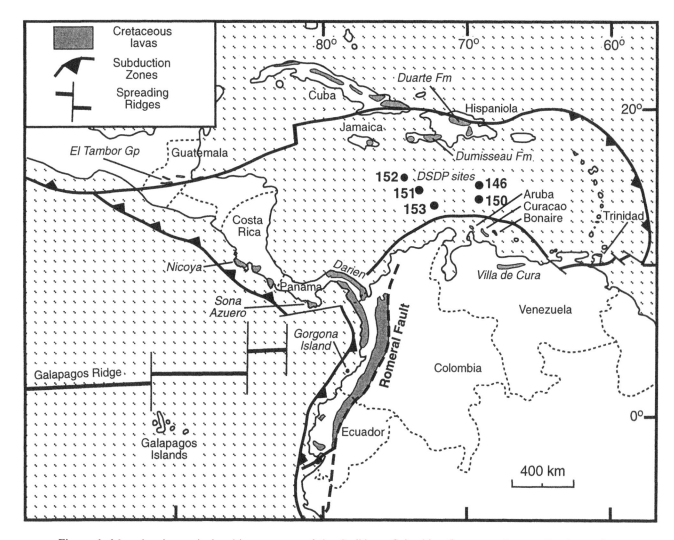

Figure 1. Map showing main basaltic exposures of the Caribbean-Colombian Cretaceous Igneous Province, after *Donnelly et al.* [1990] and *Iturralde-Vinent* [1994]. Numbered dots: locations of DSDP sites from which basalts were recovered. Galápagos may represent the remnant hotspot marking the site of earlier plateau formation, but this is not certain.

margin of South America [e.g., *Kerr et al.*, 1996a].

Recently, it has been reported that long-wavelength magnetic anomalies over the Venezuelan and Colombian basins in the Caribbean display NE–SW and E–W linear patterns [*Hall, 1995*]. These anomalies have been interpreted as being the result of an Early Cretaceous phase of seafloor spreading at the Farallon-Pacific-Phoenix triple junction, at which the CCCIP may have formed at ~90 Ma [*Hall, 1995*].

The CCCIP is, for the most part, composed of pillow lavas and massive flows or shallow sills of basalt/diabase with occasional occurrences of more picritic lavas. The age of CCCIP has been the subject of some debate in the literature. Pre-1990 knowledge of the age of the province

was based on fossils found in associated sediments and on K-Ar dates (see Figures 2 and 3 of *Donnelly et al.* [1990]). However, it is likely that sea floor alteration and zeolite- to greenschist-facies metamorphism have resulted in Ar loss, and it is therefore doubtful if any of the K-Ar age determinations can be trusted. Although Jurassic fossil ages have been cited for parts of the CCCIP (Costa Rica, Venezuela and Puerto Rico), the tectonic relationship between the fossil-bearing sediments and the basalts is by no means clear [*Donnelly et al.*, 1990; *Donnelly*, 1994]. Sediments overlying the submerged basalts in the Caribbean Sea, drilled by DSDP Leg 15, place a biostratigraphic age limit of 88–83 Ma on the cessation of igneous activity in the province. (All ages are based on the time-scale of *Harland et al.* [1990]).

Recent ^{40}Ar-^{39}Ar dating of CCCIP basalts from, Haiti, Curaçao, Isla Gorgona in Colombia, and the Nicoya Peninsula in Costa Rica [*Sinton and Duncan*, 1992; *Sinton et al.*, 1993] has shown that the ages of all these basalts group around 87–90 Ma. These results suggest that the bulk of volcanic activity associated with the CCCIP occurred over a relatively short period. Nevertheless, as will be shown below, there still is some evidence to suggest that in several areas of the Caribbean the volcanic activity may be older, and there is an obvious need for more high-precision radiometric dating in the province.

3. FIELD RELATIONS AND GEOCHEMISTRY OF THE CCCIP

3.1. Curaçao and Aruba

The islands of Curaçao and Aruba, located 70 km north of the Venezuelan coast (Figure 1), contain some of the best-preserved sections of the CCCIP. The Curaçao lava succession is more than 5 km thick [*Klaver*, 1987], with abundant pillow lavas. The lower half of the exposed succession is mostly composed of picrites and olivine basalts. The picrites can contain as much as 31 wt% MgO [*Beets et al.*, 1982]. *Kerr et al.* [1996c] have shown that these high-MgO picrites are olivine cumulates and have calculated, using the maximum Fo content of the olivines, that the primary magmas of the Curaçao picrites contained 16–18 wt% MgO. The upper exposed half of the Curaçao lava succession consists of more-evolved, plagioclase- and clinopyroxene-phyric pillow basalts, with dolerite sills and occasional reworked hyaloclastites.

There is only one known intercalation of pelagic sediments within the Curaçao lava succession [*Klaver*, 1987]. This observation implies that the Curaçao lavas were erupted over a relatively short time span. Such rapid emplacement is consistent with a mantle plume origin for the Curaçao lava succession. The pelagic sediments within the top half of the lava succession consist of siliceous shales and limestones [*Klaver*, 1987]. *Wieldmann* [1978] reported the occurrence of fossil ammonites of Mid-Albian (100–105 Ma) age within these pelagic sediments. This age is significantly older than the 88–90 Ma found using ^{40}Ar-^{39}Ar methods [*C. Sinton*, pers. comm. 1994]. However, reference to *Wieldmann* [1978] reveals that the preservation state of the Curaçao ammonites is "rather poor, due to low grade metamorphism and deformation" and "very few (ammonites) are complete, but even then distorted. Generally only small fragments are preserved." Given this information, it is conceivable that, due to misidentification, the ammonite ages are incorrect and the

age of the Curaçao lava succession is, as the ^{40}Ar-^{39}Ar data suggest, 88–90 Ma.

Kerr et al. [1996c] have shown that the picrites of the Curaçao succession are related to the basalts by simple fractional crystallisation involving olivine, clinopyroxene, and plagioclase (Figure 2) and that for the most part the Curaçao lavas possess nearly constant Nd isotopic ratios (Figure 3b), and incompatible trace element ratios (Figure 3a) with chondritic rare-earth element (REE) patterns. These features suggest that either the inferred plume source region of the Curaçao lavas was homogeneous or that melts from a heterogeneous source became homogenised during melting or in magma chambers during ascent to the surface. The elevated ^{87}Sr/^{86}Sr of the Curaçao basalts can be explained by contamination of picritic magmas with highly altered oceanic crust along with concomitant fractional crystallisation [*Kerr et al.* 1996c].

The Aruba lava succession has been less intensely studied than the Curaçao lavas. The basaltic sequence is about 3 km thick [*Beets et al.*,1984] and is intruded and metamorphosed by the 85–90 Ma [*Priem et al.*, 1986] tonalitic/gabbroic Aruba batholith. The lava succession comprises pillow basalts and dolerite sills, with intercalations of volcaniclastic sediments (containing Turonian [90.5–88.5 Ma] ammonites) and some well sorted conglomerates [*Beets et al.*, 1984]. The chemistry of the Aruba lavas is similar to that of the Curaçao basalts, except for the presence of more evolved ferrobasalts. Plagioclase and clinopyroxene are the main fractionating phases. The MgO content of the lavas varies from 10 wt% for the basalts (which occur nearer the base of the exposed section) to 4 wt% for the ferrobasalts (located nearer the top of the lava pile [*Beets et al.*, 1984]). As they have chondritic REE patterns and similar incompatible element trace element ratios (Figure 4a) to the Curaçao lavas, this suggests that they were derived from a similar, or the same, mantle source region [*Beets et al.*, 1984; *Klaver*, 1987].

Beets et al. [1984] initially concluded that the Curaçao and Aruba lava successions were part of an island arc succession. However, as *Klaver* [1987] and *Kerr et al.* [1996c] have pointed out, the presence of (non-boninitic) high-MgO lavas, the absence of a subduction trace element signature, and the chemical similarity to Pacific oceanic plateau lavas (Figure 3a) provide strong evidence for a mantle plume origin for these lavas.

3.2. Venezuela

Although the coastal borderlands of Venezuela preserve some basaltic and ultramafic rocks [*Oxburgh*, 1966; *Beets et al.*, 1984], which are possibly part of the CCCIP

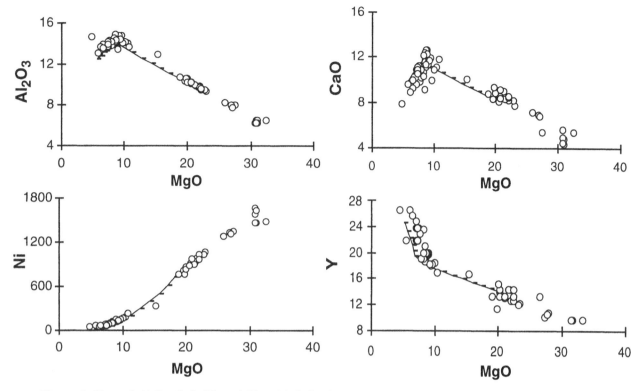

Figure 2. Plots of Al₂O₃, CaO, Ni, and Y vs MgO for Curaçao basalts and picrites. Also shown are calculated fractionation trends, with ticks at 4 wt% crystallisation intervals. After *Kerr et al.* [1996c].

[*Donnelly et al.*, 1990], many of the rocks have been extensively tectonised and metamorphosed to high-pressure/low-temperature equivalents [*Beets et al.*, 1984]. As a result of this tectonism and metamorphism, the age of these basic rock associations from the Venezuelan coastal region has not been fully resolved. As reviewed by *Beets et al.* [1984], K-Ar ages so far obtained range from 107 to 65 Ma. It is generally accepted that these basaltic and ultramafic volcanic rocks are an allochthonous terrane obducted from the north in the Late Cretaceous or Early Tertiary [*Beets et al.*, 1984]. Palaeomagnetic studies have shown that the whole belt was rotated 90° clockwise during obduction [*Stearns et al.*, 1982].

The Villa de Cura Group is a 4- to 5-km-thick, 250-km-long belt of metamorphic volcanic and volcaniclastic rocks, 50 km south of the Venezuelan coast. This approximately E-W trending unit is composed of both mafic and more siliceous rocks. *Donnelly et al.* [1990] subdivided these volcanic rocks into two chemical groups: a more evolved series of subduction-related lavas and tuffs, and a series of basalts and ultramafic rocks with transitional (T-)MORB characteristics. However, extensive tectonisation means that the interrelationships between these two groups are unclear. The mafic, T-MORB lavas are found chiefly in two formations within the Villa de

Cura Group: the Tiara and the El Carmen formations.

The basalts of the Tiara Formation are underlain by sediments containing Albian (97–112 Ma) radiolaria [*Beck et al.*, 1984; *Donnelly et al.*, 1990], thus placing a maximum age of 112 Ma on the overlying basalts. The lavas of the Tiara and the El Carmen Formations mostly possess relatively flat chondrite-normalised [*Sun and McDonough*, 1989] REE patterns (with La_N/Sm_N < 1 and Tb_N/Yb_N >1, although one of the Tiara lavas has La_N/Yb_N = 5 [*Beets et al.*, 1984]). Some of the lavas from the 1200-m-thick El Carmen Formation contain up to 12 wt% MgO and have been classified as 'picrites' by *Beets et al.* [1984], though they are more correctly classified as high-MgO picritic basalts. *Donnelly* [1989] suggested that the El Carmen formation may be correlated with the Curaçao picrites. Furthermore, *Donnelly et al.* [1990] noted that basalts of the Tiara and El Carmen formations have a very similar chemistry to the picrites and basalts of Curaçao and Aruba.

Several small ultramafic units also crop out in close proximity to the Villa de Cura Group, and it is highly likely that they are genetically related to the spatially associated basalts [*Donnelly et al.*, 1990]. One of the most noteworthy of these exposures is the Tinaquillo peridotite [e.g., *MacKenzie*, 1960]. A small exposure of dunite,

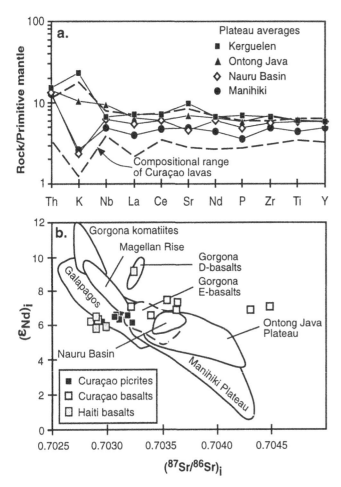

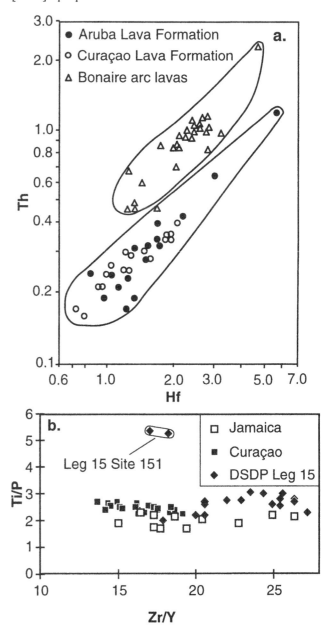

Figure 3. (a). Comparison of trace element compositions of Curaçao basalts and picrites with average compositions of other ocean plateaus: Kerguelen, Ontong Java, Nauru Basin, and Manihiki Plateau (normalised to primitive mantle values of *Sun and McDonough* [1989]), (b) Plot of $(\varepsilon_{Nd})_i$ vs $(^{87}Sr/^{86}Sr)_i$ for lavas from the CCCIP along with lavas from Cretaceous Pacific oceanic plateaus. Data sources referenced by *Kerr et al.* [1996c].

gabbro, dolerite, and peridotite on the Paraguaná peninsula has been described by *Martín-Bellizia and de Arozena* [1972].

3.3. Trinidad

The Sans Souci formation in northeastern Trinidad is represented by a 1000-m-thick sequence of basaltic pyroclastic rocks, basalts, gabbros and minor terrigenous sedimentary rocks [*Wadge and MacDonald*, 1985]. Compositionally, these lavas resemble normal (N-) MORB tholeiites, and (with one exception) the analysed basalts have low levels (<10 × primitive mantle values) of incompatible trace elements [*Wadge and MacDonald*,

1985]. A K-Ar whole rock age determination on a Sans Souci basalt yielded an age of 87±4 Ma [*Wadge and MacDonald*, 1985]. Although this places the formation within the age range of other Caribbean oceanic basalts, the altered nature of the Sans Souci basalts requires that the age be treated with caution. *Wadge and MacDonald* [1985] proposed that the Sans Souci Formation is

Figure 4. (a) Plot of Th vs Hf comparing tholeiitic plateau lavas from Aruba and Curaçao with Cretaceous arc lavas from Bonaire; after *Beets et al.* [1984]. (b) Plot of Ti/P vs Zr/Y for basalts from Jamaica [*Wadge et al.* 1982], Curaçao [*Kerr et al.*, 1996b] and Caribbean DSDP Leg 15 (G. F. Marriner, unpublished data).

separated from the rest of Trinidad by a series of strike-slip faults, and they further suggested that the basalts are part of the CCCIP that was juxtaposed along the northern margin of South America in Late Cretaceous times.

3.4. Jamaica

The 40-km² Bath-Dunrobin Formation, eastern Jamaica, consists of an inlier of highly deformed basalts, dolerites, and isotropic gabbros [*Wadge et al.*, 1982]. Thin sedimentary intercalations in this ~2.5-km-thick ophiolitic assemblage contain mostly Campanian (83–74 Ma) fossils, although older, possibly reworked, fossils have been found within the succession [*Wadge et al.*, 1982]. The chemistry of the Bath-Dunrobin basalts as reported by *Wadge et al.* [1982] is MORB-like and very similar to the Aruba and Curaçao basalts, as well as to some of the basalts from DSDP Leg 15 (Figure 4b). Despite these similarities *Wadge et al.* [1982] suggested oceanic crust comprising the Bath-Dunrobin ophiolite had been obducted from the north after formation in the Yucatan Basin. However, as *Donnelly* [1994] pointed out, there is no strong evidence that the Bath-Dunrobin Formation was not derived from the south, and thus it is probably part of the CCCIP.

3.5. DSDP Leg 15 Drill Sites

Drilling during DSDP Leg 15 encountered basalts or dolerites at five sites (Figure 1) in the central Caribbean [*Donnelly et al.*, 1973]. As noted above, the biostratigraphic age of the basalts and dolerites at sites 146, 150, 151, and 153 is late Turonian (88 Ma); however, at the most westerly site (152), an early Campanian (83 Ma) age has been assigned [*Donnelly et al.*, 1973]. This result suggests that, in the west of the province, volcanism continued for up to 5 m.y. after activity had ceased in the east. The basalts are all plagioclase-phyric tholeiites [*Bence et al.*, 1975]. *Donnelly et al.* [1973] and *Bence et al.* [1975] reported major and rare earth element data for the Leg 15 basalts, and we have recently analysed a suite of samples from the five DSDP Leg 15 sites for a range of major and trace elements. The new chemical data are presented in Figure 5.

Both *Donnelly et al.* [1973] and *Bence et al.* [1975] subdivided the lavas and sills into two chemical groups: those with slightly depleted light REE patterns (La_N/Sm_N <1) and TiO_2 < 1.8 wt% (Holes 146, 150, 152, and 153); and those with La_N/Sm_N >2 and TiO_2 >2.0 wt% (Hole 151) (Figure 5b). These two groups can also be observed in the new data set (Figure 5). The new trace element data also reveal that the basalts from Site 151 possess higher Nb/Zr

and Ti/Y ratios than the rest of Leg 15 basalts (Figure 5e), consistent with derivation from a mantle source region with garnet stable in the residue.

Bence et al. [1975] suggested that the basalts from Holes 146, 150, 152, and 153 could be related to each other by fractional crystallisation of olivine and plagioclase, reflected in decreasing Ca and Ni with lower $MgO/[MgO + Fe_2O_3(t)]$. However, lower incompatible element contents and lower Nb/Zr ratios reveal that the basalts from Site 152 cannot be related to those from sites 146, 150, and 153 by fractional crystallisation alone. The Site 152 basalts may therefore be the result of either more extensive melting of the same source region which produced the 146, 150 and 153 basalts or melting of a separate mantle source that was slightly more depleted in incompatible elements.

Bence et al. [1975] proposed that the basalts from DSDP Leg 15 were derived from two different parental liquids. However, the new trace element data (Figure 5e) suggest that three different parental liquids may have been involved. As will be shown below, such variation has also been found at other sites throughout the province.

3.6. Hispaniola

Hispaniola contains two main belts of Cretaceous basaltic rocks: the Dumisseau Formation in the southwest and the more central Duarte and Siete Cabezas Formations. Smaller, poorly studied, ophiolitic bodies are found along the north coast of the island [*Donnelly et al.*, 1990]. The Dumisseau Formation in southwestern Haiti consists of a 1500-m-thick sequence of interbedded pillowed and massive basalts, dolerites, pelagic limestones, turbidites and siliceous siltstones. Gabbroic intrusions and dolerite dykes occasionally cut the sequence. *Maurrasse et al.* [1979] reported palaeontological evidence that the sediments associated with the upper basalts of the sequence are early Campanian (74 Ma) to late Santonian (83 Ma) in age, and fossils from the lower stratigraphic levels are mostly of Coniacian and Turonian (86–90 Ma) age. These latter ages are more in accord with recent ^{40}Ar-^{39}Ar dating of basalts of the Dumisseau Formation, which span the age range 87–90 Ma [*Sinton and Duncan*, 1992].

Sen et al. [1988], in a comprehensive assessment of the chemistry of the Dumisseau basalts, showed that the major element compositions of the lavas are very similar to those of many of the basalts drilled during DSDP Leg 15. Most of the samples are basaltic (<10 wt% MgO); however, a picrite (22.7 wt% MgO) was also discovered near the base of the succession. Because the olivines in this picrite are not highly magnesian (Fo_{87}), it is likely the rock is an

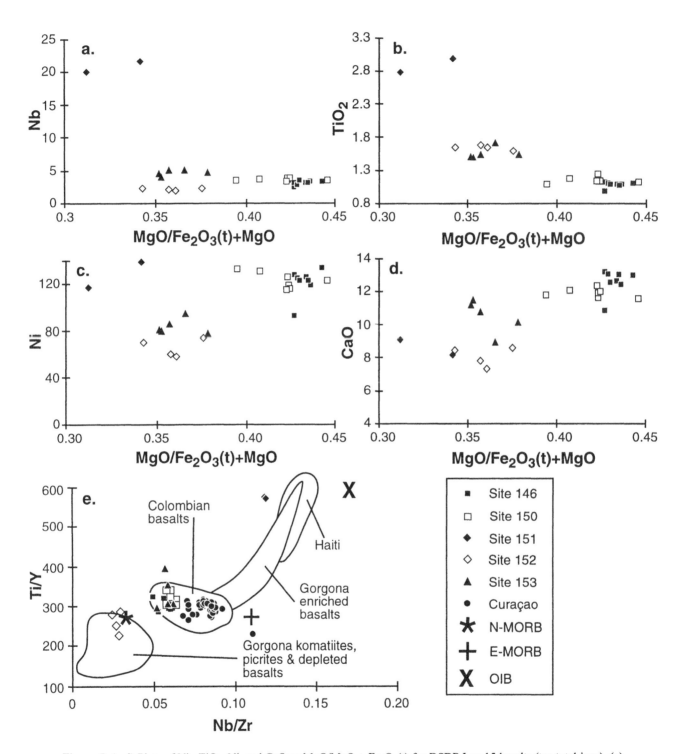

Figure 5. (a-d) Plots of Nb, TiO$_2$, Ni, and CaO vs MgO/MgO + Fe$_2$O$_3$(t) for DSDP Leg 15 basalts (t = total iron), (e) Ti/Y vs Nb/Zr showing data points for Leg 15 basalts (G. F. Marriner, unpubl. data), Curaçao lavas [*Kerr et al.*, 1996c], along with data fields for lavas from Gorgona Island [*Kerr et al.*, 1996b], western Colombia [*Nivia*, 1987] and the Dumisseau Formation, Haiti [*Sen et al.*, 1988]. Plotted compositions for average normal (N-MORB) and enriched (E-MORB) mid-ocean ridge basalt and for average ocean island basalt (OIB) after *Sun and McDonough* [1989].

olivine cumulate. Unlike the volcanic arc successions (the Los Ranchos Formation) in northern Haiti [*Donnelly et al.* 1990], the trace element characteristics of the Dumisseau basalts indicate that they are of ocean-floor affinity.

Most of the basalts are similar to those found at Site 151 in that they possess >2.0 wt% TiO_2 and La_N/Sm_N >1 (Figure 6); however, *Sen et al.* [1988] discovered several basalts at the base of the succession that have TiO_2 < 1.8 wt% and La_N/Sm_N < 1, values that are similar to those in basalts from other Leg 15 sites. This evidence supports the contention that the Dumisseau basalts are an obducted section of the CCCIP [*Sen et al.*, 1988]. Consideration of radiogenic isotope and incompatible trace element ratios supports a heterogeneous source region for the Dumisseau basalts containing at least two components [*Sen et al.*, 1988], one with enriched and one with depleted incompatible trace element contents (relative to Bulk Earth). *Sen et al.* [1988] also noted that the Sr, Nd, and Pb isotopic compositions of the Dumisseau basalts overlap with those of the Galápagos hotspot (Figure 3b) and they used this evidence to support a Pacific plume-related origin for the Caribbean Plate.

The Duarte Formation in central Hispaniola is composed predominantly of schists and amphibolites [*Donnelly et al.*, 1990]. *Donnelly and Rogers* [1978] suggested that the unmetamorphosed basalts and cherts (containing Santonian (83–86.5 Ma) radiolarians; *Donnelly et al.* [1990]) of the Siete Cabezas Formation represent an oceanic basalt sequence. The presence of both high- and low-TiO_2 basalts in the Siete Cabezas and Duarte Formations [*Donnelly and Rogers*, 1980; *Donnelly et al.*, 1990] that are similar to basalts from both DSDP Leg 15 and southern Haiti, led *Donnelly et al.* [1990] to propose that these rocks are a 'metamorphosed and tectonised allochthon of Caribbean Cretaceous basaltic rocks.'

3.7. Central America

Several igneous assemblages are exposed along the Pacific coast of Costa Rica on the Nicoya and Santa Elena peninsulas (Figure 1) and are collectively known as the Nicoya Complex. The Nicoya Complex has been studied intensively by numerous authors (see *Donnelly et al.* [1990] for a review). It consists of a lower section composed mostly of massive basalts, small gabbroic and plagiogranite plutons, with some pillow basalts and very little sediment, and an upper section consisting of massive and pillowed basalt and picritic basalts, intrusions, and minor sediments [*Donnelly et al.*, 1990]. The whole complex is strongly tectonised and *Bourgois et al.* [1984] suggested that it was emplaced in a series of nappes.

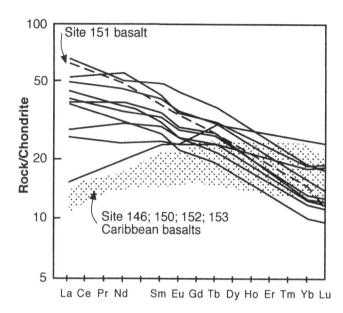

Figure 6. Chondrite-normalised rare earth element patterns for Dumisseau lavas (solid lines) [after *Sen et al.*, 1988] in comparison with basalts from DSDP Leg 15, Sites 146, 150, 152, and 153 (shaded area) and basalt from DSDP Site 151 (dashed line). DSDP data from G. F. Marriner (unpublished).

Most previous workers [e.g., *Bourgois et al.*, 1984; *Wildberg*, 1984] have subscribed to the view that the lower part of the Nicoya Complex is stratigraphically overlain by a sequence of Jurassic cherts. However, this view has recently been challenged by *Donnelly* [1994], who has pointed out that the lower part of the Nicoya Complex consists mostly of intrusive rocks and nowhere lies directly below a depositional contact with the chert, thus suggesting that the Nicoya complex may intrude the Jurassic cherts. Recent ^{40}Ar-^{39}Ar dates on several Nicoya basalts [*Sinton et al.*, 1993] have yielded early Coniacian (86.5–88.5 Ma) ages; these new age data support the view of *Donnelly* [1994] that the Nicoya Complex is significantly younger than Jurassic.

Wildberg [1984] and *Berrangé and Thorpe* [1988] presented comprehensive geochemical data for the Nicoya and Santa Elena complexes. The basalts and picritic basalts were classified as low-potassium quartz and olivine tholeiites (5-8 wt% MgO), having mostly flat REE patterns and low levels of incompatible trace elements (Figure 7). In many respects their chemistry is similar to that of the basalts from Curaçao and from DSDP Leg 15 Sites. *Wildberg* [1984] suggested that the Nicoya Complex was composed mostly of non–subduction-related oceanic crust, and he linked some of the volcanism in the upper Nicoya Complex to that of the Caribbean oceanic plateau. *Wildberg* [1984] proposed that the remaining volcanic rocks in the upper Nicoya Complex were generated in a primitive island arc, although this interpretation has been questioned [*Donnelly et al.*, 1990; *Donnelly*, 1994] .

Berrangé and Thorpe [1988] suggested that the Nicoya complex was an obducted segment of oceanic crust that had formed in a back-arc basin. However, a multi-element (primitive mantle normalised) diagram, using Berrangé and Thorpe's data (Figure 7), shows that the Nicoya basalts do not have subduction-related characteristics (i.e., they have primitive mantle normalised ratios K_N/Nb_N and $La_N/Nb_N \ll 1$) and are very similar to oceanic plateau basalts from Curaçao. Thus, from a consideration of basalt compositions and new ^{40}Ar-^{39}Ar ages, it would appear that the Nicoya Complex represents another obducted portion of the CCCIP.

Metamorphosed mafic rocks (the El Tambor Group) crop out in several places in eastern Guatemala [e.g., *Donnelly et al.*, 1990] and these have also been sampled during drilling off the western coast of Guatemala (DSDP legs 67 and 84). Biostratigraphical ages for the El Tambor Group range from lowermost Cretaceous to Cenomanian [*Donnelly et al.*, 1990]. Although *Donnelly et al.* [1990] suggested that the metamorphosed mafic rocks of Guatemala were part of the CCCIP, their older ages compared with most of the province allow the possibility they could represent oceanic crust in existence before the CCCIP.

Wildberg [1984] noted the presence of basalts very similar to those of the Nicoya Complex in the Azuero and Sona peninsulas in western Panama. However, little is known of the chemistry of these basalts. Much of eastern Panama east of the canal (the Darien region) is composed of Pre-Tertiary basaltic rocks [*Escalante*, 1990]. Cherts containing Campanian radiolaria overlie the basalts around the Golfo de San Miguel [*Bandy and Casey*, 1973], thus placing a younger age limit on the complex. *Goosens et al.* [1977] showed that basalts from the Golfo de San Miguel have chemical signatures similar to ocean basalts; they are probably part of the CCCIP.

3.8. *Cuba*

The geology of Cuba is complex and appears to represent a series of accreted terranes of both continental and oceanic material, most of which are younger than Jurassic [*Iturralde-Vinent*, 1994]. Although extensive outcrops of Upper Cretaceous volcanic and intrusive rocks can be found in Cuba, many of these include andesites, dacites, rhyolites, tuffs, and tholeiitic to calc-alkaline granitoids, all with arc-like chemistry [*Iturralde-Vinent*, 1994]. Underlying these arc lavas is an ophiolitic

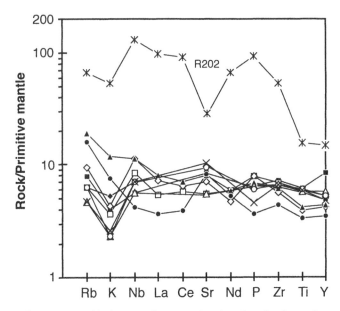

Figure 7. Multi-element diagram, showing data for lavas from the Nicoya Complex, Costa Rica. Sample R202 is significantly enriched in incompatible elements compared with other basalts from the Complex (data from *Berrangé and Thorpe* [1988]).

sequence, the basalts of which may be as young as Turonian. *Iturralde-Vinent* [1994] has interpreted this ophiolite as having formed in a back-arc basin; however, sparse and incomplete chemical analyses of these basalts mean that this theory cannot be rigorously tested. Alternatively, the ophiolite could represent another part of the CCCIP.

3.9. *Colombia and Ecuador*

Although it is generally accepted that many of the outcrops of Cretaceous basalt in the Caribbean and Central America are part of one large igneous province, it is not widely realised that this province also extends into western Colombia and Ecuador [*Marriner and Millward*, 1984; *Millward et al.*, 1984; *Kerr et al.*, 1996a]. The Romeral fault (Figure 8) marks a major terrane boundary in western Colombia. To the east of the fault, Bouguer gravity anomalies are strongly negative (−220 mgal) [*Case et al.*, 1973], confirming geological observations that the basement is composed of continental crust. In contrast, to the west of the Romeral fault, Bouguer anomalies are strongly positive (+135 mgal) [*Case et al.*, 1971]. Thus, the western Colombian-Ecuadorian Andes are composed of high-density material and, as will be shown below, this material is undoubtedly oceanic in origin.

The mafic rocks of western Colombia crop out in three accreted belts which trend approximately N–S; they are,

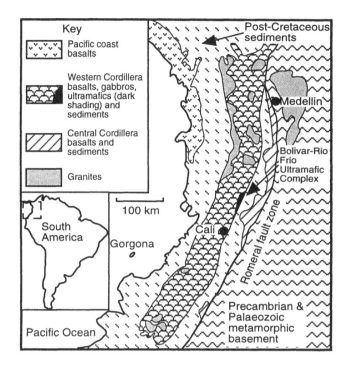

Figure 8. Simplified map of northwestern South America (after *Nivia* [1987]), showing the three separate accreted terranes containing Cretaceous basalts.

from east to west, the Central, Western, and Pacific (Serranía de Baudó) Cordilleras (Figure 8). The three cordilleras are composed of fault-bounded slices of pillowed and massive basalt, and dolerite sills [see *Goosens et al.*, 1977; *Marriner and Millward*, 1984; *Nivia*, 1987]. These major thrust/fault slices are occasionally up to 15 km in width, although more often they are <10 km wide. Several occurrences of picrites have been reported among the volcanic assemblages of the Central Cordillera [*Spadea et al.*, 1989]. These picrites are geochemically similar to the ultramafic lavas from Gorgona Island [*Aitken and Echeverría*, 1984; *Kerr et al.*, 1996b].

Fault-bounded lenses of metasediments (slates, siltstones, cherts, and sandstone turbidites), dipping steeply eastwards, sometimes separate the basaltic slices, particularly in the Western Cordillera where the volcanic sequence is thickest. Some gabbroic plugs and batholiths have intruded the basalts, along with several ultramafic complexes; however, very few dikes cut the basaltic sequences. We note one exception, around the region of Vijes, 30 km north of Cali in southern Colombia, where felsite dykes cut the basaltic sequence. These felsites also are found in breccia layers and are mixed with basalt. *Kerr et al.* [1996a] have suggested that these felsites may be analogous to Icelandic felsites, which display a similar

chemistry and also occur as mixed magmas and breccias [cf. *McGarvie*, 1984].

The volcanic succession on the island of Gorgona (Figure 8) appears to be a continuation of the basaltic sequences of Pacific Cordillera [e.g., *McGeary and Ben Avraham*, 1986] and as such has been proposed to be part of the CCCIP [*Donnelly et al.*, 1990; *Storey et al.*, 1991; *Kerr et al.*, 1996b]. Gorgona has stimulated interest in recent years as the location of the only known Phanerozoic spinifex-textured komatiites [e.g., *Aitken and Echeverría*, 1984; *Kerr et al.*, 1996b]. Picrites, picritic tuff breccias, basalts, and gabbros are also present on this small island (2.5 × 8 km). The geochemical signatures of these volcanic rocks imply that their mantle source region was markedly heterogeneous, containing components both depleted and enriched relative to chondrites (Figures 9a and 9b). The occurrence of picrites and komatiites with calculated parental magma MgO contents of 18–21 wt% requires temperatures significantly above ambient mantle and more consistent with decompression in a mantle plume [e.g., *Storey et al.*, 1991; *Kerr et al.*, 1996b].

There has been considerable discussion in the literature as to the origin of the Cretaceous basalts of the Central, Western, and Pacific Cordilleras. Most authors agree that the basalt successions are allochthonous and have been accreted onto the margin of northwestern South America in the Late Cretaceous or Early Tertiary [e.g., *McCourt et al.*, 1984; *Millward et al.*, 1984; *Bourgois et al.*, 1987]. However, whereas *Barrero* [1979] and *McCourt et al.* [1984] advocated an island-arc-related origin, *Millward et al.* [1984], *Nivia* [1987], and *Kerr et al.* [1996a] have proposed that the Cretaceous basic igneous rocks of Colombia are the remnants of an oceanic plateau and have demonstrated that the geochemistry of the Cretaceous Colombian basalts is inconsistent with an arc-related origin for the rocks of the province. Figure 10a [from *Kerr et al.*, 1996a] shows that the Cretaceous basalts of Colombia are markedly different both from Recent arc-derived volcanic rocks from Colombia and from Cretaceous arc volcanic rocks from Bonaire, Netherlands Antilles. Indeed, the geochemistry of the basalts of the Central and Western Cordilleras (Figure 10b) is very similar to that of Pacific oceanic plateaus and of the basalts of Curaçao, Netherlands Antilles (Figure 3a).

The relative ages of basalts from the three cordilleras are also a matter of debate, because the altered nature and low potassium contents of the lavas mean that K-Ar ages, although abundant in the literature (compiled by *Nivia* [1987]), span quite a wide range and are of questionable reliability. Two proposals have been made regarding the relative ages and the timing of the accretion of each of the

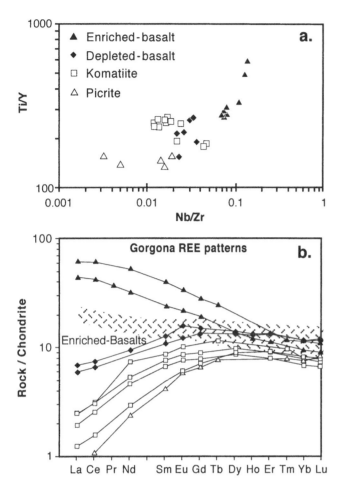

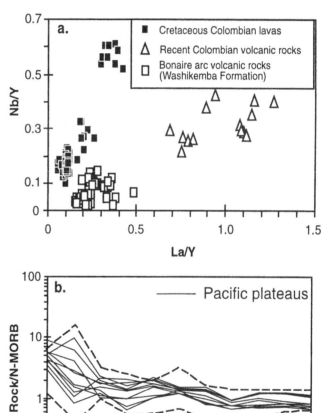

mineral ages for the Buga batholith which range from 69–113 Ma. As a result, they concluded, somewhat arbitrarily, that the Buga batholith was 99 Ma and therefore that the basalts of the Amaime formation had to be older than this. However, because the Amaime formation sensu stricto is only in faulted contact with the Buga batholith, the evidence that the Amaime basalts are older than 100 Ma is not unequivocal. The age of Central Cordillera basalts has still to be resolved satisfactorily. It is, however, possible that these basalts may be older than Turonian and so might be part of an earlier oceanic plateau which was also obducted onto the Colombian continental margin.

Figure 9. Summary of critical features of Gorgona basalts and komatiites, emphasising range of 'enriched' and 'depleted' compositions. (a) Ti/Y vs Nb/Zr. (b) Chondrite-normalised rare earth patterns (normalising values after *Sun and McDonough* [1989]). The shaded area in (b) encompasses most of the more-enriched Gorgona basalts. After *Kerr et al.* [1996b].

three volcanic belts: (a) the three belts represent a single magmatic province which developed in the late Cretaceous and has subsequently accreted onto the northwestern margin of South America [*Goosens et al.*, 1977; *Spadea et al.*, 1989], and (b) the basalts of each cordillera represent three different volcanic provinces, which successively accreted on to the continental margin [*Aspden et al.*, 1987; *McCourt et al.*, 1984].

The basalts of the Central Cordillera (the Amaime Formation) have not been dated radiometrically, nor have any fossils been found in the intercalated sediments. The Buga tonalitic batholith intrudes some amphibolites which have been assumed to be part of the same basaltic sequence as the Amaime Formation [*McCourt et al.*, 1984]. *McCourt et al.* [1984] reported K-Ar and Rb-Sr

Figure 10. (a) Plot of Nb/Y vs La/Y showing Cretaceous Amaime and Volcanic Series basalts from Colombia [*Nivia*, 1987], Cretaceous arc-related basalts from Bonaire and Recent volcanics from Colombia [*Marriner and Millward*, 1984]. (b) Comparison of multi-element patterns of Colombian basalts (dashed field) with basalts from Pacific oceanic plateaus (solid lines); data from *Nivia* [1987] and G. F. Marriner and A. D. Saunders (unpublished data). N-MORB normalisation values from *Sun and McDonough* [1989].

The age of the Western Cordillera is constrained better by Turonian to Coniacian (86–90 Ma) age ammonites and microfossils found in sediments intercalated with the basalts [*Barrero*, 1979; *Bourgois et al.*, 1985]. The biostratigraphical evidence from intercalated sediments suggests that the basalts of the Pacific Cordillera are Coniacian to Santonian (83–88 Ma) in age [*Gansser*, 1973]. Recent ^{40}Ar-^{39}Ar dating of basalts and gabbros from Gorgona Island, the proposed off-shore continuation of the Pacific Cordillera, has yielded ages of 86–88 *Ma [Sinton et al.*, 1993]. It appears then that the mafic rocks of the Western Cordillera and of Serranía de Baudó are the same age (Turonian-Coniacian) as most of the basalts in the Caribbean region. This, in conjunction with evidence of their similar chemistry (Figure 10b) and tectonic considerations, suggests that these Colombian basalts are obducted segments of the Caribbean oceanic plateau.

The Bolívar–Rio Frio Ultramafic Complex in west central Colombia (Figure 8) is a 30-km-long, 5- to 10-km-wide body that trends NNE–SSW. This mafic-ultramafic complex forms part of the easternmost portion of the Western Cordillera, with fault slices of basalt and sediment to the west. To the south, the complex is composed primarily of rhythmically layered, and in places tectonised, norites, olivine norites, and gabbronorites, which underlie a horizon of isotropic gabbronorites [*Barrero*, 1979; *Bourgois et al.*, 1982; *Nivia*, 1996]. The gabbros of the Bolívar Complex have incompatible trace element compositions that are similar to those of the basalts of the Western Cordillera [*Nivia*, 1996], suggesting that the Bolívar Complex could be genetically related to the basaltic plateau sequences.

Toward its northern end, the complex is composed of thick layers (~10 m) of serpentinised dunite separated by thinner (up to 1 m) bands of lherzolites, olivine websterites, and olivine gabbro norites [*Nivia*, 1996]. This rock association was originally interpreted by *Barrero* [1979] as a concentrically zoned body. Alternatively, *Nivia et al.* [1992] presented evidence that the complex is composed of faulted imbricated blocks of oceanic Layer 3, with the ultramafic rocks representing its lower layers. These ultramafic rocks are overlain successively by cumulus gabbros and an upper horizon of isotropic gabbros. Following our field and petrological studies on the complex, we concur with the conclusions of *Nivia et al.* [1992] and suggest that the Bolívar Complex represents the lower layers of the obducted Caribbean-Colombian oceanic plateau. These field observations on the Western Cordillera, especially the better exposures near Bolívar, have enabled us to construct a hypothetical cross section through the Caribbean-Colombian oceanic plateau (Figure

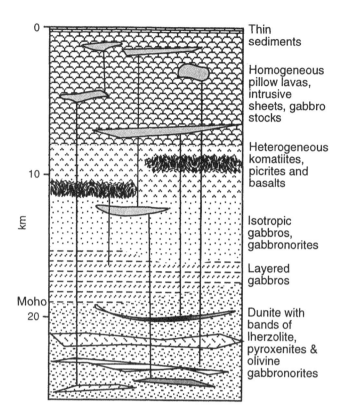

Figure 11. Hypothetical cross-section through the Caribbean oceanic plateau, based on field and petrological studies of the Bolívar Complex and other accreted segments of the CCCIP.

11). This cross section is consistent with a recent geophysical model for Ontong Java [*Farnetani et al.*, 1996] and is similar to models of Archean oceanic plateau structure [cf. *Kent et al.* 1996].

Several points may be made in relation to Figure 11. Firstly, the dominance of horizontal basalt flows and intrusive sheets and sills, as opposed to the sheeted dykes of a "typical" ophiolite, suggests that the magma supply greatly exceeded the extension necessary to accommodate it. Secondly, the layered gabbros were probably formed by the accumulation of crystals at the base of magma chambers, whereas the isotropic gabbros were produced by the in situ crystallisation of magma chamber liquids. Finally, the dunites and lherzolites possess cumulate textures that may represent part of the upper mantle sequence of the plateau.

The Bolívar Complex is intruded by a suite of plagioclase-hornblende pegmatite veins, which can be up to 2 m thick. As well as plagioclase and hornblende phenocrysts, the veins contain variable amounts of quartz and dunite, peridotite and gabbro xenoliths. Unlike the gabbros of the Bolívar Complex, the veins have a subduction-related geochemical signature (La/Nb >>1); they therefore appear to be unrelated to the formation of the oceanic plateau. *Nivia* [1996] has proposed that the veins were formed during fluid-induced partial melting at the base of the plateau, as it was obducted onto the continental margin.

4. COMPOSITIONAL VARIATION WITHIN THE CARIBBEAN–COLOMBIAN OCEANIC PLATEAU

The preceding discussion emphasises that the Caribbean oceanic plateau comprises a wide diversity of basaltic magma compositions. It is logical to group the lavas and intrusive rocks on the basis of chemical differences that cannot be accounted for by fractional crystallisation and must therefore result from mantle heterogeneities or be caused by differing partial melting processes. Three such types may broadly be identified within the CCCIP (Figures 5e and 12):
(a) tholeiitic lavas and sills having flat chondrite-normalised REE patterns ($La_N/Nd_N = 0.9-1.1$; $Sm_N/Yb_N = 1-1.5$) and $Ti/Y = 250-325$ and $Nb/Zr = 0.05-0.1$. Examples are the basalts and picrites of the Curaçao lava succession and the basalts of western Colombia. However, despite the chondritic REE patterns, the Curaçao lavas possess positive $(\varepsilon_{Nd})_i$ values (+6 to +7), which indicate that their mantle source region has had a long-term history of light REE depletion, and has only been enriched in the light REE relatively recently [*Kerr et al.*, 1996c]. In terms of their trace elements, these basalts are compositionally indistinguishable from those recovered from Pacific oceanic plateaus (Figure 3).
(b) A second group is represented by tholeiitic light-REE-depleted ($La_N/Nd_N < 0.9$) lavas and intrusive rocks, with $Ti/Y = 120-280$ and $Nb/Zr < 0.05$, and $(\varepsilon_{Nd})_i >+8$. The picrites, komatiites, and light-REE-depleted basalts of Gorgona Island; several of the Dumisseau lavas; and basalts from DSDP Leg 15 (Site 152) are the main representatives of this magma type. The Gorgona basalts are less light-REE-depleted than the komatiites and picrites, thus precluding a simple fractional-crystallisation genetic relationship. *Kerr et al.* [1996b] have proposed that the depleted basalts may have mixed with some more enriched Gorgona magmas en route to the surface.
(c) Mildly alkalic basalts with $La_N/Nd_N >1.1$; $Sm_N/Yb_N >2$; $Ti/Y >350$ and $Nb/Zr >0.1$. These basalts are chiefly found in Hispaniola, Costa Rica, Gorgona, and DSDP Leg 15, Site 151. It is significant that the areas where these basalts occur appear to be at or near the edge of the CCCIP, suggesting that they may represent lesser degrees of melting at the cooler periphery of the plume head.

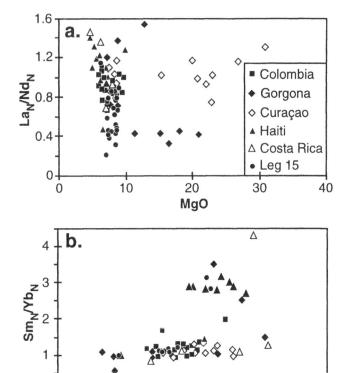

Figure 12. Summary of compositions of Colombian/Caribbean Cretaceous plateau basalts, (a) La_N/Nd_N vs %MgO; (b) La_N/Nd_N vs Sm_N/Yb_N.

Alternatively these basalts may represent late-stage melts from the plume as it cooled down.

Several authors have suggested that the CCCIP represents the melting of the initial 'plume head' phase of the Galápagos hotspot [e.g., *Duncan and Hargraves*, 1984; *Hill*, 1993]. Although there is significant evidence to suggest that the voluminous magmatism of the province was caused by the melting of a large plume head, there is little direct evidence to support the conclusion that it was the Galápagos hotspot [*Kerr et al.*, 1996b]. The geochemical evidence for the CCCIP being formed by the ancestral Galápagos plume head is also equivocal. Figure 3b shows that the Sr and Nd isotopic composition of present Galápagos lavas overlaps substantially with the composition of lavas from the CCCIP. However, given that the present-day Galápagos lavas cover quite a wide range from $\varepsilon_{Nd} = +10$ to +4 and $^{87}Sr/^{86}Sr = 0.7025$ to 0.7035, this overlap may just be coincidental; thus, whereas the geochemical evidence is not inconsistent with a Galápagos hotspot origin for the CCCIP, it is not proof of such a link.

Whether the CCCIP was formed by the Galápagos hotspot or another plume, it is obvious from the chemical diversity of rocks exposed at the surface that the plume source region from which the CCCIP was derived was markedly heterogeneous. Because most of the range of chemical and isotopic compositions seen within the CCCIP has been found within the small (8 × 3 km) island of Gorgona, this heterogeneity must be on a relatively small scale [*Kerr et al.*, 1996b]. The evidence suggests that there are essentially two plume components making up this heterogeneity, one with depleted and another with more-enriched incompatible trace element contents and isotopic ratios (relative to Bulk Earth). *Kerr et al.* [1995] and N. T. Arndt et al. (ms. submitted to EPSL, 1996) have proposed that the plume that produced the CCCIP consisted of enriched and more fusible streaks set in a depleted and more refractory matrix, which was also an integral part of the plume. The enriched plume components probably only account for <10% of the plume source region (N. T. Arndt et al., ms. submitted to EPSL, 1996); nevertheless, these components, because of their enriched nature, can exert a significant control on the chemistry of the erupted magmas.

Previous drilling into the upper sections of several Pacific oceanic plateaus has yielded basalts with predominantly flat incompatible trace element patterns (e.g., Figures 3a and 10b). In terms of isotopic ratios, lavas from individual plateaus have a more restricted range than the CCCIP. However, when taken together they display a greater degree of isotopic heterogeneity (Figure 3b). The CCCIP—whose deeper sections have been exposed by obduction and erosion—is characterised by basalts and picrites that possess a relatively wide range of enriched and depleted trace element compositions and of isotopic ratios (e.g., Figures 5, 6, 9, and 12). Is it possible then that the lower parts of oceanic plateaus have more heterogeneous basalts and more picrites than the upper sections? If so, then what mechanisms could be responsible for the increasing chemical homogeneity and more evolved basaltic compositions towards the top of a plateau?

The picrites of Curaçao occur near the base of the exposed lava succession [*Kerr et al.*, 1996c]. Additionally, many continental flood basalt provinces (e.g., West Greenland, the Siberian Traps, and the Deccan Traps) preserve sequences that have picrites near their base [see *Campbell and Griffiths*, 1992; *Kerr et al.*, 1995, for reviews]. Although there are exceptions (e.g., Deccan), these picrites mostly tend to be more incompatible-element depleted than their associated basalts [*Kerr et al.*, 1995]. Thus, the evidence from continental flood basalt provinces and the CCCIP suggests that the early stages of large

igneous province formation may be characterised by picrites and basalts with heterogeneous chemical and isotopic signatures.

Consideration should be given as to whether these features could be attributed to magmatic plumbing. In the initial stages of volcanic activity, magma chambers will probably be relatively small and poorly developed within the cool and rigid lithosphere. There is thus a greater chance that individual isotopically distinct picritic and basaltic magma batches will be able to erupt without being 'trapped' and homogenised in magma chambers. Moreover, they may interact and become contaminated with altered oceanic crust. In contrast, as the large igneous province continues to develop, magma chambers will become larger and longer-lived in a crust/lithosphere that is rheologically weaker. Also, in the later stages of magmatism the crust will be substantially thicker, making it more likely that denser picritic magmas will pond and fractionate. Magma chamber trapping of picritic melts would result in olivine fractionation, so reducing the chance of picrite eruption. Thus, although chemically and isotopically distinctive picrites and basalts are probably still formed during melt generation in these later stages of plateau development, there is more likelihood of such magma batches being trapped by large magma chambers and undergoing extensive mixing and homogenisation, thereby masking any mantle-derived heterogeneities. Figures 12 and 5e support these magma mixing proposals in that the three types (outlined at the beginning of this section) do not define distinct groups but rather form a broad continuum from depleted lavas to the more-enriched lavas.

An alternative possibility is that the mantle plume source region may have become more uniform in composition and cooler, so producing basaltic primary melts rather than picritic ones. Because plume heads should entrain surrounding material during ascent through the lower and, to a lesser extent, the upper mantle [e.g., *Griffiths and Campbell*, 1990], they may evolve toward an approximate average mantle composition [*Stein and Hofmann*, 1994]. However, during the initial stage of plume impact, particularly as the rapidly uprising hot axial zone of the plume impinges beneath the lithosphere, there may be extensive interaction with the upper asthenosphere and lower lithosphere, picking up any heterogeneities in that region. Finally, because plumes are a dynamic melt-generating system, there may be significant compositional heterogeneities resulting from the different depths of continued melt extraction (N. T. Arndt et al., ms. submitted to EPSL, 1996), but these will only be evident in the early stages of plateau development, before the development of large magma chambers that trap individually distinct magma batches.

There are several possible reasons why the deeper levels of other obducted plateaus and, in particular, Ontong Java are not exposed. The exposed sections of other oceanic plateaus (particularly Ontong Java [e.g., *Neal et al.*, this volume]) are composed almost entirely of pillowed and massive basalts with intrusive dolerite sheets, representing upper crustal levels (< 4 km). The reason for this difference with the CCCIP may relate to the age of the plateau at the time of imbrication and accretion. Ontong Java formed in two major plateau-building episodes, at ~120 and ~90 Ma [*Mahoney et al.*, 1993] and was obducted over the North Solomon Trench at 20 Ma [*Kroenke*, 1974; *Wells*, 1989], i.e., at least 70 m.y. after its formation. In contrast, the Caribbean–Colombian oceanic plateau appears to have been obducted in the Late Cretaceous (Early Tertiary?) [*Burke*, 1988], probably no more than 15 to 25 m.y. after formation. Consequently, the Caribbean-Colombian plateau may still have been relatively warm and more buoyant than the much older and cooler plateaus such as Ontong Java and Wrangellia [*Lassiter et al.*, 1995] at the time of obduction. This additional thermal buoyancy of the CCCIP may be one reason why deeper sections of this plateau were obducted. Another reason why deep crustal sequences have not (yet) been exposed in the Solomon Islands may result from the greater crustal thickness of the Ontong Java (25–42 km; [*Cooper et al.*, 1986; *Furomoto et al.*, 1976]) as compared to the Caribbean–Colombian oceanic plateau (8–20 km [*Edgar et al.*, 1971]).

The plume model of *Campbell et al.* [1989] predicts that the picritic and thus the hottest magmas associated with the initial plume head phase of magmatism will be restricted to a relatively narrow region (150 km diameter) at the axis of the plume, over the so-called tail, with the production of lower MgO magmas (< 12 wt%) in the cooler outer zones of the plume head. Although the pre-tectonic geometry of the province is difficult to determine, the presence of 88–90 Ma picritic and picritic basalt lavas throughout the CCCIP (Gorgona, Costa Rica, Curaçao, and in southern Colombia) would appear to require anomalously hot mantle (>200 °C above ambient upper mantle temperatures) within a large part of the proposed plume head and therefore seems to be inconsistent with standard plume models [e.g., *Campbell et al.*, 1989; *Saunders et al.*, this volume].

One possible explanation is that the plume head was not the conventionally assumed mushroom shape. Both *Sleep et al.* [1988] and *White* [1992] suggested that plume heads can also take the form of tabular upwelling blobs of hot material. If so, the hot ascending mantle might occupy a broad area, not just a narrow axial conduit.

The tectonic setting and structure of the lithosphere above the ascending plume head could also play an important role [cf. *Thompson and Gibson*, 1993] in determining whether or not hot mantle capable of generating picrites is present below a region. In section 2, we noted that the CCCIP may have formed at a spreading-ridge triple junction. The presence of a triple junction in the lithosphere above the impinging plume head could potentially provide a way for hot plume material to become channelled rapidly along the spreading ridges to the more distal parts of the plume head, for instance, propagating eastwards into the Caribbean region. This may have allowed eruption of picrites far from the perceived centre of the plume.

5. PLATEAU OBDUCTION

The abundant mafic volcanic sequences throughout the length of western Colombia are arranged in coast-parallel lenses alternating with graywacke-shale metasediments [see *Nivia*, 1996; also *Bourgois et al.*, 1982] that are mostly strongly deformed and vary in metamorphic grade from chlorite-pumpellyite facies phyllites to schists and occasional migmatitic gneisses. Along the eastern margin of the mafic volcanic belt, marked by the Romeral Fault, local exposures of blueschists and ultramafic rocks (now talcose and serpentinised) form lenses within the metasediments. These relationships are best accounted for by subduction-accretion processes at an active margin (lateral crustal accretion). Indeed, deep crustal xenoliths brought up in the Mercaderes volcanic diatreme [*Weber et al.*, 1995], which lies just east of the eastern margin of the mafic volcanic zone in southwestern Colombia, include mixed high-grade mafic and metasedimentary rock types that are consistent with subduction-accretion. The essence of subduction-accretion is that continent-derived graywacke-shale sediment is tectonically mixed with up-standing mafic structures on the ocean floor (arcs, aseismic ridges, plateaus, etc.) that are scraped off the subducting ocean slab [cf., *Sengör and Okuogullari*, 1991; *von Heune and Scholl*, 1993]. It is interesting to consider the fate of oceanic plateaus in this environment. Because subduction of old, cold lithosphere is the dominant driving force of plate tectonics [e.g., *Carlson et al.*, 1983], it is assumed that warm, buoyant plateaus will behave differently when they reach an active margin [*Cloos*, 1993; *Saunders et al.*, 1996]. To what extent are they obducted or underplated [*Kimura and Mukai*, 1991]?

Undoubtedly, the sheer areal extent of the mafic

volcanic complex in Colombia and the Caribbean implies that significant amounts were not subducted. However, this material is dominantly hydrothermally altered basalts and basaltic sheets, and it is only locally that the deeper, fresher parts of the plateau crust are seen (e.g., the Bolívar Ultramafic Complex). Decollément zones are important in accretionary terranes [e.g., *Moore*, 1989]; *Kimura and Ludden* [1995] have made the point that the decollément zone in normal ocean crust probably occurs at the base of the hydrothermal circulation cells, where rheologically weak altered rock is underlain by fresh dolerites and gabbros. The upper zones are peeled off and imbricated with the developing sedimentary accretionary wedge (as observed in western Colombia) whereas the fresher basalts and gabbros are subducted or underplated.

Of course, another, deeper, rheologically weak zone lies beneath the Moho in oceanic lithosphere, which is essentially temperature-dependent (>1000 °C [*Nicolas and Violette*, 1982]), and is usually the zone along which ophiolite complexes are obducted and along which there is normally metamorphism and melting as the hot ophiolitic sheet is thrusted over continental margin sediments [*Searle and Malpas*, 1982]. It is probable that the Bolívar Complex with its layered and banded peridotites, pyroxenites, and gabbros represents the deeper, hotter part of the plateau that was underthrusted by the deformed metasediments of the accretionary wedge. If so, the abundant hornblende-pyroxene pegmatites may result from hydrous partial melting of the peridotites and gabbros as fluids expelled from the underthrusted sediments entered the base of the thrust sheet. Caution is necessary because it has been shown that ophiolites (e.g., Oman ophiolite) have an abundance of gabbroic veins just beneath the Moho [*Boudier and Nicolas*, 1995], which is roughly where they occur at Bolívar. In contrast, whereas the former are 'dry' gabbroic veins, a majority of the veins at Bolívar are spectacular coarse hornblende-anorthosite pegmatites, transitional to hornblende-biotite tonalites, and some (the earlier) are quite strongly deformed. Such veins are found wherever these deep sections are exposed in Colombia [*Bourgois et al.*, 1982; *Tistl et al.*, 1994]. If they are associated with fluid entry into the base of the imbricating plateau, they potentially offer a means of dating the time of obduction and/or subcretion of an oceanic plateau at a continental margin. However, a K-Ar age of 108 ± 16 Ma on hornblende pegmatites from Bolívar [*Barrero*, 1979] and K-Ar hornblende ages clustering around ~20 Ma for the Condoto Complex southwest of Medellín [*Tistl et al.*, 1994] have been interpreted as primary intrusive emplacement ages rather than tectonic emplacement ages and need to be supplemented by ^{40}Ar-^{39}Ar ages to resolve these alternatives.

Figure 13 illustrates a possible tectonic model for plateau emplacement, with imbrication at two levels. Plateaus have a potentially longer residence time at the Earth's surface than normal oceanic crust [*Kroenke*, 1974; *Cloos*, 1993; *Saunders et al.*, 1996] and may be more difficult to subduct. However, the evidence from the Caribbean-Colombian plateau suggests that (in comparison with supra-subduction zone ophiolites) it is the upper part that is imbricated and obducted, whereas the deeper zones are subducted or tectonically underplated beneath active continental margins and only occasionally obducted. This begins to resolve the apparent paradox, identified by *Burke* [1988], that obducted plateaus are rather rare in the geological record: mostly it may be only the upper, volcanic sections of plateaus that are imbricated and obducted, as may be the case with some Precambrian greenstone belts [e.g., *Kusky and Kidd*, 1992]. What happens to the deeper parts of plateaus is a matter for speculation, but if underplated beneath continental margins and melted, they could make a major contribution to crustal growth, adding material with relatively primitive isotopic characteristics [e.g., *Tarney and Jones*, 1994; *Abbott and Mooney*, 1995].

6. ARC-DERIVED BASALTS ASSOCIATED WITH THE CCCIP

The margins of the Caribbean are marked by numerous occurrences of arc-derived lavas and equivalent intrusive rocks [e.g., *Donnelly et al.*, 1990; *Donnelly*, 1994]. The lavas are typically island arc tholeiites (called the primitive island arc series by *Donnelly et al.* [1990]), and the main exposures are found in Hispaniola, Cuba, Puerto Rico, Venezuela, Ecuador, Bonaire, parts of the Lesser Antilles, and the Quebradagrande complex east of the Romeral fault in Colombia. Nowhere are these arc lavas found to be interlayered with the Caribbean oceanic plateau basalts, so whereas there is a close spatial relationship, the nature of the relationship needs to be clarified. The arc-derived lavas and the basalts of the CCCIP overlap in age, but the island arc rocks appear to be slightly older (Cenomanian-Albian; 90–112 Ma) than the plateau-derived basalts. *Donnelly* [1994] proposed that the island arc basalts formed on top of the plateau at its margin. However, until this is confirmed by careful dating, we consider it more likely that the island arc tholeiites are the volcanic products associated with the pre-plateau-collision oceanic/continental arc in the Caribbean–Colombian region and have been accreted onto the edge of the plateau during the early stages of obduction.

Numerous calc-alkaline tonalitic batholiths of Late Cretaceous age (< 90 Ma) with some associated calc-

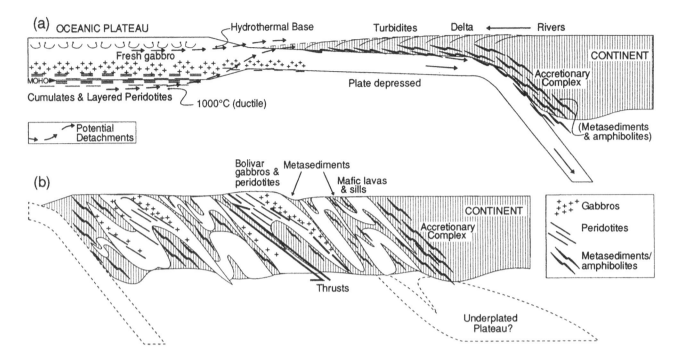

Figure 13. Schematic obduction-imbrication model for the Caribbean-Colombian plateau. (a) Thick (>15 km) plateau crust formed in close proximity to the continental margin of northwestern South America. The margin was accretionary, comprising scraped-off turbidite metasediments, amphibolites, etc. Two potential detachment zones in the approaching plateau are shown, one at the base of the hydrothermal circulation zone, the other much deeper, within the mantle lithosphere below the plateau. (b) When the plateau reached the subduction zone, only the deeper zones of the plateau were subducted. The upper layers are scraped-off and imbricated along the two possible detachments and accreted onto the continental margin. See text for further details.

alkaline extrusives are found around the Caribbean margins and in Central America [*Donnelly et al.*, 1990] and appear to largely postdate the island arc tholeiite event. The inception of this calc-alkaline phase of magmatism appears to coincide both spatially and temporally with the formation of the CCCIP, perhaps penecontemporaneously. *Donnelly et al.* [1990] suggested that the change from island arc tholeiites to calc-alkaline magmatism may be related to 'a major change in plate movements.' On the basis of significant new chemical data from Hispaniola, *Lebrón and Perfit* [1994] suggested that the transition from island arc tholeiites to calc-alkaline volcanism in the Caribbean was caused by the partial subduction of the buoyant Caribbean oceanic plateau choking the subduction zone, thus promoting reversal of subduction polarity. Our observations on Aruba suggest a close link between tonalite emplacement and the deformation/imbrication of the plateau against the South American continental margin. The age similarity between the formation of the CCCIP and the inception of voluminous tonalitic magmatism is intriguing and suggests that, if the model of *Lebrón and Perfit* [1994] is correct, the CCCIP probably formed in

relatively close proximity to the western margin of northern South America. One possibility is that if the ascending plume was channelled along an eastward-propagating rift linked to the separating North and South American Plates as a result of Atlantic opening, then the time gap between the formation of the (hot) oceanic plateau and its interaction with the westward-moving continents could be quite short.

7. CONCLUSIONS

1. The CCCIP represents the dismembered remnants of an oceanic plateau that formed in the eastern Pacific Ocean at ~87–90 Ma. Shortly after its formation the northern half of the plateau was pushed between North and South America to form the bulk of the Caribbean Plate. As the southern part of the plateau approached the continental margin of northwestern South America, it was too buoyant to totally subduct and so its topmost parts accreted onto the continental margin.

2. The lavas that occur nearer the base of the exposed CCCIP tend to be (a) more picritic and (b) mostly

compositionally more heterogeneous than basalts from nearer the top of the succession, which possess near-chondritic REE patterns and are quite uniform in composition. This observation might be explained by the development of larger magma chambers in the later stages of plateau evolution, which could act as a trap for picritic magmas in which heterogeneous picritic magmas could both fractionate and become homogenised. The inferred poorly developed magma chambers in the earlier stages of plateau evolution mean that heterogeneous lavas and picrites are more likely to reach the surface.

3. The basalts, picrites, and komatiites of the province require a markedly heterogeneous plume source region, containing at least two components.

4. The gabbros, norites, gabbronorites, dunites, olivine websterites, and lherzolites of the Bolívar Ultramafic Complex provide a glimpse of the composition of the deeper, subvolcanic sections of an oceanic plateau. From the Bolívar Complex and other accreted terranes in Colombia, a probable section through the deeper parts of an oceanic plateau has been constructed. From the top, this section consists of homogeneous basalts, followed by heterogeneous picrites and komatiites, and isotropic gabbro. This isotropic gabbro lies above layered gabbro, with dunite containing bands of lherzolite, olivine websterite, and olivine gabbro norites at still lower levels.

5. Island arc tholeiites (both of similar age and older than the CCCIP) are found around the margins of the Caribbean and to the east of the accreted plateau sequences in Colombia. It is likely that these represent the remnants of the pre-plateau–collision volcanic arc. The presence of primitive tonalitic batholiths (< 90 Ma) closely associated with the Caribbean Plateau and plateau imbrication may record a possible reversal of subduction polarity caused by the attempted obduction of the plateau.

6. Continued detailed study of the CCCIP could tell us much more about the internal structure of oceanic plateaus, as well as the fate of plateaus and their interaction with continental margins. The results should have an important bearing on understanding Precambrian greenstone belts and primitive crustal growth mechanisms, as well as on the evolution of the Caribbean region itself. There is a great need for detailed, reliable geochronology in the region.

Acknowledgments. We are grateful to Leon Pors, Carmabi Institute, Curaçao, Armando Curet and Vivi Ruiz, Aruba, and Manuel Itturalde-Vinente, Havana, Cuba, for all their help with the fieldwork, and Dirk Beets, Gerard Klaver, and Nick Arndt for much advice and discussion. Rosemary Hickey-Vargas, Bob Duncan, and John Mahoney are thanked for constructive criticism of the manuscript. This work was supported by Grants GR9/583A & GR3/8934 from the Natural Environment Research Council.

REFERENCES

Abbott, D., and W. Mooney, The structural and geochemical evolution of the continental crust - support for the oceanic plateau model of continental growth, *Rev. Geophys., 33*, 231-242, 1995.

Aitken, B. G., and L. M. Echeverría, Petrology and geochemistry of komatiites from Gorgona Island Colombia, *Contrib. Mineral. Petrol., 86*, 94-105, 1984.

Aspden, J. A., W. J. McCourt, and M. Brook, Geometrical control of subduction-related magmatism: the Mesozoic and Cenozoic plutonic history of western Colombia, *J. Geol. Soc. London, 144*, 893-905, 1987.

Bandy, O. L., and R. E. Casey, Reflector horizons and paleobathymetric history, eastern Panama, *Geol. Soc. Am. Bull., 84*, 3081-3086, 1973.

Barrero, D., Geology of the Central western Cordillera, west of Buga and Roldanillo, Colombia, *Publ. Especial, Ingeominas, 4*, 1-75, 1979.

Beck, C. M., D. Girard, and P. DeWeber, Le "Volcano-sédimentaire du Rio Guare": Un élément de la nappe ophiolitique de Lomo de Hierro, chaône Caraôbe Vénézuélienne, *Paris, Academy of Science, Comptes Rendus de Seances (D), 299*, 337-342, 1984.

Beets, D. J., G. Th. Klaver, F. F. Beunk, G. Kieft, and P. Maaskant, Picrites as parental magma of MORB-type tholeiites, *Nature, 296*, 341-343, 1982.

Beets, D. J., W. V. Maresch, G. Th. Klaver, A. Mottana, R. Bocchio, F. F. Beunk, and H. P. Monen, Magmatic rock series and high-pressure metamorphism as constraints on the tectonic history of the southern Caribbean, *Geol. Soc. Am. Mem., 162*, 95-130, 1984.

Ben Avraham, Z., A. Nur, D. Jones, and A. Cox, Continental accretion: from oceanic plateaus to allochthonous terranes, *Science, 213*, 47-54, 1981.

Bence, A. E., J. J. Papike, and R. A. Ayuso, The petrology of submarine basalts from the central Caribbean: DSDP Leg 15, *J. Geophys. Res., 80*, 4775-4804, 1975.

Berrangé, J. P., and R. S. Thorpe, The geology, geochemistry and emplacement of the Cretaceous-Tertiary ophiolitic Nicoya complex of the Osa Peninsula, southern Costa Rica, *Tectonophysics, 147*, 193-220, 1988.

Boudier, F., and A. Nicolas, Nature of the Moho transition zone in the Oman ophiolite, *J. Petrol., 36*, 777-796, 1995.

Bourgois, J., B. Calle, J. Tournon, and J-F. Toussaint, The Andean ophiolitic megastructures on the Buga-Buenaventura transverse (Western Cordillera–Valle Colombia), *Tectonophysics, 82*, 207-229, 1982.

Bourgois, J., J. Azema, P. O. Baumgartner, J. Tournon, A. Desmet, and J. Aubouin, The geologic history of the Caribbean-Cocos plate boundary with special reference to the Nicoya complex (Costa Rica) and DSDP results (Legs 67 and 84 off Guatemala): a synthesis, *Tectonophysics, 108*, 1-32, 1984.

Bourgois, J., J.-F. Toussaint, H. Gonzales, A. Orrego, and B. Calle, Les ophiolites des Andes de Colombia, Evolution structural et signification geodinamic, in *Geodynamic des*

Caraibbes Symposium, edited by A. Mascle, pp. 475-493, Technip-Paris, 1985.

Bourgois, J., J-F. Toussaint, H. Gonzales, J. Azéma, B. Calle, A. Desmet, L. A. Murcia, A. P. Acevedo, E. Parra, and J. Tournon, Geological history of the Cretaceous ophiolitic complexes of northwestern South America (Colombian Andes), *Tectonophysics, 143*, 307-327, 1987.

Burke, K., Tectonic evolution of the Caribbean, *Ann. Rev. Earth Planet. Sci., 16*, 201-230, 1988.

Burke, K., C. Cooper, J. F. Dewey, P. Mann, and J. L. Pindell, Caribbean tectonics and relative plate motions, *Geol. Soc. Am. Mem., 162*, 31-63, 1984.

Burke., K., P. J. Fox, and M. C. Sengör, Buoyant ocean floor and the origin of the Caribbean, *J. Geophys. Res., 83*, 3949-3954, 1978.

Campbell, I. H., R. W. Griffiths, and R I. Hill, Melting in an Archean mantle plume: heads it's basalts, tails it's komatiites, *Nature, 339*, 697-699, 1989.

Campbell, I. H., and R. W. Griffiths, The changing nature of mantle hotspots through time: Implications for the chemical evolution of the mantle, *J. Geol., 92*, 497-523, 1992.

Carlson, R. L., T. W. C. Hilde, and S. Uyeda, The driving mechanism of plate tectonics: relation to age of the lithosphere at trenches, *Geophys. Res. Lett., 10*, 297-300, 1983.

Case, J. E., S. L. G. Duran, A. Lopez, and W. R. Moore, Tectonic investigations on western Colombia and eastern Panama, *Geol. Soc. Am. Bull., 82*, 2685-2712, 1971.

Case, J. E., J. Barnes, G. Paris, I. H. Gonzales, and A. Vina, Trans-Andean geophysical profile, southern Colombia, *Geol. Soc. Am. Bull., 84*, 2895-2904, 1973.

Case, J. E., W. D. MacDonald, and P. J. Fox, Caribbean crustal provinces; seismic and gravity evidence, in *The Geology of North America, vol. H: The Caribbean Region*, edited by G. Dengo and J. E. Case, pp. 15-36, Geological Society of America, Boulder, CO, 1990.

Cloos, M., Lithospheric buoyancy and collision orogenesis: Subduction of oceanic plateaus, continental margins, island arcs, spreading ridges and sea mounts, *Geol. Soc. Am. Bull., 105*, 715-737, 1993.

Coffin, M. F., and O. Eldholm, Large igneous provinces: crustal structure, dimensions and external consequences, *Rev. Geophys, 32*, 1-36, 1994.

Cooper, A. K., M. S. Marlow, and T. R. Bruns, Deep structure of the central and southern Solomon Islands region: implications for tectonic origin, *Circum-Pacific Council for Energy and Mineral Resources, Earth Science Series, 4*, 137-175, 1986.

Donnelly, T. W., Late Cretaceous basalts from the Caribbean, a possible flood basalt province of vast size, *Eos Trans. AGU, 54*, 1973.

Donnelly, T. W., Geologic history of the Caribbean and Central America, in *The Geology of North America. vol. A: An Overview*, edited by A. W. Bally and A. R. Palmer, pp. 299-321, Geological Society of America, Boulder, CO, 1989.

Donnelly, T. W., The Caribbean Cretaceous basalt association: A vast igneous province that includes the Nicoya Complex of Costa Rica, *Profil, (University of Stuttgart) 7*, 17-45, 1994.

Donnelly, T. W., D. Beets, M. J. Carr, T. Jackson, G. Klaver, J.

Lewis, R. Maury, H. Schellenkens, A. L. Smith, G. Wadge, and D. Westercamp, History and tectonic setting of Caribbean magmatism, in *The Geology of North America, Vol. H: The Caribbean Region*, edited by G. Dengo and J. E. Case, pp. 339-374, Geological Society of America, Boulder, CO, 1990.

Donnelly, T. W., W. Melson, R. Kay, and J. J. W. Rogers, Basalts and dolerites of late Cretaceous age from the central Caribbean, *Init. Repts. Deep Sea Drill. Proj., 15*, 989-1012, 1973.

Donnelly, T. W., and J. J. W. Rogers, The distribution of igneous rocks throughout the Caribbean, *Geol. Mijnbouw, 57*, 151-162, 1978.

Duncan, R. A and R. B. Hargraves, Plate tectonic evolution of the Caribbean region in the mantle reference frame, *in The Caribbean-South American Plate Boundary and Regional Tectonics, Mem. 162*, edited by W. E. Bonini, R. B. Hargraves, and R. Shagam, pp. 81-93, Geological Society of America, Boulder, CO, 1984.

Edgar, N. T., J. I. Ewing, and J. Hennion, Seismic refraction and reflection in the Caribbean Sea, *Am. Assoc. Petrol. Geol., 55*, 833-870, 1971.

Escalante, G., The Geology of Southern Central America and western Colombia, in: *The Geology of North America, Vol. H: The Caribbean Region*, edited by G. Dengo and J. E. Case, pp. 201-230, Geological Society of America, Boulder, CO, 1990.

Farnetani, C. G., M. A. Richards, and M. S. Ghiorso, Petrological models of magma evolution and deep crustal structure beneath hotspots and flood basalt provinces, *Earth Planet. Sci. Lett.* (in press)

Furumoto, A. S., J. P. Webb, M. E. Odegard, and D. M. Hussong, Seismic studies of the Ontong Java Plateau, 1970, *Tectonophysics, 34*, 71-90, 1976.

Gansser, A., Facts and theories on the Andes, *J. Geol. Soc. London, 129*, 93-131, 1973.

Goossens, P. J., W. I. Rose, and D. Flores, Geochemistry of the tholeiites of the basic igneous complex of NW South America, *Geol. Soc. Am. Bull., 88*, 1711-1720, 1977.

Griffiths, R. W., and I. H. Campbell, Stirring and structure in mantle starting plumes, *Earth Planet. Sci. Lett., 99*, 66-78, 1990.

Hall, S. A., Oceanic basement of the Caribbean basins, *Geol. Soc. Am. Abstracts with Programs, 27*, p. A153, 1995.

Harland, W. B., A. L. Armstrong, A. V. Cox, L. E. Craig, A. G. Smith and D. G. Smith, *A Geologic Time Scale 1989*, 263 pp., Cambridge Univ. Press, Cambridge, 1990.

Hill, R. I., Mantle plumes and continental tectonics, *Lithos, 30*, 193-206, 1993.

Iturralde-Vinent, M. A., Cuban geology: A new plate tectonic synthesis, *J. Petroleum Geol., 17*, 39-70, 1994.

Kent, R. W., B. S. Hardarson, A. D. Saunders, and M. Storey, Plateaus ancient and modern: Geochemical and sedimentological perspectives on Archaean oceanic magmatism, *Lithos, 37*, 129-142, 1996.

Kerr, A. C., A. D. Saunders, J. Tarney, N. H. Berry, and V. L. Hards, Depleted mantle plume geochemical signatures; no paradox for plume theories, *Geology, 23*, 843-846, 1995.

Kerr, A. C., J. Tarney, G. F. Marriner, A. Nivia, A. D. Saunders, and G. Th. Klaver, The geochemistry and tectonic setting of late Cretaceous Caribbean and Colombian volcanism, *J. S. Am. Earth Sci.*, 1996a (in press).

Kerr, A. C., G. F. Marriner, N. T. Arndt, J. Tarney, A. Nivia, A. D. Saunders, and Duncan, R. A., The petrogenesis of Gorgona komatiites, picrites and basalts: new field, petrographic and geochemical constraints, *Lithos, 37*, 245-260, 1996b.

Kerr, A. C., J. Tarney, G. F. Marriner, G. Th. Klaver, A. D. Saunders, and M. F. Thirlwall, The geochemistry and petrogenesis of the late-Cretaceous picrites and basalts of Curaçao, Netherlands Antilles: a remnant of an oceanic plateau, *Contrib. Mineral. Petrol., 124*, 29-43, 1996c.

Kimura, G., and A. Mukai, Underplated units in an accretionary complex: Melange of the Shimanto Belt of eastern Shikoku, south-west Japan, *Tectonics, 10*, 31-50, 1991.

Kimura, G., and J. Ludden, Peeling oceanic crust in subduction zones, *Geology, 23*, 217-220, 1995.

Klaver, G. Th., The Curaçao lava formation an ophiolitic analogue of the anomalous thick layer 2B of the mid-Cretaceous oceanic plateaus in the western Pacific and central Caribbean, Ph.D. Thesis, University of Amsterdam, The Netherlands, 1987.

Kroenke, L. W., Origin of continents through development and coalescence of oceanic flood basalt plateaus. *Eos Trans. AGU, 55*, p. 443, 1974.

Kusky, T. M., and W. S. F. Kidd, Remnants of an Archean oceanic plateau, Belingwe greenstone belt, Zimbabwe, *Geology, 43*, 43-46, 1992.

Ladd, J. W., T. L. Holcombe, G. K. Westbrook, and N. T. Edgar, Caribbean margin geology: active margins of the plate boundary, in *The Geology of North America, Vol. H: The Caribbean Region*, edited by G. Dengo and J. E. Case, pp. 261-290, Geological Society of America, Boulder, CO., 1990.

Lassiter, J. C., D. J. DePaolo, and J. J. Mahoney, Geochemistry of the Wrangellia flood basalt province: Implications for the role of continental and oceanic lithosphere in flood basalt genesis. *J. Petrol., 36*, 983-1009, 1995.

Lebrón, M. C., and M. R. Perfit, Petrochemistry and tectonic significance of Cretaceous island-arc rocks, Cordillera Central, Dominican Republic, *Tectonophysics, 229*, 69-100, 1994.

MacKenzie, D. B., High temperature alpine-type peridotite from Venezuela, *Geol. Soc. Am. Bull., 71*, 303-318, 1960.

Mahoney, J. J., M. Storey, R. A. Duncan, K. J. Spencer, and M. Pringle, Geochemistry and age of the Ontong Java Plateau, in *The Mesozoic Pacific: Geology, Tectonics, and Volcanism, Geophys. Monogr. Ser.*, vol. 77, edited by M. S. Pringle, W. W. Sager, W. V. Sliter, and S. Stein, pp. 233-261, AGU, Washington, D. C., 1993.

Marriner, G. F., and D. Millward, The petrology and geochemistry of Cretaceous to Recent volcanism in Colombia: The magmatic history of an accretionary plate margin, *J. Geol. Soc. London, 141*, 473-486, 1984.

Martín Bellizia, C., and J. M. de Arozena, Complejo ultramáfico zonado de Tausabana-El Rodeo, gabro zonado de Siraba-Capuana y complejo sub-volcánico estratificado de Santa Ana, Paraguaná, Estado Falcón. *Trans. 6th Caribbean Geol. Conf., Margarita*, pp. 337-356, 1972.

Maurrasse, F., J. Husler, G. Georges, R. Schmitt, and P. Damond, Upraised Caribbean sea floor below acoustic reflector "B" at the southern peninsula of Haiti. *Geol. Mijnbouw, 58*, 71-83, 1979.

McCourt, W. J., J. A. Aspden, and M. Brook, New geological and geochronological data from the Colombian Andes: continental growth by multiple accretion, *J. Geol. Soc. London, 141*, 831-845, 1984.

McGarvie D. W., Torfajokull: a volcano dominated by magma mixing, *Geology, 12*, 685-688, 1984

McGeary, S., and Z. Ben-Avraham, The accretion of Gorgona Island, Colombia: multichannel seismic evidence, in *Tectonostratigraphic Terranes of the Circum-Pacific Region*, edited by D. G. Howell, pp. 543-554, Circum-Pacific Council for Energy and Mineral Resources, Houston, TX, 1986.

Meyerhoff, A. A., and H. A. Meyerhoff, Continental drift. IV: The Caribbean 'Plate', *J. Geol., 80*, 34-60, 1972.

Millward, D., G. F. Marriner, and A. D. Saunders, Cretaceous tholeiitic rocks from the western Cordillera of Colombia, *J. Geol. Soc. London, 141*, 847-860, 1984.

Moore, J. C., Tectonics and hydrogeology of accretionary prisms: role of the decollement zone, *J. Struct. Geol., 11*, 95-106, 1989.

Nicolas, A., and J. F. Violette, Mantle flow at oceanic spreading centres: models derived from ophiolites, *Tectonophysics, 81*, 319-379, 1982.

Nivia, A. Geochemistry and origin of the Amaime and Volcanic sequences, Southwestern Colombia, M. Phil. Dissertation, University of Leicester, Leicester, England, 163 pp., 1987.

Nivia, A., The Bolivar mafic-ultramafic complex, SW Colombia: the base of an obducted oceanic plateau, *J. S. Am. Earth Sci.* (in press).

Nivia, A., N. Galvis, and M. Maya, *Geologia de la Plancha 242 - Zarzal*. Ingeominas, Bogota, 163 pp., 1992.

Nur, A., and Z. Ben Avraham, Oceanic plateaus, the fragmentation of continents, and mountain building, *J. Geophys. Res., 87*, 3644-3661, 1982.

Oxburgh, E. R., Geology and metamorphism of Cretaceous rocks in Eastern Carabobo State, Venezuelan Coast Ranges, in *Caribbean Geological Investigations, Mem. 98*, edited by H. H. Hess, pp. 241-310, Geological Society of America, Boulder, CO, 1966.

Pindell, J. L., Geological arguments suggesting a Pacific origin for the Caribbean plate, in *Transactions of the Twelfth Caribbean Geological Conference, St. Crib, Virgin Islands*, edited by D. K. Larue and G. Draper, pp. 1-4, Miami Geological Society, Miami, FL, 1990.

Pindell, J. L., S. C. Cande, W. C. Pitman, D. B. Rowley, J. F. Dewey, J. LaBrecque, and W. Haxby, A plate-kinematic framework for models of Caribbean evolution, *Tectonophysics, 155*, 121-138, 1988.

Pindell, J L., and S. F. Barrett, Geologic evolution of the Caribbean region: A plate tectonic perspective, in *The Geology of North America, Vol. H: The Caribbean Region*, edited by G. Dengo and J. E. Case, pp. 405-432, Geological Society of America, Boulder, CO, 1990.

Priem, H. N. A., D. J. Beets, N. A. I. M. Boelrijk, and E. H. Hebeda, On the age of the late Cretaceous tonalitic/gabbroic batholith on Aruba, Netherlands Antilles (southern Caribbean borderland). *Geol. Mijnbouw, 65*, 247-265, 1986.

Saunders, A. D., Geochemistry of basalts from the Nauru Basin DSDP Legs 61 and 89: Implications for the origin of oceanic flood basalts, *Init. Repts. Deep Sea Drill. Proj., 89*, 499-517, 1985.

Saunders, A. D., J. Tarney, A. C. Kerr, and R. W. Kent, The formation and fate of large igneous provinces, *Lithos, 37*, 81-95, 1996.

Schubert, C., *Historical Geology of the Antillean-Caribbean Region*, Hafner, New York, 811 pp., 1935.

Searle, M. P., and J. Malpas, Petrochemistry and origin of sub-ophiolite metamorphic and related rocks in the Oman mountains, *J. Geol. Soc. London, 139*, 5-248, 1982.

Sen, G., R. Hickey-Vargas, D. G. Waggoner, and F. Maurrasse, Geochemistry of basalts from the Dumisseau Formation, southern Haiti: implications for the origin of the Caribbean sea crust, *Earth Planet. Sci. Lett., 87*, 423-437, 1988.

Sengör, A .M. C., and A. H. Okuogullari, The role of accretionary wedges in the growth of continents: Asiatic examples from Argand to plate tectonics, *Ecol. Geol. Helv., 84*, 535-588, 1991.

Sinton, C. W., and R. A. Duncan, Temporal evolution of the Caribbean Cretaceous basalt province, results of ^{40}Ar-^{39}Ar dating, *Eos Trans. AGU, 73*, 532, 1992.

Sinton, C. W., R. A. Duncan, and M. Storey, ^{40}Ar-^{39}Ar ages from Gorgona Island, Colombia and the Nicoya Peninsula, Costa Rica (abstract), *Eos Trans. AGU, 74*, 553, 1993.

Sleep, N. H., M. A. Richards, and B. H. Hager, Onset of mantle plumes in the presence of pre-existing convection, *J. Geophys. Res., 93*, 7672-7689, 1988.

Spadea, P., A. Espinosa, and A. Orrego, High-Mg extrusive rocks from the Romeral Zone ophiolites in the southwestern Colombian Andes, *Chem. Geol., 77*, 303-321, 1989.

Stearns, C., F. J. Mauk, and R. Van der Voo, Late Cretaceous to early Tertiary paleomagnetism of Aruba and Bonaire, Netherlands Antilles, *J. Geophys. Res., 87*, 1127-1141, 1982.

Stein, M., and A. W. Hofmann, Mantle plumes and episodic crustal growth. *Nature, 372*, 63-68, 1994.

Storey, M., J. J. Mahoney, L. W. Kroenke, and A. D. Saunders, Are oceanic plateaus sites of komatiite formation? *Geology, 19*, 376-379, 1991.

Sun, S.-s, and W. F. McDonough, Chemical and isotopic systematics of oceanic basalts: Implications for mantle compositions and processes, in *Magmatism in the Ocean Basins, Spec. Publ. 42*, edited by A. D. Saunders and M. J. Norry, pp. 313-345, The Geological Society, London, 1989.

Tarney, J., and C. E. Jones, Trace element geochemistry of orogenic igneous rocks and crustal growth models, *J. Geol. Soc. Lond., 151*, 855-868, 1994.

Thompson, R. N., and S. A. Gibson, Sub-continental mantle plumes, hotspots, and pre-existing thinspots, *J. Geol. Soc. Lond., 148*, 973-977, 1993.

Tistl, M., K. P. Burgath, A. Höhndorf, H. Kreuzer, R. Muñoz, and R. Salinas, Origin and emplacement of Tertiary ultramafic complexes in northwest Colombia: Evidence from geochemistry and K-Ar, Sm-Nd and Rb-Sr isotopes, *Earth Planet. Sci. Lett., 126*, 41-59, 1994.

von Huene, R., and D. W. Scholl, The return of sialic material to the mantle indicated by terrigenous material subducted at convergent margins, *Tectonophysics, 219*, 163-175, 1993.

Wadge, G., and R. MacDonald, Cretaceous tholeiites of the northern continental margin of South America: The Sans Souci Formation of Trinidad, *J. Geol. Soc. London, 142*, 297-308, 1985.

Wadge, G., T. A. Jackson, M. C. Isaacs, and T. E. Smith, The ophiolitic Bath-Dunrobin Formation, Jamaica: Significance for Cretaceous plate evolution in the northwestern Caribbean, *J. Geol. Soc. London, 139*, 321-333, 1982.

Weber, M., J. Tarney, and A. C. Kerr, Crustal and mantle xenoliths from volcanic breccias, SW Colombia: Implications for the crustal structure of the northwestern Andes, *Eos Trans. AGU, 76*, p. F376, 1995.

Wells, R. E., Origin of the oceanic basement of the Solomon Islands arc and its relationship to the Ontong Java Plateau – Insights from Cenozoic plate motion models, *Tectonophysics, 165*, 219-235, 1989.

Wieldmann, J., Ammonites from the Curaçao Lava Formation, Curaçao, Caribbean, *Geol. Mijnbouw, 57*, 361-364, 1978.

White, R. S., Magmatism during and after continental break-up, in *Magmatism and the Causes of Continental Break-up, Spec. Publ. 68*, edited by B. C. Storey, T. Alabaster, and R. J. Pankhurst, p. 1-16, The Geological Society, London, 1992.

Wildberg, H., Der Nicoya-Komplex, Costa Rica, Zentralamerika: Magmatismus und Genese eines polygenetischen Ophiolith-Komplexes, *Munstersche Forschungen zur Geologie und Palaontologie, 62*, 1-123, 1984.

Rajmahal Basalts, Eastern India: Mantle Sources and Melt Distribution at a Volcanic Rifted Margin

W. Kent and A. D. Saunders

Department of Geology, University of Leicester, Leicester LE17RH, United Kingdom

P. D. Kempton

NERC Isotope Geosciences Laboratory, Kingsley Dunham Centre, Keyworth, Nottingham NG12 5GG, United Kingdom

N. C. Ghose

Department of Geology, Patna University, Patna 800 005 Bihar, India

Basalts in the Rajmahal Hills represent the edge of a >200-m-thick sequence of predominantly tholeiitic lavas erupted in the Bengal Basin, eastern India. The $^{40}Ar/^{39}Ar$ ages of Rajmahal lavas and dikes cluster at 116–113 Ma. These ages indicate that volcanism in eastern India occurred within a longer period of igneous activity in west Australia (130–100 Ma) and was contemporaneous with the final stages of basaltic volcanism at Ocean Drilling Program Sites 749 and 750 on the central Kerguelen Plateau. Pb-Nd-Sr isotope ratios of the least crustally contaminated Rajmahal basalts differ from those of most Kerguelen Plateau lavas and appear to reflect an Indian mid-ocean ridge basalt (MORB)-like source with $^{206}Pb/^{204}Pb$ ~17.9, $^{207}Pb/^{204}Pb$ ~15.5, $\varepsilon_{Nd(t)}$ >+5, and $^{87}Sr/^{86}Sr_{(t)}$ <=0.7037. The isotopic similarities with Indian MORB are best explained if Rajmahal basalts were generated by decompressional melting of asthenosphere welling up passively beneath the rifted margin of eastern India. Upwelling could have been initiated and driven by viscous coupling of MORB-source mantle to flow in the conduit of the Kerguelen plume, located about 1000 km to the south of eastern India at 116 Ma. The MORB-source mantle lay at the edge of the thermal halo defining the plume boundary and may have been heated by the plume. We infer that the Rajmahal basalts are examples of lavas which, though associated spatially and temporally with the magmatic products of hot spot activity, were derived from compositionally 'normal' asthenosphere.

1. INTRODUCTION

The volume of melt generated during continental break-up is enhanced significantly where lithospheric rifting occurs above unusually hot asthenosphere (100–200°C hotter than ambient mantle), in the presence of a mantle

Large Igneous Provinces: Continental, Oceanic, and Planetary Flood Volcanism
Geophysical Monograph 100
Copyright 1997 by the American Geophysical Union

plume [e.g., *Foucher et al.*, 1982; *White and McKenzie*, 1989, 1995]. Differences in the composition and thickness of igneous crust across volcanic rifted margins may reflect variations in the rate and duration of lithospheric extension, mantle temperature (distance of the plume center from the developing rift system), mantle flow rate within the plume conduit, and the ease with which melt is intruded laterally in the crust [e.g., *Pederson and Ro*, 1992; *Keen et al.*, 1994]. These variations can be modeled quantitatively only in igneous provinces where exposure is excellent, the age and chemical composition of rifted margin basalts are known, and seismic and borehole data are available to estimate the timing of lithospheric extension, uplift and subsidence. Unfortunately, there are very few provinces where all of these criteria are met satisfactorily. For example, tholeiitic lavas on the rifted margins of continents which once formed parts of eastern Gondwana (India, Australia) generally are well-exposed, but are poorly sampled and thus not fully understood. One of the least-known volcanic rifted margins is that of eastern India, where the Early Cretaceous Rajmahal basalts crop out on the western margin of the Bengal Basin. In this paper, we present a comprehensive chemical and Pb-Nd-Sr isotopic study of the Rajmahal basalts and their associated dikes. Our results are used to assess the petrogenesis of the basalts, the nature of mantle sources contributing to volcanism, and controls on the distribution of Cretaceous igneous rocks along the eastern Indian rifted margin.

Evidence from Cretaceous plate tectonic reconstructions [e.g., *Davies et al.*, 1989; *Royer and Coffin*, 1992] suggests that the Rajmahal basalts form part of a large igneous province that includes rifted margin basalts in southwest Australia (Bunbury, Naturaliste Plateau), and lavas forming the central and southern Kerguelen Plateau (Figure 1). On the basis of chemical and Pb-Nd-Sr isotopic data, this igneous province has been attributed to the activity of the Kerguelen hot spot [e.g., *Mahoney et al.*, 1983, 1995; *Storey et al.*, 1989, 1992; *Kent*, 1991; *Salters et al.*, 1992; *Frey et al.*, 1996]. However, this link has been challenged on the basis of alternative plate reconstructions, suggesting that the Rajmahal basalts have a connection with the offshore 85°E Ridge (Figure 2) [*Curray and Munasinghe*, 1991]. In these reconstructions, the plume responsible for Rajmahal and 85°E Ridge volcanism is the Crozet plume, currently located beneath the Crozet Plateau at ~46.2°S. A Crozet plume source for the 85°E Ridge was tested by *Müller et al.* [1993] using a revised model for global plate motions relative to hot spots. *Müller et al.* found no evidence for Early Cretaceous magmatism above the Crozet plume, but matched the southern part of the 85°E Ridge (to 10°N) with a probably now inactive plume

located beneath the eastern Conrad Rise (~53.4°S; Figure 2). The origin of the 85°E Ridge north of 10°N was not discussed, but *Müller et al.* argued that the Rajmahal lavas were not directly produced by the Kerguelen hot spot. In their reconstruction for 118 Ma, the Kerguelen plume was predicted to lie 1000±400 km to the south of eastern India. A further challenge to the Kerguelen plume link has come from *Anderson et al.* [1992], who proposed that Cretaceous lavas in and around the eastern Indian Ocean are manifestations of decompressional melting above a 'hot cell', or large region of unusually hot upper mantle. This hot cell is one of several low-velocity anomalies in the convecting mantle inferred on the basis of seismic tomographic experiments. The cell was suggested by *Anderson et al.* to be much larger than individual present-day Indian Ocean plumes, which are 'embedded' in the low-velocity anomaly.

In addition to disagreements over which mantle plume, if any, is responsible for Cretaceous igneous activity on the eastern Indian margin, there is controversy over the extent of plume involvement in Rajmahal magmatism. In the absence of a clear 'hot spot' signature in Rajmahal basalts, *Mahoney et al.* [1983] proposed that the Kerguelen plume furnished heat, but not material, to Rajmahal volcanism. However, a decade later, ODP (Ocean Drilling Program) Leg 120 drilled volcanic basement on the central Kerguelen Plateau with Nd-Sr isotopic ratios and incompatible trace element abundances not dissimilar to those of the Rajmahal basalts [*Salters et al.*, 1992; *Storey et al.*, 1992]. These chemical and isotopic similarities were dismissed rather summarily by *Curray and Munasinghe* [1992], who favored a completely separate origin for the two features. This is in part due to the difficulty of reconciling the chemical data (which suggest some kind of link) with the apparent distance of eastern India from the center of the Kerguelen plume on Cretaceous plate reconstructions [e.g., *Müller et al.*, 1993] (note that these reconstructions assume an axisymmetric plume head located at ~49°S). Given that plate tectonic reconstructions by *Davies et al.* [1989] and *Royer and Coffin* [1992] imply a plume in close proximity to the eastern Indian margin, this difficulty is possibly an artifact of the particular hot spot reference frame employed in the *Müller et al.* [1993] reconstruction. Regardless of which plate reconstructions are employed, the chemical-isotopic measurements from Rajmahal and Kerguelen Plateau basalts require further explanation. Are the similarities merely accidental, or were both sets of basalts produced from the same mantle source? To answer this question, we need to consider the Rajmahal basalts in a local context (i.e., in terms of the Mesozoic evolution of the eastern Indian margin) and a regional

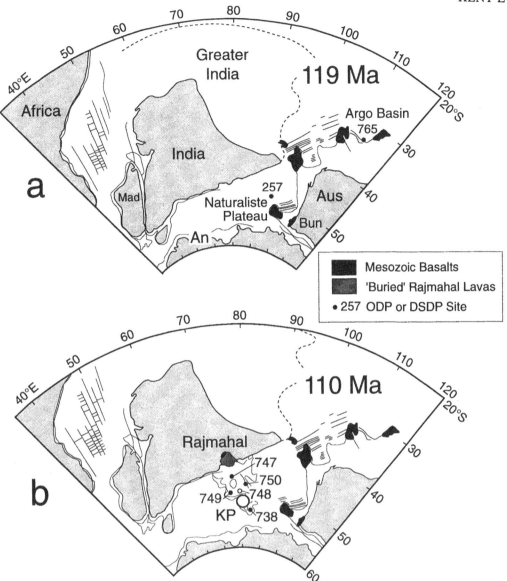

Figure 1. (a) Plate tectonic reconstruction of part of eastern Gondwana at 119 Ma [modified from *Royer and Coffin*, 1992] showing locations of Mesozoic basalts along the west Australian margin. Selected DSDP and ODP sites are shown by dots. Bathymetric contours for the Kerguelen Plateau are shown at 2000 m and 4000 m, and seafloor magnetic anomalies by thin lines. The inferred northern part of the Indian continent ('Greater India') is shown by dashed lines. Abbreviations: An - East Antarctica, Aus - Western Australia, Bun - Bunbury basalts, Mad - Madagascar. (b) Plate tectonic reconstruction of part of eastern Gondwana at 110 Ma [modified from *Royer and Coffin*, 1992] showing locations of the Rahmahal-Sylhet igneous province and ODP Leg 119 and Leg 120 sites on the Kerguelen Plateau (KP). The presumed center of the Kerguelen plume [after *Mahoney et al.*, 1995] is shown by an open circle.

context (in regard to their relationship to other Cretaceous igneous rocks in and around the eastern Indian Ocean).

2. GEOLOGICAL BACKGROUND

2.1. *The Rajmahal-Sylhet Igneous Province*

The Rajmahal Hills (approximately 24°15' to 25°15'N, 87°20' to 87°45'E) extend over an area of ~4100 km² in the Santhal Parganas district of eastern Bihar, some 300 km to the northwest of Calcutta (Figure 3). In any given traverse, up to ten basalt flows are encountered, dipping gently (≤3°) towards the east-southeast. Field studies in the Rajmahal Hills show that the exposed local thickness of the volcanic succession nowhere exceeds 230 m [e.g., *Ball*, 1877; *Kent*, 1992; *Kent et al.*, 1996]. Therefore, reports of at least twenty-five basalt flows with a cumulative

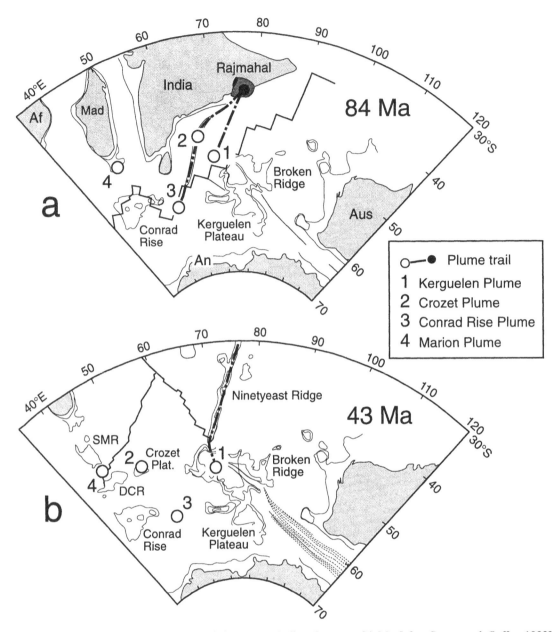

Figure 2. (a) Plate tectonic reconstruction of the eastern Indian Ocean at 84 Ma [after *Royer and Coffin,* 1992] illustrating possible plume tracks proposed by *Curray and Munasinghe* [1991] and *Müller et al.* [1993]. Locations of hot spots (open circles) are inferred from the distribution of Cenozoic seamount and ocean island volcanism. The proposed trail linking the Rajmahal basalts to the Crozet plume follows the line of the 85°E Ridge. The trail of the putative Conrad Rise plume follows the 85°E Ridge to ~10°N. Bathymetric contours for the Kerguelen Plateau and Broken Ridge are shown at intervals of 1000 m (range from 4000 to 1000 m). 'Af' denotes Africa; other abbreviations as in Figure 1. (b) Plate reconstruction of the eastern Indian Ocean at 43 Ma [after *Royer and Coffin,* 1992] showing the southern part of the Kerguelen plume track (Ninetyeast Ridge, the northern Kerguelen Plateau, and Broken Ridge) and the location of the Crozet Plateau. Bathymetric contours for the Kerguelen Plateau and Broken Ridge as in Figure 2a. The Conrad Rise, southern Madagascar Rise (SMR), and Del Cano Rise (DCR) are oceanic plateaus that may have formed above the Marion hot spot.

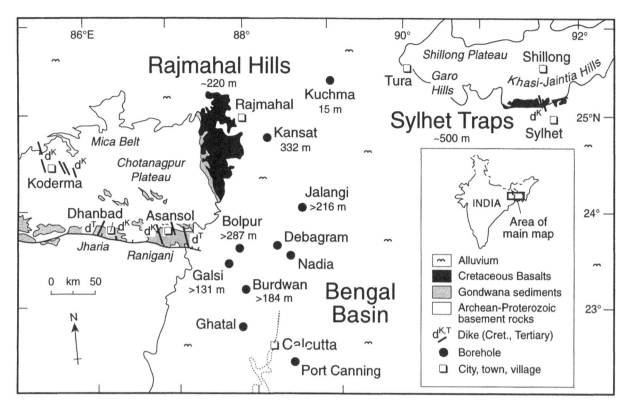

Figure 3. Map of the Rahmahal-Sylhet igneous province, showing principal outcrops of Early Cretaceous basalt, and location of exploration wells drilled in the western Bengal Basin. The minimum thickness of the Rajmahal lavas is indicated for each drill site [after *Biswas*, 1959, 1963; *Khan and Azad*, 1963], whereas thickness estimates for lavas in the Rajmahal Hills and Sylhet region are averages based on field observations [*Kent*, 1992; *Shukla*, 1992; *Kent et al.*, 1996]. Basaltic dikes are shown schematically.

thickness of ~600 m [e.g., *Baksi et al.*, 1987; *Baksi*, 1995] are incorrect. At localities such as at Sahibganj in the northern Rajmahal Hills (Figure 4), the lavas exhibit bole horizons (oxidised flow tops or tuffaceous material), attesting to their eruption in a subaerial environment. However, such features are comparatively rare, possibly due to rapid eruption of the flows.

Although basaltic lavas form >95% of the total exposure in the Rajmahal Hills, small (24–28 km²) rhyolitic lava flows occupy paleovalleys cut in the basalts at Taljhari and Berhait (see Figure 4 for locations) [e.g., *Raja Rao and Purushottam*, 1963; *Deshmukh et al.*, 1964]. Silicic tuffs and bentonites are relatively abundant, occurring at intervals throughout the lava succession. The tuffs and bentonites appear to be particularly common in the northwestern part of the Rajmahal Hills, where individual deposits are usually <1 m thick, but can locally reach thicknesses of up to 14 m [e.g., *Sengupta*, 1988]. These deposits often bear the remains of plants, whose exquisite preservation suggests that they were deposited in shallow

ephemeral lakes on the surface of the basalt flows [*Sengupta*, 1988].

On the western margin of the Rajmahal Hills, the basalts overlie unconformably sedimentary rocks of the Gondwana Supergroup (Early Permian–Early Cretaceous) (Figures 3 and 4). The Cretaceous lavas and Gondwana sediments occupy a small, broadly north-south-trending sub-basin on the western flank of the larger Bengal Basin [e.g., *Sengupta*, 1966; *Mukhopadhyay et al.*, 1986]. This sub-basin is one of several Permo-Triassic basins developed on the eastern flank of the Chotanagpur Plateau, a cratonised mobile belt consisting of >1.1-g.y.-old metamorphic rocks (Figure 3) [*Ghose*, 1983; *Mazumdar*, 1988]. The basement rocks are faulted down to the east, where hydrocarbon exploration wells and geophysical surveys carried out during the 1950s [e.g., *Biswas*, 1959, 1963; *Khan and Azad*, 1963; *Sengupta*, 1966] indicate that the Rajmahal basalts underlie much of the western Bengal Basin. The basalts have a maximum apparent thickness of 332 m in the Kansat borehole, West Bengal (Figure 3)

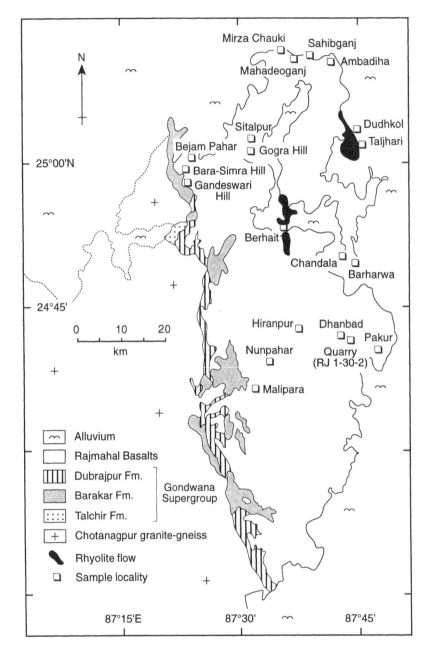

Figure 4. Map showing the distribution of volcanic and sedimentary rocks in the Rajmahal Hills and surrounding areas, and locations of sampling sites. Outline of Rajmahal basalts based on *Ball* [1877]; rhyolite flows based on *Raja Rao and Purushottam* [1963] and *Deshmukh et al.* [1964].

[*Mukhopadhyay et al.,* 1986]. To the east-northeast of the Rajmahal Hills, the lavas thin to just 15 m at the Kuchma borehole [*Khan and Azad,* 1963]. The southern limit of the lava pile is not known with any degree of certainty, due to burial of the Rajmahal basalts beneath the huge Bengal Fan, a Cretaceous–Recent sequence of fluvial sediments up to 22 km thick. Nonetheless, *Curray and Munasinghe* [1992] ventured to suggest that the lava sequence pinches out over a gravity high at ~22°19'N, 88°39'E, corresponding approximately to the margin of the Indian shield. This is consistent with seismic reflection profiles across the western Bengal Basin, which show that the depth to the top of the basalt pile increases from ~3.1 km below sea level at Ghatal to >4 km below sea level at Port Canning (Figure 3) [*Sengupta*, 1966].

The Rajmahal basalts reappear at 25°13'N, 91°21'E, where lavas equivalent in age and chemical composition to rocks in the Rajmahal Hills occupy a 60 × 4 km east-west-trending strip to the southwest of Shillong, Meghalaya (Figure 3) [e.g., *Talukdar and Murthy*, 1970; *Pantulu et al.*, 1992; *Baksi*, 1995]. These basalts, known as the Sylhet Traps, form part of an uplifted block on the northern margin of the Sylhet Trough, Bangladesh [*Johnson and Alam*, 1990]. Although not proven by drilling, they are almost certainly contiguous with basalts in the western Bengal Basin. On this basis, we infer that the Rahmahal-Sylhet igneous province spanned at least 5° of longitude (roughly 87° to 92°E), equivalent to an area of about 2 × 10^5 km². Continuation of the province east of 89°E poses difficulties for *Curray and Munasinghe's* [1991] Crozet hot spot hypothesis for the Rajmahal basalts. The hypothesis rests, in part, on an inferred link between the Rajmahal Hills (~87°E) and the 85°E Ridge, a structure of unknown origin lying offshore of the rifted margin of eastern India (Figure 2). On the other hand, the Sylhet lavas give credence to a hot spot track linking the Rahmahal-Sylhet province to the northernmost (>82-m.y.-old) section of the Ninetyeast Ridge, a >5000-km-long submarine volcanic edifice believed to have formed above the 'tail' of the Kerguelen plume [e.g., *Duncan*, 1991]. As emphasised by *Curray and Munasinghe* [1991, 1992], the northernmost Ninetyeast Ridge shows an eastward curvature beneath the sediments of the Bengal Fan, such that were it to continue northwards, its landfall would be in the vicinity of eastern Bangladesh (Figure 3). Thus, the Sylhet lavas lie exactly in the location predicted by northward extrapolation of the Kerguelen plume track. This could be pure coincidence. On balance, however, it seems unwise to regard the 85°E Ridge as part of the Rajmahal-Sylhet igneous province until more is known about the age of this edifice and its structural relationship to the Indian rifted margin.

Intriguingly, the cumulative thickness of the Sylhet lavas is significantly greater than that of the volcanic succession exposed in the Rajmahal Hills. Between eleven and thirty-five tholeiitic basalt flows with a total thickness of about 300 m are exposed in the West Khasi Hills, south of Shillong (Figure 3) [see *Pantulu et al.*, 1992; *Shukla*, 1992]. The tholeiites are overlain locally by two flows of alkali basalt and approximately 150 m of rhyolitic lavas and tuffs [*Talukdar*, 1967]. A second sequence of tholeiites, some 90–150 m thick, overlies the tuffs [*Talukdar and Murthy,* 1970]. The basalts are separated by

bole horizons, generally 0.1–0.4 m thick, but occasionally up to 14 m thick [*Shukla*, 1992]. The Sylhet lavas have a gentle southerly dip (2-3°), except where overthrust by Precambrian metamorphic rocks of the Shillong Group (northerly dip of up to 50° immediately beneath thrust planes). Basaltic dikes, ranging in thickness from 0.3 to 7.0 m, are abundant within the lava pile and show chilled margins against the lavas [*Talukdar and Murthy*, 1970]. Field evidence noted by *Talukdar and Murthy* suggests that at least some of the dikes were intruded along an east-west-trending monocline formed by folding of the Sylhet flows.

2.2. Age of the Rajmahal Basalts

Radiometric age data for the Rajmahal basalts have been obtained by several workers and are summarised in Figure 5. Most recently, three partially altered tholeiites from the Rajmahal Hills and an alkali basalt from the Bengal Basin were analysed by *Baksi* [1995] using $^{40}Ar/^{39}Ar$ step-heating methods (see also *Dalrymple and Lanphere*

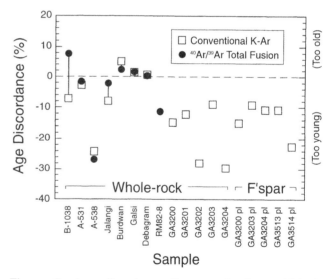

Figure 5. Age discordance diagram, showing published conventional K-Ar and $^{40}Ar/^{39}Ar$ total-fusion ages for Rajmahal whole-rock samples and feldspar separates [*McDougall and McElhinny*, 1970; *Baksi et al.*, 1987, *Baksi*, 1995] as a percentage of the best estimate of the age of the basalts (116.2±0.6 Ma [*Pringle et al.*, 1994]). Whole-rock apparent ages are shown on the left and plagioclase apparent ages on the right. Note that ages obtained by conventional K-Ar are usually lower than the best age estimate, sometimes by as much as 30%.

[1974]). These samples gave isochron ages ranging from 116.9±1.1 Ma to 110.9±1.6 Ma (1σ errors), but did not yield concordant age plateaus. This lack of concordance indicates that the ages obtained are not true crystallisation ages; they reflect loss of ^{40}Ar due to alteration and/or loss of ^{39}Ar from alteration minerals (e.g., clays) during sample irradiation. In contrast, the mean ^{40}Ar/^{39}Ar plateau age of a plagioclase separate from the southwestern Rajmahal Hills (116.2±0.6 Ma, 1σ error) reported by *Pringle et al.* [1994] fulfils rigorous statistical criteria for concordance [see *Pringle*, 1993]. This age currently provides the best estimate for the timing of Rajmahal volcanism, and is used to calculate initial Nd-Sr isotopic ratios for basalts in this study. A major problem with the ^{40}Ar/^{39}Ar database for Rajmahal basalts is that the relative stratigraphic positions of the samples analysed are not known.

In addition to ^{40}Ar/^{39}Ar ages obtained for lavas from the Rajmahal Hills, a tholeiitic lava and basaltic dike from the Khasi Hills (Sylhet) have been dated by K-Ar methods [see *Baksi et al.*, 1987; *Sarkar et al.*, 1996]. These samples gave ages of 108±4 Ma and 110±4 Ma (1σ errors), respectively. These ages appear to reflect loss of radiogenic ^{40}Ar during sample alteration and should be taken as minimum estimates of the true crystallisation ages.

Further age constraints are available from paleomagnetic data [*McDougall and McElhinny*, 1970; *Klootwijk*, 1971; *Poornachandra Rao et al.*, 1993, 1996; *Sherwood and Basu Mallik*, 1996], which show the majority of samples from the Rajmahal Hills and Sylhet Traps to be normally magnetized. However, basalts towards the top of the lava pile (exact position not documented) at three sites in the western Rajmahal Hills show transitional to reversed magnetic polarity [*Klootwijk*, 1971; *Sherwood and Basu Mallik*, 1996]. This observation is surprising, given that the radiometric ages obtained for Rajmahal basalts suggest that they were erupted during the Cretaceous Normal Polarity Superchron (118–84 Ma). In an attempt to reconcile the paleomagnetic and ^{40}Ar/^{39}Ar results, *Kent et al.* [1992b] suggested that the polarity reversal recorded by Rajmahal lavas is the so-called 'ISEA' interval. The ISEA interval, equivalent to chron M-1r of *Gradstein et al.* [1994], is a brief magnetic field reversal dated by high-resolution ammonite and foraminiferal stratigraphy at 115.0±0.3 Ma, i.e., Upper Aptian [e.g., *Tarduno*, 1990; *Gradstein et al.*, 1994]. If our interpretation of the polarity reversal is correct, it provides the best minimum estimate for the age of volcanism in the Rajmahal Hills. We note also that paleomagnetic inclination data obtained by *Klootwijk* [1971] suggest that the Rajmahal basalts were erupted at ~47°S, some 2° north of the presumed present-day latitude of the Kerguelen hot spot [e.g., *Müller et al.*, 1993].

2.3. *Dike Swarms in Eastern India*

K-Ar ages for whole-rock samples of basaltic dikes cropping out 70–250 km to the west and southwest of the Rajmahal Hills (Figure 3) range from 110–75 Ma [*Agrawal and Rama*, 1976]. This age range prompted *Agrawal and Rama* to suggest temporal continuity between Rajmahal and Deccan volcanism. However, this seems most unlikely in the light of modern interpretations of plate motions in the Indian Ocean region (see for example, *Royer and Sandwell* [1989]). Moreover, conventional K-Ar whole-rock ages for partially altered Rajmahal lavas are often younger than the most reliable estimates of crystallisation age, sometimes by up to 30% (Figure 5). This suggests that ages of much less than 110 Ma reflect variable degrees of sample alteration and/or analytical problems.

A lack of temporal continuity between Rajmahal and Deccan magmatism is supported also by preliminary results from ^{40}Ar/^{39}Ar step-heating studies of three dikes from eastern India (R. W. Kent and M. S. Pringle, unpublished data). The results indicate two distinct age populations, one of Aptian age (e.g., ^{40}Ar/^{39}Ar isochron age of whole-rock sample RJ 1-13-2 = 113.4±1.0 Ma, 1σ error) and another of Paleocene age (e.g., ^{40}Ar/^{39}Ar isochron age of the Salma dike = 64.4±0.3 Ma, 1σ error). The former population of dikes occurs over a wide region to the west and southwest of the Rajmahal Hills, whereas the latter appear to be confined largely to the Jharia and Raniganj coalfields of the Damodar Valley (Table 1 and Figure 3). A reconnaissance survey of the Chotanagpur Plateau suggests that dikes of Aptian age are quite scarce or else are poorly exposed; where exposed, they occur as short en échelon topographic ridges up to 8 m wide. Paleocene dikes, distinguished from Aptian dikes by a greater abundance of brown-weathering Fe-Ti oxides, are even less numerous than the Cretaceous dikes, but are up to 40 m wide and extend for several tens of kilometers.

In addition to basaltic dikes, subhorizontal cylindrical sills of lamproite are exposed in the Jharia and Raniganj coalfields [e.g., *Middlemost et al.*, 1988; *Kent et al.*, 1992a; *Rock et al.*, 1992]. We undertook a detailed chemical-isotopic study of the lamproites in order to determine their suitability as contaminants to the Rajmahal basalts. The full results of our lamproite study will be reported elsewhere, but we include relevant isotopic data in the discussion and in certain of the illustrations. In order to determine the age of the lamproites, we carried out ^{40}Ar/^{39}Ar laser-ablation studies of phlogopite separates from two samples collected in the Jharia coalfield (R. W. Kent and S. P. Kelley, unpublished data). These samples gave weighted mean ^{40}Ar/^{39}Ar ages of 116.6±0.8 Ma and

TABLE 1. Sample Localities and Petrographic Details

Sample Number	Locality Information	Specimen Description
Early Cretaceous Dikes: Chotanagpur Plateau and Damodar Valley (DV), Bihar and West Bengal		
RJ 1-12-1	Quarry section, 4 km east of Domchanch, Koderma (Mica Belt)	Cpx & plag-phyric mg dolerite
RJ 1-13-1	Meghatari quarry, 4 km northwest of Koderma (Mica Belt)	"
RJ 1-13-2	"	"
RJ 1-13-3	Baurhi Kalan, NH31 Highway, 22 km northwest of Koderma (Mica Belt)	"
RJ 1-19-1	1 km west of Kalidaspur colliery, Raniganj Basin, eastern DV	Cpx-phyric mg dolerite
RJ 1-19-2	"	"
RJ 1-20-3	2 km northeast of Asansol, Raniganj Basin, eastern DV	Cpx-phyric mg dolerite
RJ 1-20-4	"	"
Early Tertiary Dikes: Damodar Valley, Bihar and West Bengal		
Maheshpur I	Mohuda village, 15 km west of Dhanbad, Jharia Basin, eastern DV	Aphyric mg-cg ferrodolerite
Maheshpur II	"	"
Salma	2 km northeast of Asansol, Raniganj Basin, eastern DV	Aphyric cg ferrodolerite
Early Cretaceous Basaltic Lavas: Rajmahal Hills, Santhal Parganas, Bihar*		
RJ 1-22-1	Malipara village, 2.5 km southwest of Durio	Aphyric mg basalt
RJ 1-23-1	West flank, Gandeswari Hill, 3.5 km southwest of Lalmatia	Plag-phyric fg-mg basalt
RJ 1-23-3	"	Plag-phyric mg-cg basalt
RJ 1-23-4	"	Aphyric cg basalt
RJ 1-23-5	"	Plag-phyric fg-mg basalt
RJ 1-23-7	"	Plag-phyric mg basalt
RJ 1-25-1	North flank of Bara-Simra Hill, 0.5 km southwest of Lalmatia	Plag-phyric fg basalt
RJ 1-26-2	Northwest flank of Saharpur Hill (Bejam Pahar), 2.6 km north of Lalmatia	Plag-phyric mg basalt
RJ 1-26-7	"	Plag-phyric fg basalt
RJ 1-26-9	"	"
RJ 1-27-5	Northeast flank of Gogra Hill, 15 km east-northeast of Lalmatia	Plag-phyric mg basalt
RJ 1-27-6	"	"
RJ 1-30-1	Hiranpur, 17.5 km northwest of Pakur	Aphyric fg basalt
RJ 1-30-2	Quarry, 6 km west of Pakur	"
RJ 1-30-3	Dhanbad village, 9 km west of Pakur	"
RJ 1-30-4	"	"
RJ 1-30-5	"	"
RJ 1-30-6	"	Cpx-phyric fg basalt
RJ 1-31-1	Chandala quarry, 4 km northwest of Barharwa	Aphyric fg basalt
RJ 1-31-3	"	"
RJ 1-31-4	"	Aphyric fg trachyandesite
RJ 1-31-4A	"	"
RJ 1-31-6	"	Aphyric fg basalt
RJ 1-31-7	"	"
RJ 2-1-5	Nunpahar, 1.8 km northeast of Ambajora village	Plag-phyric cg basalt
RJ 2-1-6	"	Plag-phyric mg basalt
RJ 2-2-5	Dhudkol, 2.5 km north-northwest of Taljhari	Plag-phyric mg basalt
RJ 2-2-6	"	"
RJ 2-5-1	Western flank of ridge, 500 m northeast of Sitalpur	Plag-phyric fg basalt
RJ 2-5-2	"	"
RJ 2-5-3	"	Plag-phyric mg-cg basalt

TABLE 1. (continued)

Sample Number	Locality Information	Specimen Description
Early Cretaceous Basaltic Lavas: Rajmahal Hills, Santhal Parganas, Bihar*		
RJ 2-6-1	Northern flank of ridge, 1.5 km south of Sahibganj	Plag-phyric fg basalt
RJ 2-6-2	"	Plag-phyric mg basalt
RJ 2-6-3	"	Aphyric mg basalt
RJ 2-6-4	"	Aphyric fg basalt
RJ 2-6-5	"	"
RJ 2-6-6	"	"
RJ 2-6-7	"	"
RJ 2-7-1	3 km south of Mirza Chauki railway station, 15 km west of Sahibganj	Aphyric fg basalt
RJ 2-7-2	"	"
RJ 2-7-3	"	Cpx-phyric fg-mg basalt
RJ 2-7-4	"	Aphyric fg basalt
RJ 2-7-4A	"	"
RJ 2-7-5	"	"
RJ 2-7-7A	"	"
RJ 2-7-7B	"	"
RJ 2-8-1	Quarry, 1.5 km west of road at Ambadiha, 10 km east of Sahibganj	"
RJ 2-8-2	"	Plag-phyric fg basalt
RJ 2-8-3	"	"
RJ 2-8-4	"	"
RJ 2-8-5	"	Aphyric fg basalt
RJ 2-9-1	0.5 km south of Mahadeoganj, 5 km west of Sahibganj	Aphyric mg basalt
RJ 2-9-2	"	Aphyric fg basalt
RJ 2-9-3	"	"
RJ 2-9-4	"	"
Rhyolites: Rajmahal Hills, Santhal Parganas, Bihar*		
RJ 2-1-3	Nunpahar, 1.8 km northeast of Ambajora village	Aphyric vfg rhyolitic tuff
RJ 2-2-2	Dhudkol, 2.5 km north-northwest of Taljhari	Vesicular fg rhyolite

Notes: Cpx = clinopyroxene, plag = plagioclase. (v)fg, mg, cg = (very) fine-, medium-, and coarse-grained. *For localities in the Rajmahal Hills, samples with the lowest suffix (e.g., RJ 2-9-1) correspond to the first exposed lava flow in the section.

109.1±0.7 Ma (1σ errors). However, resistance furnace heating of phlogopite separated from the first of our two Jharia lamproites (reported by *Pringle et al.* [1994]) gave a concordant $^{40}Ar/^{39}Ar$ isochron age of 113.5±0.5 Ma (1σ). Therefore, the lamproites could be the same age as Aptian basaltic dikes from the Damodar Valley and Chotanagpur Plateau, and slightly younger than lavas in the Rajmahal Hills. If so, the laser ablation analyses could reflect mixing of once-distinct argon compositions with different apparent ages, such that the ages are precise but inaccurate. Work is in progress to clarify the geological meaning of these data. Meanwhile, the single result obtained by resistance furnace heating provides the best estimate of the true crystallisation age of the Indian lamproites.

3. CRETACEOUS VOLCANISM IN AND AROUND THE EASTERN INDIAN OCEAN

Early Cretaceous igneous activity along the west Australian margin and in eastern India was preceded by differential uplift [*Kent*, 1991] and three major phases of extensional or transtensional faulting, culminating in final breakup of the two continents at ~134 Ma [e.g., *Powell et al.*, 1988; *Marshall and Lee*, 1989]. Seafloor spreading between India and Australia began at about 155 Ma, with the formation of oceanic crust in the Argo Basin off northwest Australia (Figure 1) [*Ogg et al.*, 1992]. Over a period of at least 25 m.y., and perhaps as much as 55 m.y., thick lava sequences were erupted along the margin of western

Australia (Figure 1). These lava sequences appear to be youngest in the south [e.g., *Pyle et al.*, 1995], suggesting that the main rift zone between eastern India and western Australia propagated from north to south. It is not known whether equivalent lava sequences were erupted on the eastern Indian margin; in northeast India and Myanmar (Burma), the evidence is obscured by magmatism and sedimentation related to the India-Asia collision [see *Gibling et al.*, 1994].

A key question concerns the geodynamic significance of the west Australian margin lavas. Were these rocks generated above a mantle plume(s) or do they result from some aspect of continental breakup, for example, convection-induced melting associated with uplift along the rifted margin [e.g., *Buck*, 1986]? Chemical and isotopic results from Early Cretaceous (130–100 Ma) Bunbury and Naturaliste Plateau lavas, emplaced on the southwest part of the margin, suggest similarities between at least some Australian margin lavas and recent volcanic products of the Kerguelen hot spot (e.g., in terms of La/Nb, Th/Ta, and present-day Nd-Sr isotope ratios) [*Storey et al.*, 1992; *Mahoney et al.*, 1995; *Frey et al.*, 1996]. However, these similarities do not require the Australian lavas to have a mantle plume origin. A major obstacle to interpretation is crustal contamination, which has modified the Pb-Nd-Sr isotopic compositions of these basalts. It is noteworthy that the least crustally contaminated Bunbury and Naturaliste Plateau lavas have initial $^{87}Sr/^{86}Sr$ ratios of ~0.7039 [*Mahoney et al.*, 1995; *Frey et al.*, 1996], i.e., lower (by at least 0.0004) than those of basalts erupted on the Kerguelen Archipelago, the most recent product of the Kerguelen hot spot. This suggests a contribution to southwest Australian margin volcanism from a source other than the Kerguelen plume, most probably the Indian mid-ocean ridge basalt source ($^{87}Sr/^{86}Sr$ ~0.7025–0.7048 [e.g., *Mahoney et al.*, 1992, and references therein]).

Too few chemical data exist to evaluate the possibility of a hot spot influence on magmatism farther to the north. However, the small amount of radiometric age data for lavas on the northwest Australian margin provide some insight into the magmatic development of this region. Basalts on the Scott Plateau, the northernmost sequence of lavas on the west Australian margin, have K-Ar ages essentially the same as those of the oldest Bunbury lavas (132–128 Ma [*von Rad and Exon*, 1983]). If interpreted literally, these ages suggest that volcanism on the Scott Plateau occurred about 25 m.y. after continental breakup (recall that oceanic lavas in the Argo Basin have radiometric ages of about 155 Ma). Additional evidence for post-breakup igneous activity comes from Early Cretaceous bentonites recovered during ODP Legs 122 and 123, the distribution of which implies the existence at ~130

Ma of subaerial silicic volcanic centers along the northwest Australian margin [*Thurow and von Rad*, 1992]. If volcanic activity occurred a considerable time after breakup, we must appeal to a process other than rifted margin uplift and convection-enhanced melting to explain the origin of the basalts. Identification of the process(es) involved, and the mantle source of these rocks, awaits chemical and isotopic investigations and $^{40}Ar/^{39}Ar$ dating of core samples obtained from the Australian margin.

Beginning shortly before 118 Ma (oldest basement K-Ar age; *Leclaire et al.* [1987]), a huge oceanic plateau was constructed on juvenile oceanic crust that had formed between India and Australia-Antarctica. The origin of this edifice, the Kerguelen Plateau, has proved to be perplexing. The northernmost portion underlying the Kerguelen Archipelago is likely to have been produced in two or more stages by the Kerguelen plume. By analogy with the central portion of the plateau (see below), the oldest volcanic basement on the northern section may date from about 118 Ma. Igneous crust probably was added at 43–39 Ma during interaction of the Kerguelen plume 'tail' with the Southeast Indian Ridge (Figure 2b) [e.g., *Royer and Sandwell*, 1989; *Charvis et al.*, 1995]. The central and southern portions of the Kerguelen Plateau have $^{40}Ar/^{39}Ar$ ages in the range 114–85 Ma [*Whitechurch et al.*, 1992; *Pringle et al.*, 1994]. These parts of the plateau have been attributed to the plume 'head' stage of volcanism [e.g., *Davies et al.*, 1989; *Storey et al.*, 1989, 1992; *Weis et al.*, 1989].

Basalts from the central portion of the Kerguelen Plateau (ODP Sites 749 and 750) have major and trace element compositions consistent with generation at the intersection of a plume with a spreading ridge [*Salters et al.*, 1992; *Storey et al.*, 1992]. In contrast, basalts from ODP Site 738 on the southern part of the plateau have compositions suggesting eruption through thinned continental crust [*Alibert*, 1991; *Mahoney et al.*, 1995]. Although not proven, the possibility of blocks of continental lithosphere beneath the southern and central Kerguelen Plateau should be taken seriously. Velocities of 6.6–7.1 km s^{-1} in the lower crust below the Raggatt Basin, southeast of Site 750, allude to the presence of transitional crust comparable in velocity structure to the stretched continental crust of the Rockall Basin, eastern North Atlantic [*Operto and Charvis*, 1995, 1996]. The velocity data were interpreted by *Operto and Charvis* [1995] as evidence for underplated basalt and/or extensional shear zones akin to those recorded by seismic studies on the margins of the eastern North Atlantic [e.g., *White and McKenzie*, 1989; *Reston*, 1993]. By analogy with Hatton Bank in the North Atlantic, it is possible that failed attempts at seafloor spreading prior to continental breakup allowed highly intruded crust from the rifted margins of India and East Antarctica to become detached and stranded in the oceanic realm. Although speculative at present, we note that a Hatton Bank-type origin for portions of the central Kerguelen Plateau is capable of explaining the extraordinarily large range in Pb-Nd-Sr isotope ratios shown by Kerguelen Plateau tholeiites (see Section 4).

The youngest lavas from the central Kerguelen Plateau (85 Ma; ODP Site 747) are of an age similar to ~88 Ma basalts dredged from Broken Ridge (see Figure 2), a large (1000 × 100–200 km) submarine volcanic edifice currently lying some 1800 km to the north of the Kerguelen Plateau [*Duncan*, 1991]. Broken Ridge and Site 747 lavas represent the final stages of Kerguelen Plateau volcanism, the cessation of which was marked by a phase of accelerated seafloor spreading in the eastern Indian Ocean and construction of the Ninetyeast Ridge between about 82 Ma and 38 Ma (Figure 2b) [*Duncan*, 1991]. In common with lavas from the central Kerguelen Plateau, most basalts from DSDP (Deep Sea Drilling Project) and ODP sites on the Ninetyeast Ridge have major and trace element compositions suggestive of moderate to high degrees of partial melting (10–30%) beneath juvenile oceanic lithosphere [*Frey et al.*, 1991; *Saunders et al.*, 1991; *Kent and McKenzie*, 1994]. Plate tectonic reconstructions by, for example, *Royer and Sandwell* [1989] suggest that the Ninetyeast Ridge formed close to a spreading axis. Thus, from the Early Cretaceous to the Mid-Tertiary, a period of some 75 m.y., the eastern Indian Ocean ridge system appears to have maintained a position close to the center of the Kerguelen plume. By analogy with Iceland, this was probably accomplished via a series of rift jumps or propagation episodes as the Indian plate drifted northwards over the hot spot.

4. GEOCHEMISTRY

4.1. *Samples and Analytical Methods*

The basalt and rhyolite samples analysed in this study were collected from localities listed in Table 1 and shown in Figures 3 and 4. Care was taken to avoid lava flow tops containing analcite, stilbite, chabazite, laumontite, agate, and chalcedony. Samples range from the very fresh (all our Rajmahal basalts have weight-loss-on-ignition values of <1%) to the moderately altered (trachyandesites, weight loss on ignition of 3.6–3.7%). Specimens were selected for chemical analysis on the basis of thin sections lacking zeolites and obvious signs of alteration. Chalcedony and zeolites are particularly abundant in the rhyolite samples, and could not be removed in their entirety prior to

crushing. Small (2–3 mm) chips of each sample were hand-picked and ground in an agate barrel. Agate can contain small amounts of galena, requiring a test to be performed in order to estimate the amount of Pb contamination of basalts crushed in the agate barrel. Isotopic and isotope-dilution data obtained from a set of agate-crushed Rajmahal basalt powders were compared with measurements of splits of the same powders crushed in a tungsten carbide mill. Elemental abundances and isotopic ratios of Pb from the two splits were identical within the limits of analytical error (M. Storey, pers. comm.), suggesting that Pb isotope ratios are not affected significantly by the crushing process.

Whole-rock powders were analysed for major and trace elements by X ray fluorescence (XRF) spectrometry. Analyses were carried out at the University of Leicester using two XRF spectrometers: an ARL 8420+ and a Philips PW1400. Details of XRF operating conditions and sample preparation procedures are given by *Saunders et al.* [1991] and *Storey et al.* [1992]. Our results, along with details of precision and accuracy of the data, are presented in Table 2. Following irradiation at the University of London Reactor Centre, rare-earth element abundances and concentrations of Th, Ta, Hf, and Sc for nineteen samples were determined by instrumental neutron activation analysis (NAA) at the University of Leicester. Sample preparation and analytical techniques are similar to those described by *Fitton et al.* [1997]. The NAA results appear in Table 3, together with estimates of analytical uncertainties. Ratios of Nb/Ta and Zr/Hf can be used to check for consistency between NAA and XRF data. The Nb/Ta and Zr/Hf ratios of C1 chondrite are 17.6 and 36.3, respectively [e.g., *Sun and McDonough*, 1989]. Of the nineteen samples analysed by NAA and XRF, only two (RJ 1-20-3 and 1-26-7) have Nb/Ta <13. The non-chondritic Nb/Ta ratios of these samples most probably reflect errors in the measurement of Nb. Zr/Hf ratios in our samples range from 35–41, suggesting that the Zr and Hf abundance data also are reliable.

Isotopic ratios of Nd and Sr were measured for fifteen Rajmahal samples and a Paleocene dike (Salma) at the NERC Isotope Geosciences Laboratory (NIGL), Keyworth. Lead isotopic data were obtained at NIGL (thirteen Rajmahal samples) and at the Isotope Laboratory, University of Hawaii (two samples). Analytical methods at NIGL were described by *Kempton and Hunter* [1997]; those used in Hawaii were described by *Mahoney and Spencer* [1991]. Sr and Pb were run as the metal on single Ta and single Re filaments, respectively, using a Finnigan MAT 262 multicollector mass spectrometer (NIGL) and a VG Sector multicollector mass spectrometer (University of

TABLE 2. X Ray Fluorescence Data for Rajmahal Basalts and Rhyolites and Early Tertiary Dikes

Sample RJ-	1-19-1	1-19-2	1-20-3	1-20-4	1-22-1	1-23-1	1-23-3	1-23-4	1-23-5	1-23-7	1-25-1	1-26-2	1-26-7	1-26-9	1-27-5	1-27-6	1-30-3	1-30-4
	Group I Dikes				Group I Lavas													
SiO_2	53.62	53.03	53.55	52.72	52.84	52.61	52.10	52.95	52.04	51.77	50.95	52.89	51.19	51.49	52.36	52.07	51.33	51.60
TiO_2	1.57	1.51	1.54	1.67	1.74	1.32	1.34	1.32	1.43	1.30	1.31	1.79	2.09	2.12	1.70	1.73	1.78	1.77
Al_2O_3	14.28	14.31	14.14	13.84	13.93	15.45	15.58	15.68	15.86	15.63	15.28	14.25	14.15	13.39	14.26	13.88	13.86	13.96
$Fe_2O_3^*$	11.26	11.09	11.22	11.78	11.66	10.82	10.89	10.67	11.16	10.51	10.92	12.01	12.75	13.03	12.23	11.43	12.86	12.83
MnO	0.16	0.15	0.17	0.16	0.18	0.17	0.16	0.16	0.15	0.18	0.18	0.19	0.20	0.19	0.20	0.17	0.20	0.19
MgO	6.08	6.20	6.02	6.47	5.93	6.76	6.32	6.73	5.66	6.58	6.83	6.10	6.08	5.72	6.42	5.79	6.49	6.44
CaO	9.29	9.84	9.56	9.58	10.38	10.84	10.72	10.89	10.51	11.44	11.29	9.92	10.90	10.29	10.49	10.45	11.09	11.06
Na_2O	2.88	2.75	2.74	2.77	2.70	2.37	2.40	2.35	2.43	2.40	2.59	2.86	2.89	2.67	2.67	2.98	2.58	2.67
K_2O	0.80	0.60	0.56	0.62	0.45	0.12	0.30	0.12	0.29	0.20	0.23	0.39	0.22	0.28	0.26	0.55	0.12	0.17
P_2O_5	0.18	0.17	0.19	0.20	0.19	0.12	0.13	0.12	0.14	0.14	0.12	0.19	0.23	0.23	0.18	0.21	0.17	0.17
LOI	0.57	0.39	0.33	0.26	0.49	0.59	0.67	0.59	0.78	0.32	0.70	0.28	0.35	0.45	0.24	0.58	0.00	0.24
Total	100.69	100.04	100.02	100.07	100.49	101.17	100.61	101.58	100.45	100.47	100.40	100.87	101.05	99.86	101.01	99.84	100.48	101.10
Mg. no.	51.9	52.8	51.8	52.3	50.4	55.5	53.7	55.8	50.4	55.6	55.6	50.4	48.8	46.7	51.2	50.3	50.2	50.1
Nb	4.8	3.1	4.1	4.9	4.9	2.5	3.0	2.1	2.4	2.5	2.8	5.1	4.3	6.7	4.2	5.6	4.0	4.2
Zr	106	101	109	119	121	74	80	72	82	77	72	121	114	140	114	119	107	104
Y	31	28	32	32	36	28	26	27	26	27	24	33	35	38	34	35	32	31
Sr	316	285	287	286	258	237	234	238	243	245	241	244	237	252	247	271	224	222
Rb	22	13	10	10	38	1	6	2	3	2	2	6	6	10	9	40	2	bdl
Th	1	1	1	bdl	bdl	bdl	1	1	bdl	1	1	2	bdl	2	1	bdl	bdl	bdl
Zn	87	85	87	89	89	77	76	77	82	75	73	89	90	102	87	88	91	90
Ga	21	22	24	22	25	21	22	20	22	21	22	24	23	23	22	24	23	23
Ni	32	32	31	40	63	77	75	79	72	75	81	57	70	65	65	64	72	67
Cr	215	235	214	208	166	244	253	244	240	255	247	159	183	158	188	163	213	194
V	242	231	255	256	283	249	232	250	197	244	247	279	290	284	269	278	300	290
Ba	216	318	159	167	161	77	123	77	141	126	79	116	73	103	164	173	61	68

(table continued)

TABLE 2. (continued)

Sample RJ-	Group I Lavas									Group II Dikes				Group II Lavas				
	1-30-5	2-1-5	2-1-6	2-2-5	2-6-1	2-8-2	2-8-3	2-8-4	2-8-5	1-12-1	1-13-1	1-13-2	1-13-3	1-30-1	1-30-2	1-30-6	1-31-1	1-31-3
SiO_2	51.88	52.09	53.93	51.89	51.98	51.43	52.06	51.89	51.88	53.50	53.94	53.03	53.28	54.16	54.02	54.34	54.23	54.49
TiO_2	1.79	2.18	1.70	2.18	1.75	1.76	1.77	1.76	1.77	1.84	1.77	1.73	1.81	1.74	1.72	1.56	1.86	1.84
Al_2O_3	13.69	13.37	14.11	13.31	14.44	14.07	14.21	14.45	14.35	14.20	14.48	14.65	14.23	14.06	14.25	14.49	14.03	14.69
$Fe_2O_3^*$	12.97	13.04	11.16	13.04	11.57	11.98	11.88	11.84	11.48	11.05	10.77	10.82	10.94	10.90	10.89	10.56	11.31	11.17
MnO	0.20	0.19	0.16	0.19	0.18	0.18	0.18	0.19	0.17	0.16	0.16	0.16	0.16	0.16	0.16	0.15	0.16	0.16
MgO	6.32	5.80	6.05	5.78	6.56	6.42	6.48	6.60	6.25	6.17	6.31	6.22	6.14	5.95	6.04	5.96	5.28	5.34
CaO	10.84	10.32	9.63	10.29	10.60	10.43	10.60	10.57	10.56	9.51	9.54	9.33	9.49	9.52	9.47	9.59	9.41	9.09
Na_2O	2.75	2.80	2.67	2.81	2.76	2.76	2.89	2.78	2.75	2.85	2.93	2.88	2.96	2.50	2.76	2.68	2.51	2.95
K_2O	0.18	0.36	0.86	0.36	0.20	0.22	0.21	0.20	0.31	0.84	0.89	0.91	0.81	0.88	0.80	0.68	1.00	1.10
P_2O_5	0.18	0.21	0.21	0.22	0.18	0.18	0.18	0.18	0.18	0.24	0.23	0.23	0.23	0.22	0.22	0.21	0.24	0.24
LOI	0.04	0.12	0.63	0.08	0.27	0.07	0.01	0.15	0.30	0.18	0.35	0.59	0.26	0.26	0.02	0.97	0.72	0.07
Total	100.84	100.48	101.11	100.15	100.49	99.50	100.47	100.61	100.00	100.54	101.37	100.55	100.31	100.35	100.35	101.19	100.75	101.14
Mg no.	49.4	47.1	52.0	47.0	53.1	51.7	52.2	52.7	52.1	52.8	54.0	53.5	52.9	52.2	52.6	53.0	48.3	48.9
Nb	4.3	5.8	4.9	5.8	5.7	5.1	4.9	5.3	4.6	7.1	6.1	7.2	6.6	7.3	7.5	7.0	10.3	8.7
Zr	110	128	101	127	111	114	113	109	114	142	136	140	138	156	164	157	176	182
Y	33	38	36	37	31	32	33	32	32	30	27	30	29	34	34	32	39	37
Sr	233	246	254	250	261	255	256	260	253	373	373	384	371	385	340	345	328	326
Rb	3	2	8	12	2	4	4	3	6	16	17	18	13	11	33	27	19	41
Th	3	1	2	bdl	bdl	bdl	bdl	2	1	1	2	bdl	bdl	1	2	3	4	4
Zn	91	93	91	102	92	91	88	91	93	87	85	87	84	94	91	90	94	93
Ga	24	25	22	24	23	23	23	23	22	22	21	23	25	22	24	21	23	22
Ni	55	80	60	59	80	67	68	71	74	32	32	34	32	32	25	26	29	29
Cr	143	143	196	150	247	239	231	248	229	165	173	179	170	165	172	132	98	101
V	278	288	266	306	259	256	258	263	259	233	231	232	233	254	230	201	226	223
Ba	96	107	180	126	78	96	85	82	76	218	217	219	216	319	278	257	267	347

(table continued)

TABLE 2. (continued)

Group II Lavas

Sample RJ-	1-31-4	1-31-4A	1-31-6	1-31-7	2-2-6	2-5-1	2-5-2	2-5-3	2-6-2	2-6-3	2-6-4	2-6-5	2-6-7	2-7-1	2-7-2	2-7-3	2-7-4	2-7-4A
SiO_2	54.89	55.88	54.57	54.34	51.94	53.40	53.95	52.81	53.70	53.83	54.04	53.44	52.53	54.34	53.58	53.11	52.16	54.19
TiO_2	1.89	1.64	1.83	1.70	1.34	1.68	1.70	1.40	1.72	1.71	1.71	1.74	1.78	1.74	1.69	1.40	1.35	1.75
Al_2O_3	13.45	12.48	14.03	14.24	15.79	14.27	14.32	15.56	14.16	14.26	14.32	14.39	14.22	14.31	14.77	15.20	15.81	14.15
$Fe_2O_3^*$	11.06	10.58	10.70	10.81	9.41	11.13	10.98	9.83	11.01	11.04	11.01	11.07	11.83	11.03	10.56	9.98	9.45	10.70
MnO	0.11	0.13	0.16	0.16	0.14	0.16	0.16	0.16	0.16	0.16	0.16	0.16	0.17	0.16	0.16	0.17	0.14	0.16
MgO	3.36	3.75	5.34	5.98	8.07	6.22	6.11	7.08	5.94	5.94	6.02	6.06	6.37	5.87	5.91	7.17	8.11	5.78
CaO	5.02	6.31	8.93	9.47	10.25	9.54	9.44	10.03	9.41	9.44	9.43	9.47	10.58	9.38	9.38	9.80	10.24	9.17
Na_2O	1.36	1.18	2.92	2.81	2.38	2.68	3.07	3.39	2.70	2.80	2.84	2.73	2.64	2.78	2.82	2.60	2.40	2.79
K_2O	5.11	4.29	0.94	0.77	0.52	1.01	0.75	0.52	1.02	1.03	1.00	0.76	0.34	0.84	0.78	0.50	0.52	0.91
P_2O_5	0.29	0.17	0.25	0.22	0.18	0.22	0.22	0.19	0.22	0.22	0.23	0.23	0.18	0.23	0.17	0.18	0.18	0.23
LOI	3.59	3.68	0.47	0.37	0.19	0.00	0.25	0.52	0.03	0.02	0.10	0.37	0.07	0.24	0.25	0.16	0.14	0.35
Total	100.13	100.09	100.14	100.87	100.21	100.31	100.95	101.49	100.07	100.45	100.86	100.42	100.71	100.92	100.07	100.27	100.50	100.18
Mg. no.	37.8	41.5	50.0	52.5	63.2	52.8	52.7	59.0	51.9	51.8	52.2	52.2	51.9	51.5	52.8	59.0	63.2	51.9
Nb	9.1	8.8	10.5	7.6	5.5	8.1	7.1		8.3	8.9	8.1	8.7	7.3	8.0	8.8	6.0	5.8	8.7
Zr	182	156	186	167	108	162	166		171	170	168	171	165	171	170	118	110	173
Y	66	40	37	34	23	34	34		34	34	33	36	34	34	36	26	23	35
Sr	704	661	323	326	390	352	332		331	331	332	338	335	358	339	337	391	328
Rb	97	81	36	28	5	16	44		29	29	27	26	27	29	30	14	6	33
Th	5	2	6	3	2	2	3		3	4	4	1	4	4	5	2	2	2
Zn	107	97	96	94	76	92	86		92	89	91	93	89	91	93	79	75	94
Ga	15	14	22	23	21	21	23		22	22	24	20	23	23	23	23	20	22
Ni	31	28	28	33	57	34	33		34	33	33	34	33	32	34	39	56	32
Cr	102	92	111	146	324	151	154		146	146	148	150	144	138	146	274	316	147
V	230	123	246	230	189	225	234		221	224	222	230	236	226	247	199	187	232
Ba	754	473	338	268	195	299	276		295	304	312	315	294	339	296	198	190	297

(table continued)

TABLE 2. (continued)

Sample RJ-	2-7-5	2-7-7A	2-7-7B	2-8-1	2-9-1	2-9-2	2-9-3	2-9-4	2-1-3	2-2-2	Mah I	Mah II	Salma	W-1 Meas.	W-1 Rec.	W-2 Meas.	W-2 Rec.
	Group II Lavas								Rhyolites		Early Tertiary Dikes			Standards			
SiO₂	54.19	52.30	52.22	52.99	53.34	53.68	54.37	54.06	78.71	71.40	49.25	49.19	49.55	52.57	52.46		52.44
TiO₂	1.56	1.31	1.33	1.44	1.69	1.70	1.68	1.71	0.97	0.93	2.84	2.88	2.91	1.10	1.07		1.06
Al₂O₃	14.85	16.18	15.73	15.37	14.11	14.16	14.34	14.25	8.26	13.56	12.28	12.26	12.20	15.16	15.00		15.35
Fe₂O₃*	9.89	9.39	9.36	9.92	11.04	11.06	10.94	10.54	2.99	2.26	16.28	16.40	16.44	11.19	11.11		10.74
MnO	0.15	0.13	0.14	0.15	0.16	0.16	0.16	0.16	0.04	0.02	0.22	0.22	0.22	0.17	0.17		0.16
MgO	6.71	8.13	7.98	7.04	5.98	5.93	5.91	5.75	0.48	0.27	5.50	5.53	5.57	6.61	6.62		6.37
CaO	9.74	10.21	10.15	9.69	9.42	9.37	9.21	9.82	0.67	0.55	9.94	9.90	9.95	11.09	11.00		10.87
Na₂O	2.68	2.59	2.54	2.55	2.77	2.82	2.75	2.58	0.57	1.41	2.35	2.39	2.36	2.40	2.16		2.14
K₂O	0.65	0.53	0.52	0.72	1.00	0.96	0.90	0.74	6.26	8.34	0.57	0.58	0.57	0.61	0.64		0.63
P₂O₅	0.21	0.13	0.18	0.18	0.22	0.22	0.23	0.23	0.12	0.23	0.28	0.28	0.29	0.14	0.13		0.13
LOI	0.28	0.00	0.30	0.08	0.00	0.11	0.05	0.69	1.45	1.29	0.00	0.00	0.07				
Total	100.91	100.90	100.45	100.13	99.73	100.17	100.54	100.53	100.52	100.26	99.51	99.63	100.13	101.04	100.36		
Mg. no.	57.6	63.4	63.0	58.7	52.0	51.7	51.9	52.2	24.3	19.0	40.3	40.3	40.4				
Nb	6.2	5.4	8.7	5.4	7.7	8.2	8.4	8.3	21.4	19.7	13.0	12.9	14.2	7.0	9.9	6.4	7.9
Zr	135	108	175	125	170	168	171	151	370	425	196	197	199	96	99	99	94
Y	29	23	35	28	35	35	34	31	37	72	42	43	43	23	26	22	24
Sr	359	387	331	333	334	331	331	362	226	115	217	213	205	185	186	198	194
Rb	21	8	35	22	26	29	30	27	93	238	15	15	15	21	21	23	20
Th	1	2	4	1	3	4	2	3	8	32	1	bdl	1	1	2	3	2
Zn	85	75	95	83	91	91	93	92	64	66	109	112	110	88	84	67	77
Ga	22	21	24	22	24	23	23	23	10	17	26	26	26	19	17		20
Ni	33	56	32	37	34	35	34	31	25	7	66	67	65	73	75	71	70
Cr	251	328	141	275	151	145	149	189	86	13	82	78	84	131	119	92	93
V	222	186	228	209	222	217	228	227	122	57	392	407	414	242	257	248	262
Ba	212	193	301	195	304	305	306	213	732	1045	248	293	211	162	162	182	182

Notes: Major element abundances in weight %, trace elements in ppm. $Fe_2O_3^*$ = all Fe as Fe_2O_3. LOI = weight loss on ignition. Mg. no. = 100 . atomic $Mg/(Mg+Fe^{2+})$, assuming $FeO/Fe_2O_3 = 0.15$. bdl = below detection limit. Estimated relative errors on major and minor elements are ~1%. Precision, as estimated by one standard deviation (in ppm) from the mean, is: Nb = 0.6, Zr = 0.9, Y = 0.6, Sr = 1.4, Rb = 1.0, Th = 1.0, Zn = 1.0, Ga = 1.3, Ni = 1.9, Cr = 2.1, V = 3.4, and Ba = 1.7. Accuracy can be gauged from measured values of standards W-1 and W-2 (average of five analyses, analysed as an unknown) and recommended values [Govindaraju, 1994].

TABLE 3. Neutron Activation Analysis Data for Rajmahal Basalts and Early Tertiary Dikes

Sample RJ-	La	Ce	Nd	Sm	Eu	Gd	Tb	Yb	Lu	Ta	Th	Hf	Sc
Group I Dikes													
1-19-1	7.7	18.0	12.6	3.89	1.52	5.0	0.84	2.63	0.41	0.37	1.12	3.03	34
1-20-3	7.8	17.5	12.9	3.91	1.54	4.8	0.83	2.59	0.39	0.35	1.10	2.81	36
Group I Lavas													
1-26-7	7.0	18.1	14.0	4.30	1.61	5.4	0.90	2.91	0.45	0.37	0.70	3.03	39
1-27-6	7.2	18.1	13.4	4.23	1.62	5.3	0.92	2.85	0.44	0.34	0.87	3.08	39
1-30-5	6.9	16.5	13.1	3.91	1.51	4.9	0.86	2.85	0.43	0.33	0.81	2.87	42
Group II Dikes													
1-12-1	12.0	27.3	18.7	4.78	1.79	5.3	0.91	2.56	0.38	0.45	1.27	3.79	32
1-13-1	11.5	25.8	17.5	4.64	1.71	5.1	0.82	2.43	0.36	0.42	1.18	3.57	32
Group II Lavas													
1-31-4	35.0	54.8	34.8	7.76	2.53	9.5	1.42	4.40	0.67	0.63	4.02	4.78	34
2-7-1	16.8	37.7	22.1	5.38	1.84	5.7	1.02	2.99	0.44	0.54	3.40	4.33	33
2-7-2	16.6	38.1	21.8	5.41	1.83	5.7	1.02	3.06	0.46	0.53	3.22	4.31	32
2-7-3	12.0	26.1	15.5	4.05	1.50	4.1	0.79	2.31	0.34	0.37	1.97	3.07	30
2-7-4	10.8	23.4	14.8	3.70	1.41	4.1	0.73	2.01	0.30	0.38	1.38	2.86	30
2-7-4A	18.2	39.8	23.1	5.58	1.85	5.9	1.06	3.14	0.45	0.55	3.49	4.35	30
2-7-5	12.7	22.3	17.1	4.42	1.60	5.0	0.87	2.42	0.36	0.41	1.86	3.42	33
2-7-7A	10.1	23.6	14.5	3.63	1.42	3.9	0.71	1.98	0.30	0.37	1.39	2.82	28
2-7-7B	18.1	40.1	23.4	5.52	1.87	5.6	1.06	3.12	0.45	0.56	3.57	4.41	33
2-9-3	17.8	37.6	22.8	5.32	1.77	5.7	1.00	2.95	0.47	0.51	3.08	4.21	33
Early Tertiary Dikes													
Mah I	14.9	36.7	25.2	6.40	2.14	7.2	1.34	3.73	0.58	0.88	1.99	5.06	43
Salma	15.2	37.6	25.6	6.53	2.18	7.2	1.38	3.72	0.58	0.89	2.02	5.13	43
Standards													
BOB-1 Meas.	4.7	12.4	10.0	3.00	1.20	3.8	0.73	2.52	0.40	0.41	0.40	2.41	33
BOB-1 Rec.	4.7	13.8	10.7	3.32	1.25	4.2	0.74	2.63	0.44	0.51	0.45	2.53	
BCR-1 Meas.	24.7	53.2	28.9	6.68	2.09	6.7	1.02	3.49	0.50	0.78	6.31	5.22	33
BCR-1 Rec.	24.9	53.7	28.8	6.59	1.95	6.7	1.05	3.38	0.51	0.81	5.98	4.95	33
JB-1a Meas.	36.9	65.8	26.7	5.14	1.54	4.6	0.69	2.14	0.33	1.62	9.64	3.70	29
JB-1a Rec.	38.1	66.1	25.5	5.07	1.47	4.5	0.69	2.10	0.32	2.00	8.80	3.48	29

Notes: All abundances are in ppm. Precision was determined from eight replicate analyses of ocean floor basalt standard BOB-1, except for Sc, which was determined from eight replicate analyses of the USGS basalt standard BCR-1. Precision for each element, as estimated by one standard deviation (in ppm) from the mean, is: La = 0.12, Ce = 0.32, Nd = 0.4, Sm = 0.12, Eu = 0.03, Gd = 0.3, Tb = 0.05, Yb = 0.08, Lu = 0.01, Ta = 0.02, Th = 0.02, Hf = 0.07, and Sc = 1.0. Accuracy can be assessed by comparing recommended values for BOB-1 and JB-1a [see *Saunders et al.,* 1991; *Govindaraju*, 1994] with measured values (mean of seventeen and eighteen analyses, respectively). The standard deviations of the JB-1a analyses are (in ppm): La = 1.03, Ce = 2.01, Nd = 1.0, Sm = 0.13, Eu = 0.05, Gd = 0.1, Tb = 0.06, Yb = 0.07, Lu = 0.03, Ta = 0.06, Th = 0.30, Hf = 0.12, and Sc = 1.0.

TABLE 4. Pb, Nd, and Sr Isotopic and Isotope-Dilution Data for Rajmahal Basalts and an Early Tertiary Dike

Sample RJ-		$^{143}Nd/^{144}Nd_{(m)}$	$^{87}Sr/^{86}Sr_{(m)}$	Nd	Sm	Sr	Rb	$^{87}Sr/^{86}Sr_{(t)}$	$^{143}Nd/^{144}Nd_{(t)}$	$\varepsilon Nd(t)$	$^{206}Pb/^{204}Pb$	$^{207}Pb/^{204}Pb$	$^{208}Pb/^{204}Pb$
Group I Basaltic Dikes													
1-19-1	UL	0.512637	0.70833	a12.6	a3.89	316	21.7	0.70780	0.512495	+0.1	H18.03	H15.60	H38.48
1-19-1	L		0.70821										
Group I Basaltic Lavas													
1-26-7	UL	0.512727	0.70414	14.312	4.316	237	5.6	0.70403	0.512588	+1.9	17.93	15.56	38.12
1-30-5	UL	0.512735	0.70414	12.945	4.008	233	2.8	0.70408	0.512593	+2.0	H17.93	H15.52	H38.14
Group II Basaltic Dikes													
1-12-1	UL	0.512591	0.70520	18.166	4.772	373	16.1	0.70500	0.512471	-0.4	17.99	15.56	38.15
1-13-1	UL	0.512599	0.70862	17.559	4.629	373	17.3	0.70840	0.512478	-0.2	18.01	15.57	38.20
Group II Basaltic Lavas													
1-31-4	UL	0.512252	0.70977	36.291	8.202	704	97.4	0.70911	0.512148	-6.7	17.99	15.66	39.14
2-7-1	UL	0.512316	0.70831	22.535	5.557	358	29.4	0.70791	0.512203	-5.6	17.99	15.65	39.07
2-7-2	UL	0.512365	0.70773	21.642	5.309	339	29.7	0.70731	0.512252	-4.6	17.97	15.63	38.98
2-7-3	UL	0.512396	0.70658	15.895	a4.05	337	14.2	0.70638	0.512279	-4.1	18.03	15.59	38.55
2-7-4	UL	0.512527	0.70505	14.736	3.706	391	5.7	0.70498	0.512411	-1.5	17.98	15.64	39.03
2-7-4A	UL	0.512311	0.70793	22.443	5.460	328	32.7	0.70746	0.512199	-5.7	17.99	15.58	38.63
2-7-5	UL	0.512484	0.70541	17.281	4.454	359	20.8	0.70514	0.512366	-2.4	18.02	15.59	38.51
2-7-7A	UL	0.512518	0.70509	14.952	3.774	387	7.8	0.70499	0.512403	-1.7	17.97	15.64	39.02
2-7-7B	UL	0.512305	0.70799	22.921	5.633	331	34.8	0.70749	0.512192	-5.8	17.96	15.63	38.97
2-9-3	UL	0.512344	0.70764	a22.8	a5.32	331	30.3	0.70720	0.512237	-4.9			

(table continued)

TABLE 4. (continued)

Sample RJ-		$^{143}Nd/^{144}Nd_{(m)}$	$^{87}Sr/^{86}Sr_{(m)}$	Nd	Sm	Sr	Rb	$^{87}Sr/^{86}Sr_{(t)}$	$^{143}Nd/^{144}Nd_{(t)}$	$\varepsilon_{Nd(t)}$	$^{206}Pb/^{204}Pb$	$^{207}Pb/^{204}Pb$	$^{208}Pb/^{204}Pb$
Early Tertiary Dike													
Salma	UL	0.512672	0.70501	[a]25.6	[a]6.53	205	15.0	0.70482	0.512607	+1.0	17.18	15.39	37.76
Standard													
JB-1 Meas.		0.512768	0.70410	26.72	5.20						18.33	15.52	38.52
JB-1a Av.		0.512761	0.70411	26.50	5.15						18.32	15.51	38.49

Notes: Age-corrected to t = 116 Ma (Rajmahal Basalts), t = 65 Ma (Early Tertiary Dike). UL = unleached powder; L = leached powder. Elemental abundances of Nd, Sm, Sr, and Rb are in ppm. Nd and Sm elemental abundances were determined by isotope-dilution, except where indicated ([a] = neutron activation analysis). Concentrations of Sr and Rb were measured by X-ray fluorescence. Data were obtained at the NERC Isotope Geosciences Laboratory (NIGL), except where indicated ([H] = University of Hawaii). Isotopic fractionation corrections are $^{146}Nd/^{144}Nd = 0.7219$, $^{86}Sr/^{88}Sr = 0.1194$. Multiple analyses of the Nd standard J&M gave a value of $^{143}Nd/^{144}Nd = 0.511114\pm15$ (1σ); the Sr standard NBS987 gave a value of $^{87}Sr/^{68}Sr = 0.710237\pm25$ (1σ). Minimum uncertainties are derived from external precision of standard measurements that average 20 ppm (1σ) for $^{143}Nd/^{144}Nd$ and 28 ppm (1σ) for $^{87}Sr/^{68}Sr$. Based on repeated runs of standard NBS981, the reproducibility of Pb isotope measurements is better than $\pm0.1\%$ per a.m.u. Pb isotope ratios are corrected for fractionation relative to the NBS981 standard values of *Todt et al.* [1984]. Total procedural blanks are negligible: <175 pg for Nd, <1.3 ng for Sr, <325 pg for Pb (NIGL) and <40 pg for Pb (University of Hawaii). Uncertainties on Nd and Sm abundances are estimated at ~1%. JB-1a Av. = NIGL average values for standard JB-1a. $\varepsilon_{Nd(t)} = 0$ at the present day corresponds to $^{143}Nd/^{144}Nd = 0.512638$; $\varepsilon_{Nd(t)}$ at 116 Ma corresponds to $^{143}Nd/^{144}Nd_{(t)} = 0.512490$ for $^{147}Sm/^{144}Nd = 0.1967$, and to 0.512555 at 65 Ma.

Hawaii). Both spectrometers were operated in static mode. Nd was run as the metal on a double Re-Ta filament assembly using a VG 354 mass spectrometer (NIGL) operating in dynamic mode. Isotopic ratios and isotope-dilution abundance measurements for our samples are given in Table 4, together with data for standards and procedural blanks, and estimates of analytical uncertainties.

4.2. *Leaching Experiment*

In order to test whether the high measured $^{87}Sr/^{86}Sr$ ratio of Cretaceous dike RJ 1-19-1 is an effect of subaerial alteration, an aliquant of this sample was subjected to a multistep leaching procedure. About 150–200 mg of powder was placed in a sealed beaker on a hotplate at ~120°C for 1 h in 1 ml 6M HCl and 2 ml ultrapure H_2O. The liquid, including suspended particles of the sample, was then removed by pipette and the beaker replenished with fresh HCl. The sealed beaker was placed in an ultrasonic bath. This cycle was repeated several times until the liquid remained colorless upon addition of new HCl. The final leachate was removed, the sample rinsed in ultra-pure H_2O and dried.

Table 4 shows that the measured $^{87}Sr/^{86}Sr$ ratio of the leached aliquant is only slightly lower than that of the unleached sample. This suggests that the high $^{87}Sr/^{86}Sr$ ratio of RJ 1-19-1 is not an effect of alteration, in keeping with the low weight loss on ignition (0.57%; Table 2) and lack of evidence for alteration indicated by petrographic studies. On this basis, and petrographic evidence, we interpret high relative values of $^{87}Sr/^{86}Sr$ in some other Rajmahal basalts to reflect magmatic ratios.

4.3. *Results*

All Early Cretaceous basaltic rocks from our collection are quartz-normative tholeiites, with the exception of sample RJ 2-5-3, an olivine tholeiite (Tables 1 and 2). Within the Rajmahal Hills, it is possible that we sampled the same lava flow on more than one occasion (recall that a maximum of ten basalt flows are present at any one locality). However, we believe this is unlikely, given that most flows appear to extend along strike over distances of only a few kilometers. Due to the small areal extent of individual flows, it has not proved possible to correlate the lavas between each of the various stratigraphic sections examined.

Major and Trace Elements. Cretaceous tholeiite samples analysed in this study (Tables 2 and 3) have chemical compositions similar to basalt flows from the Rajmahal

Hills studied by previous workers [e.g., *Mahoney et al.*, 1983; *Baksi et al.*, 1987; *Sarkar et al.*, 1989; *Storey et al.*, 1992; *Baksi*, 1995]. Elemental abundances of Early Cretaceous dikes from the Chotanagpur Plateau and eastern Damodar Valley, analysed here for the first time, are very similar to those of basaltic lava flows in the Rajmahal Hills. This similarity is consistent with the radiometric age data discussed above and suggests that the dikes and lava flows share a common petrogenesis. Our analyses, together with those of previous workers, indicate a wide range in major element abundances and ratios (e.g., Figure 6). The Rajmahal basalts have SiO_2 and K_2O contents (49.4–55.9 wt% and 0.11–1.10 wt%, respectively) that generally are higher than those of normal MORBs (mid-ocean ridge basalts) erupted at present-day spreading centers in the Indian Ocean (mostly <50 wt% SiO_2 and <0.2 wt% K_2O; see, for example, *Dosso et al.* [1988]). However, the lower bound on Rb abundances (range from 1 to 44 ppm) lies well within the range exhibited by normal MORBs (mean ~0.6 ppm [e.g., *Sun and McDonough*, 1989]). The upper limits of 'immobile' trace element concentrations, such as those of Zr (72–186 ppm), Ta (0.33–0.56 ppm), and Nb (2.1–10.5 ppm) exceed those of average normal MORB (approximately 74, 0.13 and 2.3 ppm, respectively [e.g., *Sun and McDonough*, 1989]). The lower bounds for Ba (61–347 ppm) and Sr (222–391 ppm) lie within the range of values for 'enriched' MORBs, which have mean values of 57 and 155 ppm, respectively [e.g., *Sun and McDonough*, 1989]. However, it is likely that abundances of Ba (and other large-ion lithophile elements) in the Rajmahal basalts reflect, in part, the mobility of these elements in low-temperature fluids circulating through the lava pile. For example, Ba/Rb ratios in our Rajmahal samples range from 4 to 70, in contrast to near-constant Ba/Rb ratios of 11–12 [*Hofmann and White*, 1983] in pristine oceanic basalts. Magmatic abundances of these elements could have been modified also during contamination of the Rajmahal basalts by continental crust (see below).

The Rajmahal basalts were divided by *Storey et al.* [1992] into two chemical groups (I and II) on the basis of different Ti/Zr and Zr/Y ratios for a given value of MgO. Figure 7 shows that these groups are apparent also in our data set. Rajmahal Group I basalts (including samples analysed by previous workers) have Ti/Zr ranging from 82–120 and a mean Zr/Y ratio of 3.2±0.7, whereas Group II basalts have Ti/Zr ratios of 45–78 and a mean Zr/Y ratio of 4.6±0.7. A relatively evolved Rajmahal lava (RM82-10) analysed by *Mahoney et al.* [1983] appears to share features of both groups (e.g., Ti/Zr = 56, Zr/Y = 3.1), as do two "Group II" lavas from the Bengal Basin with Ti/Zr

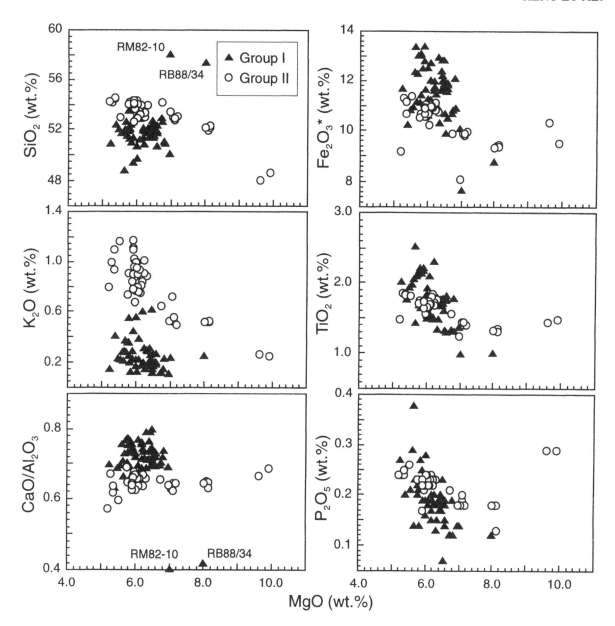

Figure 6. Major element variations for the Rajmahal basalts, versus MgO content. The Rajmahal basalts are divided into two groups (I and II) on the basis of different Ti/Zr and Zr/Y ratios (see Figure 7). In addition to our data, the plots include results obtained previously for Rajmahal lavas [*Mahoney et al.*, 1983; *Storey et al.*, 1992; *Baksi*, 1995]. Samples RM82-10 and RB88/34 have anomalously high SiO$_2$ abundances and low CaO/Al$_2$O$_3$ ratios, and are inferred to be altered.

~100 and Zr/Y >4 [*Baksi*, 1995]. We suspect that for RM82-10 these ratios are artifacts of analytical error, but at present cannot rule out a real distinction between this sample and the Rajmahal Group I and II basalts. The two anomalous Bengal Basin lavas have low Zr abundances (~90 ppm) relative to other incompatible elements and non-chondritic Zr/Hf ratios (>42); these very likely reflect analytical errors in the measurement of Zr. Two other

outliers in Figure 7 are RB88/34 [*Storey et al.*, 1992] and RJ 2-7-7B (Table 2). RB88/34 has low Zr (50 ppm) and on the basis of major element composition (e.g., Figure 6), appears to be altered. RJ 2-7-7B has low TiO$_2$ (1.33 wt%) relative to Y, possibly reflecting analytical error.

Concentrations of SiO$_2$ and K$_2$O in Rajmahal Group I basalts generally are lower than those in Group II rocks (Figure 6), whereas CaO and Sc contents are usually

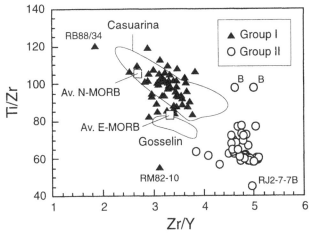

Figure 7. Plot of Ti/Zr vs. Zr/Y, comparing Rajmahal basalts with 129–123 Ma lavas from Bunbury, Western Australia, and average enriched ('E-type') and normal ('N-type') MORB. See text for an explanation of outlying Rajmahal data. 'B' denotes Bengal Basin lavas of *Baksi* [1995]. Bunbury lavas are divided on the basis of different age and chemical composition into two magma types, termed Casuarina and Gosselin [see *Frey et al.,* 1996]. MORB averages are from *Sun and McDonough* [1989].

higher in Group I basalts (9.9–11.8 vs. 8.9–10.3 wt% and 34–42 vs. 28–36 ppm, respectively). Abundances of compatible trace elements in the two groups show wide ranges: for example, Group I basalts have 31–147 ppm Ni and 105–673 ppm Cr, whereas Group II basalts have 18–210 ppm Ni and 45–790 ppm Cr (note that the ranges of values quoted above and below include published data). Rajmahal Group I basalts have low Ce/Y (0.4–1.0) and exhibit a wider range in Zr/Nb (14–34) when compared to Group II basalts (0.9–1.4 and 13–23, respectively). Group I basalts also have low La/Ta (19–22), variable Ba/Ta (165–584) and low to moderate Th/Ta (1.2–3.1). Low Nb and Ta are a feature of both Rajmahal magma types, but are more obvious in Group II basalts because of higher relative abundances of K and La on primitive-mantle-normalised trace element plots (Figure 8).

In addition to basalt samples, we analysed two silicic rocks from the northern Rajmahal Hills (Tables 1 and 2). The Taljhari rhyolite flow, represented by sample RJ 2-2-2, is a micrporphyritic pitchstone locally up to 21.5 m thick. In comparison to two amygdule-free samples of the Taljhari rhyolite analysed by *Raja Rao and Purushottam* [1963] (SiO_2 ~67.5 wt%, K_2O ~3.8 wt%), our rhyolite sample has high SiO_2 (71.4 wt%) and K_2O (8.3 wt%) (Table 2). These values, together with high abundances of Rb and Ba (238 and 1045 ppm, respectively) and the light rare-earth elements, are consistent with the presence in sample RJ 2-2-2 of abundant alkali feldspar and zeolite-

group minerals. Our second silicic sample, RJ 2-1-3, is a rhyolitic tuff with 78.7 wt% SiO_2 and 6.3 wt% K_2O. This specimen has a major element composition similar to that of a silicic tuff from the Sylhet region [see *Talukdar and Murthy,* 1970].

Samples were collected also from two 30-m-wide dikes of Paleocene age (~64 Ma) in the eastern Damodar Valley. The dike specimens (Maheshpur, abbreviated to 'Mah' in Table 2, and Salma) are characterised by low SiO_2 (49.2–

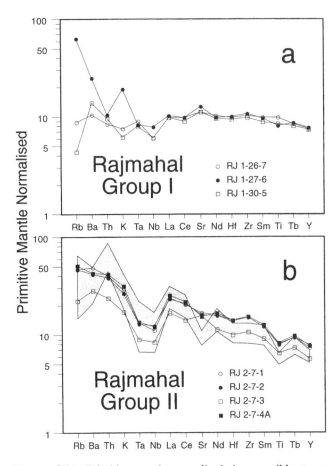

Figure 8.(a) Primitive-mantle-normalised incompatible trace element patterns for Rajmahal Group I basalts. Group I samples have moderately flat patterns (the alteration-prone large-ion lithophile elements excepted), consistent with the total melt fraction being dominated by melts generated in the spinel stability field (<100 km depth). Normalising values for primitive mantle are those of *Sun and McDonough* [1989]. (b) Primitive-mantle-normalised incompatible element patterns for Rajmahal Group II samples show features akin to average crustally contaminated basalts (shaded field) from western India (Bushe Formation, Deccan), Madagascar (East Coast tholeiites) and southern Brazil (Gramado Formation, Paraná Basin). Averages were compiled from references of *Saunders et al.* [1992]. Normalising values are from *Sun and McDonough* [1989].

49.6 wt%), high Fe_2O_3* (16.3–16.4 wt%, where Fe_2O_3* = total Fe), high TiO_2 (2.8–2.9 wt%) and high P_2O_5 (0.28–0.29 wt%). These element abundances, together with low Ni and Cr concentrations (65–67 ppm and 78–84 ppm, respectively) are consistent with extensive fractionation of tholeiitic magma at low pressure. Our chemical and isotopic results suggest that the Maheshpur and Salma dikes are very similar to ferrobasaltic lavas from the upper part of the Deccan Traps sequence at Mahabaleshwar Ghat, western India (for example, sample MB81-18 of *Mahoney et al.* [1982]). The Mahabaleshwar section lies about 1400 km to the west of the Damodar Valley, whereas the nearest exposed Deccan Traps lavas and a Mahabaleshwar-like sill [e.g., *Sen and Cohen*, 1994] lie about 240 km to the southwest of the Damodar Valley. Given the very large size of the Maheshpur and Salma dikes and the observed similarity to Mahabaleshwar ferrobasalts, it is possible that the dikes represent liquids transported eastwards across the Indian subcontinent over a distance of >1000 km.

Isotopic Variations. Figure 9 shows that the new Nd-Sr isotope ratios of Rajmahal basaltic lavas and dikes are similar to those of samples analysed previously [*Mahoney et al.*, 1983; *Storey et al.*, 1992; *Baksi*, 1995]. For our samples only, the total range in $\varepsilon_{Nd(t)}$ is +2.0 to -6.7, where t = 116 Ma. Relative to the Rajmahal lavas, the Cretaceous

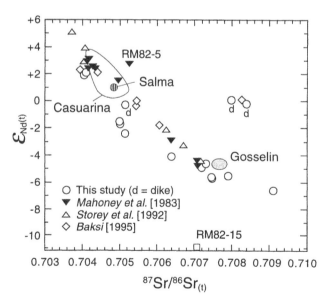

Figure 9. Initial $\varepsilon Nd(t)$ versus $^{87}Sr/^{86}Sr(t)$ for Rajmahal samples from this and previous studies, compared to lavas from Bunbury, Western Australia. RM82-15 is a rhyolitic tuff from the Rajmahal Hills with $\varepsilon Nd(t)$ of -11.1 and $^{87}Sr/^{86}Sr(t)$ of 0.70702 [*Mahoney*, 1984]. 'Salma' is a ferrobasaltic dike of Paleocene age, analysed as part of this study. Fields for the Bunbury lavas are from *Storey et al.* [1992] and *Frey et al.* [1996].

dikes show a restricted range in $\varepsilon_{Nd(t)}$ from +0.1 to -0.4. Values of $^{87}Sr/^{86}Sr_{(t)}$ for unleached rock powders lie between 0.70403 and 0.70911, extending the previous range in initial Sr isotope ratios (0.70385ε0.70828) to significantly higher values. The total range in present-day $^{206}Pb/^{204}Pb$ for our samples is 17.93ε18.03, $^{207}Pb/^{204}Pb$ = 15.52ε15.66, and $^{208}Pb/^{204}Pb$ = 38.12ε39.14.

Figure 9 reveals that the two groups of Rajmahal basalt identified by *Storey et al.* [1992] cannot easily be distinguished using Nd-Sr isotope ratios. Group I basalts are characterised by high $\varepsilon_{Nd(t)}$ (+5.1 to +0.1) and a wide range in $^{87}Sr/^{86}Sr_{(t)}$ (0.7037–0.7084), whereas Group II basalts have $\varepsilon_{Nd(t)}$ values of +0.1 to -6.7 and $^{87}Sr/^{86}Sr_{(t)}$ ranging from 0.7050 to 0.7091. In Figure 9, a Group I dike (RJ 1-19-1) and two Group II samples (RJ 1-13-1 and Jalangi, a sample from the Bengal Basin [*Baksi*, 1995]) lie above the main isotopic array. These rocks have $^{87}Sr/^{86}Sr_{(t)}$ of 0.7080–0.7084 and $\varepsilon_{Nd(t)}$ values of +0.1 to -0.2; they continue the trend from Group I sample RM82-5 [*Mahoney et al.*, 1983] towards moderately low values of ε_{Nd} and high $^{87}Sr/^{86}Sr$. A second group of samples lying offset from the main Nd-Sr isotopic array include RJ 2-7-4, 2-7-5, 2-7-7A, and a rhyolitic tuff (RM82-15 [*Mahoney*, 1984]). These rocks have moderate values of $^{87}Sr/^{86}Sr_{(t)}$ (0.7045–0.7070) and low $\varepsilon_{Nd(t)}$ (-1.5 to -11.1).

In Figures 10 and 11, Group I basalts have measured values of $^{206}Pb/^{204}Pb$, $^{207}Pb/^{204}Pb$, and $^{208}Pb/^{204}Pb$ that are very similar to those of Group II basalts (17.93–18.03 vs. 17.96–18.03, 15.52–15.60 vs. 15.56–15.66, and 38.12–38.48 vs. 38.15–39.14, respectively). Note that we do not include in our plots or discussion the Rajmahal Pb isotopic data illustrated by *Storey et al.* [1992, figures 11 to 13]. Their samples, collected independently of ours, appear to have become contaminated with Pb prior to analysis.

Figures 10 and 11 show that the new Rajmahal isotopic measurements are very similar to those of Cretaceous lavas from the southern Kerguelen Plateau (ODP Site 738), the Naturaliste Plateau, some Bunbury lavas, and some Broken Ridge lavas. In Figure 10, the Rajmahal basalts lie just outside the field of present-day MORBs from the Indian Ocean, but in Figure 11 the two data sets overlap. Lavas from islands in the southern Indian Ocean, represented by the Kerguelen Archipelago, Heard Island, and the McDonald Isles overlap with the field of Rajmahal Group II basalts, raising the possibility that at least some of our samples could have been generated by the Kerguelen hot spot (but see below). In contrast, Recent lavas from the southwest Indian Ocean, represented by alkali basalts from the Crozet Archipelago, show no overlap with the Rajmahal basalts. The Rajmahal Pb isotopic data also generally are unlike those of basalts from the ~80-m.y.-old

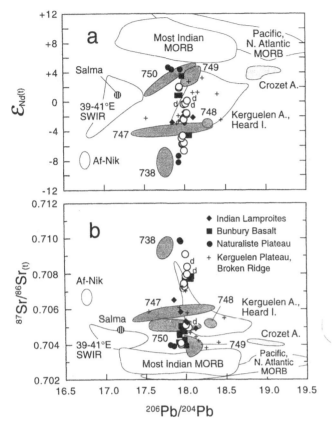

Figure 10.(a) Initial εNd(t) vs. present-day 206Pb/204Pb, and (b) 87Sr/86Sr(t) vs. present-day 206Pb/204Pb for Rajmahal basalts (data from this study, symbols as in Figure 9). The Rajmahal data are compared to our unpublished results for the Damodar Valley lamproites and data for Bunbury [*Frey et al.*, 1996] and Naturaliste Plateau lavas [*Mahoney et al.*, 1995]. We also show fields for the Kerguelen Plateau and Broken Ridge [*Alibert*, 1991; *Salters et al.*, 1992, and V. J. M. Salters, unpublished data; *Mahoney et al.*, 1995], Kerguelen Archipelago, Heard Island, and McDonald Isles [*Storey et al.*, 1988; *Gautier et al.*, 1990; *Weis et al.*, 1993; *Barling et al.*, 1994], the Afanasy-Nikitin seamount (Af-Nik) and Crozet Archipelago [*Mahoney et al.*, 1996], Indian, Pacific, and North Atlantic MORB (references of *Mahoney et al.* [1989, 1992]) and 39°-41°E Southwest Indian Ridge (SWIR) lavas [*Mahoney et al.*, 1992].

Afanasy-Nikitin seamount [see *Mahoney et al.*, 1996; *Sushchevskaya et al.*, 1997]. This seamount, together with the 85°E Ridge and Rajmahal basalts, was believed by *Curray and Munasinghe* [1991, 1992] to lie on the track of the Crozet hot spot. However, unless the isotopic composition of the Crozet plume (as represented by alkali basalts from the Crozet Archipelago) changed markedly between 116 Ma and 80 Ma, and/or components external to the plume head were involved in producing the Rajmahal basalts, the Rajmahal basalts could not have been generated above the Crozet hot spot [cf. *Kent et al.*,

1992d]. Furthermore, it is not at all obvious from the available chemical and isotopic data that Afanasy-Nikitin seamount lavas are related to the Crozet plume source at 80 Ma (see discussion by *Mahoney et al.* [1996]). Nevertheless, until samples are collected from the Crozet Plateau (the ~70-m.y.-old volcanic edifice on which the Crozet Archipelago was constructed) and the 85°E Ridge, it will not be possible to dismiss a link between the Rajmahal basalts and the Crozet hot spot.

5. DISCUSSION

5.1. *Evidence for Fractionation and Contamination of Rajmahal Magmas*

In contrast to olivine tholeiites drilled on the central Kerguelen Plateau, the Rajmahal basalts cluster along the 1

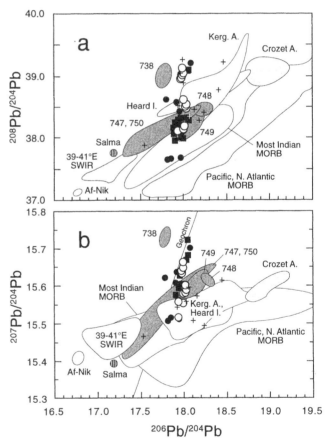

Figure 11.(a) Present-day 208Pb/204Pb vs. 206Pb/204Pb, and (b) present-day 207Pb/204Pb vs. 206Pb/204Pb for Rajmahal basalts, compared to data for other lavas erupted in and around the eastern Indian Ocean. All data are measured ratios except for those for basalts from the Afanasy-Nikitin seamount (age-corrected to 80 Ma [*Mahoney et al.*, 1996]). Symbols, abbreviations, and data sources are as in Figure 10. To avoid clutter, data for the Damodar Valley lamproites are not included.

atmosphere cotectic for olivine, clinopyroxene, plagioclase and liquid in CIPW norm space (not shown). This observation, together with our chemical results in Table 2, is consistent with fractional crystallisation at fairly shallow levels in the crust. At low pressures, plagioclase and clinopyroxene form the bulk of the crystallising assemblage in tholeiitic magmas [e.g., *Green and Ringwood*, 1967], resulting in high CaO/Al_2O_3 and low Sr and Sc concentrations in residual melts. High CaO/Al_2O_3 and Sc and low Sr contents in Rajmahal Group I basalts relative to Group II rocks (Figures 6 and 12) could imply a higher ratio of plagioclase to pyroxene in the crystallising assemblage, consonant with fractionation of Group I magmas at pressures lower than those at which Group II magmas were differentiated. This explanation was favored by *Frey et al.* [1996] as a means of accounting for compositional differences between ~129-m.y.-old and ~123-m.y.-old basalts from the Bunbury Basin, Western Australia (Casuarina- and Gosselin-type lavas, respectively; Figures 7 and 12). However, in Figure 8a, the primitive-mantle-normalised patterns of Rajmahal Group I

basalts show peaks at Sr, the opposite of what one would expect if these rocks had fractionated a large amount of plagioclase relative to Group II basalts. Furthermore, in contrast to the Bunbury lavas, Ga abundances in the two groups of Rajmahal basalt are virtually identical at different abundances of La (Figure 12a). Plagioclase segregation at low pressures should lead to Ga depletion in residual melts, in a manner analogous to the behavior of Sr [e.g., *Goodman*, 1972]. The fact that Ga is not depleted in Group II samples when compared to Group I rocks suggests that the lower CaO/Al_2O_3 and Sc and higher Sr contents of Group II basalts (Figure 12b) reflect a process other than low-pressure fractionation. One possible explanation for the differences is variable, but rather low, amounts of contamination by continental crust. Combined assimilation and fractional crystallisation would tend to result in the most evolved lavas being the most contaminated [e.g., *DePaolo*, 1981]. Figures 6, 12c, and 12d show that this is not true of the Rajmahal basalts [cf. *Mahoney et al.*, 1983]; for example, at ~7 wt% MgO, Group II rocks have SiO_2 and K_2O abundances that are

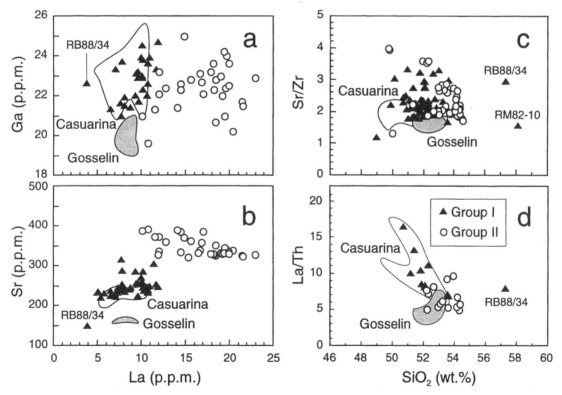

Figure 12. Plots of (a) La vs. Ga, (b) La vs. Sr, (c) Sr/Zr vs. SiO_2, and (d) La/Th vs. SiO_2 for Rajmahal basalts from this work and previous studies [*Mahoney et al.*, 1983; *Storey et al.*, 1992; *Baksi*, 1995]. Fields for Casuarina- and Gosselin-type Bunbury lavas [*Storey et al.*, 1992; *Frey et al.*, 1996] are shown for comparison. Note the locations of data for Rajmahal lavas RM82-10 and RB88/34, inferred on the basis of major element abundances (Figure 6) to be altered.

significantly higher than those of Group I basalts. This suggests that chemical differences between the two groups were established prior to extensive fractionation of low-pressure pyroxene and plagioclase.

Few chemical analyses of Indian crustal rocks and mantle xenoliths exist to identify the contaminant(s) of Rajmahal magmas, but key features of our data set allow us to discern the broad composition of the material involved. Our unpublished isotopic results for unleached samples of Damodar Valley lamproite fall roughly in the middle of the Nd-Pb and Sr-Pb isotopic arrays for Rajmahal basalts (Figure 10), allowing us to firmly rule out lamproite or the lamproite mantle source as the contaminant. This is important, because the Indian lamproites are thought to be representative of metasomatic vein material residing within the lowermost lithosphere [e.g., *Middlemost et al.*, 1988]. If such readily fusible material did not mix with Rajmahal magmas, it is most unlikely that cooler, pristine lithospheric peridotites would have acted as a contaminant (see, for example, the thermomechanical models for plume- lithosphere interactions discussed by *Arndt and Christensen* [1992]). We infer from this that the Rajmahal contaminant is crustal in origin.

The Nd-Sr isotopic ratios of Rajmahal basalts suggest three crustal contamination vectors diverging from a mantle end-member with ε_{Nd} >+5 and $^{87}Sr/^{86}Sr$ ≤0.7037 (Figure 9). The first vector, defined by data plotting above the main Nd-Sr isotopic array, points towards a crustal end-member with ε_{Nd} <-0.2 and $^{87}Sr/^{86}Sr$ >0.7084. The second vector (main array) points towards an end-member with ε_{Nd} <-7 and $^{87}Sr/^{86}Sr$ >0.7091. Rajmahal basalts defining the third vector (below the main array) were contaminated by an end-member with ε_{Nd} <-11 and $^{87}Sr/^{86}Sr$ >0.7070. These vectors could point to three separate crustal end-members, each with similar Pb isotopic characteristics ($^{206}Pb/^{204}Pb$ ~18.0, $^{207}Pb/^{204}Pb$ >15.7, and $^{208}Pb/^{204}Pb$ >39.1), or to a single contaminant that is markedly heterogeneous with respect to ε_{Nd} and $^{87}Sr/^{86}Sr$. In the absence of radiogenic isotope data for eastern Indian crust, we cannot distinguish easily between these possibilities.

If the amount of crust assimilated by Rajmahal magmas was ≤5%, as suggested by limited effects on major element (e.g., SiO_2, K_2O) concentrations, the contaminant(s) may have had very low ε_{Nd} (probably much less than -20) and high $^{87}Sr/^{86}Sr$ (>>0.710). Archean granitoids in western India have these characteristics, with ε_{Nd} of about -40, $^{87}Sr/^{86}Sr$ of about 0.820, and unusually high Sr contents, sometimes in excess of 500 ppm [*Lightfoot*, 1985]. An ancient granitic contaminant of this type, or small-volume

partial melt derived therefrom, is in keeping with slightly high SiO_2 contents and high abundances of Sr in Rajmahal Group II basalts (Figures 6, 8b, and 12b). Primitive-mantle-normalised trace element patterns (e.g., troughs at Nb and Ti in Figure 8b) and Pb isotopic ratios of Group II basalts (in particular, high $^{207}Pb/^{204}Pb$ relative to $^{206}Pb/^{204}Pb$) also are consistent with minor assimilation of old silicic crust. A likely candidate for the contaminant is the Proterozoic Chotanagpur (Bengal) granite-gneiss, which forms the bulk of surface exposures on the eastern Chotanagpur Plateau [*Ghose*, 1983]. Simple calculations suggest that in the presence of a free fluid phase, granite-gneiss would undergo partial melting if heated from an original temperature of ~300°C (appropriate to the crust of the Indian shield at a depth of ~16 km [e.g., *Negi et al.*, 1986]) to a temperature in excess of 800°C. Assimilation of small melt-fractions of Chotanagpur granite-gneiss could have occurred, for example, at the margins of dikes supplying the Rajmahal lavas. Modeling of the contamination process awaits trace element and isotopic measurements on samples of the granite-gneiss.

5.2. Depth and Degree of Partial Melting

In light of suggestions of a petrogenetic link between Rajmahal and Kerguelen Plateau basalts [e.g., *Davies et al.*, 1989; *Storey et al.*, 1992], it is pertinent to consider the melting conditions under which both suites of rocks were produced. We adopt the fractional melting inversion of *McKenzie and O'Nions* [1991, 1995], incorporating the modifications proposed by *White et al.* [1992], as a means of placing quantitative bounds on melting conditions. The inversion utilises averaged concentrations of the rare-earth elements to estimate the melt distribution as a function of depth, the total integrated melt fraction, and the total melt thickness, equivalent to the thickness of basaltic crust produced. Concentrations of the major and minor elements are then predicted from the results of the rare-earth element inversion [e.g., *White and McKenzie*, 1995]. Assumptions are made with regard to starting composition (i.e., the mantle source is specified as having a chemical composition that is primitive or depleted in light rare-earth elements relative to the bulk earth), source mineralogy, phase proportions entering the melt, and mineral-melt partition coefficients. Only data for samples with ≥6 wt% MgO were used. The inversions were run assuming that the garnet-spinel transition occurs between 80 and 100 km depth, equivalent to a mantle potential temperature of 1500°C (the potential temperature is the temperature that mantle peridotite would have if it ascended adiabatically to the surface without melting).

Figure 13a shows rare-earth element measurements for the Rajmahal basalts (average of four Group I and seven Group II samples) normalised to estimated concentrations in the bulk earth. Observed mean concentrations for the Rajmahal basalts are shown by heavy dots, with error bars indicating one standard deviation, estimated from the variance of the data. The heavy solid line shows the best fit to the data obtained by inversion. The mantle source producing the best fit is primitive mantle (single-stage melting of a depleted source composition cannot produce a good fit to the light rare-earth element abundances of crustally contaminated basalts; see below and *Brodie et al.* [1994]). The fit is least good for Eu, but for all rare-earth elements is well within the error bars. Figures 13b and 13c

illustrate the observed (dots) and predicted (solid lines) concentrations of major and other trace elements using the melt distribution with depth inferred from the inversion fit. Figure 13d shows the partial melt distribution required to produce this fit, with the dashed line illustrating the inferred original melt distribution after correction for 15.5% olivine fractionation (the correction is based on the misfit between calculated and observed average MgO and total Fe as FeO; see *McKenzie and O'Nions* [1991]).

An effect of including Group II basalts in the inversion model is that although the amount of crust added to these basalts is inferred to be modest (≤5%), the contaminant imparts a weak light rare-earth element/heavy rare-earth element fractionation signature that is not a result of

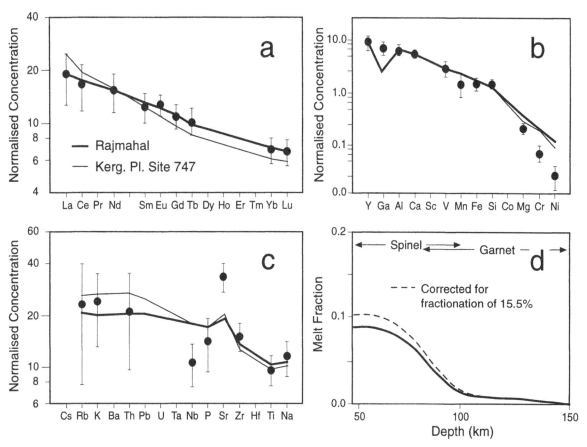

Figure 13.(a) Observed and calculated rare-earth element concentrations, normalised to bulk earth, for Rajmahal basalts with ≥6 wt% MgO (average of neutron activation data from this study) [cf. *McKenzie and O'Nions*, 1991, 1995]. Mean values for each element are shown by dots. Vertical bars illustrate one standard deviation, estimated from the variance of the data. The heavy line shows the best fit to the data obtained by fractional-melting inversion. Results obtained for Kerguelen Plateau ODP Site 747 are shown for comparison. (b) and (c) Predicted concentrations of major and other trace elements using the fractionation-corrected primary melt distribution (panel d) inferred from the rare-earth element inversion. (d) Uncorrected (heavy line) and corrected (short dashed line) melt distributions for Rajmahal basalts, estimated from the rare-earth element inversion fit. The mantle source used in the inversions is primitive mantle [see *McKenzie and O'Nions*, 1991, 1995].

melting in the mantle source (compare Figures 8a and 8b). This is expressed in the melt distribution model (Figure 13d) by a low-amplitude tail at depth >100 km, which may also include small melt-fractions generated in the garnet stability field; there is no significant effect on the melt distribution at shallower levels. If the tail is excluded, the resultant fit to the highly incompatible trace element concentrations shows a marked deterioration. This result suggests that the inversion model for Rajmahal basalts is relatively insensitive to minor amounts of crustal contamination (the effect is similar to adding further low-degree melts from the garnet stability field [cf. *Brodie et al.*, 1994]), and that inclusion of Group II samples in the model does not obscure conditions of melt generation within the asthenosphere.

The upper limit of melting in the inversion is ~50 km, typical of basalts generated beneath stretched and thinned continental lithosphere (see, for example, figures 13 to 17 of *White and McKenzie* [1995]). The lower limit of melting is not clearly defined, due to uncertainty over the amount of melt generated in the garnet stability field. The maximum melt fraction estimated from the inversion is about 10% (Figure 13d) or about two-fifths less than *McKenzie and O'Nions* [1991] estimate for the average extent of melting for normal MORB. The total melt thickness generated, after fractionation correction, is 5.1 km, or some 2 km less than the average thickness of oceanic crust [e.g., *White et al.*, 1992]. The melt thickness estimate is much greater than the observed thickness of the Rajmahal-Sylhet lava pile. This discrepancy could arise in several different ways. Some of the melt could have been redistributed by lateral flow, consistent with the occurrence of dikes to the west and southwest of the Rajmahal Hills and on the Shillong Plateau (Figure 3). Alternatively, basalt could have been removed over the last 116 m.y. by erosion. Neither of these processes are likely to account for the large discrepancy noted above. However, it is possible that melt was trapped at the Moho or in the lower crust. This inference is reasonable because the Moho probably acts as a density filter for basaltic melts which have a density of ~2.8 kg m^{-3} [e.g., *Huppert and Sparks*, 1980]. Unfortunately, a lack of modern seismic profiles across the eastern Indian margin means that at present we cannot test the hypothesis that igneous underplating has contributed to the ~35-km-thick crust beneath the southern Bengal Basin [e.g., *Brune and Singh*, 1986].

In order to place the Rajmahal results in a regional context, we used the inversion scheme also to model ODP data for the Kerguelen Plateau [*Storey et al.*, 1992] and data for Loranchet Peninsula tholeiites, northern Kerguelen Archipelago [*Storey et al.*, 1988]. Inversion results for

ODP sites on the Ninetyeast Ridge were reported by *Kent and McKenzie* [1994]. All of these lavas were probably erupted in close proximity to the Kerguelen hot spot; as noted above, the Rajmahal basalts possibly were erupted ~1000 km away from the center of this hot spot. Our modeling assumes that basalts sampled at the ODP sites (i.e., the uppermost flows) are representative of the entire igneous crustal section in terms of their rare-earth element concentrations. This is unlikely to be true, but the inversions do at least provide a quantitative means of comparing available data.

Figure 14 shows the melt distributions obtained for Kerguelen Plateau and Loranchet Peninsula basalts, corrected for the effects of olivine fractionation. Details of the fractionation correction applied to lavas from each locality are given in the caption to Figure 14. The estimated total melt fraction and total melt thickness are

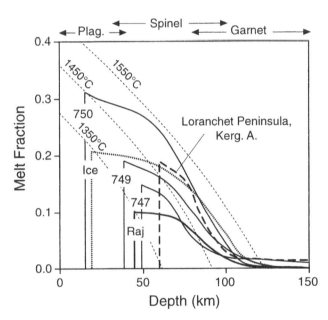

Figure 14. Partial melt distributions inferred from rare-earth element inversions of data for tholeiitic basalts from the eastern Indian Ocean, compared to Icelandic lavas. Kerguelen Plateau melt distribution curves are numbered according to ODP site. 'Raj' is the fractionation-corrected Rajmahal melt distribution from Figure 13d, whereas 'Ice' is that obtained for Icelandic basalts by *McKenzie and O'Nions* [1991]. Dotted lines show melt distribution curves predicted for isentropic decompression paths of mantle at potential temperatures of 1350°C, 1450°C, and 1550°C (see also fig. 10 of *White and McKenzie* [1995]). Corrections for olivine fractionation are 9.6% (ODP Site 747), 13.3% (Site 749), 35.9% (Site 750), and 11.8% (Loranchet Peninsula, Kerguelen Archipelago). Data sources: Rajmahal - this study; Kerguelen Plateau - *Storey et al.* [1992]; Loranchet Peninsula - *Storey et al.* [1988].

greatest at Kerguelen Plateau Site 750 (~32% and 26.2 km, respectively, using a primitive mantle source). Results for this site are comparable to those obtained for enriched MORB-like basalts from Iceland (~30% and 22.3 km, respectively, for melting of a primitive mantle source; see *McKenzie and O'Nions* [1991]) and for Cretaceous lavas from oceanic plateaus in the western Pacific (e.g., the estimated total melt fraction and total melt thickness for the Ontong Java Plateau are 28–30% and 17–24 km, respectively, depending on the source concentrations used in the inversions; see *Mahoney et al.* [1993]). The melt fractions at Kerguelen Plateau Sites 749 and 747 are somewhat smaller than for Site 750 (18% and 14%, respectively, for a primitive mantle source). Site 749 has an 'Icelandic'-type melt distribution, with the top of the melt column at a depth of ~35 km and the base at ~120 km. The total melt thickness is 11.6 km. Note that the thickness estimates for igneous crust at Sites 749 and 750 differ considerably. However, because the rare-earth element data used in the models are not necessarily representative of the entire crustal section at each site, crustal thicknesses estimated by inversion may not be close to true thicknesses. Indeed, seismic reflection and refraction data (references of *Operto and Charvis* [1995, 1996]) currently provide no obvious indication of large variations in crustal thickness between ODP sites.

It is interesting that the results for Site 749 are not dissimilar to those we obtain for tholeiites from the Loranchet Peninsula (Figure 14). These lavas are believed to be representative of the shield-building stage of subaerial volcanism on the Kerguelen Archipelago [e.g., *Storey et al.,* 1988]. The total melt fraction and total melt thickness calculated for the Loranchet tholeiites (~18% and 6.9 km, respectively) are similar to those of normal MORBs. As with tholeiites from the Kerguelen Plateau, melting begins at a depth of about 120 km, or some 50 km deeper than would be expected for ambient mantle welling up at mid-ocean ridges [cf. *White et al.*, 1992]. Deep melting is to be expected if the mantle were hotter than normal, i.e., if melting occurred in a plume (potential temperatures of 1450–1550°C [*White and McKenzie*, 1989, 1995]).

Comparison of the melt distribution curves inferred for Kerguelen Plateau ODP sites with melt distributions predicted from isentropic decompression of asthenospheric mantle (curves defined by short dashes in Figure 14) suggests a mantle potential temperature at Site 750 of about 1450°C. We infer from this that Site 750 lay close to the center of the Kerguelen plume at 114 Ma; the shallow top of the melting column suggests a position beneath the axis of an oceanic ridge, in good agreement with plate

reconstructions (Figure 1b). Melt distributions for basalts from Sites 749 and 747 again suggest high mantle potential temperatures (~1400°C) but a lower degree of melting, consonant with the presence of thicker lithosphere above the plume. These sites probably lay slightly off-axis (recall that lavas from Site 749 are about 114 m.y. old, whereas basalts from Site 747 are about 85 m.y. old). Figure 14 shows that Eocene tholeiites from the Loranchet Peninsula also have a melt distribution consistent with their being produced by melting of anomalously hot (~1460°C) mantle. The melt distribution inferred from inversion of Rajmahal basalt data (Figures 13d and 14) suggests a mantle potential temperature of about 1350°C (i.e., 40° to 50°C higher than the potential temperature required to generate normal MORBs [cf. *McKenzie and O'Nions,* 1991]). A potential temperature of 1350°C is consonant with eruption of the Rajmahal basalts on the periphery of the thermal field of the Kerguelen plume [cf. *Müller et al.,* 1993].

To summarise, the Rajmahal lavas appear to represent moderate degree (~10%) partial melts generated beneath thinned lithosphere. A near-ridge, steady-state plume ('tail') origin is suggested for Kerguelen Plateau Site 749 and 750 basalts, which have melt distributions akin to those of Recent Icelandic lavas.

5.3. *Mantle Sources*

With the exception of alteration- and/or contamination-prone incompatible elements, the primitive-mantle-normalised trace element patterns of Rajmahal Group I basalts (Figure 8a) are relatively flat. Flat patterns result from the total melt fraction being dominated by melts generated in the spinel stability field, and are characteristic of basement samples from the central Kerguelen Plateau [*Salters et al.,* 1992] and lavas from oceanic plateaus in the western Pacific (Nauru Basin, Ontong Java, Manihiki) [e.g., *Saunders,* 1986; *Mahoney et al.,* 1993]. Despite their similar trace element patterns, a few Rajmahal Group I basalts have low Th/Ta when compared to Kerguelen Plateau lavas (i.e., they are more MORB-like; see Figure 15). The same is true for the least-contaminated Casuarina-type lavas from Bunbury, Western Australia. Data for these rocks overlap the MORB field in Figure 15a. In contrast, a small number of Rajmahal Group I basalts, all Rajmahal Group II basalts, and the Gosselin-type Bunbury basalts have high Th/Ta and Ba/Ta. A likely explanation is that Ba and Th have been added to these rocks by crustal contamination of magmas. The same probably is true of Kerguelen Plateau basalts with high La/Ta, Th/Ta, and Ba/Ta [cf. *Storey et al.,* 1989, 1992; *Mahoney et al.,* 1995].

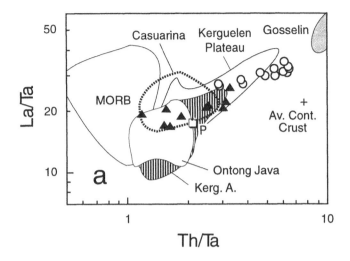

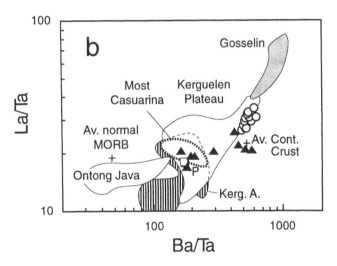

Figure 15.(a) Plot of Th/Ta vs. La/Ta for Rajmahal basalts. Note the trend of the Rajmahal Group I basalts and Casuarina (Bunbury) lavas towards the MORB field, and the similar Th/Ta ratios of average crust and Gosselin-type Bunbury basalts. Symbols, data sources, and averages as in Figure 12. The MORB field is from references of *Mahoney et al.* [1993]. Average continental crust is from *Wedepohl* [1994] and primitive mantle (denoted by 'P') is from *Sun and McDonough* [1989]. (b) Plot of Ba/Ta vs. La/Ta, showing that Rajmahal Group II basalts overlap the field of Kerguelen Plateau lavas, whereas the least crustally contaminated Rajmahal Group I basalts are intermediate between lavas from the Kerguelen- and Ontong Java plateaus. Most Casuarina-type lavas also lie in this region on the diagram, whereas Gosselin-type lavas are characterised by higher La/Ta and Ba/Ta. Symbols as in Figure 12. Data sources: Bunbury - *Storey et al.* [1992], *Frey et al.* [1996]; Kerguelen Plateau - *Salters et al.* [1992], *Mahoney et al.* [1995]; Ontong Java Plateau - *Mahoney et al.* [1993]. Average continental crust is from *Wedepohl* [1994]; average N-type MORB and primitive mantle (shown as 'P') are from *Sun and McDonough* [1989].

This explanation requires that rafts of continental lithosphere were incorporated into the Kerguelen Plateau at an early stage in its development (cf. Section 3).

In Figure 15b, Kerguelen Plateau lavas form a broadly triangular field, suggestive of mixing between three end-members with different La/Ta and Ba/Ta ratios. The high La/Ta, high Ba/Ta end-member, most apparent in Rajmahal Group II basalts, Kerguelen Plateau Site 738 basalts, and Gosselin-type Bunbury lavas, is likely to be Proterozoic crust [cf. *Frey et al.*, 1996]. The second (low La/Ta, moderate Ba/Ta) end-member is expressed in lavas from the Kerguelen Archipelago and is probably a component residing within the Kerguelen plume. The identity of the third (low La/Ta, low Ba/Ta) end-member is less clear. Clues can be gathered from Figure 15a, where the least crustally contaminated Kerguelen Plateau basalts (Site 749) have La/Ta ranging from 17 to 21 and Th/Ta from 1.6 to 1.7. The corresponding values for average normal MORB are 19 and 0.9, respectively [e.g., *Sun and McDonough*, 1989]. We infer from this that the third end-member in Kerguelen Plateau lavas is probably the Indian MORB source.

Are the Rajmahal basalts simply normal-MORB-like melts that have assimilated variable amounts of continental crust (reflected in the division into Groups I and II), or do they contain a plume component? In order to answer this question, we compare our results with the Pb-Nd-Sr isotopic ranges of Cretaceous-Tertiary flood basalts in, and on the margins of, the eastern Indian Ocean, and the ranges of Indian MORB, Kerguelen Archipelago, and Afanasy-Nikitin seamount lavas (Table 5). Analysed Rajmahal, Bunbury, and Naturaliste Plateau basalts generally do not have the moderate $^{206}Pb/^{204}Pb$ values (18.06–18.27) thought by *Weis et al.* [1993] to correspond to the 'pure Kerguelen plume', nor the high measured $^{206}Pb/^{204}Pb$ values (18.04–19.11) of Ninetyeast Ridge lavas [*Saunders et al.*, 1991; *Weis and Frey*, 1991; *Frey and Weis*, 1995]; age-correcting the data would move them even further away, by -0.1 to -0.2 in $^{206}Pb/^{204}Pb$, assuming a $^{238}U/^{204}Pb$ ratio of 5. Indo-Australian margin basalts interpreted above to be least affected by crustal contamination also have measured $^{207}Pb/^{204}Pb$ and $^{208}Pb/^{204}Pb$ values lower than those of the putative 'pure Kerguelen plume' (by up to 0.05 in $^{207}Pb/^{204}Pb$, and up to 1.03 in $^{208}Pb/^{204}Pb$; Table 5). In view of these Pb isotopic differences and the incompatible element differences noted above, we believe it unlikely that the Rajmahal, Bunbury, and Naturaliste Plateau basalts contain a component derived from the Kerguelen plume. We cannot rule out completely this possibility because the Kerguelen plume, like Hawaii, Iceland, and other large plumes, appears to be

TABLE 5. Pb, Nd, and Sr Isotopic Ranges for Selected Basalts from the Eastern Indian Ocean and Continental Margins

Province	Age (Ma)	$^{206}Pb/^{204}Pb$	$^{207}Pb/^{204}Pb$	$^{208}Pb/^{204}Pb$	$\varepsilon_{Nd(t)}$	$^{87}Sr/^{86}Sr_{(t)}$
Bunbury (Casuarina)	129	17.91-18.00	15.49-15.62	37.72-38.26	+3.5 to +0.9	0.70392-0.70506
Bunbury (Gosselin)	123	18.00-18.05	15.61-15.72	38.04-39.08	*-4.6	0.70763-0.70786
Rajmahal (Group I)	116	17.93-18.03	15.52-15.60	38.12-38.48	+5.1 to +0.1	0.70372-0.70839
Rajmahal (Group II)	116	17.96-18.03	15.56-15.66	38.15-39.14	+0.1 to -6.7	0.70498-0.70911
Naturaliste Plateau	?100	17.79-18.09	15.50-15.70	37.65-39.22	+4.9 to -12.9	0.70393-0.71302
Kerguelen Plateau	?114-85	17.38-18.42	15.43-15.75	37.80-39.14	+5.2 to -9.4	0.70351-0.70984
Broken Ridge	89-63	17.98-18.47	15.50-15.65	38.40-39.27	+3.4 to -2.7	0.70391-0.70734
Ninetyeast Ridge	82-43	18.04-19.11	15.50-15.66	38.44-39.08	+5.7 to 0.0	0.70382-0.70600
Afanasy-Nikitin Seamount	?80	16.77-18.12	15.41-15.61	37.06-37.08	+5.5 to -8.0	0.70368-0.70662
Kerguelen Archipelago	39-0	17.99-18.61	15.48-15.59	38.29-39.21	+5.7 to -2.9	0.70427-0.70598
Indian MORB	155-0	16.87-22.85	15.37-15.91	37.08-39.33	+11.3 to -4.0	0.70242-0.70562

Notes: *Epsilon Nd values calculated for two Gosselin-type samples are identical. Pb isotopic values are measured ratios, with the exception of data for the Afanasy-Nikitin seamount (age-corrected to 80 Ma). Isotopic values of Nd and Sr are age-corrected assuming the ages given in the table, with the exception of data for the Naturaliste Plateau (corrected to 110 Ma) and Kerguelen Plateau ODP Site 738 (present-day values) [see *Mahoney et al.*, 1995].

Sources for isotopic data: Bunbury - *Frey et al.* [1996]; M. Storey, unpublished data. Rajmahal - *Mahoney et al.* [1983]; *Storey et al.* [1992]; *Baksi* [1995]; this study. Naturaliste Plateau - *Mahoney et al.* [1995]. Kerguelen Plateau - *Davies et al.* [1989]; *Weis et al.* [1989]; *Alibert* [1991]; *Salters et al.* [1992], and V. J. M. Salters, unpublished data; *Mahoney et al.* [1995]. Broken Ridge - *Mahoney et al.* [1995]. Ninetyeast Ridge - *Mahoney et al.* [1983]; *Saunders et al.* [1991]; *Weis and Frey* [1991]; *Frey and Weis* [1995]. Afanasy-Nikitin seamount - *Mahoney et al.* [1996]; *Suschevskaya et al.* [1997]. Kerguelen Archipelago - *White and Hofmann* [1982]; *Storey et al.* [1988]; *Gautier et al.* [1990]; *Weis et al.* [1993]. Indian MORB - *Mahoney et al.* [1989, 1992], and references therein; *Weis and Frey* [1996].

compositionally heterogeneous [e.g., *Frey and Weis*, 1995].

Table 5 shows that analysed Rajmahal, Bunbury, and Naturaliste Plateau basalts do not have $^{206}Pb/^{204}Pb \ll 17.8$, indicating that they do not contain the low $^{206}Pb/^{204}Pb$ material found in some Late Jurassic–Early Cretaceous Indian MORBs [*Ludden and Dionne*, 1992; *Weis and Frey*, 1996] and some Kerguelen Plateau lavas (Sites 747 and 750 [*Salters et al.*, 1992; V. J. M. Salters, unpublished data]). This low $^{206}Pb/^{204}Pb$ material is believed to be ancient Gondwanan lithosphere incorporated into the shallow asthenosphere prior to, or during continental breakup [*Mahoney et al.*, 1989, 1992; *Storey et al.*, 1989, 1992]. Importantly, the absence of this material in analysed Rajmahal basalts does not rule out the Indian MORB-source as the source of Rajmahal basalts; many Indian

MORBs have $^{206}Pb/^{204}Pb$ values >17.8 (see Table 5) and the least crustally contaminated of our Rajmahal samples have Pb isotope ratios that are remarkably similar to those of average present-day Indian MORB (e.g., Figure 11). Therefore, the source of Rajmahal basalts is likely to be compositionally 'normal' (MORB-source) asthenosphere.

5.4. Origin of the Rahmahal-Sylhet Igneous Province

On the basis of the data and discussion presented above, it is possible to account for the Rajmahal and Sylhet basalts by decompressional melting of moderately hot (potential temperature about 1350°C) normal-MORB-type mantle, followed by assimilation of granitic crust and low-pressure fractional crystallisation. We envision that prior to the Early Cretaceous the Kerguelen hot spot was established beneath eastern Gondwana [e.g., *Kent*, 1991; *Kent et al.,* 1992c; *Arne*, 1994]. By 130 Ma, India had broken away from Australia–Antarctica and drifted northwards relative to the hot spot [e.g., *Müller et al.*, 1993]. Rajmahal-Sylhet volcanism occurred at about 116 Ma by melting on the periphery of a steady-state plume ('tail'). Partial melting of the ridge-centered plume tail produced the volcanic sections of the southern and central Kerguelen Plateau and Broken Ridge. Upwelling and partial melting beneath the rifted margin of eastern India may have been initiated and driven by viscous coupling of the MORB mantle to flow in the conduit of the Kerguelen plume (Figure 16).

Based on the results of numerical modeling of steady-state plumes, *Hauri et al.* [1994] proposed a similar explanation for magmatism north of the Hawaiian ridge, in which the flow of ambient mantle lying at the thermal boundary of a mantle plume becomes coupled to, but not entrained in, flow within the plume boundary. In *Hauri et al.*'s model, the ascent velocity of mantle material at the thermal boundary of the plume is of the order of 1–10 cm/y, or roughly the same as upwelling velocities inferred beneath oceanic ridges [e.g., *Sparks and Parmentier*, 1991]. With one significant modification (conductive heating of asthenospheric mantle at the edge of the plume), this model can account for the occurrence of MORB-like tholeiitic basalts on the eastern Indian margin. The location of Sylhet lavas on a northward extrapolation of the trend of

the Ninetyeast Ridge (Section 2.1) is therefore likely to be coincidental. An origin for the Rajmahal basalts outside the Kerguelen plume accommodates *Müller et al.*'s [1993] observation that the rifted volcanic margin of eastern India was about 1000 km away from the center of the Kerguelen hot spot at 116 Ma, without recourse to a large 'hot cell' [*Anderson et al.*, 1992] beneath eastern Gondwana.

Our preferred origin for the Rajmahal-Sylhet igneous province also provides an explanation for the occurrence of Aptian (c. 113–110 Ma) basaltic dikes to the west and southwest of the Rajmahal Hills, and to the south of Shillong (Figure 3). The Rajmahal and Sylhet dikes attest to magma generation or melt transport up to 300 km inland from the rifted margin, consistent with a thermal flux smaller than that inferred for the Deccan igneous province (recall our suggestion that magma in the ~64-m.y.-old Salma dike may have travelled >1000 km inland from the rifted margin of western India). The small number of crystallisation ages obtained thus far for Rajmahal dikes suggest that dike emplacement could have occurred in eastern India ~2–3 m.y. after eruption of the Rajmahal lavas. At present, we have no evidence for dikes older than 113 Ma; however, by analogy with basin evolution on the southwest Australian margin [e.g., *Marshall and Lee*, 1989], extension or transtension in the Bengal Basin probably commenced in the Late Jurassic and continued well into the Early Cretaceous. Hence, we infer that

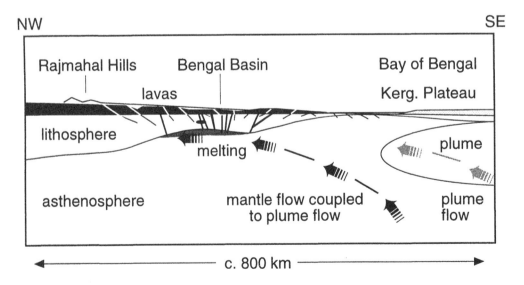

Figure 16. Cartoon illustrating an origin for the Rajmahal basalts by passive decompressional melting of normal MORB-type mantle beneath the stretched and thinned lithosphere of the Bengal Basin. Upwelling is suggested to have occurred due to viscous coupling of flow in the MORB-source mantle at the boundary of the Kerguelen plume to flow in the conduit of the plume. The center of the plume was about 1000 km to the south of eastern India at the time of Rajmahal volcanism, assuming a plate tectonic reconstruction similar to that proposed by *Müller et al.* [1993].

eruption of the Rajmahal-Sylhet lavas was preceded and accompanied by stretching and thinning of the Indian crust. Lithospheric thinning was most probably a product of extension associated with the final stages of continental breakup, rather than convective activity associated with the Kerguelen plume.

The Bunbury and Naturaliste Plateau lavas could have an origin comparable to that suggested here for the Rajmahal basalts, assuming that their Pb isotopic ratios largely reflect contamination by continental crust. The greater distance of the Scott and Exmouth plateaus from the presumed center of the Kerguelen plume makes it unlikely that lavas on these plateaus were produced by coupled flow of MORB-source mantle to flow in the 'tail' of the Kerguelen plume. However, other west Australian margin basalts, notably those on the Wallaby Plateau [*Colwell et al.*, 1994], could have been generated by these means. Unfortunately, the Wallaby Plateau lavas are not well-characterised in terms of age and chemical and isotopic composition. Therefore, the relationship of these rocks to the Kerguelen Plateau remains uncertain. If MORB-source mantle was induced to well up along the rifted continental margins by viscous coupling to flow in the Kerguelen plume, the Wallaby Plateau basalts are predicted to have relatively MORB-like isotopic signatures. Closer to the plume center, the composition of erupted magmas ought to reflect more clearly the plume isotopic signature. However, as we noted above, this simple picture is complicated by the likelihood that stranded blocks of continental lithosphere underlie parts of the central and southern Kerguelen Plateau. Moreover, there is substantial disagreement over what constitutes the 'pure Kerguelen plume' isotopic signature [e.g., *Class et al.*, 1996; *Frey and Weis*, 1996]. Until such questions are resolved, perhaps by drilling of volcanic basement on the northern Kerguelen Plateau, it will not be possible to understand fully the link between rifted margin volcanism and oceanic plateau formation in the early Indian Ocean.

Our model for the origin of Rajmahal-Sylhet basalts leads us to question whether the development of the Early Cretaceous Indian Ocean province was similar to that of other large igneous provinces. The best-known continental margin flood basalts are those of the Tertiary North Atlantic province (total volume ~6.6 × 10^6 km³ [*Eldholm and Grue,* 1994]). This igneous province is usually attributed to the initiation of the Iceland hot spot [e.g., references of *Fitton et al.,* 1997; *Saunders et al.,* this volume], whereas we account for most of the Kerguelen Plateau, Broken Ridge, and related continental margin volcanism by the activity of a plume tail. Key features of the North Atlantic province include (1) the occurrence of

62–56-m.y.-old basalt flows and associated intrusive rocks along the strike of the rifted margins for distances of more than 2000 km [e.g., *Saunders et al.,* this volume]; (2) an abundance of basaltic intrusive centers on the opposing continental margins; and (3) the existence of post-56-Ma linear basaltic ridges (the Greenland-Iceland and Faeroes-Iceland ridges) leading away from the rifted margins to the Icelandic Plateau. In comparison, the ~130–100-m.y.-old Indo-Australian margin lavas appear to be confined to isolated zones (Figure 1), each zone ranging from 200 to 800 km in length. As in the Tertiary North Atlantic igneous province, some of these zones correspond to pre-existing sedimentary basins (e.g., the Bengal Basin), but others (e.g., Naturaliste Plateau) clearly do not. Basaltic intrusive complexes are unknown, either onshore or offshore of eastern India and western Australia. Ridges of anomalously thick oceanic crust akin to those in the North Atlantic extend offshore from the eastern Indian and East Antarctic rifted margins, and appear, at least in the case of the northern Ninetyeast Ridge, to be younger than the Early Cretaceous igneous province (the Gaussberg Ridge, lying to the south of the Kerguelen Plateau, has not been studied in detail and is of uncertain age and origin).

The absence of exposed intrusive centers in eastern India and southwest Australia is puzzling, but could be an artifact of limited syn- and post-rift uplift of the continental margins, with a consequent lack of deep erosion and exposure of intrusive bodies. Alternatively, the magmatic plumbing system could have been different from that on the North Atlantic margins, preventing the ponding of basaltic magma in the upper crust. At present, there are very few constraints. The acquisition of modern seismic profiles across the eastern Indian margin would greatly improve our knowledge of the crustal structure and subsidence history of the Bengal Basin, and facilitate a detailed comparison with other rifted margins.

Even from a brief comparison such as this, it is evident that development of the eastern Indian and southwest Australian volcanic rifted margins was considerably more protracted than that of the North Atlantic margins, where the bulk of igneous activity was confined to a period of ~7 m.y. [*Saunders et al.,* this volume]. Importantly, some west Australian margin lavas, including the Bunbury basalts, were erupted at least 9–15 m.y. before the main episode of Kerguelen Plateau volcanism. Construction of the volcanic sections of the central and southern Kerguelen Plateau took place over a period of at least 29 m.y., suggesting an established, steady-state plume like that beneath present-day Iceland. Each of these observations fits with a non-plume-head origin for the Early Cretaceous Indian Ocean igneous province.

6. CONCLUSIONS

The Rajmahal basalts of eastern India are derived from MORB-like magmas that probably formed at the boundary of a mantle hot spot. Most plate tectonic reconstructions suggest that this hot spot was the Kerguelen plume. The $^{40}Ar/^{39}Ar$ ages of lava samples from the Rajmahal Hills and Bengal Basin, and dikes from the Chotanagpur Plateau and Damodar Valley (mostly 116–113 Ma) point to a short period of igneous activity on the eastern Indian margin, coincident with volcanism on the central and southern Kerguelen Plateau and within a longer period of magmatism along the continental margin of western Australia (~130–100 Ma). There is evidence, in the form of a small number of radiometric ages for continental basalts and old seafloor, for a broadly north-south progression of plateau-type volcanism on the west Australian rifted margin, suggesting that plate tectonic developments were the principal control on the timing and siting of melt generation along the main Indo-Australian rift zone (including the location of Rajmahal and Kerguelen Plateau volcanism).

The Rajmahal basalts are best explained as moderate-degree partial melts produced beneath thinned continental lithosphere. In this respect, they are quite unlike contemporaneous high-degree melts generated at Kerguelen Plateau ODP Sites 749 and 750, the elemental compositions of which are consistent with a near-ridge, plume tail origin. We infer from inversion modeling of Site 750 data that the mantle potential temperature close to the center of the Early Cretaceous Kerguelen hot spot was about 1450°C. The potential temperature of asthenosphere beneath the Indian margin (about 1000 km from the plume center at the time of Rajmahal volcanism) was probably about 1350°C.

Isotopically, the closest marine analogs to the least crustally contaminated Rajmahal basalts are present-day Indian MORBs. Our chemical and isotopic results suggest that most Rajmahal basalts, including dikes emplaced to the west and southwest of the Rajmahal Hills, assimilated small melt-fractions derived from crust similar to the Chotanagpur granite-gneiss. After contamination, Rajmahal parent magmas underwent differentiation in the upper crust. Volumetrically small occurrences of silicic volcanic and pyroclastic rocks attest to more-extreme differentiation of tholeiitic liquids in areas where conditions allowed small high-level magma chambers to become established.

Finally, we note that because they appear to have been derived from a MORB-like source, the Rajmahal basalts cannot be used to estimate the Pb-Nd-Sr isotopic composition of the Kerguelen mantle plume at 116 Ma [contra *Class et al.*, 1993]. The fact that they have been so used highlights the dangers of assuming that all basalts associated with a large igneous province necessarily have a deep-seated plume source. Careful study of other provinces may yet reveal that mantle upwelling coupled to flow within plume conduits is a potent means of producing large volumes of basalt with a MORB-like chemical and isotopic signature, well away from the plume center.

Acknowledgments. We are grateful to B. Kumar, D. Mukherjee, and B. K. Trivedi for assistance in the field, and Nick Marsh for supervising XRF work. John Mahoney generously provided Pb isotopic data for two of our Rajmahal samples. Dan McKenzie gave valuable help with the inversion modeling. We thank Vincent Salters and Mick Storey for making available to us their unpublished isotopic data for Indian Ocean basalts. Reviewers Fred Frey and Doug Pyle gave constructive and insightful comments on the first draft of the paper; Fred Frey and John Mahoney provided helpful comments on the second draft. Research in eastern India was supported by the NERC (grant GT4/89/GS/55 to R. W. Kent) and the University of Leicester (research associateship to R. W. Kent).

REFERENCES

Agrawal, J. K., and F. A. Rama, Chronology of Mesozoic volcanics of India, *Proc. Indian Acad. Sci., 84*, 157-179, 1976.

Alibert, C., Mineralogy and geochemistry of a basalt from Leg 119, Site 738: Implications for the tectonic history of the most southern part of the Kerguelen Plateau, *Proc. Ocean Drill. Prog., Sci. Results, 119*, 293-298, 1991.

Anderson, D. L., Y.-S. Zhang, and T. Tanimoto, Plume heads, continental lithosphere, flood basalt and tomography, in *Magmatism and the Causes of Continental Break-up, Spec. Publ., 68*, edited by B. C. Storey, T. Alabaster, and R. J. Pankhurst, pp. 99-124, The Geological Society, London, 1992.

Arndt, N. T., and U. Christensen, The rôle of lithospheric mantle in continental flood volcanism: Thermal and geochemical constraints, *J. Geophys. Res., 97*, 10,967-10,981, 1992.

Arne, D. C., Phanerozoic exhumation history of northern Prince Charles Mountains (East Antarctica), *Antarct. Science, 6*, 69-84, 1994.

Baksi, A. K., Petrogenesis and timing of volcanism in the Rajmahal flood basalt province, northeastern India, *Chem. Geol., 121*, 73-90, 1995.

Baksi, A. K., T. R. Barman, D. K. Paul, and E. Farrar, Widespread Early Cretaceous flood basalt volcanism in eastern India: Geochemical data from the Rajmahal-Bengal - Sylhet Traps, *Chem. Geol., 63*, 133-141, 1987.

Ball, V., Geology of the Rajmahal Hills, *Mem. Geol. Surv. India, 13*, 155-248, 1877.

Biswas, B., Subsurface geology of West Bengal, India, in *Proc. Symp. Dev. Petroleum Resources, United Nations Economic*

Commission for Asia and the Far East, Mineral Resources Development Series, 10, edited by Anon., pp. 159-161, United Nations, New York, 1959.

Biswas, B., Results of exploration for petroleum in the western part of the Bengal Basin, India, in *Proc. 2nd Symp. Dev. Petroleum Resources, United Nations Economic Commission for Asia and the Far East, Mineral Resources Development Series, 18*, edited by Anon., pp. 241-250, United Nations, New York, 1963.

Brodie, J., D. Latin, and N. White, Rare-earth element inversion for melt distribution: Sensitivity and application, *J. Petrol., 35*, 1155-1174, 1994.

Brune, J. N., and D. D. Singh, Continent-like crustal thickness beneath the Bay of Bengal sediments, *Bull. Seismol. Soc. Am., 76*, 191-203, 1986.

Buck, W. R., Small-scale convection induced by passive rifting: The cause of uplift of rift shoulders, *Earth Planet. Sci. Lett., 77*, 362-372, 1986.

Charvis, P., M. Recq, S. Operto, and D. Brefort, Deep structure of the northern Kerguelen Plateau and hotspot-related activity, *Geophys. J. Int., 122*, 899-924, 1995.

Class, C., S. L. Goldstein, S. J. G. Galer, and D. Weis, Young formation age of a mantle plume source, *Nature, 362*, 715-721, 1993.

Class, C., S. L. Goldstein, and S. J. G. Galer, Discussion of 'Temporal evolution of the Kerguelen plume: Geochemical evidence from ~38 to 82 Ma lavas forming the Ninetyeast Ridge', *Contrib. Mineral. Petrol., 124*, 98-103, 1996.

Colwell, J. B., P. A. Symonds, and A. J. Crawford, The nature of the Wallaby (Cuvier) Plateau and other igneous provinces of the west Australian margin, *J. Aust. Geol. Geophys., 15*, 137-156, 1994.

Curray, J. R., and T. Munasinghe, Origin of the Rajmahal Traps and 85°E Ridge: Preliminary reconstructions of the trace of the Crozet hotspot, *Geology, 19*, 1237-1240, 1991.

Curray, J. R., and T. Munasinghe, Reply to Comment by Kent *et al.*, *Geology, 20*, 958-959, 1992.

Dalrymple, G. B., and M. A. Lanphere, $^{40}Ar/^{39}Ar$ spectra of some undisturbed terrestrial samples, *Geochim. Cosmochim. Acta, 38*, 715-738, 1974.

Davies, H. L., S.-S. Sun, F. A. Frey, I. Gautier, M. T. McCulloch, R. C. Price, Y. Bassias, C. T. Klootwijk, and L. Leclaire, Basalt basement from the Kerguelen Plateau and the trail of a DUPAL plume, *Contrib. Mineral. Petrol., 103*, 457-469, 1989.

DePaolo, D. J., Trace element and isotopic effects of combined wallrock assimilation and fractional crystallization, *Earth Planet. Sci. Lett., 53*, 189-202, 1981.

Deshmukh, S. S., S. C. Chakraborty, and M. V. N. Murthy, The origin of the shapes and sizes of the amygdules in the pitchstone and basalt flows from Taljhari and Berhait, Santhal Parganas, Bihar, *Rec. Geol. Surv. India, 93*, 45-50, 1964.

Dosso, L., H. Bougault, P. Beuzart, J.-Y. Calvez, and J.-L. Joron, The geochemical structure of the Southeast Indian Ridge, *Earth Planet. Sci. Lett., 88*, 47-59, 1988.

Duncan, R. A., The age distribution of volcanism along aseismic ridges in the eastern Indian Ocean, *Proc. Ocean Drill. Prog., Sci. Results, 121*, 507-517, 1991.

Eldholm, O., and K. Grue, North Atlantic volcanic margins: Dimensions and production rates, *J. Geophys. Res., 99*, 2955-2968, 1994.

Fitton, J. G., A. D. Saunders, L. M. Larsen, B. S. Hardarson, and M. J. Norry, Volcanic rocks from the East Greenland margin at 63°N: Composition, petrogenesis and mantle sources, *Proc. Ocean Drill. Prog., Sci. Results, 152*, in press, 1997.

Foucher, J.-P., X. Le Pichon, and J.-C. Sibuet, The ocean-continent transition in the uniform stretching model: Rôle of partial melting in the mantle, *Phil. Trans. R. Soc. London, Ser. A, 305*, 27-43, 1982.

Frey, F. A., and D. Weis, Temporal evolution of the Kerguelen plume: Geochemical evidence from ~38 to 82 Ma lavas forming the Ninetyeast Ridge, *Contrib. Mineral. Petrol., 121*, 18-28, 1995.

Frey, F. A., and D. Weis, Reply to Discussion by Class *et al.*, *Contrib. Mineral. Petrol., 124*, 104-110, 1996.

Frey, F. A., W. B. Jones, H. Davies, and D. Weis, Geochemical and petrologic data for basalts from Sites 756, 757, and 758: Implications for the origin and evolution of the Ninetyeast Ridge, *Proc. Ocean Drill. Prog., Sci. Results, 121*, 611-659, 1991.

Frey, F. A., N. J. McNaughton, D. R. Nelson, J. R. deLaeter, and R. A. Duncan, Petrogenesis of the Bunbury Basalt, Western Australia: Interaction between the Kerguelen plume and Gondwana lithosphere? *Earth Planet. Sci. Lett., 141*, 163-183, 1996.

Gautier, I., D. Weis, J.-P. Mennessier, P. Vidal, A. Giret, and M. Loubet, Petrology and geochemistry of the Kerguelen Archipelago basalts (South Indian Ocean): Evolution of the mantle sources from ridge to intraplate position, *Earth Planet. Sci. Lett., 100*, 59-76, 1990.

Ghose, N. C., Geology, tectonics, and evolution of the Chotanagpur granite-gneiss complex, eastern India, in *Structure and tectonics of Precambrian rocks of India, Recent Researches in Geology, 10*, edited by S. Sinha Roy, pp. 211-247, Hindusthan Publishing Co., New Delhi, 1983.

Gibling, M. R., F. M. Gradstein, I. L. Kristiansen, J. Nagy, M. Sarti, and J. Wiedmann, Early Cretaceous strata of the Nepal Himalayas: Conjugate margins and rift volcanism during Gondwana break-up, *J. Geol. Soc. London, 151*, 269-290, 1994.

Goodman, R. J., The distribution of Ga and Rb in coexisting groundmass and phenocryst phases of some basic volcanic rocks, *Geochim. Cosmochim. Acta, 36*, 303-317, 1972.

Govindaraju, K., 1994 compilation of working values and sample descriptions for 383 geostandards, *Geostandards Newsletter, 18*, 1-158, 1994.

Gradstein, F. M., F. P. Agterberg, J. G. Ogg, J. Hardenbol, P. van Veen, J. Thierry, and Z. Huang, A Mesozoic timescale, *J. Geophys. Res., 99*, 24,051-24,074, 1994.

Green, D. H., and A. E. Ringwood, The genesis of basaltic magmas, *Contrib. Mineral. Petrol., 15*, 103-190, 1967.

Hauri, E. H., J. A. Whitehead, and S. R. Hart, Fluid dynamic and geochemical aspects of entrainment in mantle plumes, *J. Geophys. Res., 99*, 24,275-24,300, 1994.

Hofmann, A. W., and W. M. White, Ba, Rb, and Cs in the Earth's mantle, *Zeitsch. Naturforsch. 38*, 256-266, 1983.

Huppert, H. E., and R. S. J. Sparks, The fluid dynamics of a basaltic magma chamber replenished by influx of hot, dense ultrabasic magma, *Contrib. Mineral. Petrol., 75*, 279-289, 1980.

Johnson, S. Y., and A. M. N. Alam, Sedimentation and tectonics of the Sylhet Trough, northeastern Bangladesh, *U.S. Geol. Surv. Open File Rep., 90-313*, 42 pp., 1990.

Keen C. E., R. C. Courtney, S. A. Dehler, and M.-C. Williamson, Decompression melting at rifted margins: Comparison of model predictions with the distribution of igneous rocks on the eastern Canadian margin, *Earth Planet. Sci. Lett., 121*, 403-416, 1994.

Kempton, P. D., and A. G. Hunter, A Sr-Nd-Pb-O isotope study of plutonic rocks from MARK, ODP Leg 153: Implications for mantle heterogeneity and magma chamber processes, *Proc. Ocean Drill. Prog., Sci. Results, 153*, in press, 1997.

Kent, R. W., Lithospheric uplift in eastern Gondwana: Evidence for a long-lived mantle plume system? *Geology, 19*, 19-23, 1991.

Kent, R. W., Plume - lithosphere interaction: Petrology of Rajmahal continental flood basalts and associated lamproites, northeast India, Ph.D. thesis, University of Leicester, 1992.

Kent, R. W., and D. P. McKenzie, Rare-earth element inversion models for basalts associated with the Kerguelen mantle plume, *Mineral. Mag., 58A*, 471-472, 1994.

Kent, R. W., N. C. Ghose, P. R. Paul, M. J. Hassan, and A. D. Saunders, Coal-magma interaction: An integrated model for the emplacement of cylindrical intrusions, *Geol. Mag., 129*, 753-762, 1992a.

Kent, R. W., N. C. Ghose, A. D. Saunders, and M. Storey, Basalt stratigraphy of the Rajmahal Hills and adjacent areas, Bihar, northeast India (abstract), in *Mesozoic Magmatism of the Eastern Margin of India, Programme and Abstracts*, edited by N. C. Ghose, pp. 2-4, Patna University, Patna, 1992b.

Kent, R. W., M. Storey, and A. D. Saunders, Large igneous provinces: Sites of plume impact or plume incubation? *Geology, 20*, 891-894, 1992c.

Kent, R. W., M. Storey, A. D. Saunders, N. C. Ghose, and P. D. Kempton, Comment on 'Origin of the Rajmahal Traps and 85°E Ridge: Preliminary reconstructions of the Crozet hotspot' by J. R. Curray and T. Munasinghe, *Geology, 20*, 957-958, 1992d.

Kent, R. W., A. D. Saunders, M. Storey, and N. C. Ghose, Petrology of Early Cretaceous flood basalts and dykes along the rifted volcanic margin of eastern India, *J. Southeast Asian Earth Sci., 13*, 95-111, 1996.

Khan, A. H., and J. Azad, The geology of Pakistan gas fields, in *Proc. 2nd Symp. Dev. Petroleum Resources, United Nations Economic Commission for Asia and the Far East, Mineral Resources Development Series, 18*, pp. 275-282, United Nations, New York, 1963.

Klootwijk, C. T., Paleomagnetism of the Upper Gondwana Rajmahal Traps, northeast India, *Tectonophysics, 12*, 449-467, 1971.

Leclaire, L., Y. Bassias, M. Denis-Clocchiatti, H. Davies, I. Gautier, B. Gensous, P.-J. Giannesini, J. Ségoufin, M. Tesson, and J. Wannesson, Lower Cretaceous basalt and sediments from the Kerguelen Plateau, *Geo-Marine Lett., 7*, 169-176, 1987.

Lightfoot, P. C., Isotope and trace element geochemistry of the South Deccan lavas, India, Ph.D. thesis, The Open University, 1985.

Ludden, J. N., and B. Dionne, The geochemistry of oceanic crust at the onset of rifting in the Indian Ocean, *Proc. Ocean Drill. Prog., Sci. Results, 123*, 791-799, 1992.

Mahoney, J. J., Isotopic and chemical studies of the Deccan and Rajmahal Traps, India: Mantle sources and petrogenesis, Ph.D. thesis, University of California, San Diego (Scripps), 1984.

Mahoney, J. J., and K. J. Spencer, Isotopic evidence for the origin of the Manihiki and Ontong Java plateaus, *Earth Planet. Sci. Lett., 104*, 196-210, 1991.

Mahoney, J. J., J. D. Macdougall, G. W. Lugmair, A. V. Murali, M. Sankar Das, and K. Gopalan, Origin of the Deccan Trap flows at Mahabaleshwar inferred from Nd and Sr isotopic and chemical evidence, *Earth Planet. Sci. Lett., 60*, 47-60, 1982.

Mahoney, J. J., J. D. Macdougall, G. W. Lugmair, and K. Gopalan, Kerguelen hot spot source for Rajmahal Traps and Ninetyeast Ridge? *Nature, 303*, 385-389, 1983.

Mahoney, J. J., J. H. Natland, W. M. White, R. Poreda, S. H. Bloomer, R. L. Fisher, and A. N. Baxter, Isotopic and geochemical provinces of the western Indian Ocean spreading centers, *J. Geophys. Res., 94*, 4033-4052, 1989.

Mahoney, J. J., A. P. le Roex, Z. Peng, R. L. Fisher, and J. H. Natland, Southwestern limits of Indian Ocean Ridge mantle and the origin of low $^{206}Pb/^{204}Pb$ mid-ocean ridge basalt: Isotope systematics of the Central Southwest Indian Ridge (17°-50°E), *J. Geophys. Res., 97*, 19,771-19,790, 1992.

Mahoney, J. J., M. Storey, R. A. Duncan, K. J. Spencer, and M. Pringle, Geochemistry and age of the Ontong Java Plateau, in *The Mesozoic Pacific: Geology, Tectonics and Volcanism, Geophys. Monogr. Ser.*, vol. 77, edited by M. S. Pringle, W. Sager, W. Sliter, and S. Stein, pp. 233-261, AGU, Washington, D.C., 1993.

Mahoney, J. J., W. B. Jones, F. A. Frey, V. J. M. Salters, D. G. Pyle, and H. L. Davies, Geochemical characteristics of lavas from Broken Ridge, the Naturaliste Plateau and southernmost Kerguelen Plateau: Cretaceous plateau volcanism in the southeast Indian Ocean, *Chem. Geol., 120*, 315-345, 1995.

Mahoney, J. J., W. M. White, B. G. J. Upton, C. R. Neal, and R. A. Scrutton, Beyond EM-1: Lavas from Afanasy - Nikitin Rise and the Crozet Archipelago, Indian Ocean, *Geology, 24*, 615-618, 1996.

Marshall, J. F., and C. S. Lee, Basin framework and resource potential of the Abrolhos sub-basin, in *Aust. Bureau Mineral. Res., Geol. Geophys., Yearbook 1988-1989*, edited by K. H. Wolf and A. G. L. Paine, pp. 63-67, Australia, Bureau of Mineral Resources, Canberra, 1989.

Mazumdar, S. K., Crustal evolution of the Chotanagpur Gneissic Complex and the Mica Belt of Bihar, *Mem. Geol. Soc. India, 8*, 49-83, 1988.

McDougall, I., and M. W. McElhinny, The Rajmahal Traps of India - K-Ar ages and paleomagnetism, *Earth Planet. Sci. Lett., 9*, 371-378, 1970.

McKenzie, D., and R. K. O'Nions, Partial melt distributions from inversion of rare-earth element concentrations, *J. Petrol., 32*, 1021-1091, 1991.

McKenzie, D., and R. K. O'Nions, The source regions of ocean island basalts, *J. Petrol., 36*, 133-159, 1995.

Middlemost, E. A. K., D. K. Paul, and I. R. Fletcher, Minette-lamproite association from the Indian Gondwanas, *Lithos, 22*, 31-42, 1988.

Mukhopadhyay, M., R. K. Verma, and M. H. Ashraf, Gravity field and structures of the Rajmahal Hills: Examples of the Paleo-Mesozoic continental margin in eastern India, *Tectonophysics, 131*, 353-367, 1986.

Müller, R. D., J.-Y. Royer, and L. A. Lawver, Revised plate motions relative to the hot spots from combined Atlantic and Indian Ocean hot spot tracks, *Geology, 21*, 275-278, 1993.

Negi, J. G., O. P. Pandey, and P. K. Agrawal, Super-mobility of hot Indian lithosphere, *Tectonophysics, 131*, 147-156, 1986.

Ogg, J. G., K. Kodama, and B. P. Wallick, Lower Cretaceous magnetostratigraphy and paleolatitudes off NW Australia (ODP Site 765 and DSDP Site 261, Argo Abyssal Plain, and ODP Site 766, Gascoyne Abyssal Plain), *Proc. Ocean Drill. Prog., Sci. Results, 123*, 523-548, 1992.

Operto, S., and P. Charvis, Kerguelen Plateau: A volcanic passive margin fragment? *Geology, 23*, 137-140, 1995.

Operto, S., and P. Charvis, Deep-structure of the southern Kerguelen Plateau (southern Indian Ocean) from ocean-bottom seismometer wide-angle seismic data, *J. Geophys. Res., 101*, 25,077-25,103, 1996.

Pantulu, G., J. D. Macdougall, K. Gopalan, and P. Krishnamurthy, Isotopic and chemical compositions of Sylhet Traps basalts: Links to the Rajmahal Traps and Kerguelen hot spot (abstract), *Eos Trans. AGU, 73*, 14, Fall Meeting Suppl., 328, 1992.

Pederson, T., and H. E. Ro, Finite-duration extension and decompression melting, *Earth Planet. Sci. Lett., 113*, 15-22, 1992.

Poornachandra Rao, G. V. S., J. Mallikharjuna Rao, and M. V. Subba Rao, Palaeomagnetic and geochemical characteristics of the Rajmahal Traps, eastern India, *J. Southeast Asian Earth Sci., 13*, 113-122, 1996.

Poornachandra Rao, G. V. S., M. V. Subba Rao, and J. Mallikharjuna Rao, Geochemical and palaeomagnetic signatures in Sylhet Trap flood basalt volcanism: Preliminary results, *Geol. Mag., 130*, 163-184, 1993.

Powell, C. McA., S. R. Roots, and J. J. Veevers, Pre-break-up continental extension in East Gondwanaland and the early opening of the eastern Indian Ocean, *Tectonophysics, 155*, 261-283, 1988.

Pringle, M. S., Age-progressive volcanism in the Musicians seamounts: A test of the hot spot hypothesis for the Late Cretaceous Pacific, in *The Mesozoic Pacific: Geology, Tectonics and Volcanism, Geophys. Monogr. Ser.*, vol. 77, edited by M. S. Pringle, W. Sager, W. Sliter, and S. Stein, pp. 187-215, AGU, Washington, D. C., 1993.

Pringle, M. S., M. Storey, and J. Wijbrans, [40]Ar/[39]Ar geochronology of Mid-Cretaceous Indian Ocean basalts: Constraints on the origin of large flood basalt provinces (abstract), *Eos Trans. AGU, 75*, 44, Fall Meeting Suppl., 728, 1994.

Pyle, D. G., D. M. Christie, J. J. Mahoney, and R. A. Duncan, Geochemistry and geochronology of ancient Southeast Indian Ocean and Southwest Pacific Ocean seafloor, *J. Geophys. Res., 100*, 22,261-22,282, 1995.

Raja Rao, C. S., and A. Purushottam, Pitchstone flows in the Rajmahal Hills, Santhal Parganas, Bihar, *Rec. Geol. Surv. India, 91*, 341-348, 1963.

Reston, T. J., Evidence for extensional shear zones in the mantle, offshore Britain, and their implications for the extension of the continental lithosphere, *Tectonics, 12*, 492-506, 1993.

Rock, N. M. S., B. J. Griffin, A. D. Edgar, D. K. Paul, and J. M. Hergt, A spectrum of potentially diamondiferous lamproites and minettes from the Jharia coalfield, eastern India, *J. Volcanol. Geotherm. Res., 50*, 55-83, 1992.

Royer, J.-Y. and M. F. Coffin., Jurassic to Eocene plate tectonic reconstructions in the Kerguelen Plateau region, *Proc. Ocean Drill. Prog., Sci. Results, 120*, 917-928, 1992.

Royer, J.-Y., and D. T. Sandwell, Evolution of the eastern Indian Ocean since the Late Cretaceous: Constraints from GEOSAT altimetry, *J. Geophys. Res., 94*, 13,755-13,782, 1989.

Salters, V. J. M., M. Storey, J. H. Sevigny, and H. Whitechurch, Trace element and isotopic characteristics of Kerguelen-Heard Plateau basalts, *Proc. Ocean Drill. Prog., Sci. Results, 120*, 55-62, 1992.

Sarkar, A., A. K. Datta, B. C. Poddar, V. K. Kollapuri, B. K. Bhattacharyya, and R. Sanwal, Geochronological studies of Mesozoic igneous rocks from eastern India, *J. Southeast Asian Earth Sci., 13*, 77-81, 1996.

Sarkar, S. S., S. K. Nag, and S. Basu Mallik, The origin of andesite from Rajmahal Traps, eastern India: A quantitative evaluation of a fractional crystallisation model, *J. Volcanol. Geotherm. Res., 37*, 365-378, 1989.

Saunders, A. D., Geochemistry of basalts from the Nauru Basin, Deep Sea Drilling Project Legs 61 and 89: Implications for the origin of oceanic flood basalts, *Init. Repts. Deep Sea Drill. Proj., 89*, 499-518, 1986.

Saunders, A. D., M. Storey, I. L. Gibson, P. Leat, J. Hergt, and R. N. Thompson, Chemical and isotopic constraints on the origin of basalts from Ninetyeast Ridge, Indian Ocean: Results from DSDP Legs 22 and 26 and ODP Leg 121, *Proc. Ocean Drill. Prog., Sci. Results, 121*, 559-590, 1991.

Saunders, A. D., M. Storey, R. W. Kent, and M. J. Norry, Consequences of plume-lithosphere interactions, in *Magmatism and the Causes of Continental Break-up, Spec. Publ. 68*, edited by B. C. Storey, T. Alabaster, and R. J. Pankhurst, pp. 40-60, The Geological Society, London, 1992.

Sen, G., and T. H. Cohen, Deccan intrusion, crustal extension, doming, and the size of the Deccan - Réunion plume head, in *Volcanism*, edited by K. V. Subbarao, pp. 201-216, Wiley Eastern, New Delhi, 1994.

Sengupta, S., Geological and geophysical studies in western part of Bengal Basin, India, *Bull. Am. Assoc. Petrol. Geol., 50,* 1001-1017, 1966.

Sengupta, S., Upper Gondwana stratigraphy and palaeobotany of Rajmahal Hills, Bihar, India, *Geol. Surv. India Monogr., 98, (Palaeontologica Indica),* 180 pp., GSI, Calcutta, 1988.

Sherwood, G. J., and S. Basu Mallik, A palaeomagnetic and rock magnetic study of the northern Rajmahal volcanics, Bihar, India, *J. Southeast Asian Earth Sci., 13,* 123-131, 1996.

Shukla, R., Study of Sylhet Trap flows in Balat area, Khasi Hills District, Meghalaya (abstract), in *Mesozoic Magmatism of the Eastern Margin of India, Programme and Abstracts,* edited by N. C. Ghose, pp. 13, Patna University, Patna, 1992.

Sparks, D. W., and E. M. Parmentier, Melt extraction from the mantle beneath spreading centers, *Earth Planet. Sci. Lett., 105,* 368-277, 1991.

Storey, M., A. D. Saunders, J. Tarney, I. L. Gibson, M. J. Norry, M. F. Thirlwall, P. Leat, R. N. Thompson, and M. A. Menzies, Contamination of Indian Ocean asthenosphere by the Kerguelen - Heard mantle plume, *Nature, 338,* 574 -576, 1989.

Storey, M., A. D. Saunders, J. Tarney, P. Leat, M. F. Thirlwall, R. N. Thompson, M. A. Menzies, and G. Marriner, Geochemical evidence for plume-mantle interactions beneath Kerguelen and Heard Islands, Indian Ocean, *Nature, 336,* 371-374, 1988.

Storey, M., R. W. Kent, A. D. Saunders, J. Hergt, V. J. M. Salters, H. Whitechurch, J. H. Sevigny, M. F. Thirlwall, P. Leat, N. C. Ghose, and M. Gifford, Lower Cretaceous volcanic rocks on continental margins and their relationship to the Kerguelen Plateau, *Proc. Ocean Drill. Prog., Sci. Results, 120,* 33-53, 1992.

Sun, S.-S., and W. F. McDonough, Chemical and isotopic systematics of oceanic basalts: Implications for mantle composition and processes, in *Magmatism in the Ocean Basins, Spec. Publ., 42,* edited by A. D. Saunders and M. J. Norry, pp. 313-345, The Geological Society, London, 1989.

Sushchevskaya, N. M., G. V. Ofchinnikova, A. Y. Borisova, B. V. Belyaszsky, J. Vasilyeva, and L. K. Levsky, Geochemical heterogeneity of Afanasy-Nikitin Rise magmatism, northeast Indian Ocean (in Russian), *Petrologia,* in press, 1997.

Talukdar, S. C., Rhyolite and alkali basalt from the Sylhet Traps, Khasi Hills, Assam, *Curr. Sci., 36,* 238-239, 1967.

Talukdar, S. C., and M. V. N. Murthy, The Sylhet Traps, their tectonic history, and their bearing on problems of Indian flood basalt provinces, *Bull. Volcanol., 35,* 602-618, 1970.

Tarduno, J. A., Brief polarity reversal during the Cretaceous Normal Polarity Superchron, *Geology, 18,* 683-686, 1990.

Thurow, J., and U. von Rad, Bentonites as tracers of earliest Cretaceous post-break-up volcanism off northwestern Australia (Legs 122 and 123), *Proc. Ocean Drill. Prog., Sci. Results, 123,* 89-110, 1992.

Todt, W., R. A. Cliff, A. Hanser, and A. W. Hofmann, ^{202}Pb + ^{205}Pb double spike for lead isotopic analyses, *Terra Cognita, 4,* 209, 1984.

von Rad, U., and N. F. Exon, Mesozoic sedimentary and volcanic evolution of the starved passive continental margin off northwest Australia, in *Studies in Continental Margin Geology, Mem. 34,* edited by J. S. Watkins, C. L. Drake, and R. E. Sheridan, pp. 253-281, American Association of Petroleum Geologists, Tulsa, 1983.

Wedepohl, K. H., The composition of the continental crust, *Mineral. Mag., 58A,* 959-960, 1994.

Weis, D., and F. A. Frey, Isotope geochemistry of Ninetyeast Ridge basalts: Sr, Nd, and Pb evidence for the involvement of the Kerguelen hot spot, *Proc. Ocean Drill. Prog., Sci. Results, 121,* 591-610, 1991.

Weis, D., and F. A. Frey, Rôle of the Kerguelen plume in generating the eastern Indian Ocean sea floor, *J. Geophys. Res., 101,* 13,831-13,849, 1996.

Weis, D., Y. Bassias, I. Gautier, and J.-P. Mennessier, DUPAL anomaly in existence 115 Ma ago: Evidence from isotopic study of the Kerguelen Plateau (south Indian Ocean), *Geochim. Cosmochim. Acta, 53,* 2125-2131, 1989.

Weis, D., F. A. Frey, H. Leyrit, and I. Gautier, Kerguelen Archipelago revisited: Geochemical and isotopic study of the Southeast Province lavas, *Earth Planet. Sci. Lett., 118,* 101-119, 1993.

White, R. S., and D. McKenzie, Magmatism at rift zones: The generation of volcanic continental margins and flood basalts, *J. Geophys. Res., 94,* 7686-7729, 1989.

White, R. S., and D. McKenzie, Mantle plumes and flood basalts, *J. Geophys. Res., 100,* 17,543-17,585, 1995.

White, R. S., D. McKenzie, and R. K. O'Nions, Oceanic crustal thickness from seismic measurements and rare-earth element inversions, *J. Geophys. Res., 97,* 19,683-19,715, 1992.

White, W. M., and A. W. Hofmann, Sr and Nd isotope geochemistry of oceanic basalts and mantle evolution, *Nature, 296,* 821-825, 1982.

Whitechurch, H., R. Montigny, J. Sevigny, M. Storey, and V. J. M. Salters, 1992, K-Ar and ^{40}Ar/^{39}Ar ages of central Kerguelen Plateau basalts, *Proc. Ocean Drill. Prog., Sci. Results 120,* 71-78, 1992.

N. C. Ghose, Department of Geology, Patna University, Patna 800 005, Bihar, India.

P. D. Kempton, NERC Isotope Geosciences Laboratory, Kingsley Dunham Centre, Keyworth, Nottingham NG12 5GG, United Kingdom.

R. W. Kent and A. D. Saunders, Department of Geology, University of Leicester, University Road, Leicester LE1 7RH, United Kingdom. E-Mail: kwr@le.ac.uk

The Ontong Java Plateau

Clive R. Neal[1], John J. Mahoney[2], Loren W. Kroenke[2], Robert A. Duncan[3], and Michael G. Petterson[4]

The Alaska-size Ontong Java Plateau (OJP) in the southwest Pacific is the largest of the world's large igneous provinces and formed entirely in an oceanic environment. Limited sampling of the upper levels of basaltic basement reveals strongly bimodal ages of ≈122 Ma and ≈90 Ma. Geochemical signatures indicate two isotopically distinct, ocean-island-like mantle-source types for the 122 Ma basalts and that the 90 Ma source was almost identical to one of the 122 Ma sources, strongly suggesting that a single mantle plume caused both eruptive events. In the 125–90 Ma period, the OJP appears to have been located near the Pacific Plate Euler poles and thus moved little relative to a postulated hotspot at about 42°S, 159°W; the early phase of emplacement probably also occurred near a spreading center, but substantial volumes were emplaced off-axis. The eastern lobe of the plateau appears to have been rifted shortly after 90 Ma. Incompatible and major element data are consistent with 20–30% polybaric partial melting of a peridotite source, beginning in the garnet stability field and continuing in the spinel field. Most existing basaltic basement samples appear to have experienced 30–45% of crystal fractionation; the resulting cumulates would be wehrlitic to pyroxenitic in composition, with an average density of ≈3.25 g cm^{-3}. We conclude that these cumulates form much of the plateau's high velocity (≈7.6 km s^{-1}) basal crustal layer. The relatively high density of this layer appears to have prevented emergence of much of the OJP above sea level despite a total crustal thickness exceeding 30 km. Although the deepest levels of the crust could be eclogitized (further increasing density), post-emplacement subsidence of the plateau has probably been tempered by the presence of a roughly 85-km-thick melt-depleted mantle root with a relatively low density (≤3.25 g cm^{-3}) for mantle material. Estimates of partial melting deduced from the apparent thickness of the mantle root imply that the OJP formed by 17–31% partial melting, in excellent agreement with geochemical modeling.

[1]Department of Civil Engineering and Geological Sciences, University of Notre Dame, Notre Dame, Indiana

[2]School of Ocean and Earth Science and Technology, University of Hawaii at Manoa, Honolulu, Hawaii

[3]College of Oceanic and Atmospheric Sciences, Oregon State University, Corvallis, Oregon

[4]British Geological Survey, Muchison House, West Mains Road, Edinburgh, United Kingdom

Large Igneous Provinces: Continental, Oceanic, and Planetary Flood Volcanism
Geophysical Monograph 100
Copyright 1997 by the American Geophysical Union

1. INTRODUCTION

Oceanic plateaus are now generally recognized as the counterparts of continental flood basalts and thick volcanic sequences found at many passive plate margins. Collectively, such areas have been termed large igneous provinces or LIPs [e.g., *Coffin and Eldholm*, 1991]. Since the mid-1980s, a consensus has grown among geochemists and geodynamicists that most oceanic plateaus are created at hotspots; more recently, the larger plateaus have been ascribed to the initial "plume-head" stage of hotspot development [e.g., *Richards et al.*, 1991; *Saunders et al.*, 1992; *Kent et al.*, 1992]. Many laboratory experiments, along with numerical modeling, suggest that mantle plume initiation caused by a temperature instability at a boundary layer results in a large, inflated "plume-head" trailed by a narrow feeder conduit or "tail" [e.g., *Whitehead and Luther*, 1975; *Olson and Singer*, 1985; *Griffiths and Campbell*, 1990; *Neavel and Johnson*, 1991]. In plume-head models, emplacement of the LIP occurs in a geologically short period of only a few million years or less [e.g., *Richards et al.*, 1989]. This is a consequence of extensive decompressional melting when the rising plume head, with temperatures argued to average up to 300°C above those of ambient upper mantle, approaches and spreads out at the base of the lithosphere, where it may reach a diameter of 2000 km or more [e.g., *Campbell and Griffiths*, 1990; *Hill et al.*, 1992; *Ribe et al.*, 1995; *Kincaid et al.*, 1996]. For the largest plateaus, the enormous size of the source required suggests an origin of the plume in the lower mantle, possibly at the core-mantle boundary; a plume head originating at the 660-km mantle discontinuity, for example, would not reach sufficient size to account for the volume of magma required to generate the larger LIPs [e.g., *Coffin and Eldholm*, 1993].

Several major oceanic plateaus are located in the Pacific Ocean basin, including Hess Rise, Shatsky Rise, the Mid-Pacific Mountains, the Manihiki Plateau, and the world's largest, the Ontong Java Plateau (OJP) (Figure 1). Unlike their continental and continental-margin analogs, plateaus formed in intraoceanic settings offer a means of studying LIP sources and structure and testing models of LIP origin without the often considerable complications caused by the presence of continental lithosphere. Unfortunately, the crustal basements of oceanic plateaus remain very poorly sampled, in general, because they are largely submerged and buried beneath thick covers of marine sediments. Of the plateaus in the Pacific, the OJP is by far the best-sampled, with basement penetrations at Deep Sea Drilling Project (DSDP) Site 289 (9 m), and Ocean Drilling Program (ODP) sites 803 (26 m) and 807 (149 m). Moreover, tectonically uplifted portions of OJP basement are exposed above sea level in the eastern Solomon Islands of Santa Isabel, Ramos, Malaita, and Ulawa (Figures 2 and 3). Some of the thickest exposures on these islands recently have been sampled in detail; studies are still underway, but the results now available provide a much better, though often surprising, picture of the OJP than was available only a few years ago. This paper presents a review of geochemical, geochronological, and geophysical knowledge about the plateau and additionally reports new results of a detailed geochemical investigation of the top 2–3 km of volcanic basement exposed on the island of Malaita. Particular attention is given to the origin and effect of "hidden" cumulates in the crust of the plateau.

2. PHYSICAL FEATURES AND GROSS STRUCTURE OF THE OJP

The Alaska-size OJP covers an area of more than 1.5×10^6 km^2 and consists of two parts: the main, or high, plateau in the west and north and the eastern lobe or salient (Figure 2). The plateau surface rises to depths of ≈ 1700 m in the central region of the high plateau but lies generally between 2 and 3 km depth. The OJP is bounded to the northwest by the Lyra Basin, to the north by the East Mariana Basin, by the Nauru Basin to the northeast and the Ellice Basin to the southeast (Figure 1). Along its southern and southwestern boundaries the OJP has collided with the Solomon Islands arc and it now lies at the junction between the Pacific and Australian plates, extending eastward to the Vitiaz arc-trench system.

Although much of the surface is relatively smooth, the top of the plateau is punctuated by several sizable seamounts, including Ontong Java atoll (the largest atoll in the world), and physiography around the margins of the plateau is complicated [e.g., *Kroenke*, 1972; *Berger et al.*, 1993]. In many areas, the basement crust is covered with pelagic sediments >1 km thick, which are thickest where the plateau is at its most shallow [e.g., *Mayer et al.*, 1991]. Rough basement topography on the margins of the OJP has been interpreted as extensional horst and graben structures that predate much of the sediment cover [e.g., *Ewing et al.*, 1968; *Kroenke et al.*, 1971; *Kroenke*, 1972; *Berger and Johnson*, 1976; *Hagen et al.*, 1993].

Crustal thickness on much of the high plateau is considerable even in comparison to other plateaus. Non-seismic methods (e.g., gravity data) provide a lower limit of 25 km [*Cooper et al.*, 1986; *Sandwell and Renkin*, 1988], whereas seismic and combined seismic-gravity evidence indicates crustal thicknesses in the 30–43 km range, with an estimated average around 36 km [e.g., *Furumoto et al.*, 1970, 1976; *Murauchi et al.*, 1973; *Hussong et al.*, 1979; *Miura et al.*, 1996; *Richardson and Okal*, 1996; *Gladczenko et al.*, 1997]. A 36-km thickness translates to a volume of $> 5 \times 10^7$ km^3 [*Mahoney*, 1987; *Coffin and Eldholm*, 1993]. The maximum extent of OJP-related volcanism may be even greater, as the Early Cretaceous lavas filling the Nauru Basin to the northeast of the OJP and similar lavas in the East Mariana and Pigafetta basins to the north have been proposed to be closely related to the OJP [e.g., *Castillo et al.*, 1991, 1994; *Gladczenko et al.*, 1997 and references therein]. Indeed, by analogy with some continental flood basalt provinces [see *Ernst and Buchan*, this volume], the lavas in these basins might reflect one or more giant radiating dike swarms associated with emplacement of the OJP.

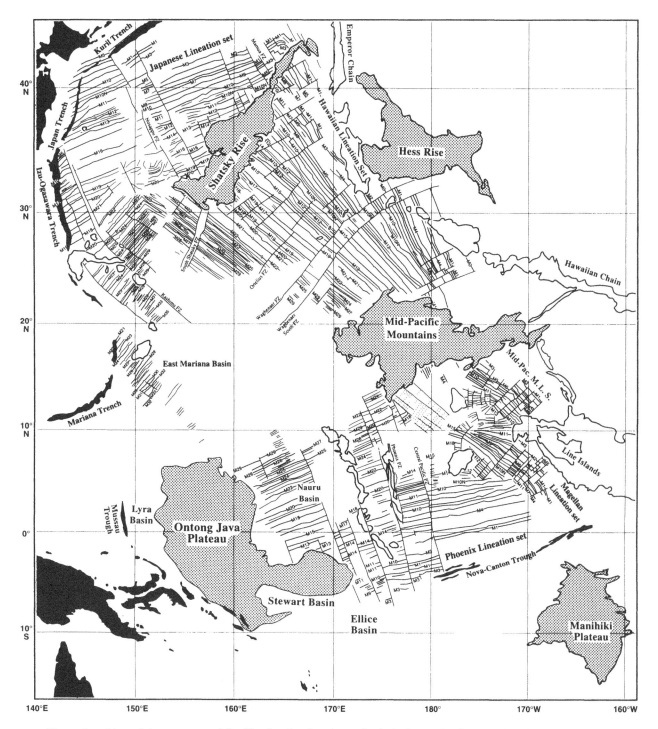

Figure 1. Map of the west-central Pacific showing locations of principal oceanic plateaus (shaded) and magnetic anomaly lineations [after *Nakanishi et al.*, 1992].

Results of recently completed seismic studies of crustal structure have, as yet, been presented only in abstract form [e.g., *Miura et al.*, 1996]. Nevertheless, a simple crustal model can be constructed from pre-1995 seismic refraction and sonobuoy wide-angle reflection/refraction measurements, despite the limitations of the data (Figure 2). Broadly, the OJP appears to have a crustal seismic structure resembling that of normal Pacific oceanic crust,

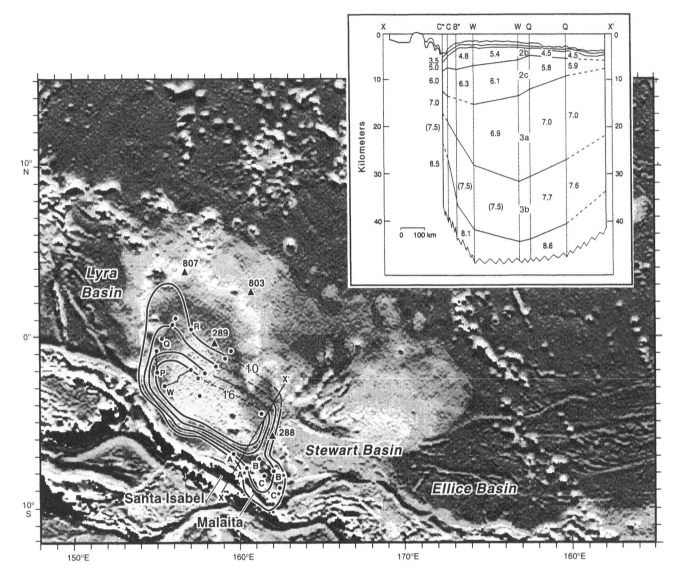

Figure 2. E-topo 5 bathymetric map of the western equatorial Pacific Basin (centered on the OJP and Nauru Basin, shaded by satellite-derived gravity fabric illuminated from the north [after *Smith and Sandwell*, 1995a,b]). The locations of drill sites 807, 803, 289, and 288 (the latter did not reach basement) are shown as triangles. Contours represent the depth to the top of Layer 3A in the high plateau (contour interval = 2 km). Deep crustal seismic refraction lines are labeled A-A*, B-B*, and C-C* [from *Furumoto et al.*, 1970] and P, Q, and R [from *Furumoto et al.*, 1976]. Sonobuoy refraction lines [from *Hussong et al.*, 1979] are represented by isolated dots. Inset is composite crustal cross-section X-X'. P-wave velocities (in km s^{-1}) are from *Furumoto et al.* [1970, 1976]; velocities in parentheses have been inserted following *Hussong et al.* [1979].

but each layer is abnormally thickened by up to a factor of five [*Hussong et al.*, 1979]. Figure 2 shows a composite section (XX') across the high plateau which depicts a thick crustal lens, more than 40 km thick where it underpins the center of the OJP. Such a crustal configuration is consistent with a structure-contour map showing the top of Layer 3A on the high plateau also presented in Figure 2.

This is based, in part, on data from the refraction surveys of *Furumoto et al.* [1970, 1976], but uses only the most reliable, upper crustal data, all from first arrivals; it also uses the later sonobuoy measurements summarized by *Hussong et al.* [1979], which showed Layer 2B velocities to be remarkably uniform. The upper crustal section includes the water column, sediments (containing some

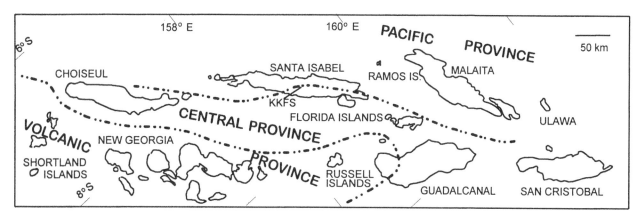

Figure 3. Division of the Solomon Islands chain into geologically distinct regions of the Pacific, Volcanic, and Central Provinces [e.g., *Coleman*, 1966, 1976; *Coleman and Packham*, 1976].

high-velocity horizons), Layer 2B (composed of dense, high velocity basalts), and Layer 2C (dolerites?). Depths to the top of Layer 3A (with P-wave velocities in the 6.9–7.0 km s^{-1} range, appropriate for high-level gabbros or possibly mafic granulites [e.g., *Rudnick and Jackson*, 1995]) range from less than 10 km around the plateau's edges to more than 16 km in a broad central region of the high plateau. The general shape is an elongate, ESE-WNW-trending depression, in much of which Layer 2 is over 12 km thick.

On the central and western parts of the high plateau, Layer 3B velocities are established to be rather high at 7.6–7.7 km s^{-1} (Figure 2). This range is appropriate for granular gabbros [*Schaefer and Neal*, 1994; *Farnetani et al.*, 1995, 1996] or garnet granulites [*Nixon and Coleman*, 1978], although *Houtz and Ewing* [1976] considered such velocities to be compatible with deep oceanic basalts or serpentinized peridotites. High sub-Moho velocities of 8.4–8.6 km s^{-1}, appropriate for eclogite [*Neal and Taylor*, 1989; *Rudnick and Jackson*, 1995; *Saunders et al.*, 1996] were detected in the northwest and southwest portions of the plateau.

3. TECTONIC SETTING OF OJP EMPLACEMENT

The original plate tectonic setting of the OJP is open to some question because well-defined magnetic anomaly lineations do not appear to be present on the plateau. Possible spreading-ridge or fracture-zone fabric is subdued and difficult to interpret unambiguously, and most pre-OJP crust to the west and south of the plateau has been subducted. However, block-faulting structures along the eastern margin of the high plateau were interpreted by *Andrews and Packham* [1975] and *Hussong et al.* [1979] to be parts of fracture zones trending roughly NNE. This

feature, together with the OJP's crustal velocity structure, possible very low-amplitude magnetic anomalies across the northern half of the high plateau, and the very sparse age data for plateau basement lavas available at the time, led *Hussong et al.* [1979] to propose an origin of the OJP at an unusually active, WNW-trending, slow-spreading ridge over a period of several tens of millions of years. An age progression across the plateau was implicit in this interpretation. Variations on this hypothesis proposed that the OJP was formed at a migrating triple junction or transform during a period of ridge jumping and heavy volcanism [*Winterer*, 1976; *Hilde et al.*, 1977; *Taylor*, 1978]; however, explanations for why volcanism was heavy were lacking. On the basis of preliminary isotopic and elemental data, together with the geophysical evidence, *Mahoney* [1987] proposed that the OJP was formed by a large ridge-centered or near-ridge plume, and discussed the possibility that this plume was the early Louisville hotspot (now located at ≈53°S [*Wessel and Kroenke*, 1997]).

Subsequent major and trace element data for basement lavas from drill sites on the OJP and from outcrops in Malaita and Santa Isabel (summarized below) were found to be consistent with the plateau having formed in the vicinity of a spreading center (or at least on thin lithosphere) by high fractions of partial melting [*Mahoney et al.*, 1993; *Tejada et al.*, 1996a]. *Tarduno et al.* [1991], *Richards et al.* [1991], and *Mahoney and Spencer* [1991] all favored the Louisville hotspot as the plume involved in the origin of the OJP; however, following *Richards et al.* [1989], they attributed the plateau to a cataclysmic outpouring of magma associated with the initial, plume-head stage of the hotspot, probably in the early Aptian. *Mahoney and Spencer* [1991] argued that, even if not initially surfacing near a spreading axis, plume heads would tend to attract ridges because of their expected control on rift propagation. Recently, *Winterer and Nakanishi* [1995] also inferred a near-ridge plume origin for the OJP; however, they interpreted bathymetry and the fabric in the new satellite-derived gravity map of *Smith and Sandwell* [1995a,b] to indicate an orientation of fracture zones and spreading axis nearly perpendicular to that suggested by *Hussong et al.* [1979].

The plateau geometry inferred by *Winterer and Nakanishi* [1995] appears difficult to reconcile with the nearby ENE-WSW M-series Nauru magnetic lineations on the east side of the OJP [e.g., *Nakanishi et al.,* 1992] (Figure 1). Moreover, *Taylor* [1978] reported M-series magnetic anomaly lineations in the nearby Lyra Basin on the west side of the OJP roughly paralleling those in the Nauru Basin on the east. A more recent aeromagnetic survey in the Lyra Basin also revealed roughly ENE-WSW-oriented M-series magnetic anomaly lineations (B. Taylor, personal communication, 1995). Our interpretation of the combined bathymetry and gravity map of *Smith and Sandwell* [1995a,b] is that the data are consistent with a NNE-trending fracture-zone fabric on the high plateau.

Rather than providing evidence for either a single, brief, cataclysmic emplacement event or a basement age progression, recent ^{40}Ar-^{39}Ar ages for OJP basement lavas yield an intriguing, strongly bimodal distribution. The ages of lavas from Sites 289, 807, and 803 on the high plateau, as well as for basement lavas from Malaita, Ramos, and Santa Isabel, suggest that most of the plateau may have formed in two relatively brief episodes, the first at 122±3 Ma, the second at 90±4 Ma (errors indicate total ranges rather than weighted means) [*Mahoney et al.,* 1993; *Tejada et al.,* 1996a,b; *Parkinson et al.,* 1996]. Thus, much of the fabric interpreted from bathymetry and satellite-derived gravity data is likely to represent preexisting oceanic crust covered by widespread plateau-basalt eruptions, consistent with pre-122 Ma, southward-younging M-series magnetic anomaly lineations in the Nauru and Lyra basins. Although a spreading center may well have been present, at least in the 122 Ma phase of eruptions, substantial magmatism must have occurred well beyond the immediate vicinity of a ridge axis. Indeed, available geophysical evidence weakly favors emplacement of most OJP lavas in an off-ridge location [*Coffin and Gahagan,* 1995]. As sampling of the plateau is still very limited, the relative crustal volumes of the ≈122 and ≈90 Ma episodes, and thus emplacement rates, are as yet unclear. However, *Tejada et al.* [1996a,b] and *Kroenke and Mahoney* [1996] suggested that the 122 Ma episode was significantly larger than the 90 Ma event, hypothesizing that the 122 Ma event generally corresponded to the construction of the main,

high plateau and that the eastern salient was the main focus of activity at 90 Ma.

4. THE OJP AND CRETACEOUS PLATE MOTIONS

Formation of the larger oceanic plateaus in the Pacific appears to be associated temporally with major changes in Pacific Plate motion. Four such changes appear to have occurred in the Late Jurassic to Early Cretaceous, a time when the plate was relatively small and being guided by the motion of adjoining plates [*Kroenke and Sager,* 1993]. These changes, at about 140, 125, 110, and 100 Ma, appear to have occurred near the times of formation of the Shatsky Rise, OJP (early event) and Manihiki Plateau, western and northern Hess Rise, and central Hess Rise, respectively, along or near the divergent boundaries of the Pacific Plate [*Kroenke and Sager,* 1993]. The most pronounced change in motion occurred at ≈125 Ma, between magnetic anomalies M1 and M0 near the Barremian-Aptian boundary [e.g., *Steiner and Wallick,* 1992], and was probably concomitant with the cessation of southwestward subduction beneath northeastern Gondwana and the beginning of northwestward subduction beneath Eurasia. This event can be noted, for example, in the reversal in magnetic anomaly lineation pattern from M3-M1 to M1-M3 east of the OJP near the west end of the Nova-Canton Trough (Figure 1; see also *Nakanishi and Winterer* [1996]). Associated with this change was the formation of a large portion of the OJP (i.e., the 122±3 Ma event) and the Manihiki Plateau (R. A. Duncan, unpubl. data, 1993). Unfortunately, existing data do not allow resolution of whether OJP eruptions followed or at least partly preceded the change. Indeed, ^{40}Ar-^{39}Ar dates for OJP lavas provide some of the best estimates of the minimum age of magnetic reversal M0 and the Barremian-Aptian boundary [*Pringle et al.,* 1992; *Mahoney et al.,* 1993].

Rough pre-100 Ma Pacific Plate motions can now be determined back to ≈145 Ma, using (1) the probable age progression along the Shatsky Rise estimated by *Sager and Han* [1993]; (2) paleomagnetic evidence of a change from a southward to a northward component of plate motion at ≈125 Ma [*Steiner and Wallick,* 1992] and limited latitudinal movement until about 90 Ma at the latitude of the Mid-Pacific Mountains [*Tarduno and Sager,* 1995]; (3) recently mapped magnetic anomaly lineations in the west-central Pacific [*Nakanishi et al.,* 1992]; (4) charts of central and western Pacific seamounts [*Mammerickx and Smith,* 1985]; and (5) the new satellite-derived gravity maps of *Smith and Sandwell* [1995a,b]. Figure 4 (bottom) shows the site of the OJP at approximately 125 Ma on the Pacific Plate, far from continental influences, centered at about 42°S, 159°W. From about 125 Ma until approximately 100 Ma the OJP appears to have been positioned very close to the Pacific Plate Euler poles, and thus moved relatively little. At approximately 100 Ma, plate motion changed from a northwestward to a more

northward trajectory, and from about 100 Ma to approximately 85 Ma the OJP moved steadily northward with the plate [*Yan and Kroenke*, 1993]. The plate reconstruction model of *Yan and Kroenke* [1993] shows that at 90 Ma, the age of the second major eruptive event, the eastern margin of the OJP passes approximately over the position occupied by the central high plateau at 125 Ma (Figure 4, middle).

The location of the plateau in the 125 Ma reconstruction of Figure 4 is at least roughly consistent with paleomagnetic data for basement and basal sediments at Site 289, which yield a paleolatitude of 30–35°S [*Hammond et al.*, 1975]; a similar paleolatitude is indicated by basement lavas at Site 807, although basement there may have been tilted around the time of emplacement [*Mayer and Tarduno*, 1993]. No hotspot is known to exist today in the vicinity of the triangle shown in Figure 4, but this area is one of the least surveyed in the world's oceans. This location is ≈1800 km distant from the Louisville hotspot, which, as noted above, has been suggested by several workers to be linked to the OJP and which has a well-marked seamount trail (the Louisville Ridge or Seamount Chain) going back to ≈70 Ma. Older portions of the Louisville hotspot trail, if they existed, have been destroyed by the Vitiaz-Tonga trench system [cf. *Mahoney and Spencer*, 1991]. To accommodate a Louisville hotspot origin for the OJP, true polar wander of ≈10–15° since 125 Ma must be invoked [e.g., *Mayer and Tarduno*, 1993]. Also, isotopic and elemental data for the 122 and 90 Ma OJP lavas are closely similar to each other, consistent with a single plume being responsible for both eruptive episodes, but they are distinct, particularly in Pb isotopic ratios, from the 0–70 Ma lavas of the Louisville Seamount Chain (see Section 6.3.1). At present, identification of the hotspot associated with the OJP remains elusive.

Between about 90 and 85 Ma, a major change in Australian and Antarctic Plate motions took place as

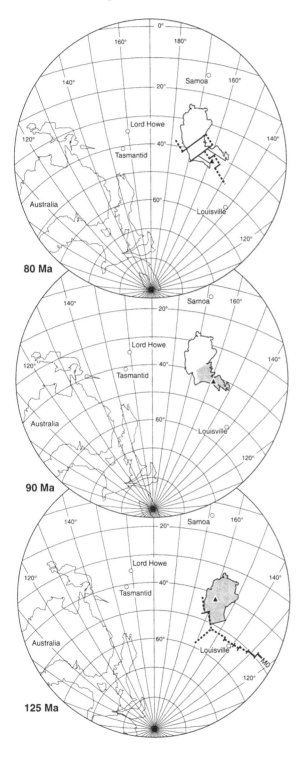

Figure 4. Bottom: approximate location of the OJP high plateau at 125 Ma. Shading represents postulated extent of 122 Ma volcanism. Several present-day hotspots are shown as circles. The triangle represents the inferred location of the OJP plume center beneath the crest of the high plateau. Heavy lines indicate a possible spreading ridge arrangement (schematic except at the eastern end). Middle: location of the OJP at 90 Ma; note that the eastern lobe of the plateau is now above the inferred plume center. Here, shading represents the postulated region of heaviest 90 Ma volcanism. Top: reconstruction for 80 Ma showing rifting of the eastern lobe (schematic). The 125 Ma reconstruction uses the 125–110 and 110–100 Ma Pacific Plate Euler poles of *Kroenke and Wessel* [1997]; the 90 and 80 Ma reconstructions are after *Yan and Kroenke* [1993].

spreading began in the Tasman Basin and in the Southwest Pacific Basin, south of Campbell Plateau [*Cande et al.*, 1989]. Following the ≈90 Ma eruptive episode on the OJP, post-emplacement rifting and seafloor spreading may have occurred for up to several million years within the plateau's eastern salient, in conjunction with spreading in the Ellice Basin to the east (Figure 4, top). A rifted character is suggested by the recent satellite-derived gravity data of *Smith and Sandwell* [1995a,b], which show that the northern and southern ridges bounding the bathymetric low in the eastern salient resemble conjugate ridges, with slight gravity lows on their southern and northern sides, respectively (Figure 2). A similar rift-margin fabric extends eastward across the adjoining Ellice Basin. Magnetic lineations have been detected in the Ellice Basin trending roughly east-west (D. Handschumacher, unpubl. data) but have not yet been identified; however, *Duncan* [1985] obtained an ^{40}Ar-^{39}Ar date of 82.6±1.2 Ma for a mid-ocean-ridge-type basalt from the eastern end of the Ellice Basin. Note that rifting of the eastern salient and spreading in the Ellice Basin soon after the ≈90 Ma OJP episode could explain why the OJP is not connected physically to a post-plateau chain of seamounts (the "plume-tail" stage of hotspot development).

5. RELATIONSHIP TO THE SOLOMON ISLANDS

Along the plateau's southern and southwestern margins, basement topography reflects faulting and deformation caused by the collision of the OJP with the Solomon Islands arc [e.g., *Kroenke*, 1972; *Kroenke et al.*, 1986, 1991; *Auzende et al.*, 1996]. The OJP arrived at the old Solomon arc during the early Neogene in a "soft" docking without significant deformation, with southwest-directed subduction ending around 27–23 Ma [e.g., *Coleman and Kroenke*, 1981; *Cooper and Taylor*, 1985]. Initiation of northeast-directed subduction beneath the plateau occurred progressively from the ESE because collision of the OJP with the old arc was diachronous. Collision moved WNW along the arc throughout the Neogene as a result of the general westward motion of the OJP, with subduction beginning on the plateau's south side at 10–5 Ma and continuing throughout the Pliocene along the arc [*Mann et al.*, 1996]. Subduction during the quiescent period between 27–23 Ma and 10–5 Ma appears to have been occurring in the Tonga and Trobriand trenches instead. The collision of the Woodlark Basin spreading-ridge complex with the southwest-facing San Cristobal forearc pushed the arc northeastward into the OJP and produced the Malaita Anticlinorium, an extensive fold belt embracing the eastern Solomon islands of Malaita, Ulawa, Ramos, and the northern half of Santa Isabel [*Kroenke*, 1972; *Kroenke et*

al., 1986; *Petterson*, 1995; *Petterson et al.*, 1997]. These islands appear to represent the tops of several-km-thick "tectonic flakes" of OJP crust thrusted onto the old forearc-backarc region, probably during the Pliocene [e.g., *Kroenke*, 1972; *Petterson et al.*, 1997]. Farther south, a large thrust-sheet of OJP crust forms the seafloor south of Malaita [*Auzende et al.*, 1996], possibly extending to and even including part of Makira (San Cristobal) [*Petterson et al.*, 1995; *Birkhold-VanDyke et al.*, 1996].

The Solomon Islands group has been subdivided for several decades into three geological regions or provinces (Figure 3) [e.g., *Coleman*, 1966, 1976; *Hackman*, 1973; *Coleman and Packham*, 1976]. The eastern region, termed the Pacific Province, appears to be an uplifted, overthrusted, largely unmetamorphosed portion of the OJP, as noted above [additional references include *Andrews and Packham*, 1975; *Hughes and Turner*, 1977; *Coleman et al.*, 1978; *Coleman and Kroenke*, 1981; *Ramsay*, 1982; *Hopson*, 1988; *Petterson*, 1995; *Tejada et al.*, 1996a]. Adjacent to the Pacific Province on the southwest is the Central Province, which contains variably metamorphosed Cretaceous and Early Tertiary seafloor and remnants of the northeast-facing arc sequence that grew during the Early to Middle Tertiary above the then southwest-plunging Pacific Plate (prior to the arrival of the OJP from the east). The boundary between the Pacific and Central provinces is generally submerged, but lies above sea level on Santa Isabel, where it forms a fault zone termed the Kaipito-Korighole fault system [e.g., *Hawkins and Barron*, 1991; *Tejada et al.*, 1996a]. Along the southwestern flank of the Central Province is the Volcanic Province, an island arc sequence composed of volcanic and intrusive rocks and active volcanoes; the age of this province appears to be <4 Ma [e.g., *Petterson*, 1995]. Significantly, the plateau appears to be more or less unsubductable [*Cloos*, 1993; *Abbott and Mooney*, 1995]. However, the post-Miocene removal of a portion of the lower OJP between Santa Isabel and Makira is evident from recent seismic surveys [*Mann et al.*, 1996; *Phinney et al.*, 1996; *Cowley et al.*, 1996].

6. GEOCHEMISTRY AND AGE OF BASEMENT

6.1. *Submarine Drillholes*

The only detailed basement stratigraphy for the entire plateau, prior to the recent and ongoing work in Malaita and Santa Isabel, came from the 149-m-thick section of ODP Site 807 and the 26 m section of Site 803 [*Mahoney et al.*, 1993]. At Site 807, the section has been divided into several units (A, C-G), each of which is a packet of low-K, tholeiitic pillow lavas and massive flows, except for Unit

F, which consists of a single 28-m-thick flow, and Unit B which is a 50-cm-thick interlava limestone [*Kroenke et al.*, 1991; *Mahoney et al.*, 1993]. Unit A (46 m thick) is isotopically (Figure 5) and chemically (Figure 6 and Table 1) distinct from Units C-G in having lower $^{206}Pb/^{204}Pb$ (18.3–18.4 vs. 18.6–18.7) and initial ε_{Nd} (+4.8 – +5.4 vs. +5.9 – +6.3), higher initial $^{87}Sr/^{86}Sr$ (0.7040–0.7041 vs. 0.70339–0.70345), and slightly higher ratios of highly incompatible to moderately incompatible elements. Consistent with a plume-head origin, the isotopic values of both groups fall within the range of those for "Dupal-type" hotspot islands (although few present-day islands have values closely similar to those of the OJP lavas). Except for elements especially susceptible to seawater alteration (e.g., Rb, K, Cs), primitive-mantle-normalized incompatible element patterns are quite flat (Figure 6), although some of the most incompatible elements are relatively depleted (e.g., Ba and Th); Li and P are also depleted whereas Mo is relatively enriched (note that rock powders were ground in alumina). Major and trace element compositions suggest both groups reflect high degrees of melting (probably 20–30%, with Units C-G representing slightly higher degrees of melting than Unit A). The single flow sampled at DSDP Site 289 is nearly identical to the Units C-G lavas. Although differing somewhat in major element composition, the Site 803 lavas are closely similar isotopically to the Units C-G basalts, with identical $^{206}Pb/^{204}Pb$ and initial ε_{Nd} and only slightly higher initial $^{87}Sr/^{86}Sr$ (0.7036–0.7037) (Figure 5). Their incompatible element signature is also similar to that of Units C-G, although they are slightly enriched in the more highly incompatible elements (Figure 6 and Table 1).

The ^{40}Ar-^{39}Ar ages for Unit A and Units C-G lavas, as well as the Site 289 basalt, are identical, within errors, at ≈122 Ma [*Mahoney et al.*, 1993]. However, the Site 803 lavas yielded significantly younger ages of ≈90 Ma; sediments above basement at Site 803 could not be used to confirm this age as the site is located in a depression which appears to contain derived sediments [*Kroenke et al.*, 1993; *Mahoney et al.*, 1993; *Sliter and Leckie*, 1993]. Foraminiferal ages of these sediments vary from Barremian to Lower Aptian [*Sliter and Leckie*, 1993].

6.2. Santa Isabel

The ≈90 Ma event was confirmed by ^{40}Ar-^{39}Ar results for the tholeiitic Sigana Basalts, which form the basement of the northern portion (Pacific Province) of Santa Isabel and Ramos Island. Both the ≈122 Ma and ≈90 Ma groups are present, with the 90 Ma lavas being particularly abundant in the middle and southeastern parts of the island

[*Tejada et al.*, 1996a; *Parkinson et al.*, 1996]. The Nd-Pb-Sr isotopic characteristics of the Sigana Basalts closely resemble those at Site 807 and Site 803. Both Unit-A-like (hereafter "A-type") and Units-C-G-like (hereafter "C-G-type") isotopic compositions are present and, as at Site 803, the ≈90 Ma lavas are similar to the C-G-type [*Tejada et al.*, 1996a]. Average incompatible element signatures are again very similar to those at the ODP sites (Figure 6), although a wider range of abundances is present among the Sigana Basalts, including several lavas that appear to be highly differentiated basalts [*Tejada et al.*, 1996a; *Parkinson et al.*, 1996].

In addition to the low-K tholeiitic Sigana Basalts, rare 90 Ma alkalic dikes, termed the Sigana Alkalic Suite [*Tejada et al.*, 1996a], are present within the Sigana Basalt exposures on Santa Isabel. Despite being indistinguishable in age from the younger group of tholeiites, they have high-$^{206}Pb/^{204}Pb$, "HIMU" or "Mangaia Group"-type isotopic signatures quite distinct from those of the tholeiites (Figure 5).

6.3. Malaita

6.3.1. OJP Basement. Following reconnaissance studies by *Rickwood* [1957] and *Pudsey-Dawson* [1960], *Hughes and Turner* [1976, 1977] mapped the southern part of Malaita, where they termed the basement section the Older Basalts. Recently, the basement lavas throughout the island have been renamed the Malaita Volcanic Group [*Petterson*, 1995]. The ^{40}Ar-^{39}Ar ages of samples collected by Hughes and Turner showed them to be 122 Ma [*Mahoney et al.*, 1993]. All are low-K tholeiites of essentially A-type composition, although rocks from the much more extensive exposures in southern Malaita exhibit somewhat greater chemical and isotopic variation than the 46 m of Unit A flows at Site 807 (e.g., Figure 5) [*Mahoney*, 1987; *Mahoney and Spencer*, 1991; *Tejada et al.*, 1996a].

The thickest exposures of basement crust (reaching 3–4 km in stratigraphic thickness [*Petterson*, 1995; *Tejada et al.*, 1996b; *Petterson et al.*, 1997]) are in the central part of Malaita, which remained unsampled until recently. Basement in central and northern Malaita is exposed in the cores of NW-SE-trending anticlinal to periclinal structures formed in response to the collision of the OJP with the Australian Plate [e.g., *Petterson et al.*, 1997]; the major outcrops of Malaita Volcanic Group rocks are in the Kwaio, Kwara'ae, Fateleka, and Tombaita areas (Figure 7). These areas are dominated by submarine basalt flows which range from <1 m to 60 m in thickness, with most between 4 and 12 m thick; dikes are rare in most locations.

TABLE 1. Compositions of Basalts from Malaita, ODP Leg 130 Sites, and Average Nauru Basalt

	Standard Reference Materials				Malaita			ODP Leg 130			Nauru Basin
	BIR-1		BHVO-1		A-Type	C-G-Type	ML475	Site 807 A-Type	Site 807 C-G-Type	Site 803	
	Publ.	Meas.	Publ.	Meas.	(n=48)	(n=21)					
SiO_2					49.92	49.25	49.68	48.24	49.88	49.09	49.7
TiO_2					1.59	1.20	0.73	1.61	1.15	1.37	1.2
Al_2O_3				14.31	14.47	14.00	14.29	14.24	15.46	13.9	
Fe_2O_3					13.71	12.69	9.84	13.52	12.26	11.59	4.0
MnO					0.20	0.21	0.16	0.20	0.20	0.16	0.2
MgO					7.11	7.75	9.99	6.66	7.64	6.25	6.8
CaO					11.29	12.35	14.52	12.11	12.13	12.22	11.4
Na_2O					2.23	2.05	1.48	2.39	2.20	2.32	2.4
K_2O					0.14	0.14	0.08	0.29	0.15	0.47	0.1
P_2O_5				0.14	0.11	0.06	0.14	0.10	0.15	0.1	
Li	[R]3.4	3.95	[P]4.6	4.12	6.42	6.71	3.51				
Be	[P]0.58	0.14	[P]1.1	0.97	0.65	0.47	0.25				
K					1139	1129	664.0				
Sc	[R]44	56.9	[R]31.8	36.4	44.0	46.5	57.9	46.5	49.3	48.9	47.4
V	[R]313	476	[R]317	369	370	314	217	325	347	312	
Cr	[R]382.0	399.8	[R]289.0	340.3	96.2	133.0	446.3	156	153	252	245
Co	[R]51.4	53.7	[R]45.0	49.2	51.0	50.9	38.8				50.0
Ni	[R]166	163	[R]121	119	70.5	98.1	133	96	94	110	115
Cu	[R]126	113.4	[R]136	135.7	107.1	140.8	128.0				
Zn	[R]71	71	[R]105	106	99	80	47				
Ga	[P]16.0	15.0	[R]21.0	25.1	19.4	17.0	13.1				
Rb	[I]1	.0148	[R]11	9.93	1.72	1.32	0.91				
Sr	[R]108	111	[R]403	425	145	153	135	188	113	172	110
Y	[R]16	15.7	[R]27.6	26.4	28.4	22.6	13.5	30.2	25.0	27.6	22.0
Zr	[P]22	15.9	[R]179	182	94.1	69.3	29.9	98.6	64.4	79.4	49.5
Nb	[P]2	0.7	[R]19	20.3	5.44	3.76	1.57	5.89	3.50	4.45	3
Mo	[I]0.5	0.10	[R]1.02	1.004	0.69	0.54	0.18	0.75	0.79	0.50	
Cs	[P]0.45	0.03	[I]0.13	0.16	0.02	0.01	0.01	0.19	0.04	0.18	
Ba	[P]7.7	6.94	[R]139	135.6	34.1	23.6	11.5	24.1	13.6	16.5	
La	[P]0.88	0.65	[R]15.8	15.9	5.59	3.66	1.61	5.79	3.38	4.31	3.4
Ce	[I]2.5	2.01	[R]39	40.6	15.72	10.55	4.80	14.39	8.99		9.5
Pr	[I]0.5	0.35	[R]5.7	5.22	2.32	1.57	0.72	2.29	1.46	1.78	
Nd	[P]2.5	2.40	[R]25.2	25.05	11.64	8.26	3.78	10.79	7.10	8.46	6.3
Sm	[R]1.1	1.04	[R]6.2	6.22	3.56	2.67	1.30	3.36	2.39	2.88	2.6
Eu	[R]0.54	0.51	[R]2.1	2.11	1.30	0.97	0.51	1.29	0.92	1.11	1.0
Gd	[P]1.9	1.77	[R]6.4	6.42	4.46	3.40	1.74	4.31	3.26	3.73	3.3
Tb	[R]0.41	0.34	[R]1.0	0.94	0.79	0.62	0.32		0.64	0.73	0.8
Dy	[R]2.4	2.43	[R]5.2	5.23	4.93	3.90	2.05	4.94	4.06	4.45	
Ho	[P]0.5	0.50	[I]0.99	0.98	1.05	0.83	0.43	1.06	0.90	0.97	
Er	[P]1.8	1.67	[R]2.4	2.51	3.01	2.37	1.26	2.99	2.46	2.74	
Tm	[P]0.27	0.24	[R]0.33	0.32	0.43	0.34	0.17				
Yb	[R]1.7	1.60	[R]2.02	2.07	2.90	2.28	1.16	2.72	2.40	2.59	2.6
Lu	[R]0.26	0.27	[R]0.29	0.28	0.43	0.34	0.17	0.41	0.35	0.38	0.4
Hf	[R]0.58	0.58	[R]4.38	4.32	2.51	1.91	0.87	2.54	1.65	2.07	1.7
Ta	[I]0.0006	0.04	[R]1.23	1.34	0.36	0.27	0.10	0.38	0.21	0.30	0.2
W	[I]0.20	0.23	[P]0.27	0.38	0.06	0.04					
Pb	[P]3.2	3.11	[P]2	2.04	1.30	0.81	0.40	0.72	0.67		
Th	[I]0.89	0.02	[R]1.08	1.23	0.53	0.33	0.12	0.52	0.28	0.35	0.3
U	[I]0.025	0.027	[R]0.42	0.43	0.14	0.09	0.03	0.14	0.12	0.21	

Our ongoing work and that of colleagues A. Saunders and T. Babbs (Univ. of Leicester, UK) reveals that broadly C-G-type lavas are present below an A-type cap some 600 m thick ([*Tejada et al.*, 1996b] and C. R. Neal et al.; M. L. Tejada et al.; T. Babbs et al., manuscripts in preparation, 1997). The range of chemical variation is again greater than in the 149 m of flows at Site 807 (Figures 8a–d, 9), but A- and C-G-type basalts can clearly be distinguished in incompatible element (e.g., Figure 6), major element (Figures 8c and 9d), and isotopic (Figure 5) diagrams. Also, all dated basement samples from central Malaita belong to the 122 Ma event [*Tejada et al.*, 1996b]. Thus, no 90 Ma lavas have been found on the island, and A-type lavas form a section more than 13 times thicker than at Site 807, some 1600 km away to the north. The transition between A-type and C-G-type groups is again sharp, but unlike Site 807, where an interlava limestone unit (Unit B) separates the two types, no corresponding sedimentary layer is seen in Malaita. These results suggest that the part of the OJP now exposed on Malaita was more proximal to, or possibly downslope from, the main locus of A-type eruptive activity than Site 807. Moreover, the combined data for Malaita, Santa Isabel, and the drill sites demonstrate that both A- and C-G-type lavas were erupted over a considerable portion of the plateau [cf. *Tejada et al.*, 1996a,b].

For each type, a voluminous, well-mixed mantle source (relative to the scale of melting) is indicated. In the context of a zoned plume-head model, one type may better represent the plume-source composition and the other better reflect average entrained mantle, or the plume source itself may have contained both types; the C-G-type is closer to some estimates of average lower mantle isotopic composition [*Tejada et al.*, 1996a,b]. Intriguingly, lavas of the much more poorly sampled Manihiki Plateau to the east of the OJP (Figure 1) are very close in age to the ≈122 Ma OJP event (R. A. Duncan, unpubl. data, 1993) and also define two isotopic groups, one of which is quite similar to the OJP A-type (Figure 5a,b), possibly suggesting that the sources of the OJP and Manihiki Plateau were related [*Mahoney and Spencer*, 1991; *Mahoney et al.*, 1993]. Indeed, *Coffin and Eldholm* [1993] speculated that the Manihiki Plateau may be part of

a "greater" OJP event, along with lavas filling the Nauru and East Mariana basins. Isotopic and elemental data (e.g., Figures 8 and 9) for the basin-filling basalts partially overlap with those for C-G-type OJP lavas [e.g., *Tokuyama and Batiza*, 1986; *Floyd*, 1986; *Castillo et al.*, 1986, 1994; *Saunders*, 1986; *Mahoney*, 1987]. However, their ^{40}Ar-^{39}Ar ages, at 111–115 Ma, are significantly different from either the 122 or 90 Ma events on the OJP, and *Castillo et al.* [1994] proposed that they reflect OJP source mantle that had flowed northward and mixed variably with ambient MORB-type asthenosphere (recall that the OJP itself probably moved very little during this period; see Section 4 and Figure 4).

The similarity in isotopic and incompatible element ratios of the 90 Ma OJP tholeiites to the 122 Ma C-G-type strongly suggests both groups shared a common plume mantle source. As noted above, the plateau may have been located at a roughly similar geographic position at these two times; however, the lack (so far) of any OJP lavas with intermediate ages argues against a smooth plume-head to plume-tail transition after the 122 Ma event. One possibility suggested by simple experimental and theoretical modeling is that the strong bimodality in age of OJP lavas represents a separation of the starting-plume head from its conduit (tail) as it rose through the 660-km mantle discontinuity, followed by formation of a second, smaller head which rose to the base of the lithosphere some 30 m.y. after the first [*Bercovici and Mahoney*, 1994]. Alternatively, plume material may have collected beneath the plateau more or less continuously after 122 Ma, but did not erupt to the surface in significant amounts until a change in the stress field of the OJP around 90 Ma provided the necessary pathways for melt egress [*Tejada et al.*, 1996a; *Ito and Clift*, 1996].

A quite different model was proposed by *Larson and Kincaid* [1996]. They suggested that the 122 Ma event reflected upward advection of a region of the 660-km boundary layer in response to accelerated subduction of slab material into the deep mantle. Ascent of this boundary layer would increase temperatures in the upper mantle above the peridotite solidus and thus cause voluminous plateau magmatism. At about the same time, increased amounts of slab material would encounter the

---

Note that all Fe is listed as Fe_2O_3. Data for Malaitan basalts from this work, for ODP basalts from *Mahoney et al.* [1993], and the average composition for the Nauru Basin basalts are calculated from *Batiza* [1986], *Tokuyama and Batiza* [1986], *Saunders* [1986], *Castillo et al.* [1986, 1991], and *Floyd* [1986, 1989]. The published compositions of standard reference basalts BIR-1 and BHVO-1 [*Govindaraju*, 1989] are shown along with our ICP-MS (inductively coupled plasma mass-spectrometric) determinations (Meas.) performed during the analyses of the Malaitan basalts. Superscripts on published values: R = recommended values (i.e., most accurately determined abundances); P = provisional abundances; I = data are for information only (least accurately determined abundances). Major element abundances are in wt%; all others are in parts per million.

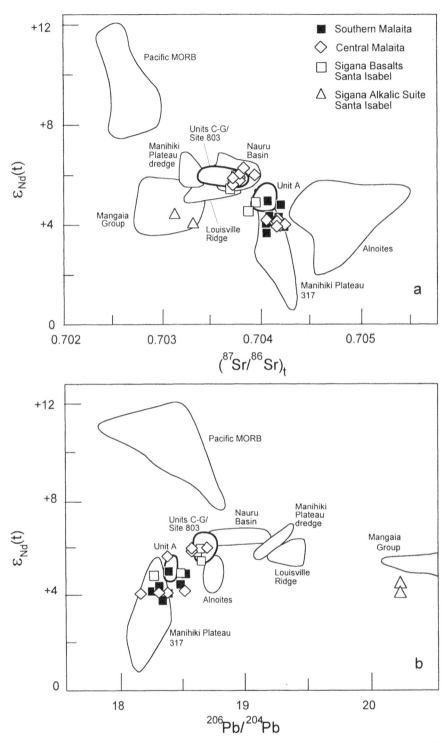

Figure 5. Initial $\varepsilon_{Nd}(t)$ vs. $(^{87}Sr/^{86}Sr)_t$ (a) and present-day $^{206}Pb/^{204}Pb$ (b) for basement lavas of Malaita and Santa Isabel and 90 Ma alkalic dikes of Santa Isabel [*Mahoney*, 1987; *Mahoney and Spencer*, 1991; *Tejada et al.*, 1996a; M. Tejada et al., unpubl. data, 1995]. Fields for A-type and C-G-type/Site 803 drillhole lavas are shown by heavy outlines, those for Malaitan alnöites, Manihiki Plateau, Louisville Ridge, Pacific MORB, and Mangaia Group islands of the South Pacific by light outlines [see *Mahoney et al.*, 1993; *Tejada et al.*, 1996a, for data sources].

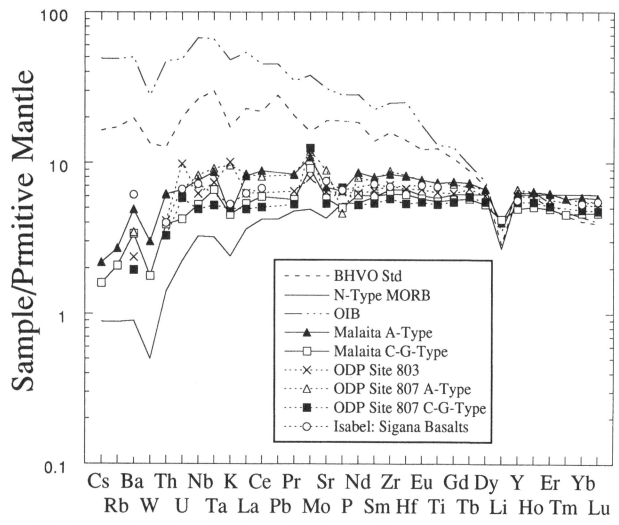

Figure 6. Primitive-mantle-normalized element profiles for average of OJP basalts of Malaita (this study), Sites 803 and 807 [*Mahoney et al.*, 1993], and Santa Isabel [*Tejada et al.*, 1996a]). Average N-type MORB and OIB patterns and primitive-mantle normalizing values are from *Sun and McDonough* [1989]. BHVO-1 standard values are from *Govindaraju* [1989].

core-mantle boundary, and the resulting instability would produce diapiric upwelling argued to reach the surface as a plume head some 20–30 m.y. after the initial magmatic event. This model accounts for the bimodality in the OJP basement ages but implies significant isotopic and chemical differences between the 122 Ma magmatism, which would have an upper mantle and/or 660-km boundary-layer source, and the 90 Ma event, which would have a source originating at the core-mantle boundary. Thus, the model does not explain the marked compositional similarity between the 122 Ma C-G-type and the 90 Ma basalts.

As noted in Section 4, aside from problems associated with pre-90 Ma plate reconstructions, attempts to establish

a geochemical link between the OJP and the Louisville hotspot have been unsuccessful. The Nd-Pb-Sr isotopic range defined by ≈70–0 Ma seamount lavas along the ≈4000-km-long Louisville Ridge (see Figure 5) is restricted and values vary little with rock type or age, age of underlying oceanic crust, or depth of melting, indicating a long-lived, isotopically homogeneous source [*Cheng et al.*, 1987; *Hawkins et al.*, 1987]. Although initial ε_{Nd} and $^{87}Sr/^{86}Sr$ values for the Louisville Ridge overlap with those of the 122 and 90 Ma OJP C-G-type basalts (Figure 5), the difference in Pb isotope ratios is substantial (e.g., 0.4 to 1.0 in $^{206}Pb/^{204}Pb$). It is highly unlikely to be the result of radiogenic ingrowth in the plume source between 90 Ma and 70 Ma, but could represent a major compositional

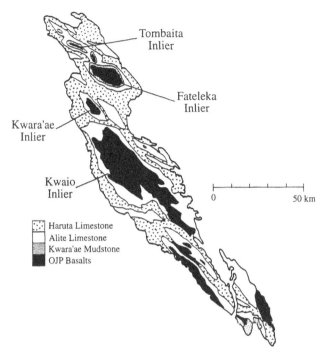

Figure 7. Simplified geological map of Malaita defining the OJP basement inliers [after *Petterson*, 1995].

change from the plume head to plume tail [*Mahoney et al.*, 1993] (as postulated to be a common occurrence by *Campbell and Griffiths* [1990]). Note, however, that the change would have to have occurred during the 20 m.y. period for which the record just happens to be missing [*Mahoney et al.*, 1993].

6.3.2. *Late-Stage Volcanism on Malaita.* Two younger volcanic events are recorded in the pelagic sedimentary section overlying basement on Malaita. The first consists of flows and sills of alkalic basalts cropping out irregularly within Eocene limestones in southern and northern Malaita and the nearby island of Ulawa. Those in southern Malaita were called the Younger Basalts by *Hughes and Turner* [1977], a term superseded by the name Maramasike Volcanic Formation, which includes both the northern and southern exposures [*Petterson*, 1995]. Although they appear at the same stratigraphic level, their thickness varies considerably from place to place, locally reaching a maximum of ≈500 m. An ^{40}Ar-^{39}Ar age of 44.2±0.2 Ma was obtained by *Tejada et al.* [1996a] for one of these basalts. Their Nd-Sr-Pb isotopic ratios are distinct from those of the basement tholeiites, and they may reflect the passage of the OJP over the Samoan hotspot [*Tejada et al.*, 1996a]. Probable counterparts are seen in seismic reflection records of the high plateau north of Malaita, which show numerous sill-like intrusions within the sedimentary section [*Kroenke*, 1972]; moreover, at least

some of the large seamounts atop the plateau may be related to Maramasike volcanism.

The second late-stage volcanic event is recorded in central Malaita as several intrusions of alnöite, a rare, ultramafic alkalic magma of deep mantle origin with affinities to kimberlite. With an age of 34 Ma [*Davis*, 1977], the alnöites significantly postdate the Maramasike Volcanic Formation. They appear to correspond to low degrees of melting of sublithospheric mantle [e.g., *Nixon and Boyd*, 1979; *Nixon et al.*, 1980; *Neal*, 1985, 1988] during flexural extension of the OJP as it overrode the outer-trench high prior to colliding with the old North Solomon trench [*Coleman and Kroenke*, 1981; *Petterson*, 1995]. Diatreme-like intrusions, probably correlative with the alnöites on Malaita, are evident in seismic reflection records over about a third of the high plateau [*Kroenke*, 1972; *Nixon*, 1980]. The alnöites contain a rich and varied suite of mantle xenoliths and xenocrysts, which have been studied extensively [e.g., *Nixon and Coleman*, 1978; *Nixon and Boyd*, 1979; *Neal*, 1988, 1995; *Neal and Davidson*, 1989]. Most of the xenoliths appear to represent lithospheric mantle and indicate that this part of the plateau (near the southern edge of the OJP) had a pre-collision lithospheric thickness of about 120 km [*Nixon and Boyd*, 1979]. Isotopically, most of the xenoliths are distinct from the OJP basement tholeiites, with generally higher initial $^{87}Sr/^{86}Sr$ and more variable ε_{Nd} values (Figure 5a) [*Neal*, 1985, 1988; *Neal and Davidson*, 1989], indicating that the lithospheric mantle of the plateau has been variably modified since the plateau formed.

7. EVALUATION OF OJP SOURCE REGION AND MELTING CHARACTERISTICS

7.1. *A Core-Mantle Boundary Origin for the OJP Plume?*

The ultimate origin of the OJP (i.e., the source of the plume) is difficult to evaluate because the presumed plume head arguably contained both plume-source and entrained-mantle components. However, geochemical signatures of the plume source could persist and be detectable in erupted magmas if such signatures were (1) of sufficient magnitude and markedly distinct from those of ambient mantle, and (2) not obliterated by melting or magmatic differentiation processes. Physical considerations imply that plume heads giving rise to the largest LIPs may originate at the core-mantle boundary [e.g., *Campbell and Griffiths*, 1990; *Coffin and Eldholm*, 1991, 1994]. A key chemical "fingerprint" of a core-mantle boundary origin may be the unusual enrichment of siderophilic trace elements [e.g., *Walker et al.*, 1995]; such enrichment may occur by

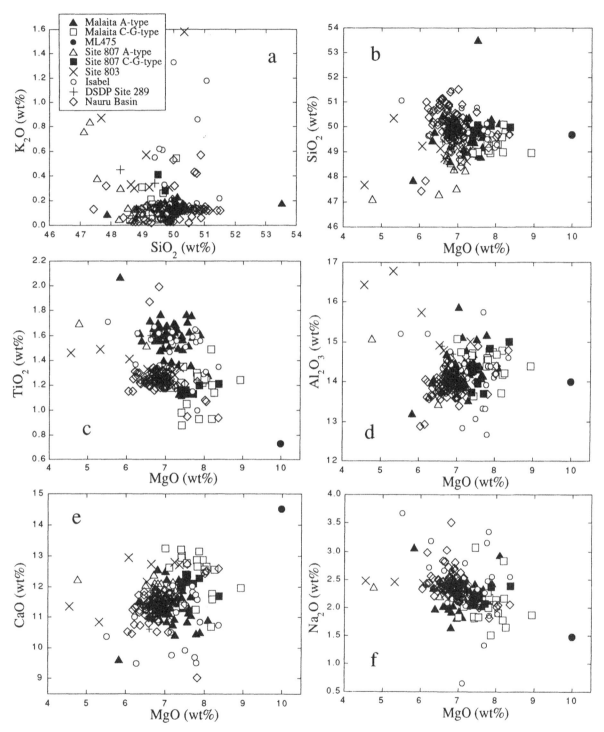

Figure 8. Major element variations in the Malaitan OJP basalts (this study) compared with those of DSDP Site 289 [*Stoeser*, 1975], ODP Leg 130 [*Mahoney et al.*, 1993], Santa Isabel [*Tejada et al.*, 1996a], and the Nauru Basin [*Batiza*, 1986; *Tokuyama and Batiza*, 1986; *Saunders*, 1986; *Castillo et al.*, 1986, 1991; *Floyd*, 1986, 1989]. (a) Classification of these basalts generally as low-K tholeiites, with K_2O values > 0.2 wt% probably resulting from low-temperature alteration; (b-f) major element variations using MgO as the fractionation index.

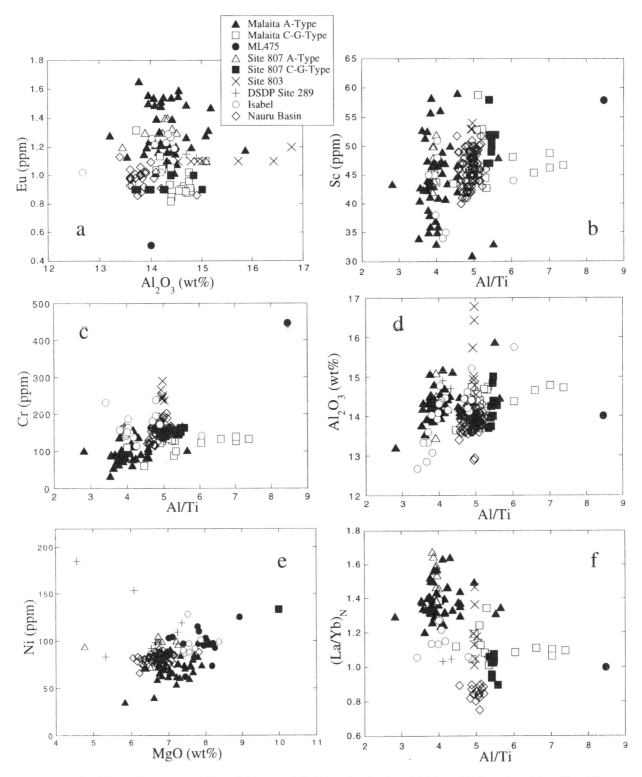

Figure 9.   Trace-element comparison of A-type and C-G-type basalts from Malaita with those from Sites 807, 803 and 289, Nauru Basin, and Santa Isabel (data sources as in Figure 8).

entrainment of small amounts of outermost core material during the initial formation of the plume or by chemical diffusion across the boundary [e.g., *Kellogg and King*, 1993]. Unlike major siderophiles (Fe, Ni, Co, etc.), which are affected strongly by silicate melting and differentiation processes, many siderophilic trace elements, such as Mo, W, Au, and Re, as well as those of the platinum group (Rh, Ru, Os, Ir, Pt, Pd), are generally incompatible in silicate minerals [e.g., *Newsom and Palme*, 1984; *Jones and Drake*, 1986; *Fleet et al.*, 1991, 1993; *Walker et al.*, 1993]. Thus, large degrees of partial melting and basaltic-type magmatic differentiation should not significantly alter the ratios of these elements to other incompatible (lithophile) elements.

Jain et al. [1995] reported the presence of a significant enrichment in W and Mo for OJP basalts from the ODP drill sites and Malaita. These ICP-MS data were interpreted as supporting a core-mantle boundary origin for the OJP plume source. However, our recent reanalysis of the samples, using a more aggressive (between-sample) wash procedure [*McGinnis et al.*, 1996, 1997] demonstrates that the W enrichments are an artifact of instrumental memory from previous samples and standards. In fact, the OJP basalts actually exhibit slightly negative anomalies on primitive-mantle-normalized element diagrams (Figure 6), similar to those seen for MORB and ocean island basalts.

In contrast to W, however, a persistent positive Mo anomaly is still evident in primitive-mantle-normalized element patterns (Figure 6). The observed enrichment of Mo in OJP basalts is consistent with a siderophile-element-enriched source region. Preliminary analyses of platinum group elements and Au in OJP basalts also appear consistent with a siderophile-enriched source [e.g., *Jain et al.*, 1996], but more detailed study is required before any definitive statements can be made.

7.2. *Partial Melting of the Plume Head Peridotite Source*

Significant amounts of partial melting of an adiabatically rising plume head should occur when it intersects the peridotite solidus in the upper mantle. One noteworthy characteristic of the OJP basement lavas is illustrated by Figure 10, showing Zr/Nb vs. Zr/Y. These elements would be highly incompatible in the fractionating assemblage (see Section 8.1) and their ratios thus insensitive to different amounts of differentiation. Relative to southern East Pacific Rise MORB, for example, the OJP basalts display a very restricted range of Zr/Nb, consistent with high degrees of partial melting (Nb being highly incompatible and Zr moderately incompatible during melting); however,

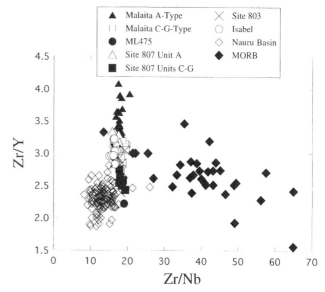

Figure 10. Variations in Zr/Y and Zr/Nb exhibited by the OJP and Nauru Basin basalts (data sources as in Figures 6 and 8) and southern East Pacific Rise MORB [*Mahoney et al.*, 1994].

the Zr/Y ratio varies by nearly a factor of two (Y being moderately incompatible in spinel peridotite but compatible in garnet). Similarly, the variations in Sm/Yb are comparable to those in La/Yb (see Figure 11). These features suggest that melting encompassed both the garnet and spinel peridotite stability fields [cf. *Mahoney et al.*, 1993; *Farnetani et al.*, 1996].

Simple forward modeling can be used to investigate some features of the mantle source and melting process that supplied the A- and C-G-type lavas. Given their generally flat primitive-mantle-normalized element patterns (Figure 6), we took a source with primitive-mantle incompatible element abundances as one possible "end-member". On the other hand, the initial ε_{Nd} values of A-type basalts cluster around +4 to +5, and those of C-G-type lavas around +6 (Figure 5a). These values are roughly halfway between the presumed primitive mantle value of 0 and the average MORB value of about +10, suggesting that a mantle source with mixed elemental characteristics could also be appropriate. For our peridotite source modeling, melting was assumed to initiate in the garnet stability field, with the bulk of melting occurring in the spinel stability field as the plume head impinged on the lithosphere. Initial source mineralogy was estimated from the modal mineralogy of garnet, garnet-spinel, and spinel peridotites studied by *Neal* [1985, 1988] to be 60% olivine, 20% orthopyroxene, 10% clinopyroxene, and 10% garnet, melting in the proportions of 15:30:25:30. After an initial phase of batch or fractional melting in the garnet stability field (minimum of 5%, maximum of 15%), the mode and

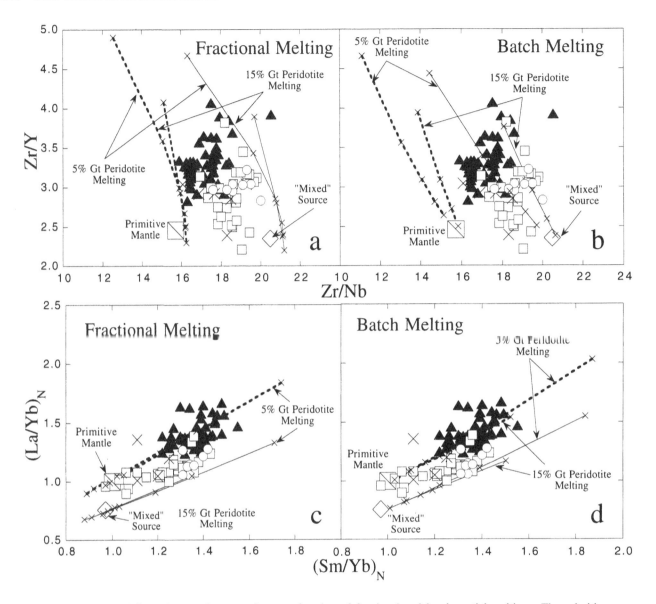

Figure 11. Modeling of trace element ratios as a function of fractional and batch partial melting. The primitive mantle source is taken from *Sun and McDonough* [1989] and the "mixed" source is a 50:50 mixture of primitive mantle and MORB source [*McKenzie and O'Nions*, 1991]. *McKenzie and O'Nions* [1991] reported concentrations for the rare earth elements in their estimated MORB source, but not Zr, Nb, or Y. In estimating concentrations of these elements we assume: (i) a Nb/La ratio of 1, producing a Nb concentration of 0.206 ppm; (ii) a Zr/Nb ratio of 32 which, on the basis of 0.206 ppm Nb, yields a Zr abundance of 6.6 ppm; and (iii) Y abundance is 10 times that of Yb, or 3.47 ppm. The melting trajectories are for hybrid melts, and incorporate melting in both the garnet and spinel peridotite stability fields. Two melting paths are shown for each source: one which incorporates a 5% partial melt and the second a 15% partial melt of the source in the garnet peridotite stability field. To each of these melts are progressively added 5% increments of melt derived from the spinel peridotite stability field after phase transformation facilitated by the rising plume head. The first cross on each melt path (i.e., that farthest from the source composition) represents a 5% or 15% melt from garnet peridotite plus a 5% melt from spinel peridotite, resulting in a minimum partial melt represented on each of these paths of 10% and 20%, respectively. The higher the degree of partial melting, the more the melt composition converges on that of the source. Partition coefficients used are from *Green et al.* [1989] and *Horn et al.* [1994]. Subscript N means chondrite-normalized.

composition of the residue were calculated and subsequent melting was calculated for a spinel peridotite, assuming all remaining garnet was converted to spinel. Melts generated from garnet and spinel peridotite were then mixed to give calculated parental melt compositions. The starting points for the model batch and fractional melting trajectories in Figure 11 represent the minimum- and maximum-degree partial melts generated in the garnet stability field (5% and 15%, respectively) which have been mixed with a 5% melt from spinel peridotite, assuming either a primitive or mixed mantle source (50% MORB source, 50% primitive mantle). The crosses on the trajectories represent mixing, in 5% increments, of these initial compositions with subsequent melts generated in the spinel stability field and formed from the residue of the initial melting. Note that because the later melts are derived from a residue after a phase transformation (and hence a change in bulk partition coefficients) the melt trajectories for neither batch nor fractional melting lead to the initial mantle source compositions as the degree of partial melting is increased.

The partial melting trajectories in Figure 11 imply that the source(s) for the OJP lavas contained a greater proportion of primitive-type mantle than MORB mantle, as the bulk of the data plot between the model trajectories; the model sources and their calculated melt trajectories do, however, generally bracket the basalt data. For simplicity, we have not considered more complicated melting processes (e.g., zone refining) which may have operated to some extent, at least in the garnet peridotite stability field, during the rise of the plume head. From our modeling, the calculated minimum degree of partial melting for the OJP basalts is 15–25% (generally for A-type basalts) and the maximum is 20–35% (generally for C-G-type basalts). However, caution is required in interpreting such melting estimates because these ranges are a product of variation in the model parameters used. The variation is produced by the choice of fractional or batch melting models, the amount of melt derived from the garnet peridotite stability field, and the type of source melted.

The data for A- and C-G-type lavas form two groups with only limited overlap (Figure 11). Although these ranges are small, we believe these groups are real in view of the high precision of the ICP-MS data (errors on individual data points would generally be within the symbol plotted). Given the isotopic differences between the two groups, the small but consistent differences in incompatible element ratios could reflect trace element differences inherent in the respective mantle sources. However, they may also reflect different amounts of partial melting followed by open-system evolution, with the A-type representing lower degrees of melting [cf. *Mahoney et*

al., 1993; *Tejada et al.*, 1996a]. Major elements provide some additional insight. Figure 9d shows the Al/Ti ratio vs. Al_2O_3; despite variable amounts of plagioclase fractionation/accumulation, the OJP basalts exhibit a fairly restricted range in Al_2O_3 (excepting Site 803, the bulk of the samples contain 13.5–14.5 wt%) and again define A- and C-G-type groups in terms of Al/Ti ratio. Changes in Al/Ti will be controlled by the degree of partial melting and, as Al is buffered, decreasing the degree of partial melting will decrease the Al/Ti ratio by increasing the relative abundance of moderately incompatible Ti. Either the individual sources for A-type and C-G-type basalts had distinct mineralogies (not likely from trace element evidence) or the A-type lavas represent a somewhat lower degree of partial melting than the C-G-type basalts.

8. CRYSTAL FRACTIONATION AND THE HIDDEN CUMULATES

8.1. *Fractionation*

Xenoliths are present in some of the 122 Ma C-G-type basement basalts in central Malaita, particularly at deeper stratigraphic levels in the Kwaio inlier. These xenoliths are gabbroic (mainly plagioclase and clinopyroxene, with minor olivine and spinel) and anorthositic. They provide graphic evidence for crystal fractionation, consistent with major-element data for the Malaitan and other OJP basalts, which show that all A-type and C-G-type lavas are evolved; for example, whole-rock molar Mg-number (= Mg/[Mg+Fe^{2+}], assuming that 75% of the total Fe is Fe^{2+}) is 0.50 to 0.29. Preliminary study indicates plagioclase is the predominant phase in the gabbroic xenoliths, whereas the basalt major-element compositions suggest that fractionation of plagioclase was comparatively minor: although Eu may undergo a slight decrease toward the top of the A-type (1.3 ppm → 1.1 ppm) and C-G-type (1.2 ppm →0.9 ppm) portions of the stratigraphic section, Sr is not depleted significantly (e.g., Figure 6), and Al_2O_3 shows a restricted range (most samples have 13.5–14.5 wt%) and does not correlate with Eu or Sr. Plagioclase phenocrysts are fairly common in the lavas, but clinopyroxene is the dominant phenocryst and appears to have been a major fractionating phase. For example, Sc abundances range from 31–59 ppm and Cr contents from 25–200 ppm for A-type and C-G-type (high-MgO sample ML475, with Cr = 446 ppm, exhibits a cumulate texture), with two basalts from Santa Isabel and the 90 Ma samples from Site 803 containing between 200 and 300 ppm Cr (Figure 9b,c). Olivine phenocrysts are rare and MgO-Ni correlations poor, but the generally low Ni contents of the lavas (Figure

9e) indicate olivine was removed, probably as an early fractionating phase. In general, correlations of major elements with typical fractionation indices, such as MgO and Mg-number, are rather poor; in part, this likely reflects a lack of strong olivine control on the analyzed basalts at this evolutionary stage, although effects of seawater-alteration on the lavas and open-system magma plumbing networks [cf. *Mahoney et al.*, 1993; *Tejada et al.*, 1996a,b] cannot be discounted.

As primary OJP magma compositions are unknown, it is difficult to quantify the amount of crystal fractionation; however, estimates can be obtained by combining petrographic and experimental petrologic data. Partial melting appears to have been initiated in the garnet stability field and concluded in the spinel stability field, as discussed above. We estimated the major-element composition of OJP parental magmas by using the experimental results (Table 2) of *Falloon and Green* [1988] for melts derived from garnet peridotite (30 kbar) and *Hirose and Kushiro* [1993] for melts from spinel peridotite (20 kbar). Rather than a single composition, we calculated a range of potential parental compositions assuming 5%, 10%, and 15% partial melting in the garnet stability field, and that the calculated melts then mix with partial melts from the spinel stability field in 5% increments, the total from both fields not exceeding 30% of partial melting (the maximum amount suggested by *Mahoney et al.* [1993] and *Tejada et al.* [1996a]). For example, 5% of garnet peridotite melt mixed with 20% of spinel peridotite melt corresponds to 25% of total melting with a 1:4 proportion of garnet and spinel peridotite end-members. This approach yields picritic melts with high MgO contents averaging 16 wt%.

TABLE 2. Melt Compositions Derived From Garnet- and Spinel-Peridotite Stability Fields and Used in Major Element Modeling

	Garnet Peridotite Melt[1]	Spinel Peridotite Melt[2]
SiO_2	45.83	47.47
TiO_2	1.08	0.75
Al_2O_3	11.70	15.53
Cr_2O_3	0.42	0.21
FeO	9.04	8.51
MgO	19.99	13.94
CaO	10.66	11.11

[1] Melt composition from MYP-90-40 at 35 kbar and 1600°C [from *Falloon and Green*, 1988].
[2] Melt composition from KLB-1 at 20 kbar and 1375°C [from *Hirose and Kushiro*, 1993].

From calculated parental magma compositions corresponding to 15–30% total melting, we modeled a broad crystallization pathway (Figure 12). The liquidus phases used and their relative proportions were derived from petrography (e.g., the range of clinopyroxene and plagioclase compositions used were those measured by electron microprobe for phenocrysts in the Malaita flows; C. R. Neal, unpubl. data, 1995) and experimental petrology (e.g., the known contraction of the olivine and plagioclase liquidus fields and expansion of the pyroxene fields during crystallization of tholeiitic magma under pressure, but <20 kbar [e.g., *Yoder and Tilley*, 1962; *O'Hara and Yoder*, 1967; *BVSP*, 1981]). *Farnetani et al.* [1996] modeled the crystallization of three experimentally derived melts from spinel peridotites under conditions approximating those expected for melts from a plume head. The main crystallizing phases predicted were olivine, clinopyroxene, spinel, plagioclase, and orthopyroxene. With increasing depth of crystallization the cumulate was predicted to change from olivine-orthopyroxene gabbro, through troctolite and/or leucogabbro, to melanogabbro, pyroxenite, or clinopyroxene-norite. In our modeling, five stages of crystallization were assumed, each crystallizing a volume equal to 10% of the starting magma: Stage 1 = 100% olivine (Fo_{90}); Stage 2 = 70% olivine (Fo_{85}) + 30% spinel; Stage 3 = 50% olivine (Fo_{80}), 50% clinopyroxene; Stage 4 = 95% clinopyroxene, 5% plagioclase; Stage 5 = 90% clinopyroxene, 5% plagioclase, 5% orthopyroxene. Figure 12 shows the calculated liquid evolution trends in a plot of MgO/TiO_2 vs. CaO/Al_2O_3 for several different model parental melts; it can be seen that compositions similar to those of most of the OJP basalts are generated primarily during Stage 4 and at the beginning of Stage 5. This result is consistent with the paucity of olivine (and orthopyroxene) phenocrysts in the lavas, and the predominance of clinopyroxene and plagioclase fractionation which has overprinted evidence of the earlier olivine removal (e.g., Figure 9e). The modeling indicates that most OJP lavas result from 30–45% of crystal fractionation and that the xenoliths seen in some flows in central Malaita correspond to relatively late stages of this process (Stage 4 or 5).

8.2. *The Hidden Cumulates*

The cumulate assemblage derived from magmatic evolution similar to that modeled above would consist primarily of early olivine and later clinopyroxene, with lesser amounts of plagioclase, orthopyroxene, and spinel. Corresponding rock types could include dunite, pyroxenite, and gabbro [cf. *Farnetani et al.*, 1996]. An

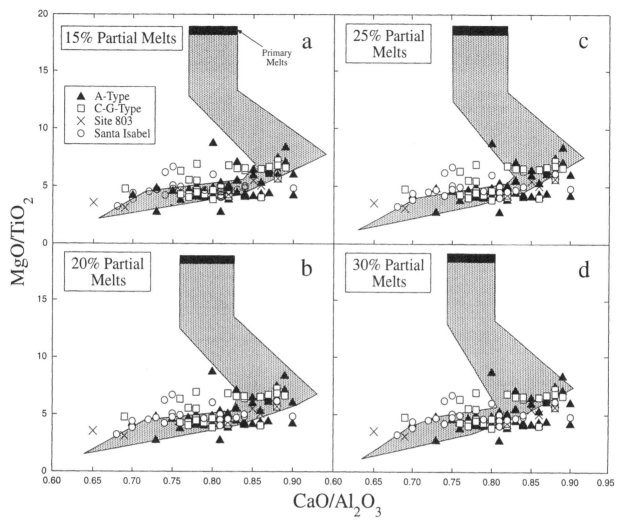

Figure 12. Modeling of the MgO/TiO$_2$ and CaO/Al$_2$O$_3$ variation in OJP basalts by crystal fractionation. Parental melts are estimated from the experimental results of *Falloon and Green* [1988] for melts derived from garnet peridotite (30 kbar) and *Hirose and Kushiro* [1993] for melts from spinel peridotite (20 kbar). These compositions are combined in 5% increments to give the total amount of partial melt at the top of each graph, with the maximum amount of melt from garnet peridotite not exceeding 15% and the resulting parent always being a hybrid of the two melt compositions. Fractional crystallization occurs in five stages, each crystallizing a volume equal to 10% of the starting magma: Stage 1 = 100% olivine (Fo$_{90}$); Stage 2 = 70% olivine (Fo$_{85}$) + 30% spinel; Stage 3 = 50% olivine (Fo$_{80}$), 50% clinopyroxene; Stage 4 = 95% clinopyroxene, 5% plagioclase; Stage 5 = 90% clinopyroxene, 5% plagioclase, 5% orthopyroxene.

MgO content of 30.1 wt% is calculated for the bulk cumulate by using the estimated mean amount of fractionation and corresponding total model fractionating assemblage. Assuming (1) that, as above, the average parental magma MgO = 16 wt%, (2) that the 7.0 wt% average MgO content of OJP samples is representative of erupted liquids on the plateau as a whole, and (3) that the 30.1 wt% MgO value is representative of the bulk cumulate pile, a rough estimate of the total amount of cumulates extracted from the magmas can be made by mass balance [e.g., *Cox*, 1993]:

$$C_P = C_L X_L + C_C X_C$$

where C_P, C_L, and C_C are the concentrations of an element, in this case MgO, in the parental magma, erupted liquid, and cumulate, respectively; and X_L and X_C are the respective mass fractions of erupted liquid and cumulates.

The relative mass of cumulates to erupted liquids is

$$X_C/X_L = (C_L - C_P) / (C_P - C_C)$$

The ratio X_C/X_L gives an estimate of the proportion of crystal cumulates necessary to lower the MgO content of the primary magma from 16.0 wt% to the average 7.0 wt%. Using the above values, the ratio is 0.64; the total, combined fractionating assemblage consists of 49.1% olivine, 40% Ca-rich clinopyroxene, 5.5% spinel, 5.0% plagioclase, and 0.4% orthopyroxene. This assemblage corresponds to a wehrlitic to pyroxenitic cumulate of density ≈3.25 g cm⁻³. In general agreement, *Cooper et al.* [1986] suggested a density for lower OJP crust of 3.0–3.25 g cm⁻³, whereas *Gladczenko et al.* [1997] estimated a lower crustal density of 3.25–3.30 g cm⁻³; both studies utilized gravity data. Conversion of these densities to P-wave velocities (V_P) indicates that the lower OJP crust should exhibit $V_P \geq 7.8$ km s⁻¹, which is not the case (see Figure 2). However, as noted by *Fountain et al.* [1994], reduced average V_P can be obtained by having a heterogeneous lower crust. This may be achieved by the presence of a mixture of gabbro and pyroxenite/wehrlite or basalt and pyroxenite/wehrlite in the lower OJP crust, or their metamorphic derivatives of granulite and possibly eclogite [cf. *Fountain et al.*, 1994].

As discussed earlier, the OJP may partly have formed at a mid-ocean ridge but part, and perhaps much, of the plateau must have been emplaced off-axis. For simplicity, we have taken two extremes: (1) emplacement of the entire plateau at a mid-ocean ridge (no preexisting oceanic crust) and (2) emplacement "off-axis" with a 7.1-km preexisting oceanic crust [*White et al.*, 1992; *Coffin and Eldholm*, 1993]. In addition, minimum and maximum average thicknesses of the OJP are taken as 25 km and 36 km, respectively (see Section 2).

Results of our "hidden cumulate" modeling are as follows. For an off-axis emplacement (presence of a 7.1-km-thick preexisting oceanic crust), cumulate thicknesses vary between 7.0 km (assuming a 25-km total OJP thickness) and 11.3 km (assuming a 36-km total OJP thickness), whereas corresponding erupted basalt sequences would be 10.9 km and 17.6 km, respectively. For an on-axis emplacement (no preexisting oceanic crust), cumulate thicknesses are 9.8 km and 14.0 km, with corresponding basaltic sequences of 15.2 and 22.0 km, respectively. For comparison, recall that the thickness of seismic Layer 2 is estimated at >12 km over much of the high plateau (Figure 2).

Seismic refraction studies of the high plateau identified a basal crustal layer with high P-wave velocity of 7.5–7.7 km s⁻¹ (Figure 2) [*Furumoto et al.*, 1976]. The thickness of this seismic layer has been estimated to be 9–16 km [*Furumoto et al.*, 1976; *Hussong et al.*, 1979; *Carlson et al.*, 1981; *Coffin and Eldholm*, 1993]. Over the years, the layer has been ascribed to a variety of possible rock types from, for example, garnet granulites to cumulate gabbros to underplated basalt magma (i.e., non-cumulate gabbros) [*Houtz and Ewing*, 1976; *Nixon and Coleman*, 1978; *Neal and Taylor*, 1989; *Carlson et al.*, 1981; *Schaefer and Neal*, 1994; *Farnetani et al.*, 1996; *Ito and Clift*, 1996]. Compositions between gabbro and pyroxenite would have P-wave velocities between 7.4 and 8.1 km s⁻¹ at pressures of ≈10 kbar [e.g., *Carmichael*, 1989]. Therefore, as noted above, mixtures of gabbroic and pyroxenitic cumulates produced by OJP magmatism, or their metamorphic derivatives, are likely to be responsible for much of the basal high-velocity crustal layer.

Significant amounts of intruded and underplated material undoubtedly contribute to the thickness of the OJP [*Tejada et al.*, 1996a; *Ito and Clift*, 1996] as, probably, does some buried preexisting oceanic crust, thus complicating such simple crustal models. Moreover, the relative amounts of magmatism (volcanism, intrusion, and underplating) associated with the 122 and 90 Ma events are unknown and, as noted earlier, the possibility of some magmatism occurring throughout the 30 m.y. period between the two eruptive events cannot be discounted at present. Intrusion and underplating associated with the seamount and Maramasike volcanism on the plateau, possibly caused by the passage of the OJP over the Samoan hotspot (which took place from ≈60–30 Ma [e.g., *Yan and Kroenke*, 1993]), was probably relatively minor but presently is impossible to quantify. A further caveat is that the OJP lavas studied to date represent only the upper levels (≤4 km) of the eruptive section and may not be representative of earlier, deeper levels, which may be more picritic in composition [e.g., *Storey et al.*, 1991].

9. UPLIFT AND SUBSIDENCE OF THE OJP

9.1. Paleodepths

The OJP is roughly isostatically compensated [e.g., *Sandwell and MacKenzie*, 1989] and most of it stands 2–3 km above the surrounding seafloor today (although still some 1700 meters or more below sea level). Subsidence rates for normal seafloor of similar age [e.g., *Stein and Stein*, 1992] suggest that the plateau should have subsided some 3 km since its formation. Remarkably, however, although there are indications that the north-central region of the high plateau has subsided a few hundred meters more than the northeastern margin since the Cretaceous [*Berger et al.*, 1992], little evidence is present for any significant amounts (i.e., >1 km) of post-emplacement subsidence at the locations where basement has been sampled. Rather, normal faults at the northern and eastern margins of the OJP indicate that the adjoining abyssal seafloor in these regions has subsided somewhat relative to the plateau [e.g., *Kroenke*, 1972; DSDP Leg 30 unpubl. reflection profiles; *Kroenke et al.*, 1986; *Hagen et al.*, 1993]. We hypothesize that the less than expected subsidence of the high plateau could reflect (1) partial thermal and magmatic "rejuvenation" at ≈90 Ma and during the OJP's ≈60-30 Ma passage [e.g., *Yan and Kroenke*, 1993] over the Samoan hotspot; (2) the near-neutral buoyancy of the "hidden cumulates" (3.25 g cm⁻³ is generally intermediate between the density of oceanic crust and upper mantle, whether fertile or melt-depleted) that would have fractionated from OJP magmas while the plateau was elevated in response to the physical presence of the plume head (see below); (3) the buoying effect of the melt-depleted (and therefore less dense) upper mantle keel beneath the OJP; and/or (4) the "flexural bulge" of the OJP along the plateau's southwest margin (a response to collision with the Australian Plate [e.g., *Coffin et al.*, 1996]), which would help maintain the elevation of the plateau, at least in this region. Also, *Ito and Clift* [1996] recently suggested that, rather than just a 90 Ma rejuvenation, subsidence may have been tempered by continued underplating of the plateau for ≈30 m.y. between the eruptive events.

If it is assumed for modeling purposes that post-emplacement subsidence of the plateau was, in fact, similar to that of normal oceanic crust, a rough "prediction" of original depth of the OJP's surface can be obtained for each drill site on the plateau. This approach was used to good effect in estimating the uplift associated with the formation of the Kerguelen Plateau [*Coffin*, 1992]. Application to the OJP yields results which are crude but at least generally consistent with emplacement of the basalts well below sea level (e.g., Site 288 = -482 meters, assuming a geophysically determined sediment thickness of 1090 m; Site 803 = -930 meters; Site 807 = -520 meters). However, a large discrepancy exists for Site 289 in that an emplacement of basement above sea level (+202 meters) is indicated, which is in direct contrast with lithological evidence (i.e., bathyal Aptian limestone above basement, lack of evidence for erosional surfaces, lack of vesicularity or oxidation in the basalt recovered, etc.). This indicates that the assumptions used (i.e., sediment and structural conditions at each site reflect regional conditions, complications arising from sediment

redeposition and submarine erosion can be ignored, and, most importantly, that subsidence followed the normal seafloor time-depth curve) are not valid in this case. In view of other considerations outlined below, we interpret these results as further evidence that the OJP did not subside as expected of normal oceanic crust. Therefore, these calculations yield only minimum emplacement depths.

Strong evidence for non-emergence of the OJP can be found in the basement lavas in Santa Isabel and Malaita, as well as at the drill sites. Lithological and chemical evidence suggests emplacement near or even below the calcite compensation depth (CCD) in some cases [*Hughes and Turner*, 1977; *Kroenke et al.*, 1991; *Hawkins and Barron*, 1991; *Tarduno*, 1992; *Saunders et al.*, 1993]. Field evidence from the basement sections of Malaita, on the southern flank of the OJP, demonstrates that very little interflow sediment is present; that which exists is siliceous rather than calcareous and the sediments overlying basement are pelagic mudstones, suggesting that emplacement occurred below the ≈122 Ma CCD in this area [*Saunders et al.*, 1993; *Neal et al.*, 1994, 1995]. An Aptian bathyal limestone lies above basement at Site 289 [*Andrews and Packham*, 1975], whereas at Site 807, on the northern flank of the plateau, a thin limestone is intercalated between the A-type and C-G-type basalts, indicating late-stage eruption depth was at least temporarily above the CCD; however, the sediments immediately above basement are not calcareous [*Kroenke et al.*, 1991]. The depth of the CCD during the Cretaceous is poorly known and fluctuated widely, but a rough estimate for 122 Ma is ≈2–3 km [e.g., *Arthur et al.*, 1986]. From CO_2 contents in Site 807 glasses, *Michael and Cornell* [1996] estimated that eruption depths of lavas at this site were in the 1000 m (A-type lavas) to 2500 m (C-G-type) range; note that many of the lavas may have flowed considerable distances from their point of eruption, so that the eruption depth and final emplacement depth may not be the same. However, these depths and the presence of limestones at Sites 807 and 289 may imply that the axis of the OJP plume head, above which maximum dynamic uplift would be expected, was situated closer to the northern end of the high plateau at 122 Ma (i.e., the large, as yet unsampled, domal region of shallower depths to the southwest of Site 289 [cf. *Tejada et al.*, 1996a]). Moreover, because the section of A-type basalts is much thicker at Malaita than at Site 807, the principal site(s) of A-type eruptive activity may not have coincided with the plume axis.

In conclusion, the combination of constructional magmatism and dynamic uplift from impingement of a plume head was apparently insufficient to raise the basement surface of much of the OJP to shallow depths, although parts may have been shallow, particularly the as yet unsampled crestal regions of the high plateau and eastern salient. This situation contrasts with those for Wrangellia [e.g., *Richards et al.*, 1991; *Lassiter et al.*, 1995] and the Kerguelen Plateau [*Coffin*, 1992], which contain evidence of extensive emergence and subaerial erosion in basal sediment and basalt stratigraphy. However, Wrangellia was built upon an island arc and the Kerguelen Plateau formed in a young, narrow ocean basin and may contain fragments of continental lithosphere [e.g., *Storey et al.*, 1992; *Mahoney et al.*, 1995; *Hassler and Shimizu*, 1995; *Operto and Charvis*, 1996].

9.2. Estimates of Initial Dynamic Uplift

Impingement of a plume on the lithosphere will produce temporary surface uplift, even if the plume surfaces beneath a mid-ocean ridge. Rough estimates of the amount of such uplift can be obtained by application of results from experimental and theoretical studies [e.g., *Olson and Nam*, 1986; *Griffiths et al.*, 1989; *Hill*, 1991; *van Keken et al.*, 1993; *Ribe*, 1996]. Such estimates require that the volume of the plume head be known or assumed; some limits on this volume can be obtained from the estimated volume of the plateau and the degree of partial melting involved in producing the basalts.

For modeling purposes, we use a minimum OJP areal value of 1.5×10^6 km^2. (*Coffin and Eldholm* [1994] gave a larger estimate of 1.86×10^6 km^2, including possible OJP-related sequences in the Nauru, East Mariana, and Lyra basins, but here we consider only the plateau itself, as defined by the 4-km depth contour.) A maximum volume of 5.4×10^7 km^3 results from using the maximum estimated average crustal thickness of 36 km, together with the assumption of no preexisting oceanic crust (Section 2). A minimum volume of 2.7×10^7 km^3 is obtained by assuming the OJP formed completely on preexisting oceanic crust and taking the estimated average crustal thickness formed by the OJP plume to be 17.9 km (25 km minus 7.1 km of preexisting crust). Assuming further that all of the OJP formed in the 122 Ma event and that the average degree of partial melting was as low as 15% (typical of values estimated for MORB), then an upper limit on the plume-head volume would be 3.6×10^8 km^3; for a spherical head, this corresponds to a diameter of 883 km. A "pseudominimum" plume-head volume of 9.0×10^7 km^3, with a spherical diameter of 555 km, is obtained by assuming 30% partial melting and using the lower estimate of OJP volume [cf. *Coffin and Eldholm*, 1993]; of

course, if significant amounts of OJP crust formed at 90 Ma, in the 122–90 Ma period, or after 90 Ma, then this minimum estimate would be lower still. Two examples of dynamic uplift modeling are presented below.

9.2.1. *Model 1.* Experiments by *Griffiths et al.* [1989] followed the evolution of surface topography as a buoyant droplet rose through a viscous liquid. Maximum surface uplift was observed when the leading edge of the diapir was 0.2 diapir diameters beneath the surface and was quantified as

$$H_{max} = \frac{0.25(1 - \lambda)(\Delta \rho D_o)}{\rho_m}$$

where: H_{max} = maximum uplift; $\lambda = (\rho_m - \rho_l)/(\rho_m - \rho_D)$; ρ_m = upper mantle density (3.3 g cm^{-3}); ρ_l = lithospheric density (3.15 g cm^{-3}); ρ_D = plume head density (varied between 3.1 and 3.15 g cm^{-3} resulting in λ values between 0.75 and 1.0; see Figure 13); $D_\rho = \rho_m - \rho_D$; D_o = diameter of spherical plume head.

As expected, the maximum amount of uplift is predicted from the largest plume head (Figure 13). However, this approach implies that maximum dynamic uplift associated with OJP formation would barely exceed 3 km and could be much less.

9.2.2 *Model 2.* *Hill* [1991] quantified dynamic uplift above a spreading (flattening) plume head and demonstrated that the amount of uplift is sensitive to the initial thermal structure of the overlying material; this is manifested by the temperature difference between the plume head and ambient mantle (DT).

$$E = \alpha \, \Delta T \, L$$

Here, E = amount of surface uplift; α = coefficient of thermal expansion (3×10^{-5} °C^{-1}); ΔT = average excess temperature anomaly (in °C); L = thickness of the anomalously hot layer (flattened plume head).

In Figure 14, the same estimates of partial melting and plateau and plume-head volume were used as in Model 1. The difference is that the spherical plume head is considered to have flattened to a disc with a diameter of either 1000, 2000, or 3000 km; the smaller the diameter of the flattened plume head, the greater its thickness and, therefore, the greater the surface uplift. Also included are results from assuming a maximum and minimum ΔT (300 and 100°C, respectively) between plume and ambient mantle. Maximum surface uplift of 4.1 km is obtained for a flattened plume head of minimum diameter, $\Delta T = 300$°C, and partial melting of 15%. If the average degree of partial melting is 20–30%, in the range of values considered most

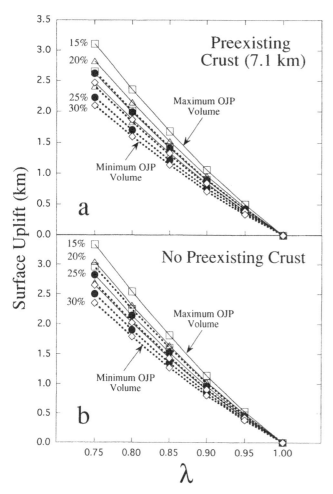

Figure 13. Estimates of surface uplift resulting from impingement of the 122 Ma OJP plume head upon the lithosphere using the model of *Griffiths et al.* [1989]. The maximum volume of the plume head is calculated by assuming a minimum and maximum average thickness of the OJP of 25 and 36 km, respectively. The size of the plume head is directly proportional to the amount of surface uplift. Calculations in (a) assume that the OJP was built upon 7.1 km of preexisting oceanic crust (off-ridge setting), whereas (b) assumes that the total thickness of the OJP was generated by the plume event (ridge setting). Dashed and solid lines represent the minimum and maximum OJP volume, respectively. The percentages represent the assumed amounts of partial melting that generated OJP magmas (i.e., the lower the assumed degree of melting, the larger the plume head required):   15% = open square; 20% = open triangle; 25% = filled circle; 30% = open diamond.

likely for OJP lavas [e.g., *Mahoney et al.*, 1993; *Tejada et al.*, 1996a; Section 7.2, this paper], then the maximum surface uplift would be between 1.0 and 3.0 km, similar to the estimate from the *Griffiths et al.* [1989] approach. (Note that these results are also similar to the minimum

uplift values for plateaus estimated by *Farnetani and Richards* [1994], who assumed a greater ΔT (350°C) and a large spherical plume head diameter (800 km)). Again, uplift could be much less.

9.3. *Effect of Added Crust*

Although by itself dynamic uplift appears incapable of raising much of the OJP from ridge or abyssal depths to above sea level, addition of as much as 36 km of new crust might be assumed to cause widespread emergence. However, overwhelming evidence to the contrary is seen at all presently sampled locations. Thus, on average, the bulk density of the OJP must be slightly greater than that of normal ocean crust simply thickened by a factor of up to five or so. For example, mass balance calculations suggest that up to 4 km of surface uplift would result from simply overthickening oceanic crust (density of 2.9 g cm^{-3}). If the density is increased to 3.08 g cm^{-3} (the average density of the OJP we calculate on the basis of P-wave velocities), the surface uplift is reduced to between only 1.8 and 3.0 km, depending upon the thickness assumed for the plateau. Indeed, average crustal densities (estimated from published P-wave velocities [e.g., *Bjarnason et al.*, 1993; *Operto and Charvis*, 1996; *White et al.*, 1996] using the methods of *Bott* [1982]) indicate that bulk crustal density increases from Kerguelen (≈ 2.8 g cm^{-3}) to Iceland (≈ 2.9 g cm^{-3}) to the OJP (≈ 3.1 g cm^{-3}). Although Iceland has an average density greater than that of Kerguelen, it is emergent because of a greater crustal thickness (22–35 km [e.g., *Bjarnason et al.*, 1993; *White et al.*, 1996] vs. 20-25 km [e.g., *Operto and Charvis*, 1996]) and because of the dynamic uplift afforded by the plume beneath the Mid-Atlantic Ridge. Much of the Kerguelen Plateau was also originally at shallow depths or subaerial [*Coffin*, 1992] because of plume-driven dynamic uplift and crustal thickening; as mentioned earlier, it may contain some continental lithospheric material, unlike Iceland.

We conclude that the key to the initial depth of OJP emplacement is largely in the high-velocity basal crustal layer, argued here to have been formed from hidden cumulates plus intruded and underplated gabbros. Also, the extreme thickness of the OJP provides conditions conducive to granulite or, possibly, eclogite development [*Fountain et al.*, 1994; *Rudnick and Jackson*, 1995; *Saunders et al.*, 1996], both of which could contribute to tempering OJP uplift. Mafic granulite xenoliths (including a garnet-bearing variety) are indeed present in the Malaitan alnöites (P. H. Nixon, pers. comm., 1996) at the southern edge of the plateau, and the high P-wave velocities of 8.4–8.6 km/s just below the Moho in the northwestern and

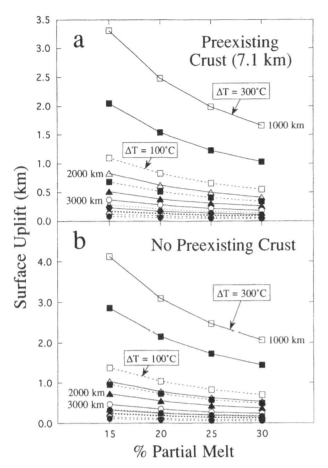

Figure 14. Surface uplift estimated for impingement of the 122 Ma OJP plume head upon the lithosphere using the model of *Hill* [1991] and varying the amount of assumed partial melting between a minimum of 15% and a maximum of 30%. Maximum and minimum plume head sizes are calculated as in Figure 13. Values of ΔT are between the plume head and ambient upper mantle; 100°C is the assumed minimum (dashed lines) and 300°C is the assumed maximum (solid lines). Solid symbols = OJP is 25 km thick; open symbols = OJP is 36 km thick. Assumed flattened plume head diameters: squares = 1000 km; triangles = 2000 km; circles = 3000 km. (a) Assumes that the OJP was built upon 7.1 km of pre-existing oceanic crust (off-ridge setting), whereas (b) assumes that the total thickness of the OJP was generated by the plume event (ridge setting).

southwestern regions of the OJP might represent eclogitized lowermost crust.

9.4. *Melt-Depleted Mantle Root to OJP*

Elevation of the OJP relative to the surrounding ocean floor appears to be persistent as the effects of temporary uplift have long since dissipated. The presence of a relatively dense lower crustal layer beneath much of the

high plateau (attributed largely to cumulates) and possible high-density eclogitic material indicated by higher than normal mantle P-wave velocities just below the Moho in parts of the high plateau (see below) would argue for subsidence of the plateau, at least in these areas. In order to maintain plateau elevation and prevent delamination of the high density material, it seems likely that a region of relatively buoyant, melt-depleted mantle resides beneath the plateau, most probably the residue from plateau formation. Degree of partial melting has been estimated from the composition of the OJP basalts to be between 20 and 30% [*Mahoney et al.*, 1993; *Tejada et al.*, 1996a; Section 7.2, this paper]. Such high degrees of partial melting, coupled with the volume of the OJP, require that a significant melt-depleted zone or root be present (barring convective removal) in the upper mantle beneath the plateau. This mantle root would cause a decrease in seismic velocities beneath the OJP and this has indeed been reported recently by *Richardson and Okal* [1996].

An estimate of the depleted mantle root thickness can be made from the mass-balance approach of *Marks and Sandwell* [1991], who used the ratio of geoid height to topography to develop a two-layer (a layer of overthickened crust and a low-density, depleted mantle root) Airy compensation model:

$$(\rho_c - \rho_w)h = (\rho_m - \rho_c)r + (\rho_m - \rho_l)t$$

Here, h = elevation of plateau above the ocean floor (2.5 km); r = thickened layer at the base of the crust, as defined by *Marks and Sandwell* [1991] (taken as thickness of the cumulate layer in our model); t = thickened depleted mantle layer; ρ_c = OJP cumulate density (3.25 g cm^{-3}); ρ_w = seawater density (1.025 g cm^{-3}); ρ_m = mantle density (3.3 g cm^{-3}); ρ_l = melt-depleted mantle density (3.24 g cm^{-3}). The thickness of the depleted mantle layer, t, can be calculated:

$$t = \frac{(\rho_c - \rho_w)}{(\rho_m - \rho_l)}(1 - \delta)h$$

The fraction of Airy compensation is represented by δ and corresponds to the compensation associated with the cumulate layer at the base of the OJP. As cumulate thickness depends on the total material added by the OJP plume event(s), r varies between 7 and 14 km (Section 8.2). In order to maintain the mass balance, δ varies between 0.06 and 0.12, respectively. This approach yields a thickness of the melt-depleted mantle of between 82 and 87 km.

As a rough check on the 82- to 87-km root thickness, the average degree of partial melting that created it can be estimated by dividing the thickness of the OJP crust by the combined thickness of the OJP crust plus depleted mantle root, and the result compared with the degree of partial melting estimated from geochemical evidence. The results imply that the degree of partial melting required to generate the depleted mantle root was between 17% (assuming addition of 17.9 km of material to preexisting 7.1 km thick ocean crust) and 31% (assuming addition of 36 km of crust with no preexisting ocean crust). These values are in excellent agreement with estimates of partial melting from geochemical data [*Mahoney et al.*, 1993; *Tejada et al.*, 1996a; Section 7.2, this paper]. Furthermore, our estimates of the depleted mantle root thickness compare well with the observed region of decreased seismic velocity beneath the OJP [*Richardson and Okal*, 1996; W.P. Richardson, pers. comm., 1996].

10. SUBDUCTIBILITY OF THE OJP

Assuming that subcrustal lithosphere was not compositionally different from surrounding asthenosphere, *Cloos* [1993] concluded from buoyancy analysis that oceanic plateaus >30 km thick would be virtually unsubductible. *Abbott and Mooney* [1995], taking into account the additional buoying effect of a melt-depleted lithospheric mantle root (depleted in both garnet and Fe), argued that oceanic crust thicker than about 25 km should be unsubductible. A critical factor in subductibility is the extent to which density increases when lower parts of a plateau's crust encounter the high pressure region of a subduction zone; the density increase resulting from transformation of lower crust to eclogite can promote subduction of at least the lower portions of plateau crust. *Saunders et al.* [1996] demonstrated that if average OJP crustal thickness is close to 36 km the basal levels of the OJP would be in the eclogite stability field. However, they argued that conversion of lower crust from granulite to eclogite facies would be long-delayed in the absence of small amounts of water (kinetically important for crystal nucleation and transport of ions) and directed stress [e.g., *Austrheim*, 1987]. They suggested that the stress field resulting from arrival of a plateau at a subduction zone and the addition of fluids escaping off a slab subsequently sinking beneath a plateau (following reversal of subduction direction after collision) would promote eclogite formation and cause eventual subduction of at least the lower plateau crust. Ultimately, only those parts of upper crust that had overridden the original forearc (e.g., Malaita, Santa Isabel) might be preserved as identifiable remnants of a plateau.

In the case of the OJP, northeastward subduction began only at 10–5 Ma (see Section 5). Although producing a steeply descending slab beneath the OJP [e.g., *Cooper and Taylor*, 1987], there has been little time for slab-derived fluids to affect more than the edge of the OJP and substantial eclogitization of lower crust by this mechanism has probably not yet occurred (though it may eventually). Nevertheless, *Petterson* [1995], *Mann et al.* [1996], and *Petterson et al.* [1997] have argued that geophysical evidence is consistent with underthrusting of the leading edge of the OJP beneath the Solomon block, even as the upper levels are overriding it. Although superficially resembling true subduction, this process is interpreted by *Petterson et al.* [1997] as a mechanical "wedge effect" of the old forearc splitting the OJP crust and forcing its lower levels downward; as such, it should be of limited magnitude and duration.

Is any significant amount of eclogite present, then, in the OJP? Earlier, we noted that sub-Moho P-wave velocities of 8.4–8.6 km s^{-1} in parts of the high plateau (e.g., Figure 2) might signal the presence of eclogite immediately below the Moho. The regions where these velocities were measured are generally far from the Solomon arc and thus from slab-derived fluids and subduction-related stress. Eclogitization in these (and possibly other) regions, if confirmed by future deep-crustal seismic studies, must have occurred largely in the absence of such factors, following or during emplacement of the plateau's thick crust. Whereas substantial amounts of eclogite could eventually lead to delamination of the plateau's lithosphere in locations far from a subduction zone [e.g., *Saunders et al.*, 1996], the buoying effect of a low-density, melt-depleted root would tend to counteract delamination. For the present, we emphasize that the possibility of an eclogite layer should be considered carefully in future models of OJP structure, subsidence, and eventual fate.

11. FUTURE RESEARCH AVENUES

To evaluate the OJP's origin, history, and consequences more fully, a number of fundamental questions remain to be answered. To a large extent they require (a) a much more comprehensive sampling on and near the OJP of basement and later-stage igneous rocks, and (b) state-of-the-art geophysical studies of the plateau's crustal structure and its relation to surrounding areas. Further detailed sampling is required, for example, to assess the relative contributions of the 122 Ma and 90 Ma events, how much—if any—volcanism occurred in the period between 122 and 90 Ma, and to document how representative the limited range of geochemical variation in the few presently sampled sites is of the plateau's upper crust as a whole. The answers, in turn, would throw light on the size and composition of the plume head(s) thought to be responsible for the OJP and the relation of plateau emplacement to plate-motion changes. The crests of the high plateau and eastern salient, which may correspond to the locations of the plume axis at 122 and 90 Ma, respectively, are likely to be important for sampling the range of geochemical compositions present. Similarly, drilling in block-faulted areas and possible dipping-reflector sequences [Kroenke, 1972] could provide a means of sampling relatively deep into the volcanic pile, possibly even into the upper levels of the cumulate sequence in some places (particularly using a ship with riser capability). Such sampling might well reveal whether early, highly magnesian lavas, predicted to exist in plateau crust and similar to those seen in stratigraphically lower parts of the Caribbean Plateau's eruptive sequence [e.g., Storey et al., 1991; Kerr et al., this volume], are present. Drilling of seismically imaged diatreme-like structures would determine if some are true kimberlites and could yield lower crustal and upper mantle xenoliths similar to those from the Malaitan alnöites, which would be critical for understanding the nature of the lower crust and the refractory/eclogitized upper mantle. Systematic analysis of highly siderophilic trace elements in both future and existing sample collections will be important for testing the hypothesis of a core-mantle boundary origin for the plume source, and for resolving which of the geochemical types seen thus far may better represent the plume source (vs. entrained mantle).

Detailed seismic studies are essential to establish a clearer picture of crustal structure (both upper and lower) and thickness, the composition of the lower crust, as well as the importance of magmatic underplating and the deep-level interaction between the southern OJP and the Solomon block. Coupled with paleodepth data from sediments at sites on and adjacent to the OJP (including the crestal regions of the high plateau and eastern salient, which may have been at relatively shallow levels during emplacement), such information is required for a better understanding of uplift and subsidence history—and of the probability that the OJP will ultimately become accreted to continental crust or at least partially subducted. Evidence is now available that lower crust at the southern edge of the plateau is underthrusting to the southwest [Mann et al., 1996; Phinney et al., 1996; Cowley et al., 1996; Petterson et al., 1997]. Whether this process, which presently appears minor, could evolve into true subduction may depend largely on the extent of lower crustal eclogitization and on the rate at which it may be occurring at present. Future seismic investigations should help evaluate the amount of eclogitization.

Geophysical (including magnetic) studies will also help to decipher the original setting of OJP emplacement, particularly its proximity to a spreading center and which parts of the plateau may have been formed essentially on-axis. Refinement of plate reconstructions for the Early Cretaceous Pacific is needed to better determine the original location of the plateau and its presumed parent plume head.

The global impacts of OJP volcanism on the surface environment, marine and atmospheric chemistry, and the biosphere are very poorly understood at present but may have been significant. For example, widespread early Aptian and early Turonian periods of marine black shale deposition (anoxic events) are well-documented [e.g., Sliter, 1989] and may correspond, at least in part, to OJP emplacement. Likewise, excursions in seawater $^{87}Sr/^{86}Sr$ to lower values appear to have occurred in the Aptian and Turonian [Jones et al., 1995]. Moreover, recent research on terrestrial plant fragments and coal seams suggests a sudden increase in atmospheric $^{13}C/^{12}C$ at the beginning of the Aptian that may largely be attributable to OJP volcanism [Gröcke et al., 1997]. In conjunction with dating and geochemical work on OJP basement rocks, further studies of Cretaceous sediments deposited on land, at different depths on pre-OJP edifices in the oceans (the Shatsky Rise, for example), and adjacent to the plateau should help reveal the OJP's role in such environmental changes.

Acknowledgments: Many thanks go to the Ministry of Energy, Water, and Mineral Resources (Geology Mapping) of the Solomon Islands; without their logistical support, constant guidance, diplomacy in the bush, and enthusiasm for field work, this study would not have been possible. Michael Storey, Tadeusz Gladczenko, and co-editor Mike Coffin provided critical reviews which greatly improved this contribution. Thanks also to S. Schaefer, J. Rigert, and J. Westerink for initial direction in cumulate and uplift modeling, and to N. Hulbirt, P. Wessel, and M. Tejada. This study was supported by NSF grants EAR 93-02471 and ECS92-14596 to CRN, EAR 92-19664 to JJM, and EAR 93-02472 to RAD.

REFERENCES

Abbott, D., and W. Mooney, The structural and geochemical evolution of the continental crust: Support for the oceanic plateau model of continental growth, *Rev. Geophys. Suppl.*, US Nat. Rept. to IUGG 1991-1994, 231-242, 1995.

Andrews, J. E., and G. H. Packham et al., *Init. Repts. Deep Sea Drill. Proj.*, 30, 753 pp., 1975.

Arthur, M. A., W. E., Dean, and S. O. Schlanger, Variations in the global carbon cycle during the Cretaceous related to climate, volcanism, and changes in atmospheric CO_2, in *The Carbon Cycle and Atmospheric CO_2: Natural Variations Archean to Present. Geophys. Monogr. Ser.*, vol. 32, edited by

E. T. Sundquist and W. S. Broecker, pp. 504-529, AGU Washington, D.C., 1986.

Austrheim, H., Eclogitization of lower crustal granulites by fluid migration through shear zones, *Earth Planet. Sci. Lett., 81,* 221-232, 1987.

Auzende, J-M., L. W. Kroenke, J-Y. Collot, Y. Lafoy, and B. Pelletier, Compressive tectonism along the eastern margin of Malaita island (Solomon Islands), *Mar. Geophys. Res., 18,* 289-304, 1996.

BVSP - Basaltic Volcanism Study Project, *Basaltic Volcanism on the Terrestrial Planets,* 1286 pp., Pergamon Press, 1981.

Batiza, R., Trace-element characteristics of Leg 61 basalts, *Init. Repts. Deep Sea Drill. Proj., 61,* 689-695, 1986.

Bercovici, D., and J. J. Mahoney J. J., Double flood-basalt events and the separation of mantle-plume heads at the 660 km discontinuity, *Science, 266,* 1367-1369, 1994.

Berger, W. H., and T. C. Johnson, Deep sea carbonates: Dissolution and mass wasting on Ontong Java Plateau, *Science, 192,* 785-787, 1976.

Berger, W. H., L. W. Kroenke, L. A. Mayer, J. Backman, T. R. Janacek, L. Krissek, M. Leckie, and M. Lyle, The record of Ontong Java Plateau: Main results of ODP Leg 130, *Geol. Soc. Am. Bull., 104,* 954-972, 1992.

Berger, W. H., R. M. Leckie, T. R. Janacek, R. Stax, and T. Takayama, Neogene carbonate sedimentation on Ontong Java Plateau: Highlights and open questions, *Proc. Ocean Drill. Prog., Sci. Results, 130,* 711-744, 1993.

Birkhold-VanDyke, A. L., C. R. Neal, J. C. Jain, J. J. Mahoney, and R. A. Duncan, Multi-stage growth for the Ontong Java Plateau (OJP)? A progress report from San Cristobal (Makira), Solomon Islands (abstract), *Eos Trans. AGU, 77,* Fall Meeting Suppl., p. 714, 1996.

Bjarnason, I. T., W. Menke, Ó. G. Flóvenz, and D. Caress, Tomographic image of the mid-Atlantic plate boundary in southwestern Iceland, *J. Geophys. Res., 98,* 6607-6622, 1993.

Bott, M. H. P., *The Interior of the Earth: Its Structure, Constitution, and Evolution,* 382 pp., Elsevier, Amsterdam, 1982.

Campbell, I. H., and R. W. Griffiths, Implications of mantle plume structure for the evolution of flood basalts, *Earth Planet. Sci. Lett., 99,* 79-93, 1990.

Cande, S. C., J. L. Labrecque, R. L. Larson, W. C. Pitman III, X. Golovchenko, and W. F. Haxby, *Magnetic Lineations of the World's Ocean Basins (chart),* American Association of Petroleum Geologists, Tulsa, Okla., 1989.

Carlson, R. L., N. I. Christensen, and R. P. Moore, Anomalous crustal structures in ocean basins: Continental fragments and oceanic plateaus, *Earth Planet. Sci. Lett., 51,* 171-180, 1981.

Carmichael, R. S., *Practical Handbook of Physical Properties of Rocks and Minerals,* 741 pp., CRC Press, Boca Raton, Fla, 1989.

Castillo, P., R. Batiza, and R. J. Stern, Petrology and geochemistry of Nauru Basin igneous complex: large-volume, off-ridge eruptions of MORB-like basalt during the Cretaceous, *Init. Repts. Deep Sea Drill. Proj., 89,* 555-576, 1986.

Castillo, P., R. W. Carlson, and R. Batiza, Origin of the Nauru Basin igneous complex: Sr, Nd, and Pb isotope and REE constraints, *Earth Planet. Sci. Lett., 103,* 200-213, 1991.

Castillo, P., M. S. Pringle, and R. W. Carlson, East Mariana Basin tholeiites: Jurassic ocean crust or Cretaceous rift basalts related to the Ontong Java plume? *Earth Planet. Sci. Lett., 123,* 139-154, 1994.

Cheng, Q., K.-H. Park, J. D. Macdougall, A. Zindler, G. W. Lugmair, J. Hawkins, P. Lonsdale, and H. Staudigel, Isotopic evidence for a hotspot origin of the Louisville seamount chain, in *Seamounts, Islands, and Atolls, Geophys. Monogr. Ser.,* vol. 43, edited by B. Keating, P. Fryer, R. Batiza, and G. Boehlert, pp. 283-296, AGU, Washington, D.C., 1987.

Cloos, M., Lithospheric buoyancy and collisional orogenesis: subduction of oceanic plateaus, continental margins, island arcs, spreading ridges, and seamounts, *Geol. Soc. Am. Bull., 105,* 715-737, 1993.

Coffin, M. F., Emplacement and subsidence of Indian Ridge plateaus and submarine plateaus, in *Synthesis of Results from Scientific Drilling in the Indian Ocean, Geophys. Monogr. Ser.,* vol. 70, edited by R. A. Duncan, D. K. Rea, R. B. Kidd, U. von Rad, and J. K. Weissel, pp. 115-125, AGU, Washington, D.C., 1992.

Coffin, M. F., and O. Eldholm, *Large Igneous Provinces:* JOI/USSAC Workshop Rept. University of Texas at Austin Institute for Geophysics *Tech. Rept. No. 114,* 79 pp., 1991.

Coffin, M. F., and O. Eldholm, Scratching the surface: Estimating the dimensions of large igneous provinces, *Geology, 21,* 515-518, 1993.

Coffin, M. F., and O. Eldholm, Large igneous provinces: crustal structure, dimensions, and external consequences, *Rev. Geophys., 32,* 1-36, 1994.

Coffin, M. F., and L. M. Gahagan, Ontong Java and Kerguelen Plateaux: Cretaceous Icelands? *J. Geol. Soc. Lond., 152,* 1047-1052, 1995.

Coffin, M. F., P. Mann, T. Shipley, E. Phinney, and S. Cowley, Structure and stratigraphy of the southern Ontong Java Plateau (abstract), *Eos Trans. AGU, 77,* Fall Meeting Suppl., p. 712, 1996.

Coleman, P. J., The Solomon Islands as an island arc, *Nature, 211,* 1249-1251, 1966.

Coleman, P. J., A re-evaluation of the Solomon Islands as an arc system, *CCOP/SOPAC Tech. Bull., 2,* 134-140, 1976.

Coleman, P. J., and L. W. Kroenke, Subduction without volcanism in the Solomon Islands arc, *Geo-Mar. Lett., 1,* 129-134, 1981.

Coleman, P. J., and G. H. Packham, The Melanesian borderlands and India-Pacific plates boundary, *Earth Sci. Rev., 12,* 197-233, 1976.

Coleman, P. J., B. McGowran, and R. W. Ramsay, New early Tertiary ages for basal pelagites, northeast Santa Isabel, Solomon Islands (central southwest flank, Ontong Java Plateau), *Bull. Aust. Soc. Explor. Geophys., 9,* 110-114, 1978.

Cooper, P., and B. Taylor, Polarity reversal in the Solomon Islands arc, *Nature, 314,* 428-430, 1985.

Cooper, P., and B. Taylor, The spatial distribution of earthquakes, focal mechanisms, and subducted lithosphere in the Solomon Islands, in *Marine Geology, Geophysics, and Geochemistry of the Woodlark Basin-Solomon Islands, Circum-Pacific Council for Energy and Mineral Resources, Earth Sci. Ser.*, vol. 7, edited by B. Taylor and N. F. Exon, pp. 67-88, Circum-Pacific Council for Energy and Mineral Resources, Houston, Tex., 1987.

Cooper, A. K., M. S. Marlow, and T. R. Bruns, Deep structure of the central and southern Solomon Islands region: Implications for tectonic origin, in *Geology and Offshore Resources of Pacific Island Arcs - Central and Western Solomon Islands, Circum-Pacific Council for Energy & Mineral Resources, Earth Sci. Ser.*, vol. 4, edited by J. G. Vedder, K. S. Pound, and S. Q. Boundy, pp. 157-175, Circum-Pacific Council for Energy and Mineral Resources, Houston, Tex., 1986.

Cowley, S., P. Mann, M. Coffin, and T. Shipley, Folds and unconformities within the Central Solomons Trough: A record of back-thrusting related to post-Miocene shallow subduction of the Ontong Java Plateau (abstract), *Eos Trans. AGU, 77*, Fall Meeting Suppl., p. 712, 1996.

Cox, K. G., Continental magmatic underplating, *Philos. Trans. R. Soc. Lond., A 342*, 155-166, 1993.

Davis, G. L., The ages and uranium contents of zircon from kimberlites and associated rocks (abstract), *Extended Abstr. 2nd Intl. Kimberlite Conf.*, Santa Fe, 1977.

Duncan, R. A., Radiometric ages from volcanic rocks along the New Hebrides-Samoa lineament, in *Geological Investigations of the Northern Melanesian Borderland, Circum-Pacific Council for Energy and Mineral Resources, Earth Sci. Ser.*, vol. 3, edited by T. M. Brocher, pp. 67-76, Circum-Pacific Council for Energy and Mineral Resources, Houston, Tex., 1985.

Ewing, J., M. Ewing, T. Aitken, and W. J. Ludwig, North Pacific sediment layers measured by seismic profiling, in *The Crust and Upper Mantle of the Pacific Area, Geophys. Monogr. Ser.*, vol. 12, edited by L. Knopoff, C. L. Drake, and P. J. Hart, pp. 147-173, AGU, Washington, D.C., 1968.

Falloon, T. J., and D. H. Green, Anhydrous partial melting of peridotite from 8 to 35 kb and the petrogenesis of MORB, *J. Petrol. Spec. Lithosphere Issue*, pp. 379-414, 1988.

Farnetani, C. G., and M. A. Richards, Numerical investigations of the mantle plume initiation model for flood basalt events, *J. Geophys. Res., 99*, 13,813-13,833, 1994.

Farnetani, C. G., M. A. Richards, and M. S. Ghiorso M. S., Modeling crustal structure in provinces affected by plume volcanism (abstract), *Eos Trans. AGU, 76*, Fall Meeting Suppl., p. 591-592, 1995.

Farnetani, C. G., M. A. Richards, and M. S. Ghiorso, Petrological models of magma evolution and deep crustal structure beneath hotspots and flood basalt provinces, *Earth Planet. Sci. Lett., 143*, 81-96, 1996.

Fleet, M. E., W. E. Stone, and J. H. Crocket, Partitioning of palladium, iridium, and platinum between sulfide liquid and basalt melt: Effects of melt composition, concentration, and oxygen fugacity, *Geochim. Cosmochim. Acta, 55*, 2545-2554, 1991.

Fleet, M. E., S. L. Chryssoulis, W. E. Stone, and C. G. Weisner, Partitioning of platinum-group elements and Au in the Fe-Ni-Cu-S system: Experiments on the fractional crystallization of sulfide melt, *Contrib. Mineral. Petrol., 115*, 36-44, 1993.

Floyd, P. A., Petrology and geochemistry of oceanic intraplate sheet-flow basalts, Nauru Basin, Deep Sea Drilling Project, Leg 89, *Init. Repts. Deep Sea Drill. Proj., 89*, 471-497, 1986.

Floyd, P. A., Geochemical features of intraplate oceanic plateau basalts, in *Magmatism in the Ocean Basins, Spec. Publ. 42*, edited by A. D. Saunders and M. J. Norry, pp. 215-230, The Geological Society, London, 1989.

Fountain, D. M., T. M. Boundy, H. Austrheim, and P. Rey, Eclogite-facies shear zones—deep crustal reflectors? *Tectonophysics, 232*, 411-424, 1994.

Furumoto, A. S., D. M. Hussong, J. F. Campbell, G. H. Sutton, A. Malahoff, J. C. Rose, and G. P. Woollard, Crustal and upper mantle structure of the Solomon Islands as revealed by seismic refraction survey of November-December 1966, *Pac. Sci., 24*, 315-332, 1970.

Furumoto, A. S., J. P. Webb, M. E. Odegard, and D. M. Hussong, Seismic studies on the Ontong Java Plateau, 1970, *Tectonophysics, 34*, 71-90, 1976.

Gladczenko, T. P., M. Coffin, and O. Eldholm, Crustal structure of the Ontong Java Plateau: Modeling of new gravity and existing seismic data, *J. Geophys. Res.*, in revision, 1997.

Govindaraju, K., 1989 compilation of working values and sample description for 272 geostandards, *Geostand. Newsl., 13*, 1-114, 1989.

Green, T. H., S. H. Sie, C. G. Ryan, and D. R. Cousens, Proton microprobe-determined partitioning of Nb, Ta, Zr, Sr, and Y between garnet, clinopyroxene, and basaltic magma at high pressure and temperature, *Chem. Geol., 74*, 201-216, 1989.

Griffiths, R. W., and I. H. Campbell, Stirring and structure in mantle starting plumes, *Earth Planet. Sci. Lett., 99*, 66-78, 1990.

Griffiths, R. W., M. Gurnis, and G. Eitelberg, Holographic measurements of surface topography in laboratory models of mantle hotspots, *Geophys. J., 96*, 477-495, 1989.

Grӧcke, D. R., A. Constantine, and M. I. Bird, Palaeo-environmental change recorded in stable carbon-isotopes in Early Cretaceous plant fragments, southeastern Australia, *Aust. J. Botany*, in press, 1997.

Hackman, B. D., The Solomon Islands fractured arc, in *The Western Pacific: Island Arcs, Marginal Seas, Geochemistry*, edited by P. J. Coleman, pp. 179-191, University of Western Australia Press, Nedlands, 1973.

Hagen, R. A., L. A. Mayer, D. C. Mosher, L. W. Kroenke, T. H. Shipley, and E. L. Winterer, Basement structure of the northern Ontong Java Plateau, *Proc. Ocean Drill. Prog., Sci. Results, 130*, 23-31, 1993.

Hammond, S. R., L. W. Kroenke, and F. Theyer, Northward motion of the Ontong Java Plateau between 110 and 30 m.y.: A paleomagnetic investigation of DSDP Site 289. *Init. Repts. Deep Sea Drill. Proj., 30*, 415-418, 1975.

Hassler, D. R., and N. Shimizu, Old peridotite xenoliths from the Kerguelen Islands (abstract), *Eos Trans. AGU, 76*, Fall Meeting Suppl., 693-694, 1995.

Hawkins, M. P., and A. J. M Barron, The geology and mineral

resources of Santa Isabel.   Solomon Islands Ministry of Natural Resources, *Geol. Div. Rept. J25*, 114 pp., 1991.

Hawkins, J. W., P. F. Lonsdale, and R. Batiza, Petrologic evolution of the Louisville seamount chain, in *Seamounts, Islands, and Atolls, Geophys. Monogr. Ser.*, vol. 43, edited by P. Fryer, B. Keating, R. Batiza, and G. Boehlert, pp. 235-253, AGU, Washington, D.C., 1987.

Hilde, T. W. C., S. Uyeda, and L. W. Kroenke, Evolution of the western Pacific and its margin, *Tectonophysics, 38*, 145-165, 1977.

Hill, R. I., Starting plumes and continental break-up, *Earth Planet. Sci. Lett., 104*, 398-416, 1991.

Hill, R. I., I. H. Campbell, G. F. Davies, and R. W. Griffiths, Mantle plumes and continental tectonics, *Science, 256*, 186-193, 1992.

Hirose, K., and I. Kushiro, Partial melting of dry peridotites at high pressures: Determination of compositions of melts segregated from peridotite using aggregates of diamond, *Earth Planet. Sci. Lett., 114*, 477-489, 1993.

Hopson, P. M., The geology of east central Santa Isabel. Solomon Islands Ministry of Natural Resources, *Geol. Div. Rept. J23*, 1988.

Horn, I., S. F. Foley, S. E. Jackson, and G. A. Jenner, Experimentally determined partitioning of high field strength and selected transition elements between spinel and basaltic melt, *Chem. Geol., 117*, 193-218, 1994.

Houtz, R. and J. Ewing, Upper crustal structure as a function of plate age, *J. Geophys. Res., 81*, 2490-2498, 1976.

Hughes, G. W., and C. C. Turner, Geology of Southern Malaita, *Honiara Min. Nat. Res. Geol. Survey Div., Bull. No. 2*, 80 pp., 1976.

Hughes, G. W., and C. C. Turner, Upraised Pacific ocean floor, southern Malaita, Solomon Islands, *Geol. Soc. Am. Bull., 88*, 412-424, 1977.

Hussong, D. M., L. K. Wipperman, and L. W. Kroenke, The crustal structure of the Ontong Java and Manihiki oceanic plateaus, *J. Geophys. Res., 84*, 6003-6010, 1979.

Ito, G., and P. Clift, Evidence for multi-staged accretion of the Manihiki and Ontong Java Plateaus from their vertical tectonic histories (abstract), *Eos Trans. AGU, 77*, Fall Meeting Suppl., p. 714, 1996.

Jain, J. C., J. A. O'Neill Jr., C. R. Neal, J. J. Mahoney, and M. G. Petterson, Siderophile elements in large igneous provinces (LIPs): Origin of the Ontong Java Plateau at the core-mantle boundary? (abstract), *Eos Trans. AGU, 76*, Fall Meeting Suppl., p. 700, 1995.

Jain, J. C., C. R. Neal, J. A. O'Neill Jr., and J. J. Mahoney, Origin of the Ontong Java Plateau (OJP) at the core-mantle boundary: Platinum group element (PGE) and gold (Au) evidence (abstract), *Eos Trans. AGU, 77*, Fall Meeting Suppl., p. 714, 1996.

Jones, C. E., H. C. Jenkyns, A. L. Coe, and S. P Hesselbo, Strontium isotopic variations in Jurassic and Cretaceous seawater, *Geochim. Cosmochim. Acta, 58*, 3061-3075, 1995.

Jones, J. H., and M. J. Drake, Geochemical constraints on core formation in the Earth, *Nature, 322*, 221-228, 1986.

Kellogg, L. H., and S. D. King, Effect of mantle plumes on the

growth of D" by reaction between the core and mantle, *Geophys. Res. Lett., 20*, 379-382, 1993.

Kent, R. W., M. Storey, and A. D. Saunders, Large igneous provinces: Sites of plume impact or plume incubation? *Geology, 20*, 891-894, 1992.

Kincaid, C., D. W. Sparks, and R. Detrick, The relative importance of plate-driven and buoyancy-driven flow at mid-ocean ridges, *J. Geophys. Res., 101*, 16,177-16,193, 1996.

Kroenke, L. W., Geology of the Ontong Java Plateau, *Hawaii Inst. Geophys. Rept., HIG-72-5*, 119 pp., 1972.

Kroenke, L. W., and J. J. Mahoney, Rifting of the Ontong Java Plateau's eastern salient and seafloor spreading in the Ellice Basin: Relation to the 90 Myr eruptive episode on the plateau (abstract), *Eos Trans. AGU, 77*, Fall Meeting Suppl., p. 713, 1996

Kroenke, L. W., and W. Sager, The formation of oceanic plateaus on the Pacific Plate (abstract), *Eos Trans. AGU, 74*, Fall Meeting Suppl., p. 555, 1993.

Kroenke, L. W., and P. Wessel, Pacific plate motion between 125 and 90 Ma and the formation of the Ontong Java Plateau (abstract), *Chapman Conf. on Global Plate Motions*, Marshall, CA, 1997.

Kroenke, L. W., R. Moberly Jr., and G. R. Heath, Lithologic interpretation of continuous reflection profiling, Deep Sea Drilling Project, Leg 7, *Init. Repts. Deep Sea Drill. Proj., 7*, 1161-1227, 1971.

Kroenke, L. W., J. Resig, and P.A. Cooper, Tectonics of the southeastern Solomon Islands: Formation of the Malaita Anticlinorium, in *Geology and Offshore Resources of Pacific Island Arcs—Central and Western Solomon Islands, Earth Sci. Ser.*, vol. 4, edited by J. G. Vedder, K. S. Pound, and S. Q. Boundy, pp. 109-116, Circum-Pacific Council for Energy and Mineral Resources, Houston, Tex., 1986.

Kroenke, L. W., W. Berger,, T. R. Janacek, et al., *Proc. Ocean Drill. Prog., Init. Repts. 130*, 1240 pp., 1991.

Kroenke, L. W., W. Berger, and Leg 130 Shipboard Scientific Party, *Proc. Ocean Drill. Prog., Sci. Results, 130*, 867 pp., 1993.

Larson, R. L., and C. Kincaid, Onset of mid-Cretaceous volcanism by elevation of the 670 km thermal boundary layer, *Geology, 24*, 551-554, 1996.

Lassiter, J. C., D. J. DePaolo, and J. J. Mahoney, Geochemistry of the Wrangellia flood basalt province: Implications for the role of continental and oceanic lithosphere in flood basalt genesis, *J. Petrol., 36*, 983-997, 1995.

Mahoney, J. J., An isotopic survey of Pacific oceanic plateaus: Implications for their nature origin, in *Seamounts, Islands, and Atolls, Geophys. Monogr. Ser.*, vol. 43, edited by B. Keating, P. Fryer, R. Batiza, and G. Boehlert, pp. 207-220, AGU, Washington, D.C., 1987.

Mahoney, J. J., and K. J. Spencer, Isotopic evidence for the origin of the Manihiki and Ontong Java oceanic plateaus, *Earth Planet. Sci. Lett., 104*, 196-210, 1991.

Mahoney, J. J., W. B. Jones, F. A. Frey, V. J. M. Salters, D. G. Pyle, and H. L. Davies, Geochemical characteristics of lavas from Broken Ridge, the Naturaliste Plateau, and southernmost

Kerguelen Plateau: Cretaceous plateau volcanism in the southeast Indian Ocean, *Chem. Geol., 120*, 315-345, 1995.

Mahoney, J. J., J. M. Sinton, M. D. Kurz, J. D. Macdougall, K. J. Spencer, and G. W. Lugmair, Isotope and trace element characteristics of a super-fast spreading ridge: East Pacific Rise, 13-23°S, *Earth Planet. Sci. Lett., 121*, 173-193, 1994.

Mahoney, J. J., M. Storey, R. A. Duncan, K. J. Spencer, and M. Pringle, Geochemistry and age of the Ontong Java Plateau, in *The Mesozoic Pacific: Geology, Tectonics, and Volcanism, Geophys. Monogr. Ser.,* vol. 77, edited by M. S. Pringle, W. W. Sager, W. V. Sliter, and S. Stein, pp. 233-262, AGU, Washington, D.C., 1993.

Mammerickx, J., and S. M. Smith, Bathymetry of the north-central Pacific, *Map and Chart Series MC-52*, The Geological Society of America, Boulder, Colo. 1985.

Mann, P., L. Gahagan, M. Coffin, T. Shipley, S. Cowley, and E. Phinney, Regional tectonic effects resulting from the progressive East-to-West collision of the Ontong Java Plateau with the Melanesian Arc System (abstract), *Eos Trans. AGU, 77*, Fall Meeting Suppl., p. 712, 1996.

Marks, K. M., and D. T. Sandwell, Analysis of geoid height versus topography for oceanic plateaus and swells using nonbiased linear regression, *J. Geophys. Res., 96*, 8045-8055, 1991.

Mayer, L. A., and J. A. Tarduno, Paleomagnetic investigation of the igneous sequence, Site 807, Ontong Java Plateau, and a discussion of Pacific true polar wander, *Proc. Ocean Drill. Prog., Sci. Results, 130*, 51-59, 1993.

Mayer, L. A., T. H. Shipley, E. L. Winterer, D. Mosher, and R. A. Hagen, Seabeam and seismic reflection surveys on the Ontong Java Plateau, *Proc. Ocean Drill. Prog., Sci. Results, 130*, 45-75, 1991.

McGinnis, C. E., C. R. Neal, and J. C. Jain, Development of an analytical technique for the accurate & precise determination of the high field strength elements (HFSEs), Cs, & Mo by ICP-MS with geological applications (abstract), *Eos Trans. AGU, 77*, Fall Meeting Suppl., p. 772, 1996.

McGinnis, C. E., C. R. Neal, and J. C. Jain, Analytical technique for the accurate and precise determination of the high field strength elements (HFSEs) by ICP-MS with geological applications, *Geostand. Newsl.,* in revision, 1997.

McKenzie, D., and R. K. O'Nions, Partial melt distributions from inversion of rare earth element concentrations, *J. Petrol., 32*, 1021-1091, 1991.

Michael, P. J., and W. C. Cornell, H_2O, CO_2, Cl, and S contents in 122 Ma glasses from Ontong Java Plateau, ODP 807C: Implications for mantle and crustal processes (abstract), *Eos Trans. AGU, 77*, Fall Meeting Suppl., p. 714, 1996.

Miura, S., M. Shinohara, N. Takahashi, E. Araki, A. Taira, K. Suyehiro, M. Coffin, T. Shipley, and P. Mann, OBS crustal structure of Ontong Java Plateau converging into Solomon Island arc (abstract), *Eos Trans. AGU, 77*, Fall Meeting Suppl., p. 713, 1996.

Murauchi, S., W. J. Ludwig, N. Den, H. Hotta, T. Asanuma, T. Yoshi, A. Kubotera, and K. Hagiwara, Seismic refraction measurements on the Ontong Java Plateau northeast of New Ireland, *J. Geophys. Res., 78*, 8653-8663, 1973.

Nakanishi, M., and E. L. Winterer, Tectonic events of the Pacific Plate related to the formation of the Ontong Java Plateau (abstract), *Eos Trans. AGU, 77*, Fall Meeting Suppl., p. 713, 1996.

Nakanishi, M., K. Tamaki, and K. Kobayashi, Magnetic anomaly lineations from late Jurassic to early Cretaceous in the west-central Pacific Ocean, *Geophys. J. Int., 109*, 701-719, 1992.

Neal, C. R., Mantle studies in the western Pacific and kimberlite-type intrusives, Unpubl. Ph.D. Thesis, University of Leeds, UK, 365 pp., 1985.

Neal, C. R., The origin and composition of metasomatic fluids and amphiboles beneath Malaita, Solomon Islands, *J. Petrol., 29*, 149-179, 1988.

Neal, C. R., The relationship between megacrysts and their host magma and identification of the mantle source region (abstract), *Eos Trans. AGU, 76*, Fall Meeting Suppl., p. 664, 1995.

Neal, C. R., and J. P. Davidson, An unmetasomatized source for the Malaitan alnöite (Solomon Islands): Petrogenesis involving zone refining, megacryst fractionation, and assimilation of oceanic lithosphere, *Geochim. Cosmochim. Acta, 53*, 1975-1990, 1989.

Neal, C. R., and L. A. Taylor, Negative Ce anomaly in a peridotite xenolith from Malaita, Solomon Islands: Evidence for crustal recycling into the mantle or mantle metasomatism?, *Geochim. Cosmochim. Acta, 53*, 1035-1040, 1989.

Neal, C. R., J. J. Mahoney, A. D. Saunders, and T. L. Babbs, Trace element characteristics of a 3.5-4 km thick section of OJP basalts, Malaita, Solomon Islands (abstract), *Eos Trans. AGU, 75*, Fall Meeting Suppl., p. 727, 1994.

Neal, C. R., S. Schaefer, J. J. Mahoney, and M. G. Petterson, Uplift associated with large igneous province (LIP) petrogenesis: A case study of the Ontong Java Plateau, SW Pacific (abstract), *Abstracts with Programs, 27 No. 6, Geol. Soc. Am.,* p. A48, 1995.

Neavel, K. E., and A. M. Johnson, Entrainment in composition-ally buoyant plumes, *Tectonophysics, 200*, 1-15, 1991.

Newsom, H. E., and M. J. Palme, The depletion of siderophile elements in the Earth's mantle: New evidence from molybdenum and tungsten, *Earth Planet. Sci. Lett., 69*, 354-364, 1984.

Nixon, P. H., Kimberlites in the south-west Pacific, *Nature 287*, 718-720, 1980.

Nixon, P. H., and F. R. Boyd, Garnet-bearing lherzolites and discrete nodule suites from the Malaita alnöite, Solomon Islands, SW Pacific, and their bearing on oceanic mantle composition and geotherm, in *The Mantle Sample: Inclusions in Kimberlites and Other Volcanics*, edited by F. R. Boyd and H. O. A. Meyer, pp. 400-423, AGU, Washington, D.C., 1979.

Nixon, P. H., and P. J. Coleman, Garnet-bearing lherzolites and discrete nodule suites from the Malaita alnöite, Solomon Islands, and their bearing on the nature and origin of the Ontong Java Plateau, *Bull. Aust. Soc. Explor. Geophys., 9*, 103-106, 1978.

Nixon, P. H., R. H. Mitchell, and N. W. Rogers, Petrogenesis of alnöitic rocks from Malaita, Solomon Islands, Melanesia, *Mineral. Mag., 43*, 587-596, 1980.

O'Hara, M. J., and H. S. Yoder Jr., Formation and fractionation of basic magmas at high pressures, *Scott. J. Geol., 3*, 67-117, 1967

Olson, P., and H. A. Singer, Creeping plumes, *J. Fluid Mech., 158*, 511-531, 1985.

Olson, P., and I. S. Nam, Formation of seafloor swells by mantle plumes, *J. Geophys. Res., 91*, 7181-7191, 1986.

Operto, S., and P. Charvis, Deep structure of the southern Kerguelen Plateau (southern Indian Ocean) from ocean bottom seismometer wide-angle seismic data, *J. Geophys. Res., 101*, 25,077-25,103, 1996.

Parkinson, I. J., R. J. Arculus, E. McPherson, and R. A. Duncan, Geochemistry, tectonics and the peridotites of the northeastern Solomon Islands (abstract), *Eos Trans. AGU, 76*, Fall Meeting Suppl., p. 642, 1996.

Petterson, M. G., The geology of north and central Malaita, Solomon Islands (including implications of geological research on Makira, Savo Island, Guadalcanal, and Choiseul between 1992 & 1995), *Geological Mem., 1/95*, Water and Mineral Resources Division, Ministry of Energy, Water, and Mineral Resources, Honiara, Solomon Islands, 1995.

Petterson, M. G., C. R. Neal, A. D. Saunders, T. L. Babbs, J. J. Mahoney, and R. A. Duncan, Speculations regarding the evolution of the Ontong Java Plateau (abstract), *Eos Trans. AGU, 76*, Fall Meeting Suppl., p. 693, 1995.

Petterson, M. G., C. R. Neal, J. J. Mahoney, L. W. Kroenke, A. D. Saunders, T. L. Babbs, R. A. Duncan, D. Tolia, B. McGrail, and M. Barron, Structure and deformation of north and central Malaita, Solomon Islands: Tectonic implications for the Ontong Java Plateau - Solomon arc collision and for the fate of oceanic plateaus, *Tectonophysics*, in revision, 1997.

Phinney, E., P. Mann, M. Coffin, and T. Shipley, Along-strike variations in the style of oceanic plateau accretion within the Malaita accretionary prism, Solomon Islands (abstract), *Eos Trans. AGU, 77*, Fall Meeting Suppl., p. 712, 1996.

Pringle, M. S., J. D. Obradovich, and R. A. Duncan, Estimated ages for magnetic anomaly M0 and interval ISEA, and a minimum estimate for the duration of the Aptian (abstract), *Eos Trans. AGU, 73*, Fall Meeting Suppl., p. 633, 1992.

Pudsey-Dawson, P.A., South Malaita geological reconnaisances, 1957-1958, *British Solomon Islands Geol. Rec., 1957-1958*, 27-31, 1960.

Ramsay, W. R. H., Crustal strain phenomena in the Solomon Islands: Constraints from field evidence and the relationship to the India-Pacific plates boundary, *Tectonophyics, 87*, 109-126, 1982.

Ribe, N. M., The dynamics of plume-ridge interaction 2. Off-ridge plumes, *J. Geophys. Res., 101*, 16,195-16,204, 1996.

Ribe, N. M., U. R. Christensen, and J. Theißing, The dynamics of plume-ridge interaction, 1: Ridge-centered plumes, *Earth Planet. Sci. Lett., 134*, 155-168, 1995.

Richards, M. A., R. A. Duncan, and V. E. Courtillot, Flood basalts and hot-spot tracks: Plume heads and tails, *Science, 246*, 103-107, 1989.

Richards, M. A., D. L. Jones, R. A. Duncan, and D. J. DePaolo, A mantle plume initiation model for the Wrangellia flood basalt and other oceanic plateaus, *Science, 254*, 263-267, 1991.

Richardson, W. P., and E. Okal, Crustal and upper mantle structure of the Ontong Java Plateau from surface waves: Results from the Micronesian PASSCAL experiment (abstract), *Eos Trans. AGU, 77*, Fall Meeting Suppl., p. 713, 1996.

Rickwood, F. K., Geology of the island of Malaita, in *Geological Reconnaissance of Parts of the Central Islands of the B.S.I.P.: Colonial Geol. and Min. Res., 6*, 300-306, 1957.

Rudnick, R. L., and I. Jackson, Measured and calculated elastic wave speeds in partially equilibrated mafic granulite xenoliths: Implications for the properties of an underplated lower continental crust, *J. Geophys. Res. 100*, 10,211-10,218, 1995.

Sager, W. W., and H.-C. Han, Rapid formation of the Shatsky Rise oceanic plateau inferred from its magnetic anomaly, *Nature, 364*, 610-613, 1993.

Sandwell, D. T., and K. R. MacKenzie, Geoid height versus topography for oceanic plateaus and swells, *J. Geophys. Res., 84*, 7403-7418, 1989.

Sandwell, D. T., and M. L. Renkin, Compensation of swells and plateaus in the North Pacific: No direct evidence for mantle convection, *J. Geophys. Res., 93*, 2775-2783, 1988.

Saunders, A. D., Geochemistry of basalts from the Nauru Basin, Deep Sea Drilling Project Legs 61 and 89, *Init. Repts. Deep Sea Drill. Proj., 89*, 499-518, 1986.

Saunders, A. D., M. Storey, R. W. Kent, and M. J. Norry, Consequences of plume-lithosphere interactions, in *Magmatism and the Causes of Continental Break-Up, Spec. Publ. 68*, edited by B. C. Storey, T. Alabaster, and R. J. Pankhurst, pp. 41-59, The Geological Society, London, 1992.

Saunders, A. D., T. L. Babbs, M. J. Norry, M. G. Petterson, B. A. McGrail, J. J. Mahoney, and C. R. Neal, Depth of emplacement of oceanic plateau basaltic lavas, Ontong Java Plateau and Malaita, Solomon Islands: Implications for the formation of oceanic LIPs? (abstract), *Eos Trans. AGU, 74*, Fall Meeting Suppl., p. 552, 1993.

Saunders, A. D., J. Tarney, A. C. Kerr, and R. W. Kent, The formation and fate of large oceanic igneous provinces, *Lithos, 37*, 81-89, 1996.

Schaefer, S., and C. R. Neal, An estimate of uplift associated with the Ontong Java Plateau and the importance of cumulates (abstract), *Eos Trans. AGU, 75*, Fall Meeting Suppl., p. 711, 1994.

Sliter, W. V., Aptian anoxia in the Pacific Basin, *Geology, 17*, 909-912, 1989.

Sliter, W. V., and R. M. Leckie, Cretaceous planktonic foraminifers and depositional environments from the Ontong Java Plateau with emphasis on Sites 803 and 807, *Proc. Ocean Drill. Prog., Sci. Results, 130*, 63-84, 1993.

Smith, W. H. F., and D. T Sandwell, Oceanographic "pseudogravity" in marine gravity fields derived from declassified Geosat and ERS-1 altimetry (abstract), *Eos Trans. AGU, 76*, Fall Meeting Suppl., p. 151, 1995a.

Smith, W. H. F., and D. T. Sandwell, Marine gravity field from declassified Geosat and ERS-1 altimetry (abstract), *Eos Trans. AGU, 76*, Fall Meeting Suppl., p. 152, 1995b.

Stein, C. A., and S. Stein, A model for the global variation in oceanic depth and heat flow with lithospheric age, *Nature, 359*, 123-129, 1992.

Steiner, M. B., and B. P. Wallick, Jurassic to Paleocene paleolatitudes of the Pacific Plate derived from the paleomagnetism of the sedimentary sequences at Sites 800, 801, and 802, *Proc. Ocean Drill. Prog., Sci. Results, 129*, 431-446, 1992.

Stoeser, D. B., Igneous rocks from Leg 30 of the Deep Sea Drilling Project, *Init. Repts. Deep Sea Drill. Proj., 30*, 410-444, 1975.

Storey, M., J. J. Mahoney, L. W. Kroenke, and A. D. Saunders, Are oceanic plateaus sites of komatiite formation? *Geology, 19*, 376-379, 1991.

Storey, M., R. Kent, A. D. Saunders, V. J. Salters, J. Hergt, H. Whitechurch, J. H. Sevigny, M. F. Thirlwall, P. Leat, N C. Ghose, and M. Gifford, Lower Cretaceous volcanic rocks along continental margins and their relationship to the Kerguelen Plateau, *Proc. Ocean Drill. Prog., Sci. Results, 120*, 33-54, 1992.

Sun, S.-s., and W. F. McDonough, Chemical and isotopic systematics of oceanic basalts: Implications for mantle composition and processes, in *Magmatism in the Ocean Basins, Spec. Publ. 42*, edited by A. D. Saunders and M. J. Norry, pp. 313-345, Geological Society of London, 1989.

Tarduno, J. A., Vertical and horizontal tectonics of the Cretaceous Ontong Java Plateau (abstract), *Eos Trans. AGU, 73*, Fall Meeting Suppl., p. 532, 1992.

Tarduno, J. A., and W. W. Sager, Polar standstill of the mid-Cretaceous Pacific plate and its geodynamic Implications, *Science, 269*, 5226-5228, 1995.

Tarduno, J. A., W. V. Sliter, L. W. Kroenke, M. Leckie, J. J. Mahoney, R. J. Musgrave, M. Storey, and E. L. Winterer, Rapid formation of the Ontong Java Plateau by Aptian mantle plume volcanism, *Science, 254*, 399-403, 1991.

Taylor, B., Mesozoic magnetic anomalies in the Lyra Basin (abstract), *Eos Trans. AGU, 59*, Fall Meeting Suppl., p. 320, 1978.

Taylor, S. R., *Planetary Science: A Lunar Perspective*, 481 pp., Lunar and Planetary Institute, Houston, Tex., 1982.

Tejada, M. L. G., J. J. Mahoney, R. A. Duncan, and M. P. Hawkins, Age and geochemistry of basement and alkalic rocks of Malaita and Santa Isabel, Solomon Islands, southern margin of the Ontong Java Plateau, *J. Petrol., 37*, 361-394, 1996a.

Tejada, M. L. G., J. J. Mahoney, R. A. Duncan, and C. R. Neal, Geochemistry and age of the Ontong Java Plateau (OJP) crust in central Malaita, Solomon Islands (abstract), *Eos Trans. AGU, 77*, Fall Meeting Suppl., p. 715, 1996b.

Tokuyama, H., and R. Batiza, Chemical composition of igneous rocks and origin of the sill and pillow basalt complex of Nauru Basin, Southwest Pacific, *Init. Repts. Deep Sea Drill. Proj., 61*, 673-687, 1986.

van Keken, P. E., D. A. Yuen, and A. P. van den Berg, The effects of shallow rheological boundaries in the upper mantle on inducing shorter time scales of diapiric flows, *Geophys. Res. Lett., 20*, 1927-1930, 1993.

Walker, D., L. Norby, and J. H. Jones, Superheating effects on metal-silicate partitioning of siderophile elements, *Science, 262*, 1858-1861, 1993.

Walker, R. J., J. W. Morgan, and M. F. Horan, [187]Os enrichment in some plumes: Evidence for core-mantle interaction? *Science, 269*, 819-822, 1995.

Wessel, P., and L. W. Kroenke, Relocating Pacific hotspots and refining absolute plate motions using a new geometric technique, *Nature*, in press, 1997.

White, R. S., J. H. McBride, P. K. H. Maguire, B. Brandsdóttir, W. Menke, T. A. Minshull, K. R. Richardson, J. R. Smallwood, R. K. Staples, and the FIRE Working Group, Seismic images of crust beneath Iceland contribute to long-standing debate, *Eos Trans. AGU, 77*, 197-201, 1996.

White, R. S., D. McKenzie, and R. K. O'Nions, Oceanic crustal thickness from seismic measurements and rare earth element inversions, *J. Geophys. Res., 97*, 19,683-19,715, 1992.

Whitehead, J. A., and D. S. Luther, Dynamics of laboratory diapir and plume models, *J. Geophys. Res., 80*, 705-717, 1975.

Winterer, E. L., Bathymetry and regional tectonic setting of the Line Islands chain, *Init. Repts. Deep Sea Drill. Proj., 33*, 731-748, 1976.

Winterer, E. L., and M. Nakanishi, Evidence for a plume-augmented, abandoned spreading center on Ontong Java Plateau (abstract), *Eos Trans. AGU, 76*, Fall Meeting Suppl., p. 617, 1995.

Yan, C. Y., and L. W. Kroenke, A plate-tectonic reconstruction of the southwest Pacific, 100-0 Ma, *Proc. Ocean Drill. Prog., Sci. Results, 130*, 697-710, 1993.

Yoder, H. S., Jr., and C. E. Tilley, Origin of basalt magmas: An experimental study of natural and synthetic rock systems, *J. Petrol., 3*, 342-532, 1962.

R. A. Duncan, College of Oceanic and Atmospheric Sciences, Ocean Administration Building 104, Oregon State University, Corvallis, OR 97331.

L. W. Kroenke and J. J. Mahoney, School of Ocean and Earth Science and Technology, University of Hawaii, Honolulu, HI 96822.

C. R. Neal, Department of Civil Engineering and Geological Sciences, University of Notre Dame, Notre Dame, IN 46556.

M. G. Petterson, British Geological Survey, Murchison House, West Mains Road, Edinburgh, EH9 3LA, United Kingdom.

The Paraná-Etendeka Province

David W. Peate

Department of Earth Sciences, The Open University, Milton Keynes, United Kingdom

Stratigraphic data and ^{40}Ar-^{39}Ar ages for the Early Cretaceous Paraná-Etendeka flood basalts indicate that the main magmatic episode lasted for several m.y. (129–134 Ma) and was linked to the northward opening of the South Atlantic Ocean, but with some earlier magmatism (135–138 Ma) found inland far from the eventual oceanic rift. The regional distribution of distinct high-Ti/Y (Urubici, Pitanga, Paranapanema, Ribeira) and low-Ti/Y (Gramado, Esmeralda) magma types in the lavas and associated dyke swarms implies that magma generation occurred over a wide area and involved different mantle sources. Low MgO contents (3-7 wt%) indicate extensive fractional crystallisation, and upper crustal assimilation was important in the evolution of the Gramado magmas. However, Paraná basalts that are considered to be uncontaminated by crust have trace element and isotope characteristics (e.g. Nb/La < 0.8; ε_{Nd_i} < 0) and major element features that appear to require mantle sources distinct from those of oceanic basalts. The minor, late-stage, Esmeralda magma type is an exception, requiring a component from incompatible-element-depleted asthenosphere. The role of the Tristan mantle plume appears to have been largely passive, with conductive heating facilitating mobilisation of old lithospheric material. Significant rhyolitic eruptions (>1000 km^3) that can be correlated across the Atlantic Ocean accompanied the final magmatic phase in the southeast Paraná and the Etendeka. The flood basalts post-date most estimates for the Jurassic-Cretaceous boundary, ruling out any link to a faunal extinction.

INTRODUCTION

The extensive Paraná lava field in central South America and the minor Etendeka remnant in Namibia once formed a single magmatic province [*Erlank et al.*, 1984; *Bellieni et al.*, 1984a] that was associated with the opening of the South Atlantic Ocean during the Early Cretaceous. This Paraná-Etendeka province ranks as one of the largest extant continental large igneous provinces (LIPs), with a preserved volume in excess of 1×10^6 km^3 [*Cordani and*

Large Igneous Provinces: Continental, Oceanic, and Planetary Flood Volcanism
Geophysical Monograph 100

Vandoros, 1967]. The Rio Grande Rise and Walvis Ridge in the South Atlantic Ocean are interpreted as representing the fossil trace of the Tristan mantle plume and they connect the flood basalt exposures of the Paraná and the Etendeka, respectively, to the present magmatic activity of the plume found on the islands of the Tristan da Cunha group and Gough (Figure 1) [*O'Connor and Duncan*, 1990; *Gallagher and Hawkesworth*, 1994]. This association has bolstered models that stress an important role for the Tristan plume in generating the flood basalts [e.g., *Morgan*, 1981; *White and McKenzie*, 1989; *Peate et al.*, 1990].

Before the 1960s, numerous papers on the petrography and general geological setting of the Paraná lavas were

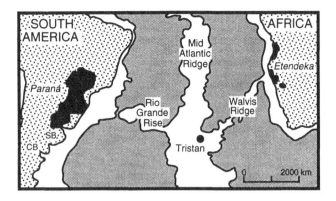

Figure 1. Map of the South Atlantic region [after *White and McKenzie*, 1989]. Shallow features, between 4 km water depth and sea level, are highlighted in white. The magmatic effects of the Tristan mantle plume, whose present location is marked by the Tristan da Cunha island group, can be traced back along the Rio Grande Rise and the Walvis Ridge to the flood basalt exposures of the Paraná and the Etendeka, respectively. SB = Salado basin, CB = Colorado basin.

published, primarily by Brazilian scientists, and these were reviewed by *Cordani and Vandoros* [1967]. Even the earliest studies [e.g., *Baker*, 1923] realised the true extent of the Paraná lava field. The continental drift debate of the 1960s saw a renewed interest in the Paraná and Etendeka lavas in the context of the breakup history of Gondwana. K-Ar dating [reviewed by *Rocha-Campos et al.*, 1988] and palaeomagnetic studies [reviewed by *Ernesto and Pacca*, 1988] confirmed the close link between the Paraná and Etendeka lavas. These studies demonstrated that they were both Early Cretaceous in age and thus not contemporaneous with the Early Jurassic Karoo and Ferrar flood basalt provinces elsewhere in Gondwana. Several geochemical studies into the petrogenesis of the Paraná magmatism began during the 1980s, involving collaborations between Brazilian, Italian, British, and American universities [e.g., *Bellieni et al.*, 1984a; *Mantovani et al.*, 1985a; *Fodor et al.*, 1985a; *Hawkesworth et al.*, 1986]. The principal results of the joint Italian/Brazilian group are summarised by *Piccirillo et al.* [1988], and all the data are collated in a book [*Piccirillo and Melfi*, 1988] which contains extensive reviews on many aspects of Paraná magmatism, including petrography, mineral chemistry, major and trace element and isotope geochemistry, K-Ar dating, and palaeomagnetism. *Erlank et al.* [1984] gave the first comprehensive description of the Etendeka magmatism and summarised the chronological, compositional, and palaeomagnetic data that showed that the Etendeka sequences once formed the eastern edge of the Paraná basin. These sequences represent a small fragment of the Paraná-Etendeka province, but the

excellent exposure has led to significant insights, particularly into the physical volcanology of the rhyolite eruptions [e.g., *Milner et al.*, 1992].

Almost a decade has passed since the last major reviews of the Paraná-Etendeka province were published. In this time, there has been a surge of interest globally in large igneous provinces, driven in part by the development of several new geodynamical models for their generation [e.g., *White and McKenzie*, 1989; *Richards et al.*, 1989]. A wealth of new information has been obtained on the Paraná-Etendeka province on a diverse range of topics such as detailed lava stratigraphy [e.g., *Peate et al.*, 1990, 1992; *Milner et al.*, 1995b], geochemical and petrogenetic modelling [e.g., *Turner and Hawkesworth*, 1995; *Peate and Hawkesworth*, 1996], precise ^{40}Ar-^{39}Ar dating [e.g., *Renne et al.*, 1992; *Turner et al.*, 1994], fission-track chronology of continental margin development [*Gallagher et al.*, 1994], and the nature of intrusive magmatism [*Piccirillo et al.*, 1990; *Regelous*, 1993]. Thus, it is appropriate to review what has been learned recently about the Paraná-Etendeka province and to see how suitable the various current plume-related geodynamic models are in explaining the principal features of the province.

GEOLOGICAL BACKGROUND

Spatial Extent of Paraná-Etendeka Magmatism

The Paraná lava field covers an area of at least 1.2×10^6 km^2 over southern Brazil, Uruguay, eastern Paraguay and northern Argentina [*Cordani and Vandoros*, 1967], and its present extent is bounded by the margin of the underlying Paraná sedimentary basin (Figure 2). The three-dimensional structure is well known by surface mapping and from numerous oil exploration boreholes (Figure 3) [*Zalan et al.*, 1987; *Peate et al.*, 1992]. The thickest preserved accumulation of lavas (1.5–1.7 km) is in the north, coincident with the deepest part of the Paraná basin. The extent of the lavas in northern Argentina is uncertain because of overlying Quaternary sediments, but borehole data indicate significant lava thicknesses (>1 km [*Leinz et al.*, 1968]). The average lava thickness over the province is estimated at about 0.7 km [*Leinz et al.*, 1968]. The Paraná lavas are spectacularly exposed along the coastal Serra Geral escarpment in southeast Brazil: the Serra Geral Formation is the formal stratigraphic name for the Paraná lavas in Brazil. As the lavas have different stratigraphic names in the neighbouring countries (Argentina: Posadas Member, Curuzú Cuatiá Formation: Paraguay: Alto Paraná Formation: Uruguay: Arapey Formation), we will use the general term Paraná province to encompass all these

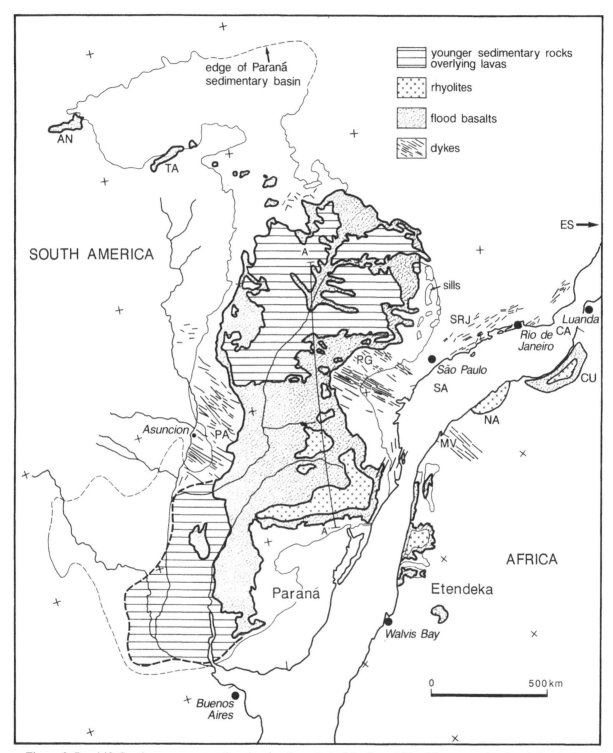

Figure 2. Pre-drift Gondwana reconstruction showing the extent of the Paraná-Etendeka magmatism in relation to the Paraná sedimentary basin and proto-Atlantic rift [*de Wit et al.*, 1988; *Peate et al.*, 1992]. Dykes are concentrated in four areas: PG = Ponta Grossa, SRJ = São Paulo - Rio de Janeiro coast, PA = eastern Paraguay, and in the Etendeka (not shown). SA = Santos basin, CA = Campos basin, ES = Esprito Santo basin, NA = Namibe basin, CU = Cuanza basin, MV = Morro Vermelho. Recent palaeomagnetic data and ^{40}Ar-^{39}Ar ages [*Montes-Lauer et al.*, 1995] indicate that the isolated basalt exposures close to the Brazil-Bolivia border (AN = Anari, TA = Tapirapuã) are Early Jurassic in age and not related to the Paraná-Etendeka magmatism as previously suggested. The + shows present-day latitude and longitude at 5° intervals.

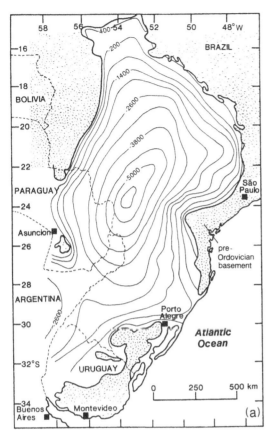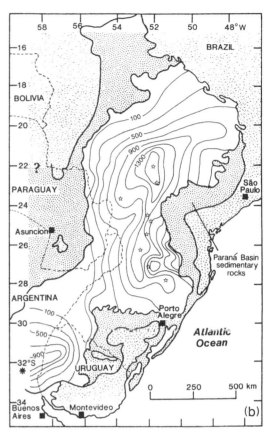

Figure 3. (a) Depth to basement beneath the Paraná sedimentary basin (contours at 600-m intervals), (b) Isopach map of Paraná lavas (contours at 200-m intervals); from *Peate et al.* [1992].

formations and the associated intrusive magmatism.

The majority of analysed samples in the literature are from the Serra Geral region, where sections up to 0.8 km thick are found. Inland from the escarpment, significant topographic relief is rarely developed on the plateau, exceptions being the deeply incised valleys of a few westward-flowing rivers. Most of the Paraná lava field is thus devoid of suitable sections for detailed flow-by-flow stratigraphic studies. Although the lava pile is greater than 1 km thick over much of the central Paraná, surface samples can only provide information about the uppermost few hundred metres. Access to borehole samples provided *Peate et al.* [1992] with a unique opportunity to look at the otherwise inaccessible deeper levels of the lava pile.

On the African Plate, the Etendeka lavas are scattered over an area of 0.8×10^5 km^2 in northwestern Namibia [*Erlank et al.*, 1984]. The coastal regions of Angola have been less well studied, but Early Cretaceous tholeiitic basalts and rhyolites are found, at least as far north as Luanda, in the onshore Namibe (NA on Figure 2) and Cuanza (CU on Figure 2) basins [*Piccirillo et al.*, 1990; *Alberti et al.*, 1992]. These Angolan lavas probably cover a similar area to the Etendeka lavas.

Significant dyke swarms are found along the Brazilian coast between São Paulo and Rio de Janeiro, in eastern Paraguay, and south of the Etendeka lava field. This distribution suggests that the lavas originally covered a much greater area of at least 2.0×10^6 km^2, now reduced by erosion. Modelling of fission track data by *Gallagher et al.* [1994] indicated that as much as 3 km of material could have been eroded from the Brazilian coastal plain since continental breakup, although not all of this would have been lavas.

Extensive Early Cretaceous lavas are found in offshore basins along the Brazilian coast [*Chang et al.*, 1992]. Lava thicknesses of 600 m have been drilled in the Campos basin (200 km east of Rio de Janeiro: CA on Figure 2) which covers an area of 10^5 km^2. *Mizusaki et al.* [1992] showed that these mainly subaerially erupted lavas were more similar in composition to the Paraná basalts than to mid-ocean ridge basalts (MORB), and they considered the Campos basin lavas to be an eastern extension of the northern Paraná lava field. *Fodor and Vetter* [1984] found similar 'Paraná-like' basalt lavas up to 150 km offshore in the Espirito Santo basin (500 km northeast of Rio de Janeiro: ES on Figure 2) and in the Santos basin (200 km southwest

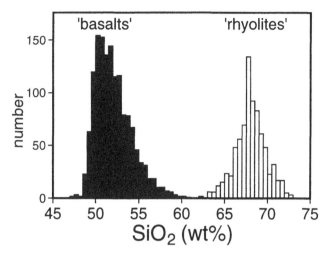

Figure 4. SiO$_2$ histogram illustrating the bimodal composition of the Paraná-Etendeka lavas. The silica gap between 61 and 63 wt% forms a natural division into 'basalts' (black shading) and 'rhyolites' (white shading). The histogram distribution does not reflect the relative erupted volumes of basalt and rhyolite magma due to the bias of sampling towards the coastal margins where rhyolites are more common. Terminology for the silicic rocks is complicated because their compositions straddle the boundaries of several fields on many classification diagrams. As a simplification, *Erlank et al.* [1984] used the term 'quartz latites' for all the Etendeka silicic rocks, whereas *Bellieni et al.* [1986] preferred the term 'rhyolite' to encompass all the compositions of the Paraná silicic rocks.

of Rio de Janeiro: SA on Figure 2). The São Paulo plateau, which extends for about 400 km southeast from the Santos basin, appears to be underlain by attentuated continental crust, and dredged samples also show affinities with the Paraná lavas [*Fodor and Vetter*, 1984]. Significant quantities of Early Cretaceous volcanic material have also been found by seismic studies throughout the Namibian continental shelf but no samples have yet been analysed [*Light et al.*, 1992].

Paraná-Etendeka magmatism is strongly bimodal (Figure 4), and the virtual absence of samples with 60–64 wt% SiO$_2$, except locally in the Etendeka, produces a natural division of the lavas into what are loosely termed 'basalts' and 'rhyolites'. The lava pile is dominated by tholeiitic basalts (>90%), but significant quantities of rhyolites are found along the Brazilian continental margin and in the Etendeka.

Regional Geology

The basement beneath the Paraná basin consists of several Archaean to Early Proterozoic cratonic blocks surrounded by Mid-Late Proterozoic mobile belts (Figure 5 [e.g., *Mantovani et al.*, 1991]). Archaean rocks are found

in the Luis Alves craton beneath the southern Paraná basin and in the San Francisco craton to the northeast of the basin. Early Proterozoic rocks are found in the Curitiba massif on the northern margin of the Luis Alves craton, and in the Transamazonian massif around the west and northwest margins of the Paraná basin. The Mid-Proterozoic Ribeira belt underlies the central and northeastern parts of the basin. The last major pre-Paraná event across the region was the Brasiliano or Pan-African orogeny at 750-500 Ma. The Brasiliano mobile belts are found beneath the eastern and northwestern Paraná basin. In Namibia, the basement geology beneath the Etendeka

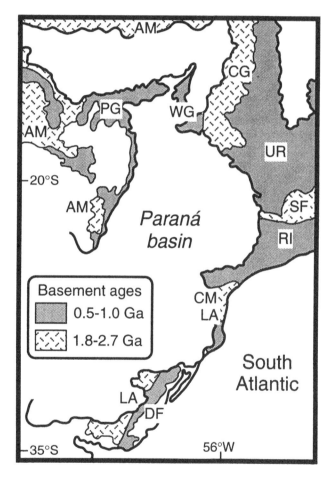

Figure 5. Simplified map of the basement geology of the Paraná region of South America, showing the Archaean cratons and Early Proterozoic massifs (speckled shading: LA = Luis Alves craton, SF = São Francisco craton, CM = Curitiba massif, AM = Transamazonian massif, CG = Central Goias massif) and the Mid-Late Proterozoic mobile belts (grey shading: RI = Ribeira belt, WG = Western Goias belt, PG = Paraguai belt, DF = Dom Feliciano belt) [adapted from *Mantovani et al.*, 1991]. The Palaeozoic Paraná sedimentary basin coincides roughly with the present outline of the Paraná flood basalts, and obscures the underlying basement geology.

lavas is formed mainly by rocks of the Pan-African Damara Sequence, which partially overlie Proterozoic rocks (2.1–1.7 Ga) on the southwest corner of the Congo craton (Figure 6b [*Milner et al.*, 1995a]).

The Paraná basin was established in the Late Ordovician as an intracratonic sedimentary basin. It is aligned roughly northeast-southwest, paralleling the Brasiliano structural trends in the underlying basement, and at its deepest point reaches over 5 km below sea level (Figure 3) [*Zalan et al.*, 1987; *Peate et al.*, 1992]. The earliest, Palaeozoic, sediments are largely marine siliciclastics, but the Mesozoic sequences are exclusively continental (lacustrine and fluvial) sediments that culminate in the aeolian sandstones of the Jurassic Botucatu Formation. The Early Cretaceous lavas were erupted subaerially directly onto the Botucatu sandstones over virtually all of the Paraná-Etendeka province, and they only overstep onto older strata (Paraná basin and pre-Ordovician basement) along the northeast margin of the Paraná lava field and in parts of the Etendeka region. The aeolian sandstones persist as intercalations, up to 160 m thick, within the Paraná and the Etendeka lava sequences, and the more northerly Paraná lavas are capped by similar aeolian sandstones of the Caiuá Formation [*Rocha-Campos et al.*, 1988]. Thus, climatic conditions were arid and desert-like throughout the period of lava eruption. This is consistent with the lack of palaeosol development within the Paraná lavas. It might also explain why development of entablature-style joint patterns are absent, because such features in the Columbia River basalts of the USA have been attributed to the effects of percolating water during cooling of the flows [*Long and Wood*, 1986].

Age of Paraná-Etendeka Magmatism

Stratigraphic and sparse fossil evidence can only constrain the age of the Paraná-Etendeka lavas to between Upper Triassic and Upper Cretaceous [*Rocha-Campos et al.*, 1988]. Over 200 K-Ar ages have been determined on Paraná-Etendeka samples (reviewed by *Erlank et al.* [1984] and *Rocha-Campos et al.* [1988]). K-Ar ages range from nearly 400 Ma to less than 100 Ma, clearly indicating problems with both excess radiogenic argon and post-crystallisation argon loss. Most samples (> 70%) yield ages in the range 115–135 Ma, with a strongly defined mode at 127 Ma. There is no significant age difference between the extrusive and intrusive phases of magmatism but the technique is not sufficiently precise to assess any systematic variation in the age of the magmatism across the province or to estimate the duration of magmatism. There have been few detailed Rb-Sr studies of the Paraná-Etendeka lavas,

partly due to the lack of sufficiently porphyritic samples with suitable mineral phases. *Mantovani et al.* [1985b] reported a combined Rb-Sr mineral isochron age of 135.5±3.2 Ma from three Chapecó rhyolite samples. In the Etendeka, an Awahab rhyolite unit gave a Rb-Sr mineral isochron age of 129.1±3.6 Ma [*Milner et al.*, 1995a].

Determining the age of the Paraná-Etendeka magmatism is critical for evaluating the temporal relationship between the volcanism and the opening of the South Atlantic Ocean. This, in turn, would help to better constrain recent geodynamic models that attempt to link continental rifting, the mantle plume, and flood basalt production. Improved estimates for the duration of magmatism would allow eruption rates to be inferred, which would have implications for the thermal mechanism behind the generation of the flood basalt magmas. Recent dating efforts have concentrated on the potential of the ^{40}Ar-^{39}Ar technique to obtain more precise age estimates [*Baksi et al.*, 1991; *Hawkesworth et al.*, 1992; *Renne et al.*, 1992, 1996a, 1996b; *Turner et al.*, 1994; *Stewart et al.*, 1996].

Renne et al. [1992], in a study of the lavas of southern Brazil, concluded that the Paraná magmatism began at 133±1 Ma and lasted less than a million years, consistent with the data of *Hawkesworth et al.* [1992] for this region. From this result, *Renne et al.* [1992] inferred a mean eruption rate of ~1.5 km^3 yr^{-1} similar in magnitude to that estimated for the Deccan province of India. By contrast, in a study covering the full areal extent of the province, *Turner et al.* [1994] suggested that the Paraná lavas were erupted over a longer interval (~10 m.y.) between 137 Ma and 127 Ma, with a mean eruption rate of ~0.1 km^3 yr^{-1}, an order of magnitude less than the estimate of *Renne et al.* [1992]. Additional work by the Open University group [*Stewart et al.*, 1996] confirmed their earlier conclusion of a significant duration of magmatism. As this result conflicts with the prevailing prejudice that most flood basalt provinces were erupted in less than a few million years, it has sparked debate, particularly about the validity of the different analytical techniques used. *Turner et al.* [1994] compared their preferred technique of laser spot heating with laser stepped heating for two whole-rock samples. One sample gave indistinguishable ages of 136 Ma from both techniques, whereas the other sample gave a poor correlation of spot analyses but produced a seemingly good, although probably meaningless, plateau age. *Renne et al.* [1996a] separated plagioclase grains from samples from a single region and analysed them in two different labs using different methods (Berkeley: laser stepped heating: Nice: furnace stepped heating) and they gave concordant age ranges. *Stewart et al.* [1996] used a sample from a Chapecó rhyolite unit to evaluate different ^{40}Ar-

[39]Ar techniques. Laser spot heating of the groundmass gave an isochron age of 131.8±1.4 Ma, which is indistinguishable from the plateau age obtained from furnace step heating of a plagioclase separate analysed in Canberra (131.6±0.2 Ma), whereas laser step heating of the plagioclase separate gave a slightly older age (133.5±1.2 Ma isochron age).

It should be stressed that the conclusions of *Renne et al.* [1992] were based on samples from just the coastal Serra Geral escarpment, which represents a small fraction of the overall province. Recent analyses by *Renne et al.* [1996a] on the Ponta Grossa dolerites (Figure 2) that are inferred to be the feeders of the northern Paraná lavas gave distinctly younger ages (129.2±0.4 to 131.4±0.4 Ma) than these workers had measured for the southern lavas, and a few had ages as young as 120 Ma. Thus, *Renne et al.* [1996a] revised their estimate of the duration of the main pulse of magmatism to about 3 m.y.. The older ages (135 to 138 Ma: n=9) of *Turner et al.* [1994] and *Stewart et al.* [1996] are found only in flows and dykes in the northern and western margins of the Paraná lava field and at the base of some of the central Paraná boreholes, parts of the province not yet sampled by the Berkeley group. Thus, based on the available evidence, there are no valid reasons to discount these older ages, and it should be emphasised that where both groups have analysed samples from the same area, the ages are in good agreement. In the correlated sequences of southeast Brazil and the Etendeka, the Berkeley group have measured ages of 131.9±0.5 Ma to 132.9±0.6 Ma (n=9) [*Renne et al.*, 1992, 1996b] and the Open University group found similar ages of 131.2±1.1 Ma to 132.9±2.8 Ma (n=4) [*Turner et al.*, 1994] with the exception of the highest flow on the escarpment that is younger (129.4±1.3 Ma). A similar situation is found with the Ponta Grossa dolerites, although *Turner at al.* [1994] did find two dykes with slightly older ages (134 Ma). In summary, the majority of analysed samples give [40]Ar-[39]Ar ages between 129 Ma and 134 Ma, but there is also good evidence for earlier magmatism inland (135–138 Ma [*Turner et al.*, 1994; *Stewart et al.*, 1996]) and for younger magmatism persisting along the coast (120–128 Ma [*Turner et al.*, 1992; *Renne et al.*, 1996a; *Stewart et al.*, 1996]).

Contemporaneous Alkaline Magmatism

Several alkalic complexes, broadly contemporaneous with the flood basalt volcanism, were emplaced around the margin of the Paraná basin (Figure 6) and cover a wide compositional spectrum including carbonatites, alkali gabbros, phonolites, syenites, and granites [*Ulbrich and Gomes*, 1981; *Milner et al.*, 1995a]. In South America, the

largest group of complexes is in the Ponta Grossa region and comprises at least six discrete centres, including the extensively studied Jacupiranga carbonatite [e.g., *Huang et al.*, 1995] which has a [40]Ar-[39]Ar age of 132 Ma [*Renne et al.*, 1993]. Other complexes of similar age occur in southern Brazil (Anitapolis, 131 Ma [*Renne et al.*, 1993]), eastern Paraguay (127 Ma [*Renne et al.*, 1993]), and in Uruguay (Mariscala, 133 Ma [*Stewart et al.*, 1996]). In Africa, this Early Cretaceous alkaline magmatism was concentrated in the Damaraland province of northern

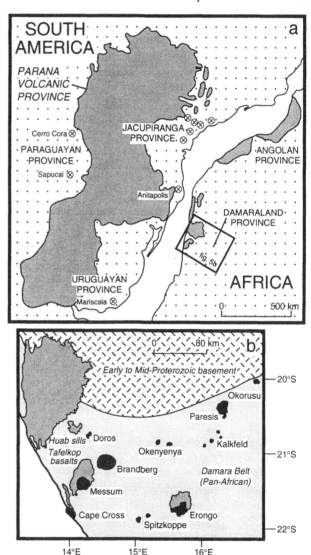

Figure 6. (a) Location of alkalic magmatism (circles with crosses) contemporaneous with the Paraná-Etendeka flood volcanism (dark grey shading); (b) detailed map of the Damaraland complexes (black shading) in relation to the Etendeka lava field (dark grey shading) and basement rocks (Early to Mid-Proterozoic; speckled shading: Pan-African; light grey shading) [from *Milner et al.*, 1995a].

Namibia and in Angola [*Marsh*, 1973]. The Damaraland complexes form a northeast-trending linear feature along the southern margin of the Etendeka lava field, extending from the coast to 350 km inland (Figure 6b). They represent high-level volcanic intrusions and many are inferred to be caldera-collapse structures. Intrusive relationships have previously been used to suggest that the alkalic magmatism largely post-dated the main flood basalt eruptions, but *Milner et al.* [1995a] concluded that magmatic activity in the alkalic complexes was contemporaneous with the onset of flood basalt volcanism, a relationship also seen in the Deccan and Siberian flood basalt provinces. *Milner et al.* [1992] have proposed from detailed mapping and geochemical correlations that the oldest Etendeka rhyolite units were erupted from the Messum complex. New dating (^{40}Ar-^{39}Ar, Rb/Sr) indicates that the Damaraland complexes range in age from 137 Ma to 124 Ma. Many were active over a long interval (Messum 132 to 127 Ma, Okenyenya 129 to 123 Ma), and Cape Cross and Paresis have ^{40}Ar-^{39}Ar ages of 137 to 135 Ma that are earlier than the main phase of Etendeka volcanism [*Milner et al.*, 1995a; *Renne et al.*, 1995b].

BASALT MAGMA TYPES

In terms of petrography, most of the Paraná lava pile can be viewed as a homogeneous sequence of virtually aphyric tholeiitic basalts [*Comin-Chiaramonti et al.*, 1988], but significant compositional variations exist. Initial classifications divided the basalts into a low-Ti group largely restricted to the south of the province, and a high-Ti group dominant in the north [*Bellieni et al.*, 1984a; *Mantovani et al.*, 1985a]. As more data became available, the original choice of 2 wt% TiO_2 to divide high- from low-Ti flows appeared arbitrary and was not governed by any natural division in the distribution of TiO_2 contents in the basalts (Figure 7a). Furthermore, based on other compositional criteria, high-Ti flows found in the south are distinct from the main group in the north [*Bellieni et al.*, 1984a].

Peate et al. [1992] attempted to clarify the status of different compositional groups within the Paraná-Etendeka lavas, with the aim of using these groups both to look at the internal stratigraphy of the province as a whole and as a means to simplify petrogenetic modelling. Six magma types were distinguished on the basis of major and trace element abundances and ratios. Analyses of representative samples of each magma type are listed in Table 1. Compositional criteria that allow flows to be assigned to a particular magma type were deliberately selected from elements routinely analysed by X-ray fluorescence (XRF) so as to be as widely applicable as possible (Table 2). More

reliance was placed on elements such as the high-field-strength (HFS) elements that are generally immobile during alteration processes. *Peate et al.* [1992] found that ~90% of available basaltic lava analyses (> 2000 with loss on ignition (LOI) <2.5 wt%) could be classified into one of the magma types. Some problems inevitably arose because the analyses were produced in several different laboratories.

While it is useful to maintain the distinction between low-Ti and high-Ti magma types when discussing petrogenesis, it is unwise to rely on the abundance of a single element for classification purposes. Thus, the magma types defined by *Peate et al.* [1992] are grouped into low-Ti (Gramado, Esmeralda) and high-Ti (Urubici, Pitanga, Paranapanema, Ribeira) varieties on the basis of a wide range of similar compositional characteristics. The Gramado and Esmeralda magma types are low-Ti magmas in the sense used by *Hergt et al.* [1991] to distinguish a compositionally distinctive group of magmas found throughout the Mesozoic flood basalt provinces of Gondwana that have low Ti/Y values (<310) similar to or less than MORB. Even though the Paranapanema and Ribeira magma types overlap in Ti contents with the low-Ti Gramado and Esmeralda magma types, they will be referred to as high-Ti types because of their high Ti/Y values (>310) and other compositional similarities with the other high-Ti magma types (Figure 7b).

The reliability of compositional data on the borehole samples had to be demonstrated by *Peate et al.* [1992] as only rock chippings were available. Petrographic study indicated that, in a given hole, the chips came from a range of levels within at least one flow and probably sampled several flows. Although samples were hand-picked in an effort to select chips from a single lithological unit and to avoid altered fragments, analyses showed that most are affected by secondary processes and have, for example, markedly higher Na contents than surface lavas. However, there is good agreement between the borehole data and analyses of surface lava samples for relatively immobile elements such as Ti, Zr, and Y, and thus, using these elements, the borehole samples can be reliably classified into different magma types and can be used to establish the regional distribution of magma types [*Peate et al.*, 1992].

Composition and Distribution of Low-Ti Magma Types

The Gramado magma type has a distinctive trace element signature relative to the high-Ti types, with a greater relative enrichment of large-ion-lithophile (LIL) over HFS elements and light rare-earth elements (LREE), and a prominent negative Ti-anomaly on a primitive-mantle-normalised diagram (Figure 8). The Esmeralda pattern is similar to the Gramado, except with generally lower incompatible trace element abundances, and it also has a lesser degree of LREE enrichment. Esmeralda

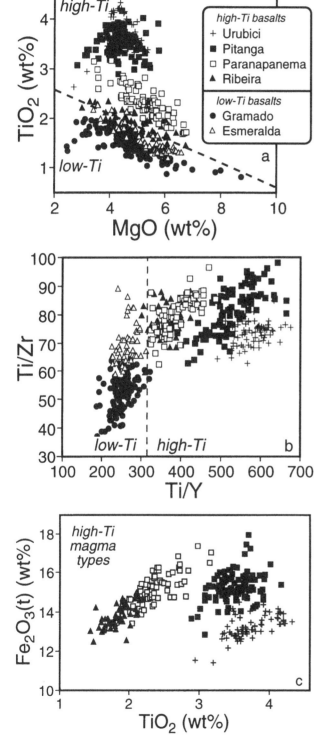

magmas have higher Ti/Zr (>60) than Gramado magmas (Figure 7b). Both of these low-Ti magma types show significant but variable depletion of Nb and Ta relative to La (Nb/La 0.5–0.8). Gramado samples have high $^{87}Sr/^{86}Sr_i$ (0.7075–0.7167) and low ε_{Ndi} (-8 to -3) relative to bulk Earth values (Figure 9), whereas Esmeralda samples form an almost linear array displaced from the Gramado field to lower $^{87}Sr/^{86}Sr_i$ (0.7046–0.7086) and higher ε_{Ndi} (-4 to +3). They both have more radiogenic Pb isotope compositions ($^{206}Pb/^{204}Pb$>18.2) than the high-Ti magma types, but Esmeralda samples have a restricted isotopic range ($^{206}Pb/^{204}Pb$=18.6–18.8) compared to Gramado lavas ($^{206}Pb/^{204}Pb$=18.4–19.1: Figure 10).

In southern Brazil, Gramado flows are mainly restricted to the coastal Serra Geral escarpment, whereas Esmeralda flows often locally cap the lava pile and are more common on the central plateau. The general pattern of surface distribution of these two magma types is consistent with the stratigraphical relationship inferred from the central Paraná boreholes and Serra Geral road sections, with the Esmeralda magmas forming a younger unit above the Gramado magmas [*Peate et al.*, 1992]. The southernmost Paraná lavas, in Argentina and Uruguay, appear to comprise only Gramado flows, although this may simply reflect the paucity of samples. In the Etendeka, the Tafelberg basalts that form most of the main lava field are compositionally equivalent to the Gramado magma type. Gramado flows are estimated to comprise up to a third of the preserved Paraná-Etendeka lava pile, with Esmeralda flows making up perhaps 5 to 10%.

Composition and Distribution of High-Ti Magma Types

The 'Northern' Magma Types. Although the Ribeira, Paranapanema, and Pitanga magma types span a wide range in TiO$_2$ contents (1.5–4.1 wt%), they are discussed together because they share many compositional features. They have similar primitive-mantle-normalised trace element patterns (Figure 8), differing only in the degree of enrichment (e.g., Ti/Y of Ribeira ~360, Paranapanema ~410, Pitanga ~530). Negative Nb-Ta anomalies relative to

Figure 7. (a) MgO vs. TiO$_2$: this shows the evolved nature and compositional variety of the Paraná basalts. The dashed line marks the approximate division between low-Ti and high-Ti magma types. (b) Ti/Y vs. Ti/Zr: low-Ti magma types (Gramado, Esmeralda) are distinguished from high-Ti magma types (Urubici, Pitanga, Paranapanema, Ribeira) by low Ti/Y (<310). Esmeralda magmas have higher Ti/Zr (>60) than Gramado magmas. (c) TiO$_2$ vs. total Fe (Fe$_2$O$_3$(t)), showing the different high-Ti basalt magma types. Data sources: *Petrini et al.* [1987], *Hawkesworth et al.* [1988], *Mantovani and Hawkesworth* [1990], *Peate* [1990], *Peate and Hawkesworth* [1996].

TABLE 1. Selected Compositional Characteristics of Paraná Basalt Bagma Types and Previous Nomenclature

Magma type	Characteristics	TiO_2	Ti/Y	Ti/Zr	$^{87}Sr/^{86}Sr_i$	Previous nomenclature
Gramado	low Ti/Y & Ti (south)	0.7-1.9	< 310	< 70	0.7075-0.7167	LTi(S), LPT, II, *Tafelberg, Albin*
Esmeralda	low Ti/Y & Ti (south)	1.1-2.3	< 310	> 60	0.7046-0.7086	LTi(S), LPT, I
Ribeira	high Ti/Y, low Ti (north)	1.5-2.3	> 310	> 65	0.7055-0.7060	LTi(N), LPT
Paranapanema	high Ti/Y & Ti (north)	1.7-3.2	> 330	> 65	0.7055-0.7063	HTi(N), IPT, III
Pitanga	high Ti/Y & Ti (north)	> 2.9	> 350	> 60	0.7055-0.7060	HTi(N), HPT
Urubici	high Ti/Y & Ti (south)	> 3.3	> 500	> 57	0.7048-0.7065	HTi(S), HPT, *Khumib*

See *Peate et al.* [1992] for additional details. Equivalent types in the Etendeka are in italics [*Duncan et al.*, 1988].

the LREE and LILE are developed to the same extent in all three magma types (Nb/La ~0.64). They also have a restricted range in Sr-, Nd- and Pb-isotope composition ($^{87}Sr/^{86}Sr_i$ = 0.7055–0.7063, εNd_i = -1.6 to -3.6, $^{206}Pb/^{204}Pb$ = 17.81–18.12) relative to the other Paraná magma types, and have been grouped together as the 'Northern' basalts on Figures 9 and 10.

These three magma types also have a close spatial association. They are found throughout the northern lava field, and down the western flank as far south as northern Argentina [*Peate et al.*, 1992]. Together, they comprise approximately half of the total preserved lava volume, with the Pitanga and Paranapanema magmas probably having roughly equal volumes each of ~20% of the total, and the Ribeira flows making up ~5%. Paranapanema lavas cover a wide area of the northern Paraná region, centred on the Paraná River, with Pitanga lavas concentrated along the northeast and eastern margins. This outcrop pattern is consistent with the borehole data, where Paranapanema samples overlie Pitanga samples. Ribeira samples have a similar surface distribution [*Petrini et al.*, 1987] to the Paranapanema samples in the northern Paraná. However, Ribeira samples were found only in one of the central Paraná boreholes, where they were below Pitanga and Paranapanema flows (Figure 11) [*Peate et al.*, 1992].

The Urubici Magma Type. Like the Pitanga magma type, this has a high TiO_2 content (>3 wt%). It shares many incompatible trace element features with the Pitanga magma type but at higher abundances (Figure 8). It is readily distinguished by its high Sr (>550 ppm) and low $Fe_2O_3(t)$ (<14.5 wt%: Figure 7c), and its greater heavy rare earth element (HREE) fractionation (Tb/Yb Urubici ~0.58, Pitanga ~0.38; Figure 12). Urubici samples partially overlap the 'Northern' basalts in Sr-Nd-Pb isotope

composition but have a wider range that extends to less radiogenic Sr and Pb values ($^{87}Sr/^{86}Sr_i$=0.7048–0.7065, $^{206}Pb/^{204}Pb$=17.46–18.25: Figures 9 and 10). Data on coexisting Ti-magnetite and ilmenite indicate lower oxygen fugacity conditions for Urubici magmas (between the quartz-fayalite-magnetite and magnetite-wustite buffers) than for Paranapanema and Pitanga magmas (between the nickel-nickel oxide and quartz-fayalite-magnetite buffers) [*Bellieni et al.*, 1984a].

The surface extent of Urubici flows is restricted to a small strip (~100 km × 350 km) along the northeast flank of the lava field in southern Brazil. Throughout this area they are interbedded with Gramado flows [*Peate et al.*, 1992]. Urubici samples are also found near the base of some of the central Paraná boreholes [*Peate et al.*, 1992]. *Duncan et al.* [1988] found high-Ti flows (Khumib magma type) in the formerly adjacent Etendeka lavas of Namibia, north of 20°S (Figure 6b) that are compositionally equivalent to the Urubici magma type [*Peate et al.*, 1992]. If Africa and South America are juxtaposed to their pre-Atlantic Ocean positions, then the Urubici and Khumib flows are directly adjacent. The coincidence of the southern limit of their outcrop indicates that the Gondwana plate reconstruction of *de Wit et al.* [1988] provides a reasonable fit to within 100 to 200 km for this region. The Urubici (and Khumib) magma type is volumetrically much less significant than the other high-Ti magma types, comprising less than 5% of the total preserved lava volume.

INTRUSIVE MAGMATISM

Ponta Grossa Dyke Swarm

A prominent swarm of northwest-southeast-striking dykes intrudes the Precambrian basement and Paraná basin

TABLE 2. Representative Analyses of Paraná-Etendeka Basalt Magma Types and Rhyolite Subgroups

sample name group	B448 ESM	DSM 06 ESM	DUP 30 GRA	DUP 38 GRA	FEG 92-23 AG	MM 90-90 CDS	FEG 92-92 SM	B980 RIB	PAR 06 PMA	CB 1110 PIT	DUP 35 URU	MM 90-88 GU	PRG 86 GI	MM 90-13 OU
		Low-Ti Basalts				Palmas Rhyolites			High-Ti Basalts				Chapecó Rhyolites	
SiO_2	49.10	51.10	50.89	56.14	66.95	66.88	70.67	50.07	50.13	51.27	53.01	64.27	65.57	65.70
TiO_2	1.52	1.37	0.95	1.68	1.08	0.93	0.69	1.57	2.19	3.35	3.76	1.48	1.24	1.22
Al_2O_3	15.19	13.77	14.92	13.19	12.90	12.79	12.54	15.51	13.56	12.70	12.86	12.92	14.56	12.92
Fe_2O_3	13.11	13.45	10.34	13.54	6.61	6.14	4.92	13.16	14.64	14.96	12.68	7.58	7.58	6.52
MnO	0.20	0.21	0.17	0.19	0.11	0.11	0.06	0.20	0.21	0.23	0.19	0.15	0.04	0.13
MgO	6.39	6.13	7.99	3.26	1.51	1.77	0.59	5.60	5.53	4.34	4.34	1.40	0.85	0.13
Na_2O	11.52	10.73	11.61	6.89	3.47	2.89	1.35	10.62	9.94	8.56	8.30	2.94	1.46	3.26
CaO	2.63	2.55	2.44	2.81	3.33	2.80	2.67	2.50	2.54	3.02	2.57	3.41	2.92	3.26
K_2O	0.17	0.54	0.51	2.57	3.47	3.96	5.15	0.58	0.99	1.05	1.70	4.15	5.34	4.35
P_2O_5	0.16	0.16	0.16	0.24	0.32	0.27	0.20	0.20	0.26	0.54	0.58	0.46	0.45	0.32
Sc	37	42	40	32	18	18	16	44	38	33	29	12	-	12
V	-	323	221	317	93	90	28	-	397	405	343	66	-	46
Cr	267	94	307	54	12	19	7	114	126	46	75	12	1	44
Co	46	50	44	41	19	18	13	46	52	34	36	18	-	16
Ni	94	58	99	16	8	8	6	93	60	20	58	7	7	5
Cu	-	169	99	61	90	57	21	-	173	140	267	11	-	18
Zn	-	89	72	105	76	74	82	-	92	132	107	129	-	106
Ga	-	21	17	22	18	18	17	-	21	26	26	23	-	24
Rb	9	19	10	86	148	174	226	8	24	30	30	103	117	131
Sr	154	163	216	211	145	131	85	273	380	459	764	401	275	321
Y	28	29	23	44	48	40	73	25	31	40	39	71	69	66
Zr	78	100	92	220	279	258	342	110	154	252	307	638	555	595
Nb	-	6	9	16	22	21	26	-	13	23	27	51	49	48
Ba	148	163	243	461	548	681	721	272	344	465	600	1002	927	1046
La	4.6	8.35	10.6	32.2	40.1	40.1	54.5	14.0	21.5	33.4	42.5	75.3	86	75.2
Ce	13.0	20.8	22.9	66.3	86.9	84.7	111	30.8	44.0	76.3	90.4	163	155	159
Nd	-	14.5	12.8	35.9	40.8	40.0	52.2	19.3	25.3	48.0	54.3	83.9	82	77.7
Sm	3.1	3.97	3.18	7.94	8.53	8.03	11.1	4.1	5.25	11.6	11.6	16.6	-	15.0
Eu	1.0	1.41	1.11	1.92	1.86	1.60	1.91	1.48	1.79	2.94	3.53	4.33	-	3.62
Tb	0.8	0.90	0.63	1.29	1.39	1.20	1.78	0.69	0.82	1.34	1.54	2.29	-	1.89
Yb	2.3	2.92	2.09	4.03	4.30	3.79	6.19	2.2	2.65	3.53	2.99	5.84	-	5.23
Lu	0.3	0.47	0.35	0.65	0.66	0.54	0.97	0.45	0.40	0.59	0.44	0.86	-	0.71
Hf	2.0	2.71	2.15	5.84	7.11	6.49	8.61	2.9	3.92	7.13	7.97	15.0	-	13.9
Ta	0.2	0.37	0.50	1.12	1.76	1.65	2.23	0.61	0.87	1.58	1.92	3.27	-	3.09
Th	0.74	1.75	2.04	9.00	13.0	13.7	18.2	1.74	2.60	3.67	4.25	8.98	-	12.6
U	0.21	0.71	-	2.23	4.58	5.10	5.81	0.24	0.87	1.00	1.34	2.15	-	2.84
εSr_i	0.9	20.1	42.2	98.9	146[a]	209	292	18.3	19.7	13.3	7.1	18.7	23.4	47.3
εNd_i	2.2	0.7	-3.8	-6.3	-6.9[a]	-7.2	-7.3	-3.2	-2.8	-2.2	-3.4	-5.1	-5.4	-5.1

[a] isotope data from sample FEG92-39.

Data from: *Petrini et al.* [1987]; *Peate* [1990]; *Alberti et al.* [1992]; *Garland et al.* [1995]; *Peate and Hawkesworth* [1996]. All samples have L.O.I. (loss on ignition) <1.6wt%, except PRG86 (3.1wt%). Basalt magma types: ESM=Esmeralda; GRA=Gramado; RIB=Ribeira; PMA=Paranapanema; PIT=Pitanga; URU=Urubici. Rhyolite sub-groups: AG=Anita Garibaldi; CDS=Caxias do Sul; SM=Santa Maria; GU=Guarapuava; GI=Giraul; OU=Ourinhos. εSr_i and εNd_i values calculated at 130 Ma relative to Bulk Earth (present-day values: $^{87}Rb/^{86}Sr=0.0847$; $^{87}Sr/^{86}Sr=0.7047$; $^{147}Sm/^{144}Nd=0.1967$; $^{143}Nd/^{144}Nd=0.51264$). Isotope ratios normalised to NBS 987 ($^{87}Sr/^{86}Sr=0.71025$) and J&M Nd ($^{143}Nd/^{144}Nd=0.51185$).

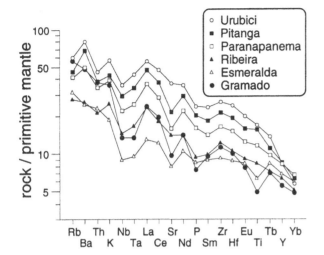

Figure 8. Primitive-mantle-normalised diagram for average least evolved (>4.5 wt% MgO) samples of each Paraná basalt magma type. Normalising values from *Sun and McDonough* [1989]. Data sources: *Peate* [1990], *Peate and Hawkesworth* [1996].

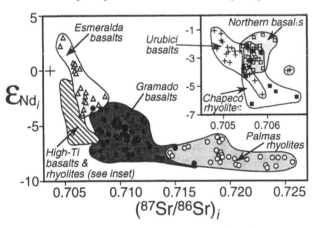

Figure 9. Initial Sr- and Nd-isotopic composition of Paraná magmas (at 130 Ma). Data for the low-Ti magmas are plotted on the main diagram, and the inset is an expanded view to highlight features of the high-Ti magmas that have a more limited isotopic variation. 'Northern Basalts' group combines the Pitanga, Paranapanema and Ribeira magma types. Cross marks bulk Earth estimate. Data sources: *Cordani et al.* [1988], *Mantovani and Hawkesworth* [1990], *Peate* [1990], *Peate and Hawkesworth* [1996].

sediments in the Ponta Grossa region, and rare dykes also crosscut the lava pile (Figures 2 and 13) [*Piccirillo et al.*, 1990]. Most of these dolerites have similar major and trace element compositions to the Paranapanema lavas, with a few similar to the Pitanga lavas [*Regelous*, 1993]. A comparison of mean palaeomagnetic poles [*Ernesto and Pacca*, 1988] suggested that the dolerites were emplaced after the eruption of the main preserved portion of the Paraná lavas, and so *Piccirillo et al.* [1990] proposed that the dykes were

the feeders to lavas erupted towards the continental margin and subsequently eroded. However, ^{40}Ar-^{39}Ar ages [*Turner et al.*, 1994; *Renne et al.*, 1996a] indicate that many of the dolerites are of similar age to the surface lavas, and thus were probably feeders to the northern Paraná lava pile, although there are a few coast-parallel dykes that give younger ages (120–125 Ma). *Raposo and Ernesto* [1995] showed from a study of magnetic susceptibility anisotropy that many dykes preserve evidence for lateral flow. Fabrics indicating a more vertical flow direction are more common in the southeast part of the swarm, perhaps indicating a source close to the proto-Atlantic rift. East-west-trending doleritic dykes near Morro Vermelho (MV on Figure 2) in southern Angola may represent a continuation of the Ponta Grossa dyke swarm onto the African Plate prior to continental separation [*Piccirillo et al.*, 1990].

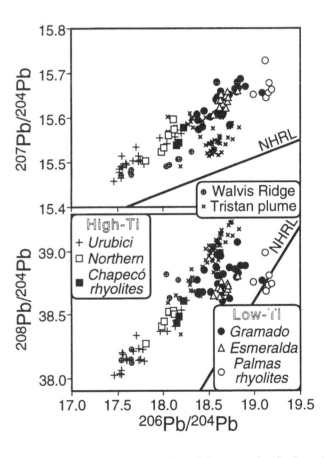

Figure 10. Variation in present-day Pb isotope ratios for Paraná basalts and rhyolites. Data for samples of recent Tristan plume activity (Tristan da Cunha, Gough, Inaccessible) and from the Walvis Ridge plume trace are plotted for reference [*Sun*, 1980; *Richardson et al.*, 1982; *le Roex et al.*, 1990; *Cliff et al.*, 1991]. Symbols and data sources for Paraná lavas as for Figure 9.

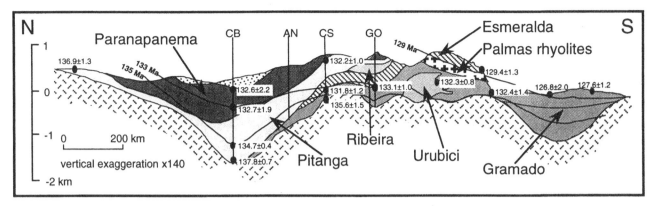

Figure 11. Schematic N-S cross section through the Paraná lavas [*Peate et al.*, 1992; *Stewart et al.*, 1996]. The timelines for 135 Ma, 133 Ma and 129 Ma (constructed from ^{40}Ar-^{39}Ar ages of *Stewart et al.* [1996]) cut across the stratigraphic units defined by the compositional magma types of *Peate et al.* [1992]. This would indicate that the magma types are diachronous.

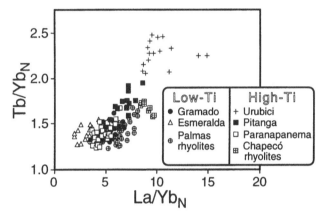

Figure 12. Tb/Yb$_N$ vs La/Yb$_N$ for Paraná basalts and rhyolites (N denotes chondrite-normalised). For the low-Ti magmas, Tb/Yb$_N$ remains fairly constant over a range in La/Yb$_N$ from 1 to 8. For the high-Ti basalt magma types, the Paranapanema magmas overlap with the low-Ti Gramado magmas, whereas the Pitanga and Urubici magmas are displaced to successively higher Tb/Yb$_N$ and La/Yb$_N$. Data sources for Paraná lavas as for Figure 9.

São Paulo - Rio de Janeiro Coastal Dyke Swarm

Numerous northeast-southwest-trending dykes, all with high-Ti compositions (Ti/Y >310 [*Comin-Chiaramonti et al.*, 1983]), intrude the Precambrian basement rocks along the coast between São Paulo and Rio de Janeiro States (Figure 2). The dyke compositions define two distinct groups, termed the Paraiba and Ubatuba magma types by *Regelous* [1993]. ^{40}Ar-^{39}Ar ages indicate that the dykes were emplaced between 129 and 133 Ma, overlapping in age with the Ponta Grossa dykes and the northern Paraná lavas [*Turner et al.,* 1994]. The Paraiba dolerites have 50–53 wt% SiO$_2$ and 3.2–4.1 wt% TiO$_2$ whereas the Ubatuba dolerites have relatively evolved compositions with 54–58

wt% SiO$_2$ and 2.2–2.6 wt% TiO$_2$, and the two types cannot be related by any simple petrogenetic process [*Hawkesworth et al.*, 1992]. *Regelous* [1993] concluded that neither the Paraiba nor the Ubatuba dolerites had extrusive equivalents in the Paraná lava pile to the west. However, it is possible that they fed the formerly adjacent basaltic lavas that crop out in northwestern Angola south of Luanda (Figure 2). These Angolan lavas have K-Ar ages of 124-145 Ma [*Piccirillo et al.*, 1990] but no compositional data are available for comparison with the dolerites.

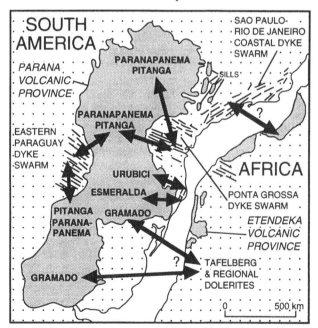

Figure 13. Location of principal dyke swarms associated with the Paraná-Etendeka magmatism. Arrows indicate possible links between dykes and surface lavas of similar age and composition [modified from *Regelous*, 1993].

Other Paraná Intrusive Magmatism

A regional airborne magnetic survey [*Druecker and Gay*, 1987] revealed a significant swarm of northwest-southeast-trending dykes that covers most of eastern Paraguay to the west of the Paraná lava field (Figure 2). Few of these dykes have been sampled but they seem to be similar in composition to the Pitanga and Paranapanema magma types. One dyke gave an old ^{40}Ar-^{39}Ar age (137 Ma), similar to other high-Ti lava flows along the western margin [*Stewart et al.*, 1996].

Intrusive magmatism is less common in southern Brazil. Occasional dykes and sills are found along the Serra Geral escarpment, and along the coastal margin where most strike approximately parallel to the coast. A few dykes have Gramado-type compositions, with Urubici-type dykes more common near the coast. On the escarpment near São Joaquim, numerous dykes and sill-like intrusions within the lava pile, including a 100-m-wide dyke intruding the Paraná basin sediments, all share a remarkably homogeneous, Esmeralda-like composition (εNd_i ~+0.5, $^{87}Sr/^{86}Sr_i$ ~0.7059, Ce/Sm_N ~1.3, where N denotes chondrite-normalised [*Peate and Hawkesworth*, 1996]).

Boreholes have revealed numerous sills, 2 to 200 m in thickness, that intrude the Paraná basin sediments, mainly within the Palaeozoic strata. Some sills crop out at the surface, north of São Paulo (Figure 2). The sills can reach a combined thickness of over 1000 m in the northern parts of the basin [*Bellieni et al.*, 1984b]. The sills are all tholeiitic and are compositionally similar to the Paraná lavas [*Bellieni et al.*, 1984b; *Regelous*, 1993]. *Peate et al.* [1990] showed that the distribution of compositional types among the sills mirrored the distribution of lava magma types at the surface, with Gramado-like sills in the south and Pitanga and Paranapanema sills in the north.

Etendeka Intrusive Magmatism

Doleritic dykes and sills are abundant within a 100-km-wide zone along the Namibian coast from the main Etendeka lava field to at least as far south as Walvis Bay (Figure 2), principally intruding the Damara basement and Karoo basin sediments [*Marsh et al.*, 1991]. The limited geochronological and palaeomagnetic data indicate that they are of similar age to the Etendeka lavas [*Erlank et al.*, 1984]. *Erlank et al.* [1984] identified three suites of dolerites: the relatively widespread Tafelberg and Regional dolerites, and the Horingbaai dolerites. The Tafelberg dolerites have identical petrographical and compositional attributes to the Tafelberg lavas, the Etendeka equivalent of the Gramado lavas in the Paraná. The Regional dolerites

are more mafic than the Tafelberg dolerites, but otherwise share many geochemical attributes with them. The MORB-like Horingbaai dolerites (εNd_i ~+8 [*Hawkesworth et al.*, 1984]) form thin dykes and sills that intrude basement rocks and the overlying Etendeka lava pile in a restricted portion of the coastal area near Cape Cross (Figure 6b). *Erlank et al.* [1984] presented ^{40}Ar-^{39}Ar ages of ~128 Ma for the Horingbaai dolerites.

Duncan et al. [1989] recognised four geochemically distinct types within the Huab sill complex, a suite of sills that intrude basement rocks and Karoo sediments in the Huab River valley (Figure 6b). One type is similar to the Tafelberg magmas, but the other three types are closest in composition to the Horingbaai dolerites. Field evidence suggests that they also postdate the main preserved lava pile. Early Cretaceous dykes have also been reported from the coastal region near the South African-Namibian border and farther south near Cape Town [*Reid*, 1990].

RELATIONSHIP OF MAGMATISM TO OPENING OF THE SOUTH ATLANTIC OCEAN

Asymmetry of Lava Distribution Relative to Proto-Atlantic Rift

Several ideas have been proposed to explain the pronounced asymmetrical distribution of the lavas relative to the eventual location of continental separation. *O'Connor and Duncan* [1990] argued that the lava distribution reflects the position of the Tristan plume at the time of magmatism, which they placed beneath the central Paraná at 130 Ma. *Thompson and Gibson* [1991] considered that interaction of the plume with the base of the lithosphere might play an important role in the location of flood basalt volcanism. They suggested that the Paraná Basin was a lithospheric 'thin-spot' where material, diverted laterally from the main plume axis to the east, could decompress to produce magmas. *White and McKenzie* [1989] linked the generation of flood basalts to rifting above a mantle plume and inferred that the lavas would be erupted from the rift zone. For the Paraná-Etendeka province, pre-existing topographic relief near the Etendeka (inferred from the overstepping of the lavas onto basement) was considered by *White and McKenzie* [1989] sufficient to prevent widespread flow of lavas eastwards onto the African Plate. Another explanation for the asymmetry is that rifting occurred via simple shear rather than pure shear [*Peate*, 1990]. *Harry and Sawyer* [1992] showed that the development of a horizontal pressure gradient in the lower crust during the early stages of extension could provide a mechanism for the lateral flow of magma over 100-200 km from the rift zone to beneath the Paraná province, thus accounting for the asymmetric lava distribution.

Compositional and age correlations between dykes and lavas give an indication of where the lava flows were erupted from. In most places the dykes are similar in composition to the local lava flows (Figure 13). Regional compositional similarities between sills and the overlying lavas also suggest that the lavas were erupted in the same general region where they now crop out. This emphasises that the different magma types do not represent the temporal evolution of a single mantle source region and that magma generation occurred over an area comparable to the areal extent of the lavas. The absence of any old dykes (along-strike equivalents of the eastern Paraguay dykes: 136–138 Ma) along the Brazilian coast suggests that at least some of the early magmas were generated and erupted at considerable distances west of the eventual Atlantic rift, and that the Paraguayan dykes had not transported magma laterally from this rift zone. The magma types were erupted in different parts of the province, perhaps at different times, as the result of a complex interplay between plume-lithosphere interaction and lithospheric extension.

Implications of Regional Stratigraphy

Stratigraphical studies can reveal the internal structure and sequential development of the lava pile. This knowledge is critical to the understanding of how magmatic sources and processes varied during the evolution of the province, and it also provides a means of determining any shifts in the principal locus of magmatism that might be linked to regional tectonic processes.

The distribution of different compositions within the lava pile discussed above indicates that flows of each basalt magma type tend to form a relatively coherent lithostratigraphical unit. By analogy with detailed stratigraphical sequences through other flood basalt provinces such as the Deccan and Columbia River, *Peate et al.* [1990, 1992] made the reasonable assumption that these units could also be considered as chronostratigraphical units. From this assumption, it appears that the internal structure of the Paraná lava pile comprises an overlapping sequence of units dipping towards the north, which suggests a northward-migrating source for the magmatism (Figure 11). *Peate et al.* [1990] suggested that this migration occurred in response to the northward propagation of rifting during the initial opening of the South Atlantic Ocean. The ^{40}Ar-^{39}Ar data of *Renne et al.* [1992, 1996a] are consistent with such a model, with the southern Brazil and the Etendeka lavas erupted at 132-

133 Ma and the northern lavas (as represented by the Ponta Grossa dolerites) erupted at 129-131 Ma. *Turner et al.* [1994], however, suggested that the earliest magmatism was inland, in the west and north, and then became concentrated near the northward-propagating South Atlantic rift.

It is important to realise that the Paraná magma types distinguished by *Peate et al.* [1992] are defined solely on compositional characteristics. They are not stratigraphically defined units as is the case in the subdivision of several other flood basalt provinces (e.g., Siberia, Deccan) where continuous sections through the lavas that can be correlated on a regional scale are more common. The ^{40}Ar-^{39}Ar data of *Turner et al.* [1994] and *Stewart et al.* [1996] suggest, in fact, that the magma types are diachronous, with different magma types being erupted simultaneously in different places and over a long period of time. This is illustrated in the cross section (Figure 11) where the time lines clearly crosscut the magma type boundaries. The oldest ^{40}Ar-^{39}Ar Paraná ages are from Paranapanema and Pitanga samples in the north and west of the lava field that, based on the borehole stratigraphy, should actually be younger than the low-Ti magmas of southeast Brazil. Between 133 and 132 Ma, Paranapanema flows were being erupted in the northern half of the province at the same time as Gramado and Urubici flows in the southern half. New data (S.P. Turner, unpubl. data) from a cored borehole in northwest Uruguay also cast doubt on the chronological significance of the geochemical stratigraphy as Paranapanema lavas overlie Gramado lavas in the centre of the province but underlie them in the south.

Duration of Magmatism

It is important for modelling the geodynamic processes involved in flood basalt generation to know the duration of magmatic activity and whether or not the eruption rate varied during this interval. Although the Paraná-Etendeka lavas appear to have been erupted over quite a significant time interval (~3 m.y. [*Renne et al.*, 1996a]; ~10 m.y. [*Stewart et al.*, 1996]), improved estimates for the duration of magmatism are clearly needed. The complete stratigraphic interval of magmatism must be sampled, and we also need to ascertain the volume of magma erupted as a function of time in order to assess variations in eruption rate. If the magma types are not chronostratigraphic then these objectives are difficult to achieve without a large regional coverage of precise ^{40}Ar-^{39}Ar ages. *Stewart et al.* [1996] suggest that eruption rates increased with time, with a less voluminous phase at 138–133 Ma followed by the main pulse at 133–129 Ma.

Palaeomagnetism potentially offers an independent

means of establishing the duration of the Paraná-Etendeka magmatism because the frequency of geomagnetic field reversals was high during the Early Cretaceous [e.g., *Harland et al.*, 1990]. Palaeomagnetic data are available from twenty stratigraphic sections in the Paraná [*Ernesto and Pacca*, 1988], and each section records up to four polarity reversals. The main drawback to establishing the total number of reversals spanned by the magmatism has been the apparent lack of suitable stratigraphic markers to correlate polarity intervals between sampled sections. *Milner et al.* [1995b] used a lithostratigraphic correlation based on compositionally distinct rhyolite units to demonstrate that magmatism in the southeastern Paraná spanned at least ten different polarity reversals. Between 135 Ma and 130 Ma, the average length of a polarity interval was 0.24 m.y., which suggests a duration of about 2.4 m.y. for eruption of the 1 km of lava in this region. This figure compares with estimates from the ^{40}Ar-^{39}Ar dating studies of <1 m.y. [*Renne et al.*, 1992] and ~3 m.y. [*Turner et al.*, 1994; *Stewart et al.*, 1996] for this part of the province.

Future studies combining magnetostratigraphy, lava composition, precise ^{40}Ar-^{39}Ar ages and cored borehole samples offer the best hope of enhancing our knowledge of the sequential development of the Paraná-Etendeka province. About a quarter of the total erupted volume lies in northern Argentina and Uruguay and yet little is known about the composition and chronology of these lava sequences as they are largely buried under younger sedimentary rocks. *Turner et al.* [1994] measured young ages (~128 Ma) on two samples, but it is uncertain whether these ages are representative of the whole lava pile in this region or whether they belong to a volumetrically minor later phase of activity as is found further north on the Serra Geral escarpment (128–129 Ma) and in coast-parallel dykes in the Ponta Grossa region (120–125 Ma [*Renne et al.*, 1996a]).

Rifting History of the South Atlantic Region

The age for the onset of sea-floor spreading within the southern South Atlantic Ocean apparently decreased northwards [*Austin and Uchupi*, 1982]. The oldest seafloor off the southern African coast near Cape Town has been assigned either to magnetic anomaly M13 or M9 (137 Ma or 130 Ma), whereas the earliest recognisable magnetic anomaly at the latitude of the Paraná-Etendeka province is M4 (~127 Ma: references of *Renne et al.* [1992]). The age of the earliest oceanic crust can only place a lower limit on the age of continental rifting in a particular region. However, studies of the tectonic and subsidence histories

of both onshore and offshore basins also suggest that the onset of rifting was earlier in the south [*Chang et al.*, 1992; *Light et al.*, 1992]. Rifting began in southern Argentina and the tip of South Africa at 200–220 Ma, and in northern Argentina (Colorado basin: Figure 1) at ~170 Ma [*Light et al.*, 1992]. The earliest extension in the offshore basins along the Brazilian margin was even younger, at ~140 Ma [*Chang et al.*, 1992].

The geometrical relationship of the Ponta Grossa dyke swarm with the Brazilian coastline and associated coast-parallel dyke swarms is reminiscent of the classic triple-junction geometry [*Burke and Dewey*, 1973] linked to domal uplift and rifting above a hotspot. For the plume-impact model of flood basalt generation [e.g., *Richards et al.*, 1989], the triple junction cannot mark the site of plume impact because the earliest magmatism was either to the northwest [*Turner et al.*, 1994] or to the south [*Renne et al.*, 1996a]. Furthermore, dykes are not oriented radially about the location of the inferred plume head, but have a dominantly northwest-southeast trend over a wide area of the province (Ponta Grossa, eastern Paraguay), a trend that can be linked to reactivation of Proterozoic basement structures. Dyke orientations are determined to a large extent by the regional stress field at the time of emplacement. The Salado and Colorado basins in northern Argentina (Figure 1), and the inferred Paraná-Chaco basin deformation zone all show a similar northwest-southeast orientation. These features are all consistent with models that require significant internal deformation and clockwise rotation within central South America during continental breakup as a result of the Central Atlantic region remaining closed as the South Atlantic Ocean opened [*Nürnberg and Müller*, 1991; *Turner et al.*, 1994]. Further work is required to decipher the exact relationship between magmatism and extension during the development of the Paraná-Etendeka province and the opening of the South Atlantic Ocean. The submerged lavas in the offshore basins are poorly studied and yet form a key piece in this puzzle.

The exact location of the Tristan plume axis during the Paraná-Etendeka magmatism is still uncertain, although *VanDecar et al.* [1995] have presented intriguing evidence for a 'fossil' plume conduit in the sublithospheric upper mantle beneath the northeast Paraná lava field. They found a low velocity anomaly, roughly cylindrical in form, 300 km across and extending from 200 km to at least 600 km depth, which they inferred to be the thermal, and possibly chemical, remnant of the original plume conduit that supplied the Tristan plume head. If this interpretation is correct, it implies that the upper mantle and lithosphere have remained coupled since the breakup of Gondwana. The long-term presence of a thermal anomaly beneath the northern Paraná will also have implications for post-Paraná magmatism and uplift in this region.

SHALLOW-LEVEL PROCESSES IN PARANÁ-ETENDEKA BASALT PETROGENESIS

Fractional Crystallisation

The majority of Paraná-Etendeka lavas have low MgO contents (3.0–6.5 wt%: Figure 7a) and low compatible element contents (<100 ppm Ni). Samples with more mafic compositions (MgO 6.5–9.0 wt%) are volumetrically insignificant (<<2%), and still have compositions (Mg# <65: Mg# is atomic ratio $100 \cdot Mg/(Mg+0.85Fe_{total})$) that are far removed from the expected composition of melts in equilibrium with mantle peridotite. These characteristics indicate that all the magmas have undergone extensive crystal fractionation, presumably in sill complexes near the base of the crust [e.g., *Cox*, 1980]. Only one picrite sample (MgO 15 wt%) has so far been found and this is from near the base of the Etendeka basalts [*Gibson et al.*, 1997]. Data for the Paraná-Etendeka basalts plot on the experimentally determined 1-atm cotectic (ol+pl+cpx+liq) in a CIPW normative Di-Ol-Hy-Qz/Ne diagram [*Thompson et al.*, 1983], which implies at least a final stage of equilibration and crystallisation at low pressures in near-surface magma reservoirs. The major-element trends within each magma type are controlled primarily by fractional crystallisation of a predominantly gabbroic crystal extract (olivine + clinopyroxene + plagioclase ± Fe-Ti oxide [*Cox*, 1980; *Bellieni et al.*, 1984a; *Peate*, 1990]).

Crustal Assimilation

The low-Ti basalts in the southern Paraná and the Etendeka show significant variations in Sr-, Nd-, Pb-, and O-isotope compositions and in highly incompatible element ratios, indicating the importance of open-system behaviour in their evolution [*Erlank et al.*, 1984; *Mantovani et al.*, 1985a; *Fodor et al.*, 1985b; *Petrini et al.*, 1987; *Mantovani and Hawkesworth*, 1990; *Peate and Hawkesworth*, 1996]. An increase in $^{87}Sr/^{86}Sr_i$ with increasing SiO_2 in the Gramado basalts is consistent with an assimilation and fractional crystallisation (AFC) style of crustal contamination, where the amount of assimilation is linked thermally to the extent of crystallisation (Figure 14). Whole-rock $\delta^{18}O$ data show a similar increase [*Fodor et al.*, 1985b] but these have been compromised by water-rock alteration [*Harris et al.*, 1989; *Iacumin et al.*, 1991]. Data on mineral separates indicate that magmatic $\delta^{18}O$ values were higher (+6.3‰ to +8.3‰) than typical mantle

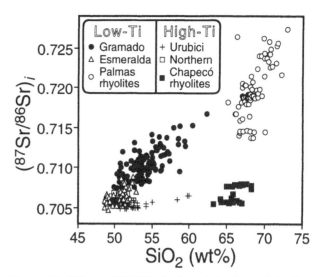

Figure 14. SiO_2 vs. $^{87}Sr/^{86}Sr_i$ for Paraná basalts and rhyolites. The effects of crustal assimilation are clearly evident in the low-Ti Gramado magma type which shows a positive correlation. The broadness of the Gramado array is attributed to regional variations in the $^{87}Sr/^{86}Sr_i$ of the pre-AFC parental magma [*Peate and Hawkesworth*, 1996]. Data sources as for Figure 7, plus *Garland et al.* [1995].

values (~+5.5‰), indicative of some crustal input [*Harris et al.*, 1989]. The contamination process was complex and cannot be represented by a single, progressively contaminated liquid line of descent, even on a local scale where significant temporal variations in the rate of contamination are evident [*Mantovani and Hawkesworth*, 1990; *Peate and Hawkesworth*, 1996].

Despite the widespread consensus for the role of contamination, disagreement exists over the composition of the 'uncontaminated' parental low-Ti magma(s). Many studies have suggested that the temporal change from Gramado to Esmeralda flows simply reflects a relatively abrupt decrease in the extent of crustal contamination of a common parental magma that had εNd_i >+4 [e.g., *Petrini et al.*, 1987]. However, in detail, it is difficult to reconcile all the compositional data with such a model. For example, Gramado samples show a positive correlation between $^{87}Sr/^{86}Sr_i$ and Th/Ta, consistent with the addition of crustal material with high $^{87}Sr/^{86}Sr_i$ and Th/Ta. Esmeralda magmas have similar Th/Ta to the less contaminated Gramado magmas but at lower $^{87}Sr/^{86}Sr_i$. Thus, *Peate and Hawkesworth* [1996] have argued that the Gramado and Esmeralda magma types evolved from distinct parental magmas. The parental Gramado magma had Nb/La <0.8 and εNd_i ~ -3, and subtle regional variations in $^{87}Sr/^{86}Sr_i$ (0.707-0.709) prior to AFC are the principal cause of the broadness of the Gramado array on Figure 14 [*Peate and Hawkesworth*, 1996]. The Esmeralda magma type shows significant ranges in $^{87}Sr/^{86}Sr_i$ and εNd_i that are reasonably correlated with incompatible element ratios such as La/Nd and Nb/La (e.g., Figure 15), but not with elemental abundances. These variations can be explained by mixing between an incompatible-element-depleted melt with εNd_i >+4 and a contaminated Gramado magma at relatively shallow levels followed by fractional crystallisation [*Peate and Hawkesworth*, 1996].

Mineral separates from high-Ti Khumib (=Urubici) samples in the Etendeka have $\delta^{18}O$ values of ~+5.9‰, indicating minimal crustal interaction [*Harris et al.*, 1989]. However, crustal assimilation does appear to have influenced the composition of some Urubici magmas, despite their relatively high incompatible element contents. Small variations in $^{87}Sr/^{86}Sr_i$ (0.7047–0.7065) are positively correlated with SiO_2 (50–60 wt%) and Th/Ta (2.1–4.0), indicative of an AFC style of contamination (Figure 14). The effects of crustal contamination appear to have been limited in the 'Northern' magma types (Pitanga, Paranapanema, Ribeira) as they all share the same, restricted, variability in Sr-Nd-Pb isotopic composition (Figures 7 and 8), and it is notable in this context that these magma types together comprise perhaps half of the Paraná erupted magma volume.

Crystal fractionation (± crustal assimilation), however, is not a viable means of explaining the variety of basaltic magma types in the Paraná, given the wide range in TiO_2 and incompatible trace element contents at a similar MgO content (e.g., 1.0 to 4.0 wt% TiO_2 at 5 wt% MgO: Figure 7a). These differences must have a deeper origin, reflecting differences in the melting process and/or multiple sources.

MANTLE ORIGINS OF THE PARANÁ-ETENDEKA BASALTS

Plume Models for Flood Basalt Generation

The immense volume of magma associated with a flood basalt province such as the Paraná-Etendeka requires a large thermal anomaly within the mantle. This is generally attributed to the presence of a mantle plume [e.g., *Morgan*, 1981], especially given the spatial and temporal connections often observed between flood basalt outcrops and the subsequent hotspot traces in the ocean basins (Figure 1). In most models, melt is generated relatively rapidly by decompression of plume mantle; the details differ as to how decompression occurs, whether in response to lithospheric extension [e.g., *White and McKenzie*, 1989], thermal erosion and thinning of the lithosphere by the plume [e.g., *Yuen and Fleitout*, 1985], or the sudden arrival of a large head of plume material during the initiation of a mantle

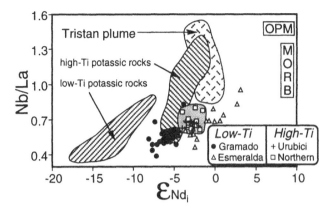

Figure 15. εNd_i vs. Nb/La diagram to illustrate the low εNd_i and Nb/La characteristics of the Paraná basalts, which are distinct from the present-day Tristan plume and MORB compositions. The positive correlation of εNd_i and Nb/La for the Esmeralda magma type indicates the addition of a MORB-like component. Data on Brazilian mafic potassic magmas which represent small-degree lithospheric mantle melts are also plotted to assess their potential role in Paraná magmatism [*Gibson et al.*, 1996]. The minor Tafelkop basalts in Etendeka are compositionally similar to the Tristan plume [*Milner and le Roex*, 1996]. Symbols and data sources for Paraná lavas as for Figure 7. Tristan plume data from *le Roex et al.* [1990] and *Cliff et al.* [1991]. OPM - average oceanic plume magma [*Gibson et al.*, 1996].

plume [e.g., *Richards et al.*, 1989]. The predictions of such plume decompression models fit the observations of some flood basalt provinces very well: e.g., in the Deccan, eruption rates were apparently very high (~1 km³ yr⁻¹), the trace element and isotope characteristics of some relatively uncontaminated basalts are similar to those of recent volcanics from the Réunion plume, and they also have major element characteristics (high Fe, low Si) consistent with melting of fertile peridotite at depth [*Peng and Mahoney*, 1995; *Turner et al.*, 1996].

However, the features of some flood basalt provinces, notably the Paraná-Etendeka, are not easily reconcilable with models that involve just decompressional melting of plume mantle [e.g., *Turner et al.*, 1996]. Paraná basalts that are considered not to have been contaminated with continental crust have certain trace element and isotopic features (i.e., low Nb/La <0.8, low εNd_i <0: Figure 15) that are not commonly observed in oceanic basalts (MORB or OIB) as well as major element compositions that require mantle sources distinct from those of typical oceanic basalts [*Hergt et al.*, 1991; *Hawkesworth et al.*, 1992; *Turner and Hawkesworth*, 1995]. It is also possible that the overall eruption rate was significantly less than in the Deccan [*Turner et al.*, 1994]. One major difference might be the influence of the lithosphere overlying the mantle plume, and hotly debated issues at present are the potential contribution of lithospheric material to flood basalt magmatism and whether the lithospheric mantle can melt to a sufficient extent to produce the observed volumes of flood basalt melts [e.g., *Hawkesworth et al.*, 1988, 1992; *Ellam and Cox*, 1991; *Arndt and Christensen*, 1992; *Saunders et al.*, 1992; *White and McKenzie*, 1995].

Plume Involvement in Paraná-Etendeka Magmatism

Compositional evidence for the involvement of the Tristan plume in the Paraná-Etendeka magmatism is minimal. The only Paraná-Etendeka magmas that have similar incompatible-trace-element and isotopic characteristics to the present-day Tristan plume, as represented by lavas from the islands of the Tristan da Cunha group and Gough, are the Tafelkop basalts and the Okenyenya alkaline intrusions, both in the Etendeka (Figure 6b) [*Milner and le Roex*, 1996]. The relatively minor Tafelkop basalts are found at the base of the lava sequences in a restricted area north of Messum, and they have high Ti/Y (> 500) and Nb/La (~1.0), and Sr-Nd-Pb isotope compositions that fall within the field for the Tristan plume [*Milner and le Roex*, 1996]. A few tholeiitic basalts with similar Nb/La (~1.0) to the Tafelkop basalts have recently been found in southern Uruguay near Mariscala (Figure 6a) [*Kirstein et al.*, 1997]. The composition of the Tristan plume has thus remained broadly constant for the last ~135 m.y., and the suggestion of *Class et al.* [1993] that differences in isotopic compositions between the older Walvis Ridge basalts and the recent Tristan magmatism result primarily from radioactive decay in the plume source can be ruled out. However, it remains difficult to distinguish the Tristan plume signature in proposed mixing arrays involving the Paraná-Etendeka lavas.

Hawkesworth et al. [1992] described some late-stage dykes along the Brazilian coast between São Paulo and Rio de Janeiro that also had similar trace element and isotope characteristics to the modern Tristan plume. However, recent work has shown that these dykes, called the São Sebastião dolerites, are in fact much younger than the Paraná-Etendeka event, with a ^{40}Ar-^{39}Ar age of 81 Ma [*Regelous*, 1993].

The only Paraná basalts that unequivocally contain a recognisably asthenospheric component are the late-stage Esmeralda magmas, and yet they represent a volumetrically insignificant portion of the total flood basalt magmatism [*Peate and Hawkesworth*, 1996]. Esmeralda magmas show significant positive correlations of ε_{Ndi} with Nb/La (Figure 15) and Sm/Ce. They indicate mixing with an incompatible-element-depleted component with a trace element signature similar to MORB (Ce/Sm$_N$ <1, Nb/La >0.9) and high ε_{Ndi} >+4 [*Peate and Hawkesworth*, 1996].

This component though, is unlike the composition of the present Tristan plume as it has higher ε_{Ndi} (Figure 15) and lower ^{208}Pb/^{204}Pb (Figure 10).

Lithospheric Involvement in Paraná-Etendeka Magmatism

An intriguing feature of the main basaltic volcanism is that high-Ti and low-Ti magma types form distinct geographical regions in the northwest and southeast of the province, respectively, which must represent either variable degrees of melting beneath the province or spatial heterogeneities within the lithosphere. *Fodor* [1987] argued that both types could be generated from a common mantle source if the degree of melting was controlled by proximity to the underlying Tristan mantle plume, with high-degree low-Ti melts generated over the plume axis in the south and low-degree high-Ti melts produced on the plume periphery to the north. *Arndt et al.* [1993] instead suggested that lithospheric thickness was a more important factor because it determined the extent to which the underlying asthenosphere could decompress and melt. Beneath a thick lithospheric cap, melting is restricted and occurs in the presence of residual garnet, leading to incompatible-element-enriched basalts with high Ti/Y and Tb/Yb similar to high-Ti basalts (Figures 7 and 12). For thinner lithosphere, melting should be more extensive and occur within the spinel peridotite field, producing basalts with lower concentrations of incompatible elements and with low Ti/Y and Tb/Yb, similar to low-Ti basalts. Any incompatible-trace-element or isotopic differences between the high-Ti and low-Ti types or relative to typical asthensopheric melts were simply dismissed by *Fodor* [1987] and *Arndt et al.* [1993] as crustal contamination effects. However, it is difficult to explain the isotopic variations of the Paraná lavas in this manner. If the Paraná mantle source is assumed to have a similar composition to the high ε_{Ndi} component seen in the Esmeralda magma type, then the trace element and isotope differences between the low-Ti Gramado magma type and the high-Ti magma types would require the coincidence of different degrees of melting with assimilation of crust of different composition. Furthermore, the simultaneous eruption of high-Ti Urubici (Tb/Yb$_N$ ~2.3) and low-Ti Gramado (Tb/Yb$_N$ ~1.3) lavas on the Serra Geral escarpment would imply significant variations in lithospheric thickness over a relatively small distance according to the model of *Arndt et al.* [1993].

The 'Northern' magma types (Pitanga, Paranapanema, Ribeira) show good evidence for a gradual increase in the degree of melting of a similar mantle source at progressively shallower depths. Paranapanema lavas, which locally overlie Pitanga lavas across most of the northwest Paraná, have lower incompatible element contents and lower Ti/Y and Tb/Yb values, but have a similar Sr-Nd-Pb isotopic composition [*Peate*, 1990]. Major element differences (lower $Fe_{8.0}$ and higher $Si_{8.0}$: fractionation-corrected to MgO 8.0 wt%) also suggest that the Paranapanema magmas were generated at lower pressures than the Pitanga magmas [*Garland et al.*, 1996]. *Gallagher and Hawkesworth* [1994] argued that these lavas were erupted through thick lithosphere because seismic data indicate that the lithospheric thickness beneath the northeastern parts of the Paraná province is presently 150-200 km, although it is uncertain to what extent the lithosphere might have rethickened since the Paraná event by conductive cooling. Extension across the Ponta Grossa structure was probably less than 20% [*Ussami et al.* 1991]. Thus there is little evidence that significant extension occurred in this region that could have produced extensive decompressional melting of asthenospheric mantle. It might be assumed from the large volume and similar compositional characteristics of the 'Northern' magma types that they represent the best estimate for the composition of the dominant component in the early stages of the Tristan plume, and yet this component was apparently not sampled subsequently by plume magmatism after continental breakup. For example, Nb/La remains constant at ~0.64 in all the 'Northern' magma types, a value which is lower than in MORB or Tristan plume magmas (Figure 15). Instead, it seems that a source within the lithospheric mantle is necessary.

Similar high-Ti and low-Ti domains can be distinguished in the other Mesozoic Gondwanan flood basalt provinces (Karoo, Ferrar). A boundary between the two types can be traced across the former Gondwanan supercontinent and has led to speculation that it represents a major compositional discontinuity within the lithospheric mantle [e.g., *Cox*, 1988; *Erlank et al.*, 1988; *Hawkesworth et al.*, 1988; *Hergt et al.*, 1991]. Supporting evidence for such a lithospheric mantle boundary comes from mafic potassic rocks erupted around the margins of the Paraná lava field during the Cretaceous. These can be divided into two compositionally and geographically distinct groups: a high-Ti group found along the northeast margin and a low-Ti group found adjacent to the central Paraná lavas along the coast in the east and in Paraguay to the west [*Gibson et al.*, 1996]. Such magmas are widely accepted as representing small-degree melts of the lithospheric mantle and so *Gibson et al.* [1996] concluded that the spatial distribution of the high- and low-Ti mafic potassic rocks was analogous to the broad high-Ti/low-Ti provinciality seen in the Paraná lavas and reflected a major

compositional boundary in the lithospheric mantle. They modelled the low- and high-Ti flood basalts as large-degree asthenospheric melts that had mixed with between 20% and 50% of the low- and high-Ti mafic, potassic, lithosphere-derived melts, respectively, and then undergone variable degrees of crustal assimilation. However, there are several problems for such a model. To explain the observed range in Ti/Y (300-700) within the high-Ti 'Northern' magma types (Figure 7b) would require the addition of between 10% and 50% of the high-Ti mafic potassic melt, and yet these magma types have relatively homogeneous Sr-Nd-Pb isotopic compositions that show no systematic variations with Ti/Y. Furthermore, most of the high-Ti mafic potassic rocks have unsuitable compositions (higher Nb/La, similar ε_{Nd_i}: Figure 15) to account for the high-Ti flood basalts by mixing with either MORB or a plume-derived melt. Ti/Y is too high (>400) in the low-Ti mafic potassic rocks for them to represent a suitable low-degree melt contaminant in the low-Ti Gramado basalts and, in addition, the already low TiO_2 contents of primitive Gramado basalts would require any asthenospheric melt to have an extremely Ti-depleted composition. *Ellam and Cox* [1991] had similar difficulties in extending their lamproite-mixing model for the high-Ti Nuanetsi picrites to the low-Ti magma types elsewhere in the Karoo province of southern Africa.

Instead, *Hergt et al.* [1991] argued that the distinctive compositional features of the Gondwana low-Ti flood basalts were inherited from an unusual, melt-depleted mantle source that had been modified by the addition of subducted sediment material. The major element compositions of the Gramado magmas indicate an origin from a mantle source distinct from that of oceanic basalts [*Hergt et al.*, 1991; *Turner and Hawkesworth*, 1995]. $Fe_{8.0}$ contents of peridotite melts depend both on pressure and source composition. Although Gramado samples with low $^{87}Sr/^{86}Sr_i$ (0.707-0.709) have similar $Fe_{8.0}$ (~10.2) to recent South Atlantic MORB, it is unlikely that both were generated at similarly shallow depths. Thus, *Hergt et al.* [1991] and *Turner and Hawkesworth* [1995] concluded that the Gramado magmas originated in an Fe-depleted source, presumably due to previous melt extraction events, argued to be in the continental mantle lithosphere.

Uncontaminated Urubici and Khumib magmas are characterised by low $^{206}Pb/^{204}Pb$ (17.4-17.8) coupled with high $^{207}Pb/^{204}Pb$ and $^{208}Pb/^{204}Pb$, typical features of the oceanic DUPAL isotopic province [*Hawkesworth et al.* 1986; *Peate*, 1990; *Milner and le Roex*, 1996]. Similar Pb isotope characteristics, and broadly similar Sr and Nd isotope compositions, are seen in the nearby contemporaneous Jacupiranga carbonatites and pyroxenites, and oxygen isotope data confirm that they are not the result of crustal contamination [*Toyoda et al.*, 1994; *Huang et al.*, 1995]. Thus, by inference, it appears that the low $^{206}Pb/^{204}Pb$ characteristic of the Urubici magmas is a mantle feature. *Hawkesworth et al.* [1986] showed that there was a close compositional link between the Urubici magmas and 70 Ma samples from DSDP Site 525A on the Walvis Ridge that have Nb/La ~0.8, $^{206}Pb/^{204}Pb$ ~17.6 and ϵNd_i ~-3 [*Richardson et al.*, 1982]. It is possible that this low $^{206}Pb/^{204}Pb$ material is an intrinsic part of the Tristan plume that is not presently being sampled by the active plume magmatism on Tristan da Cunha or Gough. However, if the Urubici basalts are derived from the lithospheric mantle, then the presence of basalts with similar trace element and isotopic features erupted within the oceanic part of the plume track sixty million years later might result from material having been thermally eroded from the base of the lithospheric mantle and entrained in the asthenosphere [*Hawkesworth et al.*, 1986; *Milner and le Roex*, 1996]. *Mahoney et al.* [1992] proposed a similar model to explain low $^{206}Pb/^{204}Pb$ rocks in the Madagascar province and along part of the Southwest Indian Ridge (39°-41°E) that lies 400 km from the Marion plume.

Lithospheric Mantle Melting Model

Although geochemical evidence for the participation of lithospheric mantle material in some flood basalt magmas is now widely accepted, the extent of this involvement has yet to be agreed [e.g., *Hawkesworth et al.*, 1988; *Ellam and Cox*, 1991; *Mahoney et al.*, 1992; *Saunders et al.*, 1992; *White and McKenzie*, 1995]. Plume decompression models for flood basalt magmatism predict minimal melt generation (<<5%) within the lithospheric mantle, and then only in the initial stages. These models might be inappropriate for some flood basalt provinces because the models assume rapid eruption rates [e.g., *White and McKenzie*, 1989] and an anhydrous solidus for the lithospheric mantle [*Arndt and Christensen*, 1992]. Melt generation within the lithospheric mantle will be facilitated by the presence of just 0.3 wt% $H_2O \pm CO_2$, which allows melting at the lower temperature volatile-enriched solidus [*Gallagher and Hawkesworth*, 1992]. *Turner et al.* [1996] developed a conductive heating model in which melting can take place solely within the lithospheric mantle, driven by conduction of heat from a mantle plume incubating beneath a lithospheric cap [cf. *Saunders et al.*, 1992]. The important controlling variables are the plume potential temperature, the thickness of the mechanical boundary layer, and the duration of heating. Because the time scales of conductive heat transfer are long, this model predicts a

protracted period of magmatism. On a time scale of ~10 m.y., melting can occur within the lithospheric mantle over a broad area (comparable to the size of the province) to produce 1-3 km of magma, without any melt contribution from the underlying plume, provided that the lithosphere is >100 km thick. If the lithospheric thickness is reduced to <100 km by extension or thermal erosion, then decompressional melting will commence within the plume, leading to higher eruption rates and the eruption of magmas dominated by sub-lithospheric melts. *Turner et al.* [1994] discussed the evidence that, during the onset of the Paraná-Etendeka magmatism, the amount of extension was too low to permit decompressional melting of the asthenosphere as the lithosphere was probably at least 150 km thick. From the geochemical arguments marshalled above, it is clear that plume mantle similar to that supplying recent Tristan da Cunha or Gough magmas or MORB asthenosphere had a minimal role in Paraná-Etendeka magmatism. Thus, the conductive heating model can account for many of the observed distinctive characteristics of the Paraná-Etendeka flood basalts. It is only in the late stages of Paraná magmatism, during eruption of the Esmeralda magma type, that lithospheric extension proceeded to such an extent to allow decompressional melting of the asthenosphere. Figure 16 presents a cartoon summary of the progressive evolution of basaltic magmatism within the Paraná province as viewed by *Garland et al.* [1996].

In the Deccan province, large volumes of basalts appear to have been derived by decompressional melting of asthenosphere associated with the Réunion mantle plume [e.g., *Peng and Mahoney*, 1995], whereas in the Paraná-Etendeka province, the Tristan plume appears to have played a largely passive role, solely providing heat by conduction to produce extensive melting of lithospheric mantle [*Turner et al.*, 1996].

RHYOLITE MAGMA TYPES AND PETROGENESIS

The Paraná-Etendeka rhyolites cover an area of at least 17,000 km^2 and typically form the uppermost units of the lava sequences. They can be divided into two distinct groups on the basis of petrography and geochemical composition (Figures 14 and 17), similar to the high-Ti/low-Ti division of the basalts [*Bellieni et al.*, 1986; *Harris et al.*, 1990]. The high-Ti rhyolites (called Chapecó rhyolites in the Paraná) are highly plagioclase-phyric, with high incompatible element abundances (e.g., Zr >500 ppm), low $\delta^{18}O$ (~+6.5‰ in pyroxene) and similar $^{87}Sr/^{86}Sr_i$ to the high-Ti basalts (0.705–0.708). In contrast, the volumetrically dominant low-Ti rhyolites (called Palmas rhyolites in the

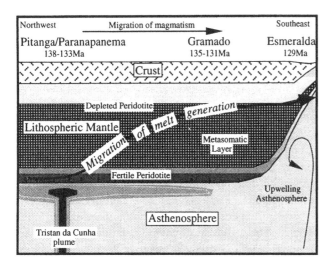

Figure 16. Cartoon section illustrating a model for the temporal evolution of magma generation within the Paraná [*Garland et al.*, 1996]. The oldest basalt unit (Pitanga) was derived from the greatest depth in the lithosphere (90-120 km), melting relatively fertile, anhydrous peridotite. The magma source became gradually shallower with time, with the Gramado basalt unit reflecting shallow-level magma generation (< 60 km), melting relatively refractory and hydrous peridotite. Asthenospheric material only becomes clearly recognisable in the youngest unit (Esmeralda), once extension has proceeded to the extent that decompression of the asthenosphere can take place.

Paraná) are virtually aphyric (<5% phenocrysts) with low incompatible element contents (e.g., Zr <400 ppm), high $\delta^{18}O$ (~+10‰ in pyroxene) and high $^{87}Sr/^{86}Sr_i$ of 0.714–0.727. Analyses of selected Paraná rhyolites are listed in Table 2. Calculated eruption temperatures for both types are unusually high for silicic magmas (950–1100°C [*Bellieni et al.*, 1986; *Garland et al.*, 1995]).

The rhyolites are restricted to the later stages of magmatism and, unlike the basalts which crop out almost exclusively on the South American Plate, the rhyolites are concentrated along the present-day continental margins, which indicates a close link with the rifting of the South Atlantic Ocean. This fact, together with the pronounced 'silica gap' between rhyolites and basalts (Figure 4), led many workers to assume that the rhyolites were the products of crustal anatexis [e.g., *Erlank et al.*, 1984; *Hawkesworth et al.*, 1988]. *Harris et al.* [1990] linked the regional distribution of the high- and low-Ti rhyolites (Figure 18) to differences in lower crustal source regions, with the high-Ti rhyolites generated from Archaean lower crust and the low-Ti rhyolites derived from melting Late Proterozoic mobile belt material. However, the coincidence of basalt and rhyolite provinciality (low-Ti Gramado basalts and Palmas rhyolites in the south; high-Ti Pitanga basalts and Chapecó rhyolites in the north) led to suggestions that the rhyolites

were linked to the basalts either by extensive fractional crystallisation and crustal assimilation or by melting of underplated basalts [*Bellieni et al.*, 1986; *Garland et al.*, 1995].

There are some minor outcrops of rhyolites near Mariscala in southern Uruguay (Figure 6a). Recent work by *Kirstein et al.* [1997] has shown that they are compositionally distinct from the Palmas and Chapecó rhyolites, and are interpreted as lower crustal melts. They have been

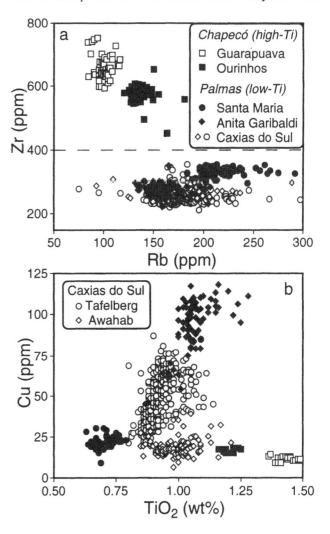

Figure 17. (a) Rb vs. Zr: 400 ppm Zr discriminates between high-Ti Chapecó and low-Ti Palmas rhyolite types, and the Chapecó subgroups are resolved by different Rb/Zr (Ourinhos >0.2; Guarapuava <0.2), (b) TiO_2 vs. Cu, illustrating compositional differences between the low-Ti rhyolite subgroups. The Caxias do Sul subgroup is divided into the Tafelberg and Awahab magma systems which lie, respectively, above and below an erosional disconformity in the field [*Milner et al.*, 1995b]. Data from *Hawkesworth et al.* [1988], *Peate* [1990], *Mantovani and Hawkesworth* [1990], *Whittingham* [1991], *Garland et al.* [1995], and S.C. Milner and A.R. Duncan (unpubl. data).

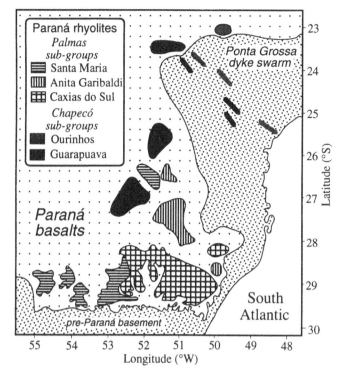

Figure 18. Distribution of rhyolite subgroups within the Paraná province [*Peate et al.,* 1992; *Garland et al.,* 1995].

dated at ~128 Ma by ^{40}Ar-^{39}Ar and they also preserve classic ignimbritic textures [*Kirstein et al.,* 1997].

High-Ti Rhyolite Magma Types

Peate et al. [1992] divided the Chapecó rhyolites into the Ourinhos subgroup (Rb/Zr >0.2, ^{87}Sr/$^{86}Sr_i$=0.7076–0.7080) and Guarapuava subgroup (Rb/Zr <0.2, ^{87}Sr/$^{86}Sr_i$=0.7055–0.7060). The Ourinhos subgroup forms a limited outcrop in the northeast of the Paraná, whereas the Guarapuava subgroup is more extensive, forming a series of separate outcrops along the eastern margin of the central Paraná lava field (Figure 18). Small petrographic differences among the Guarapuava outcrops suggest that they represent different flow units. In the African part of the province, high-Ti rhyolites have been reported from the northernmost part of the Etendeka province (Sarusas rhyolites [*Milner,* 1988]) and along the southern Angolan coast (Giraul rhyolites [*Alberti et al.,* 1992]). Although the published geochemical data are limited, both Ourinhos- and Guarapuava-like compositions are found within the Sarusus rhyolites [*Harris et al.,* 1990], but the Giraul rhyolites may represent a third subgroup as they have ^{87}Sr/$^{86}Sr_i$ of 0.7062–0.7074, intermediate between the Ourinhos and Guarapuava subgroups. *Renne et al.* [1996b]

measured a ^{40}Ar-^{39}Ar age of 132.0±0.4 Ma for the Giraul rhyolites, which is similar to ^{40}Ar-^{39}Ar ages obtained by *Turner et al.* [1994] for the Guarapuava (131.8±1.8 Ma) and Ourinhos (131.7±0.8 Ma and 128.7±1.1 Ma) rhyolites. Palaeomagnetic data, though, indicate a difference in the relative ages of the Ourinhos (reversed polarity) and Guarapuava (normal polarity) subgroups [*Ernesto and Pacca,* 1988].

Petrogenesis of the High-Ti Rhyolites

Although lower crustal granulites have been suggested as a potential source material for the high-Ti rhyolites [*Bellieni et al.,* 1986; *Harris et al.,* 1990], the close coincidence between many compositional features of the high-Ti basalts and Chapecó rhyolites make the basalts a plausible parental material. *Garland et al.* [1995] showed from a comparison of trace element and isotope compositions that, of the three high-Ti basalt magma types, only the Pitanga basalts are a suitable parent to the Chapecó rhyolites (e.g., Pitanga basalts and Chapecó rhyolites have similar Tb/Yb: Figure 12). The Chapecó rhyolites can be modelled satisfactorily either by extensive fractional crystallisation (~70%) of a parental Pitanga magma or by about 30% partial melting of a Pitanga basalt source [*Bellieni et al.,* 1986; *Garland et al.,* 1995], but the existence of significant gaps in both SiO_2 and incompatible element contents between the Pitanga basalts and Chapecó rhyolites (Figure 19) argues strongly in favour of a partial melting origin. Melting of underplated basalt, which ponded at the base of the crust because of the density contrast at the Moho [e.g., *Cox,* 1980], probably occurs during the final stages of continental breakup as the crust thins and decompresses. A similar model has been proposed to explain the voluminous high-Ti Lebombo rhyolites in the Karoo province [*Cleverly et al.,* 1984].

The Ourinhos subgroup has higher ^{87}Sr/$^{86}Sr_i$ (0.7078 vs. 0.7055) and SiO_2 and lower TiO_2 than the more voluminous Guarapuava subgroup. These differences can be explained by a localised AFC process starting with a Guarapuava magma, involving ~10% upper crustal assimilation coupled with ~20% fractionation of an assemblage primarily of plagioclase and clinopyroxene plus Fe-Ti oxides, apatite and zircon [*Garland et al.,* 1995]. Normative compositions indicate that Chapecó rhyolite magmas last equilibrated at pressures of 5–15 kb [*Garland et al.,* 1995]. Thus it appears that the magmas were extracted rapidly from lower crustal depths, without further significant differentiation of the magma (except locally for the Ourinhos subgroup). The Chapecó rhyolites preserve no textural evidence of pyroclastic activity, and each flow

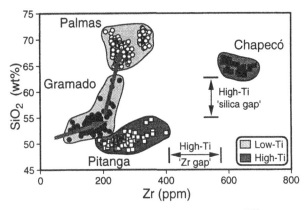

Figure 19. Zr vs. SiO$_2$ diagram highlighting the different petrogenetic origins of the high- and low-Ti rhyolites. Data sources as in Figure 16.

unit has a limited spatial extent, being <40 km long and <100 m thick [*Bellieni et al.*, 1986]. A few dykes of Chapecó composition are found within the Ponta Grossa dyke swarm (Figure 18), and so *Garland et al.* [1995] suggested that the Chapecó rhyolites are dyke-fed lava flows.

Low-Ti Rhyolite Magma Types

The Palmas rhyolites are divided into three compositional subgroups (Figure 17b) [*Peate et al.*, 1992; *Garland et al.*, 1995]. The Caxias do Sul subgroup is the largest volumetrically and is found in the southeast of the lava field near the Serra Geral escarpment where the flow units reach a combined thickness of up to 400 m (Figure 18). The Anita Garibaldi subgroup lies to the north of the Caxias do Sul subgroup, along the eastern margin of the central Paraná lava field, and field evidence shows that it is younger than the Caxias do Sul subgroup. The Santa Maria subgroup is found in two widely separated exposures: along the southern escarpment to the west of the Caxias do Sul subgroup which it overlies, and in the central Paraná where it caps the highest exposures and is thus inferred to be younger than the Anita Garibaldi subgroup as well.

The excellent exposure in the Etendeka allowed *Milner et al.* [1992] to look at the low-Ti rhyolite units on a flow-by-flow basis. Individual units are remarkably homogeneous and show no compositional zonation. Most are similar in composition to the Caxias do Sul subgroup, with the exceptions of the uppermost unit (similar to the Anita Garibaldi subgroup) and some units to the north in the Sarusas region (similar to the Santa Maria subgroup) [*Milner*, 1988]. Each unit or group of successive units is geochemically distinctive on the basis of petrography and element abundances (especially Ti, Fe and Cu), and have

been correlated throughout the Etendeka. *Whittingham* [1991] did a similar study of the Caxias do Sul subgroup in southeast Brazil, distinguishing seven distinct units, and showed that stratigraphic compositional changes mirrored those seen in the Etendeka rhyolites. *Milner et al.* [1995b] have demonstrated a direct correlation of some individual units between these two areas that are now on opposite sides of the South Atlantic Ocean.

The upper and lower parts of the Etendeka sequences are separated by an erosional disconformity that can be traced across into the southern Paraná and thus provides another reference point for reconstructions of this part of Gondwana. Rhyolites below this disconformity have similar petrography and compositions and are only found in the south of both areas. Small, stratigraphically controlled differences between units suggest that the units sequentially tapped a single, evolving magma body, referred to as the Awahab magma system [*Milner et al.*, 1995b], and they all appear to have been erupted from the Messum structure (Figure 6b), where compositionally similar quartz-monzonite intrusions are found [*Milner et al.*, 1992]. The rhyolites overlying the disconformity apparently tapped a different magma body, the Tafelberg magma system, whose eruption site is unknown. One of these units reaches its maximum thickness close to the Namibian coast, suggesting a source now offshore. The Awahab magma system comprises at least five individual eruptive units, with an estimated total volume of >8,600 km^3, dispersed over an area of >33,000 km^2, and the Tafelberg magma system is similar in terms of number of erupted units and magma volume [*Milner et al.*, 1995b]. Units of the Awahab magma system, which stratigraphically are the oldest low-Ti rhyolites, give ^{40}Ar-^{39}Ar ages of 131.9±0.6 Ma to 132.1±0.4 Ma [*Renne et al.*, 1996b], and units of the overlying Tafelberg magma system give ^{40}Ar-^{39}Ar ages of 132.8±1.1 Ma [*Renne et al.*, 1992] and 130.3±1.2 Ma [*Stewart et al.*, 1996].

Individual silicic units can have volumes >1,000 km^3, unusually high aspect ratios (1:200–1:2,000) and can have travelled lateral distances >300 km [*Milner et al.*, 1992, 1995b]. As with similar examples from the geological record, the eruptive mechanism of these units has proven controversial [e.g., *Henry and Wolff*, 1992]. *Whittingham* [1991] considered them to be extensive lava flows, mainly due to the lack of textural evidence for an explosive or ignimbritic origin. *Bellieni et al.* [1986] and *Milner et al.* [1992], on the other hand, interpreted them as rheoignimbrites, in which high temperatures maintained within an ignimbrite flow cause rewelding of particles and a final period of flow that produces many lava-like features. *Milner et al.* [1992] found rare pyroclastic textures preserved in flow margins, and *Garland et al.* [1995] found a thin, distal pyroclastic fall deposit related to the Santa Maria subgroup. However, *Manley* [1995] issued a note of caution for such an interpretation by suggesting that certain pyroclastic textures can be formed locally within a lava flow.

Petrogenesis of the Low-Ti Rhyolites

If the Palmas rhyolites are crustal melts, the basement rocks exposed in southern Brazil cannot represent a suitable source material because they have lower Nd isotope ratios than the rhyolites (Proterozoic mobile belt ϵNd_{130Ma}= -8 to -19; Archaean craton material ϵNd_{130Ma}= -21 to -32; Palmas rhyolites ϵNd_{130Ma}= -6 to -8). The relatively radiogenic Sr and Pb isotopic composition ($^{87}Sr/^{86}Sr_i$=0.714 to 0.728; $^{206}Pb/^{204}Pb$ >19.0) of the rhyolites would suggest a relatively incompatible-element-enriched, upper crustal source, but their high eruption temperatures (>1000°C) and major element composition require a fairly basic source akin to lower crustal granulites or basalts.

Several workers have instead investigated possible fractionation links between the Gramado basalts and the Palmas rhyolites [*Bellieni et al.*, 1986; *Garland et al.*, 1995]. Any crystal fractionation must be accompanied by assimilation of upper crustal material because the rhyolites continue the trend towards high $^{87}Sr/^{86}Sr$, $^{206}Pb/^{204}Pb$ and $\delta^{18}O$ seen within the basalts (e.g., Figure 14), and modelling of an AFC process can produce a good fit to the major and trace element data. The entry of significant amounts of magnetite into the calculated crystallising assemblage at ~56% SiO_2 rapidly enhances the SiO_2 content in the liquid over a small extent of fractionation and cooling and can account for the observed silica gap. It is notable that, unlike the Chapecó-Pitanga case, highly incompatible elements such as Zr show a nearly continuous trend through the Gramado basalts to Palmas rhyolites, consistent with a liquid line of descent (Figure 19). Normative compositions suggest that the Palmas rhyolites equilibrated at lower pressures (<5 kb) than the Chapecó rhyolites, indicative of shallow ponding [*Garland et al.*, 1995]. Compositional differences between the different Palmas subgroups can be explained by regional and temporal variations in the extent of assimilation and exact nature of the fractionating crystal assemblage. The Palmas rhyolites appear to have evolved from a parental basalt composition by fractional crystallisation and crustal assimilation in a stable, shallow-level magma chamber, and were then emplaced explosively at the surface as rheoignimbrites [*Garland et al.*, 1995].

PARANA-ETENDEKA FLOOD BASALTS AND MASS EXTINCTIONS ?

The possibility of a causal link between flood basalt eruptions and major faunal extinctions has been much debated in the literature [e.g., *Rampino et al.*, 1988]. The Permian-Triassic and the Cretaceous-Tertiary boundaries mark two of the most significant mass extinction events in Earth history. *Renne et al.* [1995] demonstrated that eruption of the Siberian flood basalt province coincided, within uncertainty, with the Permian-Triassic boundary. There is also a close temporal relationship between the Cretaceous-Tertiary boundary and the Deccan flood basalts, although whether they were contemporaneous is still debated [e.g., *Venkatesan et al.*, 1993; *Féraud and Courtillot*, 1994]. *Rampino et al.* [1988] correlated the eruption of the Paraná-Etendeka flood basalts to the mass extinction event at the Jurassic-Cretaceous boundary. This extinction event at the end of the Tithonian is one of the eight mass extinction events recognised by *Raup and Sepkoski* [1984], when 37% of marine genera (mostly ammonites, bivalves, and corals) disappeared.

Most estimates place the Jurassic-Cretaceous boundary at between 135 Ma and 145 Ma [e.g., *Renne et al.*, 1992], which is older than the ^{40}Ar-^{39}Ar ages of most of the Paraná-Etendeka lavas. Therefore, any temporal correlation between flood basalt eruption and mass extinction is, in this case, unlikely [*Hawkesworth et al.*, 1992; *Renne et al.*, 1992], although better constraints on the age of the Jurassic-Cretaceous boundary are clearly required. It is not possible to locate precisely the Jurassic-Cretaceous boundary within the Paraná basin sequences because of the essentially unfossiliferous nature of the local sediments. One solution might be to look at more distant and better dated sequences around the periphery of the Paraná basin, perhaps in northwestern Argentina, where distal tephra deposits from the large rhyolitic eruptions in the Etendeka and southern Brazil might be found that could provide suitable marker horizons denoting the time of at least part of the Paraná-Etendeka magmatism.

Rampino et al. [1988] suggested that the environmental impact of large flood basalt events could be significant. Individual flows can have erupted volumes of at least 700 km³ (Rosa flow, Columbia River province), which would release large quantities of sulphur dioxide and carbon dioxide into the atmosphere. The Paraná eruptions were of similar magnitude, in terms of the total volume of magma erupted, to the Deccan and Siberian lava sequences. If, indeed, the flood basalt eruptions in the Deccan and Siberia were responsible for significant faunal extinctions, perhaps in response to climatic changes caused by the repeated addition of volcanogenic gases to the atmosphere over a relatively short time span (<1 m.y.), then the apparently muted environmental effect of the Paraná eruptions might reflect a slower average eruption rate, thus giving the biosphere more time to recover between successive eruptions.

Acknowledgments. I thank Chris Hawkesworth and Marta Mantovani for initiating my interest in the Paraná-Etendeka region and for their guidance and support over the last decade. Anton le Roex and Asish Basu provided constructive reviews, and I thank John Mahoney for his editorial comments and extreme patience. Andy Duncan and Simon Milner kindly provided unpublished data from the Etendeka. Discussions on flood basalts with Chris Hawkesworth, Marta Mantovani, Nick Rogers, Janet Hergt, Simon Turner, Frances Garland, Sally Gibson, Mukul Sharma, Kerry Gallagher, Linda Kirstein and Marcel Regelous have been stimulating. This review was written while I was a research fellow in the Institute for Environmental Physics at the University of Heidelberg, Germany, funded by the Heidelberg Academy of Sciences, and I would like to thank Augusto Mangini and the group for their hospitality.

REFERENCES

Alberti, A., E. M. Piccirillo, G. Bellieni, L. Civetta, P. Comin-Chiaramonti, and E. A. A. Morais, Mesozoic acid volcanics from Southern Angola: petrology, Sr-Nd isotope characteristics, and correlation with the acid stratoid volcanic suites of the Paraná basin (south-eastern Brazil), *Eur. J. Mineral., 4,* 597-604, 1992.

Arndt, N. T., and U. C. Christensen, The role of lithospheric mantle in the generation of continental flood basalts, *J. Geophys. Res., 97,* 10967-10981, 1992.

Arndt, N. T., G. K. Czamanske, J. L. Wooden and V. A. Fedorenko, Mantle and crustal contributions to continental flood volcanism, *Tectonophysics, 223,* 39-52, 1993.

Austin, J. A., and E. Uchupi, Continental-oceanic crustal transition off southwest Africa, *Bull. Am. Ass. Petrol. Geol., 66,* 1328-1347, 1982.

Baker, C. L., The lava fields of the Paraná basin, South America, *J. Geol., 31,* 66-79, 1923.

Baksi, A. K., R. V. Fodor, and E. Farrar, Preliminary results of ⁴⁰Ar-³⁹Ar dating studies on rocks from the Serra Geral flood basalt province and the Brazilian continental margin (abstract), *Eos Trans. AGU, 72,* 300, 1991.

Bellieni G., P. Comin-Chiaramonti, L. S. Marques, A. J. Melfi, E. M. Piccirillo, A. J. R. Nardy, and A. Roisenberg, High- and low-Ti flood basalts from the Paraná plateau (Brazil): petrology and geochemical aspects bearing on their mantle origin. *N. Jb. Miner. Abh., 150,* 272-306, 1984a.

Bellieni, G., P. Comin-Chiaramonti, L. S. Marques, A. J. Melfi, E. M. Piccirillo, and D. Stolfa, Low-pressure evolution of basaltic sills from boreholes in the Paraná basin, Brazil, *Tschermaks Min. Pet. Mitt., 33,* 25-47, 1984b.

Bellieni, G., P. Comin-Chiaramonti, L. S. Marques, A. J. Melfi, A. J. R. Nardy, C. Papatrechas, E. M. Piccirillo, A. J. R. Nardy, A. Roisenberg, and D. Stolfa, Petrogenetic aspects of acid and basaltic lavas from the Paraná plateau (Brazil):

mineralogical and petrochemical aspects, *J. Petrol., 27,* 915-944, 1986.

Burke, K., and J. F. Dewey, Plume-generated triple junctions: key indicators in applying plate tectonics to old rocks, *J. Geol., 81,* 406-433, 1973.

Chang, H. K., R. O. Kowsmann, A. M. F. Figueiredo, and A. A. Bender, Tectonics and stratigraphy of the east Brazil rift system: an overview, *Tectonophysics, 213,* 97-138, 1992.

Class, C., S. L. Goldstein, S. G. Galer, and D. Weis, Young formation age of a mantle plume source, *Nature, 362,* 715-721, 1993.

Cleverly, R. W., P.J. Betton, and J. W. Bristow, Geochemistry and petrogenesis of the Lebombo rhyolites, *Geol. Soc. S. Afr. Spec. Publ., 13,* 171-194, 1984.

Cliff, R. A., P. E. Baker, and N. J. Mateer, Geochemistry of Inaccessible Island volcanics, *Chem. Geol., 92,* 251-260, 1991.

Comin-Chiaramonti, P., C. B. Gomes, E. M. Piccirillo, and G. Rivalenti, High TiO₂ dykes in the coastline of São Paulo and Rio de Janeiro States, *N. Jahrb. Miner. Abh., 146,* 133-150, 1983.

Comin-Chiaramonti, P., G. Bellieni, E. M. Piccirillo, and A. J. Melfi, Classification and petrography of continental stratoid volcanics and related intrusives from the Paraná basin (Brazil), in *Mesozoic Flood Volcanism from the Paraná Basin (Brazil): Petrogenetic and Geophysical Aspects,* edited by E. M. Piccirillo, and A. J. Melfi, pp. 47-72, IAG-USP, São Paulo, 1988.

Cordani, U. G., L. Civetta, M. S. M. Mantovani, R. Petrini, K. Kawashita, C. J. Hawkesworth, P. N. Taylor, A. Longinelli, G. Cavazzini, and E. M. Piccirillo, Isotope geochemistry of flood volcanics from the Paraná basin (Brazil), in *The Mesozoic Flood Volcanism of the Paraná Basin: Petrogenetic and Geophysical Aspects,* edited by E. M. Piccirillo, and A. J. Melfi, pp. 157-178, IAG-USP, São Paulo, 1988.

Cordani, U. G., and P. Vandoros, Basaltic rocks of the Paraná basin, in *Problems in Brazilian Gondwana Geology,* edited by J. J. Bigarella, R. D. Becker, and J. D. Pinto, pp. 207-231, 1967.

Cox, K. G., A model for flood basalt volcanism, *J. Petrol., 21,* 629-650, 1980.

Cox, K.G., The Karoo Province, in *Continental Flood Basalts,* edited by J. D. McDougall, pp. 239-271, Kluwer, 1988.

de Wit, M., M. Jeffery, H. Bergh and L. Nicolaysen, Geological map of sectors of Gondwana reconstructed to their disposition at c. 150 Ma, Bernard Price Inst. Geophys. Res., Univ. of Witwatersrand, Johannesburg (scale 1:10,000,000), 1988.

Druecker, M. D., and S. P. Gay, Mafic dyke swarms associated with Mesozoic rifting in eastern Paraguay, in *Mafic Dyke Swarms, Spec. Pap. 34,* edited by H. C. Halls and W. F. Fahrig, pp. 187-193, Geological Association of Canada, Toronto, ON, 1987.

Duncan, A. R., J. S. Marsh, S. C. Milner, and A. J. Erlank, Distribution and petrogenesis of the basic rocks of the Etendeka Formation of northwestern Namibia, in *Geochemical Evolution of the Continental Crust,* pp. 10-19, Poços de Caldas, Brazil, 1988.

Duncan, A. R., S. R. Newton, C. van den Berg, and D. L. Reid, Geochemistry and petrology of dolerite sills in the Huab River valley, Damaraland, north-western Namibia, *Communs. geol. Surv. Namibia, 5,* 5-17, 1989.

Ellam, R. M., and K. G. Cox, An interpretation of Karoo picrite basalts in terms of interaction between asthenospheric magmas and the mantle lithosphere, *Earth Planet. Sci. Lett.*, *105*, 330-342, 1991.

Erlank, A. J., J. S. Marsh, A. R. Duncan, R. M. Miller, C. J. Hawkesworth, P. J. Betton, and D. C. Rex, Geochemistry and petrogenesis of the Etendeka volcanic rocks from South West Africa/Namibia, *Geol. Soc. S. Afr. Spec. Publ.*, *13*, 195-246, 1984.

Erlank, A. J., A. R. Duncan, J. S. Marsh, R. J. Sweeney, C. J. Hawkesworth, S. C. Milner, R. M. Miller, and N. W. Rogers, A laterally extensive geochemical discontinuity in the sub-continental Gondwana lithosphere, in *Geochemical Evolution of the Continental Crust*, pp. 1-10, Poços de Caldas, Brazil, 1988.

Ernesto, M., and I. G. Pacca, Palaeomagnetism of the Paraná basin flood volcanics, southern Brazil, in *Mesozoic Flood Volcanism from the Paraná Basin (Brazil): Petrogenetic and Geophysical Aspects*, edited by E. M. Piccirillo, and A. J. Melfi, pp. 229-255, IAG-USP, São Paulo, 1988.

Féraud, G. and V. Courtillot, Comment on: 'Did Deccan volcanism pre-date the Cretaceous-Tertiary transition ?', *Earth Planet. Sci. Lett.*, *122*, 259-262, 1994.

Fodor, R. V., and Vetter, S. K., Rift-zone magmatism: petrology of basaltic rocks transitional from CFB to MORB, southeastern Brazil margin, *Contrib. Mineral. Petrol.*, *88*, 307-321, 1984.

Fodor, R. V., C. Corwin, and A. Roisenberg, Petrology of Serra Geral (Paraná) continental flood basalts, southern Brazil: crustal contamination, source material and South Atlantic magmatism *Contrib. Mineral. Petrol.*, *91* 54-65, 1985a.

Fodor, R. V., C. Corwin, and A. N. Sial, Crustal signatures in the Serra Geral flood basalt province, southern Brazil: O- and Sr-isotope evidence, *Geology*, *13*, 763-765, 1985b.

Fodor, R. V., Low- and high-TiO$_2$ flood basalts of southern Brazil: origin from picritic parentage and a common mantle source, *Earth Planet. Sci. Lett.*, *84*, 423-430, 1987.

Gallagher, K., and C. J. Hawkesworth, Dehydration melting and the generation of continental flood basalts, *Nature, 358*, 57-59 1992.

Gallagher, K., and C. J. Hawkesworth, Mantle plumes, continental magmatism and asymmetry in the South Atlantic, *Earth Planet. Sci. Lett.*, *123*, 105-117, 1994.

Gallagher, K., C. J. Hawkesworth, and M. S. M. Mantovani, The denudation history of the onshore continental margin of SE Brazil inferred from apatite fission track data, *J. Geophys. Res.*, *99*, 18117-18145, 1994.

Garland, F. E., C. J. Hawkesworth, and M. S. M. Mantovani, Description and petrogenesis of the Paraná rhyolites, *J. Petrol.*, *36*, 1193-1227, 1995.

Garland, F. E., S. P. Turner and C. J. Hawkesworth, Shifts in the source of Paraná basalts through time, *Lithos, 37*, 223-243, 1996.

Gibson, S. A., R. N. Thompson, A. P. Dickin, and O. H. Leonardos, Mafic potassic magmatic key to plume-lithosphere interactions and continental flood-basalts, *Earth Planet. Sci. Lett.*, 141, 325-341, 1996.

Gibson, S. A., R. N. Thompson, A. P. Dickin, J. G. Mitchell, and S. C. Milner, Temporal variation in magma sources related to the impact of the Tristan Plume *J. Conf. Abs.*, 2, 32, 1997.

Harland, W. B., R. L. Armstrong, A. V. Cox, L. E. Craig and D. G. Smith, *A Geological Time Scale 1989*, Cambridge University Press, Cambridge, 1990.

Harris, C., H. S. Smith, S. C. Milner, A. J. Erlank, A. R. Duncan, J. S. Marsh, and N. P. Ikin, Oxygen isotope geochemistry of the Mesozoic volcanics of the Etendeka Formation, Namibia, *Contrib. Mineral. Petrol.*, *102*, 454-461, 1989.

Harris, C., A. M. Whittingham, S. C. Milner and R. A. Armstrong, Oxygen isotope geochemistry of the silicia volcanic rocks of the Etendeka-Paraná province: source constraints, *Geology*, *18*, 1119-1121, 1990.

Harry, D. L., and D. S. Sawyer, Basaltic volcanism, mantle plumes and the mechanics of rifting: the Paraná flood basalt province of South America, *Geology, 20*, 207-210, 1992.

Hawkesworth, C. J., J. S. Marsh, A. R. Duncan, A. J. Erlank, and M. J. Norry, The role of continental lithosphere in the generation of the Karoo volcanic rocks: evidence from combined Nd- and Sr-isotope studies, *Geol. Soc. S. Afr. Spec. Publ.*, *13*, 341-354, 1984.

Hawkesworth, C. J., M. S. M. Mantovani, P. N. Taylor, and Z. Palacz, Evidence from the Paraná of south Brazil for a continental contribution to DUPAL basalts, *Nature, 322*, 356-359, 1986.

Hawkesworth, C. J., M. S. M. Mantovani, and D. W. Peate, Lithosphere remobilisation during Paraná CFB magmatism, in *Oceanic and Continental Lithosphere; Similarities and Differences*, edited by M. A. Menzies and K. Cox, pp. 205-223, Journal of Petrology, Oxford, 1988.

Hawkesworth, C. J., K. Gallagher, S. Kelley, M. S. M. Mantovani, D. W. Peate, M. Regelous, and N. W. Rogers, Paraná magmatism and the opening of the South Atlantic, in *Magmatism and the Causes of Continental Break-up, Spec. Publ. 68*, edited by B. Storey, A. Alabaster, and R. Pankhurst, pp. 221-240, The Geological Society, London, 1992.

Henry, C. D., and J. A. Wolff, Distinguishing strongly rheomorphic tuffs from extensive silicic lavas, *Bull. Volcanol., 54*, 171-186, 1992.

Hergt, J. M., D. W. Peate, and C. J. Hawkesworth, The petrogenesis of Mesozoic Gondwana low-Ti flood basalts, *Earth Planet. Sci. Lett.*, *105*, 134-148, 1991.

Huang, Y. M., C. J. Hawkesworth, P. W. van Calsteren, and F. McDermott, Geochemical characteristics and origin of the Jacupiranga carbonatites, Brazil, *Chem. Geol.*, *119*, 79-99, 1995.

Iacumin, P., E. M. Piccirillo, and A. Longinelli, Oxygen isotopic composition of Lower Cretaceous tholeiites and Precambrian basement rocks from the Paraná basin (Brazil): the role of water-rock interaction, *Chem. Geol. (Isotope Geoscience), 86*, 225-237, 1991.

Kirstein, L. A., C. J. Hawkesworth, and S. P. Turner, CFB magmatism in southern Uruaguay: marginal to a mantle plume ? *J. Conf. Abs.*, 2, 44, 1997.

Leinz, V., A. Bartorelli, and C. A. Isotta, Contribuição ao estudo do magmatismo basáltico Mesozóic da bacia do Paraná, *An. Acad. bras. Ciênc., 40*, 167-181, 1968.

le Roex, A. P., R. A. Cliff, and B. J. I. Adair, Tristan da Cunha, South Atlantic: geochemistry and petrogenesis of a basanite-phonolite lava series, *J. Petrol., 31*, 779-812, 1990.

Light, M. P. R., M. P. Maslanyj, and N. L. Banks, New geophysical evidence for extensional tectonics on the divergent margin offshore Namibia, in *Magmatism and the Causes of Continental Break-up, Spec. Publ. 68*, edited by B. Storey, A.

Alabaster, and R. Pankhurst, pp. 257-270, The Geological Society, London, 1992

Long, P. E., and B. J. Wood, Structures, textures and cooling histories of Columbia River basalt flows, *Bull. Geol. Soc. Am.*, *97*, 1144-1155, 1986.

Mahoney, J. J., A. P. le Roex, Z. Peng, R. L. Fisher, and J.H. Natland, Southwestern limits of Indian Ocean ridge mantle and the origin of low $^{206}Pb/^{204}Pb$ mid-ocean ridge basalt: isotope systematics of the central Southwest Indian Ridge (17°-50°E), *J. Geophys. Res.*, *97*, 19,771-19,790, 1992.

Manley, C. R., How voluminous rhyolite lavas mimic rheomorphic ignimbrites: eruptive styles, emplacement conditions, and formation of tuff-like features, *Geology*, *23*, 349-352, 1995.

Mantovani, M. S. M., L. S. Marques, M. A. De Sousa, L. Civetta, L. Atalla, and F. Innocenti, Trace element and strontium isotope constraints on the origin and evolution of Paraná continental flood basalts of Santa Catarina state, southern Brazil, *J. Petrol.*, *26*, 187-209, 1985a.

Mantovani, M. S. M., U. G. Cordani, and A. Roisenberg, Geoquímica isotópica em vulcânicas ácidas da Bacia do Paraná e implicações genéticas associadas, *Rev. Bras. Geoc.*, *15*, 61-65, 1985b.

Mantovani, M. S. M., and C. J. Hawkesworth, An inversion approach to assimilation and fractional crystallisation processes, *Contrib. Mineral. Petrol.*, *105*, 289-302, 1990.

Mantovani, M. S. M., A. C. B. C. Vasconcellos, W. Shukowsky, E. J. Milani, M. Basei, S. J. Hurter, and S. R. C. de Frietas, The Brusque transect (SA20) from the Dom Feliciano belt to the Amazon Craton: explanatory pamphlet, *Global Geosciences Transect Project*, 1991.

Marsh, J. S., Relationships between transform directions and alkaline igneous rock lineaments in Africa and South America, *Earth Planet. Sci. Lett.*, *18*, 317-323, 1973.

Marsh, J. S., A. J. Erlank, and A. R. Duncan, Preliminary geochemical data for dolerite dykes and sills of the southern part of the Etendeka Igneous Province, *Communs. geol. Surv. Namibia*, *7*, 71-73, 1991.

Milner, S. C., The geology and geochemistry of the Etendeka Formation quartz latites, Namibia, Ph.D. thesis, University of Cape Town, Cape Town, South Africa, 1988.

Milner, S. C., A. R. Duncan, and A. Ewart, Quartz latite rheoignimbrite flows of the Etendeka Formation, northwestern Namibia, *Bull. Volcanol.*, *54*, 200-219, 1992.

Milner, S. C., A. P. le Roex, and J. M. O'Connor, Age of Mesozoic igneous rocks in northwestern Namibia, and their relationship to continental break-up, *J. Geol. Soc. Lond.*, *152*, 97-104, 1995a.

Milner, S. C., A. R. Duncan, A. M. Whittingham, and A. Ewart, Trans-Atlantic correlation of eruptive sequences and individual silicic units within the Paraná-Etendeka igneous province, *J. Volcanol. Geotherm. Res.*, *69*, 137-157, 1995b.

Milner, S.C., and A. P. le Roex, Isotope characteristics of the Okenyenya igneous complex, northwestern Namibia: constraints on the composition of the early Tristan plume and the origin of the EM1 mantle component, *Earth Planet. Sci. Lett.*, *141*, 277-291, 1996.

Mizusaki, A. M. P., R. Petrini, G. Bellieni, P. Comin-Chiramonti, J. Dias, A. De Min, and E. M. Piccirillo, Basalt magmatism along the passive continental margin of SE Brazil (Campos basin), *Contrib. Mineral. Petrol.*, *111*, 143-160, 1992.

Montes-Lauer, C. L., I. G. Pacca, A. J. Melfi, E. M. Piccirillo, G. Bellieni, R. Petrini, and R. Rizzieri, The Anari and Tapirapuã Jurassic Formations, western Brazil: palaeo-magnetism, geochemistry and geochronology, *Earth Planet. Sci. Lett.*, *128*, 357-371, 1995.

Morgan, W. J., Hotspot tracks and the opening of the Atlantic and Indian Ocean, in *The Sea: vol. 7; The Oceanic Lithosphere*, edited by C. Emiliani, pp. 443-487, Wiley, New York, 1981.

Nürnburg, D., and R. D. Müller, The tectonic evolution of the South Atlantic from Late Jurassic to present, *Tectonophysics*, *191*, 27-53, 1991.

O'Connor, J. M., and R. A. Duncan, Evolution of the Walvis Ridge - Rio Grande Rise hotspot system: implications for African and South American plate motions over plumes, *J. Geophys. Res.*, *95*, 17,474-17,502, 1990.

Peate, D. W., Stratigraphy and petrogenesis of the Paraná continental flood basalts, southern Brazil, Ph.D. thesis, The Open University, Milton Keynes, 1990.

Peate, D. W., C. J. Hawkesworth, M. S. M. Mantovani, and W. Shukovsky, Mantle plumes and flood basalt stratigraphy in the Paraná, South America, *Geology*, *18*, 1223-1226, 1990.

Peate, D. W., C. J. Hawkesworth, and M. S. M. Mantovani, Chemical stratigraphy of the Paraná lavas (South America): classification of magma types and their spatial distribution, *Bull. Volcanol.*, *55*, 119-139, 1992.

Peate, D. W., and C. J. Hawkesworth, Lithospheric to asthenospheric transition in Low-Ti flood basalts from southern Paraná, Brazil, *Chem. Geol.*, *127*, 1-24, 1996.

Peng, Z. X., and J. J. Mahoney, Drillhole lavas from the northwestern Deccan Traps, and the evolution of Réunion hotspot mantle, *Earth Planet. Sci. Lett.*, *134*, 169-185, 1995.

Petrini, R., L. Civetta, E. M. Piccirillo, G. Bellieni, P. Comin-Chiaramonti, L. S. Marques, and A. J. Melfi, Mantle heterogeneity and crustal contamination in the genesis of low-Ti continental flood basalts from the Paraná plateau (Brazil): Sr-Nd isotopes and geochemical evidence, *J. Petrol.*, *28*, 701-726, 1987.

Piccirillo, E. M., and A. J. Melfi (Eds), *The Mesozoic Flood Volcanism of the Paraná Basin: Petrogenetic and Geophysical Aspects*, 600 pp., IAG-USP, São Paulo, 1988.

Piccirillo, E. M., A. J. Melfi, P. Comin-Chiaramonti, G. Bellieni, M. Ernesto, L. S. Marques, A. J. R. Nardy, I. G. Pacca, A. Roisenberg, and D. Stolfa, Continental flood volcanism from the Paraná basin (Brazil), in *Continental Flood Basalts*, edited by J. D. McDougall, pp. 195-238, Kluwer, 1988.

Piccirillo, E. M., G. Bellieni, H. Cavazzini, P. Comin-Chiaramonti, R. Petrini, A. J. Melfi, J. P. P. Pinese, P. Zantadeschi, and A. de Min, Lower Cretaceous tholeiitic dyke swarms in the Ponta Grossa Arch (South East Brazil): petrology, Sr-Nd isotopes, and genetic relationships from Paraná flood volcanics, *Chem. Geol.*, *89*, 19-48, 1990.

Rampino, M. R., S. Self, and R. B. Stothers, Volcanic winters, *Ann. Rev. Earth Planet. Sci.*, *16*, 73-99, 1988.

Raposo, M. I. B., and M. Ernesto, Anisotropy of magnetic susceptibility in the Ponta Grossa dyke swarm (Brazil) and its relationship with magma flow direction, *Phys. Earth Planet. Int.*, *87*, 183-196, 1995.

Raup, D. M., and J. J. Sepkoski, Periodicity of extinctions in the geologic past, *Proc. Natl. Acad. Sci. USA*, *81*, 801-805, 1984.

Regelous, M., Geochemistry of dolerites from the Paraná flood basalt province, southern Brazil, Ph.D. thesis, The Open University, Milton Keynes, 1993.

Reid, D. L., The Cape Peninsula dolerite dyke swarm, South Africa, in *Mafic Dykes and Emplacement Mechanisms*, edited by A. J. Parker, P. C. Rickwood and D. H. Tucker, pp. 325-334, Balkema, Rotterdam, 1990.

Renne, P. R., M. Ernesto, I. G. Pacca, R. S. Coe, J. M. Glen, M. Prévot, and M. Perrin, The age of Paraná flood volcanism, rifting of Gondwanaland, and the Jurassic-Cretaceous boundary, *Science*, *258*, 975-979, 1992.

Renne, P. R., D.F. Mertz, M. Ernesto, L. Marques, W. Teixeira, H. H. Ens, and M.A. Richards, Geochronologic constraints on magmatic and tectonic evolution of the Paraná province (abstract), *Eos Trans. AGU*, *74*, 553, 1993.

Renne, P. R., Z. Zichao, M. A. Richards, M. T. Black, and A. R. Basu, Synchrony and causal relations between Permian-Triassic boundary crises and Siberian flood volcanism, *Science*, *269*, 1413-1416, 1995.

Renne, P. R., K. Deckart, M. Ernesto, G. Féraud, and E. M. Piccirillo, Age of the Ponta Grossa dike swarm (Brazil), and implications to Paraná flood volcanism, *Earth Planet. Sci. Lett.*, *144*, 199-212, 1996a.

Renne, P. R., J. M. Glen, S. C. Milner, and A. R. Duncan, Age of Etendeka flood volcanism and associated intrusions in southwestern Africa, *Geology*, *24*, 659-662, 1996b.

Richards, M. A., R. A. Duncan, and V. E. Courtillot, Flood basalts and hotspot tracks: plume head and tails, *Science*, *246*, 103-107, 1989.

Richardson, S. H., A. J. Erlank, A. R. Duncan, and D. L. Reid, Correlated Nd, Sr and Pb isotope variation in Walvis Ridge basalts and implications for the evolution of their mantle source, *Earth Planet. Sci. Lett.*, *59*, 327-342, 1982.

Rocha-Campos, A. C., U. G. Cordani, K. Kawashita, H. M. Sonaki, and I. K. Sonaki, Age of the Paraná flood volcanism, in *Mesozoic Flood Volcanism from the Paraná Basin (Brazil): Petrogenetic and Geophysical Aspects*, edited by E. M. Piccirillo and A. J. Melfi, pp. 25-45, IAG-USP, São Paulo, 1988.

Saunders, A. D., M. Storey, R. W. Kent, and M. J. Norry, Consequences of plume-lithosphere interactions, in *Magmatism and the Causes of Continental Break-up*, *Spec. Publ.*, *68*, edited by B. Storey, A. Alabaster, and R. Pankhurst, pp. 41-60, The Geological Society, London, 1992.

Stewart, K., S. Turner, S. Kelley, C. J. Hawkesworth, L. Kirstein, and M. S. M. Mantovani, 3-D ^{40}Ar-^{39}Ar geochronology in the Paraná flood basalt province, *Earth Planet. Sci. Lett.*, *143*, 95-110, 1996.

Sun, S.-s., Lead isotopic study of young volcanic rocks from mid-ocean ridges, ocean islands, and island arcs, *Philos. Trans. R. Soc. London. Ser. A*, *297*, 409-445, 1980.

Sun, S.-s. and W. F. McDonough, Chemical and isotopic systematics of oceanic basalts: implications for mantle composition and processes, in *Magmatism in the Ocean Basins*, *Spec. Publ. 42*, edited by A. D. Saunders and M. J. Norry, pp. 313-345, The Geological Society, London, 1989.

Thompson, R. N., M. A. Morrison, A. P. Dickin, and G. L. Hendry, Continental flood basalts..... Arachnids rule OK ?, in *Continental Basalts and Mantle Xenoliths*, edited by C. J. Hawkesworth and M. J. Norry, pp. 158-185, Shiva Press, Nantwich, 1983.

Thompson, R. N., and S. A. Gibson, Subcontinental mantle plumes, hotspots and pre-existing thinspots, *J. Geol. Soc. Lond.*, *147*, 973-977, 1991.

Toyoda, K., H. Horiuchi, and M. Tokonami, Dupal anomaly of Brazilian carbonatites: geochemical correlations with hotspots in the South Atlantic and implications for the mantle source, *Earth Planet. Sci. Lett.*, *126*, 315-331, 1994.

Turner, S. P., M. Regelous, S. Kelley, C. J. Hawkesworth, and M. S. M. Mantovani, Magmatism and continental break-up in the South Atlantic: high precision ^{40}Ar-^{39}Ar geochronology, *Earth Planet. Sci. Lett.*, *121*, 333-348, 1994.

Turner, S. P., and C. J. Hawkesworth, The nature of the sub-continental mantle: constraints from the major element composition of continental flood basalts, *Chem. Geol.*, *120*, 295-314, 1995.

Turner, S. P., C. J. Hawkesworth, K. Gallagher, K. Stewart, D. W. Peate, and M. S. M. Mantovani, Mantle plumes, flood basalts and thermal models for melt generation beneath continents: assessment of a conductive heating model and application to the Paraná, *J. Geophys. Res.*, *101*, 11,503-11,518, 1996.

Ulbrich, H.G.J., and C.B. Gomes, Alkaline rocks from continental Brazil, *Earth Sci. Rev.*, *17*, 131-154, 1981.

Ussami, N., A. Kolisnyk, M. I. B. Raposo, F. J. F. Ferreira, E. C. Molina, and M. Ernesto, Detectabilidade magnética de diques do Arco de Ponta Grossa: um estudo integrado de magnetometria aérea e magnetismo de rocha, *Rev. Brasil. Geoc.*, *21*, 317-327, 1991.

VanDecar, J. C., D. E. James, and M. Assumpção, Seismic evidence for a fossil mantle plume beneath South America and implications for plate driving forces, *Nature*, *378*, 25-31, 1995.

Venkatesan, T. R., K. Pande, and K. Gopalan, Did Deccan volcanism pre-date the Cretaceous/Tertiary transition ? *Earth Planet. Sci. Lett.*, *119*, 181-189, 1993.

White, R. S., and D. P. McKenzie, Magmatism at rift zones: the generation of volcanic continental margins and flood basalts, *J. Geophys. Res.*, *94*, 7685-7729, 1989.

White, R. S., and D. P. McKenzie, Mantle plumes and flood basalts, *J. Geophys. Res.*, *100*, 17,543-17,585, 1995.

Whittingham, A., Stratigraphy and petrogenesis of the volcanic formations associated with the opening of the South Atlantic, southern Brazil, Ph.D. thesis, University of Oxford, Oxford, 1991.

Yuen, D. A. and L. Fleitout, Thinning of lithosphere by small-scale convective destabilisation, *Nature*, *313*, 125-128, 1985.

Zalan, P. V., S. Wolff, J. C. J Conceicão, M. A. M Astolfi, A. S. Veira, V. T. Appi, O. A. Zanotto and A. Marques, Tectonics and sedimentation of the Paraná Basin, *Anais do symposio do Gondwana*, 35, 1987.

David W. Peate, Department of Earth Sciences, The Open University, Walton Hall., Milton Keynes, MK7 6AA, United Kingdom. (e-mail: d.w.peate@open.ac.uk).

Stratigraphy and Age of Karoo Basalts of Lesotho and Implications for Correlations Within the Karoo Igneous Province

J. S. Marsh,[1] P. R. Hooper,[2] J. Rehacek,[2] R. A.Duncan,[3] and A. R. Duncan[4]

The Lesotho remnant contains the type succession for Karoo low-Ti basalts of central southern Africa. The $^{40}Ar/^{39}Ar$ dating indicates that the sequence was emplaced within a very short period at about 180 Ma and consists of a monotonous pile of compound basalt lava flows which lacks significant palaeosols and persistent sedimentary intercalations. We have used geochemistry to establish a stratigraphic subdivision of the lava pile. Thin units of basalt flows, the Moshesh's Ford, Golden Gate, Sani, Roma, Letele, and Wonderkop units, with diverse geochemical character and restricted geographical distribution, are present at the base of the succession. These are overlain by extensive units of compositionally more uniform basalt, the Mafika Lisiu, Maloti, Senqu and Mothae units, which build the bulk of the sequence. A single palaeomagnetic polarity reversal occurs within the lower third of the basalt succession and is consistently located within the Mafika Lisiu unit. This and the persistent and relatively uniform thickness of the stratigraphic units suggest that the pile was constructed in a uniform manner by eruption of basalt onto a generally planar surface from a widespread plexus of dykes. The stratigraphic sequence in Lesotho closely resembles that in the thinner sequence of low-Ti basalts of the Springbok Flats remnant, some 400 km to the north. A thin unit of high-Ti basalt within the upper part of the Springbok Flats sequence can be correlated with the thick high-Ti basalt suite along the rift-related Lebombo structure on the eastern margin of the Karoo province. This is the first established correlation between these two important outcrops of Karoo volcanic rocks and demonstrates that the low-Ti basalts of Lesotho and the cratonic interior are the approximate time equivalents of the lower part of the Lebombo sequence. This conclusion has important implications for models for the origin of the Karoo flood basalt province.

[1]Department of Geology, Rhodes University, South Africa

[2]Department of Geology, Washington State University, Pullman, Washington

[3]College of Oceanic and Atmospheric Sciences, Oregon State University, Corvallis, Oregon

[4]Department of Geological Sciences, University of Cape Town, Rondebosch, South Africa

Large Igneous Provinces: Continental, Oceanic, and Planetary Flood Volcanism
Geophysical Monograph 100

1. INTRODUCTION

Remnants of the erupted and intrusive products of the Karoo igneous province, one of the classic Mesozoic continental flood basalt provinces, are found throughout southern Africa. Such large continental igneous events are frequently ascribed to the rise of deep-seated mantle plumes and, in some instances, are thought to be related to continental breakup. Plumes can provide the anomalously high temperatures at shallow depth necessary for generation of the large volumes of basaltic magma over a short period, a characteristic of flood basalt provinces.

However, there is debate concerning plume-lithosphere

interactions, melt generation in the plume, and how the compositions of erupted basalts relate to specific sources in the plume and/or the overlying lithosphere. These questions can be addressed through geochemistry of the basalts but require comprehensive documentation of the compositions of the different geochemical types of basalt, their volumes, distribution, and age relationships.

Some recent authors [*Morgan*, 1981; *White and McKenzie*, 1989; *Campbell and Griffiths*, 1990] propose plume models for the origin of the Karoo province. The lack of a thorough knowledge of the ages, distribution, stratigraphy and geochemical character of all the igneous products of the Karoo province render their analyses unsatisfactory. For example, there is considerable documentation of the types of igneous materials that were erupted and their broad isotopic and geochemical features but, apart from the central section of the Lebombo monocline, very little is known about the details of the geochemical stratigraphy in the different erosional remnants, or how they correlate with one another. Ages are also poorly understood. In contrast to the short-lived nature of the activity in many other flood basalt sequences, available data for Karoo volcanism have been interpreted as indicating several pulses of activity over many tens of millions of years [*Fitch and Miller*, 1984]. The existence of a high-Ti, low-Ti provinciality has been known for decades, but its geographical structure has not been fully documented. Another outstanding problem is the lack of precise knowledge regarding the eruption sites for the volcanic sequence. Eruption site localities are more important than lava remnant distributions for understanding the emplacement of the Karoo province in relation to plume tectonics.

A number of detailed studies are currently underway to address these deficiencies. This paper describes results of a study aimed at establishing the stratigraphic sequence in the principal outcrop of the low-Ti basalt sequence, the Lesotho remnant, where the volcanic succession, more than 1.5 km thick, is spectacularly exposed in a deeply dissected, uplifted plateau. The basis for the stratigraphic subdivision is geochemical. Compositional variations within the basalt sequence are slight and we argue at length for the validity of the stratigraphic subdivision because they are proving crucial in establishing correlations amongst other low-Ti basalt remnants. We also show that the stratigraphic sequence is consistent with recent age and palaeomagnetic data. Finally, we discuss the wider implication of the new results; the insight they provide into the construction of the lava pile, and their use in allowing, for the first time, a stratigraphic correlation between the Lebombo and Lesotho remnants to be established. The comprehensive compositional characterization of the low-Ti basalt types of Lesotho provides the basis for detailed petrogenetic modelling, a topic to be addressed in a future paper.

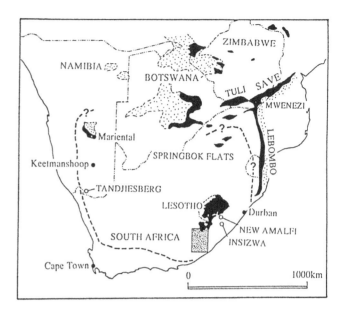

Figure 1. Map of the Karoo igneous province showing distribution of remnant basalt outcrops (black) and subsurface extent (light stippled area) and localities referred to in the text. The limit of the Central Area is indicated by the dashed line. The heavily stippled box is the area shown in detail in Figure 2.

2. OVERVIEW OF THE KAROO PROVINCE

Salient features of the province have been summarised by *Eales et al.* [1984] and *Cox* [1988] and we mention only some of the important features here. Figure 1 shows the distribution of volcanic outcrops of the province, and it is important to emphasize that these are erosional remnants of a more extensive carapace of lava which, on a wealth of circumstantial evidence, we believe to have covered much of southern Africa. These eruptions were the culmination of a long period of intracontinental sedimentation to form the Karoo sedimentary sequence and its correlatives throughout much of southern Africa and neighbouring Gondwana continents. Areas between these erosional remnants are intruded by a network of dykes and sills which are particularly well developed in the sedimentary sequences of the main and subsidary Karoo basins (Figure 2). Apart from the Botswana and Rooi Rand dyke swarms, much of the dyke network in the Karoo province is diffuse, and although systematic orientations of dykes occur in some areas, their development is not intense. Although quantification is difficult, the volume of shallow-level intrusions in the Karoo basin may have rivaled that of the erupted products. Previous geochemical studies have shown the overall similarity between the compositions of the dykes and sills and the associated basalts. By considering the intrusive and extrusive suites together, the overiding impression is one of the widespread availability of

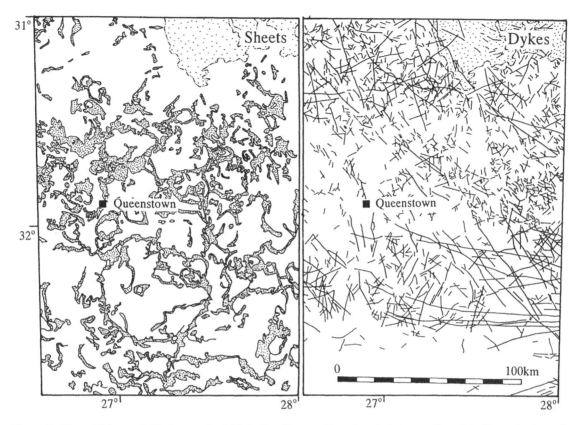

Figure 2. Map of dykes and sills (heavy stipple) intruding Karoo sedimentary strata in a portion of the Karoo basin south of Lesotho (see Figure 1 for location). Basalt outcrop shown in light stipple in top right-hand corner.

huge volumes ($1–2 \times 10^6$ km^3) of magma beneath southern Africa during the Karoo event.

To facilitate description and discussion, it is convenient to subdivide the Karoo province into subareas. *Eales et al.* [1984] identified a Central Area (Figure 1) that embraces all the Karoo volcanic remnants of the interior of southern Africa and their associated vast network of intrusive dykes and sheets. The volcanic remnants included in the Central Area are the main Lesotho remnant, the volcanic sequence near Mariental in Namibia, the Springbok Flats some 400 km north of Lesotho, and the vast outcrops and subcrops of Botswana as well as the intrusive suites associated with the main Karoo basin. We retain this grouping for convenience, with the exception of Botswana. The principal elements in grouping these different areas into a Central Area are tectonic and petrographic; i.e., rock types are overwhelmingly tholeiitic basalts of similar compositions and were emplaced in a stable cratonic environment.

Marginal to the Central Area are the thick rift-related sequences of the Tuli syncline, the Save-Mwenezi (previously Sabi-Nuanetsi) area and the 700-km-long Lebombo mono-cline. Thick and diverse assemblages of volcanic rocks were erupted in these areas, including nephelinites, tholeiitic basalts

and picrites and, along the Lebombo and Save-Mwenezi structures, a thick sequence of rhyolite. In the Mwenezi area, volcanism was also accompanied by ring complexes of granite, gabbro, syenite, and nepheline syenite.

A further north-south geochemical subdivision of the Karoo province based on a high-Ti and low-Ti provinciality within basaltic rocks was recognized by *Cox et al.* [1967]. This provinciality has been documented and discussed by *Erlank et al.* [1988] and *Sweeney and Watkeys* [1990] and is now known to occur in all Mesozoic flood basalt provinces of southern Gondwana. Although the exact boundaries between the two types are subject to debate, it is clear that they transcend Mesozoic tectonic boundaries and do not coincide with the geographical areas used here.

3. SAMPLING AND ANALYTICAL TECHNIQUES

Nine serial sections (Figure 3) were sampled in the deeply dissected and elevated basalt plateau in and adjacent to Lesotho, and in each section attempts were made to sample every flow. Detailed descriptions of sample localities are given in the Appendix. The altitude of each sample site was determined from altimeter readings (precision ±5 m) which

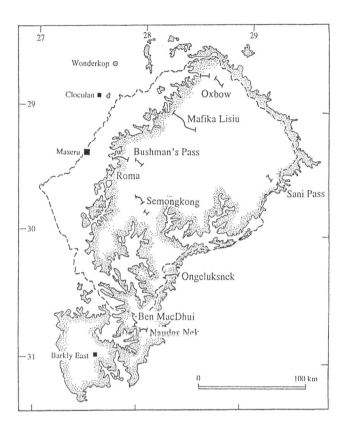

Figure 3. The basalt remnant of Lesotho and Eastern Cape showing sampling sections (labeled lines) and other localities. International boundary shown as a dashed line.

were corrected to absolute values above mean sea level by reference to several known spot heights on topographic maps. In all sections except the Bushman's Pass and Roma sections (Figure 3), several short cores were drilled at each sample site. These cores were used for whole-rock chemical analysis and palaeomagnetic studies at Washington State University (WSU) and for $^{40}Ar/^{39}Ar$ dating at Oregon State University (OSU). In addition, at the majority of sample sites, 1-2 kg whole rock samples were taken by hammer, usually within a few metres of the core sample sites. These larger samples, together with some of the core samples, were analyzed at Rhodes University (RU). Samples of Springbok Flats basalt were obtained from four borehole cores drilled through the sequence and were also analyzed at RU. At WSU major and trace elements were determined by wave-length dispersive X ray fluorescence (XRF) on 2:1 dilution fusion beads as described by *Hooper et al.* [1993]. The WSU data do not contain determinations for loss of volatiles. In addition, a subset of samples was analyzed for rare-earth elements (REE), Hf, Th, Ta, Pb, and U by inductively coupled plasma mass spectrometry, also at WSU. At RU major elements were determined by XRF using the fusion technique of *Norrish and Hutton* [1969], and trace

elements and Na were determined on undiluted pressed powder pellets as described by *Duncan et al.* [1984] and *le Roex* [1985]. Loss on ignition (LOI) and H_2O^- were determined gravimetrically. Strontium isotopes were determined at the Radiogenic Isotope Facility, University of Cape Town. Standard preparation techniques were employed and analyses were carried out on a VG Sector multicollector mass spectrometer applying a mass fractionation correction of $^{86}Sr/^{88}Sr = 0.1194$. Measured $^{87}Sr/^{86}Sr$ ratios have been normalised to a value of 0.71022 for the standard SRM-987.

Selected results are in Tables 1-3. Some 87% of the collection of 469 samples from Lesotho have been analyzed at both RU and WSU. Thus our collective data contain all the elements of two separate geochemical studies of the same sequence. Table 1 contains data that allow comparison of the WSU and RU data sets. For some elements there are inconsistent differences which are beyond simple analytical precision. These differences reflect the influence of a variety of different methods of sampling, sample preparation, and X ray analytical techniques and philosophy. Despite detailed review of our data it has not been possible to determine the causes of concentration differences. Nevertheless, use of the two sets of data, individually or in conjunction, yields consistent results in the applications described below.

4. OVERVIEW OF STRATIGRAPHIC VARIATIONS

Figure 4 shows the variation of some important compositional parameters with height through the Oxbow section, one of the most complete sections through the thickest part of the basalt sequence. This section has all of the important features shown by other profiles and, for descriptive purposes, illustrates the general vertical geochemical variations within the basalts. The main features are as follows.

1. The chemical variability in the lowermost flows, in terms of element abundances and incompatible element ratios, is large compared to the main overlying sequence.

2. Within the main sequence, there is a complex pattern of variation characterized by several sharp inflections in element abundances and interelement ratios. However, there is a general upward trend of subdued but increasing differentiation, with the uppermost flows showing a sharp trend to more evolved basaltic compositions (low magnesium number (Mg#), Ni, Cr; high Ti, P, Zr, Fe, etc.).

3. Dykes and other intrusions in the section generally have compositional characteristics of basalts higher in the sequence and can be correlated on the basis of geochemistry with known units within the basalts. However, in the Oxbow section several dykes and sheets high in the volcanic sequence have compositions which were not found amongst the basalts and may have fed flows that have been removed by erosion.

These features confirm results from earlier, more limited studies of *Cox and Hornung* [1966], *Marsh and Eales* [1984] and *Marsh* [1984]; specifically, there is a generalized trend to more evolved compositions upward in the sequence; thin basaltic units of diverse geochemistry are located at the base of the sequence; and the main part of the sequence exhibits rather limited compositional variability that is a challenge for stratigraphic subdivision. This study is aimed at a fuller exploration of the geochemical diversity amongst the lower flows and at subdividing the main sequence of flows, previously termed the Lesotho Formation by several authors [*Lock et al.*, 1974; *Marsh and Eales*, 1984].

5. THE BASIS FOR GEOCHEMICAL SUBDIVISION

The compound nature of many of the basalt flows, the presence of only one palaeomagnetic reversal near the base of the pile, and the absence of widespread sedimentary or weathering horizons leave basalt composition as the only reasonable basis for establishing a stratigraphic subdivision of the lava pile. Thus, subdivision makes use of changes in geochemical parameters, specifically element ratios of immobile high-field-strength incompatible elements and, to a lesser extent, element abundances, to indicate boundaries between sequences of flows having specific and more uniform compositional characteristics. Assessing the significance of compositional differences between adjacent samples in a stratigraphic sequence is somewhat subjective and due regard has to be given to the following:

1. The stratigraphic subdivision is only useful if it can be used by others; a repeat study should arrive at the same result. Because of the subtle nature of the variations in the sequence, we have relied on both RU and WSU data sets (i.e., effectively two separate studies) to establish the subdivisions.

2. Within-flow variation. Some flows are internally differentiated but most are not. However, all flows have undergone low-temperature alteration leading to deposition of quartz, chalcedony, and zeolite in vesiculated tops and bottoms of flows. Although we have focussed sampling on the massive, amygdale-free interior zones of the flows, our duplicate study has shown that concentrations of some elements may vary with both sample size and sample site within such zones. Our study also included multiple sampling (12 samples) of both the massive and amygdaloidal parts of a single flow. We have used results of these secondary studies and the assessment of element mobility in Karoo basalts made by *Marsh and Eales* [1984] to assist us in determining the possible stratigraphic significance of geochemical variations between adjacent flows.

3. The recognition of a given unit in its correct stratigraphic position in several sections. Stratigraphic subdivision only has meaning if individual units are persistent over large areas. Thus, the identification of a specific unit in two or more sections, in its correct relative position, is important in confirming the validity of the subdivision.

Figures 5a-c illustrate variation of some compositional parameters with height at four widely spaced sections. These diagrams illustrate the repetition of geochemical variations in several sections and the persistence of groups of flows with distinctive compositions at more or less the same stratigraphic position through all sections. This is emphasized by the dashed lines in Figure 5. We have employed diagrams like these together with more conventional variation diagrams to subdivide the succession to produce a geochemical stratigraphy. The proposed subdivision of the volcanic sequence in Lesotho is illustrated in Figure 6, showing stratigraphic columns for each of the major sections sampled by us.

Also shown in Figure 6 is the position of the palaeomagnetic reversal first established by *Van Zijl et al.* [1962] at Sani Pass and the Bushman's-Motimo Nthuze-Thaba Putsoa sequence of passes east of Maseru. We confirmed the reversal's location in Sani Pass, and a French team from Universite Sciences et Techniques du Languedoc (M. Prevot, personal communication, 1993) has confirmed the reversal and transition at Bushman's Pass. The positions of the reversal at Oxbow, Sani Pass, Ongeluksnek, Mafika Lisiu, and Ben McDhui are from our current study. The reversal provides an independent check on the validity of the geochemical stratigraphy, and this aspect will be discussed later.

The most important feature of Figure 6 is the subdivision of the main sequence of lava flows. Following *Lock et al.* [1974], *Marsh and Eales* [1984] referred informally to this main sequence as the Lesotho Formation and the basalt forming these flows as the Lesotho type. We suggest that the name Lesotho Formation be retained but that it be recognised that the formation is built of a number of basaltic magma types. We propose to abandon the term 'Lesotho magma type' in favour of the new names proposed here. Similarly, it is convenient to group the lowermost units of diverse geochemical character into a single stratigraphic unit for which we propose the name Barkly East Formation.

Below we describe the essential features of the geochemical stratigraphy and justify the definition of the different units. Most of our data are from sections concentrated in central and northern Lesotho, and we found it difficult to correlate the lower part of the succession in these northern sections with the lower part of the sequence in the two southern sections at Ongeluksnek and Ben McDhui. Thus, for convenience, the stratigraphy of the northern sections is discussed separately from that of the southern sections and correlations are drawn where appropriate. It is also convenient to discuss the lower units (Barkly East Formation), up to and including the

TABLE 1. Selected Analyses of Basalts from Units of the Barkly East Formation

	Moshesh's Ford			Golden Gate				Sani		
	BMC-1	BMC-1	ON-1	ROM-10	BUS-2	BUS-2	MLP-186	ROM-8	SP-47	SP-47
LAB**	WSU	RU	WSU	WSU	WSU	RU	WSU	WSU	WSU	RU
SiO_2	52.62	51.47	51.22	52.56	53.52	51.79	52.98	52.10	50.73	50.12
TiO_2	15.92	15.47	16.13	15.07	14.87	14.88	13.65	14.60	14.94	14.73
Al_2O_3	1.01	0.99	0.99	1.00	1.07	1.03	0.97	0.97	1.02	1.01
FeO*	9.44	9.21	9.18	10.08	9.30	9.93	10.86	9.85	9.36	9.53
MnO	0.16	0.16	0.16	0.17	0.17	0.17	0.17	0.17	0.16	0.18
MgO	10.16	9.92	10.64	10.23	10.06	10.08	8.83	10.20	11.46	11.23
CaO	6.13	6.14	6.26	6.79	5.86	6.54	7.61	8.61	6.85	6.56
Na_2O	1.03	0.99	0.38	0.69	0.71	0.65	0.92	0.62	0.53	0.46
K_2O	2.64	2.33	3.40	2.41	2.32	2.32	2.40	2.13	2.19	2.06
P_2O_5	0.196	0.191	0.166	0.132	0.141	0.143	0.146	0.155	0.167	0.159
LOI		2.46				2.21				2.59
H_2O^-		0.90				0.72				1.20
TOTAL	99.30	100.21	98.53	99.14	98.02	100.46	98.54	99.40	97.41	99.83
Ni	57	68	86	96	89	115	95	108	80	96
Cr	323	290	332	276	239	238	190	434	399	388
Sc	36	26.9	33	31	25	27.9	32	34	34	35.2
V	251	212	251	263	236	231	220	239	253	254
Ba	230	270	144	200	191	229	281	141	137	165
Rb	21	21	9	16	16	15	24	15	12	9.3
Sr	276	278	299	158	159	162	198	135	158	157
Zr	137	132	124	122	125	123	134	93	100	95
Y	27	27.4	26	27	30	30.8	27	25	24	25.7
Nb	18	20	16	4.9	4.2	3.6	7.0	4.5	3.9	2.4
Ga	19		16	21	21		23	17	18	
Cu	86	68	92	93	101	102	50	87	91	96
Zn	82	81	76	99	94	87	99	82	84	86
$^{87}Sr/^{86}Sr$			0.70594	0.70916	0.70877		0.70978	0.70877	0.70718	
$(^{87}Sr/^{86}Sr)_0$			0.70542	0.70840	0.70808		0.70890	0.70801	0.70674	

** - analyses from Washington State University (WSU) and Rhodes University (RU)

* - all Fe as FeO

Wonderkop unit, separately from the overlying main sequence units (Lesotho Formation).

5.1. *The Northern Sections*

5.1.1. *The Barkly East Formation*. Selected analyses of samples from the basaltic units of this formation are given in Table 1. Because the WSU data set is more complete, these data are used in variation diagrams illustrating the composition of these units. WSU data, together with selected duplicate analyses from the RU data set, have been used to compile Table 1. Plots of Mg# vs incompatible element concentrations for samples from the lower units (Figure 7) show strong negative correlation indicative of considerable differentiation. For all elements except P, Nb, Ta, Sr, and U, the data for all units are tightly clustered (Figure 7 b and c). However, for these five elements, variations with Mg# consist

of two or more stacked trends (Figure 7a). Thus, any plot of incompatible elements involving P, Nb, Ta, U, and Sr defines two or more parallel to subparallel or converging trends (Figures 8, 9). Within these trends, samples from different stratigraphic horizons form distinct clusters. Some of the trends do not project towards the origin of the graphs, and hence incompatible-element ratios, especially those involving the five elements identified above, should be effective discriminators between groups of samples from different stratigraphic horizons. Figure 10 shows a number of ratio-ratio plots where geochemical discrimination between different units is clearly illustrated.

In all these sections the basal unit is the Golden Gate unit, first recognized by *Marsh* [1984] in outliers of the main basaltic lava pile in the northeastern Free State (previously Orange Free State). Within the geochemical context of the basalts from Lesotho, Golden Gate basalts are enriched in Zr,

TABLE 1. (continued)

Roma		Letele		Wonderkop					
BUS-4	ROM-3	OXB-63	OXB-63	OXB-60	SP-45	MLP-183	MLP-183	R0M-11	ROM-11
WSU	WSU	WSU	RU	WSU	WSU	WSU	RU	WSU	RU
50.49	51.69	51.05	49.11	51.38	51.64	53.23	52.19	52.58	50.43
13.85	14.95	12.98	12.65	15.46	15.12	15.99	15.54	15.52	15.27
0.85	0.87	0.81	0.85	0.85	1.16	0.94	1.05	0.95	0.91
9.97	9.99	10.22	10.9	8.60	9.81	8.17	9.41	9.26	8.12
0.17	0.20	0.19	0.17	0.16	0.15	0.14	0.15	0.17	0.17
10.43	11.12	8.78	8.49	10.72	10.98	10.56	10.31	10.85	10.73
7.73	7.79	12.9	12.01	6.39	6.56	6.90	5.71	6.85	6.95
0.56	0.24	0.46	0.44	0.50	0.65	0.71	0.84	0.65	0.60
1.94	2.08	1.70	1.62	2.76	2.51	2.33	2.45	2.36	1.95
0.096	0.100	0.105	0.105	0.123	0.185	0.153	0.192	0.146	0.146
			2.97				1.92		2.36
			0.58				0.91		1.57
96.08	99.02	99.20	99.90	96.94	98.76	99.12	100.66	99.34	99.21
114	97	247	259	74	81	76	82	83	100
468	445	901	805	298	314	363	279	350	324
34	36	35	27.9	40	38	37	30.5	39	32.8
251	256	231	215	239	272	242	229	256	230
108	92	88	136	107	143	125	230	126	151
15	7	11	12	10	10	13	17	12	12
116	129	100	107	191	224	222	239	213	217
83	84	74	74	77	108	105	122	99	89
25	26	21	23	23	26	24	28.9	23	23.4
3.3	3.0	3.6	4.0	8.3	10.5	13	16	10.5	9.9
19	20	15		18	21	17		19	
96	96	47	69	87	105	89	85	101	81
80	79	85	89	72	82	74	76	110	77
0.70994		0.70887				0.70609		0.70588	
0.70893		0.70805				0.70556		0.70547	

Hf, Pb, La, Th relative to Nb, Ta, heavy REE, P, Y, U, and Ti, leading to characteristically high Zr/Nb, Zr/Y, La/Yb, Th/Nb and La/Nb and low Ti/Zr, P/Zr, and U/Pb (Figures 9 and 10). Our new data show that the flows in the upper part of the unit have higher incompatible-element abundances and La/Yb and Ti/Zr compared to the lower flows and this forms the basis for an upper and lower subdivision. This distinction cannot be made for other ratios, and collectively all flows within this unit have similar compositions which are easily distinguished from all other units. The extent of the Golden Gate unit to the south is unknown and awaits results of further studies. To the north and west, in the basalt outliers of the Free State, this unit thins and is overstepped by units higher in the sequence.

Overlying the Golden Gate flows at Bushman's Pass and Roma is the Roma unit (Figure 6). It shows the same relative enrichment amongst incompatible elements as the Golden Gate flows except for distinctively high Th/Ta and Ti/Zr, and low Zr/Y and P/Zr (Figure 10). It appears to have a limited distribution, not having been found in sections to the north of

Bushman's Pass, and its extent to the southwest is unknown.

The Sani unit Overlies the Golden Gate unit at Sani Pass. A single sample, ROM-8, from the Roma section has also been grouped into this unit. The Sani unit is similar in composition to the Golden Gate basalts but is clearly distinguished from them by high P/Zr and Ti/Zr (Figure 10). The extent of this unit along the eastern escarpment is still unknown and more sections to the north and south of Sani Pass need to be sampled. The correlation of ROM-8 with the Sani unit suggests that this unit may extend east-west across Lesotho. Although ROM-8 shares many geochemical features with the Sani Pass samples, it has much higher Th relative to other incompatible elements (Figure 9) and it is worrisome that if the Sani unit is extensive enough to reach east-west across Lesotho, it was not found in the neighbouring section at Bushman's Pass (Figure 6). However, on the evidence of ROM-8, this unit is very thin in the west and may have been missed during sampling.

The Letele unit, which is characterized by high Ti/Zr and

TABLE 2. Selected analyses (RU data) of basalts from the Main Sequence (Lesotho Formation)

	Mafika Lisiu						Mafika Lisiu - low Zr/Nb			
	MLP-155	MLP-172	OXB-43	SP-32	SP-36	ON-35	MLP-02	MLP-166	SOM-92	BMC-07
SiO_2	50.22	50.69	49.94	49.60	50.04	50.13	49.74	50.79	49.66	50.08
TiO_2	0.97	0.98	0.88	0.92	1.07	0.88	0.99	0.96	0.84	0.90
Al_2O_3	15.54	15.38	15.04	14.70	14.92	15.01	14.87	14.83	15.17	15.11
FeO*	9.83	9.60	9.44	9.71	10.04	9.76	9.38	9.79	9.01	8.49
MnO	0.17	0.17	0.16	0.17	0.18	0.17	0.16	0.18	0.15	0.15
MgO	7.36	7.01	7.29	7.48	6.59	8.56	7.33	7.87	6.39	6.33
CaO	10.21	10.96	10.34	10.61	10.53	10.78	10.15	10.32	9.86	10.79
Na_2O	2.06	2.30	2.05	1.91	2.21	2.21	2.01	2.40	2.26	2.25
K_2O	0.22	0.51	0.38	0.58	0.65	0.44	0.69	0.70	1.03	0.55
P_2O_5	0.169	0.163	0.138	0.171	0.199	0.147	0.195	0.181	0.165	0.163
LOI	3.14	1.56	2.81	3.00	2.51	1.70	2.77	1.79	3.22	3.73
H_2O^-	0.37	1.13	1.72	1.65	0.89	0.69	1.81	0.71	1.90	1.17
TOTAL	100.25	100.45	100.19	100.50	99.83	100.47	100.08	100.53	99.65	99.69
Ba	130	192	155	165	217	172	175	227	220	203
Sc	30.6	30.8	33.7	30.4	29.5	30.4	29.9	29.6	30.9	34.6
Zn	85	79	81	76	76	73	87	75	82	82
Cu	94	87	74	90	65	80	78	120	75	84
Ni	97	87	92	80	79	107	75	116	79	96
Nb	6.2	6.4	4.2	8.9	10.6	5.3	8.5	11.2	9.3	8.5
Zr	85	75	76	82	88	69	84	95	81	74
Y	25.5	23.2	23.5	24.2	25.1	22.7	25.6	24.7	24.2	23.9
Sr	251	222	176	198	231	202	202	208	187	186
Rb	2.1	10.0	8.8	10.2	12.0	8.0	10	14.2	34	9.2
Co	44	43	44	45	44	48	45	46	41	44
Cr	214	301	345	266	289	458	195	389	274	304
V	219	245	226	236	231	231	283	225	218	252
Ce	25	24	22	29	24	19	28	31	27	24
Nd	13	13	11	13	14	10	13	15	12	12
La	11	10	8	11	13	9	12	13	11	10
$^{87}Sr/^{86}Sr$	0.70657	0.70551		0.70572	0.70544	0.70529				
$(^{87}Sr/^{86}Sr)_o$	0.70611	0.70515		0.70534	0.70502	0.70497				

	Maloti						Lower Senqu					
	MLP-148	MLP-17	OXB-40	BMC-14	ON-04	BUS-31	BUS-40	ON-07	BMC-25	SP-13	MLP-31	OXB-21
SiO_2	49.82	50.60	50.72	49.58	50.61	51.63	49.86	49.38	49.20	49.64	47.42	49.77
TiO_2	0.89	0.99	0.89	0.92	1.07	0.87	1	1.07	0.92	0.94	0.93	1.03
Al_2O_3	14.81	14.81	14.55	15.46	14.92	14.51	15.4	14.35	15.17	15.21	15.73	14.78
FeO*	9.75	10.34	9.79	9.64	10.60	9.84	10.22	10.92	10.05	10.22	9.38	10.42
MnO	0.17	0.18	0.16	0.16	0.19	0.16	0.19	0.18	0.18	0.18	0.16	0.18
MgO	7.17	6.89	6.91	7.22	5.89	6.75	6.74	6.42	6.44	6.61	6.28	6.22
CaO	10.6	10.36	10.64	10.47	10.40	10.90	10.81	10.28	10.24	10.45	10.87	10.22
Na_2O	1.95	2.27	2.04	1.92	2.48	2.11	2.09	1.94	2.13	2.11	1.78	2.10
K_2O	0.62	0.64	0.54	0.45	0.64	0.37	0.54	0.56	0.65	0.56	0.46	0.80
P_2O_5	0.160	0.171	0.156	0.161	0.180	0.161	0.180	0.190	0.151	0.153	0.164	0.162
LOI	2.56	1.96	2.26	2.27	1.75	2.47	2.43	2.65	3.48	2.78	5.23	2.97
H_2O^-	2.17	1.14	1.07	1.52	0.89	0.84	1.5	1.71	1.49	1.07	0.92	1.21
TOTAL	100.67	100.33	99.73	99.78	99.62	100.61	100.96	99.65	100.09	99.92	99.31	99.86
Ba	208	225	213	192	227	190	187	200	164	195	157	245
Sc	32.6	32.2	32.8	32.5	33.7	33.8	32.3	33.7	38.5	31.3	31.9	32.9
Zn	84	88	89	86	85	82	84	93	85	82	90	85
Cu	80	88	77	87	96	89	96	101	113	54	94	85
Ni	83	82	84	92	64	93	75	68	73	79	76	67
Nb	6.8	6.7	8.7	6.8	8.2	8.0	7.1	7.8	4.9	5.3	6.1	4.8
Zr	90	99	92	83	105	93	87	98.2	80	85	87	99
Y	25.4	27.6	21.9	24.7	29.5	25.6	25.5	28	23.7	25.2	24.9	26.7
Sr	177	191	193	199	196	198	200	194	206	184	167	182
Rb	11.5	13.7	6.9	4.9	14	3.4	7.2	8.2	9.9	12	10.7	18
Co	43	44	44	46	42	46	44	44	42	45	44	43
Cr	220	194	268	209	179	279	177	151	256	193	132	223
V	235	231	227	226	246	234	248	252	273	240	243	244
Ce	30	32	25	28	31	31	26	28	21	22	29	30
Nd	14	15	14	13	16	15	15	14	11	13	13	16
La	12	12	14	9	13	12	11	13	8	12	9	13
$^{87}Sr/^{86}Sr$	0.70625		0.70615				0.70580					
$(^{87}Sr/^{86}Sr)_o$	0.70579		0.70589				0.70553					

TABLE 2. (continued)

	Upper Senqu						Mothae				Oxbow	
	MLP-115	MLP-39	OXB-11	BMC-36	SOM-04	SP-01	OXB-02	OXB-09	MLP-61	MLP-65	OXB-22	OXB-14
SiO_2	51.09	48.61	50.98	49.25	49.73	49.56	50.67	48.57	47.59	50.14	50.37	50.17
TiO_2	1.02	0.99	0.96	0.96	0.99	1.06	1.46	0.76	1.20	1.31	1.55	1.55
Al_2O_3	14.87	14.15	14.93	14.75	14.88	14.11	13.43	11.52	14.51	14.02	13.55	13.54
FeO^*	10.37	10.24	10.23	9.76	10.00	10.75	12.55	11.82	11.18	11.78	13.52	13.56
MnO	0.18	0.18	0.16	0.19	0.18	0.23	0.21	0.20	0.19	0.21	0.23	0.22
MgO	6.24	7.61	6.32	7.08	6.34	6.32	5.21	11.65	5.58	5.81	5.51	5.28
CaO	10.02	10.60	10.60	10.60	10.26	10.27	9.66	9.90	10.01	10.01	10.11	10.26
Na_2O	1.95	1.87	2.12	1.99	2.13	1.90	2.23	1.71	2.16	2.30	2.46	2.63
K_2O	0.67	0.65	0.52	0.64	1.19	0.68	0.82	0.60	0.63	0.69	0.52	0.28
P_2O_5	0.170	0.168	0.152	0.164	0.171	0.178	0.234	0.145	0.197	0.214	0.186	0.191
LOI	2.40	2.66	2.66	2.95	2.97	3.57	2.27	2.05	5.23	2.37	1.94	2.10
H_2O^-	1.9	1.92	1.05	1.62	1.02	1.38	1.28	1.08	1.53	1.28	0.44	0.47
TOTAL	101.09	99.64	100.68	99.94	99.86	100.01	100.024	100.01	100.01	100.15	100.386	100.25
Ba	201	224	185	213	247	196	279	169	200	260	190	172
Sc	34.6	34.6	34.3	34.8	34.6	35.9	35.6	37.1	38.6	35.5	34.2	33.3
Zn	87	90	93	79	79	79	97	87	99	93	106	116
Cu	102	103	101	90	99	91	150	69	110	121	210	212
Ni	55	76	73	70	62	54	52	178	57	60	62	63
Nb	6.8	6.8	7.3	5.3	6.6	8	9.1	4.7	8.6	7.5	6	6.3
Zr	97	88	91	86	92	96	132	69	109	114	116	120
Y	26.7	25.9	24.8	26.6	25.6	27.7	34	20.7	30.4	32.9	32	32
Sr	172	165	202	179	188	174	187	139	111	184	182	184
Rb	12.3	11.6	7.7	10	29	14	21	12	11.6	11.9	13.7	11.3
Co	42	46	47	46	41	41	42	70	42	43	48	48
Cr	184	298	238	244	244	235	128	427	166	158	95	92
V	259	257	233	255	249	264	278	244	286	279	356	356
Ce	28	28	24	24	30	29	38	25	34	32	31	26
Nd	16	13	14	13	14	15	19	11	16	19	16	16
La	14	11	11	13	11	13	18	10	12	14	13	11
$^{87}Sr/^{86}Sr$	0.70680						0.70679				0.70604	0.70636
$(^{87}Sr/^{86}Sr)_o$	0.70627						0.70596			0.70618	0.70548	0.70591

* - All Fe as FeO

Sample Sections: MLP - Mafika Lisiu; BMC - Ben McDhui; ON - Ongeluksnek; SOM - Semongkong
SP - Sani Pass; OXB - Oxbow; BUS - Bushman's Pass.

Th/Ta, moderate P/Zr and Zr/Nb, and low Zr/Y (Figures 9 and 10) is found from the Letele Pass northwards into the basalt outliers of the northeastern Free State. Overlying all these types is the Wonderkop unit (Figure 6) which, in contrast to underlying units, is characterized by high Nb and, to a lesser extent, P, U, and Th relative to most other incompatible elements (Figures 9 and 10). It is widespread throughout northern and central Lesotho, being present in all the northern sections, and it may have correlatives in the southern sections, as discussed later. In the basalt outliers of the Free State it oversteps underlying units to form the basal flows on Clarens sandstones at Wonderkop and Clocolan (Figure 3).

In summary, the overall geochemical stratigraphy in the lower units of the northern sections is one of a series of units (Golden Gate, Sani, Roma, and Letele units) characterized by low Nb relative to Zr and other incompatible elements. These units have variable relative enrichments of P, Ti, Y and Zr, and ratios amongst these elements serve to distinguish between these high Zr/Nb units. There is then a shift to flows with higher Nb (and consequently low Zr/Nb) which constitute the overlying and widespread Wonderkop type.

5.1.2. *The Lesotho Formation.* Selected analyses of basalts from the Lesotho Formation are in Table 2. These analyses and those in all the pertinent variation diagrams are from the RU data set. Four units build the main sequence. One of these the Maloti unit, is a thin (maximum: 120 m thick), distinctive group of flows occurring at about 2400 m in all sections, except at Sani Pass where a general upward warp in the basalt sequence results in this unit cropping out at about 2900 m. The Maloti unit effectively subdivides the Lesotho Formation into two. It marks a stage of distinct compositional change in the evolution of the main sequence basalts and is discussed first in the sections that follow. Critical compositional data for the units in the Lesotho Formation are summarised in Figures 11–13. Figure 11 is a composite diagram constructed from all the sections using the altitude of the lowest sample of the Maloti unit as a datum (i.e., for each section, the altitude of each sample is normalised to the altitude of the lowest Maloti sample). Thus, the stratigraphy in Figure 6 is directly comparable to Figure 11. The overlap in the symbols of the different units in Figure 11 is a function of the different thicknesses for some of the units, in particular the overlap in the altitude of the Mothae unit at Oxbow and Mafika Lisiu and the Senqu unit at Semongkong and Ben McDhui.

TABLE 3. Selected Analyses of Low-Ti and High Ti Basalts from the Springbok Flats

	Unit 1						Unit 2			Unit 3	
	WD4-6	WD4-8	WD4-14	LB1-2	LB1-3	RTL1-46	RTL1-39	RL1-17	RL1-11	LB1-12	RL1-24
SiO₂	50.12	50.32	50.73	48.95	50.29	50.57	48.16	47.32	49.79	48.54	48.46
TiO₂	0.87	0.95	0.90	0.82	0.87	0.91	0.86	0.93	0.96	0.94	1.00
Al₂O₃	14.55	14.45	14.86	14.31	14.78	14.18	14.53	15.12	15.07	14.75	15.18
FeO*	8.95	9.43	9.14	8.97	9.60	9.65	9.36	9.44	9.14	9.92	10.20
MnO	0.19	0.16	0.15	0.15	0.15	0.15	0.17	0.19	0.15	0.16	0.17
MgO	7.48	7.17	7.19	7.61	8.31	8.25	7.52	7.54	6.49	6.12	6.38
CaO	9.61	7.30	9.94	7.86	9.72	9.69	9.25	9.50	9.91	10.33	10.05
Na₂O	2.09	2.84	2.19	2.44	1.91	2.18	2.05	1.99	1.63	2.16	2.04
K₂O	0.71	1.06	0.76	0.99	0.75	0.63	0.45	0.58	0.72	0.65	0.75
P₂O₅	0.110	0.120	0.130	0.100	0.117	0.113	0.143	0.147	0.110	0.160	0.155
LOI	3.60	4.82	3.29	5.34	2.24	3.05	5.05	3.95	3.76	4.54	2.89
H₂O⁻	1.89	0.82	1.41	2.66	1.86	1.07	2.09	2.19	2.01	2.13	2.05
TOTAL	100.17	99.44	100.69	100.20	100.60	100.44	99.64	98.89	99.74	100.40	99.32
Ba	238	467	210	371	205	176	149	194	115	190	243
Sc	28	36	26	28	29	27	31	30	34	31	37
Zn	69	79	76	78	81	65	77	80	73	73	88
Cu	70	76	72	62	89	56	61	59	76	67	72
Ni	102	86	96	102	124	120	84	93	97	57	61
Nb	3.7	4.3	5.0	3.8	3.1	2.6	3.5	3.1	3.0	6.1	6.8
Zr	98	111	107	103	101	86	68	67	61	88	93
Y	25.4	30.4	27.7	26.2	26.4	25.0	24.0	22.8	22.2	26.4	25.6
Sr	135	225	143	213	143	142	175	152	174	209	176
Rb	16	31	19	32	17	15	7.8	8.3	15	12	16
Co	53	51	49	51	54	58	53	48	47	51	47
Cr	386	407	360	374	463	535	307	357	411	230	246
V	224	256	205	216	213	195	237	222	240	220	264
	0.71228	0.71467	0.70988	0.71762	0.71220	0.70922	0.70561	0.70569		0.70617	0.70673
	0.71140	0.71365	0.70890	0.71651	0.71132	0.70844	0.70528	0.70529		0.70575	0.70606

* - All Fe as FeO
RL1, LB1, WD4, and RTL1 refer to boreholes in Figure 18

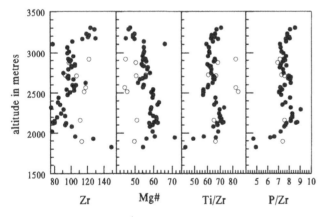

Figure 4. Variation of Zr and some interelement ratios with height in the basalts of the Oxbow section. Dots - lava flows; circles - intrusions.

Nevertheless, these slight overlaps do not distract from the effectiveness of Figure 11 in summarising the important serial geochemical features of all sections in a few simple diagrams.

5.1.2.1. *The Maloti Unit:* This unit has constant thickness of a little over 100 m and is present in all sections, including the two southern sections at Ongeluksnek and Ben McDhui.

It is characterized particularly by low Ti/Zr and to a much lesser extent by low P/Zr compared to most other Lesotho Formation basalts (Figures 11 and 13). The base of the unit is marked by an abrupt decrease in these two ratios and Zr/Nb and by a sharp increase in Zr/Y (Figure 12). Samples from the upper part of this unit may show Ti/Zr and P/Zr ratios that appear transitional between values typical of the Maloti unit and the overlying flows of the Senqu unit. Whether a persistent, but thin, transitional unit separating the Maloti and Senqu units can be unequivocally recognized must await more detailed sampling. However, the possible presence of a transitional unit does not hamper clear recognition of the top of the Maloti unit, and in this study the transitional samples have been incorporated into the Maloti unit.

5.1.2.2. *The Mafika Lisiu unit:* Between the Maloti and Wonderkop units is a sequence of flows, about 400 m thick, of rather variable composition. These variations may be a basis for futher subdivision of this unit and they are discussed below.

About 150 m below the Maloti unit is a thin unit characterized by distinctly low Zr/Nb (and high Nb) compared to other Mafika Lisiu flows, as is apparent from the RU data in Figures 11 and 12 . This group of samples is also apparent

TABLE 3. (continued)

Unit 3	Unit 4					High-Ti	
RL1-27	RTL1-05	RL1-36	RL1-43	RL1-59	RL1-60	RTL1-21	RTL1-24
47.49	49.04	49.75	49.96	48.01	49.01	51.28	46.80
1.28	1.19	1.27	1.27	1.60	1.52	3.09	3.16
15.00	14.10	14.16	14.42	13.75	13.50	14.33	14.75
11.72	11.71	11.86	11.93	12.98	12.67	10.21	10.53
0.20	0.18	0.18	0.19	0.22	0.20	0.13	0.13
5.57	6.06	5.82	5.77	6.00	5.83	3.36	3.96
10.38	10.17	10.17	10.30	10.25	9.70	7.57	8.51
2.09	2.18	2.01	2.18	2.00	1.92	2.44	2.12
0.76	0.64	0.52	0.49	0.62	0.96	2.32	1.91
0.202	0.140	0.153	0.155	0.222	0.225	0.620	0.640
2.60	2.82	2.22	1.83	2.65	2.68	3.16	4.87
1.49	1.73	1.79	1.65	1.89	1.73	1.38	2.56
98.78	99.96	99.90	100.15	100.19	99.94	99.89	99.94
258	105	182	179	222	220	1049	1034
35	38	40	40	40	43	16	16
88	92	92	91	109	108	93	96
102	155	136	131	217	175	80	123
53	81	66	63	60	53	53	55
7.0	4.2	4.9	5.0	5.7	6.9	32	33
116	89	100	99	120	126	455	460
32.0	27.9	31	32	39	41	44	46
180	158	162	170	149	144	1158	1162
12	20	7.7	5.8	10	30	45	20
47	51	50	49	53	50	44	47
174	284	218	197	154	167	69	72
281	318	301	310	358	322	203	228
0.70644	0.70642	0.70597	0.70585	0.70546	0.70635	0.70534	0.70509
0.70595	0.70548	0.70562	0.70560	0.70496	0.70481	0.70505	0.70496

in the WSU data, but their separation from other flows within this unit is less perfect. Despite their distinctive and constant Zr/Nb ratios, this group has highly variable incompatible-element ratios including P/Zr, Ti/Zr, and Zr/Y, and this is important for the possible correlation discussed in Section 5.2. Figure 6 indicates that this low-Zr/Nb group lies stratigraphically just below the Maloti unit in the Bushman's Pass section, unlike at other sections where the vertical separation is much larger. At Bushman's Pass the low-Zr/Nb group is represented by a single sample, BUS-22 from the summit of Bushman's Pass. Sample collection was resumed some 9 km to the east on the Molimo-Nthuse/Thaba Putsoa Pass at the same altitude as the summit of Bushman's Pass and, within a short interval, encountered the Maloti unit. In compiling Figure 6, no correction was made for the known slight eastward dip of the lava sequence. A 1° dip would result in a 150-m difference in elevation between the two segments of the section and would correct the apparent compression of the upper part of the Mafika Lisiu unit in the Bushman's section in Figure 6. In the Oxbow section, a single normal Mafika Lisiu basalt flow is interbedded with the low-Zr/Nb flows, but in all other northern sections the low-Zr/Nb flows appear to form a homogeneous unit. The significance of the low-Zr/Nb horizon is enhanced by the possibility that it is closely and consistently related to the palaeomagnetic polarity reversal in

the basalt sequence. Although data are sparse, the low-Zr/Nb flows also separate Mafika Lisiu basalts with different Sr-isotopic compositions. These aspects are discussed in more detail below.

In the lower part of the Mafika Lisiu unit, a group of samples with lower P/Zr, and to a lesser extent Ti/Zr, than other Mafika Lisiu basalts occurs within the interval 50 to 100 m above the base of this unit. This group could form the basis for the separation of these flows into a separate basal unit to the Lesotho Formation overlying the Wonderkop flows. However, the compositional range of this putative unit as expressed by inter-element ratios, apart from P/Zr perhaps, is large and overlaps with data for samples higher in the sequence (Figure 11). In this regard the apparently distinctive P/Zr for this group is emphasized in the WSU data by a small group of samples in the Mafika Lisiu section. This is not a feature of the RU data. In the absence of clear cut, distinctive geochemical criteria for this group its status as a separate unit relies on stratigraphy; i.e., samples from this unit have ratios that are different from samples immediately above and below but overlap with those of samples higher and lower in the sequence. Thus, we include these samples in the Mafika Lisiu unit although they are identified as separate symbols in several of the diagrams.

5.1.2.3 *The Senqu Unit*: This unit overlies the Maloti unit

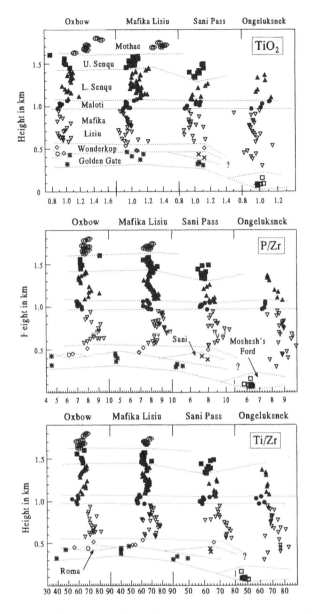

Figure 5. Vertical variations of some geochemical parameters in four sections through the Lesotho basalt sequence. Dashed lines show the correlations of different basalt units recognized on the basis of geochemistry.

and, like the Mafika Lisiu unit, has rather variable compositions which tend towards being slightly more evolved in terms of Mg# and Zr abundances compared to underlying units of the Lesotho Formation (Figures 11–13). Interelement ratios and Mg# show several abrupt changes in the succession, and in most sections the coincidence of an abrupt change (of varying magnitude) of the same sense in Ti/Zr, P/Zr and Mg# at about the same stratigraphic level forms the basis for a possible subdivision into an upper and lower Senqu

unit. However, the compositions of the basalts in these two subunits overlap completely and they cannot be distinguished independently of their stratigraphic context. Formal proposals for further subdivision of this unit must await more detailed studies.

5.1.2.4. The Mothae unit: The uppermost basalt flows in the main sequence show marked geochemical trends towards quite evolved compositions over a short stratigraphic interval. These evolved basalts, exhibiting generally lower CaO, MgO, Al_2O_3 and higher FeO, TiO_2, P_2O_5 and incompatible trace-element abundances, have been grouped into a separate unit, the Mothae unit (Figures 6, 11, and 13). The base of this unit is marked by abrupt inflections in several interelement ratios and changes in element abundances (Figures 5 and 11) in the Oxbow section, but in the Mafika Lisiu section the base is marked by an inflection in P/Zr only, with other compositional parameters varying continuously and smoothly across this boundary. Thus, although incompatible-element ratios do not distinguish this unit, Mg# and TiO_2 allow almost complete separation of these basalts from underlying units (Figures 11 and 13). We emphasize that the abrupt changes observed in the Oxbow section may be an

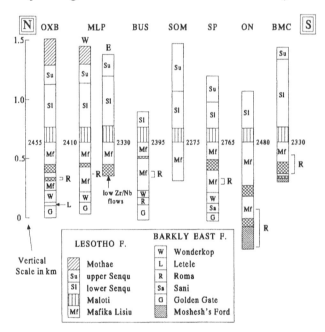

Figure 6. Summary of the stratigraphic subdivisions for each section based on geochemistry. The datum is the base of the Maloti unit which occurs in all sections. Abbreviations of sections are OXB - Oxbow; MLP - Mafika Lisiu Pass; BUS - Bushman's Pass; SOM - Semongkong; SP - Sani Pass; ON - Ongeluksnek Pass; BMC - Ben McDhui. The position of the palaeomagnetic polarity reversal is indicated by the bracket labeled R. In the ON and BMC sections the position of the reversal has not been precisely fixed due to the wide sampling interval.

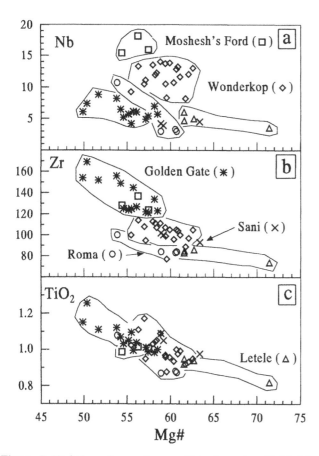

Figure 7. Variation of some incompatible elements with Mg# in units of the Barkly East Formation. The symbols indicate samples from different stratigraphic units defined on the basis of geochemistry as discussed in the text.

presumed to have fed lavas now lost due to erosion.

5.2. *The Southern Sections*

It has not been possible to correlate lower units in the two southern sections with those of the Barkly east Formation in the northern sections. As illustrated in Figure 14, at the base of both southern sections is the well-established Moshesh's Ford unit which is widespread in the region around Barkly East (Figure 3) and in the basalt outliers between Jamestown and Molteno further to the southwest [*Marsh and Eales*, 1984]. Its discovery at Ongeluksnek substantially increases

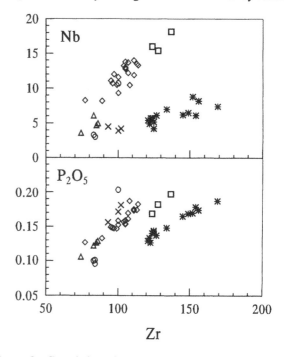

Figure 8. Covariation of some incompatible elements in basalts from the Barkly East Formation. Note subparallel and converging trends. Symbols as in Figure 7 indicate basalts from different stratigraphic units.

artefact of poor sampling density in a sequence where compositions are changing smoothly and rapidly, and the basis for a non-arbitrary definition of this unit may be confirmed only by more careful sampling.

5.1.2.5 *Intrusions*: The current study was aimed at sampling basalt lava flows in order to determine a stratigraphic subdivision for the volcanic pile. Intrusions were only sampled when encountered in the sampling sections. Generally it was possible to classify the dykes into the stratigraphic scheme described above on the basis of geochemistry, and in all cases the classification was consistent with stratigraphy; i.e., no intrusion was found to belong to a unit lower in the sequence than the flows which it intrudes. One exception to this correlation of dykes with flows is a suite of small intrusions collected at about 2500 m in the vicinity of the Oxbow Lodge on the Oxbow section. These are chemically evolved, with low Mg# and distinctively high Fe, Ti/Zr and Zr/Nb (Figures 11 and 13). They cannot be correlated with any known lava unit in Lesotho and are

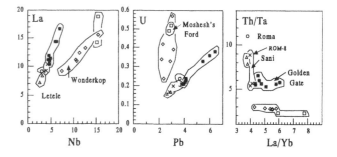

Figure 9. REE and other incompatible element variations in selected samples from units of the Barkly East Formation. Symbols as in Figure 7.

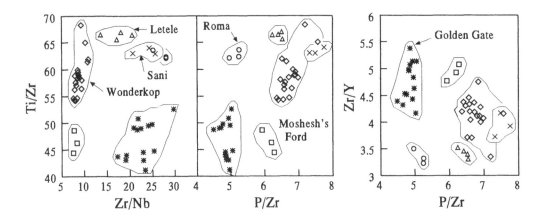

Figure 10. Plots of incompatible element ratios in basalts from units of the Barkly East Formation. Note the distinctive compositions of the different units making these plots a useful discrimination diagram. Symbols in Figure 7.

the known area over which this chemical type was erupted at the onset of basaltic activity. The northern limit to its distribution is unknown but it is not recognized from any of the northern sections. The Moshesh's Ford type is characterized by its distinctive high Nb content and by its combination of low Zr/Nb and Ti/Zr and high Zr/Y.

In the Ben McDhui section, the Moshesh's Ford unit is overlain by flows with low Zr/Nb (~ 8, which is equivalent to the Moshesh's Ford basalts), high and variable P/Zr (8.5–10) and moderate and constant Ti/Zr (~ 65). Interbedded with these is a single flow (BMC-3) with similar P/Zr and Ti/Zr, but higher Zr/Nb (~13). Above is a sequence of flows with compositional characteristics typical of normal Mafika Lisiu, Maloti, and Senqu basalts, in the correct stratigraphic order, as recognized in the northern sections. These low-Zr/Nb flows in the lower part of the southern sequence have affinities to the Wonderkop type of the northern sections on the basis of Zr/Nb, Zr/Y and to a lesser extent Ti/Zr, but not P/Zr. On the other hand, the low-Zr/Nb flows correlate perfectly in terms of geochemistry with the low-Zr/Nb Mafika Lisiu basalts.

In the lower part of the Ongeluksnek section, a sequence of flows with broadly similar compositions can also be recognized. These flows divide into a low-Zr/Nb group (Zr/Nb ~ 8) and a high-Zr/Nb group (Zr/Nb ~ 12-20). The former correlate with the low-Zr/Nb group in the Ben McDhui section and hence also with the low-Zr/Nb Mafika Lisiu flows. The other high-Zr/Nb flows are similar to the single high-Zr/Nb flow at Ben McDhui and are all compositionally identical to normal Mafika Lisiu basalts. As at Ben McDhui, the normal sequence of Mafika Lisiu, Maloti, and Senqu units is developed in the overlying sequence. Figure 14 summarises these correlations, but it should be stressed that the faulting close to our sampling section at Ongeluksnek may have influenced the relative stratigraphic position of some of the

samples and the apparent thickness of some of these units.

In summary, simple compositional correlations indicate that basalt units which form the Lesotho Formation in northern Lesotho appear to lie directly on the Moshesh's Ford unit in the southern sections. The low-Nb units which form the lowermost units in the northern sections are absent. Furthermore, provided the correlations can be sustained, the data suggest that flows of the low-Zr/Nb and normal Mafika Lisiu units interdigitate in the southern sections, in contrast to their simpler interrelationship in northern Lesotho. These conclusions are tentative and could reflect sampling inadequacies. Detailed sampling of more sections in southern Lesotho is needed to establish a reliable stratigraphy for the lower part of the sequence in this area.

5.3. Discussion

Figure 13 summarises some of the principal discriminating geochemical features of the Lesotho Formation, and Figure 15 shows the compositions of the Lesotho Formation units in relation to those of the Barkly East Formation. An important feature of Figure 13 is the minor overlap in Ti/Zr and Zr/Y between Mafika Lisiu flows and those higher in the sequence, provided the low-Zr/Nb Mafika Lisiu flows are considered separately (they are shown with a separate filled-diamond symbol in Figure 13). This underscores the significance of these low-Zr/Nb flows. The compositional variability of the Mafika Lisiu unit as a whole is also evident in comparison with the more compositionally homogeneous Senqu unit. More detailed studies in the Mafika Lisiu sequence could result in further subdivisions within the unit. The overall increasing degree of fractionation in the sequence Mafika Lisiu-Maloti-Senqu-Mothae-Oxbow is also evident from Figure 13, but the large changes in Ti/Zr and Zr/Y between the

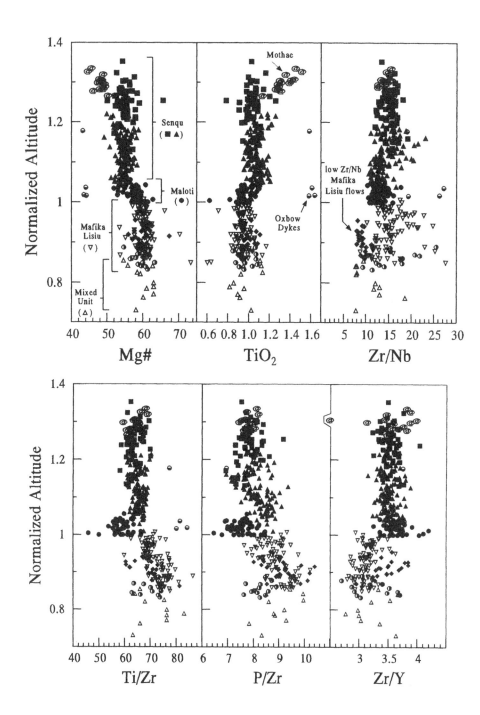

Figure 11. Composite serial section diagram showing vertical compositional variations in the Lesotho Formation. Data from all sample sections. The calculation of normalized altitude is explained in the text. The stratigraphic units defined on the basis of geochemistry are distinguished by different symbols as indicated. In the Mafika Lisiu unit the low-Zr/Nb samples are shown as solid diamonds and the possible basal subunit as half-filled circles. The status of these units is discussed in the text. For clarity, the interbedded flow section at the base of Ben McDhui and Ongeluksnek sections is identified as "Mixed Unit" (upright triangles) on the plot; see text and Figure 14 for full discussion of the stratigraphic characterization of these samples.

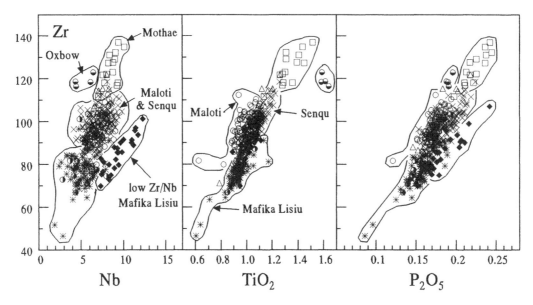

Figure 12. Covariation amongst some incompatible elements in the basalts of the Lesotho Formation. Note: symbols have been changed from those used in Figure 11 as indicated, except for the "mixed unit" samples from Bon MoDhui and Ongeluksnek, which have been classified into their geochemical type according to Figure 14. Note the tight clustering of the data and the distinctive character of the low-Zr/Nb Mafika Lisiu flows and the high Zr/Nb Oxbow intrusions.

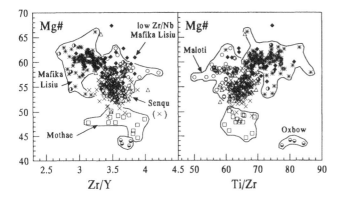

Figure 13. Interelement ratio plots for main sequence basalts. Symbols as in Figure 12.

Mafika Lisiu, Maloti, Senqu/Mothae, and Oxbow units make it unlikely that this sequence represents a normal differentiation trend involving gabbro fractionation.

The relative geochemical homogeneity of the Lesotho Formation units as a group compared to the diversity in the Barkly East Formation is emphasized in Figure 15. Basalt magma types in the Barkly East Formation are not simply related by fractionation to those of the Lesotho Formation. Interestingly, the data for the Oxbow intrusions in the Senqu unit plot on the fringes or well away from the Lesotho Formation cluster and suggest a possible return to a geochemical diversity in the late stages of eruption, similar to that which characterized the flows in the lower units.

The diagrams presented are potentially useful discrimination diagrams for stratigraphic studies. However, we emphasize that the compositional differences among the basalt units are subtle and require data of the highest quality for their recognition. Furthermore, although the general stratigraphic structure of the sequence is recognizable in both RU and WSU data sets, problems of small interlaboratory biases for some elements result in characteristic incompatible-element ratios for any one unit being slightly different in the two data sets (see Table 1 for some comparisons between the two data sets). These two effects have proved testing for the integrity of our geochemical subdivisions and make it difficult to classify any random sample, especially from the Lesotho Formation, using these diagrams. We warn against such haphazard and uncritical use of our results. As in many stratigraphic studies, an individual sample cannot be divorced from its context and, in the Lesotho succession, this requires that classification be based on results from a sequence of flows.

5.4. *Geochemical Stratigraphy: Isotopic Evidence*

Figure 16 summarises our Sr isotopic data, some of which are listed in Tables 1 and 2. In general, the results are similar to those of previous studies summarised by *Marsh and Eales* [1984] and *Bristow et al.* [1984], in that basalts of the Barkly East Formation are isotopically more diverse than those of the Lesotho Formation. This is also consistent with the trace element diversity discussed above. *Marsh et al.* [1992]

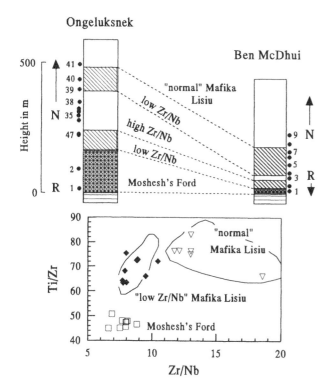

Figure 14. Detailed stratigraphic section at the base of the Ongeluksnek and Ben McDhui illustrating geochemical classification of the samples immediately overlying the Moshesh's Ford basalts. These are the "mixed unit" samples from Figure 11. The low-Zr/Nb samples (filled diamonds) correlate with the field for the low-Zr/Nb Mafika Lisiu subunit in the northern sections, whereas the interbedded high-Zr/Nb basalts (inverted triangles) correlate with the "normal" Mafika Lisiu basalts from all sections. The palaeomagnetic polarity of the samples is indicated by "R" and "N" and the numbered dots refer to sample locations and numbers.

demonstrated that Sr (and Pb) isotopic composition is an effective discriminator amongst different basalt types in the Karoo province, and it is unlikely that other isotopic systems would present a general stratigraphic picture much different from that in Figure 16. Within the Mafika Lisiu unit the Sr isotope compositions group around 0.706 and 0.705; the higher values are characteristic of samples collected in the upper part of the unit above the low Zr/Nb horizon, and the lower values are from samples below this horizon.

5.5. *Comparison with Previous Studies*

Apart from the reconnaissance study of *Marsh* [1984], the results of which have been incorporated into this investigation, the only other stratigraphic study in the Lesotho basalts is that of *Ramluckan* [1992] on an 800-m section at Sani Pass. This study differs from ours in that it terminated at Sani Top (see Appendix), within the Maloti unit, and therefore does not

include data from the uppermost part of the sequence.

Ramluckan [1992] recognized five geochemical units in the section being, from the base up, the Giants Cup, Agate Vale, Sakeng, Mkhomazana, and the Phinong units. The Giants Cup unit is about 14 m thick and, from the analyses and descriptions, appears to consist of several thin, highly altered flows; it was not sampled by us. The Agate Vale unit is equivalent to the Golden Gate unit, and the overlying Sakeng unit is equivalent to the Sani unit. The Mkhomazana unit is also very thin and may have been missed in our sample collection. These flows are strongly altered and appear to have been distinguished by *Ramluckan* [1992] on the basis of high Na_2O. We consider Na to provide a weak basis for geochemical stratigraphy and are dubious about the status of this unit. The overlying basalts form the Phinong unit, which is largely equivalent to our Mafika Lisiu unit. Although we have not made a detailed analysis of Ramluckan's data, it appears that samples at the base of his Phinong unit are equivalent to our Wonderkop unit and samples from the topmost flows classify as Maloti basalts. Also within the central part of the Phinong unit are several flows of high-Nb basalts which can be correlated to the low-Zr/Nb Mafika Lisiu subunit. In summary, the results of Ramluckan's work are quite similar to ours, although there are differences in the stratigraphic interpretation of the compositional variations.

Marsh and Eales [1984] summarised the geochemistry of a collection of basalts from a 600-m section through the lower part of the sequence at Naudes' Nek, some 12 km southeast of the Ben McDhui section. Apart from the basal Moshesh's Ford basalts, this section was regarded as being built of flows of the Lesotho magma type; in fact, the geochemical characterization of the Lesotho type of *Marsh and Eales* [1984] and *Marsh* [1987] relied on samples from the Naudes' Nek section. We have reanalyzed these samples and their compositions correlate with the Mafika Lisiu basalts A sequence of low Zr/Nb flows near the base of the section overlies Moshesh's Ford basalts as at Ben McDhui. Maloti basalts have not been encountered in the section, presumably because sampling terminated below their outcrop level. Apart from the Moshesh's Ford basalts, *Marsh and Eales* [1984] also documented a number of chemically distinct units (Kraai River, Vaalkop, Omega, and Pronksberg High-K) within the Barkly East Formation to the west around Barkly East and Molteno. None can be correlated with the units of the Barkly East Formation in the northern sections.

6. CORRELATION BETWEEN PALAEOMAGNETIC AND GEOCHEMICAL STRATIGRAPHY

We have determined the precise position of the palaeomagnetic polarity reversal at the Oxbow, Sani Pass, and

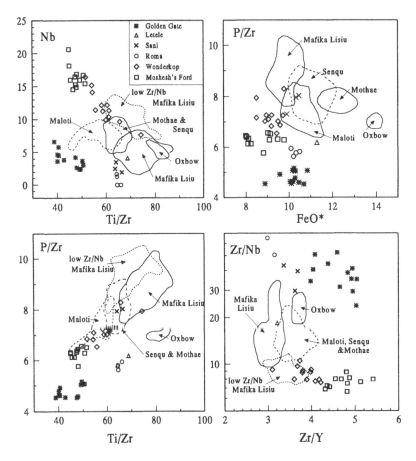

Figure 15. Variation diagrams showing the composition of samples from the Barkly East Formation in relation to composition fields for units of the Lesotho Formation.

Mafika Lisiu sections and the approximate position of the reversal at Ongeluksnek and Ben McDhui [*Rehacek*, 1995]. Although we did not determine the polarity of samples collected from the Bushman's Pass sections, we were able to correlate our geochemical sample sites with the well-established reversal described by V*an Zijl et al.* [1962]. At Semongkong and in the Mafika Lisiu (E) section, all our samples have normal polarity and the reversal presumably lies lower in the sequence than our lowest sample. The position of the reversal in each section is shown on Figure 6.

In all instances the reversal occurs within the Mafika Lisiu unit; more precisely, it occurs below the low Zr/Nb subunit at Sani Pass (95 m below), Mafika Lisiu (65 m), and Bushman's Pass (92 m). At Oxbow, Ongeluksnek, and Ben McDhui, where low Zr/Nb flows are interlayered with 'normal' Mafika Lisiu flows, the reversal appears to occur within the interbedded sequence (Figures 6 and 14). Although more detailed work is required in the two southern sections, it is clear that a consistent relationship exists between our geochemical stratigraphy and the palaeomagnetic reversal.

7. IMPLICATIONS FOR STRUCTURE AND EMPLACEMENT OF THE LAVA SEQUENCE

The stratigraphy illustrated in Figure 6 gives insight into the overall structure of the lava sequence in Lesotho. The extensive Maloti unit is particularly important in this regard. The present-day altitude of the base of this unit is shown adjacent to each section in Figure 6 and it is evident that the elevations are lower for sections lying within the basalt remnant (SOM and MLP-E) compared to those at the edges. The overall high elevation of the Sani Pass section is also a noteworthy feature. This confirms the observations made previously [e.g., *Stockley*, 1947] that the present structure of the lava remnant is one of a broad basin with slight inward dips of the flows. The overall constancy in thickness of the Maloti and other units suggests that this structure developed subsequent to eruption, but we cannot exclude the possibility of some contemporaneous subsidence.

The surface onto which the earliest basalt flows were emplaced was underlain by the Clarens sandstones of mixed

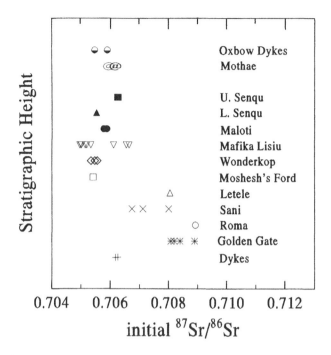

Figure 16. Summary of initial Sr isotope ratio data for the different units.

fluvial and aeolian origin. This surface exhibited mild topographic relief, with gentle slopes and occasional steep-sided valleys [*Lock et al.*, 1974]. The distribution of the thin lower units is determined in part by topography and this plausibly explains the absence of units in some of the sections. On the other hand, it is striking that the basal Golden Gate unit in northern Lesotho and the Moshesh's Ford unit in southern Lesotho show such extensive development (over areas up to 40,000 km^2). As emphasised above, sections north of latitude 30°S exhibit a reasonably continuous stratigraphy amongst the lower units, which differs completely from that in the two southern sections and the areas farther to the southwest described by *Marsh and Eales* [1984]. This lack of correlation could reflect the influence of an important topographic barrier to lower flow distribution, of geographically separated eruption sites, or of both. In this regard, it is notable that the thickness of the sequence between the base of the volcanic sequence and the Maloti unit at Ben McDhui and Ongeluksnek is variable and stands in contrast to the regular thickness of this sequence in all the northern sections. This variation is evidence of the more dramatic pre-volcanic topography in the south compared to the north, a feature also born out by the work of *Lock et al.* [1974]. Locations of eruption sites may be indicated by intrusions of specific compositional types. Intrusions of Moshesh's Ford basalts are known from Ongeluksnek and some 60 km to the southwest at Blikana [*Marsh and Eales*, 1984], but we have not found

intrusions that can be correlated with any other units of the Barkly east Formation. The problem of the lack of correlation in the units of the Barkly East Formation between north and south Lesotho awaits further detailed studies for its resolution.

The overlying, thicker geochemical units of the Lesotho Formation spread across Lesotho with more or less constant thickness. It should be pointed out that the thicknesses represented in Figure 6 have not been corrected for any dip of the strata. Several sections were sampled over horizontal distances of 15 km or greater, and a slight dip of 1° can result in over- or underestimates of thicknesses by 100 to 200 m depending on the relationship between the direction of dip and the orientation of the section. Thus, the apparent slight differences in thickness of units between different sections, particularly the northern sections in Figure 6, are not considered significant. This constancy of thickness strongly suggests that the bulk of the lava pile was built in a uniform manner (uniform eruption rates ?) on a generally planar surface. There is nothing in the stratigraphy that suggests the presence of geographically focussed lava eruption sites; even the restricted occurrence of some of the lower units seems to be a result of topographic control. Indeed, our observations and available geological maps lead us to conclude that the flows of all units were fed from a diffuse plexus of dykes spread throughout Lesotho and probably the entire Karoo basin. That the flow type in the basalt pile is overwhelmingly compound in nature may be significant in this regard. The vast subvolcanic network of dykes and sills is ample evidence of the general availability of basaltic magma in the uppermost crustal layers.

8. THE AGE OF KAROO VOLCANISM

Figure 17 summarises new ^{40}Ar/^{39}Ar age data obtained in the current study (R. Duncan, P Hooper, J. Marsh, and A. Duncan, unpublished information). A previous review of age data by *Fitch and Miller* [1984] proposed that activity in the Central Area was episodic over a period of 40 to 50 m.y. starting at 190–195 Ma, and with a main phase of emplacement at about 175 to 180 Ma. That this long span of activity should be represented in the Lesotho sequence defies geological evidence, such as the lack of significant weathering horizons and significant sedimentary units interbedded in the basalts, suggesting a short period for the emplacement of the basalt sequence. The preservation of the transition in the palaeomagnetic polarity reversal, in particular, implies a very short period for the building of the Lesotho succession. Much of the spread in the old K/Ar ages shown in Figure 17 comes from dolerites, which may be thought to preserve the time range of activity more completely compared to a lava pile which has undergone an unknown amount of erosion. On the

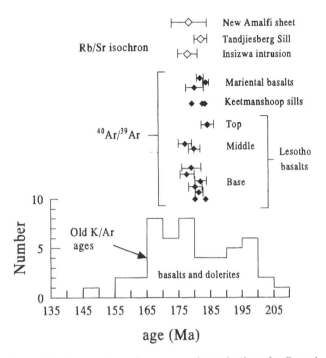

Figure 17. Comparison of recent age determinations for Central Area basalts and intrusions with previous ages largely determined by conventional K-Ar techniques (see text for references). $^{40}Ar/^{39}Ar$ ages with error bars are whole rock plateau ages and those without are total fusion ages.

other hand, dolerite intrusions are relatively undegassed in comparison to lava flows and dolerite age determinations by conventional K/Ar methods may be more susceptible to the presence of excess argon .

Our new age data for the basalts in and around Lesotho show a tight clustering around 180 Ma. Significantly, the basalts from Mariental in Namibia are identical in age to the Lesotho basalts, as are the dolerite sills from the Keetmanshoop area in southern Namibia. These new $^{40}Ar/^{39}Ar$ ages are consistent with the Rb/Sr internal isochron age obtained by *Richardson* [1984] for the Tandjiesberg sill in southern Namibia (Figure 1) and are also consistent with the geochemical evidence of a very close compositional correspondence between these Namibian basalts and dolerites and those of the main Karoo basin. Data for other differentiated intrusions are included in Figure 17; these are Rb/Sr isochron ages for the Insizwa intrusion (F. J. Kruger, personal communication, 1993) and the New Amalfi intrusion [*Williams*, 1995] (Figure 1). Thus all the new data from the Central Area show a tight clustering around 180 Ma and this small range is consistent with age data from other large flood basalt provinces which emphasize the short time spans (1–3 m.y.) for the emplacement of the bulk of their igneous products. Specifically, the latest data offer no support for an

earlier age for the commencement of basalt eruption in southern Lesotho at 190–195 Ma [*Fitch and Miller*, 1984].

An age of ca 180 Ma for the Central Area activity is consistent with the correlations between the Lesotho/ Springbok Flats remnants and the central Lebombo made below. R. Duncan (unpublished data, 1995) determined $^{40}Ar/^{39}Ar$ ages for nephelinite, picrite, basalt and rhyolite from the Lebombo. The mafic rocks yield ages of 180±2 Ma and the Jozini rhyolites slightly younger ages of about 177 Ma. These are all within error of the Rb/Sr isochron age of 179±4 Ma for the Jozini rhyolites obtained by *Allsopp et al.* [1984]. Ages of 180 Ma have also been obtained from $^{40}Ar/^{39}Ar$ measurements on plagioclase separates from the Kirwan basalts, Antarctica (R. Duncan, unpublished data, 1993), which *Harris et al.* [1990] correlated on geochemical grounds with low-Ti basalts of southern Lebombo, a correlation in keeping with their close spatial relationship in Gondwana reconstructions. There is now overwhelming evidence that the whole Karoo igneous province was emplaced within a short period at 180 Ma. Furthermore, the Kirwan basalt data and the precise U-Pb age of 181.2±0.7 Ma for the Dufek intrusion [*Minor and Mukasa*, 1995] extend the area of Gondwana influenced by Karoo-age flood basalt volcanism deep into Antarctica.

9. CORRELATIONS WITH OTHER KAROO VOLCANIC SUCCESSIONS

9.1. *Springbok Flats*

The only other major remnant in the Central Area for which a detailed geochemical stratigraphy has been established is the poorly exposed Springbok Flats remnant some 400 km due north of Lesotho (Figure 1). A detailed geochemical and isotopic investigation (J. Marsh, unpublished information) of four widely spaced borehole cores through the sequence has led to the recognition of five stratigraphic units (see Table 3 for representative analyses). One of these units consists of 30 m of high-Ti flows in the upper part of the sequence, which are compositionally and isotopically [*Marsh et al.*, 1992] identical to the low-Fe, high-Ti-Zr basalts at the base of the Sabie River Formation studied by *Sweeney et al.* [1994] in the central Lebombo. Thus, the Springbok Flats succession provides a vital correlating link between the Central Area basalts and those of the Lebombo. This is important because there are at least two palaeomagnetic polarity reversals of the sense R to N in the Lebombo succession and, as will be shown below, the geochemical correlation is critical in determining which one correlates with the reversal in the Lesotho succesion. Previous studies have been unable to determine any reliable correlation between the two sequences.

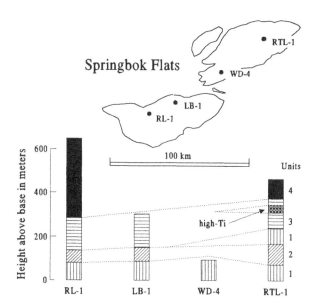

Figure 18. Map showing location of the Springbok Flats boreholes and a summary of the geochemical stratigraphy determined in each core.

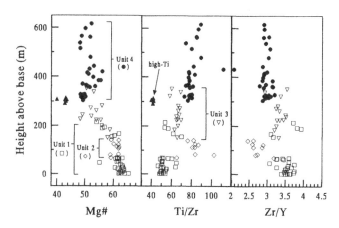

Figure 19. Compositional variations with height in the Springbok Flats succession. This is a composite diagram containing data from all cores plotted against height above the base of the lava flows.

Figure 18 illustrates the stratigraphic sequence in the Springbok Flats and Figures 19 and 20 summarise some pertinent geochemical data. Logging of borehole cores provided no geological basis (apart from the easy recognition of the macrophenocrystic high-Ti basalt flows) for the subdivision of the basalt flow sequence, and Figure 18 was constructed using geochemistry, employing the same basic criteria for the recognition of distinct geochemical units as in the current Lesotho study. Leaving aside the high-Ti basalts, the overall compositional variation in the Springbok Flats sequence is very similar to that in Lesotho; there is an overall upward trend towards evolved basalts, and major and trace element

and isotopic compositions [*Marsh et al.*, 1992] overlap almost completely, although the more evolved Fe-rich compositions are more extensively developed in the upper Springbok Flats sequence than in Lesotho. Previously, *Marsh and Eales* [1984] categorized the Springbok Flats basalts as a separate type differing from the Lesotho basalts. The few analyses of Springbok Flats basalt on which this proposal was based were very Fe-rich and comparisons were made with the lower, least-evolved basalts of the Lesotho Formation at the Naude's Nek in the southern part of the Lesotho remnant. The more complete data set provides a truer picture of the relationships and the impression gained is that the Springbok Flats section (max. preserved thickness about 700 m) represents a compressed version of the Lesotho sequence.

In detail, the basal Unit 1 in the Springbok Flats sequence shows strong geochemical affinities to the basal Golden Gate unit in northern Lesotho in having high $^{87}Sr/^{86}Sr$ and Zr/Nb, low P/Zr and Ti/Zr (Figures 20 and 21) and similar relative REE abundances. An important difference between the two units is their degree of differentiation; Unit 1 is relatively primitive with Mg# = 60–66 whereas Golden Gate basalts are considerably more evolved with Mg# = 50–58. Figure 21 indicates that Unit 1 and the Golden Gate unit have the highest Zr/Y ratios of the units in their respective successions but that

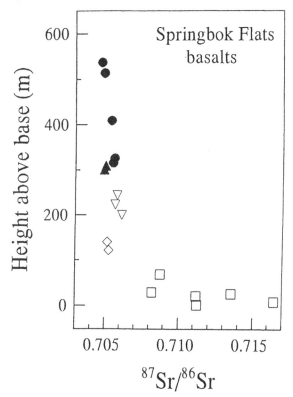

Figure 20. Vertical variation in initial Sr isotope ratios through the Springbok Flats succession. Symbols defined in Figure 19.

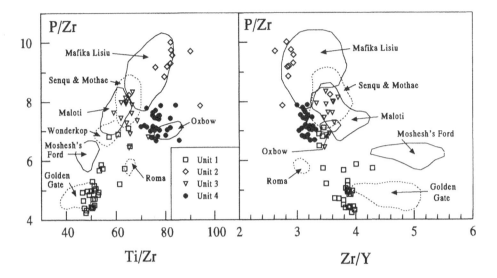

Figure 21. Comparison of some pertinent interelement ratios in Springbok Flats basalt units with those of the succession in Lesotho. Symbols defined in Figure 19.

the absolute Zr/Y ratios differ. This difference may reflect the different degree of evolution (Zr/Y ratios should increase slightly with fractional crystallization). In general, the geochemical and stratigraphic evidence for a correlation of the two is compelling, except that the Golden Gate basalts are more evolved and perhaps slightly more contaminated (as suggested by the high and variable Sr-isotope data) compared to Unit 1 basalts.

Unit 2 basalts with high P/Zr and Ti/Zr, low Zr/Y, and similar compatible- and incompatible-element abundances correlate with the Mafika Lisiu unit. Unit 3 is compositionally similar to the Senqu unit (Figure 21). The Fe-rich basalts of Unit 4 correlate with the more evolved Senqu and Mothae units in many respects, except for Ti/Zr, La/Yb and La/Sm which are similar to ratios in the Oxbow dykes. However, the Unit 4 basalts are more primitive, with lower Zr/Y, FeO*, and TiO_2 and higher Al_2O_3 and MgO compared to the Oxbow dykes, but could evolve by fractionation to the Oxbow compositions. Correlation between Unit 4 and Oxbow intrusions is also favoured by $^{87}Sr/^{86}Sr$ ratios, but the isotopic evidence is not exclusive.

In summary, the stratigraphic units identified in Lesotho have a close geochemical affinity to those in the Springbok Flats sequence. Although correlations are not perfect, it is important to emphasise that the temporal changes in geochemical character of the basalts, as reflected in the geochemical stratigraphy, are extremely similar in both areas (Figure 22). Although we hesitate to conclude that the Springbok Flats basalts flowed from the same feeders that supplied the Lesotho sequence, it is clear that the magma supply systems for both sequences evolved in an identical

fashion with time. Considering all the evidence, we suggest that the two sequences can be correlated as illustrated in Figure 23.

9.2. Lebombo

The Lebombo monocline shows considerable lateral and vertical lithological variation in the volcanic sequence as summarised in the maps by *Eales et al.* [1984]. In the northern Lebombo, sporadic nephelinites at the base are

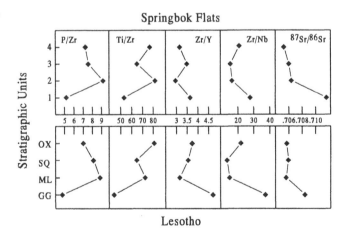

Figure 22. Comparative variation of interelement ratios and Sr-isotopic composition between units in Lesotho and Springbok Flats. Although the actual value of a ratio may not correspond precisely in the correlated units, the overall pattern of variation is striking. Abbreviations: OX - Oxbow; SQ - Senqu; ML - Mafika Lisiu; GG - Golden Gate units.

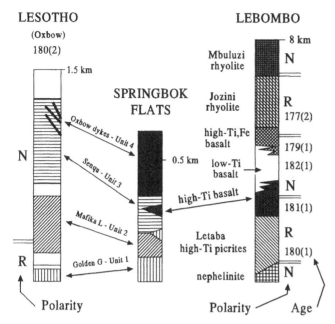

Figure 23. Summary of the proposed correlations between Lesotho, Springbok Flats, and Lebombo. The diagram is partly schematic and not to scale, particularly with regard to the composite Lebombo section. Magnetostratigraphy from this study and Hargraves (personal communication, 1995). Ages from this study. Error on ages indicated by number in brackets. Note the different thicknesses of mafic rocks preserved at each locality.

overlain, in sequence, by Letaba picrites, Sabie River high-Ti basalts, and Jozini rhyolites. Southwards, the picrites overstep the nephelinites and are, in turn, overstepped by the basalts. In the central Lebombo a wedge of low-Ti basalt separates the high-Ti basalts into a basal low-Fe variety and an overlying high-Fe variety. The low-Ti basalts thicken southwards at the expense of the high-Ti types and completely dominate the basalt sequence in the southern Lebombo. In addition, in Swaziland, the Mbuluzi rhyolites overlie the Jozini rhyolites and are in turn overlain by the little-known Movene basalts. Nowhere is the complete sequence preserved, and the succession in Figure 23 is a composite column showing the various lithologies in their correct stratigraphic order, as well as our preliminary palaeomagnetic polarity reversal stratigraphy and $^{40}Ar/^{39}Ar$ ages.

Duncan et al. [1984] demonstrated that the low-Ti basalts of the Sabie River Formation of the southern Lebombo were geochemically distinct from other Central Area basalts, specifically the basalts previously referred to as the Lesotho type. However, in the light of the new age, palaeomagnetic and geochemical data discussed here, and the work of *Sweeney et al.* [1994], this question deserves reconsideration. Firstly, the correlation established for the high-Ti basalts located at the base of the Sabie River Formation in the

Lebombo sequence (Figure 23) suggests that comparisons of low-Ti basalts should only consider those basalts lying above the correlated high-Ti basalt flows. In the Lebombo this includes all the low-Ti basalts, but in the Springbok Flats these include only the uppermost Fe-rich flows; i.e., the equivalent of the Senqu and Mothae flows and the Oxbow dykes. Figure 24 demonstrates that the Lebombo basalts are compositionally different from the Springbok Flats basalts, principally in being more evolved, i.e., richer in Fe, Ti, Zr and poorer in MgO, Al_2O_3, etc. In general terms these Lebombo basalts could have evolved from Unit 4 basalts from the Springbok Flats, but whether such evolved basalts existed in the Central Area to provide a direct correlation is uncertain. However, as *Marsh et al.* [1992] emphasized, the isotopic character of the low-Ti basalts from the two areas is strikingly different and, on available evidence, we suggest that the Lebombo magma system developed in a largely separate manner from that supplying the Central Area.

The geochemical correlations between Lesotho, Springbok Flats, and the Lebombo illustrated in Figure 23 indicate that the polarity reversal at the base of the Lesotho pile correlates with that in the low-Fe, high-Ti basalt in the lower part of the

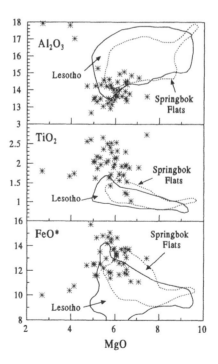

Figure 24. Comparison between Central Area low-Ti basalts (Lesotho, Springbok Flats) and those in the Lebombo (asterisks) illustrating the more evolved character of the Lebombo basalts. Data from this study and *Sweeney et al.* [1994]. Note the almost complete overlap in composition between the Lesotho basalts (data from all units) and those of the Springbok Flats.

Sabie River Formation. This correlation implies that the basaltic eruptions in the Lebombo were contemporaneous with those in Springbok Flats and Lesotho. However, the geochemical correlation shown in Figure 23 suggests that the main volume of Lebombo basalt, specifically the Sabie River low-Ti basalts, largely postdates the main low-Ti emplacement event in the Central Area.

10. CONCLUSIONS

Detailed sampling, together with precise X ray fluorescence analysis, has allowed us to construct a geochemical stratigraphy for the Karoo basalts in Lesotho. This stratigraphy is consistent with results of palaeomagnetic polarity studies and $^{40}Ar/^{39}Ar$ age determinations. Thin basaltic units with diverse geochemistry form the Barkly East Formation at the base of the sequence and, although some of these units may be found over areas approaching 40,000 km^2, others are found over rather limited areas. In the sections north of latitude 30° S there is a reasonably continuous stratigraphy amongst the lower units which differs completely from that in the sections to the south. Further investigations are needed to map out the stratigraphy in the lower part of the sequence in southern Lesotho and to determine the cause of the poor north-south correlations between the lower units.

Overlying the Barkly East Formation is the Lesotho Formation, comprising a number of thick units of constant thickness which can be traced throughout the Lesotho remnant. These units build the main bulk of the basalt pile and their flows appear to have been fed from a widespread plexus of dykes. The new age data indicate that the entire Lesotho lava pile and its correlatives up to 1300 km away (Mariental in Namibia) were emplaced at 180±2 Ma, an age that is consistent with Rb-Sr isochron ages from differentiated dolerite intrusions in the Central Area mafic rocks in the Lebombo. There is no evidence in the new data for a long period of episodic activity as proposed previously. The apparent short duration of the activity is entirely consistent with the lack of significant subaerial weathering horizons or palaeosols in the basalt sequence, with the palaeomagnetic data, and with the overall close geochemical correlation between the basalt flows and dolerite intrusions.

There is a strong geochemical similarity between the basalt sequence in Lesotho and that in the Springbok Flats some 400 km north of Lesotho. We propose specific correlations between individual units in the two sequences and suggest that the Springbok Flat succession represents a compressed version of the Lesotho sequence. Through correlation of geochemically distinctive high-Ti basalt flows high in the upper part of the Springbok Flats sequence with the high-Ti basalts at the base of the basaltic succession in the central Lebombo, it appears that the bulk of the Central Area basalts may slightly predate the basaltic volcanism (but not necessarily the nephelinitic or picritic extrusions) of the Lebombo. However, these age differences, if valid, are not resolvable by $^{40}Ar/^{39}Ar$ dating. In any event, apart from the high-Ti basalt in the Springbok Flats succession, the large geochemical, isotopic, and petrographic differences between the Lebombo volcanic suite and that of the Central Area emphasizes that the Lebombo magmatic system evolved and erupted its volcanic products separately from that in the Central Area and Lesotho in particular. The rift environment of the Lebombo provided ample scope for tapping magma sources perhaps not available beneath the stable cratonic interior and may also have played an important role in confining the erupted products in rift depressions along the Lebombo and to the east. Available data demonstrate that few of the flows erupted along the Lebombo flowed westwards across the cratonic interior of southern Africa.

APPENDIX

Sample Localities

The Ben McDhui section (BMC samples) was sampled along the jeep track that ascends Carlisle's Hoek just east of the village of Rhodes to the weather station at the base of Ben Mcdhui and from there up the slope to the summit beacon. At Ongeluksnek Pass the road follows a zone of shearing and small faults. Samples were thus collected in three traverses up a series of streams to the south of the road from the border post at the foot of the pass to the top. The uppermost samples (ON-3 to 18) were taken up the slope from the top of the road pass to the summit of an unnamed peak immediately to the south. At Sani Pass samples SP-1 to 20 were collected from the summit of the hill Kotisephola some 11 km from the top of pass, down onto the road and then down the road to Sani Flats. The remaining samples SP-21 to 51 were taken from Sani Top down the pass to the basalt-Clarens sandstone contact. The Oxbow sample suite was also collected in two sections. The upper section (OXB-1 to 25) was from the summit of the hill Mahlasela down the Mahlasela Pass to the area around the Oxbow Lodge. The remaining samples were collected from the top of the Moteng Pass down to the basalt-Clarens sandstone contact. The Mafika Lisiu sample suite comes from a double collection. The Mafika Lisiu (E) samples MLP-1 to 50 were collected starting in the bed of the Malibamatso River just north of the Village of Ha Lejone up to the road and then following the road up to the summit of the Mafika Lisiu Pass. In the western section, MLP 100 to 186 and the detailed collection MLM 1 to 17 were collected from the top of the pass down to the base of the basalts just west of

Pitseng. MLP-60 to 65 were collected up the hill immediately south of the top of the pass. Samples in the Bushman's Pass section were collected along three passes, Bushman's, Molimo Nthuse, and Thaba Putsoa that lie along the main road from Maseru eastwards to Mantsonyane. BUS-1 to 22 were taken from the bed of the Liphiring stream just east of Nazareth along the road to the top of Bushman's Pass, and samples BUS-24 to 40 were collected from just south of Ha Chalalisa to the summit of Thaba Putsoa Pass.

The Roma Section was sampled on the Semonkong road south of Roma from the Thabana-li-Mele craft centre on the Makhaleng river northwards along the road to Roma. Finally, the Semongkong section, also a two-part section, was sampled from the summit of Thaba Putsoa just west of the Ha Ramabanta–Semonkong road down onto that road (SOM-1 to 18) then along the road to the crossing of the Makhoalipana stream (SOM-19 to 40). SOM-82 to 100 were collected in the steep gully that descends from the road to the bottom of the Le Bihan Falls on the Maletsunyane River just south of Semongkong.

Acknowledgments. Funds for this research were provided by the Foundation for Research Development, South Africa (Marsh, Duncan) and the National Science Foundation (Hooper). Vehicle costs were subsidized by Rhodes University, Geological Survey of Namibia, and Gold Fields of South Africa. J.S.M. thanks the Anglo American Corporation and S. Marsh for access to the Springbok Flats cores. We thank the following colleagues for field and laboratory assistance: Mike Jackson, Phoenix Hoyle. Reviews by K. Cox and R. Fodor improved the manuscript.

REFERENCES

Allsopp, H. L., W. I. Manton, J. W. Bristow, and A. J. Erlank, Rb-Sr geochronology of Karoo felsic volcanics, *Spec. Publ. Geol. Soc. S. Afr., 13*, 273-280, 1984.

Bristow, J. W., H. L. Allsopp, A. J. Erlank, J. S. Marsh, and R. A. Armstrong, Strontium isotope characterization of Karoo volcanic rocks, *Spec. Publ. Geol. Soc. S. Afr., 13*, 295-330, 1984.

Campbell, I. H., and R. W. Griffiths, Implications of mantle plume structure for the origin of flood basalts, *Earth Planet. Sci. Lett., 99*, 79-93, 1990.

Cox, K. G., The Karoo Province, in *Continental Flood Basalts,* edited by J. D. Macdougall, pp. 239-271, Kluwer Academic, Dordrecht, 1988.

Cox, K. G., and G. Hornung, The petrology of the Karroo basalts of Basutoland, *Amer. Mineral., 51*, 1414-1432, 1966.

Cox, K. G., R. MacDonald, and G. Hornung, Geochemical and petrological provinces in the Karroo basalts of southern Africa, *Amer. Mineral., 52*, 1451-1474, 1967.

Duncan, A. R., A. J. Erlank, and J. S. Marsh, Regional geochemistry of the Karoo Igneous Province, *Spec. Publ. Geol. Soc. S. Afr., 13*, 355-388, 1984.

Eales, H. V., J. S. Marsh, and K. G. Cox, The Karoo Igneous Province: an introduction, *Spec. Publ. Geol. Soc. S. Afr., 13*, 1-26, 1984.

Erlank, A. J., A. R. Duncan, J. S. Marsh, R. S. Sweeney, C. J. Hawkesworth, S. C. Milner, R. McG. Miller, and N. W. Rogers, A laterally extensive geochemical discontinuity in the subcontinental Gondwana lithosphere (abstract), *Geochemical Evolution of the Continental Crust,* Pocos de Caldas, Brazil, 1-10, 1988.

Fitch, F. J., and J. A. Miller, Dating Karoo igneous rocks by conventional K-Ar and ^{40}Ar/^{39}Ar age spectrum methods, *Spec. Publ. Geol. Soc. S. Afr., 13*, 247-266, 1984.

Harris, C., J. S. Marsh, A. R. Duncan, and A. J. Erlank, The petrogenesis of the Kirwan basalts of Dronning Maud Land, Antarctica, *J. Petrol., 31*, 341-369, 1990.

Hooper, P. R., D. M. Johnson, and R. M. Conrey, Major and trace element analyses of rocks and minerals by automated X-ray spectrometry, *Open File Report,* 37 pp., Washington State University, Pullman, 1993.

le Roex, A. R., Geochemistry, mineralogy, and magmatic evolution of the basaltic and trachytic lavas from Gough Island, South Atlantic, *J. Petrol., 26*, 149-186, 1985.

Lock, B. E., A. L. Paverd, and T. J. Broderick, Stratigraphy of the Karoo volcanic rocks of the Barkly East district, *Trans. Geol. Soc. S. Afr., 77*, 373-374, 1974.

Marsh, J. S., Geochemistry of Karoo basalts and dolerites in the northeastern Orange Free State: recognition and origin of two new Karoo basalt magma types (abstract), *Geocongress '84,* 91-94, Geological Society of South Africa, Potchefstroom, 1984.

Marsh, J. S., Basalt geochemistry and tectonic discrimination within continental flood basalt provinces, *J. Volcanol. Geotherm. Res., 32*, 35-49, 1987.

Marsh, J. S., and H. V. Eales, Chemistry and petrogenesis of igneous rocks of the Karoo Central area, southern Africa, *Spec. Publ. Geol. Soc. S. Afr., 13*, 27-68, 1984.

Marsh, J. S., R. A. Armstrong, and R. S. Sweeney, New Pb, Sr, and Nd isotope data from the Karoo Province (abstract), *24th Congress,* 262-264, Geological Society of South Africa, Bloemfontein, 1992.

Minor, D. R., and S. B. Mukasa, A new U-Pb crystallization age and isotope geochemistry of the Dufek layered mafic intrusion; implications for the formation of the Ferrar Province (abstract), *Eos Trans AGU, 76*, 5284, 1995.

Morgan, W. J., Hotspot tracks and the opening of the Atlantic and Indian oceans, in *The Sea,* vol. 7, edited by C. Emiliani, pp. 443-487, Wiley Interscience, New York, 1981.

Norrish, K., and J. T. Hutton, An accurate X-ray spectrographic method for the analysis of a wide range of geological samples, *Geochim. Cosmochim. Acta, 33*, 431-453, 1969.

Ramluckan, V. R., The petrology and geochemistry of the Karoo sequence basaltic rocks in the Natal Drakensberg at Sani Pass, MSc thesis, University of Durban-Westville, Durban, 1992.

Rehacek, J., Chemical and paleomagnetic stratigraphy of basalts in northern Lesotho, Karoo Province, PhD thesis, Washington State University, Pullman, 1995.

Richardson, S. R., Sr, Nd and O isotope variation in an extensive Karoo dolerite sheet, southern Namibia, *Spec. Publ. Geol. Soc. S. Afr., 13*, 289-294, 1984.

Stockley, G. M., *Report on the Geology of Basutoland,* 114 pp., Government Printer, Maseru, Basutoland, 1947.

Sweeney, R. S., and M. K. Watkeys, A possible link between lithosphere architecture and Gondwana basalts, *J. African Earth Sci.,* *10,* 707-716, 1990.

Sweeney, R. S., A. R. Duncan, and A. J. Erlank, Geochemistry and petrogenesis of Central Lebombo basalts of the Karoo Igneous Province, *J. Petrol.,* *37,* 95-125, 1994.

Van Zijl, J. S. V., K. W. T. Graham, and A. L. Hales, The palaeomagnetism of the Stormberg lavas of South Africa, I: Evidence for a genuine reversal of the Earth's field in Triassic-Jurassic times, *Geophys. J. R. Astron. Soc.,* *7,* 23 -29, 1962.

White, R. S., and D. P. McKenzie, Magmatism at rift zones: the generation of volcanic continental margins and flood basalts, *J. Geophys. Res.,* *94,* 7685-7729, 1989.

Williams, C., The petrogenesis of the New Amalfi Sheet: a highly differentiated Karoo intrusion, MSc thesis, Rhodes University, Grahamstown, 1995.

_____

A. R. Duncan, Department of Geological Sciences, University of Cape Town, Rondebosch 7700, South Africa (e-mail - ard@ucthpx.uct.ac.za).

R. A. Duncan, College of Ocean and Atmospheric Sciences, Oregon State University, Corvallis, OR 97331-5503, USA (e-mail - rduncan@oce.orst.edu).

P. R Hooper and J. Rehacek, Department of Geology, Washington State University, Pullman, WA 99164-2812, USA (e-mail - prhooper@mail.wsu.edu).

J. S. Marsh, Department of Geology, Rhodes University, Grahamstown 6140, South Africa (e-mail - jsm@rock.ru.ac.za).

Siberian Traps

Mukul Sharma

The Lunatic Asylum of the Charles Arms Laboratory, Division of Geological and Planetary Sciences
California Institute of Technology, Pasadena, California

This paper examines recent attempts to estimate the timing and duration of Siberian flood volcanism and to model the petrogenesis of the Siberian Traps. The most significant findings are (1) the bulk of the Siberian lavas erupted within a period of about one million years at 250 Ma; (2) one of the earliest eruptions has a $^3He/^4He$ ratio which is 12.7 times the atmospheric value; (3) the early-stage lavas that constitute about 8% of the province display widely varying mineralogy and chemistry, with $\varepsilon_{Nd}(t)$ values ranging from -10 to +7; and (4) the voluminous late-stage lavas (~90% of the province) are remarkably homogeneous in mineralogy and chemistry, show nearly flat rare earth element patterns (Ce = 20 × primitive mantle; Ce/Yb ~ 2 × primitive mantle ratio), have higher Th/Ta and La/Ta ratios than those expected from melting of primitive mantle, and a volume-weighted $\varepsilon_{Nd}(t)$ ~ +2. The timing and duration of Siberian flood volcanism support a first-order causal link between the volcanism and the Permian-Triassic mass extinction. Assessment of the geochemical and isotopic data points to contributions from two different sources for the Siberian Traps. The early-stage lavas were derived from a mantle source with ε_{Nd} ~ +8 and containing residual garnet. The late-stage lavas were derived from a shallower mantle source with ε_{Nd} ~ +4. The magmas assimilated variable amounts of continental crust. Several major outstanding problems regarding the province remain to be solved.

1. INTRODUCTION

The intent of this paper is to summarize existing geochemical and isotopic data and their interpretation for the origin and petrogenesis of the Siberian Traps. A well-characterized, stratigraphically controlled suite of rocks with selected geochemical and isotopic data is used to show the existence of systematic temporal variations in major and trace elements and in initial Sr-, Nd-, and Pb-isotopic ratios.

These variations are then used to evaluate the existing models of the genesis of the Siberian Traps.

Several workers from the former Soviet Union have contributed significantly to understanding the geology of the Siberian Traps [e.g., *Zolotukhin and Al'mukhamedov*, 1988 and references therein]. Most of the early Soviet studies were conducted in the western and northwestern parts of the province as a result of the discovery of large Cu-Ni sulfide deposits in the Noril'sk region (Figure 1). A number of such studies were aimed at understanding the origin of the sulfide ores. Systematic geochemical work to understand the origin and petrogenesis of the Siberian Traps was undertaken by G.V. Nesterenko and co-workers [*Nesterenko et al.*, 1964, 1969a,b, 1971, 1972; *Nesterenko and Frolova*, 1965; *Balashov and Nesterenko*, 1966; *Nesterenko and Al'mukha-*

Large Igneous Provinces: Continental, Oceanic, and Planetary
Flood Volcanism
Geophysical Monograph 100

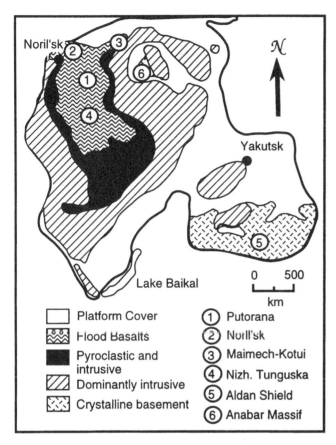

Figure 1. Schematic map of Permian-Triassic Siberian Traps (modified from *Renne and Basu* [1991]).

medov, 1966; *Nesterenko et al.*, 1972; *Nesterenko and Al'mukhamedov*, 1975; see also *Masaitis et al.*, 1966]. However, until quite recently only very sketchy outlines of the geology and geochemistry of this extensive region existed in the western scientific literature [*Lurie and Masaitis*, 1964; *Nalivkin*, 1973; *DePaolo and Wasserburg*, 1979; *Basaltic Volcanism Study Project*, 1981; *Khain*, 1985]. Because of the political changes in the Soviet Union in the mid-1980s, several groups in North America and in Europe obtained access to a vast collection of well-characterized rock samples, especially from the Noril'sk region where drillholes had been sunk to explore for Cu-Ni ores. As a result, a substantial amount of good-quality geochemical and isotopic data for the flood basalt province has been obtained in the last six years. A review of the earlier work on the stratigraphy and geochemistry of the Siberian Traps was given by *Zolotukhin and Al'mukhamedov* [1988]. In addition, several excellent papers focussing on the geology and geochemistry of the Noril'sk region have appeared recently [*Wooden et al.*, 1993; *Lightfoot and Naldrett*, 1994; *Hawkesworth et al.*, 1995; *Fedorenko et al.*, 1996].

2. GEOLOGIC SETTING

2.1. General

Details of the geologic setting of the Siberian Traps have been given by *Zolotukhin and Al'mukhamedov* [1988 and references therein], *Fedorenko* [1980, 1981], *Naldrett et al.* [1992], *Hawkesworth et al.* [1995] and *Fedorenko et al.* [1996]. The flood basalts occupy the northwestern margin of the Siberian platform, which has been a stable craton since the end of the Precambrian. It is surrounded to the north by the Taimyr Peninsula and to the west by the East European-Urals block [*Khain*, 1985]. Within the western part of the Siberian platform, early Paleozoic dolomites, limestones and argillaceous sediments are overlain by Devonian calcareous and dolomitic marls, dolomites, and sulfate-rich evaporites and early Carboniferous shallow-water limestones. These are in turn unconformably overlain by the Carboniferous to Late Permian siltstone, sandstone, conglomerate, and coal measures of the Tungusskaya Series. The Siberian Traps overlie the Tungusskaya Series sediments and are present mainly in the Tunguska basin. At present, the extrusive basaltic rocks are estimated to occupy an area of 3.4×10^5 km^2 with an average thickness of 1 km [*Lurie and Masaitis*, 1964]. *Fedorenko et al.* [1996] summarized estimates by different workers of the original volume of the Siberian Traps. They suggested that the total initial volume of the lavas and intrusions could be much higher than 2×10^6 km^3. The entire volcanic succession lacks significant interbedded sedimentary rocks or paleosols. The basement through which the lavas were erupted is not exposed in the Tunguska basin. It is inferred to be similar to Archean-Early Proterozoic granulite-granite gneisses, amphibolites, crystalline schists, quartzites and marble exposed in the Aldan shield to the southeast and in the Anabar Massif to the northeast [*Khain*, 1985]. The thickness of the lava pile in the Tunguska basin is typically greater than 3 km in the northwest and thins to the southeast, where the sequence is only a few tens of meters thick [*Zolotukhin and Al'mukhamedov*, 1988]. Individual flows are, in general, a few tens of meters thick and extend over a few tens of kilometers; however, some flows are as thick as 150 m and can be traced for several hundred kilometers [*Zolotukhin and Al'mukhamedov*, 1988].

One of the most interesting characteristics of the lava pile is the presence of large quantities of basaltic tuffs which are distributed widely but irregularly within the Tunguska basin [*Zolotukhin and Al'mukhamedov*, 1988]. The pyroclastic deposits reach a maximum thickness of 700 m in the center of the Tunguska basin and thin toward the north and west where the lavas are well developed. To the south of the Tunguska basin, the tuffs are present without accompanying lava. The intrusive facies of the traps crop out mainly at the margins of the Tunguska basin where they intrude the Precambrian basement, Paleozoic sediments, and the associated volcanic rocks. Within the Precambrian basement, the intrusions are present as thin sills or dikes; most of the intrusive bodies, however, are present as thicker differentiated and undifferentiated sills in the Devonian sediments and to a lesser extent in the associated volcanic rocks [*Hawkesworth et al.*, 1995]. Sills vary in thickness from a few meters up to 500 m and in some areas contribute to as much as 50% of the thickness of the sedimentary-volcanic succession [*Zolotukhin and Al'mukhamedov*, 1988]. Linear dikes and dike swarms, the latter extending to several hundred kilometers, are present in the northern and northeastern margins of the Tunguska basin. Although the feeder dikes for individual formations are not observed, it is believed that the bulk of the Siberian Traps lavas were erupted through large linear vent systems [*Khain*, 1985]. Recent work on the northwestern margin of the Tunguska Basin suggests that magmatic activity was focused along discrete lineaments and that during the early phase of volcanism the centers of volcanic activity switched episodically between different eruptive sites [*Hawkesworth et al.*, 1995].

The Siberian Traps can be divided broadly into four regions of fundamentally different volcanic sequences [*Zolotukhin and Al'mukhamedov*, 1988; *Sharma et al.*, 1991;

Fedorenko et al., 1996] (Figure 1): (1) Putorana, (2) Noril'sk, (3) Maimecha-Kotui, and (4) Nizhnyaya Tunguska. The great majority of the Putorana rocks are relatively homogeneous, aphyric, and polyphyric tholeiitic basalts [*Zolotukhin and Al'mukhamedov*, 1988; *Sharma et al.*, 1991]. The volcanism in the Noril'sk region is marked by widely varying rock types from picritic through tholeiitic to subalkalic basalts and basaltic andesites [*Zolotukhin and Al'mukhamedov*, 1988; *Sharma et al.*, 1991]. The volcanic rocks of the Maimecha-Kotui area are quite evolved as shown by a wide variety of rock types that include picrites, tholeiitic basalt, alkaline-olivine basalt, trachybasalt, trachyandesite, basanite, olivine nephelinite and maimechite [e.g., *Zolotukhin and Al'mukhamedov*, 1988]. The Nizhnyaya Tunguska region is remarkable as it consists dominantly of basaltic tuffs. Figure 2 gives the currently understood correlation of volcano-stratigraphic sequences present in the four regions [*Sadovnikov*, 1981; *Zolotukhin et al.*, 1986; *Zolotukhin and Al'mukhamedov*, 1988; *Sharma et al.*, 1991; *Fedorenko et al.*, 1996].

The alkalic to ultra-alkalic rocks of the Maimecha-Kotui region crop out on the western slope of the Anabar Massif, about 800 km northeast of the Noril'sk type-section (Figure 1). At present, the stratigraphic position of various suites of the Maimecha-Kotui region relative to those in the other regions is not clear (Figure 2). (Note that in this paper the terms "suite" and "formation" have been used

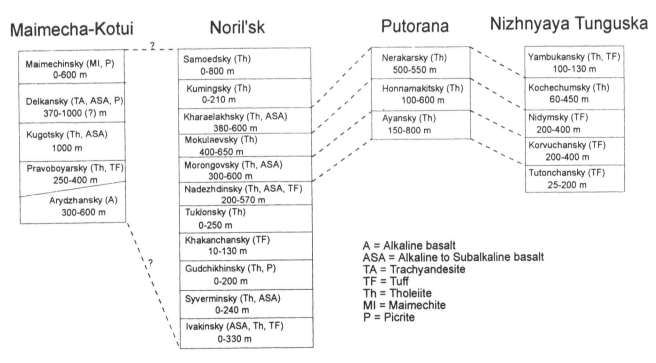

Figure 2. Composite volcano-stratigraphic section of the Siberian Traps (modified from *Zolotukhin and Al'mukhamedov* [1988], *Sharma et al.* [1991], and *Fedorenkco et al.* [1996]).

interchangeably and refer to a group of successive flows with similar geochemical characteristics; the former term is widespread in Russian literature whereas the latter has been used by *Lightfoot et al.* [1993]). ^{40}Ar-^{39}Ar studies indicate that the volcanism in the Maimecha-Kotui region was somewhat earlier than in the Noril'sk region and it may have continued throughout the main-stage eruptions (see below; cf. *Basu et al.* [1995]). The lavas of the Noril'sk region are divided into 11 suites, with the composite thickness of the entire section reaching up to 3 km [*Fedorenko*, 1981; *Zolotukhin and Al'mukhamedov*, 1988; *Fedorenko et al.*, 1989]. The Putorana stratigraphy is divided into three suites (maximum thickness = 1.8 km). These three suites and their counterparts in the Noril'sk and Nizhnyaya Tunguska regions represent the main phase of the volcanic activity in the Siberian Traps, with outpouring of > 90% of the lavas. The Nizhnyaya Tunguska stratigraphy is divided into five suites with a composite thickness of about 1 km. The present review concentrates on the geochemical data for the extrusive rocks from the Putorana and Noril'sk sections.

2.2. *Basalt Compositional Magma Types of the Noril'sk Region*

Extensive work in the Noril'sk region has shown that not every stratigraphic break is associated with a significant change in magma chemistry [*Hawkesworth et al.*, 1995]. *Naldrett et al.* [1992] reviewed the major and trace element data for the Noril'sk area and suggested that five principal magma types were involved in the generation of the lavas of the formations up to and including the Mokulaevsky (see Figure 2): (1) the Ivakinsky and Syverminsky magma type of alkalic and subalkalic affinity, (2) the Gudchikhinsky Ni-rich suite which includes picritic basalts, (3) the primitive but Ni-depleted Tuklonsky suite, which is characterized by flat rare earth element (REE) profiles and also includes picritic basalts, (4) the Lower Nadezhdinsky type, which is light-REE-enriched and has low Nd and high Sr isotope ratios, and (5) the Mokulaevsky type, which is primitive and has close similarities to the Tuklonsky type. In the Upper Nadezhdinsky and the Morongovsky there are flows with compositions transitional between the Mokulaevsky and Lower Nadezhdinsky [*Hawkesworth et al.*, 1995]. *Naldrett et al.* [1992] also identified five principal groups among the intrusive rocks of the Noril'sk area: (1) those of alkalic and subalkalic affinity, (2) Ti-rich dolerite dikes found only in the northeastern part of the Noril'sk region, (3) dolerite dikes and sills found throughout the Noril'sk region, (4) differentiated intrusions not related to the centers of mineralization, and (5) differentiated intrusions present in the vicinity of the centers of mineralization. On the basis of bulk

rock chemistry and field relations, *Naldrett et al.* [1992] and *Fedorenko* [1991] assessed whether the different intrusion types could be linked to the specific magma types in the lava pile. Utilizing an extensive chemical data-set for the Noril'sk region, *Fedorenko et al.* [1996] proposed a new classification for the rocks exposed in the region. These workers identified eight different varieties of primary mantle melts which they classified into four primitive magma types: (1) a low-Yb type with high TiO_2 (this is a characteristic of lavas from the lower to middle Morongovsky), (2) a moderate-Yb type with moderate TiO_2 (Ivakinsky-Gudchikhinsky), (3) a high-Yb type with low TiO_2 corresponding to upper Morongovsky-Samoedky, and (4) a high-Yb type with very low TiO_2 (Tuklonsky). Because of a lack of geochemical data such classification schemes have not been proposed in other areas of the Siberian Traps.

3. AGE OF SIBERIAN VOLCANISM

The age of inception and duration of Siberian flood volcanism are currently topics of active research. The motivation for establishing the precise age of inception of the volcanism stems from the possibility that the massive volcanic activity may be related to the Permo-Triassic extinction event (see below). Early geochronological investigations using the whole-rock K-Ar method suggested that the eruption of the Siberian flood basalts was concentrated between 235 and 220 Ma but lasted from 240 to 200 Ma [*Zolotukhin and Al'mukhamedov*, 1988 and references therein]. The first ^{40}Ar-^{39}Ar investigation was by *Baksi and Farrar* [1991a], who reported whole-rock incremental heating data for two lava flows from near the bottom and near the top of the volcanic succession. They concluded that the volcanism began at ~238 Ma and lasted for ~10 m.y. Subsequent refinement of laboratory techniques led *Baksi and Farrar* [1991b] to revise their estimate of the age of volcanism to 244–240 Ma. Further results from two other ^{40}Ar-^{39}Ar laboratories showed apparent inconsistencies. *Renne and Basu* [1991] analyzed whole-rock and plagioclase samples from the Ivakinsky and Nerakarsky formations (see Figure 2) using laser incremental heating, and inferred that the Siberian Traps erupted over an extremely short time interval (900,000 ☐ 800,000 years) beginning at about 248 Ma. Additional data were reported by *Dalrymple et al.* [1991, 1995], who found slightly younger ages (245-244 Ma) for the flood basalts than for biotites from intrusions cutting the basalts (249 ± 1 Ma). These workers concluded that the plagioclase samples used to date the basalts had lost ~2% of their radiogenic ^{40}Ar and that the basalts were older than 249 Ma. The discrepancy in the ^{40}Ar-^{39}Ar ages appeared to be partly related to Ar loss or gain and partly to the age of the neutron flux monitor used by the different laboratories. An additional limit on the age of inception of the volcanism was provided by ion microprobe U-Pb data for zircons from the Noril'sk I intrusion that yielded a concordia intercept date of 248.0 ± 3.7 Ma [*Campbell et al.*, 1992]. The Noril'sk I intrusion cuts through the lower third of the volcanic sequence in the Noril'sk region. Accurate dating of this intrusion would thus give a minimum age of inception of volcanism. The zircon data, although significant in providing independent corroboration of the ^{40}Ar-^{39}Ar results, were still too imprecise to resolve the apparent inconsistencies in ^{40}Ar-^{39}Ar data from the three different laboratories. Recently, *Renne* [1995] showed that (1) the ^{40}Ar-^{39}Ar basalt data from the three laboratories yield consistent ages (249–246 Ma) if normalized to the same reference monitor standard and (2) the biotites analyzed by *Dalrymple et al.* [1991, 1995] have excess ^{40}Ar. Furthermore, recalculation of the age of the reference standard with the astronomically calibrated geomagnetic polarity time scale [*Renne et al.*, 1994] gives an age of 250.0 ± 1.6 Ma for the inception of the Siberian flood basalt volcanism. This conclusion is consistent with additional ^{40}Ar-^{39}Ar data obtained by *Pringle et al.* [1995]. Also, *Kamo et al.* [1996] published a precise U-Pb zircon and baddeleyite age for the Noril'sk I leucogabbro of 251.2 ± 0.3 Ma, consistent with the adjusted ^{40}Ar-^{39}Ar results. In summary, it appears that the bulk of the Siberian Traps erupted at around 250 Ma, possibly largely within a period of about 1 m.y. The age of inception is within error of the estimated age for the Permo-Triassic boundary, 251.1 ± 3.6 Ma [*Claoué-Long et al.*, 1991, 1995] and 249.9 ± 1.5 Ma [*Renne et al.*, 1995].

The alkalic volcanism in the Maimecha-Kotui area may have been somewhat older than the main stage of flood volcanism in the Putorana and Noril'sk regions. This is suggested by ^{40}Ar-^{39}Ar dating of an olivine nephelinite from the lower part of the Maimecha-Kotui section which gave a plateau age of 253.3 ± 2.6 Ma [*Basu et al.*, 1995]. The uncertainty associated with the age of volcanism in Maimecha-Kotui is too large to establish if the volcanism in this region is older than the main pulse of volcanism (= 250.0 ± 1.6 Ma). However, *Basu et al.* [1995] pointed out that the inferred uncertainty of ± 1.6 Ma for the main pulse of volcanism is large because of averaging of data from three different labs and that in order to make an intralaboratory comparison the uncertainty may be may be reduced to ±0.3 Ma; the latter is the calculated uncertainty in the age of a sample from the Ivakinsky formation analyzed by these workers [*Renne and Basu*, 1991]. *Basu et al.* [1995] concluded that the age of 253.3 ± 2.6 Ma for the olivine nephelinite from Maimecha-Kotui is older than the initiation of the main stage of flood volcanism at 250.0 ± 0.3 Ma.

Clearly, this interpretation needs to be evaluated further by obtaining new data for the lower part of the Maimecha-Kotui area.

The Maimechinsky suite (see Figure 2) from Maimecha-Kotui has been considered to be either younger than the Samoedsky formation [*Horan et al.*, 1995; *Fedorenko et al.*, 1996] or correlated with the Nerakarsky or Samoedsky formations [*Zolotukhin and Al'mukhamedov*, 1988]. A phlogopite from an alkalic-ultrabasic intrusion (the Gulin intrusion) associated with the Maimechinsky suite gave an ^{40}Ar-^{39}Ar age of 250.4 ± 1.3 Ma [*Renne et al.*, 1994; *Basu et al.*, 1995; cf. *Dalrymple et al.*, 1995]. This result led *Basu et al.* [1995] to infer that the termination of alkaline volcanism in the Maimecha-Kotui area was synchronous with the initiation of main-stage tholeiitic volcanism to the southwest in the Putorana and Noril'sk areas. This conclusion is subject to other interpretations as ^{40}Ar-^{39}Ar dating cannot distinguish between the inception and termination of the main-stage volcanism. Further, because the Maimechinsky formation has been considered to be either younger or synchronous with the highest formations in the Noril'sk and Putorana sections, it is more likely that the termination of alkalic volcanism in the Maimecha-Kotui region was synchronous with the termination of main-stage tholeiitic volcanism elsewhere.

Additional limits on the duration of the Siberian Traps eruption are provided by paleomagnetic data from the Noril'sk area [*Lind et al.*, 1994]. These data show that the Ivakinsky formation has reversed polarity, whereas the younger suites of the sequence are characterized by normal polarity [*Lind et al.*, 1994; see also *Campbell et al.*, 1992]. This observation led *Campell et al.* [1992] to suggest that the eruption of the Siberian Traps occurred within 600,000 years, spanning a R-N interval during the Illawara Reversals, a time of rapid reversals from the latest Permian to the middle Triassic. However, recent paleomagnetic work by *Mitchell et al.* [1994] cast doubt on the above estimates of duration of the flood volcanism. If the Noril'sk stratigraphy represents accurately all stages of Siberian volcanism, then all the samples from the Putorana region should have normal polarity. *Mitchell et al.* [1994] found that three samples from the southern part of the Putorana region have a reversed polarity, suggesting that the flows from this region may be younger than their counterparts in the Noril'sk area. Alternatively, the Noril'sk area may in fact be a R-N-R sequence [*Mitchell et al.*, 1994]. Another interpretation of the data would be that the inferred correlation between the various units (see Figure 2) from the Noril'sk and the Putorana regions is not correct. A much more detailed paleomagnetic and geochemical study of the lava pile is required to differentiate among these possibilities and to obtain an accurate estimate of the duration of the volcanism. If the inferred correlation between the Nerakarsky formation from the Putorana region and the Kharaelakhsky formation from the Noril'sk area is correct (see Figure 2), the ^{40}Ar-^{39}Ar and U-Pb dating indicate that an estimated minimum of 75% of the total volume of the Siberian Traps was erupted within ~ 1 m.y.

4. SIBERIAN FLOOD VOLCANISM AND THE PERMO-TRIASSIC MASS EXTINCTION

The Permo-Triassic mass extinction was the most catastrophic in the geologic record, with as many as 90% of marine species and 70% of terrestrial vertebrate families dying out [*Erwin*, 1994]. Whether or not a relationship exists between this mass extinction and Siberian flood volcanism depends directly on the relative ages of the two events. As mentioned above, *Campbell et al.* [1992] and *Renne et al.* [1995] showed that the bulk of the Siberian flood volcanism is synchronous with the Permo-Triassic boundary. *Renne et al.* [1995] argued that the volcanism produced sufficient stratospheric sulfate aerosols to cause rapid global cooling which in turn led to marine regression. A short-lived volcanic winter was followed immediately by greenhouse conditions resulting from a buildup of volcanogenic CO_2. The environmental extrema engendered the mass extinction. Although the synchronicity of Siberian flood volcanism and the Permo-Triassic boundary appears established, the above model is not without critics. For example, *Erwin* [1994] and references therein suggested that the climatic effects of SO_2 emanations are self-limiting and may not lead to global cooling. Furthermore, Erwin argued [quoted by *Kamo et al.*, 1996] that to have been a major contributor to the mass extinction, Siberian volcanism should have somewhat preceded the extinction. The above arguments led *Kamo et al.* [1996] to suggest that Siberian flood volcanism may have been a major obstruction to early Triassic biospheric recovery but not the ultimate cause of the extinction.

5. ARE THE SIBERIAN FLOOD BASALTS THE RESULT OF A STARTING PLUME?

The production of continental flood basalts (CFBs) has been attributed to the arrival of plume heads from the core-mantle boundary [e.g., *Richards et al.*, 1989; *Hill*, 1991]. According to the starting-plume model, initial rapid and voluminous CFB eruptions are followed by decreasing eruption rates and, in some cases, the generation of a hotspot track as a lithospheric plate moves over a relatively stationary mantle plume. Alternatively, some CFBs have

been attributed to the melting of metasomatized shallow mantle caused by lithospheric stretching, with or without a plume [e.g., *Hawkesworth et al.*, 1984, 1986; *Gallagher and Hawkes-worth*, 1992]. Yet another model of CFB generation was given by *King and Anderson* [1995], who suggested that pull-apart of an asymmetric lithosphere (see below) could generate conditions conducive to CFB volcanism.

Morgan [1981] proposed that the Siberian flood basalt eruption was linked to the initiation of the Jan Mayen hotspot. According to his plate-tectonic reconstruction, the Lomonosov Ridge could be a part of the hotspot track. Figure 3 is a polar view showing lithospheric plates and the present locations of the Siberian Traps, the Lomonosov Ridge, and the Jan Mayen hotspot. A detailed plate-tectonic reconstruction of the opening of the Arctic Ocean is not available at present to test Morgan's proposal. Further, it is likely that the Lomonosov Ridge is not a segment of the conjectured hotspot track as seismic, gravity and magnetic data from the Lomonosov Ridge point to a continental structure underlying 1–2 km thick sediments [see *Jokat et*

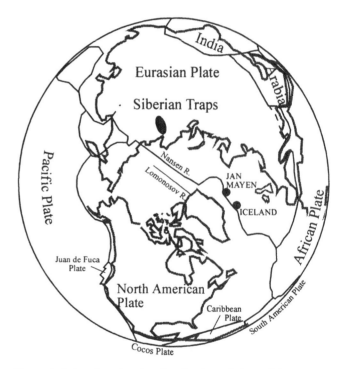

Figure 3. Polar view showing the plate boundaries and the present locations of the Siberian Traps, the Lomonosov Ridge and the Jan Mayen hotspot. The onset of volcanism on the Siberian platform 250 m.y. ago was proposed by *Morgan* [1981] to be a consequence of the arrival of a plume currently active at Jan Mayen, the Lomonosov Ridge being the postulated hotspot track left during the opening of the Arctic Ocean. However, the plate-tectonic reconstruction for 250 Ma is incomplete for lack of data, and at present no hotspot can be linked readily to the Siberian Traps.

al., 1992 and references therein]. Moreover, similarity of P-wave velocity structure determined for the Lomonosov Ridge and that measured along the outer Kara and Barents shelves, along with geomagnetic anomaly patterns in the Nansen Basin, suggest that the ridge is a fragment of continental margin detached from Siberia by normal seafloor spreading at the Nansen Ridge [*Sweeney et al.*, 1982; *Forsyth and Mair*, 1984; *Wilson*, 1985; *Jokat et al.*, 1992].

Numerical modeling of the mantle plume initiation model for CFB events suggests that large-scale uplift (0.5–4 km) should occur in the general region of the flood volcanism [e.g., *Farnetani and Richards*, 1994]. Utilizing these predictions, *Renne et al.* [1995] suggested that a mantle plume head associated with Siberian flood volcanism might have caused an uplift of 1–3 km over an area ~500 km in radius. The uplift should have reached a maximum 5–20 m.y. before the onset of the volcanism. However, there is no evidence for such uplift in the general area where the Siberian flood basalts are developed best [e.g., *Kamo et al.*, 1996]. Indeed, the Middle Carboniferous to Late Permian Tungusskaya Series which is inferred to underlie much of the Siberian Traps contains some of the largest coal-measures in the world, with few erosional breaks. These observations led *Kamo et al.* [1996] to conclude that although the tectonic conditions during Tungusskaya sedimentation were somewhat unstable, sedimentation generally compensated subsidence and the area remained close to sea level during basin filling. Further, available paleontological and geochemical data indicate that the initial volcanic eruptions in the Noril'sk region may have occurred underwater in shallow lakes or lagoons [*Fedorenko*, 1991]. Thus, the lack of uplift in the general region of flood volcanism provides no evidence for a plume underlying the Siberian platform before or during the eruptions. If the Siberian volcanism had a plume origin, then the lack of uplift in the western Siberian platform before or during the eruption of the Siberian Traps must be explained. This question requires an evaluation of models of plume-lithosphere interaction to explore the specific issue of whether or not extension in the upper mantle and crust can mitigate the need for substantial uplift [*Fedorenko et al.*, 1996]. Note that the above arguments about a lack of evidence of uplift in the western Siberian platform rely implicitly on one key inference: the Tungusskaya Series is as extensive as the Siberian Traps. This inference is based on some outcrops at the margin of the flood basalt province and has to be re-evaluated. For example, if the Tungusskaya Series is not as extensive as the Siberian Traps, it can be speculated that the sediments and associated coal-measures were deposited in grabens developed in a region that was uplifted because of plume activity.

King and Anderson [1995] suggested that during tectonic pull-apart of asymmetric lithospheres (i.e., lithospheres with two greatly varying thicknesses because of their relative ages as, for example, younger continental margins attached to Archean cratons) as much as 10^4 km^3 of magma may be generated. The magma production is accomplished via transport of mantle material from the thicker part to the thinner part of the lithosphere. According to *King and Anderson* [1995], this model can explain many features of large igneous provinces including the rapid turn-on and turn-off of the volcanism, the absence of uplift in some provinces, and the presence of many large igneous provinces at continental margins adjacent to Archean cratons. All of these features characterize the Siberian Traps. However, the volume of magma produced in Siberia is >100 times more than that achievable in *King and Anderson's* model.

Recent studies have noted that the numerous dike swarms and sheeted-dike-like sequences in the Siberian Traps indicate that lava eruption occurred in an extensional environment [*Zonenshain et al.*, 1990]. However, despite the extensional conditions and crustal thinning (8–10 km), the flood volcanism was not associated with significant lithospheric rifting [*Zonenshain et al.*, 1990]. Thus, mantle decompression resulting from rifting was probably not the primary cause of widespread melting [*Zonenshain et al.*, 1990; see also *White and McKenzie*, 1995].

Collectively, the above observations suggest that the Siberian Traps eruption cannot be linked directly either to lithospheric stretching in the absence of a plume or to hotspot initiation. Yet there appears to be a consensus supporting a plume origin among those working on the Siberian Traps [e.g., *Sharma et al.*, 1991, 1992; *Renne and Basu*, 1991; *Arndt et al.*, 1993; *Wooden et al.*, 1993; *Renne et al.*, 1995; *White and McKenzie*, 1995; *Hawkesworth et al.*, 1995; *Fedorenko et al.*, 1996]! Two pieces of evidence have engendered such a confluence of opinion: (1) the large volume (> 2×10^6 km^3) of magma emplaced and (2) the short duration of ~1 m.y. of eruption. Additional evidence connecting a mantle plume and the Siberian Traps comes from the ^{3}He/^{4}He analysis of the olivine nephelinite from the lower part of Maimecha-Kotui section which has an ^{40}Ar-^{39}Ar age of 253.3 ± 2.6 Ma [*Basu et al.*, 1995]. Olivine phenocrysts of this rock showed ^{3}He/^{4}He ratios up to 12.7 times the atmospheric ratio (R_A), much higher than the average value for mid-ocean ridge basalts (= $8 \times R_A$) [e.g., *Graham et al.*, 1992]. This observation indicates that there is a connection between Siberian flood volcanism and an undegassed mantle source [*Basu et al.*, 1995], ostensibly a plume from the lower mantle [e.g., *Porcelli and Wasserburg*, 1995]. A similar result has been reported for some early-erupted alkalic basalts associated with Deccan Traps

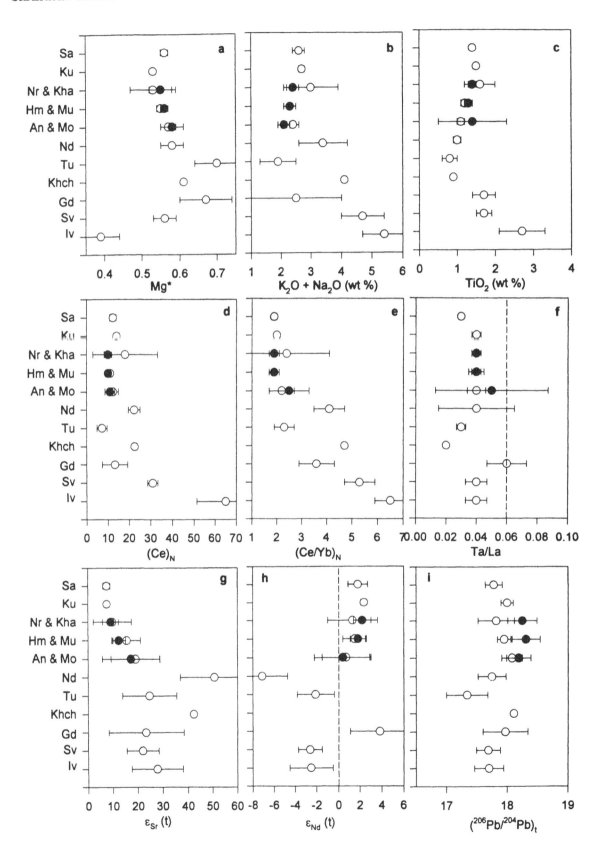

magmatism [*Basu et al.*, 1993], which was triggered by the initiation of the Réunion hotspot [e.g., *Morgan*, 1981]. In summary, the large volume of magma erupted, the short duration of eruption and, consequently, the high eruption rates (~1.7 km^3 yr^{-1} [*Renne and Basu*, 1991]), are consistent with a starting-plume origin for the Siberian Traps. Further, the high ^{3}He/^{4}He signature in an early lava also indicates a hotspot origin. However, whether the Siberian Traps and Jan Mayen hotspot are related is not clear at present.

6. TEMPORAL GEOCHEMICAL AND ISOTOPIC VARIATIONS

An assessment of temporal geochemical and isotopic variations in a flood basalt province is vital in understanding the compositional evolution of the magmas and the variations in source composition. Furthermore, data on these variations can be combined with estimates of the volume of erupted magma to establish the relative roles of various crust and mantle reservoirs and to evaluate models of flood basalt genesis. Numerous major and trace element determinations as well as Sr-, Nd-, and Pb-isotopic analyses are now available from the Siberian Traps. Although most of the samples have come from the northern and northwestern parts of the Siberian Traps (Noril'sk and northern Putorana), they are adequate to derive first-order conclusions about the evolution of magma sources. Detailed accounts of element and isotopic variations in stratigraphically controlled drill core and outcrop samples may be found in *Lightfoot et al.* [1990, 1993], *Sharma et al.* [1991, 1992], *Wooden et al.* [1993], *Hawkesworth et al.* [1995], and *Fedorenko et al.* [1996]. These workers also discussed detailed petrogenetic models for the Siberian Traps. For the following analysis, the data are obtained from the above studies. In addition, some unpublished trace and rare earth element data from the Putorana are also used (A.R. Basu and M. Sharma, unpublished data). The complete data set may be obtained from the author. Some selected data are presented in Figures 4 through 13.

Figure 4 illustrates stratigraphic variations in the average values of selected oxides and in element and isotopic ratios in Noril'sk (open circles) and Putorana (filled circles). The error bars around the average value for a stratigraphic unit reflect 1σ variations within that unit. Note that the data from the correlated formations from Noril'sk and Putorana are placed next to each other. There appears to be no significant geochemical and isotopic difference between the correlated units from the Noril'sk and the Putorana regions (Figure 4). Some notable features of the geochemical and isotopic stratigraphy are listed below.

(1) The oldest lavas of the Ivakinsky formation show the lowest Mg* = 0.39 (molar Mg* = Mg/(Mg + 0.85*Fe$_{tot}$)) and the highest TiO$_2$ content (= 2.7 wt.%); the younger, voluminous lavas from Putorana and Noril'sk display Mg* = 0.55–0.58 and TiO$_2$ = 1.2–1.4 wt.% (Figure 4a and c). None of the lavas show Mg* >0.79 that would suggest equilibration with mantle olivine [e.g., *Albarède*, 1992]. The lavas with the highest Mg* come from Gudchikhinsky suite and typically have MgO >10 wt.% [*Sharma et al.* 1991; *Wooden et al.*, 1993].

(2) The lavas become less alkaline from the base of the lava pile upward with decreasing (Ce)$_N$ and decreasing (Ce/Yb)$_N$ (here the subscript refers to chondrite-normalized values) (Figure 4b, d and e); the bulk of the Siberian Traps (Ayansky through Nerakarsky in the Putorana and Morongovsky through Kharaelakhsky in the Noril'sk region) has an average total alkali content of ~2.5, (Ce)$_N$ ~10 and (Ce/Yb)$_N$ ~2.

(3) With the exception of the Gudchikhinsky picrites, the Ta/La ratio of the bulk of the Siberian Traps is distinctly lower than the primitive mantle Ta/La ratio of 0.06 [*Sun and McDonough*, 1989] (Figure 4f).

(4) The average ε$_{Sr}$ (t) values of suites throughout the lava pile are +8 to +51 (1 ε unit = 1 part in 10,000 relative to the estimated bulk-earth isotope ratio at time t; the bulk earth is assumed to have (^{87}Sr/^{86}Sr)$_{250m.y.}$ = 0.7045; Figure 4g). Apart from the Khakanchansky basaltic tuff, the mean ε$_{Sr}$(t) values of the early lavas (Ivakinsky through Tuklonsky) range from +20 to +30. The Nadezhdinsky basalts, which erupted after the Tuklonsky picrites, display a sharp increase in the mean ε$_{Sr}$(t) value (to +51). The bulk of the Siberian Traps lavas show mean ε$_{Sr}$(t) values decreasing progressively up-section (Figure 4g).

Figure 4. Stratigraphic variations of selected major elements, trace elements, elemental ratios and isotopes in the Siberian Traps. Data from the Noril'sk and Putorana regions are plotted as open and filled circles, respectively. Y-axis: Iv = Ivakinsky, Sv = Syverminsky, Gd = Gudchikhinsky, Khch = Khakanchansky, Tu = Tuklonsky, Nd = Nadezhdinsky, An = Ayansky, Mo = Morongovsky, Hm = Honnamakitsky, Mu = Mokulaevsky, Nr = Nerakarsky, Kha = Kharaelakhsky, Ku = Kumingsky, Sa = Samoedsky. The data from An, Hm and Nr suites of the Putorana region are combined with the correlated units Mo, Mu, and Kha in the Noril'sk region. This legend will be followed for Figures 5 through 13. Only averages and 1σ variations are plotted for each of the stratigraphic units. There are no significant geochemical variations between the correlated units from the Putorana and Noril'sk regions. The subscript "N" in panels d and e refers to chondrite-normalized values. The vertical lines in panels f and h show the position of the primitive mantle [after *Sun and McDonough*, 1989].

(5) The average ε_{Nd} (t) values (calculated assuming bulk earth $(^{143}Nd/^{144}Nd)_{250m.y.} = 0.512316$) show wide variability in the lower part of the lava sequence (-7 to +4; Figure 4h). The bulk of the Siberian lavas have ε_{Nd} (t) values ~+2 (Figure 4h).

(6) The $(^{206}Pb/^{204}Pb)_t$ ratios of the entire Siberian lava pile appear to vary only modestly from 17.3 to 18.3 (Figure 4i).

Geochemical studies of flood basalt provinces associated with the breakup of Gondwana have revealed the existence of two types of basalt: a low-Ti and a high-Ti basalt [e.g., *Bellieni et al.*, 1984; *Marsh*, 1987; *Ellam and Cox*, 1991; *Hergt et al.*, 1991]. The Siberian Traps also display large variations in TiO$_2$ contents for a given Mg* (Figure 5). *Lightfoot et al.* [1993] divided the ultramafic and mafic lavas into two broad groups: a lower assemblage with high Ti and an upper assemblage with low Ti. The high-Ti suites include the Ivakinsky, Syverminsky, and Gudchikhinsky and the low-Ti suites include the rest of the formations (Figure 5). The low-Ti and high-Ti groups are not related by fractional crystallization of similar phase assemblages from a uniform parental magma as they have overlapping ranges in Mg* but different TiO$_2$ contents. The distinction between the high-Ti and low- Ti groups becomes quite clear when Gd/Yb ratio is plotted against La/Sm ratio [*Lightfoot et al.*, 1993; *Hawkesworth et al.*, 1995] (Figure 6). Figure 6 shows that,

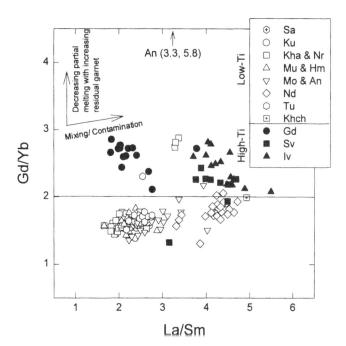

Figure 6. Variations in La/Sm versus Gd/Yb. Data from *Lightfoot et al.* [1993], *Wooden et al.* [1993], *Hawkesworth et al.* [1995], and M. Sharma and A.R. Basu (unpublished data). The arrow points to the location of one data point from the Ayansky formation which has extremely high Gd/Yb ratio and lies outside the diagram. A Gd/Yb ratio of 2.0 is chosen arbitrarily to distinguish the low-Ti and high-Ti groups [after *Lightfoot et al.*, 1993].

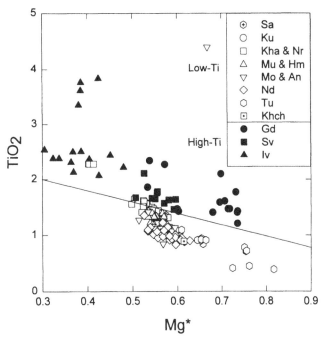

Figure 5. Plot of TiO$_2$ (wt%) versus Mg* after *Lightfoot et al.* [1993]. Mg* = molar Mg/(Mg + 0.85*Fe$_{tot}$). The distinction between the low-Ti and high-Ti basalts is indicated approximately by the solid line.

with very few exceptions, the high-Ti group is depleted in heavy REE (Yb) relative to middle REE (Gd) in comparison to the low-Ti group. Accordingly, *Lightfoot et al.* [1993] used a Gd/Yb ratio = 2 to demarcate the two groups (Figure 6). The differences in the Gd/Yb ratio may suggest different depths of melting for the two groups (see below). Further, the high La/Sm ratios (>3) of the Ivakinsky, Syverminsky, and Nadezhdinsky suites may reflect continental crustal contamination [e.g., *Lightfoot et al.*, 1993; *Hawkesworth et al.*, 1995] (Figure 6).

Melts derived from primitive mantle are expected to have a Th/Ta ratio of ~2.3 (Figure 7a) [*Sun and McDonough*, 1989; see also *Wooden et al.*, 1993]. Because both Th and Ta are highly incompatible elements, the melts are expected to retain their Th/Ta ratio during closed-system fractional crystallization and evolve along the solid line in Figure 7a. Most of the low-Ti and high-Ti basalts appear to have evolved with Th/Ta ratios that are quite different from each other (Figure 7a). Also, the Th/Ta ratios of the lavas are much higher than expected from the closed-system crystallization of primitive-mantle-derived melts. The only exceptions to this rule are some picrites from the Gudchikhinsky formation (Figure 7a). The high-Ti lavas

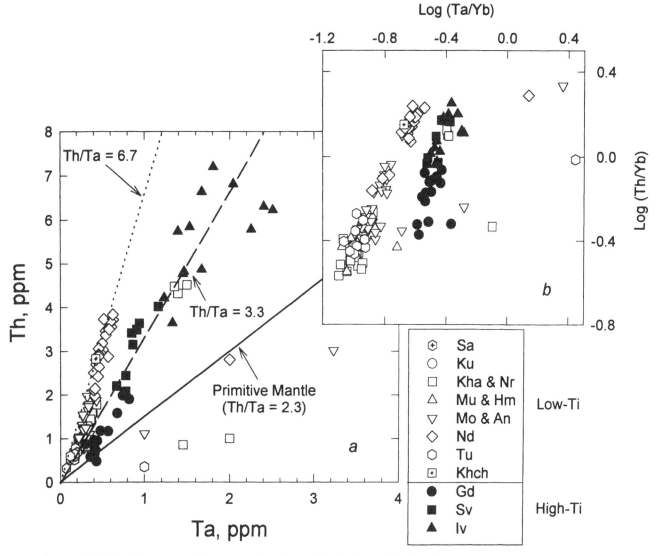

Figure 7. (a) Tantalum versus Th concentrations in the Siberian lavas. The low-Ti and high-Ti magmas evolved with different initial Th/Ta ratios. Interestingly, the Th/Ta ratios for nearly all, including the least contaminated lavas, are much higher than the primitive mantle Th/Ta ratio [e.g., *Sun and McDonough*, 1989]. (b) Log (Th/Yb) versus log (Ta/Yb). Most data for the low-Ti and high-Ti basalts plot along two different trends with slopes >45°, indicating (1) the derivation of magmas from sources with two different Th/Ta ratios and (2) that both types of basalts are contaminated by continental crust, which has high Th/Ta ratios.

display wide variations in Th/Ta ratios which are as high as 4 for some Ivakinsky and Syverminsky samples and as low as 2.3 for some Gudchikhinsky samples. On the other hand, most of the low-Ti samples have relatively constant Th/Ta ratios over a range of Th and Ta concentrations, suggesting generation of these lavas via fractional crystallization of a magma source with Th/Ta ~6.7 [cf. *Wooden et al.*, 1993]. In detail, however, the Th/Ta ratios of both high-Ti and low-Ti groups increase somewhat with increasing Ta concentration. This is illustrated in Figure 7b, which is a plot of log (Ta/Yb)

versus log (Th/Yb). On this diagram, magmas originating from a common source and evolving in a closed system define an array with a slope of 45° [*Russell and Cherniak*, 1989]. In Figure 7b, the low-Ti and high-Ti groups define two distinct arrays (with some exceptions), each with a slope >45°. As the average upper continental crust may have Th/Ta as high as 10 [e.g., *Condie*, 1993], the observed enhancement in the Th/Ta ratios of Siberian lavas is consistent with contamination of magmas with continental crust.

Incompatible element patterns in flood basalt provinces have been used extensively to assess the nature of magma sources and contaminants. Figure 8 illustrates the primitive-mantle-normalized averages of moderately to highly

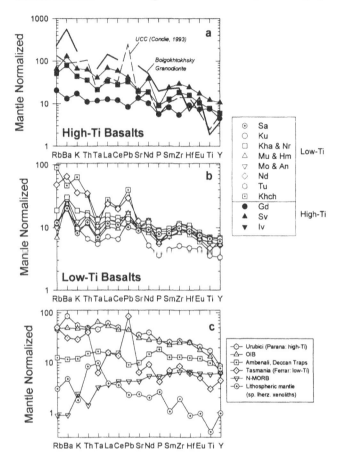

Figure 8. Primitive-mantle-normalized incompatible-element plot (primitive-mantle values from *Sun and McDonough*, 1989). For the sake of clarity, the data are presented in three different panels: (a) averages of high-Ti basalt suites, average upper continental crust (UCC) [*Condie*, 1993], and Bolgokhtokhsky granodiorite [*Hawkesworth et al.*, 1995]; (b) averages of low-Ti basalt suites; and (c) average Urubici magma type (representing the Paraná high-Ti lavas) [*Peate*, this volume], ocean island basalts (OIB) [*Sun and McDonough*, 1989], the least contaminated unit from the Deccan Traps (Ambenali Formation) [*Lightfoot et al.*, 1990; *Peng et al.*, 1994], the average of low-Ti Ferrar basalts [*Hergt et al.*, 1991], normal mid-ocean ridge basalts (N-MORB) [*Sun and McDonough*, 1989], and the average continental lithospheric mantle as sampled by spinel harzburgite xenoliths [*McDonough*, 1990]. The trace element patterns of most of the Siberian lavas are between the OIB/Paraná high-Ti basalts and Ambenali Formation patterns. Also, unlike their Gondwanan counterparts (e.g., low-Ti Ferrar basalts) most of the low-Ti Siberian lavas do not show a pronounced negative Ti anomaly. The incompatible element patterns of the Siberian Traps are not diagnostic of derivation from either a MORB-type or an OIB-type mantle source.

incompatible elements for each of the 11 units in the combined Noril'sk and Putorana sequences. The trace element patterns are compared with averages of selected formations from other flood basalt provinces as well as normal mid-ocean ridge basalts (N-MORB) [*Sun and McDonough*, 1989], ocean island basalts (OIB) [*Sun and McDonough*, 1989], and the continental lithospheric mantle as sampled by spinel harzburgite xenoliths [*McDonough*, 1990] (Figure 8c). Additionally, mantle-normalized patterns of the average upper continental crust (UCC) [*Condie*, 1993] and Bolgokhtokhsky granodiorite intrusion [*Hawkesworth et al.*, 1995] are given in Figure 8a; the latter was assumed by *Hawkesworth et al.* [1995] to reflect the composition of the local continental contaminant in the Noril'sk region. Both high-Ti and low-Ti Siberian lavas show high concentrations of large-ion-lithophile elements (Rb, Ba) relative to high-field-strength elements (Ta, P, and Ti) (Figure 8). The trace element characteristics of the high-Ti basalts appear to be intermediate between those of the Ambenali (Deccan) and Urubici (Paraná; cf. Figures 8a and 8c). None of the suites display the relatively smooth patterns shown by average OIB, the Urubici and the Ambenali, however. All Siberian low-Ti basalts show (1) enrichments in Zr and (2) depletions in Ta relative to La and (3) Ti/Zr = 32–117. These characteristics are similar to those observed in the low-Ti Gondwanan basalts. Such features have been argued to be distinctive of continental lithospheric mantle containing small amounts of subducted sediment [e.g., *Hergt et al.*, 1991]. An important distinction between low-Ti Siberian lavas and their Gondwanan counterparts is that the former show low Rb/Ba ratios (except the Nadezhdinsky)—a feature that could result from contamination with a Bolgokhtokhsky granodiorite-type melt [see also *Hawkesworth et al.*, 1995]. However, in contrast to the low-Ti Gondwanan basalts, average lithospheric mantle, average upper continental crust, or Bolgokhtokhsky granodiorite, the low-Ti Siberian lavas (except the Khakanchansky and Nadezhdinsky) do not show pronounced negative Ti anomalies (Figure 8).

Sharma et al. [1991, 1992] studied representative samples from the Noril'sk and Putorana regions and found that (1) on an ε_{Sr}-ε_{Nd} diagram, a majority of the rocks fall in the region of OIB and (2) the bulk of the Siberian Traps has $\varepsilon_{Nd}(t)$ ~+2 and $\varepsilon_{Sr}(t)$~+7. Further work by *Lightfoot et al.* [1993], *Wooden et al.* [1993] and *Hawkesworth et al.* [1995] substantiated these observations. Figure 9 shows the available Sr and Nd isotopic data. Also shown in the diagram are the fields of present-day Pacific MORB and OIB. Note that data for 18 of 26 high-Ti basalts do not fall in the OIB field. In contrast, data for most of the low-Ti basalts (the Morongovsky, Mokulaevsky, Kharaelakhsky and their Putorana counter-parts) fall in the OIB field. Significantly,

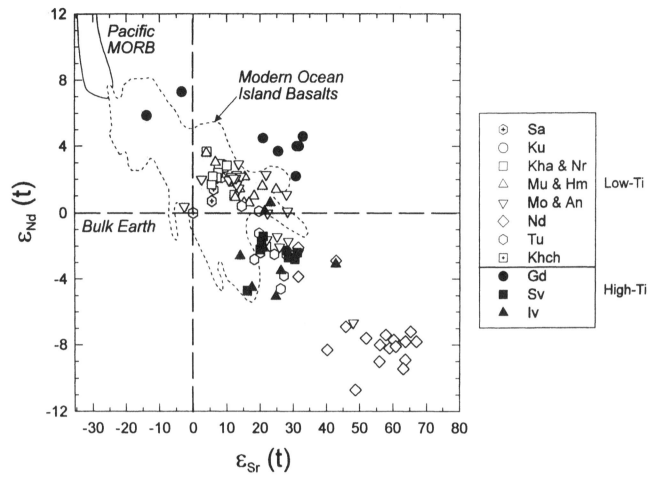

Figure 9. Plot of ε_{Sr} (t) versus ε_{Nd} (t) for the Siberian flood basalts. Data sources: *Sharma et al.* [1991, 1992]; *Lightfoot et al.* [1993]; *Wooden et al.* [1993]; *Hawkesworth et al.* [1995]. Field of modern ocean island basalts (OIB) is from *Hart and Staudigel* [1989]. The Pacific MORB field is defined by data from *Macdougall and Lugmair* [1986], *White et al.* [1987], and *Mahoney et al.* [1994].

the early, volu-metrically negligible eruptions (Ivakinsky through Nadezhdinsky) have the extreme ε_{Sr} and ε_{Nd} values. Equally remarkable is that the voluminous late-stage eruptions (Morongovsky and Ayansky through Kharaelakhsky and Nerakarsky) display a relatively narrow range of ε_{Nd} values. Another curious aspect of the data is that the isotopic extremes are confined to one or two suites and the data do not show a continuous mixing array between the extremes. For example, the highest ε_{Nd} and the lowest ε_{Sr} values come from the Gudchikhinsky picrites. Similarly, the samples from the Nadezhdinsky (and a sample from the Morongovsky and Ayansky) show the highest ε_{Sr} and the lowest ε_{Nd}.

The Pb isotopic compositions of the low-Ti Siberian basalts define a rough positive trend on the $(^{206}Pb/^{204}Pb)_t$ vs $(^{207}Pb/^{204}Pb)_t$ plot (Figure 10a). The bulk of the Siberian data lie in a cluster at the lower end of the array defined by

modern OIB and MORB. Data for all the basalts, except two of the Tuklonsky suite, fall well to the right of the geochron in Fig 10a, indicating their derivation from a mantle source with multi-stage evolution of the U-Pb system [see also *Sharma et al.*, 1992]. In comparison with the bulk of the Siberian basalts, most of the Gudchikhinsky lavas have higher $^{207}Pb/^{204}Pb$ for a given $^{206}Pb/^{204}Pb$. The rough positive trend shown by the low-Ti lavas appears to be consistent with mixing with a contaminant having a low time-integrated U/Pb ratio, possibly lower crust. Mixing with a Bolgokhtokhsky granodiorite-type melt (with $(^{206}Pb/^{204}Pb)_t$ = 16.29 and $(^{207}Pb/^{204}Pb)_t$ = 15.3) could also produce the observed trend [*Wooden et al.*, 1993]. On the $(^{206}Pb/^{204}Pb)_t$ vs $(^{208}Pb/^{204}Pb)_t$ diagram (Figure 10b), the Siberian basalt data lie slightly above the Northern Hemisphere Reference Line (NHRL) [*Hart*, 1984] and in the fields of MORB and OIB. Unlike the lavas from Paraná, Deccan, and Columbia River

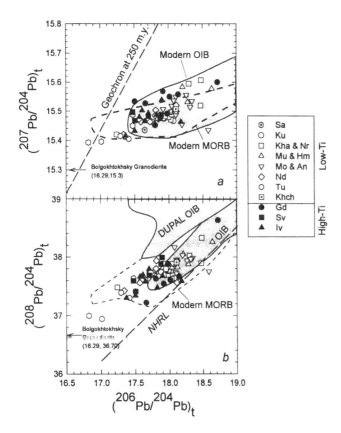

Figure 10. Plot of $(^{206}Pb/^{204}Pb)_t$ versus $(^{207}Pb/^{204}Pb)_t$ (a) and $(^{208}Pb/^{204}Pb)_t$ (b) in Siberian basalts. The data points fall in the fields of OIB and MORB and trend toward the field of DUPAL OIB. Most of the Siberian data fall to the right of the geochron at 250 Ma in panel (a), indicating a multistage U/Pb evolution of the source material. Note also that the Bolgokhtokhsky granodiorite could be a contaminant. Data sources: *Sharma et al.* [1992]; *Wooden et al.* [1993]. Fields of OIB, MORB and DUPAL OIB are after *Hart,* [1984], *Zindler and Hart* [1986], *Ito et al.* [1987], *Storey et al.* [1988], *le Roex et al.* [1990], *Barling and Goldstein* [1990], *Dosso et al.* [1991], *Mahoney et al.,* [1992, 1994], and *Woodhead and Devey* [1993]. Northern Hemisphere Reference Line (NHRL) is from *Hart* [1984].

provinces, none of the Siberian basalts fall in the modern DUPAL OIB field.

7. MANTLE SOURCES, CRYSTAL FRACTIONATION AND CONTAMINATION

The problem of assigning the roles of different mantle sources in CFBs is a non-trivial exercise as at least three mantle reservoirs may be implicated in the generation of flood basalts: (1) a MORB-type mantle, (2) an OIB-type mantle which may also be undegassed, and (3) the continental lithospheric mantle (CLM). Whereas the

compositional variations of the MORB-type and OIB-type mantle reservoirs are grossly identified [e.g., *Zindler and Hart*, 1986], the CLM reservoir appears to be very heterogeneous [e.g., *McDonough*, 1990]. The problem is further compounded as magmas may evolve and may get contaminated by continental crustal material en route to the surface. In this section, temporal and spatial variations in the chemical and isotopic composition of the Siberian Traps will be used to assess mantle sources, magma differentiation, and crustal contamination. Specifically, the following questions are significant: (1) whether the bulk of the Siberian Traps displays evidence of substantial crustal contamination, (2) whether or not a relationship exists between the chemical evolution of a lava suite and the extent to which it is contaminated, and (3) the mechanism of crustal contamination. A related issue is whether the primary magmas were picritic or basaltic [cf. *Basaltic Volcanism Study Project*, 1981].

As explained above, the low-Ti and high-Ti groups in the Siberian Traps cannot be related by fractional crystallization of similar phase assemblages from a uniform parental magma (Figure 5) [*Lightfoot et al.*, 1993]. Further, in comparison to the low-Ti group, the high-Ti group is relatively depleted in heavy REE (Figure 6), suggesting different depths of generation. *Lightfoot et al.* [1993] showed that the Gd/Yb ratios of high-Ti lavas are consistent with their generation at garnet stabilization depths. In contrast, the low-Ti basalts were likely generated by partial melting of spinel peridotite. These observations may suggest melting of a common source at two different depths to produce the Siberian lavas [e.g., *Arndt et al.*, 1993; *Wooden et al.*, 1993]. Alternatively, two different sources may have melted to produce the low-Ti and high-Ti groups [*Lightfoot et al.*, 1990, 1993; *Fedorenko et al.*, 1996]. Figure 7 shows that the least-contaminated lavas (i.e., those with the lowest Th/Ta) of the low- and high-Ti groups have two distinct Th/Ta ratios. As Ta and Th are not expected to fractionate significantly from each other during moderate to high degrees of partial melting and not at all during fractional crystallization processes, this result suggests derivation of magma from two distinct sources. The main problem with this interpretation is that the Th/Ta ratios displayed by the least-contaminated lavas of the low-Ti and high-Ti groups are both higher than the estimated primitive mantle value. Thus, even these lavas may be slightly contaminated. This observation indicates that the following two possibilities cannot be ruled out: (1) the magmas were derived from a common source but were contaminated by two different contaminants and (2) the magmas were derived from two sources and were contaminated by two different contaminants.

Figure 11 shows the variations of $^{147}Sm/^{144}Nd$ ratios with $\varepsilon_{Nd}(t)$ in the Siberian Traps and compares them to a solid line

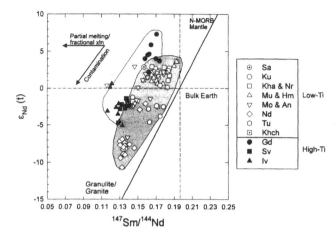

Figure 11. Relationships between $^{147}Sm/^{144}Nd$ and $\varepsilon_{Nd}(t)$ in Siberian basalts. First-order effects of partial melting/fractional crystallization (xln) and contamination can be assessed using this diagram. The high-Ti and low-Ti basalts suggest mixing between two mantle members with positive ε_{Nd} values and two continental crustal contaminants with low Sm/Nd and negative ε_{Nd}. Alternatively, melting of a single mantle source, first at high pressure and then at low pressure, could generate magmas that, on variable mixing with continental material, would produce the two trends. Data sources are the same as in Figure 9. See text for discussion and for explanation of line and arrows.

that represents mixing of N-MORB, bulk-earth-type mantle and an average sediment (not shown). The high-Ti and low-Ti groups define two broad arrays with positive slopes. The horizontal component of the arrays suggests variable degrees of partial melting of the mantle source(s) and/or fractional crystallization. In comparison to the low-Ti lavas, the high-Ti basalts, in general, could have been generated by lower degrees of partial melting of the mantle source. This observation is consistent with the Gd/Yb data in Figure 6. The positive correlation in Figure 11 would result from contamination of mantle-derived magmas (with high Sm/Nd and ε_{Nd}) with continental crustal material such as granulite (with low Sm/Nd and ε_{Nd}). The two arrays in Figure 11 allow the possibility of magma derivation from two mantle reservoirs with ε_{Nd} values of approximately +4 and +8. Further, the two arrays allow the possibility of more than one contaminant. The latter observation is consistent with the evolution of high-Ti and low-Ti groups with two distinct Th/Ta ratios (Figure 7).

Alternatively, the data could suggest derivation of the high-Ti group by lower degrees of partial melting of a source with ε_{Nd} of +8; high degrees of partial melting of such a source at shallower depths could produce the low-Ti basalts. The main difficulty with this interpretation is that the ε_{Nd} values of none of the low-Ti samples exceed +4. Indeed, the

low-Ti group has a volume-weighted ε_{Nd} (t) value of about +2.0 [*Sharma et al.*, 1992]. However, it may be argued that continental crustal contamination may have uniformly lowered the ε_{Nd} values of the low-Ti group; two features of the low-Ti lavas strongly support this argument: (1) an extremely high Th/Ta ratio of ~6.7, which is about three times higher than that of the primitive mantle, and (2) a Ta/La ratio of 0.04, which is about 30% lower than the primitive mantle value of 0.06.

The Ta/La ratio, in conjunction with the observed ε_{Nd} (t) values, points to the nature of the magma source(s) and the extent of crustal contribution in the case of the low-Ti group [see also *Hawkesworth et al.*, 1995]. To illustrate this, a simple binary mixing calculation is used for two cases: (I) basaltic magma, Ta = 0.1 ppm, Ta/La = 0.06, Nd = 9 ppm and ε_{Nd} = +4; (II) basaltic magma, Ta = 0.1 ppm, Ta/La = 0.06, Nd = 9 ppm and ε_{Nd} = +8. A tonalitic contaminant [*Condie*, 1993; see also *Wooden et al.*, 1993] is assumed, with Ta = 0.7 ppm, Ta/La = 0.02, Nd = 22 ppm and ε_{Nd} = -11. When ε_{Nd} for the magma is +4 (case I), addition of about 6% of the contaminant gives rise to a mixture with ε_{Nd} = +2.0 and Ta/La = 0.04. In contrast, when ε_{Nd} for the magma is +8 (case II), no proportion of this contaminant and postulated magma can reconcile the observed ε_{Nd} and Ta/La. The results remain the same with other likely crustal contaminants and indicate that if the magma has ε_{Nd} = +8, only contaminants with ε_{Nd} values < -37 (> 3.3 Ga continental crust) would be able to reconcile the observed ε_{Nd} and Ta/La. Recent studies in the Aldan shield suggest the presence of predominantly <3.0 Ga continental crust and extensive reworking of continental material in the mid-Proterozoic [e.g., *Nutman et al.*, 1992 and references therein]. Thus it is concluded that the source of the low-Ti magmas had an ε_{Nd} value of ~ +4. It follows that the mantle sources of the high-Ti and low-Ti groups are different. Note that the above arguments are based on forward mixing calculations that are not unique. Nonetheless, the conclusions appear to be the simplest explanation for the data.

What was the mechanism that governed the high Th/Ta ratios of the high-Ti and low-Ti magmas? Was it mixing of variable amounts of subducted shale within the lithospheric mantle [*Lightfoot et al.*, 1990, 1993; *Hawkesworth et al.*, 1995], or was it contamination with two continental crustal components with different Th/Ta ratios [*Wooden et al.*, 1993]? This issue is difficult to resolve as the trace element contents and Sr-, Nd- and Pb-isotopic compositions of subducted shale and some types of continental crust may not differ substantially from each other. Specifically, mantle containing subducted shale may have negative Ti, Ta, and Nb anomalies, high Th/Ta ratios, radiogenic Sr and Pb isotopes, and unradiogenic Nd isotopes. The contaminated

Siberian lavas display all the above characteristics except that they have relatively unradiogenic Pb isotopic compositions, pointing to a contaminant with low time-integrated U/Pb ratio as the most significant contaminant. Such a contaminant is generally considered to be the lower continental crust [see also *Wooden et al.*, 1993]. Thus, the simplest explanation of the data is that there are two different crustal contaminants for the high-Ti and the low-Ti groups. If it is assumed that the contaminant is continental crust with ε_{Nd} = -11 (which is the lowest value observed in a Nadezhdinsky sample; see Figures 9, 11) and a $^{147}Sm/^{144}Nd$ ratio of 0.11, then the minimum age of the contaminant is ~10^9 y. If it is assumed that the other contaminant is continental crust with ε_{Nd} = -5 (which is the lowest value observed in an Ivakinsky sample; see Figures 9, 11) and a $^{147}Sm/^{144}Nd$ ratio of 0.11, then the minimum age of this contaminant is ~450 Ma.

Variations of ε_{Sr} (t) and $(^{206}Pb/^{204}Pb)_t$ for the Siberian basalts are plotted in Figure 12 and show a broad negative trend with some samples from the Tuklonsky suite falling to the left of this trend. These data point to the presence of two contaminants within the low-Ti group: one with $^{206}Pb/^{204}Pb$ < 6.7, ε_{Sr} ~20, and the other with $^{206}Pb/^{204}Pb$ ≤17.2, ε_{Sr} > +70. The Pb and Sr isotopic compositions of the Bolgokhtokhsky granodiorite are consistent with it being a contaminant for the Tuklonsky lavas. However, the Sr isotopic composition of the Bolgokhtokhsky granodiorite (= 0.70634) is too low to achieve the Sr isotopic shift required of a suitable contaminant for the Nadezhdinsky suite (Figure 12) [*Wooden et al.*, 1993].

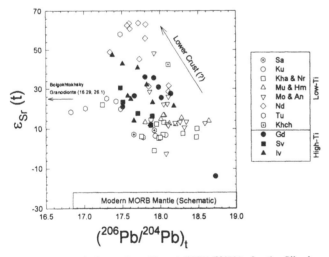

Figure 12. Variations of ε_{Sr} (t) and $(^{206}Pb/^{204}Pb)_t$ for the Siberian basalts. Note that the Bolgokhtokhsky granodiorite cannot be a dominant contaminant. The data point to a contaminant with low time-integrated U/Pb ratio, possibly lower crust [e.g., *Rudnick and Fountain*, 1995; see also *Wooden et al.*, 1993].

In summary, the trace-element and isotopic data for the Siberian Traps are consistent with derivation of magmas from two sources. The high-Ti magmas were generated by low degrees of partial melting of a mantle source with ε_{Nd} ~8. Larger degrees of partial melting of a magma source with ε_{Nd} ~+4 produced the low-Ti lavas. The two groups appear to have evolved with two distinct Th/Ta ratios much higher than the primitive mantle value, suggesting (1) the influence of two different contaminants and (2) either early contamination or buffering of Th/Ta ratios [*Wooden et al.*, 1993]. The relatively unradiogenic Pb isotopic composition of the contaminated lavas appears to favor the lower continental crust as the dominant contaminant.

Although the Mg* of the Siberian lavas varies from 0.30 to 0.70, the bulk of the rocks show a rather restricted range (~0.54 to 0.56). The Mg* is much lower than expected for the primary magmas of MORBs and OIBs, which may be as high as ~0.81 [*Albarède*, 1992]. Recent measurements of primary melt inclusions in MORB olivines found Mg* varying from 0.73 to 0.78 [*Sobolev and Chaussidon*, 1996] The Ni contents of primary melt inclusions in OIBs could be greater than 600 ppm [*Clague et al.*, 1991]. If the Siberian Traps basalts were primary, they would reflect derivation from a mantle more Fe-rich than the normal MORB source. Indeed, a deep mantle source with abnormally high Fe contents has been inferred for the plume-derived lavas at the Isle of Skye [*Scarrow and Cox*, 1995]. The Siberian lavas, in general, have Ni contents varying between 150 and 200 ppm. Thus, compared to a primary melt with Mg* = 0.78 and Ni = 600 ppm, the Siberian lavas appear to have undergone about 30% reduction in Mg* and Ni content. Taken together, the above considerations suggest that olivine fractionation, and not an unusual magma source with high Fe content, is responsible for the low Mg*. If the parental liquids of the Siberian Traps had Mg* of about 0.78, the extent of olivine fractionation needed to reduce the Mg* to 0.55 is about 33% (where olivine is with Mg* = 0.90). Note that the above arguments assume that (1) the Ni concentration of the precursors of the Siberian lavas was much higher than observed and (b) the reduction in the Ni content of the lavas resulted from olivine fractionation. These assumptions cannot be justified by the present data set as a plot of Mg* versus Ni concentration (not shown) shows no strong correlation between these two parameters for the bulk of the Siberian Traps.

Figure 13 shows the variation of ε_{Nd} (t) versus Mg*. Crustal contamination and magma evolution appear to be related in the early-erupted high-Ti group as indicated by a crude positive correlation between the Mg* and ε_{Nd} (t) values (Figure 13) [cf. *Sharma et al.*, 1991]. In contrast, no such correlation is seen for the late-erupted, voluminous low-Ti

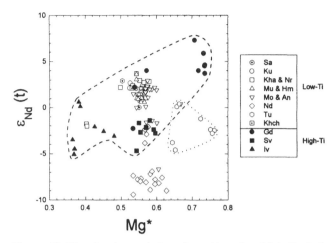

Figure 13. Plot showing variation of ε_{Nd} (t) against Mg*. Dashed line encloses all high-Ti basalt data. The Tuklonsky suite basalts (enclosed by the dotted line) are cumulates.

group: with the exception of the Tuklonsky lavas, the Mg* values for the least contaminated and the most contaminated low-Ti basalts are similar to each other [cf. *Lightfoot et al.*, 1993; *Wooden et al.*, 1993; *Hawkesworth et al.*, 1995]. *Wooden et al.* [1993] noted that the mineralogy and chemical compositions of the low-Ti basalts (except the Tuklonsky suite) suggest that the magmatic precursors of these lavas underwent variable olivine + plagioclase + clinopyroxene fractionation at crustal depths. The basalts from the Tuklonsky suite display high Mg* and low ε_{Nd} values. These lavas are most probably cumulates [*Wooden et al.*, 1993] and were evolved by mixing with low ε_{Nd} crustal melts that were also LREE-enriched [e.g., *Peng et al.*, 1994].

Major element studies in other flood basalt provinces also suggest similar scenarios, with substantial magmatic evolution through gabbroic fractionation (i.e., fractionation at less than 5 kbar of olivine, plagioclase, and clinopyroxene in crustal magma chambers) [e.g., *Krishnamurthy and Cox*, 1977; *Cox*, 1980; *Basaltic Volcanism Study Project*, 1981]. Magmatic differentiation and contamination may occur in large chambers at deep crustal levels [e.g., *Wooden et al.*, 1993; *Peng et al.*, 1994] similar to those preserved in the Ivrea Zone of northern Italy [e.g., *Voshage et al.*, 1990]. Indeed, magma ponding at or near the crust-mantle boundary (the Moho) appears to be a natural consequence of the density contrast between picritic melts and the continental lid [e.g., *Cox*, 1980; *Farnetani et al.*, 1996 and references therein]. The two types of contamination seen in the Siberian Traps have also been recognized in other flood basalt provinces: (1) the most evolved magmas are the most contaminated; examples come from the Columbia River province [*Carlson et al.*, 1981, 1983], Paraná [*Hawkesworth et al.*, 1986; *Petrini et al.*, 1987], and Ferrar Supergroup

[*Menzies and Kyle*, 1990]; and (2) the most primitive basalts (high Ni, Cr and Mg*) are the most contaminated; examples come from the Deccan Traps [e.g., *Mahoney et al.*, 1982; *Devey and Cox*, 1987; *Mahoney*, 1988] and the British Tertiary Volcanic Province [e.g., *Thompson et al.*, 1986].

8. MODELS OF GENERATION OF SIBERIAN TRAPS

8.1 *General*

As discussed earlier, there appears to be little dispute over the involvement of a mantle plume in the generation of the Siberian flood volcanism. However, there is disagreement over the composition of the plume source (relatively MORB-like mantle versus relatively "undepleted" and undegassed, possibly lower mantle source), relative roles of the plume source and CLM in the generation of the bulk of the Siberian Traps, and the extent of modification of the magmas by the continental crust. The following models have been proposed.

(1) Melting of the CLM in response to heating by a mantle plume with further modification of the magma by continental crust [*Lightfoot et al.*, 1990, 1993; *Hawkesworth et al.*, 1995]. This model is identical to that proposed for the CFBs related to the breakup of Gondwana [e.g., *Hawkesworth et al.*, 1986]. According to this model, the low-Ti basalts of the Tuklonsky and Mokulaevsky suites are the result of melting of CLM that contained sediments subducted previously. The lower-assemblage, high-Ti basalts from Gudchikhinsky suite are derived from a mantle plume similar to those postulated for ocean island basalts [*Lightfoot et al.*, 1993].

(2) Melting of a lower mantle plume (ε_{Nd} ~ +2 to +4) with minimal involvement of MORB-type mantle, CLM, or continental crust [*Sharma et al.*, 1991, 1992; *Basu et al.*, 1995; cf. *DePaolo and Wasserburg*, 1979]. According to this model, early eruptions of volumetrically insignificant lavas were dominated by contributions from the MORB-type mantle (e.g., Gudchikhinsky). The bulk of the Siberian flood basalts were derived from decompressional melting in the plume head. Magmas en route to the surface were modified slightly by the continental crust (Morongovsky through Kharaelakhsky and their Putorana counterparts).

(3) Melting of a deep mantle plume with large-scale modification of the magmas in shallow-level magma chambers. According to this model, the uniform composition of many of the low-Ti magmas reflects buffering by continental crust [*Arndt et al.*, 1993; *Wooden et al.*, 1993]. The degree of melting, which is contingent upon the thickness of the lithosphere, controls whether or not garnet is present as a residue. The low-degree melts (with residual garnet in the source) evolve to high-Ti basalts. In contrast,

the melts derived from a large degree of melting of mantle peridotite in the absence of garnet evolve to become the low-Ti basalts.

8.2. Melting of the Continental Lithospheric Mantle

Lightfoot et al. [1990, 1993] and *Hawkesworth et al.* [1995] proposed that Siberian flood volcanism was the result of melting in the CLM engendered by a hot mantle plume. The principal argument is based upon low Ta/La and Nb/La ratios shown by the upper-assemblage low-Ti basalts, a feature that these authors attributed to a CLM containing subducted sedimentary material. This model was criticized by *Arndt et al.* [1993] and *Wooden et al.* [1993], who argued that the CLM is too cold and too dry and does not have an appropriate trace-element composition to form a major source for the flood basalts. On the basis of the elevated (^{87}Sr/^{86}Sr)$_t$, low $\varepsilon_{Nd}(t)$, and unradiogenic Pb isotopic compositions of the low-Ti lavas, *Wooden et al.* [1993] concluded that the low Ta/La and Nb/La ratios resulted from contamination by old continental crust. *White and McKenzie* [1995] gave additional arguments against the CLM being a major source of the Siberian Traps: (1) if the source of heat for generating the flood basalts was a hot mantle plume, then it is very difficult to explain how rapid melting in the plume head could be suppressed; and (2) conduction of heat in the CLM is too slow to produce large volumes of magma in a short period. These arguments were examined quantitatively by *Turner et al.* [1996], who tested the feasibility of extensive melting in the lithosphere above a plume instead of within the plume itself. They concluded that the most favorable circumstances under which significant lithospheric melting would ensue without melting in the underlying plume include a >100-km-thick, volatile-enriched lithosphere (0.3 wt.% $H_2O \pm CO_2$) overlying a plume with a potential tem-perature between 1380 and 1580 °C. Under such conditions, ~1–2-km-thick flood basalts may be derived from the CLM if the eruption occurs over 10–15 m.y. A period of 10–15 m.y. is required because of the extremely low thermal conductivity of the CLM. Because the bulk of the Siberian Traps was most probably produced in ~1 m.y., it is most likely that most of the lavas were derived from melting within the plume.

8.3. Melting of a Lower Mantle Plume

On the basis of volume-weighted Nd-isotopic composition of the Siberian lavas (average $\varepsilon_{Nd}(t)$ = +1.8), which also showed a sharp cut-off at $\varepsilon_{Nd}(t)$ = 0, *Sharma et al.* [1991, 1992] proposed that 90% of the Siberian Traps lavas were derived from the partial melting of a 'nearly primitive' lower-mantle-derived plume. This suggestion is identical to

that offered by *DePaolo and Wasserburg* [1979]. *White and McKenzie* [1995] inverted the REE patterns of the lavas to show that following the initial stages of melting, during which the CLM could be involved, the subsequent bulk of the Siberian Traps basalts were plausibly formed from melting of a plume. This model is obviously too simple, however, as it relies heavily on REE patterns and Nd isotopic data, not taking into account the complex trace element patterns displayed by the lavas (Figures 4, 7, 8). In particular, the model does not explain the cause of systematic variations in Th/Ta ratios that are also much higher than expected from melting of a primitive mantle source (Figure 7). As discussed above, continental crustal contamination probably explains the high Th/Ta and Ta/La ratios. Further, the 'nearly primitive' ε_{Nd} values of ~+2 of the bulk volume of the Siberian Traps can be explained by the modest contamination of OIB-type melts with ε_{Nd} of ~ +4 (Figures 9, 11).

8.4. Buffering of Plume-Derived Magmas by Continental Crust

Wooden et al. [1993] and *Arndt et al.* [1993] invoked large-scale buffering of plume-derived magmas by continental crust in shallow-level magma chambers to account for (1) the nearly uniform major- and trace-element and isotopic compositions of the low-Ti lavas, (2) high (^{87}Sr/^{86}Sr)$_t$ ratios, and (3) negative Nb-Ta anomalies in the lavas. In formulating the model, *Wooden et al.* [1993] observed that the low-Ti lavas show significant amounts of fractional crystallization of olivine, plagioclase, and clinopyroxene, and reflect evolution in magma chambers at pressures less than 5 kbar. *Wooden et al.* [1993] modeled magma evolution in chambers that were periodically replenished and tapped, in which the magmas were continuously fractionating and assimilating crustal wall rocks. According to their model, the high-Ti and low-Ti magmas were derived from different degrees of melting of a plume source. This model is able to explain many of the features observed in the Siberian basalts as resulting from crustal contamination and magma mixing, including (1) the high (^{87}Sr/^{86}Sr)$_t$ and low Ta/La ratios (Figure 4), (2) high and nearly constant Th/Ta ratios (Figure 7), and (3) nearly uniform major element and trace element compositions of the upper-stage lavas. However, the model does not account for the existence of two isotopically distinct magma sources (see, e.g., Figure 11) for the high-Ti and low-Ti lavas.

8.5. Considerations for a General Model of the Siberian Flood Volcanism

Any general model of generation of the Siberian Traps must account for the following key observations and inferences.

(1) The bulk of the magmatism occurred rapidly, within ~1 m.y.

(2) The flood volcanism was not accompanied by significant uplift or rifting.

(3) At least some of the early alkaline ultramafic magmatism possessed $^3He/^4He$ signatures indicative of the involvement of a high ^3He source.

(4) Incompatible element patterns of the high-Ti and low-Ti groups are not diagnostic of derivation from either a MORB-type or an OIB-type mantle source.

(5) During the initial stages of magmatism, high-Ti lavas were derived from a mantle source with ε_{Nd} ~+8 and containing residual garnet.

(6) The bulk of the Siberian Traps, represented by the voluminous late-stage low-Ti lavas, was derived from a shallower mantle source with ε_{Nd} ~+4.

(7) The Siberian Traps magmas underwent variable continental crustal contamination, as indicated by high Th/Ta and low Ta/La ratios. The contaminants had low time-integrated U/Pb ratios as indicated by the unradiogenic Pb isotopic ratios of the contaminated lavas. It is possible that magma processing took place in crustal reservoirs that underwent periodic replenishment, periodic tapping, crystal fractionation, and wallrock assimilation.

At present, no model has been proposed that would satisfy all the above observations. *Fedorenko et al.* [1996] suggested that *Anderson's* [1994] model of melting of perisphere should be investigated to explain the high-Ti basalts. Recently, D. L. Anderson (The helium-lead paradox, ms. in process) has suggested that an incompatible-element-rich perisphere and not deep mantle plumes supplies the high $^3He/^4He$ to OIB magmas. This idea presents an intriguing possibility as the perisphere may also be the source of the early alkaline-ultramafic lavas from the Maimecha-Kotui area which display a high $^3He/^4He$ signature. However, the ultimate source of ^3He has to be the lower mantle which is generally considered to be the relatively undegassed mantle reservoir [see e.g. *Porcelli and Wasserburg*, 1995]. Thus one must ask the question of whether the two magma sources inferred in the Siberian Traps could be identified with perisphere (high-Ti, ε_{Nd} ~+8) and plume (low-Ti, ε_{Nd} ~+4, high $^3He/^4He$). Note that in the above discussion it is assumed that the perisphere is a viable mantle reservoir for producing large volumes of melts. The existence of this reservoir as a viable mantle source is not accepted by many geochemists.

9. WHERE DO WE GO FROM HERE?

Recent studies have shown that the Siberian Traps have the potential to provide important insights into continental flood magmatism and its role in the earth's history. The province is enormous and we are just beginning to understand the complexities [see *Fedorenko et al.*, 1996]. In the following, an outline of some of the areas for future research is provided. At present, the sampling is confined largely to the northern and northwestern parts of the province. High-quality geochemical, isotopic, and geochronological data from additional sections in the southern Putorana (e.g., Nizhnyaya Tunguska area) are needed to evaluate the proposed models of Siberian flood volcanism. Further, analyses of other rock types associated with the Siberian Traps, including kimberlites, alkalic basalts, and maimechites, are required to understand the plume-CLM interaction. An important problem intimately tied to the evolution of some of the early lavas in the Noril'sk region is their relationship to ore mineralization. Detailed sulfur isotope analysis of the lavas and the associated mineralized intrusions is needed to evaluate this issue [see *Hawkesworth et al.*, 1995; *Lightfoot and Hawkesworth*, this volume].

The extent to which the CLM and continental crust were involved in the generation of the Siberian Traps magmas should be evaluated further. To this end, significant contributions have come from (1) *Horan et al.* [1995], who discussed the Os isotopic variations in some early-erupted picrites and maimechites from the Maimecha-Kotui region, and (2) *Walker et al.* [1994], who analyzed ore-bearing intrusions from Noril'sk for Os, Nd, and Pb isotopes. However, such studies presently can be conducted only on samples with low Re/Os ratios and high enough Os abundances. This limitation leads to a focus on volumetrically insignificant flows which may not be representative of the bulk of the Siberian Traps. A detailed oxygen isotopic study on mineral separates is needed to assess the role of crustal contamination relative to CLM influences [cf. *Peng et al.*, 1994].

Another important issue is that of spatial progression of volcanism on the Siberian platform and its relationship, if any, with the Lomonosov Ridge (Figure 3). A detailed plate-tectonic reconstruction based on precise ^{40}Ar-^{39}Ar and paleomagnetic dating is needed to address this problem. In this context, some accessible portions of the Lomonosov Ridge should also be sampled.

Acknowledgments. I would like to thank A. R. Basu for many discussions on the genesis of the Siberian Traps. David Peate persuaded me to write this review paper and gave me new insights into the data. Comments from J. L. Wooden, an anonymous reviewer, and J. J. Mahoney significantly improved this paper. Many thanks to Don Porcelli, Tom Latourette, and Paul Asimow for all the suggestions. This work is supported by a grant to G. J. Wasserburg (NASA-NAGW 3337). Division contribution No. 5661 (928).

REFERENCES

Albarède, F., How deep do common basaltic magmas form and differentiate?, *J. Geophys. Res.*, *97*, 10,997-11,009, 1992.

Anderson, D. L., The sublithospheric mantle as the source of continental flood basalts; the case against the continental lithosphere and plume head reservoirs, *Earth Planet. Sci. Lett.*, *123*, 269-280, 1994.

Arndt, N. T., G. K. Czamanske, J. L. Wooden, and V. A. Fedorenko, Mantle and crustal contributions to continental flood volcanism, *Tectonophysics, 223*, 39-52, 1993.

Baksi, A. K., and E. Farrar, $^{40}Ar/^{39}Ar$ dating of the Siberian Traps, USSR: evaluation of the ages of the two major extinction events relative to episodes of flood-basalt volcanism in the USSR and the Deccan Traps, India, *Geology, 19*, 461-464, 1991a.

Baksi, A. K., and E. Farrar, $^{40}Ar/^{39}Ar$ dating of whole-rock basalts (Siberian Traps) in the Tunguska and Noril'sk area, Siberia (abstract), *Eos Trans. AGU, 72*, 570, 1991b.

Balashov, Y. A., and G. V. Nesterenko, Distribution of the rare earths in the traps of the Siberian platform, *Geochem. Int., 3*, 672-679, 1966.

Darling, J. and S. L. Goldstein, Extreme isotopic variations in Heard Island lavas and the nature of mantle reservoirs, *Nature, 348*, 59-62, 1990.

Basaltic Volcanism Study Project, *Basaltic Volcanism on the Terrestrial Planets*, 1286 pp., Pergamon Press, New York, 1981.

Basu, A. R., P. R. Renne, D. K. DasGupta, F. Teichmann, and R. J. Poreda, Early and late alkali igneous pulses and a high-^3He plume origin for the Deccan flood basalts, *Science, 261*, 902-905, 1993.

Basu, A. R., R. J. Poreda, P. R. Renne, F. Teichmann, Y. R. Vasiliev, N. V. Sobolev, and B. D. Turrin, High-^3He plume origin and temporal-spatial evolution of the Siberian flood basalts, *Science, 269*, 822-825, 1995.

Bellieni, G., P. Brotzu, P. Comin-Chiaramonti, M. Ernesto, A. Melfi, I. G. Pacca, and E. M. Piccirillo, Flood basalt to rhyolite suite in southern Paraná Plateau (Brazil): palaeomagnetism, petrogenesis and geodynamical implications, *J. Petrol., 25*, 579-618, 1984.

Campbell, I. H., G. K. Czamanske, V. A. Fedorenko, R. I. Hill, and V. Stepanov, Synchronism of the Siberian Traps and the Permian-Triassic boundary, *Science, 258*, 1760-1763, 1992.

Carlson, R. W., G. W. Lugmair, and J. D. Macdougall, Columbia River volcanism: the question of mantle heterogeneity or crustal contamination, *Geochim. Cosmochim. Acta, 45*, 2483-2499, 1981.

Carlson, R. W., G. W. Lugmair, and J. D. Macdougall, Columbia river volcanism: the question of mantle heterogeneity or crustal contamination (reply to a comment by D. J. DePaolo), *Geochim. Cosmochim. Acta, 47*, 845-846, 1983.

Clague, D. A., W. S. Weber, and J. E. Dixon, Picritic glasses from Hawaii, *Nature, 353*, 553-556, 1991.

Claoué-Long, J. C., Z. Zichao, M. Guogan, and D. Shaohua, The age of the Permian-Triassic boundary, *Earth Planet. Sci. Lett., 105*, 182-190, 1991.

Claoué-Long, J. C., W. Compston, J. Roberts, and C. M. Fanning, Two carboniferous ages: a comparison of SHRIMP zircon dating with conventional ages and $^{40}Ar/^{39}Ar$ analysis, in *Geochronology, Time Scales, and Stratigraphic Correlation, Spec. Publ. 54*, edited by W. Bergenn, J. Hardenbol and D. Kent, pp. 1-27, (SEPM) Society for Sedimentary Geology, Tulsa, OK, 1995.

Condie, K. C., Chemical composition and evolution of the upper continental crust: contrasting results from surface samples and shales, *Chem. Geol., 104*, 1-37, 1993.

Cox, K. G., A model for flood basalt vulcanism, *J. Petrol., 21*, 629-650, 1980.

Dalrymple, G. B., G. K. Czamanske, M. A. Lanphere, V. Stepanov, and V. Fedorenko, $^{40}Ar/^{39}Ar$ ages of samples from the Noril'sk-Talnakh ore-bearing intrusions and the Siberian flood basalts, Siberia (abstract), *Eos Trans. AGU, 72*, 570, 1991.

Dalrymple, G. B., G. K. Czamanske, V. A. Fedorenko, O. N. Simonov, M. A. Lanphere, and A. P. Likhachev, A reconnaissance $^{40}Ar/^{39}Ar$ geochronologic study of ore-bearing and related rocks, Siberian Russia, *Geochim. Cosmochim. Acta, 59*, 2071-2083, 1995.

DePaolo, D. J. and G. J. Wasserburg, Nd isotopes in flood basalts from the Siberian Platform and inferences about their mantle sources, *Proc. Natl. Acad. Sci. USA, 76*, 3056-3060, 1979.

Devey, C. W., and K. G. Cox, Relationships between crustal contamination and crystallisation in continental flood basalt magmas with special reference to the Deccan Traps of the Western Ghats, India, *Earth Planet. Sci. Lett., 84*, 59-68, 1987.

Dosso, L., B. B. Hanan, H. Bougault, J. G. Schilling, and J. L. Joron, Sr-Nd-Pb geochemical morphology between 10°N and 17°N on the Mid-Atlantic Ridge-A new MORB signature, *Earth Planet. Sci. Lett, 106*, 29-43, 1991.

Ellam, R. M., and K. G. Cox, An interpretation of Karoo picritic basalts in terms of interaction between asthenospheric magmas and the mantle lithosphere, *Earth Planet. Sci. Lett., 105*, 330-342, 1991.

Erwin, D. H., The Permo-Triassic extinction, *Nature, 367*, 231-236, 1994.

Farnetani, C. G., and M. A. Richards, Numerical investigations of the mantle plume initiation model for flood basalt events, *J. Geophys. Res., 99*, 13,813-13,833, 1994.

Farnetani, C. G., M. A. Richards and M. S. Ghiorso, Petrological models of magma evolution and deep crustal structure beneath hotspots and flood basalt provinces, *Earth Planet. Sci. Lett., 143*, 81-94, 1996.

Fedorenko, V. A., Petrochemical series of extrusive rocks of the Noril'sk region, *Soviet Geol. Geophys., 22*, 66-74, 1981.

Fedorenko, V. A., Tectonic control of magmatism and regularities of Ni-bearing localities on the north-western Siberian platform, *Soviet Geol. Geophys., 32*, 41-47, 1991.

Fedorenko, V. A., and O. A. Dyuzhikov, Establishing the period of late Paleozoic-early Mesozoic volcanism in the Noril'sk region of the Siberian platform, *Soviet Geol. Geophys., 21*, 114-117, 1980.

Fedorenko, V. A., V. M. Kuligin, G. C. Vitozhents, S. K. Mikhalev, and L. V. Makeeva, Rare earth elements in magmatic formations in the Noril'sk region, *Soviet Geol. Geophys., 30*, 61-71, 1989.

Fedorenko, V. A., P. C. Lightfoot, A. J. Naldrett, G. K. Czamanske, C. J. Hawkesworth, J. L. Wooden, and D. S. Ebel, Petrogenesis of the flood-basalt sequence at Noril'sk, North Central Siberia, *Inter. Geol. Rev., 38,* 99-135, 1996.

Forsyth, D. A. and J. A. Mair, Crustal structure of the Lomonosov Ridge and the Fram and Makarov Basins near the north Pole, *J. Geophys. Res., 89,* 473-481, 1984.

Gallagher, K., and C. Hawkesworth, dehydration melting and the generation of continental flood basalts, *Nature, 358,* 57-59, 1992.

Graham, D. W., W. J. Jenkins, J.-G. Schilling, G. Thompson, M. D. Kurz, and S. E. Humphris, Helium isotope geochemistry of midocean ridge basalts from the south-Atlantic, *Earth Planet. Sci. Lett., 110,* 133-147, 1992.

Hart, S. R., A large-scale isotope anomaly in the southern hemisphere mantle, *Nature, 309,* 753-756, 1984.

Hart, S. R., and H. Staudigel, Isotopic characterization and identification of recycled components, in *Crust/Mantle Recycling at Convergence Zones,* edited by S. R. Hart and L. Gülen, pp. 15-28, Kluwer Academic Publishers, Dordrecht, 1989.

Hawkesworth, C. J., J. S. Marsh, J. S. Duncan, A. J. Erlank, and M. J. Norry, The role of continental lithosphere in the generation of the Karoo volcanic rocks: evidence from combined Nd- and Sr-isotope studies, *Spec. Publ. Geol. Soc. S. Afr., 13,* 341-354, 1984.

Hawkesworth, C. J., M. S. M. Mantovani, P. N. Taylor, and Z. Palacz, Evidence from the Paraná of south Brazil for a contribution to Dupal basalts, *Nature, 322,* 356-359, 1986.

Hawkesworth, C. J., C. J. Lightfoot, V. A. Fedorenko, S. Blake, A. J. Naldrett, W. Doherty, and N. S. Gorbachev, Magma differentiation and mineralisation in the Siberian continental flood basalts, *Lithos, 34,* 61-88, 1995.

Hergt, J. M., D. W. Peate, and C. J. Hawkesworth, The petrogenesis of Mesozoic Gondwana low-Ti flood basalts, *Earth Planet. Sci. Lett., 105,* 138-148, 1991.

Hill, R. I., Starting plumes and continental break-up, *Earth Planet. Sci. Lett.,* 104, 398-416, 1991.

Horan, M. F., R. J. Walker, V. A. Fedorenko, and G. K. Czamanske, Osmium and neodymium isotopic constraints on the temporal and spatial evolution of Siberian flood basalt sources, *Geochim. Cosmochim. Acta, 59,* 5159-5168, 1995.

Ito, E., W. M. White, and C. Göpel, The O, Sr, Nd and Pb isotope chemistry of MORB, *Chem. Geol., 62,* 157-176, 1987.

Jokat, W., G. Uenzelmann-Neben, Y. Kristoffersen, and T. M. Rasmussen, Lomonosov Ridge—a double-sided continental margin, *Geology,* 20, 887-890, 1992.

Kamo, S. L., G. K. Czamanske, and T. E. Krough, A minimum U-Pb age for Siberian flood-basalt volcanism, *Geochim. Cosmochim. Acta, 60,* 3505-3511, 1996.

Khain, V. E., *Geology of the USSR,* 272 pp., Gebrüder Borntraeger, Berlin, 1985.

King, S. D., and D. L. Anderson, An alternative mechanism of flood basalt formation, *Earth Planet. Sci. Lett., 136,* 269-279, 1995.

Krishnamurthy, P., and K. G. Cox, Picrite basalts and related lavas from the Deccan Traps of Western India, *Contrib. Mineral. Petrol., 62,* 53-75, 1977.

le Roex, A. P., R. A. Cliff, and B. J. Adair, Tristan da Cunha, South Atlantic: Geochemistry and petrogenesis of a basanite-phonolite lava series, *J. Petrol., 31,* 779-812, 1990.

Lightfoot, P. C., and A. J. Naldrett, *Proceedings of the Sudbury-*

Noril'sk Symposium, 392 pp., Ontario Geological Survey, Toronto, 1994.

Lightfoot, P. C., A. J. Naldrett, N. S. Gorbachev, W. Doherty, and V. A. Fedorenko, Geochemistry of the Siberian Trap of the Noril'sk area, USSR, with implications for the relative contributions of crust and mantle to flood basalt magmatism, *Contrib. Mineral. Petrol., 104,* 631-644, 1990.

Lightfoot, P. C., C. J. Hawkesworth, J. Hergt, A. J. Naldrett, N. S. Gorbachev, V. A. Fedorenko, and W. Doherty, Remobilisation of the continental lithosphere by a mantle plume: major- trace-element, and Sr- Nd- and Pb-isotope evidence from picritic and tholeiitic lavas of the Noril'sk District, Siberian Trap, Russia, *Contrib. Mineral. Petrol., 114,* 171-188, 1993.

Lind, E. N., S. V. Kropotov, G. K. Czamanske, S. C. Gromme, and V. A. Fedorenko, Paleomagnetism of the Siberian Flood Basalts of the Noril'sk area: a constraint on eruption duration, *Int. Geol. Rev., 36,* 1139-1150, 1994.

Lurie, M. L., and V. L. Masaitis, Plateau Basalts, *Int. Geol. Cong., 22 ,* Moscow, 1-12, 1964.

Macdougall, J. D. and G. W. Lugmair, Sr and Nd isotopes in basalts from the East Pacific Rise: significance for mantle heterogeneity, *Earth Planet. Sci. Lett., 77,* 273-284, 1986.

Mahoney, J. J., Deccan Traps, in *Continental Flood Basalts,* edited by J. D. Macdougall, pp. 151-194, Kluwer Academic Publishers, Dordrecht, 1988.

Mahoney, J. J., J. D. Macdougall, G. W. Lugmair, M. Sankar Das, A. V. Murali, and K. Gopalan, Origin of the Deccan Trap flows at Mahabaleshwar inferred from Nd and Sr isotopic and chemical evidence, *Earth Planet. Sci. Lett., 60,* 47-60, 1982.

Mahoney, J. J., A. P. le Roex, Z. Peng, R. L. Fisher, J. H. Natland, Southwestern limits of Indian Ocean ridge mantle and the origin of low $^{206}Pb/^{204}Pb$ mid-ocean ridge basalt: isotope systematics of the central Southwest Indian Ridge (17°–50° E), *J. Geophys. Res., 97,* 19,771-19,790, 1992.

Mahoney, J. J., J. M. Sinton, M. D. Kurz, J. D. Macdougall, K. J. Spencer, and G. W. Lugmair, Isotope and trace element characteristics of a super-fast spreading ridge: East Pacific rise, 13–23°S, *Earth Planet. Sci. Lett., 121,* 173-193, 1994.

Marsh, J. S., Basalt geochemistry and tectonic discrimination within continental flood basalt provinces, *J. Volcanol. Geotherm. Res., 32,* 35-49, 1987.

Masaitis, V. L., I. I. Abramovich, D. A. Dodin, and A. A. Smyslov, Uranium in Siberian Platform trap rocks, *Geochem. Int., 3,* 392-405, 1966.

McDonough, W. F., Constraints on the composition of the continental lithospheric mantle, *Earth Planet. Sci. Lett., 101,* 1-18, 1990.

Menzies, M. A., and P. R. Kyle, Continental volcanism: a crust-mantle probe, in *Continental Mantle,* edited by M. A. Menzies, pp. 157-177, Oxford University Press, Oxford, 1990.

Mitchell, C., J. G. Fitton, A. I. Al'mukhamedov, and A. I. Medvedev, The age and duration of flood basalt magmatism: geochemical and palaeomagnetic constraints from the Siberian Province, *Mineral. Mag., 58A,* 617-618, 1994.

Morgan, W. J., Hotspot tracks and the opening of the Atlantic and Indian Oceans, in *The Sea, vol. 7,* edited by C. Emiliani, pp. 443-487, Wiley, New York, N.Y., 1981.

Naldrett, A. J., P. C. Lightfoot, V. A. Fedorenko, W. Doherty, and N. S. Gorbachev, Geology and geochemistry of intrusions and flood basalts of the Noril'sk region, USSR, with implications for the origin of the Ni-Cu ores, *Econ. Geol., 87*, 975-1004, 1992.

Nalivkin, D. V., *Geology of the U.S.S.R*, 855 pp., University of Toronto Press, Toronto, 1973.

Nesterenko, G. V., A. I. Al'mukhamedov, and Y. I. Belyayev, Cadmium in the differentiation of a basic magma, *Geochem. Int., 9*, 432-437, 1972.

Nesterenko, G. V., and A. I. Al'mukhamedov, Geochemical peculiarities of the trappean magma differentiation, in *Recent Contributions to Geochemistry and Analytical Chemistry*, edited by A. I. Tugarinov, pp. 233-242, John Wiley, New York, 1975.

Nesternko, G. V., and A. I. Al'mukhamedov, Titanium in the pyroxenes of differentiated traps, *Geochem. Int., 3*, 767-777, 1966.

Nesterenko, G. V., and L. P. Frolova, Lithium and rubidium in Siberian Trap rocks, *Geochem. Int., 2 (Supplement)*, 295-305, 1965.

Nesterenko, G. V., N. S. Avilova, and N. P. Smirnova, Rare elements in the traps of the Siberian platform, *Geokhimiya, 10*, 970-976, 1964.

Nesterenko, G. V., I. Y. Belyayev, and P. H. Phi, Silver in the evolution of mafic rocks, *Geochem. Int., 2*, 162-169, 1969a.

Nesterenko, G. V., Y. I. Belyayev, and F. Fam-Khung, Silver in the evolution of mafic rocks, *Geochem. Int., 6*, 119-126, 1969b.

Nesterenko, G. V., Y. B. Znamenskiy, A. I. Al'mukhamedov, and V. D. Tsykhanskiy, Nb. Ta, Zr, and Hf in trap magma differentiation, *Geochem. Int., 8*, 725-738, 1971.

Nesterenko, G. V., A. I. Al'mukhamedov, and Y. I. Belyayev, Cadmium in the differentiation of basic magma, *Geochem. Int., 6*, 669-675, 1972.

Nutman, A. P., I. V. Chernyshev, H. Baadsgaard, and A. P. Smelov, The Aldan Shield of Siberia, USSR: the age of its Archaean components and evidence for widespread reworking in the mid-Proterozoic, *Precamb. Res., 54*, 195-210, 1992.

Peng, Z. X., J. Mahoney, P. Hooper, C. Harris, and J. Beane, A role for lower continental crust in flood basalt genesis? Isotopic and incompatible element study of the lower six formations of the western Deccan Traps, *Geochim. Cosmochim. Acta, 58*, 267-288, 1994.

Petrini, R., L. Civetta, E. M. Piccirillo, G. Bellieni, P. Comin-Chiaramonti, L. S. Marques, and A. J. Melfi, Mantle heterogeneity and crustal contamination in the genesis of low-Ti continental flood basalts from the Paraná Plateau (Brazil): Sr-Nd isotope and geochemical evidence, *J. Petrol., 28*, 701-726, 1987.

Porcelli, D. and G. J. Wasserburg, Mass transfer of helium, neon, argon, and xenon through a steady-state upper mantle, *Geochim. Cosmochim. Acta, 59*, 4921-4937, 1995.

Pringle, M. S., C. Mitchell, J. G. Fitton, and M. Storey, Geochronological constraints on the origin of large igneous provinces: Examples from the Siberian and Kerguelen flood basalts, D. L. Anderson, S. R. Hart, and A. W. Hofmann, convenors, Plume 2, *Terra Nostra, 3/195*, Alfred-Wegner-Stiftung, Bonn, 120-121, 1995.

Renne, P. R., Excess [40]Ar in biotite and hornblende from the Noril'sk 1 intrusion, Siberia: implications for the age of the Siberian Traps, *Earth Planet. Sci. Lett., 131*, 165-176, 1995.

Renne, P. R., and A. R. Basu, Rapid eruption of the Siberian Traps flood basalts at the Permo-Triassic boundary, *Science, 253*, 176-179, 1991.

Renne, P. R., A. L. Deino, R. C. Walter, B. D. Turrin, C. C. Swisher, T. A. Becker, G. H. Curtis, W. D. Sharp, and A.-R. Jaouni, Intercalibration of astronomical and radio-isotopic time, *Geology, 22*, 783-786, 1994.

Renne, P. R., Z. Zichao, M. A. Richards, M. T. Black, and A. R. Basu, Synchrony and causal relations between Permian-Triassic boundary crises and Siberian Flood Volcanism, *Science, 269*, 1413-1416, 1995.

Richards, M. A., R. A. Duncan, and V. E. Courtillot, Flood basalts and hotspot tracks: Plume heads and tails, *Science, 246*, 103-107, 1989.

Rudnick, R. L. and D. M. Fountain, Nature and composition of the continental crust: a lower crustal perspective, *Rev. Geophys., 33*, 267-309, 1995.

Russell, J. K., and D. Cherniak, *Theory and application of Pearce element ratios to geochemical data analysis*, 315 pp., Geological Association of Canada Short Course, Vancouver, 1989.

Sadovnikov, G. I., Correlation and origin of the volcanogenic formations of the Tunguska basin, Northern Anabar region and Taymyr, *Izvestiya Academ. Nauk., 9*, 49-63, 1981.

Scarrow, J. H., and K. G. Cox, Basalts generated by decompressive adiabatic melting of a mantle plume: a case study from the Isle of Skye, NW Scotland, *J. Petrol., 36*, 3-22, 1995.

Sharma, M., A. R. Basu, and G. V. Nesterenko, Nd-Sr isotopes, petrochemistry, and origin of the Siberian flood basalts, USSR, *Geochim. Cosmochim. Acta, 55*, 1183-1192, 1991.

Sharma, M., A. R. Basu, and G. V. Nesternko, Temporal Sr- Nd- and Pb-isotopic variations in the Siberian flood basalts: implications for the plume-source characteristics, *Earth Planet. Sci. Lett., 113*, 365-381, 1992.

Sobolev, A. V., and M. Chaussidon, H_2O concentrations in primary melts from supra-subduction zones and mid-ocean ridges: implications for H_2O storage and recycling in the mantle, *Earth Planet. Sci. Lett., 137*, 45-55, 1996.

Storey, M., A. D. Saunders, J. Tarney, P. Leat, M. F. Thirlwall, R. N. Thompson, M. A. Menzies, and G. F. Marriner, Geochemical evidence for plume-mantle interactions beneath Kerguelen and Heard Islands, Indian Ocean, *Nature, 326*, 371-374, 1988.

Sun, S.-S., and W. F. McDonough, Chemical and isotopic systematics of oceanic basalts: implications for mantle composition and processes, in *Magmatism in the Ocean Basins, Spec. Publ. 42*, edited by A. D. Saunders and M. J. Norry, pp. 313-345, The Geological Society, London, 1989.

Sweeney, J. F., J. R. Weber and S. M. Blasco, Continental ridges in the Arctic ocean: LOREX constraints, *Tectonophysics, 89*, 217-237, 1982.

Thompson, R. N., M. A. Morrison, A. P. Dickin, I. L. Gibson, and R. S. Harmon, Two contrasting styles of interaction between basic magmas and continental crust in the British Tertiary Volcanic Province, *J. Geophys. Res., 91*, 5985-5997, 1986.

Turner, S., C. Hawkesworth, K. Gallagher, K. Stewart, D. Peate, and M. Mantovani, Mantle plumes, flood basalts, and thermal models for melt generation beneath continents: assessment of a conductive heating model and application to the Paraná, *J. Geophys. Res., 101*, 11,503-11,518, 1996.

Voshage H., A. W. Hofmann, M. Mazzucchelli, G. Rivalenti, S. Sinigoi, I. Raczek, G. Demarchi, Isotopic evidence from the Ivrea Zone for a hybrid lower crust formed by magmatic underplating, *Nature, 347*, 731-736, 1990.

Walker, R. J., J. W. Morgan, M. F. Horan, G. K. Czamanske, E. J. Krogstad, V. A. Fedorenko, and V. E. Kunilov, Re-Os isotopic evidence for an enriched-mantle source for the Noril'sk-type, ore-bearing intrusions, Siberia, *Geochim. Cosmochim. Acta, 58*, 4179-4197, 1994.

White, R. S., and D. McKenzie, Mantle plumes and flood basalts, *J. Geophys. Res., 100*, 17,543-17,585, 1995.

White, W. M., A. W. Hofmann, and H. Puchelt, Isotope geochemistry of Pacific mid-ocean ridge basalt, *J. Geophys. Res., 92*, 4881-4893, 1987.

Wilson, J. T., Room at the top of the world, *Nature, 316*, 768, 1985.

Wooden, J. L., G. K. Czamanske, V. A. Fedorenko, N. T. Arndt, C. Chauvel, R. M. Bouse, B.-S. King, R. J. Knight, and D. F. Siems, Isotopic and trace-element constraints on mantle and crustal contributions to Siberian continental flood basalts, Noril'sk area, Siberia, *Geochim. Cosmochim. Acta, 57*, 3677-3704, 1993.

Woodhead, J. D. and C. W. Devey, Geochemistry of the Pitcairn seamounts, I: source character and temporal trends, *Earth Planet. Sci. Lett., 116*, 81-99, 1993.

Zindler, A., and S. Hart, Chemical geodynamics, *Ann. Rev. Earth Planet. Sci., 14*, 493-571, 1986.

Zolotukhin, V. V., and A. I. Al'mukhamedov, Traps of the Siberian Platform, in *Continental Flood Basalts*, edited by J. D. Macdougall, pp. 273-310, Kluwer Academic Publishers, New York, 1988.

Zolotukhin, V. V., A. M. Vilensky, and O. A. Djuzhikov, *Basalts of the Siberian Platform*, 245 pp., Nauka, Novosibirsk, 1986.

Zonenshain, L. P., M. I. Kuzmin, and L. M. Natapov, Geology of the USSR: a plate tectonic synthesis, in *Geology of the USSR: a Plate Tectonic Synthesis*, edited by B. M. Page, p. 242, American Geophysical Union, Washington, D.C., 1990.

M. Sharma, The Lunatic Asylum of the Charles Arms Laboratory, Division of Geological and Planetary Sciences, Mail Code 170-25, California Institute of Technology, Pasadena, CA 91125.

Giant Radiating Dyke Swarms: Their Use in Identifying Pre-Mesozoic Large Igneous Provinces and Mantle Plumes

Richard E. Ernst

Geological Survey of Canada, Ottawa, Ontario, Canada
Department of Earth Sciences, University of Western Ontario, London, Ontario, Canada

Kenneth L. Buchan

Geological Survey of Canada, Ottawa, Ontario, Canada

The identification of large igneous provinces (LIPs) and associated mantle plumes has been confined for the most part to the Mesozoic and Cenozoic record. With few exceptions, pre-Mesozoic LIPs have been partially or totally destroyed by tectonic and erosional processes. Here, we describe a method of identifying the remnants of pre-Mesozoic LIPs and the location of paleoplumes from the convergent points of giant radiating dyke swarms. It is based on the observations from seven case histories that giant radiating dyke swarms converge towards known mantle plume centers beneath the volcanic accumulations of Mesozoic and Cenozoic LIPs. We describe three additional Mesozoic and 14 pre-Mesozoic swarms which likely focus on paleoplumes. Several other candidates are also considered.

1. INTRODUCTION

Large igneous provinces (LIPs) include continental flood basalts, volcanic passive margins, oceanic plateaus, oceanic basin flood basalts, some seamounts and submarine ridges, and represent large volumes of magma originating from processes other than normal sea-floor spreading [*Coffin and Eldholm*, 1994]. They have been a focus of much recent interest regarding their implications for the thermal structure of the mantle, specifically their relationship with mantle plumes, and their association with continental breakup [e.g., *Morgan*, 1981; *Richards et al.*, 1989; *White and McKenzie*, 1989; *Storey*, 1995; *Anderson*, 1995]. LIPs that are likely to

have a mantle plume origin include continental flood basalts, oceanic plateaus, and oceanic basin flood basalts. Characteristics suggestive of a mantle plume origin include emplacement of large volumes of magma in a short time interval [e.g., *Coffin and Eldholm*, 1994], linkage to hotspot tracks [e.g., *Morgan*, 1981; 1983], and associated topographic doming [e.g., *Cox*, 1989; *LeCheminant and Heaman*, 1989].

Coffin and Eldholm [1994] summarized the distribution of LIPs. Most are of Mesozoic and Cenozoic age. In the pre-Mesozoic record LIPs are poorly preserved, and thus their former locations have been difficult to identify. Destruction of pre-Mesozoic LIPs has resulted from many processes, including subduction, continent-continent collision, and erosion.

In this paper, we discuss the surface geometry and other characteristics of giant dyke swarms (swarms with lengths >300 km) [*Ernst et al.*, 1995b] and their importance in understanding mantle plume dynamics and the generation of

Large Igneous Provinces: Continental, Oceanic, and Planetary
Flood Volcanism
Geophysical Monograph 100

plume-related LIPs. Although the origin of most giant swarms is unclear, and small swarms have a variety of origins including those that are subduction and continent-collision related [*Ernst et al.*, 1995a and references therein], we suggest that giant swarms which have a radiating geometry are derived from mantle plumes. A giant radiating swarm converges towards the former mantle plume center and associated flood basalt (or its erosional remnant).

In presenting radiometric ages, we indicate the uncertainty level (1σ or 2σ) where available. However, in many publications the uncertainty levels are poorly documented or not reported. Provisional names for dyke swarms are indicated by a superscript pn.

2. CONTINENTAL FLOOD BASALTS AND THEIR ASSOCIATED DYKE SWARMS

2.1. *Introduction*

Continental flood basalts and associated passive margins are the only mantle-plume-related LIPs for which a significant amount of information on dyke swarms is presently available. We will focus on continental examples, although the possibility of using swarms to locate oceanic mantle plumes is discussed briefly in section 5.5.

Dyke swarms are associated with all continental flood basalts, although their full extent and distribution are often difficult to assess because of a lack of exposure. The feeder dyke system is typically hidden beneath the volcanic pile. Therefore, it is necessary to rely on differential erosion to expose portions of the feeder system. Furthermore, where dyke swarms have been emplaced at depth and did not reach the paleosurface, erosion may not have been sufficient to expose them.

The volcanic portions of continental flood basalt provinces are typically located along craton margins and, hence, are likely to be deformed and destroyed during subsequent ocean closing. *Fahrig* [1987] noted that coeval dyke swarms extend into the craton where they are better protected from later deformation. In addition, dyke swarms are less likely to be eroded than flood basalts because they lie underneath the lavas which they feed and reach to at least midcrustal depths [e.g., *Fahrig*, 1987; *Ernst et al.*, 1995a]. Thus, coeval dyke swarms may provide the best, and in many cases the only, record of old continental flood basalts.

In instances where a giant dyke swarm has been dismembered by plate tectonic processes, its primary radiating pattern may only become apparent when continents have been reconstructed. Only then can the focus of the radiating pattern be used to locate the plume center and associated

LIP. Furthermore, the criteria of arranging continental blocks so as to restore the primary geometry of the swarm may facilitate the reconstruction itself [*Ernst et al.*, 1995a].

In this section, dyke swarms associated with the major Cenozoic and Mesozoic continental flood basalts are described in order of increasing age (Figure 1 and Table 1). A consistent pattern emerges, in which most swarms converge towards the center of topographic uplift marking the location of the plume head. These examples suggest that dyke swarms associated with continental flood basalts may, at least on a regional scale, form a simple radiating pattern centered on the mantle plume. This relationship will be further explored in later sections of the paper, where giant radiating dyke swarms are utilized to establish the location of flood basalts that have been removed by erosion or deformed in later orogenies.

2.2. *Columbia River Event*

The Columbia River basalt group (Figure 2) was erupted between 17.5 and 6 Ma (K-Ar). Most of the extrusive rocks (97%) were emplaced during the first 3.5 m.y. [*Hooper*, 1988; *Tolan et al.*, 1989] and are thought to be related to the arrival of a mantle plume centered several hundred kilometers to the south, which subsequently migrated east-northeast to its present position beneath Yellowstone [e.g., *Smith and Braile*, 1993; *Parsons et al.*, 1994] (Figure 2). Activity to the south along the Nevada rift has been related to lithospheric rifting associated with the same plume [*Zoback et al.*, 1994; *Parsons et al.*, 1994].

Extensive dyke swarms are associated with this LIP (Figure 2). Flows of the Columbia River Basalt group erupted from north-northwest-trending fissure systems and dykes of the 17-15 Ma (K-Ar) Chief Joseph and Monument swarms [e.g., *Swanson et al.*, 1975; *Tolan et al.*, 1989; *Atkinson and Lambert*, 1990]. Dykes associated with the Nevada rift trend south from the plume center and are dated at 17-14 Ma (K-Ar) [*Zoback et al.*, 1994].

Other poorly dated swarms which may also belong to this event include the Steens Mountain dykes of north-northeast trend [*Walker and MacLeod*, 1991] and the dykes of the Cascade Range with a trend of about 315°. The Cascade Range[pn] dykes are related to 17 to 10 Ma basaltic and andesitic flows [unit Tbaa of *Walker and MacLeod*, 1991] and, on the basis of their age, probably belong to the Columbia River event.

The Chief Joseph, Monument, and Cascade Range[pn] dykes radiate over an angle of 45° about the site of presumed plume initiation. The Nevada rift[pn] dykes also trend towards the same center. However, the Steens Mountain dykes do not fit the radiating pattern.

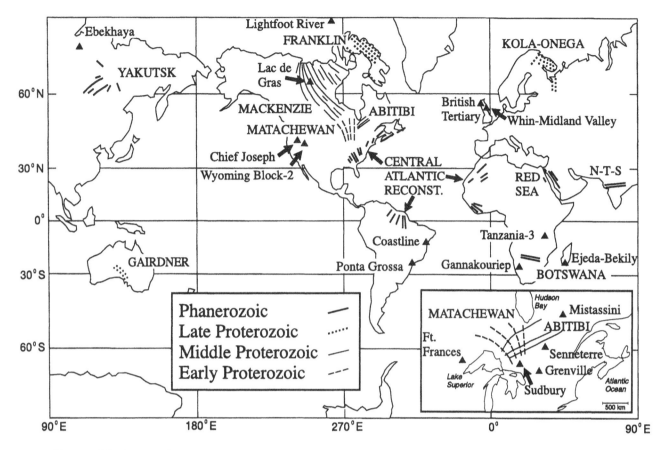

Figure 1. Distribution of key dyke swarms associated with events discussed in this paper. Swarms longer than about 500 km are labeled in uppercase. Triangles and lowercase names identify smaller swarms. N-T-S is Narmada-Tapti-Son. Swarms that are discussed as part of a larger event are Botswana (Lower Limpopo Karoo event); British Tertiary (North Atlantic Volcanic Province event); Chief Joseph (Columbia River event); Ejeda-Bekily (Madagascar event); Gairdner (Willouran event); Grenville (Central Iapetus event); Lac de Gras (Booth River event); N-T-S (Deccan event); Ponta Grossa (Paraná-Etendeka event); Red Sea (Afar event); Senneterre (Ungava Bay event); Tanzania-3 (Bukoban event) and Whin-Midland Valley (Jutland event); Wyoming Block-2 (Willouran event). The north-central portion of North America has been enlarged at the lower right to better illustrate the large number of swarms located there.

2.3. *Afar Event*

The Afar plume (Figure 3), which caused broad regional uplift [*Camp and Roobol*, 1992] and incipient rifting along the Gulf of Aden and the Red Sea, is associated with the onset of separation of Africa and Arabia [e.g., *White and McKenzie*, 1989; *Cox*, 1989]. The Yemeni and Ethiopian LIPs are the main volcanic expression of this plume, but additional components include about 20 volcanic fields (locally called 'harrats' in the Arabian Peninsula and the large dyke swarm which parallels the Red Sea [e.g., *Coleman*, 1993]. The total volume of associated magmatism is greater than 0.35×10^6 km^3 [*Mohr*, 1983; *Baker et al.*, 1996]. The magmatism of this event is discussed in three main age ranges, 30–25 Ma, 24–21 Ma, and less than 5 Ma.

The oldest stage of activity is marked by the Yemeni LIP (age range of 30–25 Ma, ^{40}Ar-^{39}Ar), the Ethiopian LIP (about 30 Ma, ^{40}Ar-^{39}Ar) and at least one harrat, Harrat Hadan (28-27 Ma, ^{40}Ar-^{39}Ar) [*Baker et al.*, 1996; *Hofmann et al.*, 1995; *Sebai et al.*, 1991a, respectively]. All of these units have a transitional to alkali basalt composition [*Coleman*, 1993; *Camp and Roobol*, 1989]. Few dykes have yet been identified that are coeval with this early activity. Exceptions are those associated with Harrat Hadan [*Sebai et al.*, 1991a], which trend towards the plume center (Figure 3).

A younger 24-21 Ma (^{40}Ar-^{39}Ar) tholeiitic stage of magmatism consists of the major coast parallel Red Sea dykes, Yemeni dykes and associated plutons [*Sebai et al.*, 1991a]. Some volcanism in Ethiopia may also be of this age based on ^{40}Ar-^{39}Ar dating [*Hofmann et al.*, 1995]. In

TABLE 1. Dyke Swarms and Large Igneous Provinces (LIPs) Associated with Plume Arrival[a]

Event (section of paper)	Plume Head[b]	Age of Plume Arrival (Ma)	Component Dyke Swarms of Radiating Pattern	Selected Coeval LIPs or Remnant LIPs
Columbia River Event (section 2.2)	Yellowstone	17-14	Chief Joseph Monument ? Cascade Range[pn] Nevada rift	Columbia River Basalt Group CFB
Afar Event (section 2.3)	Afar	~25	Red Sea Ethiopian Yemeni Harrat Hadan ? Al Ghayl[pn]	Ethiopian CFB Yemeni CFB
North Atlantic Volcanic Province Event (section 2.4)	Iceland	61-53	E. Greenland Tertiary W. Greenland Tertiary British Tertiary	NAVP mafic plutons and lavas E. Greenland PM W. Greenland PM Vøring Plateau OP
Deccan Event (section 2.5)	Réunion	~66	Narmada-Tapti-Son Panvel[pn] Mt. Girnar[pn] Cambay Graben[pn] ? Rani[pn]	Deccan CFB
Madagascar Event (section 2.6)	Marion	90-83	Ejeda-Bekily Morondava[pn] Tamatave[pn]	Madagascar CFB N. Madagascar Ridge OP
Alpha Ridge Event (section 2.9)		~100	Hazen Strait[pn] Lightfoot River[pn]	Alpha Ridge OP Queen Elizabeth Islands CFB
Paraná-Etendeka Event (section 2.7)	Tristan da Cunha	137-127	Ponta Grossa Santos-Rio de Janeiro Florianópolis ? Paraguay West Bodoquena Serro do Caiapó --------------[c]	Paraná CFB
			Etendeka Horingbaai Mehlberg Cape Pennisula	Etendeka CFB
Lower Limpopo Event (section 2.8)	Bouvet? Marion?	~184	Botswana Lebombo Orange River	Karoo CFB ? Ferrar CFB
Lower Zambesi Event (section 2.8)	–	~180	Rushinga Cholo Gorongosa	Karoo CFB ? Ferrar CFB

TABLE 1. (continued)

Event (section of paper)	Plume Head[b]	Age of Plume Arrival (Ma)	Component Dyke Swarms of Radiating Pattern	Selected Coeval LIPs or Remnant LIPs
Central Atlantic Event (section 4.1)	Fernando de Noronha? Cape Verde?	~200	ENA Charleston ---- Amapá Guyana[pn] ----- Morocco[pn] Taoudenni[pn] Liberia[pn]	Florida volcanism
Siberian Trap Event (section 2.9)	Jan Mayen?	~248	Ebekhaya Maymecha[pn]	Siberian CFB
Jutland Event (section 4.2)		~300	Whin-Midland Valley Oslo Rift Scania	Oslo rift volcanism Whin & Midland valley sills
Yakutsk Event (section 4.3)		~350	Chara-Sinsk Vilyui-Marcha Dzhardzhan[pn] Tomporuk[pn]	Dzhalkan volcanics
Central Iapetus Event (section 4.4)		~600	Long Range Grenville Adirondack ? Southern Appalachian[pn]	Tibbit Hill volcanism
? Gannakouriep Event (section 4.13)		~720	Gannakouriep	--
Franklin Event (section 4.5)		723-718	Franklin Thule	Natkusiak CFB Coronation sills
Willouran Event (section 4.6)		~800-780	Mackenzie Mountains Wyoming Province Hottah sheets ---------- Gairdner	Willouran volcanism
? Bukoban Event (section 4.13)		1000-800	Tanzania-3[d] Tanzania-4[d] Tanzania-5[d]	? Bukoban volcanism
? Coastline (Bahia) Event (section 4.13)		~1000	Salvador Ilhéus-Olivença-Camacã Itacaré	--
Kola-Onega Event (section 4.7)		~1040	Kola-Onega	--
Abitibi Event (section 4.8)		~1140	Abitibi Eye Dashwa	--

TABLE 1. (continued)

Event (section of paper)	Plume Head[b]	Age of Plume Arrival (Ma)	Component Dyke Swarms of Radiating Pattern	Selected Coeval LIPs or Remnant LIPs
Keweenawan Event (section 4.8)		1110-1085	–	Keweenawan CFB
Sudbury Dyke Event (section 4.13)		1238-1235	Sudbury	–
Mackenzie Event (section 4.9)		1270-1265	Mackenzie	Coppermine River volcanics Muskox Intrusion
? Booth River Event (section 4.13)		2030-2023	Lac de Gras	Booth River Complex
Fort Frances Event (section 4.10)		~2076	Fort Frances	? Mille Lacs Group
Ungava Bay Event (section 4.13)		~2200	Senneterre Klotz Maguire	–
Matachewan Event (section 4.11)		2490-2445	Matachewan	Volcanics and mafic plutons
Mistassini Event (section 4.12)		~2470	Mistassini	–
? McArthur Basin Event (section 4.13)		–	McArthur Basin-1[pn]	–

Superscript pn indicates provisional names.

[a] Details and references in text. Labelling of LIP types after *Coffin and Eldholm* [1994]: CFB is continental flood basalt, PM is passive margin, and OP is oceanic plateau. NAVP is North Atlantic Volcanic Province. "?" in table indicates uncertainty either in the identification of an event or in the correlation of swarms or LIPs to an event.

[b] For Mesozoic and Cenozoic LIPs, the plume is given the name of the hotspot that backtracks to the site of plume initiation.

[c] Dashed lines separate events presently located on separate continents.

[d] Tanzania 3, 4 and 5 are the Tanzania III, IV and V of *Halls et al.* [1987].

addition, an age of 24 ± 3 Ma (K-Ar) has been determined for the major episode of dyking within the Ethiopian Plateau [*Megrue et al.*, 1972; p. 84 of *Mohr and Zanettin*, 1988]. In the southeastern portion of the Yemen LIP, Yemeni dykes have ages of ~25.5 and 18.5–16 Ma (^{40}Ar-^{39}Ar) [*Zumbo et al.*, 1995a].

The character of the 24-21 Ma dykes changes across the Ad Darb fault (Figure 3) [pp. 58-60 of *Coleman*, 1993]. The Red Sea dykes north of the fault are mainly individual dykes with widths up to 100 m, whereas the Yemeni dykes south of the fault tend to be sheeted dyke complexes with more widely spaced dykes having widths of 0.5 to 15 m [*Coleman*, 1993]. The greatest concentration of Yemeni dykes is along the coastal escarpment zone, but less concentrated swarms lace the interior [*Mohr*, 1991]. The Yemeni dykes are divided into subswarms, some with very high densities of

intrusion resulting in local crustal extension up to 50%. Al Ghayl[pn] dykes trend at a high angle to the Yemeni dykes and are assigned a broad age range of Cenozoic [*Kruk*, 1980].

After correction for the opening of the Red Sea, the Red Sea and Yemeni dykes trend away from the plume center for a distance of nearly 2000 km [*Eyal and Eyal*, 1987; *Baldridge et al.*, 1991]. However, as these dykes have ages of 24–21 Ma, they postdate the main LIP activity at 30–25 Ma and their distribution must be strongly influenced by rifting along the Red Sea. Likewise, the 24 Ma dyke swarms in the Ethiopian Plateau have variable trends, which appear more related to rift faults rather than focussing on the plume center [figure 3 of *Mohr and Zanettin*, 1988].

The youngest activity, 5–0 Ma magmatism, occurs within the western portion of the Ethiopian LIP [*Zumbo et al.*, 1995b; figure 3.1 of *Coleman*, 1993]. In addition, most

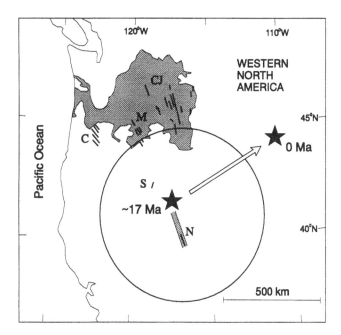

Figure 2. Dyke swarms of the 17–14 Ma Columbia River-Nevada Rift magmatic event. Swarms associated with the arrival of the Yellowstone plume at 17 Ma include the Chief Joseph (CJ), Monument (M), Steens Mountain (S) and Nevada rift[pn] (N) swarms, and possibly the Cascade Range[pn] (C) swarm [*Tolan et al.*, 1989; *Walker and MacLeod*, 1991; *Zoback et al.*, 1994]. Stars locate the position of the mantle plume at 17 Ma and its present position under Yellowstone [*Parsons et al.*, 1994]. Distribution of volcanic rocks (shaded) after *Tolan et al.* [1989] and *Zoback et al.* [1994]. Circle with a radius of ~400 km outlines the region of plume-generated uplift [*Parsons et al.*, 1994].

harrats exhibit activity in this time range [*Féraud et al.*, 1987; figure 3.1 of *Coleman*, 1993].

2.4. *North Atlantic Volcanic Province Event*

The North Atlantic Volcanic Province (Figure 4), with a total volume of more than 6.6×10^6 km³, consists of extrusive rocks, intrusive centers and associated dykes of the British Tertiary province in the British Isles; volcanic rocks that extend for 2000 km along the east coast of Greenland, in the adjacent ocean, on the Vøring Plateau (off Norway), and on the Rockall Bank; and minor accumulations in the Disko Island area of west Greenland and adjacent Baffin Island [e.g., *Dickin*, 1988; *White and McKenzie*, 1989; *Coffin and Eldholm*, 1994]. Ages from all three areas fall between about 61 and 53 Ma (mostly based on K-Ar ages) with the preponderance of magmatism occurring between 57.5 and 54.5 Ma [*White and McKenzie*, 1989, 1995 and references therein]. *White and McKenzie* [1989] proposed that volcanism in all of these areas was associated with an upwelling

mantle plume located beneath the east coast of Greenland. Plume-generated uplift occurred between 65 and 55 Ma [*Nadin and Kusznir*, 1995] and was associated with the breakup and separation of Europe and Greenland. The tail of the plume is now expressed by continuing volcanic activity in Iceland.

An alternative model was proposed by *Lawver and Müller* [1994], who computed a different track for the Iceland hotspot on the basis of studies of plate motions and other hotspots. Their 60 Ma location for the hotspot lies in west central Greenland, a distance of nearly 1000 km from that advocated by *White and McKenzie* [1989]. Furthermore, they suggested that the plume arrived much earlier, possibly at 130 Ma, beneath Ellesmere Island in northern Canada.

Dykes of the British Tertiary province have a dominant northwest trend, although more local radiating dyke sets are

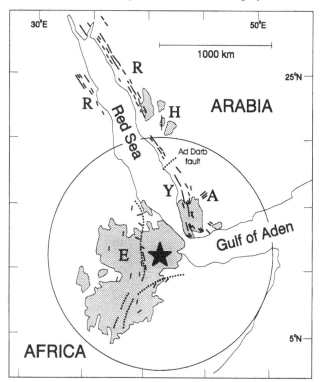

Figure 3. Dyke swarms of the ~25 Ma Afar magmatic event of Africa and Arabia in the present-day configuration of the region. Red Sea dykes (R) are after *Eyal and Eyal* [1987], Ethiopian dykes (E) after *Mohr and Zanettin* [1988], Yemeni (Y) dykes after *Mohr* [1991], Harrat Hadan (H) dykes after *Sebai et al.*, [1991], and Al Ghayl[pn] (A) dykes after *Kruk* [1980]. Shaded pattern shows the distribution of volcanic rocks [*White and McKenzie*, 1989], including the Ethiopian LIP, the Yemeni LIP near the southern tip of the Arabian Peninsula and smaller 'harrats' farther north. Circle with a radius of 1000 km depicts the probable extent of the thermal anomaly in the mantle surrounding the plume [after *White and McKenzie*, 1989]. Dotted lines mark rift-related faults in Ethiopia.

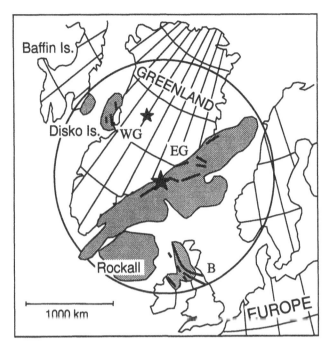

Figure 4. Dyke swarms of the ~60 Ma North Atlantic Volcanic Province. The British Tertiary Province dykes (B) are after *Speight et al.* [1982], and the east Greenland dykes (EG) and west Greenland dykes (WG) after *Nielsen* [1987]. Distribution of volcanism after *White and McKenzie* [1989] and *Coffin and Eldholm* [1994]. Continental reconstruction after *White and McKenzie* [1989]. Large star locates plume center of *White and McKenzie* [1989]; small star locates the proposed 60 Ma hot spot position of *Lawver and Müller* [1994].

associated with intrusive complexes [*Speight et al.*, 1982; *MacDonald et al.*, 1988]. Dyke swarms abound along and oriented parallel to the coast of east Greenland [*Nielsen*, 1987]. In west Greenland, a minor dyke set is associated with volcanic activity [*Nielsen*, 1987].

In a reconstruction of the North Atlantic (Figure 4), the dominant dykes of the British Tertiary Province and those of Greenland have overall trends which roughly converge towards the plume center of *White and McKenzie* [1989]. On the other hand, the dykes of Greenland do not converge towards the alternative 60 Ma hotspot location of *Lawver and Müller* [1994]. Therefore, on the basis of the radiating swarm model, we prefer the *White and McKenzie* plume location. This is also the location of the Kangerdlugssuak triple junction of *Burke and Dewey* [1973].

The British Tertiary dykes show a prominent swing in trend from southeast to east-southeast. This change in direction probably reflects the decreasing influence of plume-generated uplift and increasing influence of a regional stress component [*Ernst et al.*, 1995a,b]. Similar deviation

from a simple radiating pattern is observed in several other of the largest dyke swarms described below.

2.5. Deccan Event

The Deccan flood basalt province in India (Figure 5) has a volume of more than 8×10^6 km^3, and is associated with the separation of India from the Seychelles [e.g., *Mahoney*, 1988; *White and McKenzie*, 1989, 1995]. The majority of the magmatism was quite rapid, 65.5 ± 0.5 Ma (1σ) [whole rock ^{40}Ar-^{39}Ar from *Baksi*, 1994] and is thought to be related to the arrival of the Réunion plume [e.g. *Morgan*, 1981; *Storey*, 1995]. The present drainage pattern is consistent with the existence of a Late Cretaceous topographic swell centered above the plume center [*Cox*, 1989].

Locally, the distribution of dykes associated with the Deccan flows can be quite complex, with radial, crosscut-

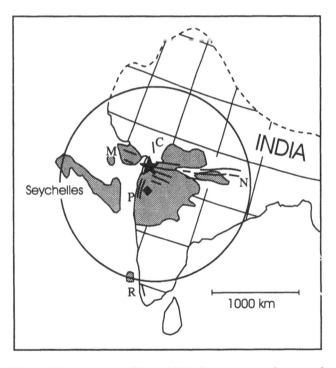

Figure 5. Dyke swarms of the ~ 66 Ma Deccan magmatic event of India and the Seychelles. Component swarms include the Narmada-Tapti-Son (N), Panvel Flexure[pn] (P), Cambay Graben[pn] (C), and Mt. Girnar[pn] (M) swarms, and probably the Rani[pn] (R) swarm. Distribution of dykes follows *Murthy* [1987] and *Radhakrishna et al.* [1994]. Distribution of flood basalts (shaded), region of anomalously hot mantle around the plume (circle) and continental reconstruction for >65 Ma are after *White and McKenzie* [1989]. Star shows location of plume center after *Bhattacharji et al.* [1994]. Diamond locates center of magmatism of *Hooper* [1990] and is near the plume center of *White and McKenzie* [1989]. The Kutch region is located just north of the Mt. Girnar[pn] swarm.

ting, curved, random, and subparallel patterns [*Auden*, 1949, *Karkare and Srivastrava*, 1990; *Sant and Karanth*, 1990; *Hooper*, 1990]. However, several regional-scale swarms are observed. The most dramatic is the east-west-trending swarm which follows the Narmada-Tapti-Son Lineament, an often-reactivated Precambrian suture zone, for more than 600 km [e.g., *Murthy*, 1987; *Deshmukh and Sehgal*, 1988]. East-trending dykes of the Mt. Girnar area could be viewed as a continuation of the Narmada-Tapti-Son swarm. The dykes associated with the Panvel flexure trend north-south [e.g., *Deshmukh and Sehgal*, 1988] and are of at least two generations: older tholeiitic pre-flexure dykes which dip 60-80° east and younger alkaline dykes which are vertical [*Dessai and Bertrand*, 1995]. The Cambay graben is paralleled by north- to northwest-trending dykes [*Bhattacharji*, 1988].

The dykes of the Narmada-Tapti-Son lineament, Cambay graben, Mt. Girnar area, and Panvel flexure show a rough convergence toward a locality which has been interpreted as a plume center mainly on the basis of rift convergence [*Bhattacharji*, 1988; *Bhattacharji et al.*, 1994] (Figure 5). This location was also identified as the Gulf of Khambat triple junction by *Burke and Dewey* [1973].

This plume center differs from the plume center suggested by *White and McKenzie* [1989] and the center of magmatism proposed by *Hooper* [1990], both of which are located a few hundred kilometers to the south (see symbol in Figure 5). *Hooper's* [1990] center of magmatism is marked by large randomly oriented feeder dykes which he considered the primary feeders for the Deccan LIP. However, *Karkare and Srivastava* [1990] and *Bhattacharji et al.* [1994] identified primary feeder dykes aligned along the Narmada-Tapti-Son zone. Furthermore, dyke frequency along the Narmada-Tapti-Son zone increases toward the plume center shown in Figure 5 [*Bhattacharji et al.*, 1994]. Additional main feeder dykes were identified by *Karkare and Srivastava* [1990] paralleling the Panvel zone, and in the Kutch region.

There clearly are complexities to the overall radiating swarm pattern. There are some crosscutting trends throughout the region and randomly oriented dykes in local areas [e.g., *Hooper*, 1990]. The complexity in the dyke distribution may reflect the superposition of an early radiating swarm, associated with plume head arrival, and later dykes whose orientation is controlled by both rifting and the changing uplift topography as the plume flattened against the lithosphere [*Griffiths and Campbell*, 1991].

There are other dyke swarms of similar age that are likely associated with the Deccan event. Deccan-age dykes of unspecified orientation are reported from the Seychelles, which were adjacent to India at that time [*Devey and Stephens*, 1991]. Also, in the Kerala region of southwest

India, the northwest-trending Rani[pn] dykes (Figure 5) have an age of 69 ±1 Ma (^{40}Ar-^{39}Ar whole rock) [*Radhakrishna et al.*, 1990, 1994].

2.6. *Madagascar Event*

Widespread Cretaceous flood basalts are present in Madagascar (Figure 6) and can be related to the track of the Marion hotspot and the breakup of Madagascar and India [*Mahoney et al.*, 1991; *Storey*, 1995; *Storey et al.*, 1995]. On the basis of numerous high-precision ^{40}Ar-^{39}Ar analyses (whole rock and feldspar) this LIP has been dated at 90–83 Ma [*Storey et al.*, 1995].

Dyke swarms associated with the volcanism are mapped in three regions [*Storey et al.*, 1995]. Together, the main dyke trends in each region form a radiating pattern which fans over about 65° and converges near the estimated 88 Ma location of the Marion hotspot (Figure 6). We interpret this focal point (large star) to mark the arrival of the Marion plume.

Because southwest India was adjacent to Madagascar at that time, *Storey et al.* [1995] suggested that some dykes of India may be related to this event. The K-Ar whole rock age, 81 ±3 Ma, for north-northwest-trending Palai[pn] dykes [e.g., *Radhakrishna et al.*, 1990; 1994] is roughly similar in age to the Madagascar LIP, and, therefore, may be related. However, they do not fit the radiating pattern defined by the majority of dykes of Madagascar.

2.7. *Paraná-Etendeka Event*

The separation of South America and Africa is associated with the arrival of the Tristan da Cunha plume at ~130 Ma. Melting in this plume head is thought to be responsible for a major LIP, the Paraná basalt province of eastern South America with a volume of >1.2 × 10⁶ km³, and the smaller, originally contiguous, Etendeka province of southwestern Africa (Figure 7) [e.g., *Storey*, 1995; *White and McKenzie*, 1989]. The age range of the Paraná event (based on the ^{40}Ar-^{39}Ar method) has been reported as 133–131 Ma [*Renne et al.*, 1993] and as 138–127 Ma [*Turner et al.*, 1994; *Stewart et al.*, 1996]. Identical ^{40}Ar-^{39}Ar ages of 133–131 Ma are also reported for the Etendeka LIP [*Renne et al.*, 1996b]. The present drainage pattern is consistent with an Early Cretaceous topographic swell centered over the plume center [*Cox*, 1989].

The most prominent dyke swarm of the Paraná event is the Ponta Grossa swarm, which is well defined over a distance of about 300 km [*Sial et al.*, 1987; *Raposo and Ernesto*, 1995] and has an age of 132-129 Ma [*Renne et al.* 1996a]. Other smaller swarms, the Santos-Rio de Janeiro and

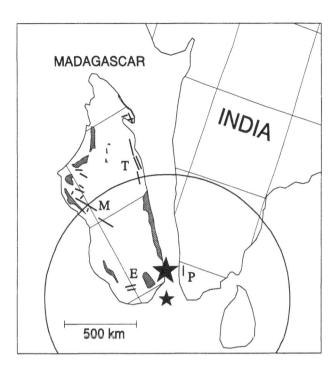

Figure 6. Dyke swarms of the ~88 Ma Madagascar magmatic event. Distribution of dykes and volcanic rocks (shaded) in Madagascar are after *Storey et al.* [1995]. E, M, and T are the Ejeda-Bekily, Morondava[pn] and Tamatave[pn] dykes, respectively, of Madagascar. The Palai[pn] (P) swarm of India [*Radhakrishna et al.,* 1994] is possibly correlative. The small star and circle locate the 88-Ma center and possible extent of the Marion plume [*Storey et al.,* 1995]. The large star locates the convergence point for the dykes. Reconstruction for ~88 Ma is after *Storey et al.* [1995].

Florianópolis swarms, are coast parallel and together with the Ponta Grossa dykes form a convergent pattern focussed towards the southeast. *Stewart et al.* [1996] reported ages of 133.3–129.4 Ma (^{40}Ar-^{39}Ar) for the Santos-Rio de Janeiro swarm, whereas *Renne et al.* [1996a] indicated that the coast parallel dykes may be younger with ages of 126–119 Ma (^{40}Ar-^{39}Ar). A thousand kilometers to the west and northwest of the plume center, respectively, are the West Bodoquena and Serro do Caiapó swarms which parallel the Ponta Grossa swarm and are approximately contemporaneous with it [*Sial et al.,* 1987]. Finally, the large northwest-trending Paraguay swarm, which is traced from aeromagnetic maps, is also thought to belong to the Paraná event [*Druecker and Gay,* 1987] although dykes seem to have a slightly older age of 138–137 Ma [*Stewart et al.,* 1996].

Studies of the anisotropy of magnetic susceptibility of the Ponta Grossa swarm indicate mainly subhorizontal magma flow, except towards its eastern end where the inferred flow is typically steeper [*Raposo and Ernesto,* 1995]. These data are consistent with a plume source to the east of the swarm

(Figure 7). However, as noted by *Renne et al.* [1996a], the actual site of plume impact may have occurred farther south because the age of the Ponta Grossa swarm postdates the main magmatism of the southern Paraná lavas to the south (133–132 Ma) by a few million years.

In Africa, dyke swarms associated with the Etendeka event are poorly defined, largely because of difficulty in distinguishing them from Karoo-age dykes (section 2.8). Nevertheless, some small swarms are evident. The Horingbaai dykes, which include both north- and northwest-trending sets [*Namibia Geological Survey,* 1988], have

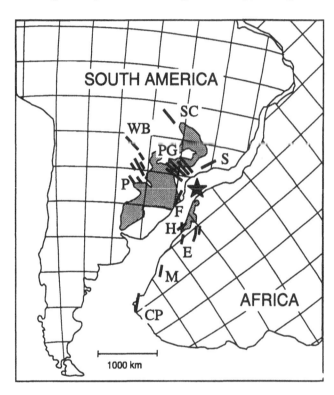

Figure 7. Dyke swarms of the ca. 130 Ma Paraná-Etendeka magmatic event of South America and Africa. In South America, the Paraná swarms include the Ponta Grossa (PG), Paraguay (P), Santos-Rio de Janeiro (S), Florianópolis (F), West Bodoquena (WB), and Serro do Caiapó (SC) swarms. In Africa, the Etendeka-aged swarms include the Etendeka (E), Horingbaai (H; north- and northwest-trending sets shown) and Cape Peninsula (CP) swarms and the single Mehlberg (M) dyke. South American dyke distributions are after *Sial et al.* [1987], *Druecker and Gay* [1987] and *Turner et al.* [1994]. Information on the Florianópolis dykes is from P. Comin-Charamonti (1995, pers. comm.) and *Piccirillo et al.* [1991]. Dykes of Africa are after *Namibia Geological Survey* [1988], *Duncan et al.* [1990], *Reid* [1990], *Reid and Rex* [1994], and *Mubu* [1995]. Distribution of volcanic rocks is after *White and McKenzie* [1989]. Continental reconstruction for ~130 Ma after *O'Connor and Duncan,* [1990]. Star locates approximate convergence point of dykes.

MORB-like chemistry and intrude the basal units of the Etendeka [*Duncan et al.*, 1990]. A coast-parallel swarm (not shown in Figure 7) was suggested by *Hawkesworth et al.* [1992], but its extent is unclear (D. Peate, pers. comm. 1995; A. Duncan, pers. comm. 1995). However, a recent aeromagnetic interpretation recognized an Etendeka swarm which is coast parallel but is located farther from the coast [*Mubu*, 1995]. Along the coast to the south are northwest-trending dykes which collectively are termed the Southern Cape dykes by *Mubu* [1995]. From south to north, they consist of the Cape Peninsula dykes with an age of 132 ±6 Ma (2σ; K-Ar whole rock) [*Reid*, 1990; *Reid et al.*, 1991a], the Cederberg dykes (not shown) [*Hunter and Reid*, 1987] of unknown age and the Mehlberg dyke with an age of 134 ±3 Ma (^{40}Ar-^{39}Ar whole rock) [*Reid and Rex*, 1994].

Most of the dyke swarms of the Paraná and Etendeka events converge towards the general region advocated for the plume centre using other criteria. Our preferred location based on the dyke pattern differs slightly from that of other authors [e.g., *White and McKenzie*, 1989; *Harry and Sawyer*, 1992] but is similar to that of the Sao Paulo triple junction of *Burke and Dewey* [1973]. The slight misfit of the Santos-Rio de Janeiro and Florianópolis subswarms may be explained by their later emplacement along the rift margin. The weak convergence of the Paraguay, Serro do Caiapó, and Cape Peninsula subswarms may reflect the deflection of these dykes into a regional stress field outside the influence of the plume-uplift [*Ernst et al.* 1995b] or that plume impact actually occurred further south [*Renne et al.*, 1996a]. The northwest-trending Horingbaai dykes fit the radiating pattern. However, the role of the north-trending Horingbaai dykes in the tectonic story remains unclear.

2.8. Karoo Event

Large parts of the Karoo magmatic province of southeastern Africa, and the Ferrar magmatic province of Antarctica were emplaced in perhaps less than 1 million years at 184 Ma, on the basis of U-Pb baddeleyite/zircon and ^{40}Ar-^{39}Ar dates [*Encarnación et al.*, 1996 and references therein]. The volumes of Karoo and Ferrar magmatism are at least 2×10^6 km^3 and 0.5×10^6 km^3, respectively. The Karoo magmatism and Ferrar magmatism in adjacent Antarctica have been related to a mantle plume (Figure 8) which was associated with the breakup of Africa and Antarctica [e.g., *White and McKenzie*, 1989]. The part of the Ferrar province located in the Transantarctic Mountains, although coeval with the Karoo has different composition and may have a subduction-related origin [*Encarnación et al.*, 1996]. The Karoo plume center of *White and McKenzie* [1989] is similar to the Lower Limpopo triple junction of *Cox* [1970] and

Burke and Dewey [1973]. The present drainage patterns in southeastern Africa are consistent with a Jurassic topographic swell centered on this plume center [*Cox*, 1989]. Possible candidate plumes are the Marion hotspot [*Morgan*, 1981; *Richards et al.*, 1989; *Duncan and Richards*, 1991] and the Bouvet hotspot [*Lawver et al.*, 1992; *Storey*, 1995].

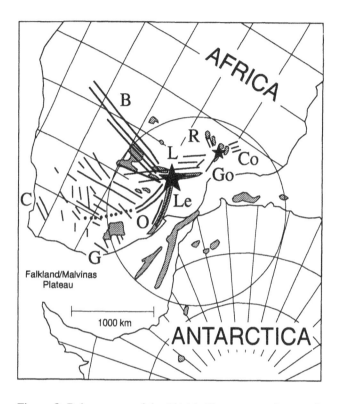

Figure 8. Dyke swarms of the 184 Ma Karoo magmatic event in southeast Africa. The Botswana dykes (B), Orange River dykes (O), and the Lebombo dykes (Le) [*Vail*, 1970; *Wilson et al.*, 1987; *Mubu*, 1995] radiate from the Lower Limpopo triple junction (large star) of [*Cox*, 1970; *Burke and Dewey*, 1973]. The possible arcuate continuation of the Orange River dykes is indicated by a dotted line [*Mubu*, 1995]. The Limpopo dykes (L) are also of Karoo age, but crosscut the Botswana swarm [*Wilson*, 1990]. The Rushinga (lower Zambesi, Imhamangombe) (R), Cholo (Shire Highlands) (Co) and Gorongoza (Go) swarms [*Woolley and Garson*, 1970; *Cox*, 1970; *Vail*, 1970; *MacDonald et al.*, 1983; *Eales et al.*, 1984; *Wilson et al.*, 1987] appear to radiate from the Lower Zambesi triple junction of *Burke and Dewey* [1973]. Numerous other swarms including the east-trending Gap (G) [*Cox*, 1970] and northwest-trending Cederberg (C) dykes [*Hunter and Reid*, 1987; *Mubu*, 1995] are not obviously correlated with any single center. The distribution of volcanic rocks, the Africa-Antarctica reconstruction for this period and the circle which encloses the region of anomalously hot mantle around the Lower Limpopo plume are after *White and McKenzie* [1989]. The Transantarctic Mountains (unlabeled) containing related Ferrar magmatism are located near the bottom of the diagram in Antarctica.

At least three prominent dyke swarms of Karoo age converge on the Lower Limpopo plume center [Figure 16.7 of *Windley*, 1984]. These include the Botswana (1500 km long), Lebombo, and Orange River swarms which form a convergent pattern spanning 100° of arc (Figure 8). Preliminary estimates using measurements of anisotropy of magnetic susceptibility indicate vertical flow at the eastern end of the Botswana swarm and dominantly horizontal flow farther west—an observation consistent with a plume source underlying the focal region [*Ernst and Duncan*, 1995].

According to *Windley* [1984; figure 16.7], other Karoo-age dyke swarms converge on the Lower Zambesi triple junction of *Burke and Dewey* [1973], about 700 km from the Lower Limpopo triple junction (Figure 8). These include the Rushinga, Cholo, and Gorongoza swarms. This focal region and the associated volcanic centers [*Macdonald et al.*, 1983] may define a second site of Karoo-age plume arrival.

There are numerous other Karoo-aged swarms which are not obviously correlated with either the Lower Limpopo or Lower Zambesi plume centers. For example, the Limpopo dykes crosscut the Botswana swarm [*Wilson*, 1990] and so are not related to the Lower Limpopo plume center. Similarly, they intersect the Gorongoza swarm and so probably are not related to the Lower Zambesi plume center either.

2.9. *Additional Cases*

Dyke swarms are associated with other LIPs, but their distribution and (or) age are as yet rather poorly documented. For example, the volcanic rocks of the Alpha Ridge oceanic plateau and nearby Queen Elizabeth Islands of the Canadian Arctic (Figure 9a) form a large igneous province which may have been produced by a plume [*Embry and Osadetz*, 1988]. Most of this volcanism is Cretaceous in age (Hauterivian to early Cenomanian). Dykes of Cretaceous age also occur over a wide area on the northern Arctic islands and *Embry and Osadetz* [1988] noted that many appear to radiate from the plume. In detail, the Hazen Strait[pn] swarm, based largely on aeromagnetic interpretation, and the Lightfoot River swarm may represent subswarms of a giant radiating dyke swarm whose focal point marks a plume center. Precise dating is necessary to test this model. The tectonic relationship of similar-age east-trending Surprise Fiord[pn] dykes that cut the area is not clear.

A second example involves the Siberian Traps (Figure 9b) which cover an area of 1.5×10^6 km^2 in the Siberian platform [*Zolotukhin and Al'Mukhamedov*, 1988]. They were apparently extruded within a period of 5 to 10 m.y. at ~248 Ma, based on ^{40}Ar-^{39}Ar dating [*Renne and Basu*, 1991; *Baksi and Farrar*, 1991]. Several dyke swarms are associated with the Siberian Traps. Two Permo-Triassic swarms, the

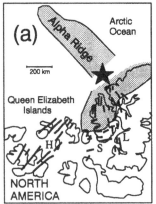 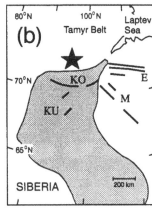

Figure 9. Additional cases. (a) Dyke swarms of the Alpha Ridge event of northern North America (modified after *Embry and Osadetz* [1988]). Dyke sets include Hazen Strait[pn] (H), Lightfoot River (L), and Surprise Fiord[pn] (S). Location of Alpha Ridge and Queen Elizabeth Islands volcanism from *Embry and Osadetz* [1988]. (b) Dyke swarms of the Siberian Traps of Siberia [*Krasnov et al*, 1966; *Blagovyeshchenskaya*, 1973; *Thomshin and Okrugin*, 1995]. Dyke sets include the Ebekhaya (E), Maymecha[pn] (M), Kureyka[pn] (KU) and Kochikha[pn] (KO). There are also deformed Permo-Triassic dykes in the Taymyr belt to the north [*Vakar*, 1962]. Distribution of Siberian Trap volcanics is after *Zolotukhin and Al'Mukhamedov* [1988].

Ebekhaya and Maymecha[pn] swarms, cut basement rocks east of the main accumulation of volcanic rocks [*Blagovye-shchenskaya*, 1973; *Tomshin and Okrugin*, 1995] and may be part of a giant radiating pattern. Other swarms do not readily fit this pattern. Kureyka[pn] dykes cut the volcanic rocks and may be younger. Kochikha[pn] dykes form an arc about the focal point of the Ebekhaya and Maymecha[pn] swarms. They could represent a giant circumferential swarm, located along the rim of a central mega-caldera, similar to the mega-caldera suggested for the Mackenzie event (see section 4.9) [*Baragar et al.*, 1996]. Alternatively, they may represent a feature associated with coronae, a type of plume collapse feature common on Venus but not previously identified on Earth (R. E. Ernst and K. L. Buchan, Identifying Venusian Coronae on Earth Using Giant Circumferential Dyke Swarms, manuscript in preparation). Circumferential swarms have been observed on a much smaller size in the Galapagos Islands [*Chadwick and Dieterich*, 1995].

3. GIANT DYKE SWARMS

Large dyke swarms were emplaced throughout the Proterozoic and the Phanerozoic (Figure 1). Several of the younger swarms are part of known continental flood basalts as outlined in the previous section. Many more swarms, especially those of pre-Mesozoic age, are not associated with known continental flood basalts. Before discussing the uses of giant dyke swarms in identifying the localities of mantle plume centers and associated LIPs (section 4), and in understanding characteristics of plume-generated LIP events (section 5), it is instructive to summarize what is known about giant dyke swarms.

3.1. General Characteristics

Of a total of more than 500 diabase dyke swarms recently catalogued in the Global Mafic Dyke GIS Database Project [*Ernst et al.*, 1996], 141 are classified as giant dyke swarms with lengths ≥300 km [*Ernst et al.*, 1995b]. Of these, 63 have lengths ≥500 km and 17 have lengths ≥1000 km. Giant dyke swarms are concentrated in North America, eastern South America, southern Africa, India, and Australia. This distribution probably reflects more complete information on dykes and their distributions in these areas than is available elsewhere.

Giant dyke swarms can be grouped into five types based on their geometry [*Ernst et al.*, 1995b] (Figure 10). Three types exhibit a radiating pattern: type I has a continuous fanning pattern, type II has a fanning pattern subdivided into separate subswarms, and type III has subswarms of subparallel dykes which radiate from a common point. Two types show a linear pattern: type IV has subparallel dykes distributed over a broad zone and type V has subparallel dykes restricted to a narrow zone. Subparallel types IV and V could, in some cases, represent distal portions of types I and III, respectively, if the segment observed is outside the region of plume-induced uplift and subject only to a regional stress field [*Ernst et al.*, 1995a,b].

Beyond a certain distance, giant radiating swarms may become subparallel, as the influence of the regional stress field exceeds those of the plume-generated uplift [*Ernst et al.*, 1995a,b]. On the other hand, near the plume center, dykes may exhibit variable trends as a result of being located on the relatively flat portion (the crest) of the topographic uplift, where there are insignificant differential horizontal stresses.

3.2. Relationship with Plumes

Recent work has indicated a direct connection between the emplacement of giant radiating dyke swarms (types I, II, and III) and the arrival of mantle plumes at the base of the lithosphere [*LeCheminant and Heaman*, 1989; *Heaman*, 1991; *Ernst et al.*, 1995a,b]. Evidence includes (1) a radiating dyke pattern suggestive of a centrally located magma source; (2) coeval volcanic and plutonic rocks in the focal

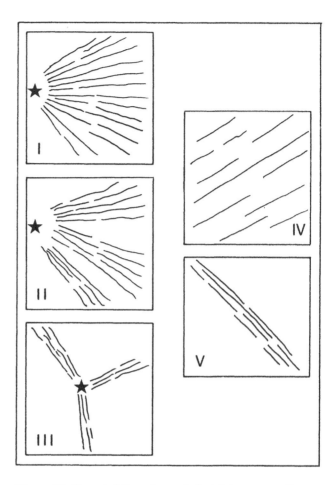

Figure 10. Characteristic patterns of giant dyke swarms. Swarms are divided into 5 types: I= continuous fanning pattern; II= fanning pattern subdivided into separate subswarms: III= subswarms of subparallel dykes which radiated from a common point; IV= subparallel dykes distributed over a broad area; V= subparallel dykes restricted to a narrow zone. Stars locate probable mantle plume centers as defined by radial dyke patterns. After *Ernst et al.* [1995b] and reprinted with permission of A. A. Balkema, Rotterdam.

region which may represent the remnants of LIPs emplaced above a mantle plume; (3) lateral magma flow in dykes beyond the focal region; (4) uplift of the focal region in response to the arrival of the mantle plume; and (5) rapid emplacement of some swarms. Each of these characteristics is elaborated below.

(1) The radiating pattern and continuity of swarms for up to 2500 km suggests that the magma source was localized in the vicinity of the focal area [*Ernst et al.*, 1995a]. In some cases the giant radiating pattern only becomes apparent after continents have been reassembled into their pre-drift configuration [e.g., *May*, 1971; *Park et al.* 1995; *Ernst et al.*, 1995b].

(2) Key evidence supporting the presence of a plume center beneath the focal region of radiating swarms is the abundance of coeval volcanic rocks and plutons in the focal region [*Fahrig*, 1987]. However, it should be noted that in some cases the volcanism was displaced up to two or three hundred kilometers from the center (section 5.1).

(3) Studies of the anisotropy of magnetic susceptibility of giant radiating dyke swarms indicate that, far from the swarm's focus, magma flow was subhorizontal [*Greenough and Hodych*, 1990; *Ernst and Baragar*, 1992; *Raposo and Ernesto*, 1995; *Ernst and Duncan*, 1995]. On the other hand, magma flow was subvertical in the focal area [*Ernst and Baragar*, 1992; *Ernst and Duncan*, 1995]. This pattern is consistent with magma injected as thin vertical blades from a mantle plume beneath the focal area and flowing laterally to great distances, as much as 2000 km or more [e.g., *Ernst et al.*, 1995a].

(4) Broad regional uplift has been documented in the focal region of some giant radiating dyke swarms [e.g., *LeCheminant and Heaman*, 1989; *Rainbird*, 1993]. Because topographic uplift is associated with the arrival of mantle plumes [e.g., *White and McKenzie*, 1989; *Cox*, 1989; *Richards et al.*, 1989; *Griffiths and Campbell*, 1991], it follows that uplifted focal regions of dyke swarms also mark the arrival of a plume head [*LeCheminant and Heaman*, 1989]. On the other hand, it is unclear at present whether plume tails [*Richards et al.*, 1989] can generate giant dyke swarms. The smaller magnitude of volcanism associated with plume tails suggests that dyke swarms associated with plume tails should be much smaller than those associated with plume heads.

(5) High precision U-Pb dating of baddeleyite from several dykes of the Mackenzie giant radiating dyke swarm (see section 4.9) has demonstrated that emplacement occurred within a few million years (1267±2 Ma; 2σ) [*LeCheminant and Heaman*, 1989]. This result is consistent with the observations from the Columbia River, Deccan, Karoo, and Siberia LIPs for rapid emplacement in less than a few million years [e.g., see refs. of *Storey*, 1995, *White and McKenzie*, 1995]. However, as discussed in section 5.3, evidence from other dyke swarms indicates that some dyke events may be episodic over intervals of up to 30 m.y.

3.3. *Origin of Type I, II, and III Swarms*

The differences between type I, II, and III patterns may reflect the distribution of high-level magma chambers and (or) the presence of coeval or pre-existing rift zones.

Baragar et al. [1996] have discussed the distribution of high-level magma chambers in the vicinity of the plume head associated with the Mackenzie swarm (Figure 11a). They suggest that magma access and loci for magma

chamber development were provided by graben collapse near the uplift core.

In this model, magma chambers grow at a number of centers distributed concentrically about the center of uplift, with each chamber responsible for the injection of a subswarm of dykes. If the chambers are close together then no gaps in the dyke distribution should be observed (type I, Figure 10). The classic type I example is the Mackenzie dyke swarm of northern North America (see section 4.9). If the chambers are widely spaced, however, dyke-poor gaps will be present between subswarms as in type II (Figure 10). The best example of a giant swarm exhibiting a type II pattern is the Matachewan swarm of North America (see section 4.11). Other type II examples include the Columbia River Basalt Group dykes (Figure 2) and the swarms associated with the Lower Limpopo center (Figure 8).

The presence of coeval or pre-existing rifts may cause a concentration of dykes in the vicinity of and parallel to the rifts. This model is best suited to type III swarms (Figure 10). In detail, *Fahrig* [1987] suggested that dykes are associated with continental breakup and emplaced parallel to (passive margin dykes) and perpendicular to (failed arm dykes) the rift margin (Figure 11b). Examples of passive-margin dykes include the Red Sea dykes of the Afar event (Figure 3), the east Greenland dykes of the North Atlantic Tertiary event (Figure 4), the Panvel dykes of the Deccan event (Figure 5), Santos-Rio de Janeiro and Florianópolis dykes of the Paraná event (Figure 7), and the Lebombo dykes of the Lower Limpopo (Karoo) event (Figure 8). Failed-arm types may include the Narmada-Tapti-Son dykes of the Deccan (Figure 5), the Botswana swarm of the Lower Limpopo event (Figure 8), the Grenville swarm of eastern North America (section 4.4), and the Yakutsk dykes of Siberia (section 4.3).

A consequence of the *Fahrig* [1987] model is that most passive-margin dykes are destroyed or deformed in a subsequent collision, whereas failed-arm dykes survive largely intact (Figure 11b). This may explain why many dyke swarms are truncated at one end by younger orogenic belts [*Buchan and Halls*, 1990].

If both widely spaced magma chambers and coeval rifts are present, then giant swarms may exhibit geometries intermediate between types I, II, and III. For example, the Central Atlantic reconstructed swarm (Figure 12 and section 4.1) appears to exhibit a rather continuous fan but also has a concentration of dykes along the Africa-South America and the North America-Africa rift margins.

4. APPLICATION OF GIANT DYKE SWARMS IN IDENTIFYING PALEOPLUMES AND PALEO-LIPS

In this section we summarize plume-related magmatic

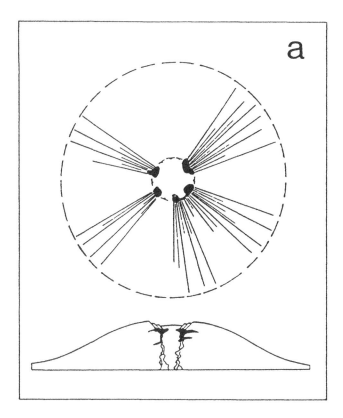

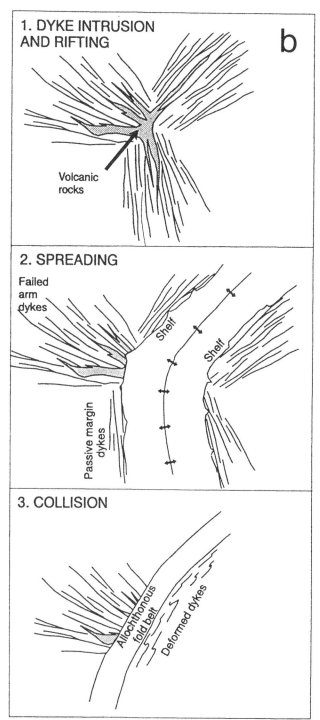

Figure 11. (a) Model of *Baragar et al.* [1996] for magmatism related to plume-induced uplift as described in the text. Reprinted with permission of Oxford University Press. (b) Model of *Fahrig* [1987] for three-stage plate tectonic cycle in the evolution of mafic continental dyke swarms. Reproduced with the permission of the Geological Association of Canada.

events that have been identified on the basis of their converging dyke patterns.

4.1. Central Atlantic Event

The Central Atlantic reconstructed swarm is an example of a dyke swarm dismembered by plate tectonic processes (Figure 12). Since being first identified by *May* [1971] in a pre-Atlantic continental reconstruction, the 270° fan has remained the classic example of a reconstructed giant radiating dyke swarm [*Oliveira et al.*, 1990; *Sebai et al.*, 1991b; *Bertrand*, 1991; *Ernst et al.*, 1995a]. Dykes of this swarm are traced up to 2800 km from the plume center. The North American portion of the swarm, known as the Eastern North America dykes, is the best studied geochemically [e.g., *McHone et al.*, 1987; *de Boer et al.* 1988; *McHone*, 1996] and the most precisely dated, at 201 ±2 Ma (2σ; U-Pb) [*Dunning and Hodych*, 1990]. Dates of 206–196 (^{40}Ar-^{39}Ar on plagioclase) have been obtained for related dykes in Morocco and Algeria, Mali, and Iberia (not shown) [*Sebai et*

al., 1991b]. The convergent point of the Central Atlantic reconstructed swarm and the interpreted location of the mantle plume is the Blake Plateau.

The plume thought most likely responsible for the Central Atlantic reconstructed swarm is the Fernando de Noronha hotspot [*Hill*, 1991]. The Cape Verde plume was suggested

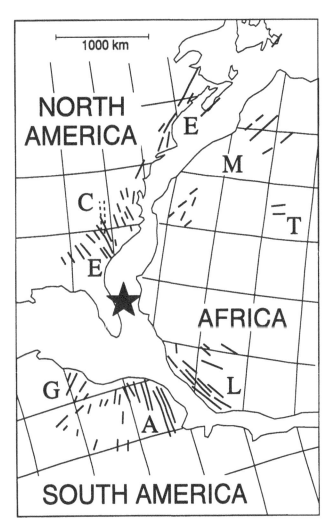

Figure 12. Dyke swarms of the 200 Ma Central Atlantic magmatic event of North America, South America, and Africa [after *May*, 1971; *McHone et al.*, 1987; *Oliveira et al.*, 1990; *Sebai et al.*, 1991b] in a reconstruction of Rowley [in *Keppie and Dallmeyer*, 1989]. In North America these consist of the Eastern North America (E) subswarm which is locally crosscut (dashed lines) by the slightly younger Charleston (C) subswarm [*Ragland et al.*, 1983; *Smith*, 1987; *de Boer et al.*, 1988]. In Africa, the dykes [e.g., *Sebai et al.*, 1991b] are divided on the basis of distribution into the Morocco[pn] (M), Liberia[pn] (L), and Taoudenni[pn] (T) subswarms. In South America numerous subswarms [*Sial et al.*, 1987; *Gibbs*, 1987; *Choudhuri et al.*, 1991] are grouped into the north-northwest trending Amapá (A) and related subswarms (Cassiporé, Jari, Cayenne, Apatoe), and in Guyana[pn] (G), the northeast-trending Cerro Bolivar, Supenaam and Tukutu River and possibly the Rio Trombetas subswarms [*Choudhuri et al.*, 1991; *Ernst et al.*, 1995b, 1996; *Sial et al.*, 1987; *Gibbs*, 1987]. The star marks the focal point of the swarms.

sedimentation in the Newark-type rift basins of eastern North America [*Hill*, 1991]. Sedimentary deposition in these basins began at ~230 Ma but was interrupted at 215–210 Ma, presumably by uplift associated with the approaching plume. Volcanism commenced at ~200 Ma. The presence of the rift basins prior to volcanism indicates that the region was under tension before the plume arrived. The plume model is supported by preliminary magnetic fabric studies from the Eastern North America dykes. These data indicate predominately subhorizontal magma flow [*Greenough and Hodych*, 1990; *Ernst et al.*, 1995c] in dykes located north of about 41°N (i.e., distal from the plume center).

Interestingly, despite the large extent of the Central Atlantic reconstructed swarm, the amount of volcanism in the focal region is minor [*Hill*, 1991]. Some associated volcanic rocks exist in the subsurface of Florida [*de Boer et al.*, 1988; *Heatherington and Mueller*, 1991] and in Morocco [*Sebei et al.*, 1991b; *Bertrand*, 1991]. Extensive volcanism also occurred along the entire east coast of North America, the so-called East Coast Margin igneous province [*Holbrook and Kelemen*, 1993]. However, these volcanic rocks have been related to the transition to oceanic rifting, rather than to an initial plume event [e.g. *Austin et al.* 1990; *Holbrook and Kelemen*, 1993; *Holbrook et al.* 1994].

4.2. Jutland Event

We define the Jutland event (Figure 13) on the basis of the convergence of three swarms: the *Whin-Midland Valley*, Oslo Rift and Scania swarms, all of age ~300 Ma. The most impressive of these is the >400-km-long, east-trending *Whin-Midland Valley* swarm in the British Isles and the adjacent North Sea [*Macdonald et al.*, 1981; *Smythe et al.*, 1995]. The large Whin Valley and Midland Valley sills [e.g., *Dunham and Strasser-King*, 1982] are associated with this swarm. In the Olso rift of Norway the oldest igneous activity consists of a swarm of north-northwest- to north-northeast-trending dykes with an age of 297±9 Ma (2F; Rb-Sr mineral isochron) [*Sundvoll and Larsen*, 1993]. In addition, northeast-trending Kongsberg dykes with an age >275 Ma may be coeval with the 300 Ma dykes [*Sundvoll and Larsen*, 1993]. The east-southeast- to southeast-trending Scania dykes of the Tornquist Line of southern Sweden [*Gorbatschev et al.*, 1987; *Sundvoll and Larsen*, 1993] have essentially the same age (294±4 Ma; K-Ar) [*Klingspor*, 1976]. According to *Gorbatschev et al.* [1987, p. 370] this swarm (Scania) continues into Poland, beneath younger sedimentary cover.

The coeval nature of the three magmatic events was noted by *Smythe et al.* [1995]. They linked the Whin-Midland Valley and Oslo Rift swarms into a single arcuate swarm explained by a "concentration of regional horizontal tensile

by *White and McKenzie* [1989] but was located far (~2000 km) from the focal point of the swarm at 200 Ma [*Morgan*, 1983]. Arrival of the plume is marked by an interruption in

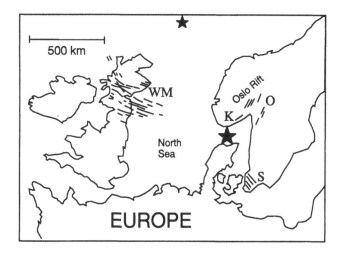

Figure 13. Dykes of the ~300 Ma Jutland event of Europe. Distribution of Whin-Midland Valley (WM) dykes after *Smythe et al.* [1995]. Oslo Rift (O), and Scania (S) dykes are after *Gorbatschev et al.* [1987] and the Kongsberg (K) dykes are after *Sundvoll and Larsen* [1993] and references therein. Star marks convergence of Whin-Midland Valley, Oslo Rift and Scania swarms. Small star locates center of arcuate swarm defined by Whin-Midland Valley and Oslo Rift dykes [*Smythe et al.*, 1995].

stresses through the Faeroe-Shetland area" to the north. They noted that the Scania dykes did not fit this model. We note that the arcuate dykes may define a giant circumferential swarm, and the Scania dykes, a radiating swarm, both associated with a plume centered to the north (see model in R. E. Ernst and K. L. Buchan, in preparation). We prefer an alternative plume model, in which the three swarms—Whin-Midland Valley, Oslo Rift, and Scania— define a giant radiating swarm focussed on a mantle plume. This model is consistent with the triple junction model (rift-rift-rift type) of *Burke and Dewey* [1973], in which convergence of the Midland Valley rift, the Oslo graben, and the Danish Trough (on the south side of the Tornquist Line) define the Jutland triple junction.

4.3. Yakutsk Event

The Devonian-age Yakutsk swarm (Figure 14) is located in the Siberian craton and fans over an angle of about 150°. Subswarms are associated with rift arms of a triple junction [*Shpount and Oleinikov*, 1987; *Gusev and Shpount* 1987]. The Vilyuy-Markha and Chara-Sinsk subswarms are located on either side of the Vilyuy rift and do not converge towards the proposed plume centre because of subsequent Devonian 'v'-shaped opening of the rift [figures 21.35 and 21.40 of *Sengör and Natal'in*, 1996]. The Tomporuk[pn] subswarm and associated Dzhalkan volcanic rocks [*Levashov*, 1979] paralleled rifting which probably resulted in separation of a

narrow strip of Paleozoic passive-margin deposits [pp. 554-556 of *Sengör and Natal'in*, 1996].

The *Yakutsk* swarm is poorly dated with most K-Ar whole rock ages falling in the range between 375 and 320 Ma. Nevertheless, the overall radiating pattern of the swarm and the association with triple junction rifting and ocean opening suggest a plume origin.

4.4. Central Iapetus Event

Several late Proterozoic dyke swarms of eastern North America may locate a mantle plume associated with the breakup of a late Proterozoic supercontinent and the formation of the Iapetus Ocean (Figure 15). Ocean opening occurred along nearly the same boundary as the subsequent Atlantic Ocean.

Three subswarms are found on the Canadian Shield. The east-trending Grenville dykes extend over a length of about 700 km and a width of 100 km and occur along the Ottawa Graben, a failed Iapetus rift zone [*Kumarapeli et al.*, 1990;

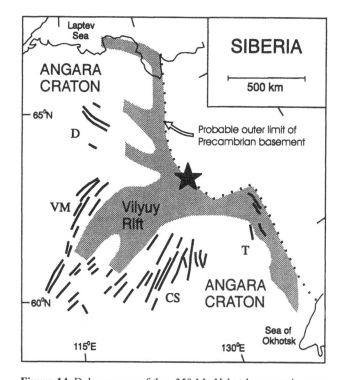

Figure 14. Dyke swarms of the ~350 Ma Yakutsk magmatic event of Siberia [after *Shpount and Oleinikov*, 1987]. CS and VM locate the Chara-Sinsk and Vilyuy-Markha subswarms [*Tomshin and Koroleva*, 1990; *Erinchek et al.*, 1995] and D and T are the Dzhardzhan and Tomporuk subswarms [*Shpount and Oleinikov*, 1987; *Ernst et al.*, 1996]. Shaded pattern locates associated paleorift [after *Shpount and Oleinikov*, 1987]. The outer limit of Precambrian basement is from *Rosen et al.* [1994]. Star locates focal point of swarm assuming some closure of the Vilyuy rift.

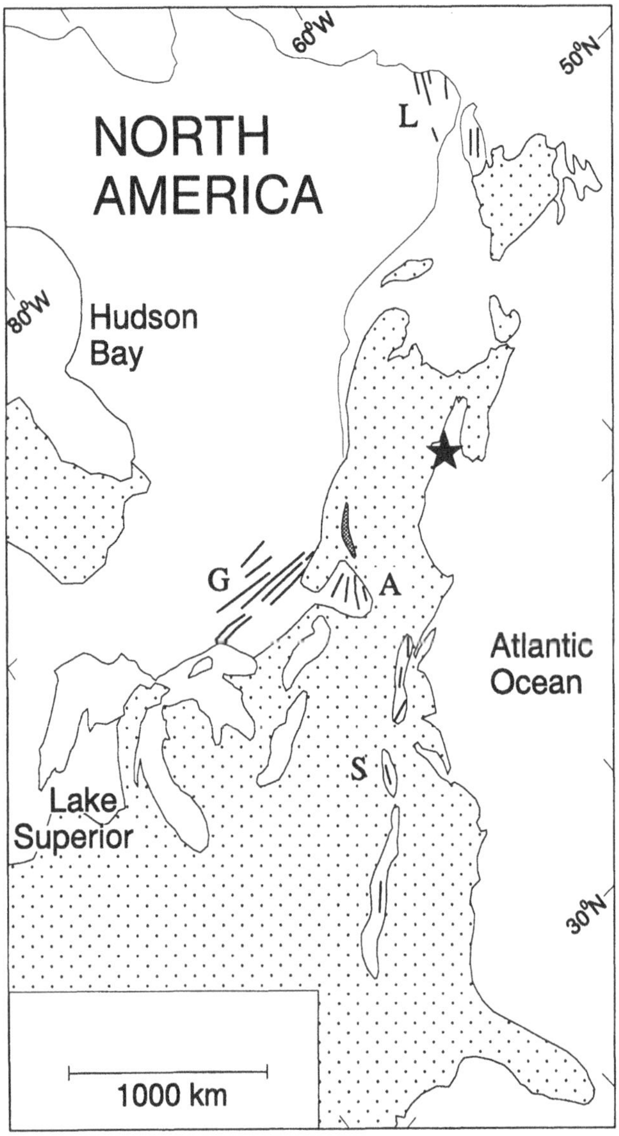

Figure 15. Dyke swarms of the ~600 Ma Central Iapetus magmatic event of eastern North America, associated with the opening of the Iapetus Ocean. Long Range dykes (L) are after *Fahrig and West* [1986], *Kamo et al.* [1989] and *Kamo and Gower* [1994]; Grenville dykes (G) are after *St. Seymour and Kumarapeli* [1995]; Adirondack dykes (A) are after *St. Seymour and Kumarapeli* [1995] and *Coish and Sinton* [1992]. Southern Appalachian[pn] dykes (S) are after *Goldberg and Butler* [1990] and are restricted to exposed basement massifs. Star locates the approximate convergent point of the swarms. Shaded area locates Tibbit Hill volcanic rocks (including outcrop areas and subsurface extensions interpreted from gravity modelling). Dotted pattern locates Phanerozoic cover rocks.

St. Seymour and Kumarapeli, 1995]. They have been dated at 590 +2/-1 Ma (2σ; U-Pb on baddeleyite) [*Kamo et al.*, 1995]. Grenville dykes converge slightly towards the east in the direction of the Sutton Mountains triple junction, which

is centered on the outcrops of Tibbit Hill metavolcanic rocks with an age of 554 ±4/-2 Ma (2σ; U-Pb on zircon) [*Kumarapeli et al.*, 1989]. The Adirondack dykes have an arcuate trend which also converges toward the Sutton Mountains triple junction [*St. Seymour and Kumarapeli*, 1995]. These dykes have K-Ar ages of 588-542 Ma [*Isachsen et al.*, 1988]. The northeast-trending Long Range dykes occur in Newfoundland and southeastern Labrador. Based on U-Pb baddeleyite and zircon geochronology, their age is 615±2 Ma (2σ) [*Kamo et al.*, 1989; *Kamo and Gower*, 1994].

South of the Canadian Shield, coast-parallel dykes are found over a length of nearly 1500 km in Precambrian basement massifs in the central and southern Appalachian Mountains. They are considered to have been part of the eastern margin of Laurentia during the late-Precambrian rifting which formed the Iapetus Ocean [*Goldberg and Butler*, 1990]. These Southern Appalachian[pn] dykes have been variably metamorphosed to low and medium grades by the Appalachian orogeny but the common northeast trend of the dykes is considered to be primary. The only isotopic date currently available for the Southern Appalachian[pn] dykes is a 734 ±26 Ma (1σ) Rb-Sr age at the south end of the swarm in the Bakersville subswarm [*Goldberg et al.*, 1986; *Goldberg and Butler*, 1990]. There is evidence for two pulses of continental rifting in the central and southern Appalachians, at 550-600 Ma and at 700–760 Ma [*Aleinikoff et al.*, 1995; *Tollo and Hutson*, 1996]. The Bakersville age suggests that at least some of the Southern Appalachian[pn] dykes belong to the older event.

The Long Range, Grenville, Adirondack, and perhaps some of the Southern Appalachian[pn] swarms belong to the younger rift event. The radiating dyke pattern may indicate that this rifting was caused by a mantle plume. The focal point of the dykes is located farther east than the Sutton Mountains triple junction even after correction of the latter for 100 km of Appalachian thrusting [*St. Seymour and Kumarapeli*, 1995]. This location is similar to that of the Montreal triple junction of *Burke and Dewey* [1973].

The age range for the Central Iapetus event is quite broad, at least 25 m.y. (615–590 Ma). If the 554 Ma Tibbit Hill volcanic rocks are also considered, then the age range is extended to about 60 m.y. However, it is possible that the Tibbit Hill rocks are not related to the arrival of a plume but instead represents rift products associated with the incipient opening of the Iapetus Ocean [*St. Seymour and Kumarapeli*, 1995].

The Baltoscandian dykes, distributed along the western margin of Baltica, are also thought to be associated with the opening of the Iapetus Ocean [*Andréasson*, 1994]. The most precisely dated of these is the Sarek subswarm in northernmost Sweden (68°N) with an age of ~606 Ma (U-Pb zircon and ^{40}Ar-^{39}Ar mica on baked wallrock) [*Svenningsen*, 1995].

During the Caledonide collision, the passive margin containing the Baltoscandian dykes shortened by hundreds of kilometers [*Andréasson*, 1994]. Therefore, their original distribution and orientation, as well as their relationship with the Central Iapetus dykes of Laurentia, remain to be determined.

4.5. *Franklin-Natkusiak Event*

Franklin-Natkusiak magmatism consists of the Natkusiak volcanic rocks, Coronation sills, Franklin dykes, Thule dykes, and other coeval sills and dykes (Figure 16). An age of 723+3/-2 Ma (2σ; U-Pb baddeleyite) [*Heaman et al.*, 1992] is based on six sills and a single dyke. Other dykes have been correlated by K-Ar dating and (or) paleomagnetism [e.g., *Fahrig et al.*, 1965; *Jones and Fahrig*, 1978; *Dawes*, 1991]. Franklin dykes extended for more than 1500 km across the Arctic islands and the adjacent mainland of North America and northwestern Greenland (Figure 16) [*Robertson and Baragar*, 1972; *Fahrig*, 1987; *Nielsen*, 1987; *Jefferson et al.*, 1994; *Dawes*, 1991; *Ernst et al.*, 1995b]. *Heaman et al.* [1992] and *Rainbird* [1993] proposed that a mantle plume located north of Victoria Island was responsible for this magmatism and they provided stratigraphic evidence for regional uplift above the plume center. A more precise location for this plume center is provided by the converging dyke pattern.

In general, the dykes associated with this event have a fanning distribution (Figure 16). The densest subswarms trend southeasterly across Baffin Island and the adjacent mainland of North America, and also across northwest Greenland after restoration to its pre-drift position [*Rowley and Lottes*, 1988]. Other, less dense subswarms include the southeast-trending dykes that are associated with Natkusiak volcanic rocks on Victoria Island, and a few north-trending dykes on the mainland to the south.

Other small swarms, thought to be of similar age, do not fit the simple radiating pattern; in particular, the south-trending dykes on northern Baffin Island and southwesterly-trending dykes on Somerset Island [*Fahrig and West*, 1986]. The explanation of these discordant trends is not known.

Sill complexes of Franklin age are abundant on Victoria Island and the mainland of North America (Figure 16).

4.6. *Willouran Event*

The Willouran magmatic event is recognized in southern Australia, where the ~800 Ma (Sm-Nd mineral isochron) [*Zhao and McCulloch*, 1993] Gairdner dyke swarm has been linked to a mantle plume beneath the coeval Willouran volcanic province [*Zhao et al.*, 1994] (Figure 17).

Recently, *Park et al.* [1995] have proposed that 780 Ma

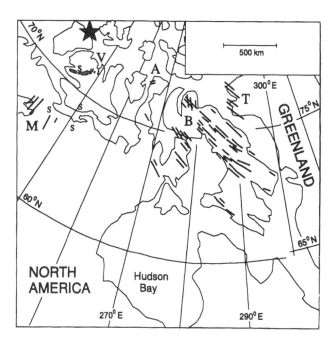

Figure 16. Dyke swarms of the ~723 Ma Franklin-Natkusiak magmatic event of North America and Greenland. Greenland is in the reconstructed position of *Rowley and Lottes* [1988]. Other Arctic islands remain in their present position relative to North America. Franklin dykes are found on Baffin Island and the adjacent mainland (B) and on Somerset Island (A; Aston dykes) [*Fahrig and West*, 1986], on Victoria Island (V) [*Rainbird et al.*, 1994a, b] and on the North American mainland (M) [*Fahrig and West*, 1986; *Baragar and Donaldson*, 1973]. Thule dykes of northwest Greenland are after *Dawes* [1991]. Sill complexes (s) of North America are from *Fahrig* [1987]. Shading locates Natkusiak volcanic rocks. Star locates plume center based on convergence of dykes.

mafic dykes and sheets in three widely separated areas of western North America (the northwestern Canadian Shield, the Mackenzie Mountains in the northern Cordillera, and the Wyoming Province) represent subswarms of the Willouran event. Ages are based on U-Pb baddeleyite dating from *LeCheminant and Heaman* [1994] and Harlan and Premo in *Park et al.* [1995].

Paleomagnetic evidence [*Park et al.*, 1995] indicates that the North American subswarms have not moved relative to one another since emplacement and that their roughly radial pattern is primary. Together with the Gairdner dykes of Australia, *Park et al.* [1995] proposed that they form a giant radiating dyke swarm in a reconstruction of Australia and Laurentia [*Borg and DePaolo*, 1994]. The Willouran volcanic province is located near the southeastern end of the Gairdner swarm [*Zhao et al.*, 1994], near the focal area of the giant radiating swarm after reconstruction. The various subswarms extend between 1200 and 2000 km from the focal point. *Park et al.* [1995] suggested that the Willouran

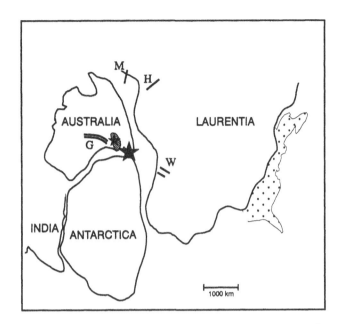

Figure 17. Dyke swarms of the ~780 Ma Willouran magmatic event in Australia and western North America [*Park et al.*, 1995 and references therein]. Continental reconstruction is that of *Borg and DePaolo* [1994]. The 780 Ma dykes of the Mackenzie Mountains (M) and the Wyoming Province (W) and inclined Hottah (H) sheets in the Canadian Shield of western North America are compared with ~800 Ma Gairdner (G) dykes of southern Australia [*Park et al.*, 1995]. The small star marks the mantle plume center proposed by *Zhao et al.* [1994] on the basis of the distribution of Willouran volcanic rocks (shaded). The large star marks the location of the proposed plume center defined on the basis of the converging dyke pattern [*Park et al.*, 1995]. Present outline of eastern North America (dotted pattern) is shown to indicate the orientation of Laurentia.

event may have been a precursor to Late Proterozoic breakup of a supercontinent which incorporated both Australia and Laurentia.

4.7. *Kola-Onega Event*

The Kola-Onega dyke swarm of Finland and Russia [*Berkovsky and Platunova*, 1987; *Gorbatschev et al.*, 1987] with a length of 500 km and a width of 600 km, radiates from the north over an angle of about 50° (Figure 18). *Mertanen et al.* [1996] have published Sm-Nd isochron ages of 1042±50, 1013±32, and 1066±34 Ma (2σ) for the Laanila, Ristijärvi, and Kautokeino dykes, respectively, all of which belong to the western portion of the swarm.

Convergence of the Kola-Onega swarm suggests a plume center to the north. Coeval diabase sills along the northern coast of the Kola Peninsula [*Sinitsyn*, 1963, cited by *Berkovsky and Platunova*, 1987 and by *Gorbatschev et al.*, 1987] may be part of a remnant LIP.

4.8. *Abitibi and Keweenawan Events*

Mafic magmatic events associated with the Mid-Continent Rift of North America occur over a span of at least 65 m.y. (Figure 19), consisting of an older 1140 Ma Abitibi event and a younger 1110–1085 Ma Keweenawan event.

The weakly fanning Abitibi dyke swarm [*Ernst and Buchan*, 1993] was emplaced at 1141±1 Ma (2σ; U-Pb baddeleyite) [*Krogh et al.*, 1987]. It consists mainly of a few very wide (up to 200 m) dykes. The largest of these is termed the Great Abitibi dyke and can be traced for nearly 700 km. On the basis of magnetic fabric studies and longitudinal compositional variations, that dyke was probably emplaced laterally from a source area to the southwest [*Ernst*, 1990; *Ernst and Bell*, 1992]. The Abitibi swarm focuses south of Lake Superior in the vicinity of the Goodman Swell, an area of reset Rb-Sr biotite ages which are thought to reflect local uplift at 1128±20 Ma [*Peterman and Sims*, 1988]. *Hutchinson et al.* [1990] proposed that the swell represents uplift associated with a mantle plume responsible for the younger Keweenawan activity, whereas

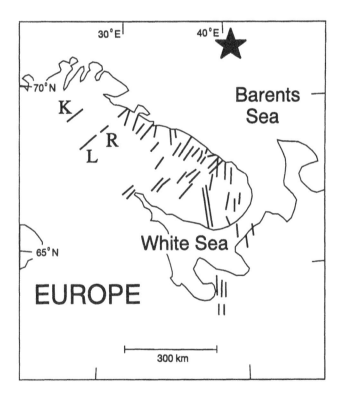

Figure 18. Dykes of the ~1000 Ma Kola-Onega magmatic event of northern Europe [*Berkovsky and Platunova*, 1987; *Gorbatschev et al.*, 1987]. L, R, and K label the Laanila, Ristijärvi, and Kautokeino dykes, respectively, for which Sm-Nd dates have been reported [*Mertanen et al.*, 1996]. Star locates the convergence point of the dykes.

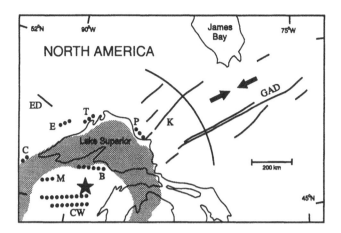

Figure 19. Dyke swarms of the 1140-1085 Ma Abitibi and Keweenawan magmatic event(s) of central North America. GAD and K label the Great Abitibi and Kipling dykes, respectively, of the 1140 Ma Abitibi dyke swarm (distribution after *Ernst and Buchan* [1993]). The minor Eye Dashwa (ED) swarm is also thought to be about 1140 Ma [*Osmani*, 1991]. Star locates center of Goodman swell [*Peterman and Sims*, 1988]. Keweenawan dykes dated as or assumed to be ~1100 Ma are marked by dotted lines [*Green et al.*, 1987] and include Pukaskwa (P), Thunder Bay (or Pigeon River) (T), Ely-Moose (E), Carlton County (C), Baraga (Marquette) (B), Mellen-Gogebic (M) and Central Wisconsin (CW) subswarms. The arc marks the maximum extent of the radial pattern in the Abitibi swarm and the arrows mark the inferred orientation of the regional stress maximum. The Mid-Continent (Keweenawan) rift is shaded.

Ernst et al. [1995a] suggested that the swell represents plume-generated uplift which could also be synchronous with emplacement of the Abitibi dyke swarm. At a distance of 600 km from the center of the swell, the Kipling dyke of the Abitibi swarm swings parallel to the Great Abitibi and other nearby Abitibi dykes. This probably marks the distance at which the influence of the regional stress field exceeded the plume-generated uplift stress (Figure 19).

To the west, the minor Eye-Dashwa swarm [*Osmani*, 1991] has K-Ar (whole rock) ages of 1132±27 and 1143±27 Ma (2σ) (samples GSC87-55 and GSC87-56 [*Hunt and Roddick*, 1987]) and a virtual geomagnetic pole (our preliminary unpublished data) similar to that of the Abitibi swarm. Hence, the Eye-Dashwa dykes may form a subswarm of the Abitibi swarm.

A second period of activity occurred ~30–65 m.y. later. Voluminous Keweenawan volcanism and sill emplacement were accompanied by a number of dyke swarms, generally aligned parallel to the rift arms [*Green et al.*, 1987]. They include the northwest-trending Pukaskwa dykes east of Lake Superior, the northeast-trending Thunder Bay (or Pigeon River), Ely-Moose and Carlton County dykes northwest and west of the lake, and the east-trending Baraga (Marquette)

dykes south of the lake. The Mellen-Gogebic and Central Wisconsin swarms may also be Keweenawan in age [*Green et al.*, 1987; *King*, 1990].

4.9. *Mackenzie Event*

One of the largest known dyke swarms on Earth is the Mackenzie swarm of northwestern North America (Figure 20). It covers an area of 2.7×10^6 km^2, extends more than 2600 km from the inferred plume center, and fans dramatically over an angle of about 100° [*Fahrig and Jones*, 1969; *Fahrig*, 1987]. Coeval flood basalts, the Coppermine River lavas, as well as the layered mafic/ultramafic Muskox intrusion, are present near the focus of the swarm, although the focus itself is hidden beneath younger cover rocks. High-precision U-Pb baddeleyite dating has demonstrated that the

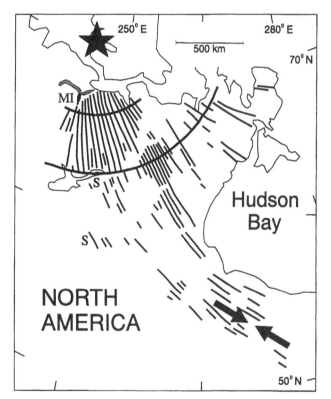

Figure 20. Dykes of the 1270–1265 Ma Mackenzie magmatic event of northern North America. Coeval Coppermine River volcanic rocks are shaded. Star marks focal point of swarm. S marks location of coeval sills [after *Fahrig and West*, 1986; *Hulbert et al.*, 1993]. There are also other related sills in the eastern Canadian Arctic islands (covered by the scale bar) [*Ernst et al.*, 1995a]. MI is Muskox Intrusion. The inner arc marks the transition from vertical flow to horizontal flow in the swarm based on magnetic fabric studies [*Ernst and Baragar*, 1992]. The outer arc marks the extent of the purely radial pattern. Arrow indicates the inferred orientation of the regional stress maximum.

Mackenzie dyke swarm, the Muskox intrusion, and several sills far from the focus region were emplaced within a period of less than 5 million years beginning at 1272 Ma, with all the dated dykes being injected at 1267±2 Ma (2σ) [*LeCheminant and Heaman*, 1989, 1991]. The Coppermine River lavas have not been dated precisely but are correlated with the Mackenzie dykes based on similar compositions, paleomagnetism, K-Ar ages, and a decrease in dyke abundance upwards in the volcanic pile, consistent with at least some of the dykes being feeders [*Gibson et al.*, 1987; *Baragar et al.*, 1996].

According to *LeCheminant and Heaman* [1989] the Mackenzie swarm was initiated by a mantle plume impinging on the lithosphere. Their conclusion is based on the radiating pattern of dykes and the stratigraphic evidence for uplift in the focal region preceding magma injection, and it is supported by magnetic fabric data indicating vertical flow in the focal region and lateral flow beyond it [*Ernst and Baragar*, 1992]. Breakup along this northern boundary of Laurentia at the time of dyke emplacement [*LeCheminant and Heaman*, 1989] is thought to have opened the Poseidon Ocean [*Fahrig*, 1987].

4.10. *Fort Frances Event*

The northwest-trending Fort Frances dyke swarm to the west of Lake Superior has a length of 300 km and a width of 400 km (Figure 21). The distribution of the Fort Frances dykes, based in part on aeromagnetic interpretation [*Chandler*, 1991], suggests a fan of about 35° [*Buchan and Halls*, 1990]. U-Pb baddeleyite ages of 2077 +4/-3 and 2076 +5/-4 Ma (2σ) were reported by *Wirth et al.* [1995] and *Buchan et al.* [1996], respectively.

Southwick and Day [1983] proposed that the Fort Frances swarm was related to a hotspot beneath the Paleoproterozoic Animikie Basin to the southeast of the exposed swarm. They also suggested that the swarm was injected to the northwest along late Archean fractures which were reactivated as the failed arm of a triple junction. The Fort Frances swarm converges to the southeast to a postulated mantle plume center, consistent with the hotspot model of *Southwick and Day* [1983]. A rift-related sequence of rocks (Mille Lacs Group) in the lower Animikie basin includes volcanic units that may be coeval with the Fort Frances swarm [*Southwick and Day*, 1983].

4.11. *Matachewan Event*

The Matachewan dyke swarm occurs over an area of 2.5×10^5 km^2 in the southern and central Superior Province

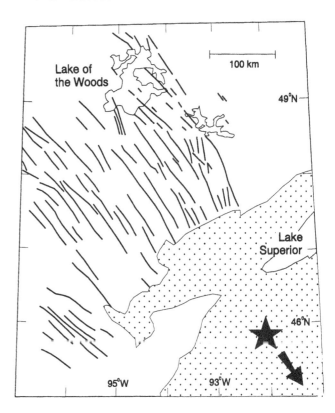

Figure 21. Dykes of the ~2076 Ma Fort Frances (or Kenora-Kabetogama) magmatic event of central North America (modified from figure 2 of *Chandler* [1991]). Younger cover rocks of the Animikie Basin are shown in dotted pattern. Star and arrow indicate direction to convergence point of swarm.

[e.g., *Halls*, 1991] (Figure 22). It is composed of three subswarms separated by areas of few dykes. After correction for later deformation [*West and Ernst*, 1991; *Bates and Halls*, 1991; *Bird et al.* 1996] each subswarm radiates from a focal region centered approximately on eastern Lake Huron.

The age of the Matachewan swarm has been defined by U-Pb baddeleyite dates of 2446±3 and 2473 +16/-9 Ma (2σ) [*Heaman*, 1995] indicating that it was emplaced over a minimum of 30 m.y. However, more data are needed to establish whether the magmatism was episodic or continuous over this interval.

A number of coeval volcanic and intrusive units are found at the southeastern end of the exposed swarm [*Krogh et al.*, 1984; *Prevec et al.*, 1995], not far from the swarm focus and presumed mantle plume center [*Halls and Bates*, 1990]. Direct evidence for magma flow direction is not seen in the Matachewan swarm, but a preferential direction of dyke-splitting (bifurcation) implies northward injection of magma away from the focal region and plume [*Halls*, 1982; *Bisson*, 1985].

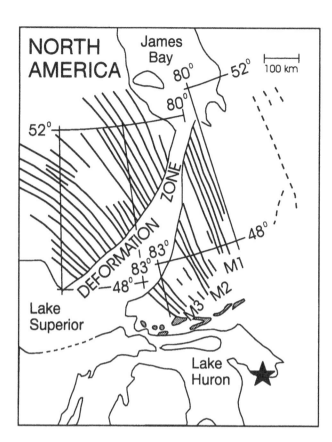

Figure 22. Dyke swarms of the 2470–2450 Ma Matachewan magmatic event of central North America [*Fahrig and West*, 1986]. after correction for a subsequent deformation event [*West and Ernst*, 1991]. Main subswarms are labelled M1, M2 and M3. Dotted lines are N-S dykes tracked on aeromagnetic maps and identified as possibly Matachewan in the region of ~48°N, 78°E based on paleomagnetic results [*Buchan et al.*, 1993]. Shaded areas in focal region represent coeval volcanic and intrusive units [*Bennett et al.*, 1991; *Prevec et al.*, 1995; *S. Prevec*, written communication, 1995].

4.12. *Mistassini Event*

The Mistassini dyke swarm of eastern North America fans over an angle of 30° (Figure 23). The focal point of the swarm and possible plume center is located southeast of the swarm within the Grenville Province. No coeval volcanics have been identified in the focal region. The swarm consists of both tholeiitic and komatiitic suites [*Fahrig et al.*, 1986]. The age of the swarm is about 2470 Ma based on U-Pb baddeleyite/zircon dating [*Heaman*, 1994].

4.13. *Additional Cases*

There are several additional examples in which pre-

Mesozoic giant dyke swarms may be associated with paleoplumes. However, in each case, dyke distribution, age, or tectonic setting is still speculative.

The Gannakouriep swarm of western South Africa appears to fan from a focus to the south (Figure 24a). It can be interpreted as (1) a radiating swarm with a fan angle of 25° which in the north swings into a regional paleostress field oriented north-northeast [*Ernst et al.*, 1995b] or (2) a linear north-northeast-trending swarm which has been reoriented into a northward trend within the deformed Gariep belt [*Ransome*, 1992]. Gannakouriep dykes are dated at 717±11 Ma (Rb-Sr mineral isochron) [*Reid et al.*, 1991b]. They may be related to Pan-African rifting that developed just prior to the onset of Gariep sedimentation [*Gresse and Scheepers*, 1993].

The north-trending Salvador, northwest-trending Itacaré, and west-trending Ilhéus-Olivença-Camacã dykes appear to define a radiating swarm which covers an angle of about 80°

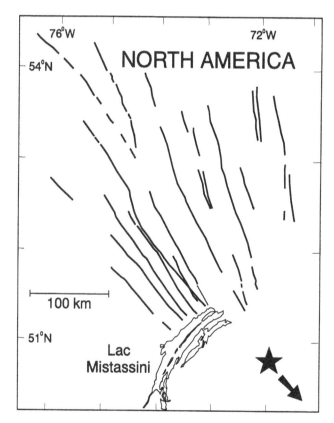

Figure 23. Dykes of the Mistassini magmatic event of northeastern North America (based on *Fahrig and West* [1986] and a new tracing from aeromagnetic shadowgram maps produced from the National Aeromagnetic Data Base by the Geological Survey of Canada Geophysical Data Center). Star and arrow indicate direction to convergence point of swarm.

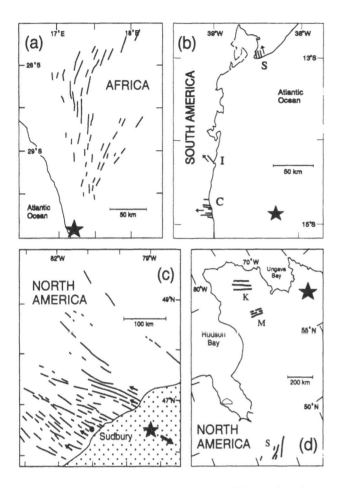

Figure 24. Additional cases. (a) Dykes of the ~720 Ma Gannakou-riep event [*Reid et al.* 1991b; *Ransome*, 1992; *Gresse and Scheepers*, 1993]. Star locates convergent point of swarm. (b) Dyke swarms of the ~1000 Ma Coastline (Bahia) magmatic event of Brazil [*Gomes et al.*, 1989; *Correa-Gomes et al.*, 1991; *Correa-Gomes and Tanner de Oliveira*, 1994a,b; *Correa-Gomes*, 1995]. Component subswarms are labelled as follows: S is Salvador, I is Itacaré and C is Ilhéus-Olivença-Camacã. Arrows indicate flow directions determined from textural observations by *Correa-Gomes and Tanner de Oliveira* [1994a,b] and *Correa-Gomes* [1995]. Star

(Figure 24b). Flow lineations and other rheological markers in the dykes of all three areas indicate that magma flow had an inclination of 35–25° upwards from a source to the east or southeast. According to *Correa-Gomes et al.* [1991] and *Correa-Gomes* [1995], these dykes are associated with a failed attempt at continental breakup between South America and Africa. Isotopic dating yields an age range of 150 m.y. This may indicate a complicated scenario involving more than one plume event. Two Salvador dykes give different ages: the first, an age of >924 Ma (^{207}Pb-^{206}Pb, 2.6% discordant) [*Heaman*, 1991], and the second, ages of

1021±8 Ma and 1003±33 (^{40}Ar-^{39}Ar host rock biotite and dyke plagioclase) [*D'Agrella-Filho et al.*, 1989]. The Ilhéus and Olivença subswarms have paleomagnetic directions that differ by about 20°, perhaps correlating with the reported age difference of 65±45 m.y. (Ilhéus dykes, 1011±24 and 1012±24 Ma (1σ), and Olivença dykes 1077±24 and 1078±18 Ma (1σ); ^{40}Ar-^{39}Ar host rock biotite and dyke plagioclase) [*D'Agrella-Filho et al.*, 1990; *Renne et al.*, 1990].

The 1238–1235 Ma (U-Pb baddeleyite) [*Krogh et al.*, 1987; *Dudàs et al.*, 1994] Sudbury dykes define a broad, roughly linear swarm trending northwest for a distance greater than 200 km in the Superior and Southern provinces of the Canadian Shield (Figure 24c). The general extent of the swarm has been defined primarily by paleomagnetic studies [e.g., *Palmer et al.*, 1977]. In addition, some Sudbury dykes can be traced southeast into the Grenville Province as increasingly deformed bodies for distances of at least 50 km and perhaps as much as 100 km [*Bethune and Davidson*, 1988; *Dudàs et al.*, 1994]. Magnetic fabric studies [*Ernst*, 1994] reveal a consistent horizontal flow fabric and a consistent sense of imbrication suggesting magma flow from the southeast. Therefore, the magma source (plume?) was likely located to the southeast, although in the absence of swarm convergence a more precise location cannot be determined.

It is suggested (K.L. Buchan et al., Paleomagnetism and Geochronology of Dykes of the Ungava, manuscript in preparation) that three widely separated Paleoproterozoic dyke swarms in northeastern North America represent remnants of a major magmatic event (Figure 24d). The Klotz, Maguire and Senneterre dykes have U-Pb ages in the range of 2230–2210 Ma ([*Buchan et al.*, 1993]; K. L. Buchan et al., ms. in preparation), and converge over an angle of ~80° towards a focal point near Ungava Bay.

Other examples which may represent radiating dyke swarms are found in Tanzania, northern Canada, and northern Australia (Figure 1). In Tanzania, three dyke swarms of undefined age radiate from the southern end of Lake Victoria over an angle of about 90°. They have been related to the nearby Bukoban volcanic rocks [*Halls et al.*, 1987] which have a poorly defined age of 1200-800 Ma. In the Slave Province of the Canadian Shield, the 2030-2023 Ma (U-Pb baddeleyite) [*LeCheminant and van Breemen*, 1994] Lac de Gras swarm fans approximately 10° from a focal point at the coeval Booth River intrusive suite in the Kilohigok basin [*LeCheminant*, 1994]. In northern Australia, the McArthur Basin dyke swarm [*Tucker and Boyd*, 1987] of unknown age fans to the south over an angle of about 50°.

5. IMPLICATIONS OF GIANT RADIATING DYKE SWARMS FOR LIPS AND PLUMES

The previous section has outlined evidence suggesting that pre-Mesozoic giant radiating dyke swarms are remnants of LIPs which, as a result of erosion, have lost most or all of their volcanic component. In this section we discuss (1) the relationship between LIP distributions and uplift topography as determined from radiating swarms, (2) a method to map the boundary of melt generation in the plume, (3) evidence from dyke swarms suggesting that some LIP events may be tens of millions of years in duration rather than the <5 million years considered typical of post-Paleozoic LIPs, (4) lateral feeding of sills distal from the plume, and (5) the possibility that some continental swarms are fed from oceanic LIPs.

5.1. Correlation of Dyke Distribution, Uplift Topography, and LIPs

Two dyke swarm parameters can be related to plume-generated uplift: the radial-linear transition in the swarm pattern and the convergence point of the dykes.

Firstly, the location of the transition between the radial and the more-distal, subparallel portions of swarms is controlled by the balance between the stresses due to plume-generated uplift and regional forces [McKenzie et al., 1992; Ernst et al., 1995b]. Therefore, this transition marks the approximate outer boundary of the topographic uplift at the time of dyke intrusion. Modeling by Griffiths and Campbell [1991] suggests that uplift topography varies during plume ascent, but that at the time of probable peak magma emplacement (about 5 m.y. after maximum uplift), the outer boundary of uplift is similar to that of the plume head. In actual cases, the uplift will be subject to additional factors such as heterogeneity in the structure and lithology of the lithosphere and may be quite complicated. However, we will take the simple model as a starting point. In the cases of the Mackenzie and Abitibi swarms (Figures 19 and 20), the transition occurs at about 1000 km from the plume center, consistent with plume modelling of Griffiths and Campbell [1991].

Secondly, the convergence point of swarms can be used to locate the position of maximum uplift. We have already discussed in a general way how the convergent dyke pattern locates a plume center. In more detail, the location of the focus can be calculated rigorously from the convergence of great circles which represent dykes within the radial portion of the swarm [Ernst et al., 1995b]. In the case of the Mackenzie swarm, which covers a large portion of the

Canadian Shield, the calculated convergence point based on 10 representative dykes distributed evenly across the swarm is 70°N, 244°E with a 95% confidence uncertainty of ±2° [Ernst et al., 1995b]. Other proposed plume centers in this paper have not been calculated rigorously.

In general, most of the volcanic rocks associated with a mantle plume are found near the center of topographic uplift. Examples include the Afar (Figure 3), North Atlantic Volcanic Province (Figure 4), and Deccan (Figure 5) events. However, in some cases, the maximum accumulation of volcanic rocks is offset hundreds of kilometers from the center of the topographic uplift. For example, the Columbia River basalt group, which resulted from the arrival of the Yellowstone plume, has a maximum accumulation of volcanic rocks nearly 500 kilometers from the plume center on the edge of the uplift (Figure 2). Because this event is only 17 m.y. old, the present uplift probably reflects the paleo-uplift.

An older example of volcanism offset from the center of topographic uplift involves the Abitibi and Keweenawan events (Figure 19) at 1140 and 1110 to 1085 Ma. The inferred center of uplift associated with these events is located to the south of Lake Superior, in the vicinity of the Goodman Swell [Peterman and Sims, 1988], a region of ~1130 Ma Rb-Sr biotite ages reset during uplift. Assuming the swell as the plume center, we note that Keweenawan rifting and major volcanism is concentrated in an arc located ~200 km away from the center of the Goodman Swell.

Two other examples of LIPs whose exposed volcanic rocks appear to be offset from the plume center include the Natkusiak volcanics of the 723 Ma Franklin event and the Coppermine River volcanic rocks of the 1270 Ma Mackenzie event. In both cases the volcanic rocks are located approximately 300 km south of the plume center (Figures 16 and 20). However, for these older events, including the Abitibi and Keweenawan events (discussed above), it is difficult to be sure that erosion has not preferentially removed the lavas on the topographic uplift.

There are a number of possible explanations for offset of volcanism from the plume center. Firstly, the observed distribution of volcanic rocks is strongly influenced by the development of rifts, and such accumulations will also be resistant to erosion. Secondly, the presence of local previously thinned lithospere, a "thinspot" in the terminology of Thompson and Gibson [1991] may facilitate accumulation of volcanic rocks in these areas. Thirdly, in the model of Baragar et al. [1996] magma chambers are located in a ring associated with central graben collapse (Figure 11a) and are displaced from the plume center by a distance of several hundred kilometers. Volcanism from such chambers would also be offset from the plume center, providing a possible explanation for the observed offset. Finally, giant circumferential dyke swarms (R. E. Ernst and K. L. Buchan, ms. in preparation) may feed lava flows offset from the plume center by hundreds of kilometers.

5.2. Location of Boundary of Plume

The pattern of magma flow in giant radiating dyke swarms can be related to melt generation in the plume head. In particular, the transition from vertical to horizontal flow with increasing distance from the focus (see magnetic fabric studies of section 3.2) is thought to mark the outer boundary of melt generation in the underlying mantle plume [*Ernst and Baragar*, 1992] and may reflect the penetration of only the central portion of the plume to the depths conducive to voluminous melt generation.

In only three swarms—the Mackenzie, Botswana, and the Ponta Grossa—has a transition between vertical flow and horizontal flow been determined from magnetic fabric studies. In the Mackenzie swarm (Figure 20), the transition occurs within 500 km of the focus [*Ernst and Baragar*, 1992]. In the case of the Botswana swarm (Figure 8), predominant vertical flow is indicated at the eastern end of the swarm, with lateral flow predominating at, and presumably beyond, a distance of ~400 km west of the plume center [*Ernst and Duncan*, 1995]. A magnetic fabric study of the Ponta Grossa swarm (Figure 7) shows dykes with steep flow giving way to dykes with horizontal flow with increasing distance from the plume center [*Raposo and Ernesto*, 1995]. The transition occurs at least 200 km from the plume center displayed in Figure 7.

As the radius of the plume head after arrival and flattening against the lithosphere (~1000 km [e.g., *White and McKenzie*, 1989]) is likely to be much larger than our estimates of this vertical to horizontal dyke-flow transition boundary (≤ 500) km, we may speculate on what additional factor(s) is impeding shallow plume penetration and associated melt generation beyond the transition distance. *Helmstaedt and Gurney* [1994] suggested that the penetration of the plume to a shallow level can be limited by a thick refractory lithospheric mantle root underlying adjacent Archean crustal areas. Furthermore, because thick lithospheric roots appear necessary for kimberlites to be diamondiferous, mapping of the boundary between shallow parts of plumes and adjacent mantle roots using magma flow trajectories in dykes could help predict the distribution of diamondiferous vs. barren kimberlites [*Helmstaedt and Gurney*, 1994; *LeCheminant et al.*, 1996].

5.3. Duration of LIP Events

Many plume-related LIPs of Mesozoic and Cenozoic age were emplaced rapidly [see summaries of *Storey*, 1995; *White and McKenzie*, 1995]. As noted in earlier sections, most of the volcanism in the Deccan, Siberian, Karoo, Columbia River, and Paraná LIPs occurred in time spans of less than a few m.y. In other cases, LIPs form over much longer periods. For example, the Icelandic plateau formed over more than 20 m.y. However, this is an example of a plume tail captured by a slow-spreading ridge [*Oskarsson et al.*, 1985].

The identification of pre-Mesozoic paleoplumes and remnant LIPs from convergent dyke swarms provides additional examples for which timing of LIP emplacement can be assessed. Unfortunately, only a few of the pre-Mesozoic LIPs (Table 1) have a sufficient number of precise dates to allow estimates of the time span of emplacement.

Some pre-Mesozoic events do appear to have occurred rapidly. For example, based on high precision U-Pb dating, both the Mackenzie and Franklin events occurred in less than 5 m.y. at 1267 and 723 Ma, respectively. On the other hand, other swarms seem to exhibit a much longer emplacement history. The Matachewan dykes and associated mafic plutons yield ages between about 2490 and 2445 Ma. However, more data are required to determine whether the magmatism was episodic or continuous over this interval. The Central Iapetus event spans at least 25 m.y. based on U-Pb ages for the 615 Ma Long Range dykes and the 590 Ma Grenville dykes. The ~1140 Ma Abitibi and ~1100 Keweenawan dykes may represent another example of repeated magmatic activity associated with a single plume.

A possible explanation for multiple plume 'impacts' from one plume has been offered by *Bercovici and Mahoney* [1994]. They suggested that separation of the plume head from its tail at the 660-km-deep boundary could result in accumulation at that boundary of a second plume head. It should be noted that the second plume head will only appear in the same place at the surface if the plate has not moved substantially in the interval.

5.4. Remote Feeding of Sills

In their mapping and modeling study of Early Tertiary volcanic centers in western North America, *Hyndman and Alt* [1987] made a convincing case that dykes of a radiating swarm may change orientation along strike and become laccoliths. *Ernst et al.* [1995a] summarized evidence supporting a similar origin for some sill complexes associated with giant radiating dyke swarms, based on the radiating patterns of these swarms and evidence for lateral flow. Three examples are given below.

Associated with the Mackenzie swarm (Figure 20) are coeval sills within (or near) older sedimentary basins at distances of 800, 1000, 1400, and 1500 km from the proposed plume center [*Ernst et al.*, 1995a]. Given that Mackenzie dykes are thought to have been fed through lateral injection beyond 500 km [*Ernst and Baragar*, 1992], it is unlikely that these sills were fed vertically, as vertical flow would require the existence of several very widely separated magma sources of identical age. Instead, it is more probable that these sills were fed from laterally emplaced Mackenzie dykes, some of which changed orientation, becoming sills within (or near) sedimentary basins [*Fahrig*, 1987; *Ernst et al.*, 1995a]. This reorientation of dykes into sills requires a modification of the stress conditions by a horizontal plane of weakness, or by the local load due to sediment accumulation.

Other examples in which coeval sills are located far from plume centers include the Franklin and Paraná events. In the case of the Franklin event (Figure 16), sills at a distance of about 800 km from the probable plume center may be fed in a similar fashion to the Mackenzie sills described above. Likewise, *White* [1992] proposed that Paraná sills within the sediments of the Paraná basin were fed laterally through dykes "rooted in the rifted area far to the east."

5.5. *Implications for Oceanic LIPS*

An important class of LIPs includes the oceanic plateaus, which may be due to mantle plume heads impinging on the base of the oceanic lithosphere. Some of the largest examples are the Ontong Java and Kerguelen plateaus [e.g., *Coffin and Eldholm*, 1994, and references therein] which may mark the arrival of mid-Cretaceous superplumes [*Larson*, 1991].

Just as giant radiating dyke swarms are generated when plumes give rise to continental flood basalts, they might also be expected to occur when plumes produce oceanic plateaus [*Ernst et al.*, 1995a]. On Venus, where giant radiating dyke swarms cut basaltic rocks, the swarms are observed by their surface grabens [*McKenzie et al.*, 1992; *Grosfils and Head*, 1994; *Ernst et al.*, 1995a]. By analogy, giant dyke swarms on Earth should radiate through oceanic crust away from oceanic mantle plumes. Many such swarms may be concentrated parallel to spreading ridges as plume heads in ocean basins are likely to "capture" spreading ridges even from a considerable distance away [*Mahoney and Spencer*, 1991]. Dykes may be difficult to

observe in the ocean basins, either directly or using geophysical techniques, because individual dykes are narrow, intrude host rock with a similar basaltic composition to that of the dykes themselves, and are typically covered by sediments. Whether such swarms would be more readily observed where they have propagated into adjacent continental crust is unclear, because their mode and depth of emplacement upon reaching the continental crust have not been studied. However, if continental swarms associated with distal oceanic plumes exist, a complete radiating dyke pattern would not be observed even after reconstructing plate positions to their configuration at the time of oceanic plume emplacement. As yet, there are no known examples of giant dyke swarms in ocean basins or on adjacent continents which can be linked to oceanic plumes.

6. DISCUSSION

Giant dyke swarms are important in extending the record of mantle plume-related events and LIPs through the Paleozoic and Proterozoic (Table 1). In order to assess plume production through time, we will consider only the magmatic events in continental areas (continental flood basalts) because the pre-Mesozoic oceanic LIP record has been removed by subduction.

Most continental flood basalt events that have been identified to date [e.g., *Coffin and Eldholm*, 1994] are of Mesozoic and Cenozoic age (Figure 25a). As described above, several of these include coeval giant radiating dyke swarms whose fanning patterns locate approximate centers of associated mantle plumes. However, giant radiating dyke swarms are also found throughout the Paleozoic and Proterozoic (Figure 25b), over an interval of time that is an order of magnitude longer than that represented by the Mesozoic and Cenozoic. As discussed in the paper, they were likely generated by mantle plumes.

A great number (more than 100) other giant swarms do not show an obvious radiating pattern [*Ernst et al.*, 1996]. Fifty-six of these which are dated are shown in Figure 25c. Many of these giant swarms may represent portions of radiating dyke swarms. If so, they are also likely to be related to mantle plumes and paleocontinental flood basalts. When all dated giant dyke swarms and known continental flood basalt events are combined (Figure 25d) they indicate a nearly continuous record of major magmatic events from the earliest Proterozoic to the present. Thus, in order to fully appreciate and understand the role of mantle plumes and continental flood basalts it is imperative that the entire record of such events through geological time be investigated. Giant radiating swarms provide the most enduring

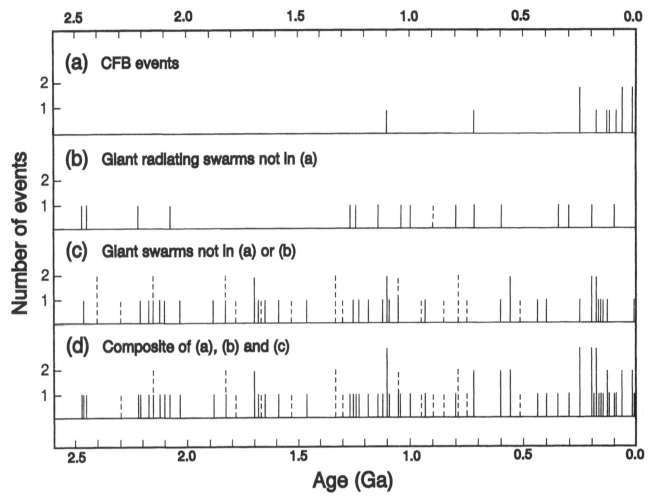

Figure 25. Bar graph showing the age distribution of events potentially associated with the arrival of a plume head at the base of the continental lithosphere. (a) continental flood basalt events after *Coffin and Eldholm* [1994], (b) giant radiating dyke swarms exclusive of those associated with the continental flood basalt events of (a) (this paper), (c) giant dyke swarms not observed to radiate (from *Ernst et al.* [1996]), and (d) composite of (a), (b) and (c). Solid bar used for events dated to within 100 m.y. Dashed bar indicates age uncertainties between 100 and 250 m.y. Events with age uncertainties greater than 250 m.y. are not shown.

record of these events.

Acknowledgments. We would like to thank Bob Baragar, Bill Davis, Henry Halls, Steve Kumarapeli, Rob Rainbird, Tony LeCheminant, and Marie-Claude Williamson for fruitful discussions regarding the relationship of dyke swarms to mantle plume events. In addition, Rob Rainbird and Garth Jackson are thanked for assistance in distinguishing the dykes associated with the Franklin-Natkusiak event, André Ciesielski and the Geological Survey of Canada Geophysical Data Center for information on the Mistassini swarm, Steve Prevec for information on the volcanic rocks and plutons associated with the Matachewan swarm, and Karl Wirth for information on the Fort Frances swarm. Colin Reeves provided a copy of the thesis by M. Mubu with some key details on dykes of the Karoo events. An internal Geological Survey of Canada review of this manuscript by Bob Baragar was appreciated. Thorough reviews by Kevin Burke, Mike Wingate, and volume co-editor John Mahoney were particularly helpful, especially in the younger LIP examples. This is Geological Survey of Canada contribution # 37095 and Canadian LITHOPROBE # 787.

REFERENCES

Aleinikoff, J. N., R. E. Zartman, M. Walter, D. W. Rankin, P. T. Lyttle, and W. C. Burton, U-Pb ages of metarhyolites of the Catoctin and Mount Rogers formations, central and southern Appalachians: evidence for two pulses of Iapetan rifting, *Am. J. Sci.*, 295, 428-454, 1995.

Anderson, D. L., Lithosphere, asthenosphere, and perisphere, *Rev. Geophys.*, 3, 125-149, 1995.

Andréasson, P. G., The Baltoscandian margin in Neoproterozoic-early Palaeozoic times. Some constraints on terrane derivation and accretion in the Arctic Scandinavian Caledonides, *Tectonophysics*, 231, 1-32, 1994.

Atkinson, S. S., and R. St. J. Lambert, The Roza Member feeder dyke system, Columbia River basalt group, USA: Compositional variation and emplacement, in *Mafic Dykes and Emplacement Mechanisms*, edited by A. J. Parker, P. C. Rickwood, and D. H. Tucker, pp. 447-459, Balkema, Rotterdam, 1990.

Auden, J. B., Dykes in western India: A discussion of their relationships with the Deccan traps, *Trans. National Inst. Sci. India*, 3, 123-157, 1949.

Austin Jr., J. A., P. L. Stoffa, J. D. Phillips, J. Oh, D. S. Sawyer, G. M. Purdy, E. Reiter, and J. Makris, Crustal structure of the Southeast Georgia embayment-Carolina trough: Preliminary results of a composite seismic image of a continental suture (?) and a volcanic passive margin, *Geology*, 18, 1023-1027, 1990.

Baker, J., L. Snee, and M. Menzies, A brief Oligocene period of flood volcanism in Yemen: implications for the duration and rate of continental flood volcanism at the Afro-Arabian triple junction, *Earth Planet. Sci. Lett.*, 138, 39-55, 1996.

Baksi, A. K., Geochronological studies on whole-rock basalts, Deccan Traps, India: evaluation of the timing of volcanism relative to the K-T boundary, *Earth Planet. Sci. Lett.*, 121, 43-56, 1994.

Baksi, A. K., and E. Farrar, $^{40}Ar/^{39}Ar$ dating of the Siberian Traps, USSR: Evaluation of the ages of the two major extinction events relative to episodes of flood-basalt volcanism in the USSR and the Deccan Traps, India, *Geology*, 19, 461-464, 1991.

Baldridge, W. S., Y. Eyal, Y. Bartov, G. Steinitz, and M. Eyal, Miocene magmatism of Sinai related to the opening of the Red Sea, *Tectonophysics*, 197, 181-201, 1991.

Baragar, W. R. A., and J. A. Donaldson, Coppermine and Dismal lakes map-areas. *Geol. Surv. Can. Paper* 71-39, 1973.

Baragar, W. R. A., R. E. Ernst, L. Hulbert, and T. Peterson, Longitudinal petrochemical variation in the Mackenzie dyke swarm, Northwestern Canadian Shield, *J. Petrol.*, 37, 317-359, 1996.

Bates, M. P., and H. C. Halls, Broad-scale Proterozoic deformation of the central Superior Province revealed by paleomagnetism of the 2.45 Ga Matachewan dyke swarm, *Can. J. Earth Sci.*, 28, 1780-1796, 1991.

Bennett, G., B. O. Dressler, and J. A. Robertson, Huronian supergroup and associated intrusive rocks, in *Geology of Ontario, Spec. Vol. 4, Part 1*, pp. 549-591, Ontario Geological Survey, Toronto, ON, 1991.

Bercovici, D., and J. Mahoney, Double flood basalts and plume head separation at the 660-kilometer discontinuity, *Science*, 266, 1367-1369, 1994.

Bertrand, H., The Mesozoic tholeiitic province of northwest Africa: a volcano-tectonic record of the early opening of Central Atlantic, in *Magmatism in Extensional Structural Settings: The Phanerozoic African Plate*, edited by A. B. Kampunzu and R. T. Lubala, pp. 148-188, Springer-Verlag, 1991.

Berkovsky, A. N., and A. P. Platunova, Dyke swarms of the East European craton: Aeromagnetic and geological evidence, in *Mafic Dyke Swarms, Spec. Pap. 34*, edited by H. C. Halls and W. F. Fahrig, pp. 373-377, Geological Association of Canada, 1987.

Bethune, K. M., and A. Davidson, Diabase dykes and the Grenville Front southwest of Sudbury, Ontario, in *Current Research, Part C, Canadian Shield, 88-1C*, pp. 151-159, Geological Survey of Canada, Ottawa, ON, 1988.

Bhattacharji, S., Propagating mafic dike swarms in the Deccan volcanics, hot spot track and interplate rifting, in *International Symposium on Mafic Dykes and Related Magmatism in Rifting and Intraplate Environments with Workshop on Mafic Dyke Magmatism in the Baltic Shield, August 8-13, 1988*, p. 11, IGCP-257 Technical Report Number One, Institute of Geology, Lund University, Sweden, 1988.

Bhattacharji, S., N. Chatterjee, J. M. Wampler, and M. Gazi, Mafic dikes in Deccan volcanics - Indicator of India intraplate rifting, crustal extension and Deccan flood basalt volcanism, in *Volcanism: Radhakrishna Volume*, edited by K. V. Subbarao, pp. 253-276, Wiley Eastern Limited, 1994.

Bird, R. T., W. R. Roest, M. Pilkington, R. E. Ernst, and K. L. Buchan, The application of digital geophysical data to the restoration of crustal deformation in the Canadian Shield, in *Current Research, Part C, Canadian Shield, Pap. 96C*, pp. 117-124, Geological Survey of Canada, Ottawa, ON, 1996.

Bisson, J. R., Quantitative analysis of width, spacing, trend and bifurcation direction of 2.6 Ga dykes in selected areas of the Central Superior Province. B.Sc. Thesis, 86 pp., Carleton University, Ottawa, 1985.

Blagovyeshchenskaya, M. N., *Geological map of Siberian platform and adjoining territories, scale 1:1,500,000*, Ministry of Geology of the USSR, 1973.

Borg, S. G., and D. J. DePaolo, Laurentia, Australia, and Antarctica as a Late Proterozoic supercontinent: Constraints from isotopic mapping, *Geology*, 22, 307-310, 1994.

Buchan, K. L., and H. C. Halls, Paleomagnetism of Proterozoic mafic dyke swarms of the Canadian shield, in *Mafic Dykes and Emplacement Mechanisms*, edited by A. J. Parker, P. C. Rickwood, and D. H. Tucker, pp. 209-230, Balkema, Rotterdam, 1990.

Buchan, K. L., J. K. Mortensen, and K. D. Card, Northeast-trending Early Proterozoic dykes of southern Superior Province: Multiple episodes of emplacement recognized from integrated paleomagnetism and U-Pb geochronology, *Can. J. Earth Sci.*, 30, 1286-1296, 1993.

Buchan, K. L., H. C. Halls, and J. K. Mortensen, Paleomagnetism, U-Pb geochronology, and geochemistry of Marathon dykes, Superior Province and comparison with the Fort Frances swarm. *Can. J. Earth Sci.*, 33, 1583-1595, 1996.

Burke, K., and J. F. Dewey, Plume-generated triple junctions: key indicators in applying plate tectonics to old rocks, *J. Geol.*, 81, 406-433, 1973.

Camp, V. E., and M. J. Roobol, The Arabian continental alkali basalt province: Part I. evolution of Harrat Rahat, Kingdom of Saudi Arabia, *Geol. Soc. Am. Bull.*, 101, 71-95, 1989.

Camp, V. E., and M. J. Roobol, Upwelling asthenosphere beneath western Arabia and its regional implications, *J. Geophys. Res.*, 97, 15,255-15,271, 1992.

Chadwick, Jr., W. W., and J. H. Dieterich, Mechanical modeling of circumferential and radial dike intrusion on Galapagos volcanoes, *J. Volcanol. Geotherm. Res.*, 66, 37-52, 1995.

Chandler, V. W., *Aeromagnetic anomaly map of Minnesota, scale 1:500,000, State Map Series Map S-17*, Minnesota Geological Survey, St. Paul, MN, 1991.

Choudhuri, A., E. P. Oliveira, and A. N. Sial, Mesozoic dyke swarms in northern Guiana and northern Brazil and the Cape Verde- Fernando de Noronha plume vortices: a synthesis, in *Extended Abstracts for the International Symposium on Mafic Dykes*, compiled by W. Teixeira, M. Ernesto, and E. P. Oliveira, pp. 17-22, São Paulo, Brazil, 1991.

Coffin, M. F., and O. Eldholm, Large igneous provinces: crustal structure, dimensions, and external consequences. *Rev. Geophys.*, 32, 1-36, 1994.

Coish, R. A., and C. W. Sinton, Geochemistry of mafic dikes in the Adirondack mountains: implications for the constitution of Late Proterozoic mantle. *Contrib. Mineral. Petrol.*, 110, 500-514, 1992.

Coleman, R. G., *Geologic Evolution of the Red Sea*, Oxford University Press, New York, 186 pp., 1993.

Correa-Gomes, L. C., The mafic dyke swarm along the coastline of Bahia State, Brazil: An attempt of continental breakup between South America and Africa 1.0 Ga ago? (abstract) *Program & Abstracts for the Third International Dyke Conference, Sept. 4-8, 1995, Jerusalem, Israel*, p. 19, 1995.

Correa-Gomes, L. C., and M. A. F. Tanner de Oliveira, *Map of mafic dyke provinces, Bahia, Brazil, scale 1:1000000*, SICT/SME/ UFBA/PPPG, 1994a.

Correa-Gomes, L. C., and M. A. F. Tanner de Oliveira, Map of Bahia state, Brazil: major provinces, temporal evolution and present knowledge. Some evidences about upper mantle behavior. *International Symposium on the Physics and Chemistry of the Upper Mantle, August 14-19, 1994* São Paulo, Brazil, 1994b.

Correa-Gomes, L. C., M. A. F. Tanner de Oliveira, H. Concieção, and M. B. Abram, Tectonic styles and chemistry of the mafic dykes in the eastern part of São Francisco craton, Bahia, Brazil, in *Extended Abstracts for the International Symposium on Mafic Dykes*, compiled by W. Teixeira, M. Ernesto, and E. P. Oliveira, pp. 66-70, São Paulo, Brazil, 1991.

Cox, K. G., Tectonics and vulcanism of the Karroo period and their bearing on the postulated fragmentation of Gondwanaland, in *African Magmatism and Tectonics*, edited by T. N. Clifford and I. G. Gass, pp. 211-235, Oliver & Boyd, Edinburgh, 1970.

Cox, K. G., The role of mantle plumes in the development of continental drainage patterns, *Nature*, 342, 873-876, 1989.

D'Agrella-Filho, M. S., I. G. Pacca, T. C. Onstott, P. R. Renne, and W. Teixeira, Paleomagnetism and geochronology of mafic dikes from the regions of Salvador, Olivença and Uauá, São Francisco craton, Brazil: present stage of the USP/Princeton University collaboration. *Boletim IG-USP Série Científica* 20, 1-8, 1989.

D'Agrella-Filho, M. S., I. G. Pacca, P. R. Renne, T. C. Onstott, and W. Teixeira, Paleomagnetism of Middle Proterozoic (1.01 to 1.08 Ga) mafic dykes in southeastern Bahia State - São Francisco Craton, Brazil, *Earth Planet. Sci. Lett.*, 101, 332-348, 1990.

Dawes, P. R., *Geological Map of Greenland, Sheet 5, Thule, scale 1:500,000*, Geological Survey of Greenland., 1991.

de Boer, J. Z., J. G. McHone, J. H. Puffer, P. C. Ragland, and D. Whittington, Mesozoic and Cenozoic magmatism, in *The Geology of North America, Volume I-2, The Atlantic Continental Margin, U.S.*, edited by R. E. Sheridan and J. A. Grow, pp. 217-241, Geological Society of America, Boulder, CO, 1988.

Deshmukh, S. S., and M. N. Sehgal, Mafic dyke swarms in Deccan volcanic province of Madhya Pradesh and Maharashtra, in *Deccan Flood Basalts, Mem. 10*, edited by K. V. Subbarao, pp. 323-340, Geological Society of India, 1988.

Dessai, A. G., and H. Bertrand, The "Panvel Flexure" along the Western Indian continental margin: an extensional fault structure related to Deccan magmatism, *Tectonophysics*, 241, 165-178, 1995.

Devey, C. W., and W. E. Stephens, Tholeiitic dykes in the Seychelles and the original spatial extent of the Deccan, *J. Geol. Soc. London*, 148, 979-983, 1991.

Dickin, A. P., The North Atlantic Tertiary Province, in *Continental Flood Basalts*, edited by J. D. Macdougall, pp. 111149, Kluwer Academic Publishers, 1988.

Druecker, M. D., and S. P. Gay Jr., Mafic dyke swarms associated with Mesozoic rifting in eastern Paraguay, South America, in *Mafic Dyke Swarms, Spec. Pap. 34*, edited by H. C. Halls and W. F. Fahrig, pp. 187-193, Geological Association of Canada, Toronto, ON, 1987.

Dudás, F. Ö., A. Davidson, and K. M. Bethune, Age of the Sudbury diabase dykes and their metamorphism in the Grenville Province, Ontario, in *Radiogenic Age and Isotopic Studies: Report 8, Current Research 1994-F*, pp. 97-106, Geological Survey of Canada, Ottawa, ON, 1994.

Duncan, R. A., and M. A. Richards, Hotspots, mantle plumes, flood basalts, and true polar wander, *Rev. Geophys.*, 29, 31-50, 1991.

Duncan, A. R., R. A. Armstrong, A. J. Erlank, J. S. Marsh, and R. T. Watkins, MORB-related dolerites associated with the final phases of Karoo flood basalt volcanism in southern Africa, in *Mafic Dykes and Emplacement Mechanisms*, edited by A. J. Parker, P. C. Rickwood, and D. H. Tucker, pp. 119-129, Balkema, Rotterdam, 1990.

Dunham, A. C., and V. E. H. Strasser-King, Late Carboniferous intrusions of northern Britain, in *Igneous Rocks of the British Isles*, edited by D. S. Sutherland, pp. 277-283, John Wiley and Sons, Ltd., 1982.

Dunning, G., and J. P. Hodych, U/Pb zircon and baddeleyite ages for the Palisades and Gettysburg sills of the northeastern United States: Implications for the age of the Triassic/Jurassic boundary, *Geology*, 18, 795-798, 1990.

Eales, H. V., J. S. Marsh, and K. G. Cox, The Karoo igneous province: An introduction, in *Petrogenesis of the Volcanic Rocks of the Karoo Province, Spec. Publ. 13*, edited by A. J. Erlank, pp. 1-26, The Geological Society of South Africa, Johannesburg, 1984.

Embry, A. F., and K. G. Osadetz, Stratigraphy and tectonic significance of Cretaceous volcanism in Queen Elizabeth Islands, Canadian Arctic Archipelago, *Can. J. Earth Sci.*, 25, 1209-1219, 1988.

Encarnación, J., T. H. Fleming, D. H. Elliot, and H. V. Eales, Synchronous emplacement of Ferrar and Karoo dolerites and the early breakup of Gondwana, *Geology*, 24, 535-538, 1996.

Erinchek, Ju. M., E. D. Milshtain, and R. S. Kontorovich, The structure of the Middle Paleozoic Viluy-Markha dyke belt interpreted from aeromagnetic data (Siberian Plate, Russia)(abstract), *Program & Abstracts for the Third International Dyke Conference, Sept. 4-8, 1995, Jerusalem, Israel*, edited by A. Agnon and G. Baer, pp. 27, 1995.

Ernst, R. E., Magma flow directions in two mafic Proterozoic dyke swarms of the Canadian Shield: As estimated using anisotropy of magnetic susceptibility data, in *Mafic Dykes and Emplacement Mechanisms*, edited by A. J. Parker, P. C. Rickwood, and D. H. Tucker, pp. 231-235, Balkema, Rotterdam, 1990.

Ernst, R. E., Mapping the magma flow pattern in the Sudbury dyke swarm using magnetic fabric analysis, in *Current Research 1994-E*, pp. 183-192, Geological Survey of Canada, Ottawa, ON, 1994.

Ernst, R. E., and W. R. A. Baragar, Evidence from magnetic fabric for the flow pattern of magma in the Mackenzie giant radiating dyke swarm, *Nature*, 356, 511-513, 1992.

Ernst, R. E., and K. Bell, Petrology of the Great Abitibi dyke, Superior Province, *J. Petrology*, 33, 423-469, 1992.

Ernst, R. E., and K. L. Buchan, Paleomagnetism of the Abitibi dyke swarm, southern Superior Province, and implications for the Logan Loop, *Can. J. Earth Sci.*, 30, 1886-1897, 1993.

Ernst, R. E., and A. R. Duncan, Magma flow in the giant Botswana dyke swarm from analysis of magnetic fabric (abstract), *Program & Abstracts for the Third International Dyke Conference, Sept. 4-8, 1995, Jerusalem, Israel*, edited by A. Agnon and G. Baer, pp. 30, 1995.

Ernst, R. E., J. W. Head, E. Parfitt, E. Grosfils, and L. Wilson, Giant radiating dyke swarms on Earth and Venus, *Earth Sci. Rev.*, 39, 1-58, 1995a.

Ernst, R. E., K. L. Buchan, and H. C. Palmer, Giant dyke swarms: characteristics, distribution and geotectonic applications, in *Physics and Chemistry of Dykes*, edited by G. Baer and A. Heimann, Balkema, Rotterdam, pp. 3-21, 1995b.

Ernst, R. E., A. G. Lindsey, and J. Z. de Boer, Flow pattern in the early Jurassic Higganum (Conn./Mass.) and contemporary Christmas Cove (Maine) dikes using magnetic fabric analysis (abstract), *1995 Abstracts with Programs, GSA Northeastern Section*, pp. 42, 1995c.

Ernst, R. E., K. L. Buchan, T. D. West, and H. C. Palmer, Diabase (dolerite) dyke swarms of the world: first edition, 1:35,000,000 map, 104 pp., *Geol. Surv. Can. Open File 3241*, 1996.

Eyal, Y., and M. Eyal, Mafic dyke swarms in the Arabian-Nubian Shield, *Isr. J. Earth Sci.*, 36, 195-211, 1987.

Fahrig, W. F., The tectonic settings of continental mafic dyke swarms: Failed arm and early passive margin in *Mafic Dyke Swarms, Spec. Pap. 34*, edited by H. C. Halls and W. F. Fahrig, pp. 331-348, Geological Association of Canada, Toronto, ON, 1987.

Fahrig, W. F., and D. L. Jones, Paleomagnetic evidence for the extent of Mackenzie Igneous Events, *Can. J. Earth Sci.*, 6, 679-688, 1969.

Fahrig, W. F., and T. D. West, *Diabase dyke swarms of the Canadian shield, Map 1627A*, Geological Survey of Canada, Ottawa, ON, 1986.

Fahrig, W. F., E. H. Gaucher, and A. Larochelle, Paleomagnetism of diabase dykes of the Canadian Shield, *Can. J. Earth Sci.*, 2, 278-298, 1965.

Fahrig, W. F., K. W. Christie, E. H. Chown, D. Janes, and N. Machado, The tectonic significance of some basic dyke swarms in the Canadian Superior Province with special reference to the geochemistry and paleomagnetism of the Mistassini swarm, Quebec, Canada, *Can. J. Earth Sci.*, 23, 238-253, 1986.

Féraud, G., G. Giannérini, and R. Campredon, Dyke swarms as paleostress indicators in areas adjacent to continental collision zones: examples from the European and northwest Arabian plates, in *Mafic Dyke Swarms, Spec. Pap. 34*, edited by H. C. Halls and W. F. Fahrig, pp. 273-278, Geological Association of Canada, Toronto, ON, 1987.

Gibbs, A. K., Contrasting styles of continental mafic intrusions in the Guiana Shield, in *Mafic Dyke Swarms, Spec. Pap. 34*, edited by H. C. Halls and W. F. Fahrig, pp. 457-465, Geological Association of Canada, Toronto, ON, 1987.

Gibson, I. L., N. S. Madhurendra, and W. F. Fahrig, The geochemistry of the Mackenzie dyke swarm, Canada in *Mafic Dyke Swarms, Spec. Pap. 34*, edited by H. C. Halls and W. F. Fahrig, pp. 109-121, Geological Association of Canada, Toronto, ON, 1987.

Goldberg, S. A., and J. R. Butler, Late Proterozoic rift-related dykes of the southern and central Appalachians, eastern USA, in *Mafic Dykes and Emplacement Mechanisms*, edited by A. J. Parker, P. C. Rickwood, and D. H. Tucker, pp. 131-144, Balkema, Rotterdam, 1990.

Goldberg, S. A., J. R. Butler, and P. D. Fullagar, The Bakersville dike swarm: geochronology and petrogenesis of Late Proterozoic basaltic magmatism in the southern Appalachian Blue Ridge, *Am. J. Sci.*, 286, 403-430, 1986.

Gomes, L. C. C., M. A. F. Tanner de Oliveira, and L. R. B. Leal, Structural features associated with mafic dikes: examples from the Atlantic coastal belt of Bahia, Brazil, *Boletim IG-USP, Série Científica*, 20, 21-24, 1989.

Gorbatschev, R., A. Lindh, Z. Solyom, I. Laitakari, K. Aro, S.B. Lobach-Zhuchenko, M. S. Markov, A. I. Ivliev, and I. Bryhni, Mafic dyke swarms of the Baltic shield, in *Mafic Dyke Swarms, Spec. Pap. 34*, edited by H. C. Halls and W. F. Fahrig, pp. 361-372, Geological Association of Canada, Toronto, ON, 1987.

Green, J. C., T. J. Bornhorst, V. W. Chandler, M. G. Mudrey, P. E. Myers, L. J. Pesonen, and J. T. Wilband, Keweenawan dykes of the Lake Superior region: Evidence for evolution of the Middle Proterozoic Midcontinent Rift of North America in *Mafic Dyke Swarms, Spec. Pap. 34*, edited by H. C. Halls and W. F. Fahrig, pp. 289-302, Geological Association of Canada, Toronto, ON, 1987.

Greenough, J. D., and J. P. Hodych, Evidence for lateral magma injection in the early Mesozoic dykes of eastern North America, in *Mafic Dykes and Emplacement Mechanisms*, edited by A. J. Parker, P. C. Rickwood, and D. H. Tucker, pp. 35-46, Balkema, Rotterdam, 1990.

Gresse, P. G., and R. Scheepers, Neoproterozoic to Cambrian (Namibian) rocks of South Africa: a geochronological and geotectonic review, *J. Afr. Earth Sci.*, 16, 375-393, 1993.

Griffiths, R. W., and I. H. Campbell, Interaction of mantle plume heads with the Earth's surface and onset of small-scale convection, *J. Geophys. Res.*, 96, 18,295-18,310, 1991.

Grosfils, E. B., and J. W. Head, The global distribution of giant radiating dike swarms on Venus: implications for the global stress state, *Geophys. Res. Lett.*, 21, 701-704, 1994.

Gupta, V. K., *Shaded image of total magnetic field of Ontario, east-central sheet, scale 1:1 000 000, Map 2586*, Ontario Geological Survey, Toronto, ON, 1991a.

Gupta, V. K., *Shaded image of total magnetic field of Ontario, southern sheet, scale 1:1 000 000, Map 2587*, Ontario Geological Survey, Toronto, ON, 1991b.

Gusev, G. S., and B. R. Shpount, Precambrian and Paleozoic rifting in northeastern Asia, *Tectonophysics*, 143, 245-252, 1987.

Halls, H. C., The importance and potential of mafic dyke swarms in studies of geodynamic processes, *Geosci. Canada*, 9, 145-154, 1982.

Halls, H. C., The Matachewan dyke swarm, Canada: An early Proterozoic magnetic field reversal, *Earth Planet. Sci. Lett.*, 105, 279-292, 1991.

Halls, H. C., and M. P. Bates, The evolution of the 2.45 Ga Matachewan dyke swarm, Canada, in *Mafic Dykes and Emplacement Mechanisms*, edited by A. J. Parker, P. C. Rickwood, and D. H. Tucker, pp. 237-249, Balkema, Rotterdam, 1990.

Halls, H. C., K. G. Burns, S. J. Bullock, and P. M. Batterham, Mafic dyke swarms of Tanzania interpreted from aeromagnetic data in *Mafic Dyke Swarms, Spec. Pap. 34*, edited by H. C. Halls and W. F. Fahrig, pp. 173-186, Geological Association of Canada, Toronto, ON, 1987.

Harry, D. L., and Sawyer, D. S., Basaltic volcanism, mantle plumes, and the mechanics of rifting: the Paraná flood basalt province of South America, *Geology*, 20, 207-210, 1992.

Hawkesworth, C. J., K. Gallagher, S. Kelley, M. Mantovani, D.W. Peate, M. Regelous, and N. W. Rogers, Paraná magmatism and the opening of the South Atlantic, in *Magmatism and the Causes of Continental Break-up, Spec. Publ. 68*, edited by B. C. Storey, T. Alabaster, and R. J. Pankhurst, pp. 221-240, Geological Society of London, 1992.

Heaman, L., U-Pb dating of giant radiating dyke swarms: potential for global correlation of mafic magmatic events, *Extended Abstracts for the International Symposium on Mafic Dykes, São Paulo, Brazil*, compiled by W. Teixeira, M. Ernesto, and E. P. Oliveira, pp. 7-9, 1991.

Heaman, L. M., 2.45 Ga global mafic magmatism: Earth's oldest superplume?, in *Eighth International Conference on Geochronology, Cosmochronology & Isotope Geology, Berkeley, California, Program with Abstracts (Berkeley, California)*, p. 132, U.S. Geol. Surv. Circular 1107, 1994.

Heaman, L. M., U-Pb dating of mafic rocks: past, present and future (abstract), *Program with Abstracts Geol. Assoc. Can./Mineral. Assoc. Can.*, 20, A43, 1995.

Heaman, L. M., A. N. LeCheminant, and R. H. Rainbird, Nature and timing of Franklin igneous events, Canada: Implications for a Late Proterozoic mantle plume and the break-up of Laurentia, *Earth Planet. Sci. Lett.*, 109, 117-131, 1992.

Heatherington, A. L., and P. A. Mueller, Geochemical evidence for Triassic rifting in southwestern Florida, *Tectonophysics*, 188, 291-302, 1991.

Helmstaedt, H.H., and J.J. Gurney, Geotectonic controls on the formation of diamonds and their kimberlitic and lamproitic host rocks: Applications to diamond exploration, in *Proceedings 5th International Kimberlite Conference, Araxá, Brazil 1991, v. 2, Diamonds: Characterization, Genesis and Exploration*, edited by H. O. A. Meyer, and O. H. Leonardos, pp. 236-250, 1994.

Hill, R. I., Starting plumes and continental break-up, *Earth Planet. Sci. Lett.*, 104, 398-416, 1991.

Hofmann, C., G. Feraud, R. Pik, C. Coulon, G. Yirgu, D. Ayalew, C. Deniel, and V. Courtillot, $^{40}Ar/^{39}Ar$ dating of Ethiopian Traps. *IUGG XXI General Assembly Abstracts Week A*, A465, 1995.

Holbrook, W. S., and P. B. Kelemen, Large igneous province on the US Atlantic margin and implications for magmatism during continental breakup, *Nature*, 364, 433-436, 1993.

Holbrook, W. S., G. M. Purdy, R. E. Sheridan, L. Glover III, M. Talwani, J. Ewing, and D. Hutchinson, Seismic structure of the U.S. Mid-Atlantic continental margin, *J. Geophys. Res.*, 99, 17,871-17,891, 1994.

Hooper, P. R., The Columbia River Basalt, in *Continental Flood Basalts*, edited by J. D. Macdougall, pp. 1-33, Kluwer Academic Publishers, 1988.

Hooper, P. R., The timing of crustal extension and the eruption of continental flood basalts, *Nature*, 345, 246-249, 1990.

Hulbert, L., B. Williamson, and R. Thériault, Geology of Middle Proterozoic MacKenzie diabase suites from Saskatchewan: An overview and their potential to host Noril'sk-type Ni-Cu-PGE mineralization, in *Summary of Investigations 1993, Misc. Report 93-4*, edited by R. Macdonald. T. I. I. Sibbald, C. T. Harper, D. F. Paterson, and P. Guliov, pp. 112-126, Saskatchewan Geological Survey, Saskatchewan Mineral Resources, Regina, SK, 1993.

Hunt, P. A., and J. C. Roddick, A compilation of K-Ar ages, report 17, in *Radiogenic Age and Isotopic Studies: Report 1. Geol. Surv. Can.*, Paper 87-2, pp. 143-210, 1987.

Hunter, D. R., and D. L. Reid, Mafic dyke swarms in southern Africa, data, in *Mafic Dyke Swarms, Spec. Pap. 34*, edited by H. C. Halls and W. F. Fahrig, pp. 445-456, Geological Association of Canada, Toronto, ON, 1987.

Hutchinson, D. R., R. S. White, W. F. Cannon, and K. J. Schulz, Keweenaw hot spot: Geophysical evidence for a 1.1 Ga mantle plume beneath the Midcontinent rift system, *J. Geophys. Res.*, 95, 10,869-10,884, 1990.

Hyndman, D. W., and D. Alt, Radial dikes, laccoliths, and gelatin models, *J. Geol.*, 95, 763-774, 1987.

Isachsen, Y. W., W. M. Kelly, C. Sinton, R. A. Coish, and M. T. Heizler, Dikes of northeast Adirondack region: introduction to their distribution, orientation, mineralogy, chronology, magnetism, chemistry and mystery, in *New York State Geological Association, 60th Ann. Meeting, Field Trip Guidebook*, edited by J.F. Olmsted, pp. 215-243, New York State, Albany, 1988.

Jefferson, C. W., L. J. Hulbert, R. H. Rainbird, G. E. M. Hall, D. C. Grégoire, and L. I. Grinenko, Mineral resource assessment of the Neoproterozoic Franklin igneous events of Arctic Canada: Comparison with the Permo-Triassic Noril'sk - Talnakh Ni-Cu-PGE deposits of Russia, *Open File 2789*, 48 pp., Geological Survey of Canada, Ottawa, ON, 1994.

Jones, D. L., and W. F. Fahrig, Paleomagnetism and age of the Aston dykes and Savage Point sills of the Boothia Uplift, Canada, *Can. J. Earth Sci.*, 15, 1605-1612, 1978.

Kamo, S. L., and C. F. Gower, Note: U-Pb baddeleyite dating clarifies age of characteristic paleomagnetic remanence of Long Range dykes, southeastern Labrador, *Atl. Geol.*, 30, 259-262, 1994.

Kamo, S. L., C. F. Gower, and T. E. Krogh, Birthdate for the Iapetus Ocean? A precise U-Pb zircon and baddeleyite age for the Long Range dikes, southeast Labrador, *Geology*, 17, 602-605, 1989.

Kamo, S. L., T. E. Krogh, and P. S. Kumarapeli, Age of the Grenville dyke swarm, Ontario-Quebec: implications for the timing of Iapetan rifting, *Can. J. Earth. Sci.*, 32, 273-280, 1995.

Karkare, S. G., and R. K. Srivastava, Regional dyke swarms related to the Deccan Trap alkaline province, India, in *Mafic Dykes and Emplacement Mechanisms*, edited by A. J. Parker, P. C. Rickwood, and D. H. Tucker, pp. 335-347, Balkema, Rotterdam, 1990.

Keppie, J. D., and R. D. Dallmeyer, *Tectonic map of Pre-Mesozoic terranes in Circum-Atlantic Phanerozoic orogens, scale 1: 5,000,000*, Nova Scotia Department of Natural Resources, Halifax, 1989.

King, E. R., Precambrian terrane of north-central Wisconsin: an aeromagnetic perspective, *Can. J. Earth Sci.*, 27, 1472-1477, 1990.

Klingspor, I., Radiometric age-determination of basalts, dolerites and related syenite in Skåne, southern Sweden, *Geol. Fören. Stockholm Förh.*, 98, 195-216, 1976.

Krasnov, I. I., M. J. Lurje, and V. L. Masaitis (editors), Geology of the Siberian Platform, Nedra, Moscow, USSR, 447 pp., 1966.

Krogh, T. E., D. W. Davis, and F. Corfu, Precise U-Pb zircon and baddeleyite ages for the Sudbury structure, in *The Geology and Ore Deposits of the Sudbury Structure, Spec. Vol. 1*, edited by E. G. Pye, A. J. Naldrett, and P. E. Giblin, pp. 431-445, Ontario Geological Survey, Toronto, ON, 1984.

Krogh, T. E., F. Corfu, D. W. Davis, G. R. Dunning, L. M. Heaman, S. L. Kamo, N. Machado, J. D. Greenough, and E. Nakamura, Precise U-Pb isotopic ages of diabase dykes and mafic to ultramafic rocks using trace amounts of baddeleyite and zircon, in *Mafic Dyke Swarms, Spec. Pap. 34*, edited by H. C. Halls and W. F. Fahrig, pp. 147-152, Geological Association of Canada, Toronto, ON, 1987.

Kruk, W., *Geological Map of the Yemen Arab Republic, Sheet Al Hazm, 1:250,000*, Federal Institute for Geosciences and Natural Resource, 1980.

Kumarapeli, S. P., G. R. Dunning, H. Pintson, and J. Shaver, Geochemistry and U-Pb zircon age of commenditic metafelsites of the Tibbit Hill Formation, Quebec Appalachians, *Can. J. Earth Sci.*, 26, 1374-1383, 1989.

Kumarapeli, S. P., K. St. Seymour, A. Fowler, and H. Pinston, The problem of the magma source of a giant radiating mafic dyke swarm in a failed arm setting, in *Mafic Dykes and Emplacement Mechanisms*, edited by A. J. Parker, P. C. Rickwood, and D. H. Tucker, pp. 163-171, Balkema, Rotterdam, 1990.

Larson, R.L., Geological consequences of superplumes, *Geology*, 19, 963-966, 1991.

Lawver, L. A., and R. D. Müller, The Iceland hotspot track, *Geology*, 22, 311-314, 1994.

Lawver, L. A., L. M. Gahagan, and M. F. Coffin, The development of paleoseaways around Antarctica, *Antarct. Res. Ser.*, 56, 7-30, 1992.

LeCheminant, A. N., *Proterozoic diabase dyke swarms; Lac de Gras and Aylmer Lake area, District of Mackenzie, Northwest Territories, scale 1: 250,000, Open File 2975*, Geological Survey of Canada, Ottawa, ON, 1994.

LeCheminant, A. N., and L. M. Heaman, Mackenzie igneous events, Canada: Middle Proterozoic hotspot magmatism associated with ocean opening, *Earth Planet. Sci. Lett.*, 96, 38-48, 1989.

LeCheminant, A. N., and L. M. Heaman, U-Pb ages for the 1.27 Ga Mackenzie igneous events, Canada: Support for a plume initiation model (abstract), in *Program with Abstracts*, 16, A73, Geological Association of Canada, Waterloo, ON, 1991.

LeCheminant, A. N., and L. M. Heaman, 779 Ma mafic magmatism in the northwestern Canadian Shield and northern Cordillera: A new regional time-marker (abstract), in *8th International Conference on Geochronology, Cosmochronology and Isotope Geology, Program with Abstracts (Berkeley, California). U.S. Geol. Surv. Circular 1107*, pp. 197, 1994.

LeCheminant, A. N., and O. van Breemen, U-Pb ages of Proterozoic dyke swarms, Lac de Gras area, N.W.T.: evidence for a progressive break-up of an Archean supercontinent, *Program with Abstracts*, 19, A62, Geological Association of Canada, Toronto, ON, 1994.

LeCheminant, A. N., L. M. Heaman, O. van Breemen, R. E. Ernst, W. R. A. Baragar, and K. L. Buchan. Mafic magmatism, mantle roots and kimberlites in the Slave craton, in *Searching for Diamonds in Canada, Open File 3228*, edited by A. N. LeCheminant, D. G. Richardson, R. N. W. DiLabio, and K. A. Richardson, pp. 161-169, Geological Survey of Canada, Ottawa, ON, pp. 161-169, 1996.

Levashov, K. K. Paleorift structure in the eastern environs of the Siberian Platform. *Int. Geol. Rev.*, 21, 188-200, 1979.

Macdonald, R., D. Gottfried, M. J. Farrington, F. W. Brown, and N. G. Skinner, Geochemistry of a continental tholeiite suite: late Palaeozoic quartz dolerite dykes of Scotland, *Trans. R. Soc. Edinburgh: Earth Sci.*, 72, 57-74, 1981.

Macdonald, R., R. Crossley, and K. S. Waterhouse, 1983. Karoo basalts of southern Malawi and their regional petrogenetic significance, *Mineral. Mag.*, 47, 281-289, 1983.

MacDonald, R., L. Wilson, R. S. Thorpe, and A. Martin, Emplacement of the Cleveland dyke: Evidence from geochemistry, mineralogy and physical modelling, *J. Petrol.*, 29, 559-583, 1988.

Mahoney, J. J., Deccan Traps, in *Continental Flood Basalts*, edited by J. D. Macdougall, pp. 151-194, Kluwer Academic Publishers, Dordrecht, 1988.

Mahoney, J. J., and K. J. Spencer, Isotopic evidence for the origin of the Manihiki and Ontong Java oceanic plateaus, *Earth Planet. Sci. Lett.*, 104, 196-210, 1991.

Mahoney, J., C. Nicollet, and C. Dupuy, Madagascar basalts: tracking oceanic and continental sources, *Earth Planet. Sci. Lett.*, 104, 350-363, 1991.

May, P. R., Pattern of Triassic-Jurassic diabase dykes around the North Atlantic in context of predrift position of the continents, *Geol. Soc. Am. Bull.*, 82, 1285-1292, 1971.

McHone, J. G., Broad-terrane Jurassic flood basalts across northeastern North America, *Geology*, 24, 319-322, 1996.

McHone, J. G., M. E. Ross, and J. D. Greenough, 1987. Mesozoic dyke swarms of eastern North America, in *Mafic Dyke Swarms, Spec. Pap. 34*, edited by H. C. Halls and W. F. Fahrig, pp. 279-288, Geological Association of Canada, Toronto, ON, 1987.

McKenzie, D., J. M. McKenzie, and R. S. Saunders, Dike emplacement on Venus and on Earth, *J. Geophys. Res.*, 97, 15,977-15,990, 1992.

Megrue, G. H., E. Norton, and D. W. Strangway, Tectonic history of the Ethiopian rift as deduced by K-Ar ages and paleomagnetic measurements of basaltic dikes, *J. Geophys. Res.*, 77, 5744-5754, 1972.

Mertanen, S., L.J. Pesonen, and H. Huhma, Palaeomagnetism and Sm-Nd ages of the Neoproterozoic diabase dykes in Laanila and Kautokeino, northern Fennoscandia, in *Precambrian Crustal Evolution in the North Atlantic Region, Spec. Publ. 112*, edited by T. S. Brewer, pp. 331-358, The Geological Society, London, 1996.

Mohr, P., Ethiopian flood basalt province, *Nature*, 303, 577-584, 1983.

Mohr, P., Structure of Yemeni Miocene dike swarms and emplacement of coeval granite plutons, *Tectonophysics*, 198, 203-221, 1991.

Mohr, P., and B. Zanettin, The Ethiopian flood basalt province, in *Continental Flood Basalts*, edited by J. D. Macdougall, pp. 63-110. Kluwer Academic Publishers, Dordrecht, 1988.

Morgan, W. J., Hotspot tracks and the opening of the Atlantic and Indian oceans, in *The Sea. volume 7*, edited by C. Emiliani, pp. 443-487. Wiley Interscience, New York, 1981.

Morgan, W. J., Hotspot tracks and the early rifting of the Atlantic, *Tectonophysics*, 94, 123-139, 1983.

Mubu, M. S., Aeromagnetic mapping and interpretation of mafic dyke swarms in southern Africa, M.Sc. thesis, Dept. of Earth Resource Surveys, International Institute for Aerospace Survey and Earth Sciences, Delft, the Netherlands, 1995.

Murthy, N. G. K., Mafic dyke swarms of the Indian shield, in *Mafic Dyke Swarms, Spec. Pap. 34*, edited by H. C. Halls and W. F. Fahrig, pp. 393-400, Geological Association of Canada, Toronto, ON, 1987.

Nadin, P. A., and N. J. Kusznir, Palaeocene uplift and Eocene subsidence in the northern North Sea basin from 2D forward and reverse stratigraphic modelling, *J. Geol. Soc. London*, 152, 833-848, 1995.

Namibia Geological Survey *(Cape Cross), 1:250,000 Geological Series*, Sheet 2013, 1988.

Nielsen, T. F. D., Mafic dyke swarms in Greenland: A review in *Mafic Dyke Swarms, Spec. Pap. 34*, edited by H. C. Halls and W. F. Fahrig, pp. 349-360, Geological Association of Canada, Toronto, ON, 1987.

O'Connor, J. M., and R. A. Duncan, Evolution of the Walvis ridge-Rio Grande rise hot spot system: implications for African and South American plate motions over plumes, *J. Geophys. Res.*, 95, 17,475-17,502, 1990.

Oliveira, E. P., J. Tarney, and X. J. Jono, Geochemistry of the Mesozoic Amapá and Jari dyke swarms, northern Brazil: Plume-related magmatism during the opening of the central Atlantic, in *Mafic Dykes and Emplacement Mechanisms*, edited by A. J. Parker, P. C. Rickwood, and D. H. Tucker, pp. 173-183, Balkema, Rotterdam, 1990.

Oskarsson, N., S. Steinthorsson, and G. E. Sigvaldason, Iceland geochemical anomaly: origin, volcanotectonics, chemical fractionation and isotope evolution of the crust, *J. Geophys. Res.*, 90, 10,011-10,025, 1985.

Osmani, L. A., Proterozoic mafic dike swarms in the Superior Province of Ontario, in *Geology of Ontario, Spec. Vol. 4, Part 1*, edited by P. C. Thurston, H. R. Williams, R. H. Sutcliffe, and G. M. Stot, pp. 661-681, Ontario Geological Survey, Toronto, ON, 1991.

Palmer, H. C., B. A. Merz, and A. Hayatsu, The Sudbury dikes of the Grenville Front region: paleomagnetism, Petrochemistry, and K-Ar age studies, *Can. J. Earth Sci.*, 14, 1867-1887, 1977.

Park, J. K., K. L. Buchan, and S. S. Harlan, A proposed giant radiating dyke swarm fragmented by the separation of Laurentia and Australia based on paleomagnetism of ca. 780 Ma mafic intrusions in western North America, *Earth Planet. Sci. Lett.*, 132, 129-139, 1995.

Parsons, T., G. A. Thompson, and N. H. Sleep, Mantle plume influence on the Neogene uplift and extension of the U.S. western Cordillera?, *Geology*, 22, 83-86, 1994.

Peterman, Z. E., and P. K. Sims, The Goodman swell: A lithospheric flexure caused by crustal loading along the midcontinent rift system, *Tectonics*, 7, 1077-1090, 1988.

Piccirillo, E. M., G. Bellieni, R. Petrini, G. Cavazzini, P. Comin-Chiaramonti, M. H. F. Macedo, A. J. Melfi, P. Zantedeschl, J. P. P. Pinese, and G. Martins, Mesozoic mafic dykes and intrusives from Brazil: petrology, geochemistry and Sr-Nd isotopes. *Extended Abstracts for the International Symposium on Mafic Dykes, São Paulo, Brazil*, compiled by W. Teixeira, M. Ernesto, and E. P. Oliveira, pp. 15-16, 1991.

Prevec, S. A., R. S. James, R. R. Keays, and D. C. Vogel, Constraints on the genesis of Huronian magmatism in the Sudbury area from radiogenic isotopic and geochemical evidence., in The Northern Margin of the Southern Province of the Canadian Shield, Program and Abstracts, *Can. Mineral.*, 33, 4, 930-932, 1995.

Radhakrishna, T., M. Joseph, P. K. Thampi, and J. G. Mitchell, Phanerozoic mafic dyke intrusions from the high grade terrain of southwestern India: K-Ar isotope and geochemical implications, in *Mafic Dykes and Emplacement Mechanisms*, edited by A. J. Parker, P. C. Rickwood, and D. H. Tucker, pp. 363-372, Balkema, Rotterdam, 1990.

Radhakrishna, T., R. D. Dallmeyer, and M. Joseph, Palaeomagnetism and $^{36}Ar/^{40}Ar$ vs. $^{39}Ar/^{40}Ar$ isotope correlation ages of dyke swarms in central Kerala, India: tectonic implications, *Earth Planet. Sci. Lett.*, 121, 213-226, 1994.

Ragland, P. C., R. D. Hatcher Jr., and D. Whittington, Juxtaposed Mesozoic dike sets from the Carolinas: A preliminary assessment, *Geology*, 11, 394-399, 1983.

Rainbird, R. H., The sedimentary record of mantle plume uplift preceding eruption of the Neoproterozoic Natkusiak flood basalt,

J. Geol., 101, 305-318, 1993.

Rainbird, R. H., D. A. Hodgson, W. Darch, and R. Lustwerk, *Bedrock and surficial geology of Northeast Minto Inlier, Victoria Island, District of Franklin, Northwest Territories, scale 1:50,000, Open File 2781,* Geological Survey of Canada, Ottawa, ON, 1994a.

Rainbird, R. H., D. A. Hodgson, and C. W. Jefferson, *Bedrock and surficial geology, Washington Islands (NTS 78 B/5) District of Franklin, Northwest Territories, scale 1:50,000, Open File 2920,* Geological Survey of Canada, Ottawa, ON, 1994b.

Ransome, I. G. D., The geodynamics, kinematics and geochemistry of the Gannakouriep dyke swarm, M.Sc. thesis, University of Cape Town, 1992.

Raposo, M. I. B., and M. Ernesto, Anisotropy of magnetic susceptibility in the Ponta Grossa dyke swarm (Brazil) and its relationship with magma flow direction, *Phys. Earth Planet. Inter.,* 87, 183-196, 1995.

Reid, D. L., The Cape Peninsula dolerite dyke swarm, South Africa, in *Mafic Dykes and Emplacement Mechanisms,* edited by A. J. Parker, P. C. Rickwood, and D. H. Tucker, pp. 325-334, Balkema, Rotterdam, 1990.

Reid, D. L., and D. C. Rex, Cretaceous dykes associated with the opening of the South Atlantic: the Mehlberg dyke, northern Richtersveld, *South Afr. J. Geol.,* 97, 135-145, 1994.

Reid, D. L., A. J. Erlank, and D. C. Rex, Age and correlation of the False Bay dolerite dyke swarm, south-western Cape, Cape Province, *South Afr. J. Geol.,* 94, 155-158, 1991a.

Reid, D. L., I. G. D. Ransome, T. C. Onstott, and C. J. Adams, Time of emplacement and metamorphism of late Precambrian mafic dykes associated with the Pan-African Gariep orogeny, southern Africa: implications for the age of the Nama group, *J. Afr. Earth Sci.,* 13, 531-541, 1991b.

Renne, P. R., and A. R. Basu, Rapid eruption of the Siberian Traps flood basalts at the Permo-Triassic boundary, *Science,* 253, 176-179, 1991.

Renne, P. R., T. C. Onstott, M. S. D'Agrella-Filho, I. G. Pacca, and W. Teixeira, $^{40}Ar/^{39}Ar$ dating of 1.0–1.1 Ga magnetizations from the São Francisco and Kalahari cratons: tectonic implications for Pan-African and Brasiliano mobile belts, *Earth Planet. Sci. Lett.,* 101, 349-366, 1990.

Renne, P. R., D. F. Mertz, M. Ernesto, L. Marques, W. Teixeira, H. H. Ens, and M. A. Richards, Geochronologic constraints on magmatic and tectonic evolution of the Paraná Province (abstract), *Eos Trans. AGU 1993 Fall Meeting Suppl,* p. 553, 1993.

Renne, P. R., K. Deckart, M. Ernesto, G. Féraud, E. M. Piccirillo, Age of the Ponta Grossa dike swarm (Brazil), and implications to Paraná flood volcanism, *Earth Planet. Sci. Lett.,* 144, 199-211, 1996a.

Renne, P. R., J. M. Glen, S. C. Milner, A. R. Duncan, Age of Etendeka flood volcanism and associated intrusions in southwestern Africa, *Geology,* 24, 659-662, 1996b.

Richards, M. A., R. A. Duncan, and V. E. Courtillot, Flood basalts and hot spot tracks: Plume heads and tails, *Science,* 246, 103-107, 1989.

Robertson, W. A., and W. R. A. Baragar, The petrology and paleomagnetism of the Coronation sills, *Can. J. Earth Sci.,* 9, 123-140, 1972.

Rowley, D. B., and A. L. Lottes, Plate-kinematic reconstructions of the North Atlantic and Arctic: Late Jurassic to present, *Tectonophysics,* 155, 73-120, 1988.

Rosen, O. M., K. C. Condie, L. M. Natapov, and A. D. Nozhkin, Archean and Early Proterozoic evolution of the Siberian craton: a preliminary assessment, in *Archean Crustal Evolution, Developments in Precambrian Geology 11,* edited by K. C. Condie, pp. 411-459, Elsevier, Amsterdam, 1994.

Sant, D. A., and R. V. Karanth, Emplacement of dyke swarms in the Lower Narmada valley, western India, in *Mafic Dykes and Emplacement Mechanisms,* edited by A. J. Parker, P. C. Rickwood, and D. H. Tucker, pp. 383-389. Balkema, Rotterdam, 1990.

Sebai, A., V. Zumbo, G. Féraud, H. Bertrand, A.G. Hussain, G. Giannérini, and R. Campredon, $^{40}Ar/^{39}Ar$ dating of alkaline and tholeiitic magmatism of Saudi Arabia related to the early Red Sea rifting, *Earth Planet. Sci. Lett.,* 104, 473-487, 1991a.

Sebai, A., G. Feraud, H. Bertrand, and J. Hanes, $^{40}Ar/^{39}Ar$ dating and geochemistry of tholeiitic magmatism related to the early opening of the Central Atlantic rift, *Earth Planet. Sci. Lett.,* 104, 455-472, 1991b.

Şengör, A. M. C., and B. A. Natal'in, Paleotectonics of Asia: fragments of a synthesis, in *The Tectonic Evolution of Asia,* edited by A. Yin and T. M. Harrison, pp. 486-640, Cambridge University Press, 1996.

Shpount, B. R., and B. V. Oleinikov, A comparison of mafic dyke swarms from the Siberian and Russian platforms, in *Mafic Dyke Swarms, Spec. Pap. 34,* edited by H. C. Halls and W. F. Fahrig, pp. 393-400, Geological Association of Canada, Toronto, ON, 1987.

Sial, A. N., E. P. Oliveira, and A. Choudhuri, Mafic dyke swarms of Brazil, in *Mafic Dyke Swarms Spec. Pap. 34,* edited by H. C. Halls and W. F. Fahrig, pp. 379-383, Geological Association of Canada, Toronto, ON, 1987.

Smith. W. A., Paleomagnetic results from a crosscutting system of northwest and north-south trending diabase dikes in the North Carolina Piedmont, *Tectonophysics,* 136, 137-150, 1987.

Smith, R. B., and L. W. Braile, The Yellowstone hotspot: physical properties, topographic and seismic signature, and space-time evolution (abstract), *Eos Trans. AGU 1993 Fall Meeting Suppl,* p. 602, 1993..

Smythe, D. K., M. J. Russell, and A. G. Skuce, Intra-continental rifting from the major late Carboniferous quartz-dolerite dyke swarm of NW Europe, *Scott. J. Geol.,* 31, 151-162, 1995.

Southwick, D. L., and W. C. Day, Geology and petrology of Proterozoic mafic dikes, north-central Minnesota and western Ontario, *Can. J. Earth Sci.,* 20, 622-638, 1983.

Speight, J. M., R. R. Skelhorn, T. Sloan, and R. J. Knaap, The dyke swarms of Scotland, in *Igneous Rocks of the British Isles,* edited by D. S. Sutherland, pp. 449-459, Wiley and Sons, London, U.K., 1982.

Stewart, K., S. Turner, S. Kelley, C. Hawkesworth, L. Kirstein, M. Mantovani, 3-D, $^{40}Ar-^{39}Ar$ geochronology in the Paraná continental flood basalt province, *Earth Planet. Sci. Lett.,* 143, 95-109, 1996.

Storey, B. C., The role of mantle plumes in continental breakup:

case histories from Gondwanaland, *Nature*, 377, 301-308, 1995.

Storey, M., J. J. Mahoney, A. D. Saunders, R. A. Duncan, S. P. Kelley, and M. F. Coffin, Timing of hot spot-related volcanism and the breakup of Madagascar and India, *Science*, 267, 852-855, 1995.

St. Seymour, K., and P. S. Kumarapeli, Geochemistry of the Grenville dyke swarm: Role of plume-source mantle in magma genesis, *Contrib. Mineral. Petrol.*, 120, 29-41, 1995.

Sundvoll, B., and B. T. Larsen, Rb-Sr and Sm-Nd relationships in dyke and sill intrusions in the Oslo Rift and related areas, *Nor. Geol. Unders. Bull.*, 425, 25-41, 1993.

Svenningsen, O.M., Extensional deformation and dyke emplacement along the Late Precambrian Baltoscandian passive margin: the Sarek dyke swarm, Arctic Swedish Caledonides (abstract), *Program & Abstracts for the Third International Dyke Conference, Sept. 4-8, 1995, Jerusalem, Israel*, edited by A. Agnon, and G. Baer, pp. 73, 1995.

Swanson, D. A., T. L. Wright, and R. T. Helz, Linear vent systems and estimated rates of magma production and eruption for the Yakima basalt on the Columbia Plateau, *Am. J. Sci.*, 275, 877-905, 1975.

Thompson, R. N., and S. A. Gibson, Subcontinental mantle plumes, hotspots and pre-existing thinspots, *J. Geol. Soc. London*, 148, 973-977, 1991.

Tolan, T. L., S. P. Reidel, M. H. Beeson, J. L. Anderson, K. R. Fecht, and D. A. Swanson, Revisions to the estimates of the areal extent and volume of the Columbia River Basalt Group, in *Volcanism and Tectonism in the Columbia River Flood-Basalt Province, Spec. Pap. 239*, edited by S. P. Reidel, and P. R. Hooper, pp. 1-20, Geological Society of America, Boulder, CO, 1989.

Tollo, R. P., and F. E. Hutson, 700 Ma rift event in the Blue Ridge province of Virginia: a unique time constraint on pre-Iapetan rifting of Laurentia, *Geology*, 24, 59-62, 1996.

Tomshin, M. D., and O. V. Koroleva, Composite dykes of the Vilyuisk paleorift system, Siberian Platform, Yakutia, USSR in *Mafic Dykes and Emplacement Mechanisms*, edited by A. J. Parker, P. C. Rickwood, and D. H. Tucker, pp. 535-540, Balkema, Rotterdam, 1990.

Tomshin, M. D., and A. V. Okrugin, Dyke swarm of alkaline basites in the north of the Siberian platform, *Program with Abstracts, Third International Dyke Conference, Jerusalem, Israel*, edited by A. Agnon, and G. Baer, p. 76, 1995.

Tucker, D. H., and D. M. Boyd, Dykes of Australia detected by airborne magnetic surveys, in *Mafic Dyke Swarms, Spec. Pap. 34*, edited by H. C. Halls and W. F. Fahrig, pp. 163-172, Geological Association of Canada, Toronto, ON, 1987.

Turner, S., M. Regelous, S. Kelley, C. Hawkesworth, and M. Mantovani, Magmatism and continental break-up in the South Atlantic: high precision ^{40}Ar-^{39}Ar geochronology, *Earth Planet. Sci. Lett.*, 121, 333-348, 1994.

Vail, J. R., Tectonic controls of dykes and related irruptive rocks in eastern Africa, in *African Magmatism and Tectonics*, edited by T. N. Clifford, and I. G. Gass, pp. 337-354, Oliver & Boyd, Edinburgh, 1970.

Vakar, B. A., Trap formations of the Taymyr, *Petrology of Eastern Siberia: Volume 1, Siberian Platform and its Northern Rim*,

edited by G. D. Afanas'ev, pp. 256-340, Academy of Sciences of the U.S.S.R, 1962.

Walker, G. W., and N. S. MacLeod, *Geologic Map of Oregon, scale 1:500,000*, U.S. Department of the Interior, U.S. Geological Survey, Washington, D.C., 1991.

West, G. F., and R. E. Ernst, Evidence from aeromagnetics on the configuration of Matachewan dykes and the tectonic evolution of the Kapuskasing Structural Zone, Ontario, Canada, *Can. J. Earth Sci.*, 28, 1797-1811, 1991.

White, R. S., Magmatism during and after continental break-up, in *Magmatism and the Causes of Continental Break-Up, Spec. Publ. 68*, edited by B. C. Storey, T. Alabaster, and R. J. Pankhurst, pp. 1-16, The Geological Society, London, 1992.

White, R. S., and D. McKenzie, Magmatism at rift zones: the generation of volcanic continental margins and flood basalts, *J. Geophys. Res.*, 94, 7685-7729, 1989.

White, R. S., and D. McKenzie, Mantle plumes and flood basalts, *J. Geophys. Res.*, 100, 17543-17585, 1995.

Wilson, J. F., A craton and its cracks: some of the behaviour of the Zimbabwe block from the Late Archean to the Mesozoic in response to horizontal movements and the significance of some of its mafic dyke fracture patterns, *J. Afr. Earth Sci*, 10, 483-501, 1990.

Wilson, E. D., D. L. Jones, and J. D. Kramers, Mafic dyke swarms of Zimbabwe, in *Mafic Dyke Swarms, Spec. Pap. 34*, edited by H. C. Halls and W. F. Fahrig, pp. 433-444, Geological Association of Canada, Toronto, ON, 1987.

Windley, B. F., *The Evolving Continents*, John Wiley and Sons, 2nd ed., 1984.

Wirth, K. R., J. D. Vervoort, and L. M. Heaman, Nd isotopic constraints on mantle and crustal contributions to 2.08 Ga diabase dykes of the southern Superior Province (abstract), *Program & Abstracts for the Third International Dyke Conference, Sept. 4-8, 1995, Jerusalem, Israel*, edited by A. Agnon, and G. Baer, p. 84, 1995.

Woolley, A. R., and M. S. Garson, Petrochemical and tectonic relationship of the Malawi carbonatite-alkaline province and the Lupata-Lebombo volcanics, in *African Magmatism and Tectonics*, edited by T. N. Clifford, and I. G. Gass, pp. 237-262, Oliver & Boyd, Edinburgh, 1970.

Zhao, J.-X., and M. T. McCulloch, Sm-Nd mineral isochron ages of Late Proterozoic dyke swarms in Australia: evidence for two distinctive events of mafic magmatism and crustal extension, *Chem. Geol.*, 109, 341-354, 1993.

Zhao, J.-X, M. T. McCulloch, and R. J. Korsch, Characterization of a plume-related ~800 Ma magmatic event and its implications for basin formation in central-southern Australia, *Earth Planet. Sci. Lett.*, 121, 349-367, 1994.

Zoback, M. L., E. H. McKee, R. J. Blackely, and G. A. Thompson, The northern Nevada rift: Regional tectono-magmatic relations and middle Miocene stress direction, *Geol. Soc. Am. Bull.*, 106, 371-382, 1994.

Zolotukhin, V. V. and A. I. Al'Mukhamedov, Traps of the Siberian platform, in *Continental Flood Basalts*, edited by J. D. Macdougall, pp. 273-310, Kluwer Academic Publishers, Dordrecht, 1988.

Zumbo, V., G. Féraud, H. Bertrand, and G. Chazot, ^{40}Ar/^{39}Ar chronology of Tertiary magmatic activity in Southern Yemen during the early Red Sea-Aden rifting, *J. Volcanol. Geotherm. Res.*, 65, 265-279, 1995a.

Zumbo, V., G. Féraud, P. Vellutini, P. Piguet, and J. Vincent, First [40]Ar/[39]Ar dating on Early Pliocene to Plio-Pleistocene magmatic events of the Afar - Republic of Djibouti, *J. Volcanol. Geotherm. Res.*, 65, 281-295, 1995b.

Kenneth L. Buchan, Geological Survey of Canada, 601 Booth St., Ottawa, Ontario, K1A 0E8 Canada, kbuchan@gsc.nrcan.gc.ca.

Richard E. Ernst, Geological Survey of Canada, 601 Booth St., Ottawa, Ontario, K1A 0E8 Canada, rernst@gsc.nrcan.gc.ca; and Department of Earth Sciences, University of Western Ontario, London, Ontario, N6A 5B7 Canada.

Plume/Lithosphere Interaction in the Generation of Continental and Oceanic Flood Basalts: Chemical and Isotopic Constraints

John C. Lassiter[1] and Donald J. DePaolo

Department of Geology and Geophysics, UC Berkeley, Berkeley, California

The plume initiation model for flood basalt genesis predicts that melt generation will occur almost entirely within the plume head, beneath the lithosphere. However, isotopic and trace element differences between continental and oceanic flood basalts (CFBs and OFBs, respectively) require the incorporation of a lithospheric component in the former. Debate persists as to whether the "continental" signatures present in many CFBs derive from contamination of (plume-derived) melts with small volumes of highly incompatible-element-enriched lithospheric components (e.g., through crustal assimilation) or substantial melt generation within the lithospheric mantle. The chemical and isotopic compositions of OFBs and CFBs are used to constrain the mantle sources and melting environment responsible for flood basalt generation and the extent of crustal assimilation. Major and trace element trends in CFBs reflect smaller extents of partial melting than in OFBs and the presence of garnet in CFB source regions. These observations are consistent with the plume initiation model, because thick continental lithosphere will inhibit ascent and melt generation in sublithospheric plumes. However, melt generation from refractory, Fe-poor lithospheric mantle is also indicated for several CFBs, including the Siberian Traps. Depth of melt generation typically decreases with time for a given province, as indicated by the removal of a garnet signature in trace element ratios (e.g., decreasing Sm/Yb). These chemical and temporal trends suggest early melt generation within hydrous but otherwise refractory lithospheric mantle, followed by mechanical erosion of the lithosphere that allows additional ascent and melt generation within the underlying plume.

---

[1]Send correspondence to current address at: Department of Terrestrial Magnetism, Carnegie Institution of Washington, 5241 Broad Branch Road, NW, Washington, DC 20015

Large Igneous Provinces: Continental, Oceanic, and Planetary Flood Volcanism
Geophysical Monograph 100
Copyright 1997 by the American Geophysical Union

1. INTRODUCTION

The association of hotspots and flood basalts has long been recognized [e.g., *Morgan*, 1971]. Fluid-dynamic models [e.g., *Griffiths*, 1986; *Griffiths and Campbell*, 1990] have produced a plausible model for flood basalts as forming when large plume heads associated with new mantle plumes ascend beneath the lithosphere. However, considerable debate persists concerning the actual source of magmas for flood basalts. Several theoretical treatments of plume/lithosphere interaction have concluded that melt generation in response to plume ascent is largely restricted to the plume itself, with little melt generation occurring within the lithospheric mantle [e.g., *Arndt and Christensen*, 1992; *Farnetani and Richards*, 1994]. However, many continental flood basalts (CFBs) have isotopic and trace element signatures very dissimilar to those found in either oceanic flood basalts (OFBs) or ocean island basalts (OIBs). Although crustal assimilation may account for some of the differences between continental and oceanic flood basalts, several studies have proposed that the continental litho-spheric mantle (CLM) is a major source of CFB melts [e.g., *Lightfoot et al.*, 1990a; *Gallagher and Hawkesworth*, 1992].

In this paper, we review the available chemical and isotopic data for lavas from several continental and oceanic flood basalt provinces and OIBs. Our purpose is to determine the role of the lithospheric mantle in the generation of flood basalts and to evaluate whether the predictions concerning melt generation inherent in the plume initiation model are borne out. For example, do CFBs, which form in regions where the lithosphere is old and thick, reflect smaller percentages of melting at greater depth than OFBs (or OIBs), as predicted by the plume initiation model? In order to establish the role of the lithospheric mantle in the generation of flood basalts, it is necessary to distinguish chemical signatures arising from crustal contamination from those generated by partial melting of the CLM. Are the distinctive chemical and isotopic signatures of CFBs explained adequately by crustal contamination alone, or is melt generation within the CLM required? We review the chemical evidence for both crustal contamination and melt generation from the CLM as well as the inherent limitations of geochemical studies for uniquely distinguishing between the two. Although every flood basalt province is unique, our goal is to illustrate several features that are common to the evolution of many continental and oceanic flood basalts. By so doing, we hope to develop a generic model of plume/lithosphere interaction that is consistent with the available observations. Finally, by discussing the limitations of traditional isotopic and trace element studies for distinguishing crustal contam-ination from CLM melting, we seek to stimulate future research capable of addressing this long-standing problem in flood basalt studies.

2. MELT FRACTIONS AND DEPTH OF MELT GENERATION

Systematic relationships between lithosphere thickness, depth of melting, and melt fraction are expected for plume-generated basalts. As hot mantle ascends in upwelling zones, it intersects its solidus at some depth and begins to partially melt. The depth at which melting begins is largely a function of the potential temperature of the ascending mantle [e.g., *McKenzie and Bickle*, 1988], although variations in composition, especially volatile content, may also result in more or less fertile source regions [e.g., *Falloon and Green*, 1989]. In extensional or rift environments, such as at mid-ocean ridges, lithosphere is thin or absent, so there is essentially no barrier to adiabatic ascent [*McKenzie and Bickle*, 1988]. The thickness of the resultant melt zone and the total amount of melt produced are directly related to potential temperature, so that average depth and average degree of melting are strongly correlated for different mid-ocean ridge segments [e.g., *Klein and Langmuir*, 1987]. In non-rifting environments, the lithosphere acts as a mechanical barrier to mantle ascent. The thickness of the melt zone and the maximum extent of partial melting are therefore limited by the lithosphere thickness [e.g., *McKenzie and O'Nions*, 1995]. Consequently, plumes rising beneath thick, rigid lithosphere should undergo less partial melting than plumes rising under thin lithosphere, and the average depth of melting will be greater [e.g., *Ellam*, 1992; *Saunders et al.*, 1992].

The depth and extent of melting in plumes can be estimated from analysis of both the major and trace element compositions of basalts. For example, with increasing depth of origin FeO^T (total Fe as FeO) increases, and with increasing melt fraction at a given pressure Na$_2$O decreases and CaO/Al$_2$O$_3$ increase [e.g., *Hirose and Kushiro*, 1993; *Baker and Stolper*, 1994]. For trace elements, incompatible element concentrations are approximately inversely pro-portional to melt fraction. Depth of melting is qualitatively indicated by the presence or absence of a residual-garnet signature in the basalts, because garnet is stable in mantle peridotite only at depths greater than ~75 km [cf. *Nickel*, 1986].

The effects of fractional crystallization on chemical composition must be removed in order to evaluate chemical signatures deriving from melt generation. The majority of lavas from both continental and oceanic flood basalt provinces have undergone significant fractionation, evidenced by low Mg#s [=100 × molar Mg/(Mg+Fe^{2+})] and MgO and Ni contents [e.g., *Mahoney*, 1988; *Wooden et al.*, 1993; *Lassiter et al.*, 1995a]. To compare the compositions of lavas that have undergone different fractionation histories, major element compositions were corrected to a constant 8 wt% MgO using best-fit linear regressions, excluding picritic samples and samples with less than 4 wt% MgO [e.g., *Langmuir et al.*, 1992]. This regression technique is similar to the regressions of *Peng et al.* [1994] and *Turner and Hawkesworth* [1995]. The main limitations are the poorly defined major element trends of many flood basalt suites, which result in large uncertainties in the regressions, and the fact that fractionation trends are only approximately linear because different phases may become saturated at different stages of fractionation. The results of our regressions are generally consistent with those of *Turner and Hawkesworth* [1995] and *Peng et al.* [1994], although some differences exist. For example, we calculate somewhat lower Fe_8 and higher Na_8 for average Deccan lavas than do *Turner and Hawkesworth* [1995] (subscript denotes oxide extrapolation to 8 wt%). More importantly, we also utilize incompatible element ratios to assess the depth and extent of melting. As we discuss below, our estimates for relative depth and degree of melting obtained by these two independent methods are largely mutually consistent.

2.1. Major Element Regressions

Figure 1 shows calculated mean Na_8 and $(Ca/Al)_8$ values plotted against Fe_8 for several continental and oceanic flood basalt suites. Also shown are average extrapolated values for two intraplate hotspots (Hawaii and Réunion) and one on-ridge hotspot (Iceland). The mid-ocean ridge basalt (MORB) array provides a reference corresponding to basalts produced where there is approximately zero-thickness lithosphere, but where large variations in mantle potential temperature result in strong correlations between depth of melting and melt fraction. If plume-source and MORB-source mantle have similar major element compositions, intraplate plume-derived melts should be shifted off the MORB array so that smaller melt fractions (evidenced by higher Na_8 and lower $(Ca/Al)_8$) are associated with a higher pressure (greater depth) of origin (higher Fe_8), as qualitatively shown in Figure 1.

Many CFBs have major element compositions consistent with the effects outlined above of melt generation beneath a lithospheric lid. With the possible exception of the Columbia River basalts (but see below) and some Gondwana basalts, all of the OFBs, CFBs, and OIBs examined possess high Fe_8 compared with most MORB (e.g., >10 wt%), suggesting derivation from mantle with anomalously high potential temperature. Lavas from the Siberian and Deccan Traps have higher Na_8 and lower $(Ca/Al)_8$ for a given Fe_8 than most MORB and plot at the edge of or outside the MORB array in Figure 1. The Wrangellia flood basalts also possess major element compositions suggestive of melt generation beneath a lithospheric lid. The Wrangellia flood basalts are unique

among the OFBs considered here because, although these flood basalts were erupted in an oceanic environment, the preexisting arc lithosphere through which the basalts

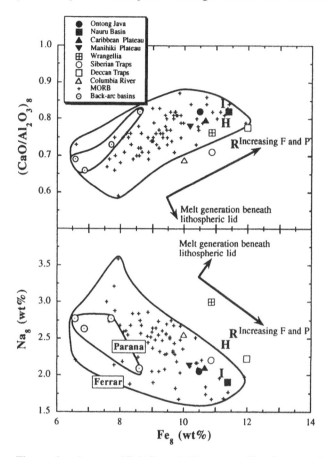

Figure 1. Average $(Ca/Al)_8$ and Na_8 versus Fe_8 for several continental and oceanic flood basalt suites and three ocean island hotspots. Data for these and subsequent figures are from the following: Ontong Java Plateau [*Mahoney*, 1987; *Mahoney and Spencer*, 1991; Mahoney et al., 1993]; Nauru Basin [*Floyd*, 1986; *Mahoney*, 1987; *Mahoney and Spencer*, 1991; *Castillo et al.*, 1991]; Caribbean Plateau [Donnelly et al., 1973; *Aitken and Etchevera*, 1984; *Sen et al.*, 1988; *Kerr et al.*, 1996]; Manihiki Plateau [*Jackson et al.*, 1976; *Clague*, 1976; *Mahoney*, 1987; *Mahoney and Spencer*, 1991]; Wrangellia [*Lassiter et al.*, 1995a]; Siberian Traps [*DePaolo and Wasserburg*, 1979; *Lightfoot et al.*, 1990b; *Sharma et al.*, 1991; *Sharma et al.*, 1992; *Lightfoot et al.*, 1993; *Wooden et al.*, 1993]; Deccan Traps [*Mahoney et al.*, 1982; *Cox and Hawkesworth*, 1985; *Mahoney*, 1988; *Lightfoot et al.*, 1990a; *Peng et al.*, 1994]; Columbia River [*Carlson et al.*, 1981; *Carlson*, 1984; *Nelson*, 1983; *Nelson*, 1989; *Hooper and Hawkesworth*, 1993]; Iceland (I) [*Hemond et al.*, 1993]; Hawaii (H) [*Rhodes*, 1996; *Yang et al.*, 1996]; Réunion (R) [*Fisk et al.*, 1988; *Albarede and Tamagnan*, 1988]. Averages for MORB and back-arc basin basalts are from *Klein and Langmuir* [1987]. Paraná and Ferrar low-Ti basalt averages recalculated from *Hergt et al.* [1991].

erupted was more similar in terms of its thickness and composition to continental lithosphere than to oceanic lithosphere. The Wrangellia flood basalts therefore represent a hybrid flood basalt province, sharing similarities to both continental and oceanic flood basalts [*Lassiter et al.*, 1995a]. That the inferred melting environments of the Wrangellia flood basalts and some CFBs are similar is thus not entirely unexpected.

Most other OFBs, including the Ontong Java Plateau and Nauru Basin, the Manihiki Plateau, and the Caribbean Basin, all show evidence of high melt fractions and depth of origin but no effect of a lithospheric lid because they plot well within the high-melt-fraction end of the MORB field. This observation is consistent with previous conclusions and with suggestions that several of these plateaus formed near ridges or triple junctions [e.g., *Mahoney*, 1987; *Mahoney et al.*, 1993]. Icelandic basalts also plot within the high-melt-fraction end of the MORB-array, as expected given Iceland's location along a spreading ridge. In contrast, lavas from the intraplate Hawaiian and Réunion hotspots have higher Na_8, and lavas from Réunion have lower $(Ca/Al)_8$ for their Fe_8 values than basalts within the MORB-array. As with many of the CFB suites, these trends suggest melt generation beneath the lithosphere within hot mantle plumes.

Because of the well-documented isotopic differences between MORB, flood basalts, and OIBs, we must consider whether the chemical trends described above reflect variations in the major element composition of the mantle sources rather than differences in melting conditions. Recent studies indicate that major element variations are present in the plume sources of some OIBs [e.g., *Hauri*, 1996]. However, the shifts away from the MORB melting array observed for several CFBs and for lavas from Wrangellia, Réunion, and Hawaii are unlikely to be the result of such variations, because most OFBs and on-ridge hotspot lavas from Iceland do not record such shifts in major element composition despite possessing similar Sr- and Nd-isotopic signatures as the Réunion, Hawaii, and Wrangellia lavas.

Crustal assimilation may also affect the extrapolated major element compositions of CFBs. For example, the low Fe_8 inferred for the Columbia River basalts appears in part to be the result of crustal contamination. Figure 2 compares the estimated Fe_8 and Na_8 for the Imnaha and the Grande Ronde suites of the Columbia River basalts. Crustal assimilation appears to have strongly affected the isotopic and trace element compositions of the Grande Ronde basalts [e.g., *Carlson et al.*, 1981; *Nelson*, 1989]. In contrast, the Imnaha basalts appear more directly derived from a mantle plume source with less crustal

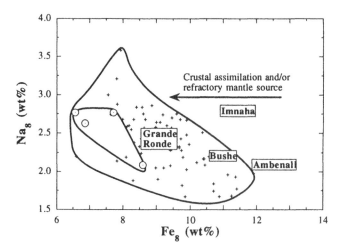

Figure 2. Typical effects of crustal assimilation on flood basalt composition are qualitatively illustrated by the lower Fe_8 values of (crustally contaminated) Bushe and Grande Ronde lavas compared with less contaminated lavas from the Ambenali and Imnaha suites of the Deccan Traps and Columbia River Basalt Province. The trend to lower Fe_8 produced by crustal contamination or melt generation from an Fe-poor source (e.g., back-arc basin lavas) is roughly opposite that expected for melt generation at high pressure beneath a litho-spheric lid. Symbols for MORB and back-arc lavas are as in Figure 1.

interaction [e.g., *Brandon and Goles*, 1988; *Hooper and Hawkesworth*, 1993]. Consistent with this conclusion, the Imnaha basalts have higher Fe_8 values than Grande Ronde basalts and have major element compositions that suggest they were generated beneath a lithospheric lid. A similar comparison can be made for the Bushe and Ambenali suites of the Deccan Traps. Although oxygen isotopic data for phenocrysts indicate that the Bushe and several other early formations in the Deccan assimilated crustal material, Ambenali lavas are interpreted to more closely record the isotopic composition of a sublithospheric source [e.g., *Cox and Hawkesworth*, 1985; *Peng et al.*, 1994]. As with the Grande Ronde basalts, Peng et al. [1994] noted that Bushe lavas have lower Fe_8 than less contaminated lavas such as the Ambenali. These examples illustrate that crustal contamination will in general lower the inferred Fe_8 content of a suite of lavas. The high Fe_8 values of CFBs such as the Deccan and Siberian Traps therefore cannot be explained by this mechanism.

Melt generation from refractory mantle sources (e.g., the CLM) will have a similar effect on Fe_8. The low Fe content of many back-arc lavas (Figure 1) suggests they are derived from a source more refractory than typical MORB-source mantle [e.g., *Falloon et al.*, 1994] and thus are similar to what we expect for melts derived from refractory lithosphere. CLM-derived melts should commonly possess lower Fe contents for two reasons. First, Fe content in primitive magmas is strongly influenced by the pressure of melt generation, with Fe increasing with increasing pressure [e.g., *Hirose and Kushiro*, 1993]. Therefore, for similar source compositions, melts from sublithospheric plumes should have higher Fe contents because of their greater depth of origin. Also, lithospheric mantle is likely to be more refractory on average than the underlying convecting mantle. Melt extraction, and the accompanying chemical buoyancy this imparts, may be important for stabilizing ancient lithospheric mantle [e.g., *Jordan*, 1978]. Mantle xenoliths typically are depleted in a basaltic component; e.g., they have lower Fe, Ca, Al, and Na relative to primitive mantle estimates [e.g., *McDonough*, 1990, 1992]. As a result, melts generated from refractory CLM should have lower Fe than melts generated under comparable conditions from asthenospheric mantle. Picrites from the Karoo confirm that CLM-derived melts can be relatively Fe-depleted. *Ellam et al.* [1992] showed that Karoo picrites with a large lithospheric-mantle-derived component (as indicated by unradiogenic Os isotopes) had lower Fe contents than picrites without a lithosphere-derived com-ponent. Thus, crustal contamination and melt generation from refractory lithospheric mantle both reduce the inferred Fe content of parental magmas and thereby obscure the signatures of melt generation beneath a lithospheric lid.

Because the major element signatures of both crustal contamination and melt generation from refractory litho-sphere are almost opposite to the signatures produced by melt generation from plumes beneath the lithosphere, the high Na_8 and low $(Ca/Al)_8$ for a given Fe_8 recorded by average lavas from the Deccan and Siberian Traps and Wrangellia relative to most MORB or OFBs likely reflect melt generation beneath the lithosphere. This conclusion is further strengthened from consideration of trace element fractionations in these lavas, as discussed below.

2.2. Trace Element Ratios

Trace element abundance ratios provide an independent means of assessing the depth and extent of partial melting responsible for generation of flood basalts. Because rare earth element (REE) partition coefficients in most common mantle phases are well established, and because the bulk solid/melt partitioning of heavy REE is dramatically different for garnet and spinel peridotite, REE profiles are particularly useful for determining the relative extents of partial melting and the mineralogy (and therefore depth) of the source of flood basalts. Partial melting of either garnet

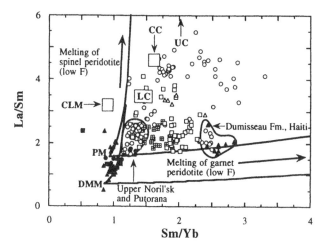

Figure 3. La/Sm and Sm/Yb values for continental and oceanic flood basalt suites. Batch melting trends for garnet and spinel peridotite calculated using the partition coefficients, modal abundances, and primitive (PM) and depleted (DMM) mantle trace element ratios of *McKenzie and O'Nions* [1991]. Arrows denote the effect of decreasing melt fraction (F). Median composition of CLM [*McDonough*, 1990], lower (LC), upper (UC), and bulk continental crust (CC) [*Taylor and McLennan*, 1985] shown for comparison. Symbols are as in Figure 1.

or spinel peridotite will preferentially enrich the lighter REE such as La, which are more incompatible in most mantle phases than middle REE such as Sm. However, the degree of enrichment of middle REE relative to heavy REE such as Yb depends greatly on whether garnet exists as a residual phase during melting, because the heavy REE are preferentially retained by garnet during melting but not by most other mantle phases [cf. compilation of distribution coefficients by *McKenzie and O'Nions*, 1991]. Further-more, fractional crystallization will produce only modest changes in La/Sm and Sm/Yb ratios compared with variations caused by changes in melt fraction or source mineralogy. For the REE, the evolved nature of many flood basalts therefore introduces fewer uncertainties than it did in our assessment of major element trends.

The enrichments in La/Sm and Sm/Yb produced by batch melting of spinel and garnet peridotite are shown in Figure 3, along with La/Sm and Sm/Yb values for lavas from several continental and oceanic flood basalt provinces. Lavas from the Ontong Java Plateau and the Nauru Basin have relatively flat REE patterns, with lower La/Sm and Sm/Yb than most continental flood basalts, and appear to represent high-degree melts [*Castillo et al.*, 1991; *Mahoney et al.*, 1993]. Lavas from the Caribbean Plateau have variable La/Sm and Sm/Yb ratios. Basalts from Curaçao and drill sites in the central Caribbean primarily have low Sm/Yb ratios similar to those of

Ontong Java and Nauru Basin lavas [*Donnelly et al.*, 1973; *Kerr et al.*, 1996], but lavas from the Dumisseau Formation, Haiti, have high Sm/Yb ratios and appear to have been generated through smaller degrees of melting of a garnet-bearing source [*Sen et al.*, 1988]. Though not shown in Figure 3, lavas from the Manihiki Plateau also have nearly flat REE patterns [*Jackson et al.*, 1976], similar to lavas from Ontong Java and Nauru Basin.

In contrast, most CFBs have higher La/Sm and Sm/Yb than OFBs. The higher La/Sm and Sm/Yb in most CFBs suggest they are generated through smaller degrees of partial melting than most OFBs (see below for discussion of the effects of crustal contamination). The Wrangellia flood basalts have lower La/Sm and Sm/Yb than most CFBs but higher than other OFBs, which indicates these basalts were generated through intermediate degrees of partial melting. As we observed for the major element compositions of the Wrangellia basalts, this is consistent with the prevolcanic history of Wrangellia, because these flood basalts erupted through preexisting are lithosphere which, given its age (minimum of ~55 to 125 Ma at the time of volcanism [*Lassiter et al.*, 1995b]), was most likely thicker than average oceanic lithosphere but thinner than ancient continental lithosphere.

Compared with the oceanic plateaus, continental flood basalts display much greater scatter in their REE profiles. However, much of this scatter arises from contamination of plume melts by melts derived from continental crust and/or lithospheric mantle. As with the major element trends, we must examine the effects of this contamination on trace element ratios in order to understand the plume melting environment. A common characteristic of this contamination in many continental flood basalt suites is a depletion in high-field-strength elements (HFSEs), particularly Nb and Ta, relative to other similarly incompatible elements. For example, CFBs commonly have sub-chondritic Nb/La and Ta/La ratios, whereas most oceanic plume-derived basalts (OFBs and OIBs) have higher than chondritic Nb/La and Ta/La [e.g., *Arndt and Christensen*, 1992]. In order to examine the conditions of melt generation within the plumes responsible for flood basalt volcanism, we have attempted to filter out the greater part of the continental contamination by only plotting in Figure 4 data for lavas with minimal Nb-Ta depletions (e.g., Nb/La or Ta/La > 0.85 × chondritic value). When the data are filtered in this manner, the scatter present in Figure 3 is reduced significantly. More importantly, although the average La/Sm of many CFB suites is diminished, the Sm/Yb values are similar to the unfiltered data. As a result, it is much more apparent from the shallow La/Sm-Sm/Yb trend of the filtered data that

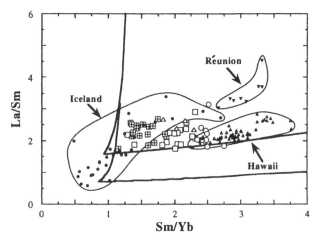

Figure 4. La/Sm-Sm/Yb values for continental flood basalt lavas (and Wrangellia) with minimal Nb-Ta depletions. Only samples with Nb/La > 0.9 or La/Ta < 20 are included. Field for Iceland lavas includes samples with > 5 wt% MgO. Batch melting curves are as in Figure 3. Symbols for Wrangellia, Siberian and Deccan Traps, and Columbia River basalts are as in Figure 1.

melting of garnet-bearing sources were important in the generation of these lavas. Because garnet is only stable in the mantle at depths greater than ~75 km [e.g., *Nickel*, 1986], the more pronounced garnet signature in the REE patterns of relatively un-contaminated CFBs compared with most OFBs is consistent with the major element indications that CFBs are primarily generated through smaller degrees of partial melting of anomalously hot mantle beneath a thick lithospheric lid.

Comparison of the major element and trace element trends for individual suites of lavas further strengthens this conclusion. For example, most Columbia River basalts have low Nb/La values and also have lower Fe_8 values than many other flood basalts (Figure 1). However, lavas from the relatively uncontaminated Imnaha suite of the Columbia River Basalt Group have high Fe_8 and Na_8 (Figure 2), which we interpret as resulting from melt generation beneath the continental lithosphere. The Imnaha lavas, represented in Figure 4, are also characterized by near-chondritic Nb/La values and have high Sm/Yb and moderate La/Sm values that are most readily explained by melting of a garnet-bearing source.

Data for three ocean island suites are also plotted in Figure 4. Lavas from the intraplate Hawaiian and Réunion hotspots are strongly enriched in Sm relative to Yb, again indicating melt generation from a garnet-bearing source. Icelandic basalts show a considerable range in La/Sm and Sm/Yb, but on average have much lower Sm/Yb ratios than Hawaiian or Réunion lavas and have REE patterns similar to those of many OFBs. Again, these results are consistent with the differences in major element compositions between Icelandic and intraplate hotspot lavas. Icelandic lavas have lower Na_8 for a given Fe_8 than lavas from either Hawaii or Réunion. Because the Hawaii, Réunion, and Iceland plumes are all characterized by similar isotopic signatures (e.g., high $^3He/^4He$ but ε_{Nd} and $^{87}Sr/^{86}Sr$ values intermediate between those found in MORB and inferred for the bulk earth [*Condomines et al.,* 1983; *Albarede and Tamagnan,* 1988; *Kurz et al.,* 1996]), the major and trace element differences between lavas from the intraplate Hawaiian and Réunion hotspots and the on-ridge Iceland hotspot are unlikely to reflect systematic differences in the compo-sitions of these plumes. The observed compositional differences between Iceland, Hawaii, and Réunion lavas are instead what are expected given the tectonic settings of these three hotspots, and are consistent with the conclusions of *Ellam* [1992], who reported a similar relationship between average Ce/Yb in ocean island basalts and the age of the lithosphere through which the basalts were erupted.

An unexpected finding of our analysis is that basalts from Réunion and to a lesser extent from Hawaii apparently represent smaller degrees of melting than either continental or oceanic flood basalts. A comparison of Réunion and Deccan basalts in Figures 1 and 4 is especially instructive, because initiation of the Réunion plume has been linked to the formation of the Deccan Traps [e.g., *Vandamme and Courtillot,* 1990]. In particular, once the effects of continental contamination are removed, for instance by comparing Réunion lavas with the relatively uncon-taminated Ambenali suite, the Réunion lavas have major element compositions that plot farther from the MORB field and have greater incompatible element fractionations, indicating derivation from smaller extents of melting [e.g., *Ellam,* 1992]. This may seem surprising, because plume "tails" should be at least as hot as plume "heads," as entrainment of ambient mantle in rising plume heads will tend to reduce the potential temperature of the head [*Griffiths and Campbell,* 1990]. Furthermore, ancient con-tinental lithosphere is typically several hundred kilometers thick, much thicker than average oceanic lithosphere (~100 km) [e.g., *Chapman and Pollack,* 1977]. As we discuss in a later section, the higher melt fractions inferred for many CFBs than for intraplate hotspot basalts may reflect the erosion of continental lithosphere that often accompanies flood basalt volcanism. Erosion of oceanic lithosphere by steady-state plume "tails" may be significantly less pronounced, perhaps because oceanic lithosphere is typically drier and therefore stronger than continental lithosphere or because the time

during which any portion of a moving lithospheric plate resides over a narrow plume conduit is too short to permit significant heating and softening of the lithospheric lid.

3. CRUSTAL ASSIMILATION OR MELT GENERATION WITHIN THE CLM?

The above discussion illustrates the strong control that the lithospheric mantle has on melting in sublithospheric plumes. This influence over the melting environment is generally consistent with the plume initiation model for flood basalt genesis, which predicts melt generation to be largely confined to the plume head beneath the lithosphere [e.g., *Arndt and Christensen,* 1992; *Farnetani and Richards,* 1994]. However, pronounced isotopic and trace element differences exist between continental and oceanic flood basalts (Figure 5). Because high $^3He/^4He$ ratios in lavas associated with several CFBs suggest a component of undegassed lower mantle origin [*Basu et al.,* 1993; *Basu et al.,* 1995; *Dodson et al.,* 1996], the compositions of deep-seated mantle plumes ascending beneath the continents should not systematically differ from those ascending beneath ocean basins. Therefore, the isotopic and trace element differences between CFBs on the one hand and OIBs and OFBs on the other require the incorporation of a "continental" component in the former. If these isotopic differences can be explained as resulting from volumetrically small additions of material to plume-derived melts, such as through crustal assimilation, then there is no problem with plume model predictions that melting is confined largely to the plume interior. However, if a substantial fraction of magma production occurs within the lithosphere, we would need to modify current models of plume/lithosphere interaction. It is important to remember that contamination of plume-derived basalts from continental crust and melt generation from the lithospheric mantle are not mutually exclusive, but that it is the magnitude of the latter that may help constrain models of plume/lithosphere interaction.

Unfortunately, resolving this most basic question on the origin of the isotopic and trace element signatures in CFBs has proven extremely difficult. Because of the heterogeneity of both continental crust and the CLM and the complexity of the processes involved in partial melting and assimilation, most geochemical studies have failed to conclusively prove or disprove the involvement of specific components. In the following discussion, we review the geochemical evidence and circumstantial arguments in support of both crustal and lithospheric-mantle sources for the "continental component" in various CFB suites.

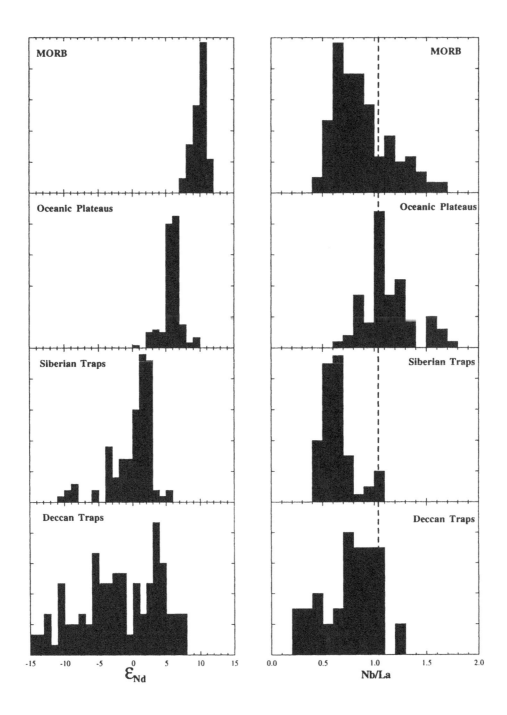

Figure 5. Representative ranges for εNd and Nb/La in different tectonic settings. Dashed line denotes chondritic Nb/La ratio *[Sun and McDonough,* 1989]. CFBs are isotopically much more heterogeneous than either OFBs or MORB. OFB ranges include samples from the Ontong Java Plateau, Nauru Basin, Manihiki Plateau, Caribbean Basin, and Wrangellia Terrane. MORB data sources: *Dosso et al.* [1993]; *Price et al.* [1986]; *Prinzhofer et al.* [1989]; *Michael* [1995].

3.1. Chemical Discrimination of Crust and CLM Components in CFBs

In order to distinguish the effects of crustal assimilation from assimilation or melting of the CLM, we must examine geochemical signatures that are clearly different in these two reservoirs. Oxygen isotope data, particularly for phenocryst phases, show promise in this regard, because O isotope variations in both asthenospheric and lithospheric mantle are small compared with the large ^{18}O-enrichments commonly found in continental crust. Fresh MORB possess well-defined δ^{18}O values of 5.7 ± 0.3‰ [Ito et al., 1987]. CLM-derived xenoliths show slightly greater O isotope heterogeneity, but for the most part also possess δ^{18}O < 6‰ [e.g., Kyser, 1986; Mattey et al., 1994]. In contrast, many CFBs have elevated δ^{18}O, suggesting they have assimilated ^{18}O-enriched continental crust. For example, phenocrysts in the Bushe and several other lower stratigraphic formations of the Deccan Traps have higher δ^{18}O values than commonly found in uncontaminated mantle-derived melts or xenoliths [Peng et al., 1994]. Furthermore, there is a rough positive correlation in Deccan lavas between δ^{18}O values and other indications of "continental" contamination such as elevated ^{87}Sr/^{86}Sr. Lavas of the Grande Ronde suite from the Columbia River Basalt Province are also characterized by elevated whole-rock δ^{18}O values [e.g., Nelson, 1983; Carlson, 1984], as are many lavas of the Paraná [e.g., Fodor et al., 1985; Harris et al., 1990]. Unfortunately, relatively few detailed studies of stable isotope systematics in CFBs have been performed. Especially needed are studies examining δ^{18}O values in both whole-rock and phenocryst phases, because O isotopes in phenocrysts are more resistant to postmagmatic alteration than are whole-rocks [e.g., Eiler et al., 1996].

Osmium isotopes also have the potential to differentiate crustal and CLM contributions to CFBs. Continental crust has high Re/Os and is characterized by extremely radiogenic Os isotopes. In contrast, numerous studies have demonstrated that the CLM is characterized by subchondritic γ_{Os} due to long-term Re depletion [e.g., Walker et al., 1989; Carlson and Irving, 1994; Pearson et al., 1995]. Therefore, assimilation of crustal and CLM components will have opposite effects on Os isotopes. Unfortunately, because Os behaves as a compatible element during fractional crystallization, Os abundances in the highly evolved lavas typically dominant in flood basalt provinces are extremely low. As a result, most Os isotopic studies of CFBs have concentrated on relatively rare picrites. However, the high Os abundances in picrites make them much less sensitive to crustal assimilation than

Os isotopes in basaltic lavas. Nevertheless, several Os isotope studies have suggested CLM additions to flood basalts rather than addition of continental crust. Osmium isotope variations in picrites from the Karoo province provide the most unambiguous evidence for a CLM component in a flood basalt sequence. These picrites display a positive correlation between Os and Nd isotopes, so that picrites with the strongest "continental" signatures (e.g., low ε_{Nd}) have the lowest γ_{Os}. Therefore, the continental component in the Karoo picrites must have been derived from the (low γ_{Os}) CLM rather than (high γ_{Os}) continental crust [Ellam et al., 1992]. Other Os isotope studies have suggested a CLM component in picrites from the North American midcontinent rift [Shirey, 1997] and the Siberian Traps [e.g., Horan et al., 1995].

Recent improvements in low-blank Re and Os analysis and sample/spike equilibration methods (thus allowing more accurate Re/Os determinations) have opened the window for Os isotopic studies of the low-MgO (and therefore low-Os) basalts that make up the bulk of most CFB provinces. Low-Os basalts are much more sensitive to crustal contamination than high-Os picrites, so that the presence or absence of radiogenic Os isotopes in these basalts places a strong constraint on the timing and extent of crustal contamination. For example, basalts of the Ferrar flood basalt province have initial γ_{Os} values only slightly higher than values typically found in oceanic plume-derived lavas. Therefore, either these basalts were not crustally contaminated to any significant degree, or crustal contamination occurred when the basalts were much more picritic so that the higher Os concentrations buffered the effects of crustal assimilation [Molzahn et al., 1996]. In contrast, the Grande Ronde basalts of the Columbia River basalt province have extremely radiogenic Os isotopes, whereas lavas of the Imnaha group in the same province do not [Chesley et al., 1996]. As we discussed earlier, the Grande Ronde basalts have been strongly influenced by crustal contamination, but the Imnaha basalts appear to be essentially unmodified plume-derived basalts. These studies illustrate the potential for Os isotopes to reveal the respective roles of crustal assimilation and CLM melting in flood basalt genesis.

In contrast to Os and O isotopes, most trace element ratios are much less diagnostic of crustal or CLM contamination. For example, many CFBs have pronounced depletions in HFSEs relative to plume-related basalts from oceanic settings (OIBs and OFBs), as indicated, for example, by low Nb/La (Figure 5; see also Arndt and Christensen [1992]). Average continental crust is also strongly depleted in HFSEs [e.g., Taylor and McLennan, 1985]. In contrast, although xenoliths derived from the

CLM have highly variable Nb/La ratios, on average they are not depleted in HFSEs [e.g., *McDonough*, 1990]. These observations have led many researchers to conclude that the low Nb/La values present in many CFB suites require crustal assimilation rather than assimilation of CLM-derived melts [e.g., *Wooden et al.*, 1993; *Arndt et al.*, 1993; *Brandon and Goles*, 1995].

Unfortunately, HFSE depletions in CFBs are not as diagnostic of crustal contamination as is commonly assumed. The conclusion that HFSE depletions are uniquely diagnostic of crustal assimilation is belied by the existence of CLM-derived, HFSE-depleted lavas. Many lamproites, for example, have Nb/La < 1 [*Mitchell and Bergman*, 1991], as do minettes from the Colorado Plateau [*Thompson et al.*, 1990]. The Karoo picrites clearly illustrate that at least some CFBs derive their HFSE depletions from the CLM rather than continental crust. These picrites possess strong negative Nb anomalies [*Ellam and Cox*, 1989, 1991] that are correlated with γ_{Os} (Figure 6). This correlation is opposite that expected for assimilation of high-γ_{Os} crustal material and therefore indicates that the extremely low Nb/La values in these picrites and associated basalts must have been derived from some component or process within the low-γ_{Os} CLM. A number of factors could explain why appropriate sources for the Karoo picrites are not well represented in global xenolith data bases. Mantle xenoliths may provide a chemically biased sampling of the CLM, or Nb/La ratios may be fractionated during partial melting to a greater extent than is commonly assumed (e.g., Nb may be retained by trace phases such as rutile at small degrees of partial melting [*Green and Pearson*, 1987], thereby decreasing Nb/La in the melt). In any event, although the origin of the HFSE depletions in the Karoo picrites is not completely understood, they nevertheless demonstrate that such depletions cannot be used to rule out the presence of CLM-derived components in other flood basalt suites.

3.2. *Melt Generation from the CLM: Evidence from the Siberian Traps*

The Siberian Traps provide an excellent example for assessing the relative importance of plume, crustal, and lithospheric mantle components in a CFB. Although this province has been studied extensively, little consensus exists concerning the nature of the mantle source(s) for these lavas or the degree to which the magmas assimilated crustal material. For example, *Sharma et al.* [1991, 1992] concluded that the bulk of the Siberian flood basalts interacted little with either the lithospheric mantle or crust and that their compositions closely resemble the

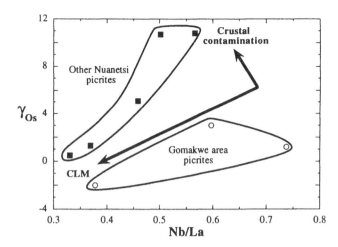

Figure 6. The positive correlation between Nb/La and γ_{Os} observed for picrites from the Karoo flood basalt province is inconsistent with the derivation of HFSE depletions (e.g., low Nb/La) through crustal assimilation, but is consistent with mixing between melts derived from a high γ_{Os}, high Nb/La plume source and a low γ_{Os}, low Nb/La (and Fe poor) source within the CLM. Picrites sampled from near the Gomakwe intrusion consistently plot along separate, subparallel trends in various element-isotope plots [*Ellam et al.*, 1992]. Data for Karoo picrites: *Ellam and Cox* [1989]; *Ellam and Cox* [1991]; *Ellam et al.* [1992].

composition of their plume source. In contrast, *Lightfoot et al.* [1993] proposed that the isotopic and trace element signatures of lavas from the Noril'sk region of the Siberian Traps reflect the incorporation of a lithospheric mantle-derived melt, whereas *Wooden et al.* [1993] proposed that the compositions of these lavas were controlled primarily by crustal assimilation and fractional crystallization in large, near-steady-state crustal magma chambers. In this section we examine each of these scenarios in turn, drawing upon data available from the literature.

The Siberian Traps can be divided into three subprovinces: the Putorana, the Meimecha-Kotui, and the Noril'sk [e.g., *Sharma et al.*, 1991]. The Putorana formations represent the volumetric bulk (~90%) of the province [e.g., *Sharma et al.*, 1992]. However, most of the detailed stratigraphic and geochemical work on the Siberian Traps has concentrated on the earlier sequences recovered in drill cores and exposures from the Noril'sk region. The upper Noril'sk formations appear to be the chemical and temporal equivalents of those of the Putorana [e.g., *Sharma et al.*, 1992] and thus reflect the composition of the bulk of the province. However, the heterogeneous lower Noril'sk formations contain lavas that represent distinct geochemical end-members. Mixtures of these end-members can explain most of the geochemical characteristics of the upper sequences. In particular, most upper-sequence lavas have isotopic and trace element ratios that, to first order, can be explained by mixtures of lavas similar to those of the Gudchikhinsky and Tuklonsky suites. The origin of these volumetrically minor lower sequences is therefore critical to our understanding of the origin of the main upper sequences.

Lavas from the Siberian Traps display considerable isotopic heterogeneity. The ε_{Nd} values, for example, range from +6 to -11 (Figure 5). However, *Sharma et al.* [1991, 1992] noted that most of this heterogeneity was restricted to lavas erupted early in the history of the province and that volumetrically most lavas were very homogeneous, with $\varepsilon_{Nd} \approx +2$. They concluded that this homogeneity requires the lavas to have been derived from a homogeneous, nearly primitive plume source, with little subsequent modification. However, the conclusion that a homogeneous lava sequence, no matter how large, requires melt generation from a homogeneous (plume) mantle source appears unjustified. For example, the isotopically homogeneous Putorana and upper Noril'sk lavas also have very homogeneous major element compositions, but these compositions are far removed from those of primary mantle melts. The Mg#s in the Putorana and upper Noril'sk vary systematically between ~50 and 60, yet primary mantle melts have much higher Mg#s (≥70) [e.g., *Roeder and Emslie*, 1970]. Obviously, some process either accompanying or following crystal fractionation acted on these lavas to produce the observed homogeneity in major element compositions, and there is no reason to believe that the same process could not also account for the observed isotopic homogeneity.

Furthermore, most Siberian Traps lavas, including the homogenous upper sequences discussed by *Sharma et al.* [1992], have many trace element signatures that are unlike those observed in other plume-related lavas. For example, most Siberian Traps lavas (including the homogeneous upper Noril'sk section) have pronounced HFSE depletions, with $(Ta/La)_N$ and $(Nb/La)_N$ <1 (subscript denotes normalization to chondritic ratio), whereas most OIBs and OFBs have superchondritic ratios (Figure 5; see also *Arndt and Christensen* [1992]). Thus, although we cannot exclude the possibility that these trace element signatures derive from a plume source, the inferred composition of the Siberian plume would necessarily be unlike that inferred for any other mantle plume.

In contrast to *Sharma et al.* [1991], *Wooden et al.* [1993] proposed that the compositions of most Siberian Traps lavas were significantly modified from that of their initial plume source, predominantly through assimilation of continental crust in large, periodically replenished and tapped crustal magma chambers. The modeling of *O'Hara*

and Mathews [1981] suggests that in such a system signatures of crustal assimilation may become decoupled from those of fractional crystallization as the magma approaches a steady-state major element composition. However, although the major element homogeneity of most Putorana and upper Noril'sk lavas, combined with their evolved nature, does require some process of homogenization subsequent to melt generation, this does not directly resolve whether contamination occurred in the lithospheric mantle or the crust. Below, we outline several observations that suggest a lithospheric-mantle origin for many of these features is more likely.

Wooden et al. [1993] proposed that crustal processing of Siberian magmas took place in large, quasi-steady-state magma chambers, in part to explain the apparent decoupling of fractional crystallization from crustal assimilation. However, Os isotope data for picrites from the lower Noril'sk section suggest that any crustal contamination must have occurred prior to significant fractional crystallization. Most researchers concur that the Gudchikhinsky picrites, which have $\varepsilon_{Nd} \approx +4$ and negligible HFSE anomalies, likely reflect the composition of the Siberian plume [e.g., *Wooden et al.*, 1993]. In contrast, the Tuklonsky picrites have $\varepsilon_{Nd} \approx -2$ and pronounced HFSE depletions, yet they possess essentially identical γ_{Os} to the Gudchikhinsky picrites [*Horan et al.*, 1995]. Although minor isotopic differences exist between the Tuklonsky picrites and basalts, their similar trace element patterns suggest similar contamination histories for both. Therefore, if the Tuklonsky basalts were derived from picritic parental liquids similar to the Tuklonsky picrites, then contamination must have occurred before significant fractional crystallization of these picritic parents. However, *Wooden et al.* [1993] suggested that the Tuklonsky picrites contain cumulus olivine and are thus derived from more evolved, low MgO (and therefore low Os) liquids similar to the Tuklonsky basalts. However, if this were the case, then the similarity of Os isotopes in the Gudchikhinsky and Tuklonsky picrites again precludes significant mixing of radiogenic continental crust with these "parental" basaltic melts, because the low Os content of evolved basalts would make Os isotopes highly sensitive to such contamination. Thus, regardless of whether the Tuklonsky basalts are derived from picritic melts or the Tuklonsky picrites are derived from basaltic melts, the "continental" isotopic and trace element signatures in these lavas must have been imparted before the basalts underwent significant fractional crystallization.

The steady-state magma chamber model also cannot easily explain the apparent decoupling between incompatible trace element ratios that we expect to be correlated if these ratios are controlled by crustal assimilation. For example, the Ta depletions in most Siberian Traps lavas were taken by *Wooden et al.* [1993] as evidence for crustal assimilation. However, if this were the case, we would expect a positive correlation between La/Ta and La/Sm, because upper continental crust or moderate-degree partial melts of the lower continental crust will in general be enriched in light REE. Figure 7 shows La/Ta-La/Sm variations for lavas of the Noril'sk region. Some Noril'sk lavas do possess the high La/Sm, high La/Ta values predicted for crustal assimilation. In particular, Nadezhdinsky suite lavas, which most researchers agree are crustally contaminated, possess highly elevated La/Sm and high La/Ta. However, most upper Noril'sk lavas possess modest La/Sm values and high La/Ta values. Significantly, the Tuklonsky picrites possess even higher La/Ta values than the bulk of the upper Noril'sk section but only slightly higher La/Sm. These lavas clearly assimilated a component very different than that contained in the Nadezhdinsky lavas. The La/Ta and La/Sm values of most upper-sequence lavas are intermediate between those of the Tuklonsky picrites and the Gudchikhinsky picrites, as are the Nd- and Pb-isotopic ratios. Therefore, the bulk of the Siberian Traps (=upper Noril'sk + Putorana) may have been generated from mixtures of primary melts similar to the Gudchikhinsky and the Tuklonsky picrites.

Assimilation of upper continental crust cannot produce the La/Sm-La/Ta variations in the Tuklonsky and upper-sequence lavas because of the high La/Sm in upper crustal rocks (Figure 7) [e.g., *Taylor and McLennan*, 1985]. It is possible that bulk assimilation of some types of low La/Sm mafic lower crust could explain this correlation, although the combination of high La/Ta and low La/Sm in lower crustal xenoliths is quite rare. However, in the more likely scenario where plume-derived lavas assimilated partial melts of the lower crust, the increase in La/Sm produced from partial melting of granulitic or amphibolitic sources again precludes such assimilation as a means to account for the Tuklonsky lavas. Finally, we note that other flood basalt sequences that are clearly crustally contaminated, such as the Grande Ronde suite of the Columbia River basalts or the Bushe in the Deccan, do not have the low La/Sm, high La/Ta values found in the Tuklonsky. We therefore consider that these correlations may reflect the incorporation of a HFSE-depleted melt derived from the CLM.

Appropriate sources for the low La/Sm-high La/Ta Tuklonsky picrites are also rare in CLM-derived mantle xenoliths. However, the CLM-derived Karoo picrites have many trace element similarities with the Tuklonsky

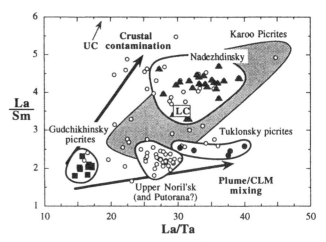

Figure 7. La/Ta-La/Sm variations in Siberian Traps lavas. Most crustal rocks possess elevated La/Sm ratios, and partial melting will further increase La/Sm. Therefore, crustal contamination will increase La/Sm as well as La/Ta. The very high La/Ta ratios of the Tuklonsky picrites combined with only modest La/Sm enrichments suggest that these picrites have not experienced significant crustal contamination, but instead contain a CLM-derived, HFSE-depleted component. Lavas with high La/Sm (e.g., La/Sm > 3), which likely have assimilated crustal material, are restricted to the early phases of volcanism in the Noril'sk region, and are also characterized by elevated δ18O [*Das Sharma et al.*, 1994]. Lavas from the upper part of the Noril'sk section have normal δ18O, and La/Ta-La/Sm values intermediate between those of the Tuklonsky and the Gudchikhinsky picrites. These lavas probably result from mixing of plume- and CLM-derived melts.

picrites. Figure 7 shows that the Karoo picrites have similar La/Ta but higher La/Sm values than the Tuklonsky picrites and upper Noril'sk lavas. However, ratios of highly to moderately incompatible elements such as La/Sm are strongly influenced by the extent of partial melting. Therefore, we can envision a scenario wherein larger extents of partial melting of a source similar to the CLM source of the Karoo picrites, perhaps accompanied by the removal of garnet as a residual phase, could produce melts similar to the Tuklonsky series lavas.

Oxygen isotopic data also do not support a crustal assimilation model for the Tuklonsky and upper-sequence lavas. Most Siberian Traps lavas, including samples from the Gudchikhinsky and Tuklonsky picrites, the upper Noril'sk, and the Putorana, have δ18O values within the range of or only slightly higher than values found in fresh MORB glasses (δ18O ≤ 6.1, with few exceptions) [*Das Sharma et al.*, 1994]. In contrast, lavas from the Nadezhdinsky and Ivakinsky suites have δ18O ≥ 6.9, suggesting that these lavas have assimilated high δ18O continental crust. Both these suites are characterized by high La/Sm and La/Ta ratios, which are also consistent with assimilation of high-La/Sm continental crust. The normal mantle δ¹⁸O found in the Tuklonsky and most other Siberian Traps lavas, in contrast, suggests that these lavas have not assimilated a significant amount of high δ¹⁸O crustal material. Although mafic lower crustal xenoliths have lower average δ¹⁸O than metasediments or average upper continental crust [e.g., *Fowler and Harmon*, 1990], mafic lower crustal components also have less pronounced HFSE depletions, thus requiring greater addition of such material to account for the high La/Ta values in the Tuklonsky picrites. The required combination of low La/Sm, normal mantle δ¹⁸O, but high La/Ta is quite rare in crustal rocks.

Correlations between major elements and isotopic compositions in Siberian Traps lavas are also difficult to reconcile with crustal assimilation models and instead suggest the presence of a CLM-derived, Fe-poor component. Values of FeO and ε_{Nd} are correlated in Siberian Traps lavas (Figure 8a). This correlation is similar to that observed in picrites from the Karoo flood basalts between ε_{Nd}, γ_{Os}, and Fe_2O_3. As we discussed earlier, the association of low γ_{Os} and low Fe_2O_3 in the Karoo picrites indicates that melts derived from the (low γ_{Os}) CLM had lower Fe than melts derived from the underlying plume [*Ellam et al.*, 1992].

The correlation between FeO and ε_{Nd} in the Siberian Traps is unlikely to reflect the shallow-level evolution of these lavas. For example, because average upper continental crust has low Fe (but also low Mg) [e.g., *Taylor and McLennan*, 1985], the observed correlation between FeO and ε_{Nd} could in principle be produced by assimilation of Fe-poor, low-ε_{Nd} upper continental crust. However, upper continental crust is characterized by high La/Sm and so is precluded by the La/Sm-La/Ta variations in the Tuklonsky and upper-sequence lavas as discussed above. Figure 8b shows the same correlation between FeO and ε_{Nd} after lavas from the crustally contaminated Nadezhdinsky suite and other high La/Sm lavas have been removed. Although bulk assimilation of lower crust is permissible, lower crust is more mafic than upper crust and has a higher Fe content [e.g., *Taylor and McLennan*, 1985; *Rudnick and Taylor*, 1987]. Therefore, a very large amount of assimilation of lower crust is required to produce the FeO-ε_{Nd} trend in Figure 8b. Furthermore, simple assimilation of low Mg, low Fe continental crust should produce a positive correlation between Fe and Mg, which is not observed. Although it is possible that some convolution of crustal assimilation coupled with magma chamber processing could account for this correlation, a simpler explanation is that these variations result from

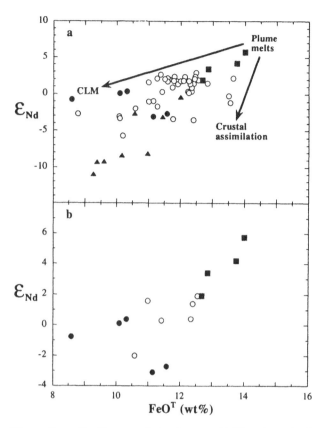

Figure 8. a. Positive correlation between FeO^T and ε_{Nd} in the Siberian Traps. b. FeO^T-ε_{Nd} correlation after removal of lavas from the Nadezhdinsky suite and other high La/Sm lavas. High δ¹⁸O and La/Sm values in Nadezhdinsky suite lavas are consistent with crustal contamination, but Tuklonsky lavas have normal mantle δ¹⁸O and low La/Sm. Symbols are as in Figure 7. See text for discussion.

mixing of Fe-poor, CLM-derived melts with Fe-rich, plume-derived melts similar to that observed for the Karoo picrites.

If the correlation between FeO and ε_{Nd} in the Siberian Traps lavas does reflect mixing of Fe-poor melts from the CLM with Fe-rich plume-derived melts, then a substantial mass fraction of CLM-derived melt must be present in the more contaminated, low ε_{Nd} lavas. Assimilation of small amounts of highly incompatible-element-enriched melts such as lamproites into tholeiitic magmas cannot reproduce the observed correlation. This has important implications for the plume impact model of flood basalt genesis, which predicts little or no melt generation from the CLM [*Arndt and Christensen*, 1992; *Farnetani and Richards*, 1994]. It is therefore important to assess the average melt fraction derived from the lithosphere rather than the plume. The Putorana formations represent ~90 vol% of the entire Siberian Traps province [*Sharma et al.*, 1991] and so

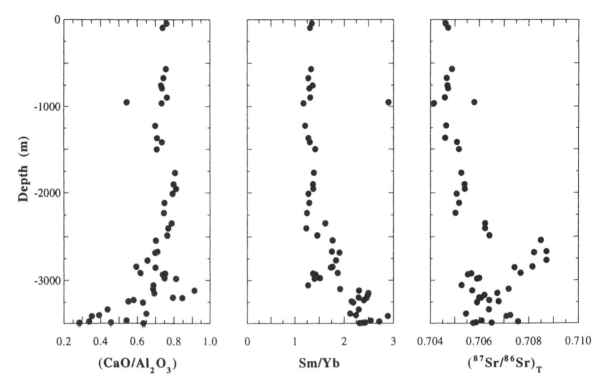

Figure 9. The decrease in Sm/Yb and increase in CaO/Al2O3 with increasing stratigraphic height observed in lavas from the Noril'sk region are consistent with a temporal increase in the degree of partial melting and removal of garnet as a residual phase within the source region. These changes in the melting environment within the mantle are temporally correlated with a general decline in $^{87}Sr/^{86}Sr$ (with the exception of lavas from the Nadezhdinsky suite), suggesting that the isotopic variations are also related to processes occurring within the mantle, e.g. a shift from CLM to plume melting. Data are from *Wooden et al.* [1993].

provide a good estimate of the average composition of the Siberian Traps. Furthermore, the Putorana lavas generally plot near the high FeO-high ε_{Nd}, "plume"-derived end of the array in Figure 8. The ε_{Nd} values of average Putorana lavas (≈+2) are only slightly lower than ε_{Nd} values of the Gud-chikhinsky picrites (≈+4), which we interpret as reflecting the uncontaminated Siberian plume. Therefore, even if this shift is due entirely to assimilation of Fe-poor CLM-derived melts with ε_{Nd} values less than or equal to ≈-2 (the value of most Tuklonsky picrites), the fraction of CLM-derived melt in these lavas is still relatively small, probably less than 10–20%. This is still a large amount of CLM-generated melt compared with predictions of the plume impact model. However, melt volumes of this magnitude may be feasible if hydrous regions with depressed solidi are involved in melt generation.

The temporal evolution of the Siberian Traps provides a final line of circumstantial evidence supporting a model of plume/lithospheric mantle interaction rather than one of crustal assimilation. Figure 9 illustrates the stratigraphic variation in lavas of the Noril'sk region. These lavas record a general decrease in Sm/Yb and an increase in CaO/Al2O3 with increasing stratigraphic height, which most likely reflects the removal of garnet as a residual phase in the source(s) of the lavas [e.g., Wooden et al., 1993] and an increase in melt fraction. The decrease in Sm/Yb, in particular, is unlikely to reflect changes in the amount or kind of crustal assimilation, and so this progression reflects a change in the melting environment. Over the same interval, however, there is a general accompanying decline in tracers of "continental" contamination, such as high La/Ta or $^{87}Sr/^{86}Sr$. Although these two trends do not entirely mirror one another (e.g., the Gudchikhinsky picrites, which have the most pronounced garnet signature of any Siberian Traps lavas, are also the least contam-inated), the overall correlation is quite clear. Because one of these trends reflects a change in melting conditions within the mantle, this correlation suggests that the accompanying decline in the strength of continental signatures also reflects a change occurring within the mantle, e.g., a decrease in the amount of melt generated from the CLM relative to melt generated from the underlying plume. The fact that the inferred CLM-derived component decreases at the same time the apparent depth of melt generation decreases further suggests that this shift was brought about in part by the mechanical erosion of the CLM, which allowed the Siberian plume to ascend to shallower levels.

In summary, no single piece of evidence precludes the possibility that the compositions of the Tuklonsky picrites or the volumetrically dominant Putorana and upper Noril'sk lavas were controlled by assimilation of very specific crustal components. However, in combination the available data do not support such a model. In contrast, melt generation from a sediment-enriched lithospheric-mantle component similar in many respects to the source of the Karoo picrites and mixing of plume- and lithosphere-derived melts provide a simpler mechanism for producing most of the isotopic and trace element characteristics of the Siberian Traps. Crustal contamination, which significantly altered the compositions of some early lavas (e.g., the Nadezhdinsky suite), was primarily of secondary importance during the main stages of volcanism.

4. COMMON CHARACTERISTICS OF FLOOD BASALTS: TOWARDS A GENERAL MODEL

4.1. Temporal Evolution

The temporal increase in melt fraction and decrease in the depth of melting inferred from major and trace element trends in the Siberian Traps is a common feature of many CFBs. For example, Kerr [1994] and Fram and Lesher [1993] inferred a temporal progression towards higher degrees of partial melting and decreasing depth of melting for basalts from the North Atlantic Tertiary Province. Basalts erupted early in the formation of the North Atlantic Tertiary Province have high Sm/Yb, requiring the presence of garnet as a residual phase in their source region. Later basalts lack this garnet signature. Kerr [1994] suggested that progressive erosion of the lithosphere beneath the North Atlantic Tertiary Province allowed the Iceland plume to ascend to shallower levels. The resulting higher degrees of melting and/or a shift towards melting of spinel peridotite led to the removal of garnet as a residual phase. Stewart and Rogers [1996] inferred a similar temporal increase in melt fraction for lavas of the southern Ethiopian rift and proposed thermomechanical erosion of the lithosphere by the Afar plume. Lavas at the base of the Keweenawan section of the North American midcontinent rift also have significantly higher Sm/Yb ratios than most lavas higher in the sequence, a pattern that is mirrored by

other sequences from the midcontinent rift [cf. Shirey et al., 1994; Nicholson et al., 1997]. Again, this temporal decline in Sm/Yb is likely the result of decreasing depth of melting with time. Finally, a similar temporal progression has been suggested for the Deccan Traps. Peng and Mahoney [1995] noted a north-south trend of decreasing ratios of highly to moderately incompatible elements (e.g., Nb/Y) in the western Deccan Traps, which they suggested might reflect progressive lithospheric thinning and an increase in the extent of partial melting for the (younger) southern Deccan relative to the (older) northern Deccan lavas. Erosion of the Indian lithosphere during Deccan volcanism is consistent with geophysical studies that indicate the modern Indian lithosphere is significantly thinner and hotter than most other shields [e.g., Negi et al., 1986].

The temporal progression from low to higher melt fractions is observed both in flood basalts associated with continental rifting (e.g., the North Atlantic Tertiary Province) and in flood basalts not associated with rifting (e.g., Siberian Traps) and is often accompanied by a switch from continental isotopic and trace element signatures to more plume-like signatures. As in the Siberian Traps, these observations suggest significant lithospheric erosion during flood basalt volcanism, often accompanied by partial melting of the CLM. The inferred lithospheric erosion occurs over time scales too short to be explained by simple conductive heating of the lithosphere. Lithospheric erosion may be accelerated if partial melting of fusible (hydrous?) domains within the lithosphere reduces lithosphere viscosity and enhances convective erosion by the plume head. This interpretation is consistent with the observation that the chemical signatures of contamination are often, though not always, strongest in lavas erupted early within a flood basalt sequence (e.g., Siberian Traps [Wooden et al., 1993]; Wrangellia [Lassiter et al., 1995a]) because progressive melting of the lithosphere should deplete any hydrous domains present.

4.2. Major Element Trends

Both the Karoo flood basalts and Siberian Traps record the incorporation of an Fe-poor, HFSE-depleted component. In the Karoo picrites, and most likely in the Siberian Traps as well, this component appears to have been derived from the lithospheric mantle. Similar correlations between major element and isotopic composition are seen in other flood basalt provinces as well. For example, the Deccan Traps also show evidence for mixing of Fe-rich plume melts with Fe-poor CLM melts. Figure 10 shows a correlation between FeO^T and

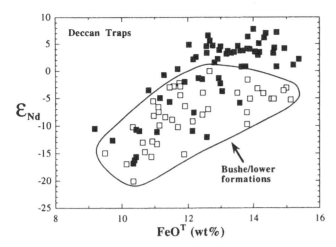

Figure 10. FeO^T-ε_{Nd} in the Deccan Traps. Crustally contaminated lavas from the Bushe and other early Deccan formations (open symbols) have lower ε_{Nd} for a given FeO^T than lavas erupted later in the sequence. This shift is similar to that observed for lavas from the crustally contaminated Nadezhdinsky suite in the Siberian Traps. However, the correlation between FeO^T and ε_{Nd} persists when these formations are removed, and does not appear to depend upon the extent of fractional crystallization.

ε_{Nd} in Deccan lavas similar to what we observe for the Siberian Traps. As with the Siberian Traps, we must again assess the degree to which the observed FeO^T-ε_{Nd} trend reflects the effects of fractional crystallization and/or crustal assimilation rather than variations in primary melt composition. In the absence of Fe-Ti oxide precipitation, fractional crystallization will usually increase Fe content in basaltic magmas. An unusual feature of the Deccan lavas is that the least fractionated lavas tend to be the most contaminated [e.g., *Mahoney*, 1988]. Therefore, the correlation between FeO^T and ε_{Nd} could simply reflect this fact. However, Deccan lavas spanning only a narrow range in MgO content display the same trend as the entire data set, regardless of whether one focuses solely upon high- or low-MgO samples. Crustal contamination also does not appear sufficient to explain the observed correlation. Crustal contamination is particularly pronounced in lavas of the Bushe and other early Deccan formations [*Cox and Hawkesworth*, 1985; *Peng et al.*, 1994]. These early Deccan lavas have consistently lower ε_{Nd} for a given FeO^T than most other Deccan lavas (Figure 10), suggesting that crustal contamination had a more pronounced effect on isotopic and trace element ratios than on major element composition. However, when these lavas are removed, the correlation between FeO^T and ε_{Nd} persists.

Many flood basalts related to the breakup of Gondwana appear to contain an even larger fraction of melt derived from the CLM. *Hergt et al.* [1991] noted that lavas from the Paraná and the Ferrar have lower Fe$_8$ than typical MORB, and they argued that these melts were generated from previously melt-depleted, but subsequently hydrated, lithospheric mantle. *Turner and Hawkesworth* [1995] reached similar conclusions using major element regressions similar to those we have utilized in our discussion of melt generation. Support for a lithospheric mantle source for the Paraná lavas comes from isotopic variations in potassic rocks from southern Brazil. Many Brazilian potassic lavas have Sr- and Nd-isotopic characteristics similar to the so-called high-Ti Paraná [*Gibson et al.*, 1995], yet some of these lavas have unradiogenic Os isotopes, which suggests a lithospheric mantle origin [*Carlson et al.*, 1997]. *Gibson et al.* [1995] also noted an apparent spatial correlation between high- and low-Ti mafic potassic volcanic rocks and the occurrence of high- and low-Ti flood basalts, suggesting that both high- and low-Ti Gondwanan basalts may contain a significant CLM-derived component. Recent Os isotope analyses of Ferrar flood basalts also support a lithospheric mantle origin over crustal contamination models for these basalts [*Molzahn et al.*, 1996].

There is less chemical evidence from the Gondwanan flood basalts for substantial melt generation within sub-lithospheric plumes than for most other flood basalt provinces. This lack of evidence for plume-generated melts may reflect a greater water content and thus a lower solidus within the pre-rifting Gondwanan lithosphere as a result of protracted subduction along the supercontinent's margins or may reflect a lower than average excess potential temperature for the Tristan and other Gondwanan plumes presumably responsible for triggering lithospheric melting. In any case, melt generation from the CLM during flood basalt genesis appears to be a common feature of many CFBs, a fact that appears at odds with traditional anhydrous plume/lithosphere interaction models.

5. CONCLUSIONS

The interplay of plume, CLM, and crustal processes affecting the chemical and isotopic compositions of flood basalts is a complex one that varies significantly from province to province. However, the preceding discussion has highlighted several common features in the evolution of many continental and oceanic flood basalts. Our interpretations of these features are summarized below.

(1) Lithosphere thickness controls flood basalt

composition by limiting adiabatic ascent of sublithospheric plumes. The presence of this "lithospheric signature" in CFBs, but not in OFBs, indicates that melting initiated within sublithospheric plumes and that substantial thinning of the lithosphere caused by extension or protracted incubation are not necessary prerequisites for flood volcanism. However, the observation that OIBs are generated from even smaller extents of melting than CFBs suggests that some thinning of continental lithosphere does occur.

(2) Melting of lithospheric mantle and mixing of CLM- and plume-derived melts can explain the isotopic and trace element trends observed in many CFB suites including the bulk of the Siberian Traps. Significant melting of the lithospheric mantle is in conflict with anhydrous plume/lithosphere interaction models, and therefore suggests melting of hydrous regions within the CLM, as proposed by *Gallagher and Hawkesworth* [1992]. However, CLM-derived melts are generally more prevalent during the early phases of volcanism, with later lavas containing a greater proportion of plume-derived melts. On average, the total proportion of melt generated from the CLM is less than 10-20% of that generated from sublithospheric sources.

(3) Average depth of melt generation decreases while melt fraction increases during the course of volcanism for several CFBs. Furthermore, this pattern is observed both in CFBs associated with extension or rifting (e.g., the North Atlantic Tertiary Province) and where extension is not observed (e.g., Siberian Traps). Combined with the short duration of volcanism measured for many CFBs, this and the accompanying decline in fraction of CLM-derived melt suggest progressive mechanical erosion of the lithosphere, possibly triggered by partial melting of discrete hydrous domains within the CLM and an accompanying decrease in lithosphere viscosity.

The plume initiation model for flood basalt genesis is generally consistent with our observations, particularly regarding the depth of initial melt generation and extent of melting in continental and oceanic flood basalts. However, the fraction of CLM-derived melt inferred from our analysis exceeds predictions of anhydrous plume/lithosphere interaction models. More sophisticated models that consider compositional differences between the CLM and sub-lithospheric plumes are necessary to account for these discrepancies.

Unambiguous differentiation of crustal and CLM-derived components in CFBs has proven to be exceedingly

difficult using traditional incompatible trace element and isotopic arguments, as any survey of the flood basalt literature will attest. We have attempted to illustrate the means by which a combination of trace element and major element arguments can help differentiate the various processes within the plume, lithospheric mantle, and crust that operate simultaneously during flood basalt genesis. However, future progress in this area will require additional data from isotopic systems capable of clearly distinguishing crustal and CLM components. Recent studies examining oxygen and osmium isotopes in flood basalts offer the best hope that many of the difficult questions concerning the role of continental crust and lithospheric mantle in flood basalt genesis may soon finally be resolved.

Acknowledgments. We thank N. Arndt and an anonymous reviewer for their comments, R. Carlson, S. Shirey, and A. Brandon for useful suggestions and discussions, and J. Mahoney for his editorial patience. This material is based upon work supported under a National Science Foundation graduate research fellowship, The Berkeley Fellowship, NSF grant EAR 9304419, an NSF postdoctoral fellowship, and by the Carnegie Institution of Washington.

REFERENCES

Aitken, B. G., and L. M. Etchevera, Petrology and geochemistry of komatiites and tholeiites from Gorgona Island, Columbia, Contrib. *Mineral. Petrol.*, 86, 94-105, 1984.

Albarede, F., and V. Tamagnan, Modeling the recent geochemical evolution of the Piton de la Fornaise volcano, Réunion Island, 1931-1986, *J. Petrol.*, 29, 997-1030, 1988.

Arndt, N. T., and U. Christensen, The role of lithospheric mantle in continental flood volcanism: Thermal and geochemical constraints, *J. Geophys. Res.*, 97, 10,967-10,981, 1992.

Arndt, N. T., G. K. Czamanske, J. L. Wooden, and V. A. Fedorenko, Mantle and crustal contributions to continental flood volcanism, *Tectonophysics*, 223, 39-52, 1993.

Baker, M. B., and E. M. Stolper, Determining the composition of high-pressure mantle melts using diamond aggregates, *Geochim. Cosmochim. Acta.*, 58, 2811-2827, 1994.

Basu, A. R., P. R. Renne, D. K. DasGupta, F. Teichmann, and R.J. Poreda, Early and late alkali igneous pulses and a high-3He plume origin for the Deccan flood basalts, *Science*, 261, 902-906, 1993.

Basu, A. R., R. J. Poreda, P. R. Renne, F. Teichmann, Y.R. Vasilev, N.V. Sobolev, and B.D. Turrin, High-3He plume origin and temporal-spatial evolution of the Siberian flood basalts, *Science*, 269, 822-825, 1995.

Brandon, A. D., and G. G. Goles, A Miocene subcontinental plume in the Pacific Northwest: Geochemical evidence, *Earth Planet. Sci. Lett.*, 88, 273-283, 1988.

Brandon, A. D., and G. G. Goles, Assessing subcontinental lithospheric mantle sources for basalts: Neogene volcanism in the Pacific Northwest, USA as a test case, *Contrib. Mineral. Petrol.*, 121, 364-379, 1995.

Carlson, R. W., G. W. Lugmair, and J. D. MacDougall, Columbia River volcanism: The question of mantle heterogeneity or crustal contamination, *Geochim. Cosmochim. Acta*, 45, 2483-2499, 1981.

Carlson, R. W., Isotopic constraints on Columbia River flood basalt genesis and the nature of the subcontinental mantle, *Geochim. Cosmochim. Acta.*, 48, 2357-2372, 1984.

Carlson, R. W., and A. J. Irving, Depletion and enrichment history of subcontinental lithospheric mantle: An Os, Sr, Nd, and Pb isotopic study of ultramafic xenoliths from the northwestern Wyoming craton, *Earth Planet. Sci. Lett.*, 110, 99-119, 1994.

Carlson, R. W., S. Esperança, and D. P. Svisero, Chemical and Os isotopic study of Cretaceous potassic rocks from southern Brazil, *Contrib. Mineral. Petrol.*, in press, 1997.

Castillo, P. R., R. W. Carlson, and R. Batiza, Origin of Nauru basin igneous complex: Sr, Nd, and Pb isotope and REE constraints, *Earth Planet. Sci. Lett.*, 103, 200-213, 1991.

Chapman, D. S., and H. N. Pollack, Regional geotherms and lithospheric thickness, *Geology*, 5, 265-268, 1977.

Chesley, J. T., J. Ruiz, and P. R. Hooper, Crust-mantle mixing: implications based on the Re-Os isotope systematics of the Columbia River Basalt Group, *EOS Trans. AGU*, 77, 832, 1996.

Clague, D. A., Petrology of basaltic and gabbroic rocks dredged from the Danger Island troughs, Manihiki Plateau, *Init. Repts. Deep Sea Drill. Proj.*, 33, 891-911, 1976.

Condomines, M., K. Grönvold, P. J. Hooker, K. Muehlenbachs, R. K. O'Nions, N. Oskarsson, and E. R. Oxburgh, Helium, oxygen, strontium, and neodymium isotopic relationships in Icelandic volcanics, *Earth Planet. Sci. Lett.*, 66, 125-136, 1983.

Cox, K. G., and C. J. Hawkesworth, Geochemical stratigraphy of the Deccan Traps at Mahabaleshwar, Western Ghats, India, with implications for open system magmatic processes, *J. Petrol.*, 26, 355-377, 1985.

Das Sharma, S., D. J. Patil, R. Murari, G. Gopalan, and G. V. Nesterenko, Oxygen isotope systematics of Siberian basalts, *J. Geol. Soc. India*, 44, 327-330, 1994.

DePaolo, D. J. and G. J. Wasserburg, Neodymium isotopes in flood basalts from the Siberian Platform and inferences about their mantle sources, *Proc. Natl. Acad. Sci.*, 76, 3056-3060, 1979.

Dodson, A., D. J. DePaolo, and B. M. Kennedy, Noble gasses in the Columbia River Basalt, *Eos Trans. AGU*, 77, 288, 1996.

Donnelly, T. W., W. Melson, R. Kay, and J. J. W. Rogers, Basalts and dolerites of late Cretaceous age from the central Caribbean, *Init. Repts. Deep Sea Drill. Proj.*, 15, 989-1011, 1973.

Dosso, L., H. Bougault, and J. -L. Joron, Geochemical morphology of the north Mid-Atlantic Ridge, 100-240N: Trace element-isotope complementarity, *Earth Planet. Sci. Lett.*, 120, 443-462, 1993.

Eiler, J. M., J. W. Valley, and E. M. Stolper, Oxygen isotope ratios in olivines from the Hawaii Scientific Drilling Project, *J. Geophys. Res.*, 101, 11807-11814, 1996.

Ellam, R. M., Lithospheric thickness as a control on basalt geochemistry, *Geology*, 20, 153-156, 1992.

Ellam, R. M., and K. G. Cox, A Proterozoic lithospheric source for Karoo magmatism: Evidence from the Nuanetsi picrites, *Earth Planet. Sci. Lett.*, 92, 207-218, 1989.

Ellam, R. M., and K. G. Cox, An interpretation of Karoo picrite basalts in terms of interaction between asthenospheric magmas and the mantle lithosphere, *Earth Planet. Sci. Lett.*, 105, 330-342, 1991.

Ellam, R. M., R. W. Carlson, and S. B. Shirey, Evidence from Re-Os isotopes for plume-lithosphere mixing in Karoo flood basalt genesis, *Nature*, 359, 718-721, 1992.

Falloon, T. J., and D. H. Green, The solidus of carbonated, fertile peridotite, *Earth Planet. Sci. Lett.*, 94, 364-370, 1989.

Falloon, T. J., D. H. Green, and A. L. Jaques, Refractory magmas in back-arc basin settings, *Mineral. Mag.*, 58A, 263-264, 1994.

Farnetani, C. G., and M. A. Richards, Numerical investigations of the mantle plume initiation model for flood basalt events, *J. Geophys. Res.*, 99, 13813-13834, 1994.

Fisk, M. R., B. G. J. Upton, and C. E. Ford, Geochemical and experimental study of the genesis of magmas of Réunion Island, Indian ocean, *J. Geophys. Res.*, 93, 4933-4950, 1988.

Floyd, P. A., Petrology and geochemistry of oceanic intraplate sheet-flow basalts, Nauru basin, Deep Sea Drilling Project leg 89, *Init. Repts. Deep Sea Drill. Proj.*, 89, 471-497, 1986.

Fodor, R. V., C. Corwin, and A. N. Sial, Crustal signatures in the Serra Geral flood-basalt province, southern Brazil: O- and Sr-isotope evidence, *Geology*, 13, 763-765, 1985.

Fowler, M., and R. S. Harmon, The oxygen isotope composition of lower crustal granulite xenoliths, *in Petrology and Geochemistry of Granulites*, edited by D. Vielzeuf, pp. 493-506, NATO/ASI Series, Vol. X, 1990.

Fram, M. S., and C. E. Lesher, Geochemical constraints on mantle melting during creation of the North Atlantic basin, *Nature*, 363, 712-715, 1993.

Gallagher, K., and C. Hawkesworth, Dehydration melting and the generation of continental flood basalts, *Nature*, 258, 57-59, 1992.

Gibson, S. A., R. N. Thompson, A. P. Dicken, and O. H. Leonardos, High-Ti and low-Ti potassic magmas: Key to plume-lithosphere interactions and continental flood-basalt genesis, *Earth Planet. Sci. Lett.*, 136, 149-165, 1995.

Green, T. H., and N. J. Pearson, An experimental study of Nb and Ta partitioning between Ti-rich minerals and silicate liquids at high pressure and temperature, *Geochim. Cosmochim. Acta*, 51, 55-62, 1987.

Griffiths, R. W., Thermals in extremely viscous fluids, including the effects of temperature-dependent viscosity, *J. Fluid Mech.*, 166, 115-138, 1986.

Griffiths, R. W., and I. H. Campbell, Stirring and structure in mantle starting plumes, *Earth Planet. Sci. Lett.*, 99, 66-78, 1990.

Harris, C., A. M. Wittingham, S. C. Milner, and R. A. Armstrong, Oxygen isotope geochemistry of the silicic volcanic rocks of the Etendeka-Paraná province: Source constraints, *Geology*, 18, 1119-1121, 1990.

Hauri, E. H., Major-element variability in the Hawaiian mantle plume, *Nature*, 382, 415-419, 1996.

Hemond, C., N. T. Arndt, U. Lichtenstein, and A. W. Hofmann, The heterogeneous Iceland plume: Nd-Sr-O isotopes and trace element constraints, *J. Geophys. Res.*, 98, 15833-15850, 1993.

Hergt, J. M., D. W. Peate, and C. J. Hawkesworth, The petrogenesis of Mesozoic Gondwana low-Ti flood basalts, *Earth Planet. Sci. Lett.*, 105, 134-148, 1991.

Hirose, K., and I. Kushiro, Partial melting of dry peridotite at high pressures: determination of compositions of melts segregated from peridotite using aggregates of diamond, *Earth Planet. Sci. Lett.*, 114, 477-489, 1993.

Hooper, P. R., and C. J. Hawkesworth, Isotopic and geochemical constraints on the origin and evolution of the Columbia River Basalt, *J. Petrol.*, 34, 1203-1246, 1993.

Horan, M. F., R. J. Walker, V. A. Fedorenko, and G. K. Czmanske, Osmium and neodymium isotopic constraints on the temporal and spatial evolution of Siberian flood basalt sources, *Geochim. Cosmochim. Acta*, 59, 5159-5168, 1995.

Ito, E., W. M. White, and C. Göpel, The O, Sr, Nd, and Pb isotope geochemistry of MORB, *Chem. Geol.*, 62, 157-176, 1987.

Jackson, E. D., K. E. Bargar, B. P. Fabbi, and C. Heropoulos, Petrology of the basaltic rocks drilled on leg 33 of the Deep Sea Drilling Project, *Init. Repts. Deep Sea Drill. Proj.*, 33, 571-630, 1976.

Jordan, T. H., Composition and development of the subcontinental tectosphere, *Nature*, 274, 544-548, 1978.

Kerr, A. C., Lithospheric thinning during the evolution of continental large igneous provinces: A case study from the North Atlantic Tertiary province, *Geology*, 22, 1027-1030, 1994.

Kerr, A. C., J. Tarney, G. F. Marriner, G. T. Klaver, A. D. Saunders, and M. F. Thirwall, The geochemistry and pertogenesis of the late-Cretaceous picrites and basalts of Curaço, Netherland Antilles: a remnant of an oceanic plateau, *Contrib. Mineral. Petrol.*, 124, 29-43, 1996.

Klein, E. M., and C. H. Langmuir, Global correlations of ocean ridge basalt chemistry with axial depth and crustal thickness, *J. Geophys. Res.*, 92, 8089-8115, 1987.

Kurz, M. D., T. C. Kenna, J. C. Lassiter, and D. J. DePaolo, Helium isotopic evolution of Mauna Kea volcano: first results from the 1-km drill core, *J. Geophys. Res.*, 101, 11781-11791, 1996.

Kyser, T. K., Stable isotope variations in the mantle, in *Stable Isotopes in High Temperature Geological Processes, Rev. Mineral.*, 16, edited by J.W. Valley, H.P. Taylor, and J.R. O'Neil, pp. 141-146, Mineralogical Society of America, Chelsea, MI, 1986.

Langmuir, C. H., E. M. Klein, and T. Plank, Petrological systematics of mid-ocean ridge basalts: constraints on melt generation beneath ocean ridges, in *Mantle Flow and Melt Generation at Mid-Ocean Ridges, Geophys. Monogr Ser.*, vol. 71, edited by J. P. Morgan, D. K. Blackman, and J. M. Sinton,

pp. 183-280, AGU, Washington, D.C., 1992.

Lassiter, J. C., D. J. DePaolo, and J. J. Mahoney, Geochemistry of the Wrangellia flood basalt province: Implications for the role of continental and oceanic lithosphere in flood basalt genesis, *J. Petrol.*, 36, 983-1009, 1995a.

Lassiter, J. C., M. T. Silk, M. A. Richards, D. J. DePaolo, C. G. Farnetani, and R. A. Duncan, Constraints on the origin of the Wrangellia flood basalt province: support for the plume impact model of flood basalt formation, *Eos Trans. AGU*, 76, 587, 1995b.

Lightfoot, P. C., C. J. Hawkesworth, C. W. Devey, N. W. Rogers, and P. W. C. Van Calsteren, Source and differentiation of Deccan Trap lavas: Implications of geochemical and mineral chemical variations, *J. Petrol.*, 31, 1165-1200, 1990a.

Lightfoot, P. C., A. J. Naldrett, N. S. Gorbachev, W. Doherty, and V. A. Fedorenko, Geochemistry of the Siberian Trap of the Noril'sk area, USSR, with implications for the relative contributions of crust and mantle to flood basalt magmatism, *Contrib. Mineral. Petrol.*, 104, 631-644, 1990b.

Lightfoot, P. C., C. J. Hawkesworth, J. Hergt, A. J. Naldrett, N. S. Gorbachev, V. A. Fedorenko, and W. Doherty, Remobilization of the continental lithosphere by a mantle plume: Major-, trace-, and Sr-, Nd-, and Pb-isotope evidence from picritic and tholeiitic lavas of the Noril'sk district, Siberian Trap, Russia, *Contrib. Mineral. Petrol.*, 114, 171-188, 1993.

Mahoney, J. J., An isotopic survey of Pacific ocean plateaus: Implications for their nature and origin, *in Seamounts, Islands, and Atolls, Geophys. Monogr. Ser.*, vol. 43, edited by B. Keating, P. Fryer, R. Batiza, and G. Boehlert, pp. 207-220, AGU, Washington, D.C., 1987.

Mahoney, J. J., Deccan Traps, in *Continental Flood Basalts*, edited by J. D. Macdougall, pp. 151-194, Kluwer Academic Publishers, Boston, 1988.

Mahoney, J. J., and K. J. Spencer, Isotopic evidence for the origin of the Manihiki and Ontong Java plateaus, *Earth Planet. Sci. Lett.*, 104, 196-210, 1991.

Mahoney, J. J., J. D. Macdougall, G. W. Lugmair, A. V. Murali, M. Sankar Das, and K. Gopalan, Origin of the Deccan Trap flows at Mahabaleshwar inferred from Nd and Sr isotopic and chemical evidence, *Earth Planet. Sci. Lett.*, 60, 47-60, 1982.

Mahoney, J. J., M. Storey, R. A. Duncan, K. J. Spencer, and M. Pringle, Geochemistry and age of the Ontong Java plateau, in *The Mesozoic Pacific: Geology, Tectonics, and Volcanism, Geophys. Monogr. Ser.*, vol. 77, edited by M. Pringle, W. Sager, W. Sliter, and S. Stein, pp. 233-261, AGU, Washington, D.C., 1993.

Mattey, D., D. Lowry, and C. Macpherson, Oxygen isotope composition of mantle peridotite, *Earth Planet Sci. Lett.*, 128, 231-241, 1994.

McDonough, W. F., Constraints on the composition of the continental lithospheric mantle, *Earth Planet. Sci. Lett.*, 101, 1-18, 1990.

McDonough, W. F., Chemical and isotopic systematics of continental lithospheric mantle, *Proc. 5th Intl. Kimberlite Conf.*, 478-485, 1992.

McKenzie, D., and M. J. Bickle, The volume and composition of melt generated by extension of the lithosphere, *J. Petrol.*, 29, 625-679, 1988.

McKenzie, D., and R. K. O'Nions, Partial melt distributions from inversion of rare earth element concentrations, *J. Petrol.*, 32, 1021-1091, 1991.

McKenzie, D., and R. K. O'Nions, The source regions of ocean island basalts, *J. Petrol.*, 36, 133-159, 1995.

Michael, P., Regionally distinctive sources of depleted MORB: Evidence from trace elements and H2O, *Earth Planet. Sci. Lett.*, 131, 301-320, 1995.

Mitchell, R. H., and S. C. Bergman, *Petrology of Lamproites*, 447 pp., Plenum Press, New York, 1991.

Molzahn, M., L. Reisberg, G. Wörner, Os, Sr, Nd, Pb, O isotope and trace element data from the Ferrar flood basalts, Antarctica: evidence for an enriched subcontinental lithosphere source, *Earth Planet. Sci. Lett.*, 144, 529-546, 1996.

Morgan, W. J., Convection plumes in the lower mantle, *Nature*, 230, 42-43, 1971.

Negi, J. G., O. P. Pandey, and P. K. Agrawal, Super-mobility of hot Indian lithosphere, *Tectonophysics*, 131, 147-156, 1986.

Nelson, D. O., Implications of oxygen-isotope data and trace-element modeling for a large-scale mixing model for the Columbia River basalt, *Geology*, 11, 248-251, 1983.

Nelson, D. O., Geochemistry of the Grande Ronde basalt of the Columbia River Basalt Group; a reevaluation of source control and assimilation effects, in *Volcanism and Tectonism in the Columbia River Flood-Basalt Province, Spec. Pap. 239*, edited by S. P. Reidel and P. R. Hooper, pp. 333-341, Geological Society of America, Boulder, CO, 1989.

Nicholson, S. W., S. B. Shirey, K. J. Schulz, and J. C. Green, Evolution of 1.1 Ga midcontinent rift basalts: rift-wide correlation and the interaction of multiple mantle sources during rift development, *Can. J. Earth Sci.* special IGCP 336 vol., in press, 1997.

Nickel, K. G., Phase equilibria in the system SiO2-MgO-Al2O3-CaO-Cr2O3 (SMACCR) and their bearing on spinel/garnet lherzolite relationships, *Neues Jahrb. Miner. Abh.*, 155, 259-287, 1986.

O'Hara, M. J., and R. E. Mathews, Geochemical evolution in an advancing, periodically replenished, periodically tapped, continuously fractionated magma chamber, *J. Geol. Soc. London*, 138, 237-277, 1981.

Pearson, D. G., R. W. Carlson, S. B. Shirey, F. R. Boyd, and P.H. Nixon, Stabilization of Archaean lithospheric mantle: A Re-Os isotope study of peridotite xenoliths from the Kaapvaal craton, *Earth Planet. Sci. Lett.*, 134, 341-357, 1995.

Peng, Z. X., J. Mahoney, P. Hooper, C. Harris, and J. Beane, A role for lower continental crust in flood basalt genesis? Isotopic and incompatible element study of the lower six formations of the western Deccan Traps, *Geochim. Cosmochim. Acta*, 58, 267-288, 1994.

Peng, Z. X., and J. J. Mahoney, Drillhole lavas from the northwestern Deccan Traps, and the evolution of Réunion hotspot mantle, *Earth Planet. Sci. Lett.*, 134, 169-185, 1995.

Price, R. C., A. K. Kennedy, M. Riggs-Sneeringer, and F. A.

Frey, Geochemistry of basalts from the Indian Ocean triple junction: Implications for the generation and evolution of Indian Ocean ridge basalts, *Earth Planet. Sci. Lett.*, 78, 379-396, 1986.

Prinzhofer, A., E. Lewin, and C. J. Allègre, Stochastic melting of the marble cake mantle: Evidence from local study of the East Pacific Rise at 12050'N, *Earth Planet. Sci. Lett.*, 92, 189-206, 1989.

Rhodes, J. M., Geochemical stratigraphy of lava flows sampled by the Hawaii Scientific Drilling Project, *J. Geophys. Res.*, 101, 11729-11746, 1996.

Roeder, P. L., and R. F. Emslie, Olivine-liquid equilibrium, *Contrib. Mineral. Petrol.*, 29, 275-289, 1970.

Rudnick, R. L., and S. R. Taylor, The composition and petrogenesis of the lower crust: a xenolith study, *J. Geophys. Res.*, 92, 13981-14005, 1987.

Saunders, A. D., M. Storey, R. W. Kent, and M. J. Norry, Consequences of plume-lithosphere interaction, in *Magmatism and the Causes of Continental Break-up, Spec. Publ. 68*, edited by B. C. Storey, T. Alabaster, and R. J. Pankhurst, pp. 41-60, Geological Society of London, 1992.

Sen, G., R. Hickey-Vargas, D. G. Waggoner, and F. Maurrasse, Geochemistry of basalts from the Dumisseau Formation, southern Haiti: implications for the origin of the Caribbean Sea crust, *Earth Planet. Sci. Lett.*, 87, 423-437, 1988.

Sharma, M., A. R. Basu, and G. V. Nesterenko, Nd-Sr isotopes, petrochemistry, and the origin of the Siberian flood basalts, USSR, *Geochim. Cosmochim. Acta*, 55, 1183-1192, 1991.

Sharma, M., A. R. Basu, and G. V. Nesterenko, Temporal Sr-, Nd-, and Pb-isotopic variations in the Siberian flood basalts: Implications for the plume-source characteristics, *Earth Planet. Sci. Lett.*, 113, 365-381, 1992.

Shirey, S. B., Re-Os isotopic compositions of Midcontinent rift system picrites: implications for plume-lithosphere interaction and enriched mantle sources, *Can. J. Earth Sci.*, in press, 1997.

Shirey, S. B., K. W. Klewin, J. H. Berg, and R. W. Carlson, Temporal changes in the sources of flood basalts: Isotopic and trace element evidence from the 1100 Ma old Keweenawan Mamainse Point Formation, Ontario, Canada, *Geochim. Cosmochim. Acta*, 58, 4475-4490, 1994.

Stewart, K., and N. Rogers, Mantle plume and lithosphere contributions to basalts from southern Ethiopia, *Earth Planet. Sci. Lett.*, 139, 195-211, 1996.

Sun, S. -s., and W. F. McDonough, Chemical and isotopic systematics of oceanic basalts: implications for mantle composition and processes, *in Magmatism in the Ocean Basins, Spec. Publ. 42*, edited by A. D. Saunders and M. J. Norry, pp. 313-345, Geological Society of London, 1989.

Taylor, S. R., and S. M. McLennan, *The Continental Crust: Its Composition and Evolution*, 312 pp., Blackwell Scientific Publications, London, 1985.

Thompson, R. N., P. T. Leat, A. P. Dickin, M. A. Morrison, G. L. Hendry, and S. A. Gibson, Strongly potassic mafic magmas from lithospheric mantle sources during continental extension and heating: Evidence from Miocene minettes of northwest Colorado, USA, *Earth Planet. Sci. Lett.*, 98, 139-153, 1990.

Turner, S., and C. Hawkesworth, The nature of the subcontinental mantle: Constraints from the major-element composition of continental flood basalts, *Chem. Geol.*, 120, 295-314, 1995.

Vandamme, D, and V. Courtillot, Latitudinal evolution of the Réunion hotspot deduced from paleomagnetic results of Leg 115, *Geophys. Res. Lett.*, 17, 1105-1108, 1990.

Walker, R. J., R. W. Carlson, S. B. Shirey, and F. R. Boyd, Os, Sr, Nd, and Pb isotope systematics of southern African peridotite xenoliths: Implications for the chemical evolution of subcontinental mantle, *Geochim. Cosmochim. Acta*, 53, 1583-1595, 1989.

Wooden, J. L., G. K. Czmanske, V. A. Fedorenko, N. T. Arndt, C. Chauvel, R. M. Bouse, B. S. W. King, R. J. Knight, and D. F. Siems, Isotopic and trace-element constraints on mantle and crustal contributions to Siberian continental flood basalts, Noril'sk area, Siberia, *Geochim. Cosmochim.* Acta, 57, 3677-3704, 1993.

Yang, H. -J., F. A. Frey, J. M. Rhodes, and M. O. Garcia, Evolution of Mauna Kea volcano: Inferences from lava compositions recovered from in the Hawaii Scientific Drilling Project, *J. Geophys. Res.*, 101, 11747-11767, 1996.

J. C. Lassiter and D. J. DePaolo, Berkeley Center for Isotope Geochemistry, Department of Geology and Geophysics, University of California, Berkeley CA 94720

Flood Basalts and Magmatic Ni, Cu, and PGE Sulphide Mineralization: Comparative Geochemistry of the Noril'sk (Siberian Traps) and West Greenland Sequences

Peter C. Lightfoot[1]

Department of Earth Sciences, The University of Toronto, Toronto, Ontario, M5S 3B1, Canada

Chris J. Hawkesworth

Department of Earth Sciences, The Open University, Milton Keynes, MK7 6AA, UK

The geological settings of the continental flood basalts at Noril'sk in the northwestern Siberian Traps and in West Greenland have a number of parallels. These include (1) a thick (0–3.5 km) sequence of tholeiitic and picritic lavas; (2) the inferred presence of a mantle plume or hot spot; (3) comagmatic intrusions, some of which are picritic; (4) faults which penetrate the upper mantle and acted as conduits for magmatism; and (5) an epicontinental setting with sulphidic sediments. The intrusions at Noril'sk are known to host giant deposits (>555 million tonnes with 2.7 wt% Ni, 3.9 wt% Cu, 3 ppm Pt, 12 ppm Pd) of sulphide mineralization, whereas those in West Greenland contain only small known showings of nickel sulphide mineralization. The lavas at both Noril'sk and West Greenland exhibit an empirical relation between crustal contamination and depletion in Ni and Cu, and at Noril'sk such lavas are highly depleted in the platinum group elements (PGE). Different styles of crustal contamination are recognized, but the observed depletion in Ni, Cu, and the PGE is similar irrespective of the nature of the contaminant. Moreover, the main trends are controlled by igneous crustal components that are likely to have had low S contents. Thus, we argue that initial sulphur saturation occurred largely in response to the increase in SiO_2 (typically 5 wt%) brought about by 20–25% crustal contamination, rather than to the addition of significant quantities of crustal-derived S. At Noril'sk, minor- and trace-element arguments indicate that the contaminated lavas contain <0.5% of shallow-level evaporite-rich sediments, and so although such sediments may have contributed significant quantities of S, they were not responsible for the crustal contamination that is associated with the distinctive Ni, Cu, and PGE depletions in the Nadezhdinsky lavas. Mass-balance considerations suggest that as little as 1% of the S available in the mantle-derived magmas and assimilated evaporite-rich sediments at Noril'sk is locked up in the known deposits. The sulphides at Noril'sk have unusually high metal contents, and so they must have interacted with large volumes of magma. In one model the intrusions acted as open-system conduits, or chonoliths, in which the sulphides were continually upgraded as more magma passed through. The earlier magmas were stripped of Ni, Cu and PGE, but as the system evolved the sulphides became saturated and, in some cases, isolated, and so the later magmas were progressively less depleted in metals. In another model the sulphides formed at much deeper levels and were emplaced into the intrusions as sulphide magmas. Whichever model proves to be correct, the lavas at Noril'sk are characterised by a progressive upward recovery in Ni, Cu, and PGE abundances (over about 700 m) as the degree of contamination falls, but those in West Greenland are not. This may in turn have implications for exploration in other large igneous provinces.

[1] Present address: Inco Limited, Field Exploration Office, Highway 17 West, Copper Cliff, Ontario, P0M 1N0, Canada

Large Igneous Provinces: Continental, Oceanic, and Planetary Flood Volcanism
Geophysical Monograph 100
Copyright 1997 by the American Geophysical Union

1. INTRODUCTION

Continental flood basalts (CFB) represent major magmatic events, and some are associated with important sulphide mineralization: for example, the Norils'k deposits in the Siberian Traps, the Duluth Complex in the Keweenawan midcontinent rift, and the Insizwa deposit in the Karoo [e.g., *Naldrett and Lightfoot*, 1993]. Different CFB are characterised by different major- and trace-element and radiogenic-isotope compositions, and there is an active debate both over the extent to which one model is applicable to all CFB, and whether melting is triggered by lithospheric extension or the emplacement of deep-seated mantle plumes [e.g., *Richards et al.*, 1989; *White and McKenzie*, 1989, 1995; *Gallagher and Hawkesworth*, 1994; *Turner et al.*, 1996]. Most CFB have relatively high inferred melt generation rates and were therefore associated with mantle hot spots, but the relative contributions from mantle plumes and both the crustal and mantle portions of the continental lithosphere vary from one CFB province to another [e.g., *Mahoney et al.*, 1982; *Macdougall*, 1988; *Hergt et al.*, 1991; *Turner and Hawkesworth*, 1995]. Such considerations are relevant to understanding the availability of primitive S-undersaturated magma [*Keays*, 1995] and the nature of the high-level processes governing the composition of the erupted magmas and the Ni, Cu, and platinum group elements (PGE) mineral potential of any comagmatic intrusions.

Crustal contamination of mantle-derived melts typically results in elevated SiO_2 and lithophile-element abundances, more "enriched" radiogenic isotope compositions, and in some cases highly contaminated magmas that also exhibit pronounced depletions in Ni, Cu, and the PGE [*Lightfoot et al.*, 1990, 1993, 1994; *Naldrett et al.*, 1992, 1995; *Brügmann et al.*, 1993; *Wooden et al.*, 1993; *Fedorenko*, 1994; *Hawkesworth et al.*, 1995]. The recognition that in some areas it was the lavas with strong crustal contamination signatures that were preferentially depleted in Ni, Cu, and PGE and hence, for example, had low Cu/Zr, led to the suggestion of a link between the formation of magmatic sulphide liquids and contamination of flood-basalt magmas by the continental crust [*Naldrett et al.*, 1992]. This suggestion encouraged *Naldrett and Lightfoot* [1993] and *Lightfoot et al.* [1994] to argue that the composition of flood-basalt magmas could be used as an indication of the likelihood that comagmatic intrusions are hosts to giant multibillion dollar Ni, Cu, and PGE sulphide mineral deposits such as those found at Noril'sk and Talnakh in Russia (see Figure 1). Recent efforts have focused more closely on an understanding of the reason for this strong empirical relation, not least because the data provide the mineral exploration industry with improved models for the location of mineral deposits. However, such models are very sensitive to whether sulphide formation is triggered simply by an increase in the silica contents of the magmas [*Irvine*, 1975] and/or by the addition of large amounts of crustal sulphur [e.g., *Ripley*, 1981]. Al-

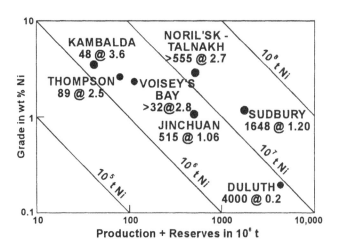

Figure 1. Plot of Ni grade against tonnage (production + reserves) for major sulphide deposits, after *Naldrett and Lightfoot* [1993]. The Noril'sk deposit is half the size of that at Sudbury, but has a much higher reserve of Ni and PGE.

ternatively, sulphide formation may be due to a physical control such as magma temperature, pressure, or fugacity of sulphur or oxygen [e.g., *Wendlandt*, 1982].

In this contribution we consider data from two CFB: the Noril'sk region of the Siberian Traps and the Qeqertarssuaq (Disko) Island and Nuussuaq Peninsula of the West Greenland CFB, because both include contaminated tholeiitic lavas with pronounced depletions in Ni and Cu [e.g., *Pedersen*, 1985a; *Naldrett et al.*, 1992, 1995; *Lightfoot et al.*, in press; P. C. Lightfoot et al., ms. in preparation]. We note that other CFB such as the Osler Volcanic Group of the Keweenawan midcontinent rift also exhibit these features [*Lightfoot et al.*, 1991]. Geological and geochemical data are used to investigate the following:

(i) whether the addition of crustal silicate material alone is sufficient to trigger the segregation of large volumes of sulphide from mafic magmas, or whether an external source of sulphur such as shale or evaporite-rich sediment is required to produce giant magmatic sulphide deposits;

(ii) the extent to which contamination by different crustal components can be recognized and then linked to siderophile and chalcophile element depletion; and

(iii) whether there are specific features of the lava chemostratigraphy that can be used to identify equilibration between sulphide and silicate liquids in the high-level conduits of CFB.

We first review the controls on the formation of magmatic sulphides, then compare aspects of the geology of the West Greenland and Siberian CFB, and ultimately use this information together with geochemical data to address the above questions.

2. CONTROLS ON THE FORMATION OF MAGMATIC SULPHIDES

A number of important controls govern the likelihood that a mafic magma will segregate and concentrate economic amounts of Ni-, Cu-, and PGE-rich sulphides.

2.1. *Source of Sulphur*

Mantle-derived magmas are likely to contain some sulphur, and S may be introduced by assimilation of basement or sediments from the continental crust. Basaltic rocks typically outgas S during subaerial eruption, and therefore the measured S contents of basaltic rocks are a poor index of those in the original magmas. Mafic intrusive rocks devoid of sulphide contain rarely more than 1000 ppm S, MORB glasses contain ~800 ppm [*Hamlyn et al.,* 1985], and the pre-eruption S contents of Columbia River flood basalts have been measured at 1900 ppm [*Thordarson,* 1995]. Crustal rocks have vastly differing sulphur contents. For example, gneisses and granitoids from West Greenland contain an average of 220 ppm S and quartzofeldspathic sediments 700 ppm S [*Shaw et al.,* 1976]; mineralized and unmineralized greenstone-belt rocks typically contain 5–25 wt% S, and <2000 ppm S, respectively (P. C. Lightfoot, unpublished data, 1996); a composite of black shales from the Nuussuaq area of West Greenland contains 0.8% S [*Pedersen,* 1979] and evaporitic sediments at Noril'sk have 19.5 wt% S (P. C. Lightfoot, unpublished data, 1996). Addition of crustal S to mafic magmas may be an important control for a number of deposits, such as at Noril'sk [*Grinenko,* 1985] and Duluth [*Ripley,* 1981].

2.2. *Sulphur Solubility*

The solubility of sulphur in mafic magmas depends on their silica content. *Irvine* [1975] demonstrated that the addition of silica-rich crustal material to mafic magmas changes the proportions of tetrahedral to octahedral sites in the magma and thereby the solubility of sulphur. Thus, on Figure 2 [from *Irvine,* 1975] the addition of SiO_2 to a homogenous, sulphur-rich liquid of composition A changes its composition to point B, where it consists of two liquids, one silicate-rich (Y) and the other sulphide-rich (X). Identification of crustal contamination in mafic magmas, and in particular the magnitude of the increase in SiO_2 due to contamination, is therefore important to understanding the causes of sulphide formation. Moreover, there are a number of CFB whose compositions have been modified significantly by crustal contamination processes and which have a smaller than expected inventory of chalcophile ele-

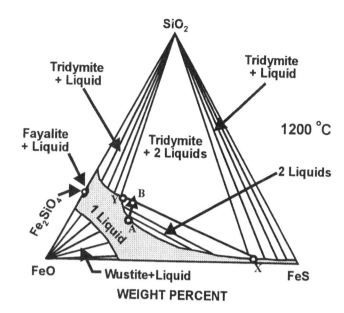

Figure 2. The 1200°C isotherm of the Fe-S-O system illustrating how the addition of SiO_2 to a homogenous sulphur-rich liquid of composition A results in a change to composition B, where it consists of two liquids, one silicate-rich (Y) and the other sulphide-rich (X). After *Irvine* [1975].

ments like Ni, Cu, and PGE [e.g., *Naldrett et al.,* 1992; *Lightfoot et al.,* 1994, in press, in preparation; *Brügmann et al.,* 1993].

Overall, the solubility of sulphur in mafic magmas is a function of magma temperature, pressure, and both fO_2 and fS_2 (where f = fugacity) [*MacLean,* 1969; *Wendlandt,* 1982]. These physical parameters are linked to composition; for example, magma temperatures depend on the degree of differentiation and fS_2 depends on the relative abundance of sulphur in the system. In this contribution we are concerned with the compositional controls on S solubility.

2.3. *S-Undersaturated Mafic Magmas*

If a silicate liquid is to generate a sulphide deposit with elevated Ni, Cu, and PGE, it is essential that the liquid was not S-saturated at an earlier stage in its evolution [*Keays,* 1995]. If magmas equilibrated with and were separated from sulphide early in their evolution, then their Ni, Cu and PGE contents would be so low that even if external sulphur was assimilated, any late-stage, higher-level sulphides would tend to be barren of economic concentrations of ore metals. Hot, primitive magmas appear more likely to be S-undersaturated because at low Fe contents the S solubility of a magma is much higher than at high Fe contents [*Wendlandt*, 1982]. The degree of partial melting also controls the amount of S in the magma, and *Keays* [1995] for example, argued that S-saturated magmas are generated at <25% partial melting.

2.4. *Partitioning of the Metals and Upgrading of the Sulphide Liquid*

Conventional sulphur-saturation models for the generation of magmatic Ni, Cu, and PGE deposits [e.g., *Naldrett*, 1989] propose that the equilibration of silicate magma with immiscible magmatic sulphide liquid results in the strong depletion of Ni, Cu, and PGE in the silicate and enrichment of these elements in the immiscible sulphide liquid. This effect is due in part to the high partition coefficients of these elements into the sulphide liquid (D_{Ni} = 500–900, D_{Cu} = 1400, D_{Pt} = 3900, D_{Pd} = 35,000 [*Peach et al.*, 1990]) but is also a function of the ratio of magma to sulphide—the R factor of *Campbell and Naldrett* [1979]. More recently, the importance of progressive upgrading of the sulphide liquid in Ni, Cu, and PGE has been addressed by *Naldrett et al.* [1995]. They and *Rad'ko* [1991] recognized that the sulphides at Noril'sk must have equilibrated with large volumes of silicate magma in the conduits of the Siberian CFB and that the sulphides had been upgraded in Ni, Cu, and PGE by continued equilibration with successive magma batches passing through the conduits. The magma, depleted in these metals, was then erupted as the lavas of the Nadezhdinsky Formation of the Siberian Traps [*Naldrett et al.*, 1992, 1995].

3. COMPARATIVE GEOLOGY AND GEOCHEMISTRY OF THE SIBERIAN TRAPS AND WEST GREENLAND CFB

3.1. *Large-Scale Tectonic Setting*

The lavas of both the Noril'sk region of the Siberian Traps and the West Greenland CFB appear to be associated with the emplacement of a mantle plume. In the case of the Siberian Traps, the 250 Ma sequence of basalts is located on the Siberian Platform to the southwest of the East Siberian lowlands, the Enisei Trough and the Khatanga Trough [*Aplonov*, 1988]. The Noril'sk region is at the northwest corner of the Siberian Traps (Figure 3), and it contains a remarkable diversity of rock types, including two different sequences of primitive picritic lavas, one of which has many geochemical traits consistent with relatively large degrees of melting of asthenospheric upper mantle [*Lightfoot et al.*, 1993; *Wooden et al.*, 1993]. *Morgan* [1981] suggested that melt generation was linked to

the Jan Mayen hot spot, whereas *Fedorenko et al.* [1996] argued that structural and sedimentological evidence does not support uplift accompanying a mantle plume.

In West Greenland, lavas in the Davis Strait region were erupted during rifting between Canada and Greenland at ~63 Ma [*Holm et al.*, 1993]. The lavas crop out in Canada on Baffin Island at Cape Dyer [*Clarke and Pedersen*, 1976; *Clarke*, 1991] and along the west coast of Greenland on Qeqertarssuaq (formerly Disko Island), and Nuussuaq (the Nuussuaq Peninsula) (Figure 4), Ubekendt Island, and the Svartenhuk Peninsula. *Holm et al.* [1993] showed that the picritic rocks are compositionally coincident with Icelandic lavas and suggested that the picritic lavas of the West Greenland CFB were derived, at least in part, from a mantle plume. *Lightfoot et al.* [in press] report new geochemical data which indicate that the lavas contain up to 14–16 wt% MgO. This contrasts with at least the mineralized intrusions at Noril'sk, which have Fo_{82} olivines that were in equilibrium with a low-Mg tholeiitic parental magma.

3.2. *Regional Setting*

The lavas of the Noril'sk region of the Siberian Traps and the West Greenland CFB were both erupted in regions of extensive Phanerozoic sedimentation. In West Greenland, the lavas were erupted onto a prograding deltaic assemblage on the margin of a Proterozoic to Archean craton [*Pedersen and Pulvertaft*, 1992], whereas at Noril'sk, they were erupted onto an epicontinental assemblage of shallow-water sediments and sabkha deposits [e.g., *Naldrett et al.*, 1992]. At Noril'sk, Ripheian dolomites, argillites, and limestones of marine origin rest unconformably on Lower Proterozoic gneisses and crystalline schists. The Ripheian sediments are overlain by extensive Devonian calcareous and dolomitic marls, dolomites, and sulphate-rich evaporites, and Lower Carboniferous shallow-water limestones. These rocks are unconformably overlain by the Middle Carboniferous to Upper Permian Tungusskaya epicontinental lagoonal and continental sediments, including siltstones, sandstones, conglomerates, and coal measures [*Smirnov*, 1966; *Glazkovsky et al.*, 1977; *Simonov*, 1994]. The overlying Siberian Traps represent $>1 \times 10^6$ km^3 of erupted magma that peaked in the Early Triassic at 248–250 Ma [*Renne and Basu*, 1991; *Campbell et al.*, 1992; *Dalrymple et al.*, 1995].

In West Greenland, the subhorizontal volcanic rocks rest unconformably on a rugged palaeosurface composed of fault scarps, basement ridges, and deeply eroded channels through the underlying Cretaceous-Tertiary sediments and Archean-Proterozoic basement [*Clarke and Pedersen*,

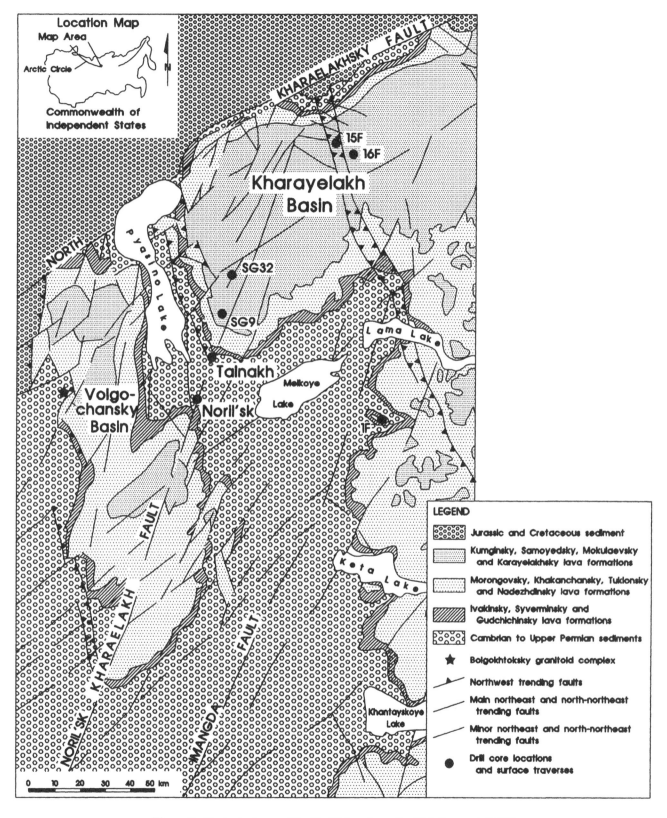

Figure 3. Geological map of the Noril'sk region, after *Lightfoot et al.* [1993].

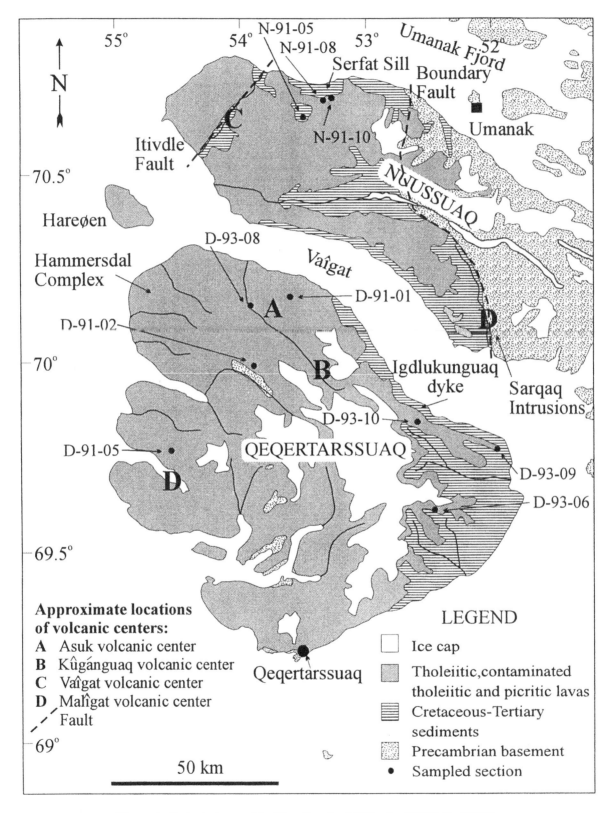

Figure 4. Geological map of the West Greenland CFB, after *Ulff-Møller* [1991].

1976]. On Qeqertarssuaq, a gneiss ridge runs north-south through the centre of the island and was an important topographic control during the eruption of lavas that acted as a barrier to the migration of some flows. Early volcanism was dominated by the development of picritic pillow lavas and hyaloclastite breccias flowing onto an epicontinental deltaic sequence [*Pedersen and Pulvertaft,* 1992]. The sediments on Nuussuaq consist of a <5-km-thick fluvial epicontinental deltaic sequence with moderate to high carbon content and associated hydrocarbon accumulations, but without exposed evaporite sequences. Deltaic sedimentation appears to have prograded from the southeast towards the northwest, and numerous gneiss inliers on Qeqertarssuaq indicate that the depth to the basement is shallower than on Nuussuaq.

3.3. *Volcanic Stratigraphy*

In the West Greenland CFB and at Noril'sk, up to 3500 m of picritic and tholeiitic lavas crop out. The sequence at Noril'sk thins significantly towards the southeast on the Putorana plateau where only a few tens of meters of lavas are present within a thick sequence of tuffs. The lava stratigraphy was established by Russian workers based on stratigraphic, petrological, and major-element criteria [e.g., *Fedorenko,* 1981], and these subdivisions have been retained in subsequent studies of the trace-element and radiogenic isotope variations. The lavas consist of a Lower Sequence of three formations, the Ivakinsky, Syverminsky, and Gudchikhinsky, overlain by an Upper Sequence of the Khakhanchansky, Tuklonsky, Nadezhdinsky, Morongovsky, Mokulaevsky, Kharaelakhsky, Kumginsky and Samoedsky Formations (Figure 5). The Ivakinsky consists dominantly of subalkalic basalts and trachybasalts, whereas the overlying Syverminsky consists of a sequence of tholeiites. The Gudchikhinsky contains a lower unit of tholeiitic basalts and an upper unit of picrites [*Lightfoot et al.,* 1990; 1993]. The rocks of the Upper Sequence are dominantly tholeiites with subordinate picrites in the upper unit of the Tuklonsky [*Lightfoot et al.,* 1993] and andesitic basalts in the Kharaelakhsky [*Hawkesworth et al.,* 1995]. The rocks of the Tuklonsky through to the Mokulaevsky are of principal concern in this manuscript. These rocks occupy a considerable portion of the stratigraphy (Tuklonsky: 2–220 m thick; Nadezhdinsky: 150–670 m thick; Morongovsky: 230–790 m thick; Mokulaevsky: 255–1095 m thick [*Lightfoot et al.,* 1994]). Picrites comprise <1% of the stratigraphic thickness in the Noril'sk area [*Fedorenko,* 1994].

The volcanic stratigraphy of the West Greenland CFB on Qeqertarssuaq (Figure 4) is subdivided into three litho-

stratigraphic units termed the Vaigat, Malîgat, and Hareöen Formations [*Hald and Pedersen,* 1975] (Figure 6). The division of the volcanic stratigraphy was based largely on lithostratigraphy and the relative positions of key marker

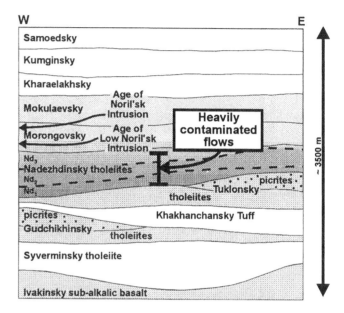

Figure 5. Stratigraphy of the Siberian Traps at Noril'sk projected onto a west-east section, after *Lightfoot et al.* [1994].

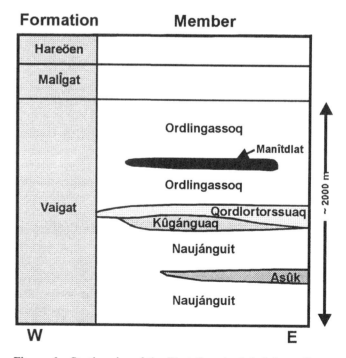

Figure 6. Stratigraphy of the West Greenland tholeiites, after *Pedersen* [1985b]. The contaminated units are the Asûk and Kûgánguaq members.

horizons, rather than geochemical criteria. Thus, the Vaigat Formation was distinguished from the overlying Malîgat Formation on the presence of picritic lavas, hyaloclastites, and a small number of basalt horizons. The Malîgat Formation consists of feldspar-phyric tholeiitic basalts, and the Hareöen Formation consists of transitional olivine-phyric basalts separated from the Malîgat by shales with coal seams. *Pedersen* [1985a] subdivided the Vaigat Formation into several members within two volcanic cycles: the lower cycle of ~900 m thickness consists dominantly of picrites (Naujánguit member) with two 50- to 100-m-thick tholeiite horizons (the Asûk and Kûgánguaq members). The tholeiitic horizons of the Asûk contain shale xenoliths and metallic iron [*Pedersen*, 1978; *Klock et al.*, 1986], whereas the Kûgánguaq member is devoid of native iron. The Asûk and Kûgánguaq members are both highly amygdaloidal, reddish brown coloured flow groups which have elevated silica contents attributed to crustal contamination [*Lightfoot et al.*, in press]. The Naujánguit flows were followed by the eruption of olivine poor tholeiitic basalts of the Qordlortorssuaq member in a waning stage of volcanic activity [*Pedersen*, 1985b]. The upper cycle of the Vaigat Formation comprises 800 m of picrites of the Ordlingassoq member, which includes three horizons of distinctive, contaminated lavas intercalated with the picrites [*Pedersen*, 1985b].

On Nuussuaq, the Vaigat Formation is exposed east of the Itivdle Fault (Figure 4) and consists of more than five packages, or subunits, of contaminated lava flows with ~1 km of picritic rocks [e.g., *Hald*, 1977a, b; *Lightfoot et al.*, in press]. These lavas have been assigned an interim nomenclature because it has yet to be established whether individual subunits can be correlated from Nuussuaq to Qeqertarssuaq. On Nuussuaq, the more basaltic, contaminated lavas are therefore in members termed B0 to B4, and the picrites are in members P0 to P6. One of the contaminated lava horizons south of the Serfat sill (Figure 4) has associated native iron mineralization. The native iron occurs as blebby to massive cumulates within the lower part of a complex flow unit at the base of the B3 unit. This native iron resembles the native iron in a large boulder resting on Vaigat Formation lavas in Stordal, central Qeqertarssuaq [e.g., *Ulff-Møller*, 1991, and references therein].

The Malîgat Formation is composed of thicker, massive basalts. Both formations occur throughout central and western Qeqertarssuaq, west of the Itivdle Fault on Nuussuaq, and also on the east side of Nuussuaq (Figure 4). On Qeqertarssuaq, the Malîgat Formation is subdivided into the Rinks Dal (~750 m thick), the Nordfjord, and the Niaqussat members [*Pedersen*, 1975]. The Hareöen Formation is ~250 m thick and consists of olivine porphyritic transi-

tional basalts and is developed only on Hareöen Island.

3.4. *Regional Structure*

The lavas at Noril'sk and in the West Greenland CFB are associated with major faults, some of which may have acted as the conduits for magma migration. At Noril'sk the epicontinental sediments and unconformable basalt sequence are cut by a major series of NNW-SSE-trending faults which appear to have played a central role in controlling the surface distribution of different portions of the lava stratigraphy [e.g., *Naldrett et al.*, 1992]. They may also have been the foci for many of the intrusions in the region, in particular those which host the major deposits of Ni, Cu, and PGE sulphide mineralization at Noril'sk and Talnakh (Figure 3). The three major, possibly mantle-penetrating faults are the Noril'sk-Kharaelakh, Imangda, and North Kharaelakhsky. In the lower part of the stratigraphy, the primitive lavas of the Gudchikhinsky are thickest along the Noril'sk-Kharaelakh and North Kharaelakh faults and were presumably erupted along these structures. In contrast, the lavas of the Tuklonsky are thickest along the Imangda Fault and were erupted well to the east of Noril'sk. Lavas of the Nadezhdinsky were erupted in the Noril'sk region and centred close to the Noril'sk-Kharaelakh Fault. *Fedorenko* [1981, 1994] and *Lightfoot et al.* [1990, 1993, 1994] discussed the thickness variations in the different formations of lava in the Noril'sk Region. *Naldrett et al.* [1992] used isopach diagrams for the thickness of these formations to demonstrate that magmatic activity switched episodically between different eruptive sites, rather than migrating steadily across the province as, for example, when a plate moves across a mantle plume [e.g., *Devey and Lightfoot*, 1986].

In West Greenland, the Boundary Fault cuts through the eastern part of Nuussuaq and marks the eastern extent of Cretaceous and Tertiary sediments where they abut basement Proterozoic gneiss [*Pulvertaft*, 1989]; the fault may have served locally as a controlling structure for the lateral containment of the flood basalts (Figure 4). The Itivdle Fault cuts across western Nuussuaq and defines the boundary of the Malîgat Formation flows on western Nuussuaq, which are downfaulted by more than 1 km relative to the main Vaigat package east of the fault. The eruption of the picritic lavas in the West Greenland CFB appears to have been controlled by major faults; for example the Vaigat Formation picritic lavas are believed to have been centred close to the Itivdle Fault (Figure 4), but some of the packages of tholeiites (e.g., the Asûk and Kûgánguaq) were erupted in separate sub-basins away from the major fault lines [*Pedersen*, 1985a]. This situation contrasts with the

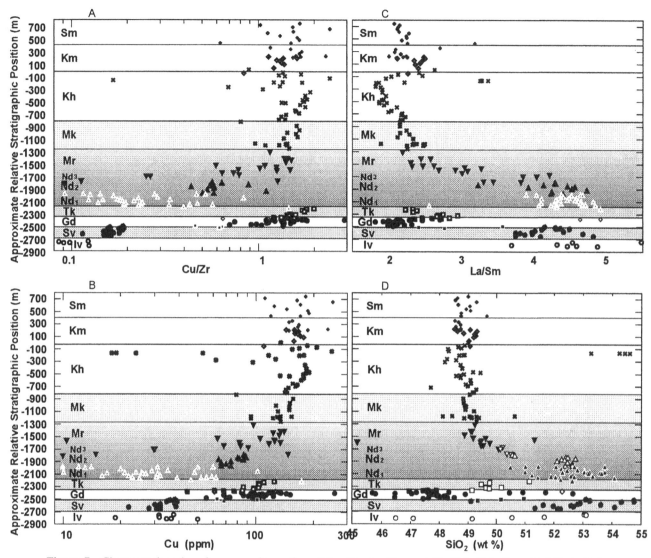

Figure 7. Chemostratigraphy of a composite section of the Siberian Traps at Noril'sk [data from *Lightfoot et al.,* 1993, 1994, ms. in preparation; *Wooden et al.,* 1993; *Hawkesworth et al.,* 1995].

Noril'sk region where much of the eruption was centred on major mantle-penetrating faults. The centre of Malîgat volcanism is poorly known and although it may have been west of Qeqertarssuaq, the presence of many Malîgat-like dolerite sills (Sarqaq dolerites) along the Boundary Fault in the Sarqaq valley (Figure 4) suggests that at least some of the Malîgat lavas were erupted close to the Boundary Fault.

3.5. *Chemostratigraphy of the Lavas of the Noril'sk Region and the West Greenland CFB*

Figure 7 shows the composite chemostratigraphy of the Siberian Traps at Noril'sk. The section has been recon-

structed from data for samples of drill core (hole SG-32) and surface outcrop (sections 1F, 15F, and 16F) (P. C. Lightfoot et al., ms. in preparation). The section of stratigraphy focused on here is that between the base of the Tuklonsky and the top of the Mokulaevsky Formations. In this stratigraphic interval there is a marked drop in Cu content at the base of the Nadezhdinsky followed by a gradual recovery in abundance levels upwards through the section to the Mokulaevsky (Figure 7a) [*Lightfoot et al.,* 1990; 1993; 1994; *Wooden et al.,* 1993]. The changes in Cu content are accompanied by similar variations in Ni and PGE contents [*Lightfoot et al.,* 1990; *Brügmann et al.,* 1993], and by a sudden increase in La/Sm at the base of the Nadezhdinsky followed by a progressive decline to-

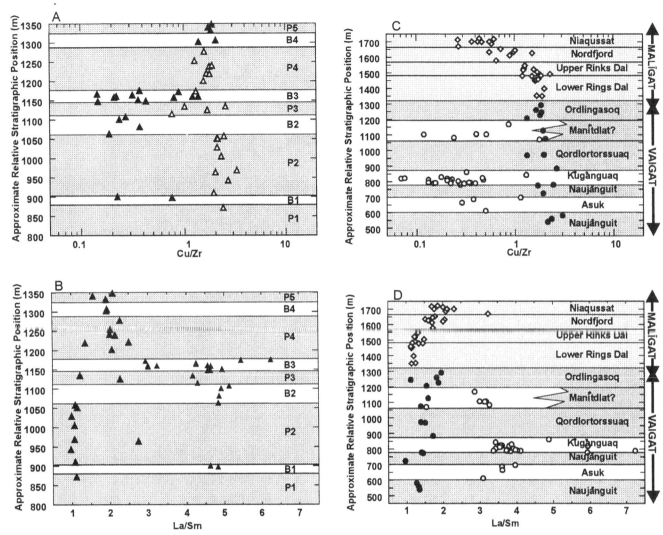

Figure 8. Chemostratigraphy of composite sections of the West Greenland CFB, with data from *Lightfoot et al.* [in press]. (a) Cu/Zr versus stratigraphic position, Nuussuaq. (b) La/Sm versus stratigraphic position, Nuussuaq. (c) Cu/Zr versus stratigraphic position on Qeqertarssuaq. (d) La/Sm versus stratigraphic position on Qeqertarssuaq.

wards the top of the Mokulaevsky (Figure 7b). This change in La/Sm is accompanied by systematic changes in SiO₂, in large ion lithophile element to high field strength element ratios (LILE/HFSE), and in radiogenic isotopic ratios and has been explained in terms of a progressive decline in the amount of contamination up-section [*Lightfoot et al.*, 1990, 1993; *Wooden et al.*, 1993; *Hawkesworth et al.*, 1995]. *Naldrett et al.* [1995] and P. C. Lightfoot et al. (ms. in preparation) ascribe the progressive changes in the metal contents of these lavas to continued equilibration of highly contaminated Tuklonsky-type magma with a sulphide liquid in the conduits of the CFB, which are now represented by the Noril'sk and Talnakh intrusions and their associated magmatic Ni, Cu, and PGE sulphides.

In the West Greenland CFB on Qeqertarssuaq, a composite stratigraphy has been developed based on the data of *Lightfoot et al.* [in press] (Figure 8). *Pedersen* [1985a,b] first demonstrated that many of the tholeiitic lavas from Qeqertarssuaq had been contaminated by crustal material and that not only was olivine in these lavas depleted in Ni but also that some of the whole-rock compositions had extremely low Ni and Cu contents. He further suggested that the strong Ni and Cu depletion was caused by the segregation of a magmatic sulphide liquid from the magma. *Lightfoot et al.* [in press] demonstrate that the depletion in Ni and Cu is associated with magmas that had elevated contents of SiO₂, light rare earth elements and LILE, similar to those in the contaminated tholeiites at Noril'sk (see Figure 9a). In the stratigraphic section shown in Figure 8, the depletion in Cu is observed to be a sudden event, much like that recorded at the base of the Nadezhdinsky in Figure 7a. The thickness of the Cu-depleted units is ~100 m, which is less than the 500-m thickness at Noril'sk (and the 1000-m sequence in the Osler Group volcanics of the Keweenawan midcontinent rift [*Lightfoot et al.,* 1991; *Naldrett and Lightfoot,* 1993]). However, there are a number of units on Qeqertarssuaq and Nuussuaq that show Cu depletion, which makes the overall volume of Cu-depleted lavas ~10^3 km^3, compared with ~5×10^3 km^3 at Noril'sk.

One of the more significant differences between the West Greenland and Noril'sk successions is that the former preserves no evidence of a gradual recovery in the Cu contents of the lavas either within or above the contaminated, low-Cu units, such as the Asûk or the Kûgánguaq members on Qeqertarssuaq. The most Cu-depleted rocks

show the same low La/Sm as the Noril'sk lavas, but again there is no gradual decrease in La/Sm upwards through either the Asûk or the Kûgánguaq members (see Figure 8). On Nuussuaq, like Qeqertarssuaq, the tholeiitic members have low Cu contents and high La/Sm ratios relative to the picritic rocks [see *Lightfoot et al.,* in press], and they also show no systematic compositional changes within individual units. Such differences in the rates of chemical change with stratigraphic height between West Greenland and Noril'sk (and the Keweenawan) presumably reflect differences in the dynamics of the magma feeder systems, which may in turn have influenced the extent to which different magma batches equilibrated with separated sulphides.

4. DISCUSSION

At issue is whether different processes of contamination

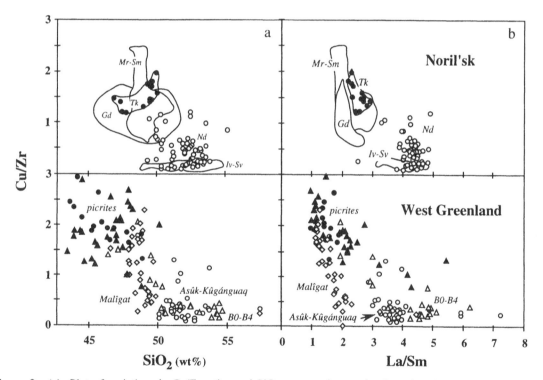

Figure 9. (a) Plot of variations in Cu/Zr ratios and SiO$_2$ contents in samples from the Siberian Traps [data from *Lightfoot et al.,* 1990, 1993, 1994, ms. in preparation; *Wooden et al.,* 1993; *Hawkesworth et al.,* 1995], and the West Greenland CFB [data from *Lightfoot et al.,* in press].
(b) Plot of Cu/Zr- La/Sm ratios for samples from the Siberian Traps [data from *Lightfoot et al.,* 1993, rns. in preparation; *Wooden et al.,* 1993; *Hawkesworth et al.,* 1995], and the West Greenland CFB [data from *Lightfoot et al.* in press]. For the Siberian lavas the following abbreviations are used for the different formations: Iv - Ivakinsky; Sv - Syverminsky; Gd - Gudchikhinsky; Tk - Tuklonsky (filled circles); Nd - Nadezhdinsky (open circles); Mr - Morongovsky; Sm - Samoedsky. For the West Greenland lavas the symbols are as follows: filled circles - Vaigat Formation picrites on Qeqertarssuaq; filled triangles - Vaigat Formation picrites on Nuussuaq; open circles - Asûk and Kûgánguaq members; open triangles - B0 to B4 rocks in the Vaigat Formation on Nuussuaq; open diamonds - Malîgat Formation.

and crustal end-members can be identified and the extent to which the sulphur and silica contents of the contaminants determine whether a magma will segregate magmatic sulphide. Further considerations are whether the geochemical variations in the lava stratigraphy provide information about the timing of sulphide saturation and any upgrading of the siderophile element deposits by, for example, continued interaction of the sulphides with successive magma batches in the CFB conduits.

4.1. Relation Between Ni-Cu-PGE Depletion and Major-Element, Trace-Element, and Isotopic Composition

The tholeiitic lavas of the Nadezhdinsky Formation have relatively low Cu contents, low Cu/Zr, and elevated SiO_2 and La/Sm (Figure 9 and *Lightfoot et al.* [1994]). The same traits are found in the tholeiitic lavas of the Asûk, Kûgánguaq, and B0–B4 units of the Vaigat Formation, and portions of the Malîgat Formation in West Greenland (these rocks will be referred to below as the CWGT, i.e., contaminated West Greenland tholeiites). The negative

correlations between Cu/Zr and both SiO_2 and La/Sm are developed better in the CWGT than in the Nadezhdinsky lavas, although this is in part because the CWGT lavas with high Cu/Zr have significantly lower SiO_2 and La/Sm than their Noril'sk counterparts. The CWGT tend to have consistently lower Cu/Zr than the Nadezhdinsky lavas, and the range of Cu/Zr in the latter suggests that the degree of equilibration with sulphide was less complete than in the CWGT. The Ivakinsky and Syverminsky lavas at Noril'sk are tholeiitic to subalkalic basalts whose low Cu/Zr ratios may reflect small degrees of melting in the presence of residual sulphide.

The shifts to high La/Sm and low Cu/Zr in the CWGT and Nadezhdinsky are accompanied by increases in $^{87}Sr/^{86}Sr_o$ and decreases in ε_{Nd} (Figure 10). The Noril'sk lavas define a broad trend of increasing La/Sm with decreasing ε_{Nd} which was modelled by *Lightfoot et al.* [1994] and *Naldrett et al.* [1995] in terms of contamination of a Tuklonsky parental magma by a crustal component with high $^{87}Sr/^{86}Sr$ and La/Sm, and low ε_{Nd}. This contamination was then followed by progressive mixing of the

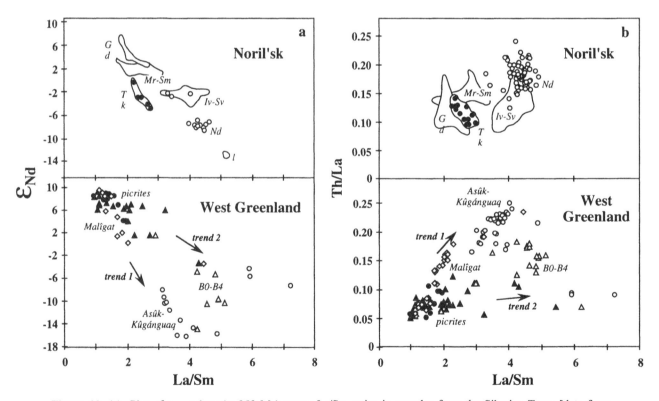

Figure 10. (a) Plot of ε_{Nd} values (at 250 Ma) versus La/Sm ratios in samples from the Siberian Traps [data from *Lightfoot et al.*, 1990, 1993, 1994, ms. in preparation; *Wooden et al.*, 1993; *Hawkesworth et al.*, 1995], and the West Greenland CFB [data from *Lightfoot et al.*, in press]. The symbols and abbreviations are as for Figure 9, except that l is a lamprophyre apparently associated with the Siberian Traps. (b) Plot of Th/La against La/Sm ratios for the Siberian Traps and West Greenland CFB, with data sources, symbols, and abbreviations as for Figure 9.

contaminated magma with Mokulaevsky magma in the conduits of the CFB system. Such a model requires the establishment of a large reservoir of contaminated Tuklonsky magma followed by progressive replenishing of this reservoir with Mokulaevsky magma [e.g., *Brügmann et al.*, 1993].

The CWGT are different from the Noril'sk lavas in that the samples with higher La/Sm ratios have a range in ε_{Nd} consistent with contributions from two La-enriched end-members (Figure 10a). Most of the data for rocks from Qeqertarssuaq plot on a steeper trend of La/Sm against ε_{Nd} than those from Nuussuaq, except that the three samples with the highest La/Sm ratios have relatively high ε_{Nd}. This indication of two La-enriched components in the lavas of the CWGT is reinforced by the plot of Th/La against La/Sm (Figure 10b), in that lavas with lower ε_{Nd} values are characterised by elevated Th/La, and rocks with higher La/Sm ratios have intermediate Th/La in addition to slightly higher ε_{Nd} values. On both diagrams, the CWGT data fan out from the fields for the Vaigat Formation picritic lavas, and there are broad increases in Th/La with decreasing ε_{Nd} in both Noril'sk and West Greenland. The least Th- and La-enriched rocks from Noril'sk still have slightly elevated Th/La and La/Sm and low ε_{Nd} values, which has led to contrasting models in which such magmas were derived largely from incompatible-element-enriched source regions in the mantle lithosphere [*Lightfoot et al.*, 1993] or from the asthenospheric mantle with additional crustal contamination [*Wooden et al.*, 1993].

4.2. Nature of the Crustal Components in the Siberian and Greenland Lavas and Intrusions

The nature of the crustal material incorporated into mantle-derived magmas is important to models of sulphide formation and specifically to whether sulphide saturation is primarily triggered by the addition of excess S, or by increases in silica content. Crustal components may include bulk sedimentary or meta-igneous rocks, or melts thereof, and in principle may be distinguished on the basis of selected minor-element and trace-element ratios. For example, upper-crustal rocks tend to have elevated Rb/Ba, Rb/Sr, and hence with time relatively high $^{87}Sr/^{86}Sr$; deep crustal melts are more likely to have been generated in the presence of residual garnet and hence to have low heavy rare earth elements and higher Sr contents, and most crustal melts have relatively low P_2O_5 and TiO_2 [e.g., *Taylor and McLennan*, 1985; *Harris et al.*, 1986; *Hawkesworth and Clarke*, 1994].

Because the shifts to increasing La/Sm in the CWGT and Nadezhdinsky lavas are accompanied by shifts to more

crustal-like Sr and Nd isotopic ratios (e.g., Figure 10a), higher SiO_2 and low Ti/Zr, they have been largely attributed to crustal contamination processes [*Lightfoot et al.*, 1990, 1993, 1994, in press; in preparation; *Naldrett et al.*, 1992, 1995; *Wooden et al.*, 1993; *Hawkesworth et al.*, 1995; *Fedorenko*, 1994]. Thus, in these rocks La/Sm may be regarded as an index of contamination, and from the above studies, the following general changes accompany the increase in La/Sm in both the CWGT and the Nadezhdinsky:

(i) increases in La/Yb, Zr/Y, Th/Nb and Th/La (e.g., Figure 10b),

(ii) reduction in Cu/Zr (Figure 9), Ti/Zr, P/Zr and Ta/La (e.g., Figure 11),

(iii) increases in $^{87}Sr/^{86}Sr_0$ and reduction in ε_{Nd} (Figure 10a),

(iv) increases in SiO_2 often with little change in Mg-number (see Section 4.3).

Such changes are all consistent with crustal contamination, but there are also differences in both the size of the changes and the element ratios between the CWGT and

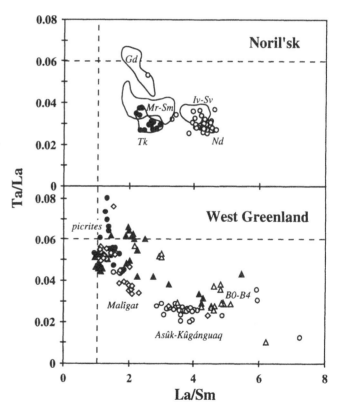

Figure 11. Plot of Ta/La against La/Sm ratios for the rocks of the Siberian Traps and West Greenland. Symbols, abbreviations and data sources as for Figure 9. The dashed lines are at the Ta/La and La/Sm ratios of primitive mantle.

Nadezhdinsky lavas that reflect both differences in their crustal contaminants and, more speculatively, in the processes of contamination and fractionation (see also section 4.3). These differences include those between average Rb/Ba and Rb/Sr in the Nadezhdinsky lavas (0.08 and 0.07 respectively [*Lightfoot et al.*, 1994]) and in the CWGT (0.16 and 0.14) and the fact that the CWGT have relatively low Sr/Zr (1.3–1.8), whereas in the Nadezhdinsky Sr/Zr and Sr are extremely variable (0.6–4.1 and 35–518 ppm) and show no systematic trends with initial ^{87}Sr/^{86}Sr (P. C. Lightfoot et al., ms. in preparation). Among the major elements, Fe_2O_3 decreases and SiO_2 contents increase more rapidly with decreasing MgO contents in the Noril'sk than in the West Greenland picrites. Moreover the Cu/Zr ratios of the CWGT are more restricted and on average lower than those in the Nadezhdinsky.

For the Nadezhdinsky, the combination of elevated Zr/Y and La/Sm, low Ti/Zr, P/Zr and Ta/La (Figure 11), and Rb/Ba similar to the Tuklonsky [*Lightfoot et al.*, 1993] is more consistent with contamination by mid- to lower crustal material than by upper crustal metasediments [e.g., *Lightfoot et al.*, 1994; *Hawkesworth et al.*, 1995]. Thus, the model of *Lightfoot et al.* [1994] and *Fedorenko* [1994] involved contamination of a primitive Tuklonsky lava with 8% granodioritic melt (similar in composition to the younger, Triassic Bolgokhtokh intrusion) and ~24% subsequent fractionation of olivine, plagioclase, and augite in the ratio 35:41:24. Other models of contamination of the Nadezhdinsky have utilised different amounts of crust [*Lightfoot et al.*, 1990; *Wooden et al.*, 1993; *Fedorenko*, 1994], but the observed changes in trace-element ratios strongly suggest that the contaminant was granodioritic in composition. Such material will typically contain <0.01 wt% S, and so it would appear that the contaminant responsible for the elevated La/Sm ratios in the Nadezhdinsky rocks was unlikely to be the source of significant amounts of additional sulphur. The bulk rocks of the Lower Talnakh intrusions have many of the isotopic and incompatible-element characteristics of the Nadezhdinsky lavas, and *Hawkesworth et al.* [1995] emphasised that these too are unlikely to be due to crustal contamination with sedimentary material. New data from P. C. Lightfoot et al. (ms. in preparation) further demonstrate that the Devonian evaporites beneath the Siberian Traps have high Sr abundances (2051–2812 ppm) and ^{87}Sr/^{86}Sr (0.7078–0.7084) and very low rare earth element contents (<10 × chondrite). Thus, evaporite contamination would result in large changes in Sr contents but have little effect on La/Sm or ε_{Nd} values.

In summary, the available evidence from Noril'sk indicates that the contamination processes responsible for the

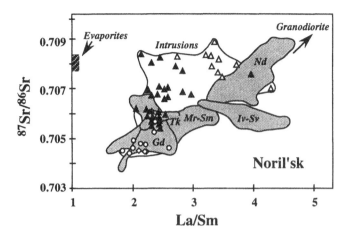

Figure 12. Plot of initial La/Sm versus Sr isotope ratios for lavas and intrusive rocks from Noril'sk, Siberia. The small field for the evaporites is simply to illustrate the observed range in Sr isotopic composition (P. C. Lightfoot et al., ms. in preparation). Symbols for the different intrusions are as follows: filled triangles - Noril'sk-type; open triangles - Lower Talnakh-type; circles - dolerites; open diamonds - differentiated, unmineralized intrusions [after *Hawkesworth et al.*, 1995]. Other data from *Lightfoot et al.* [1990, 1993, 1994].

elevated La/Sm, Th/La, and SiO_2 and the reduction in Ti/Zr, P/Zr, and Ta/La involved a granodiorite component and took place earlier than any addition of sulphur from the evaporites. These are the contamination processes associated with the observed depletion in Ni, Cu, and PGE (e.g., Figures 7, 9 and 10), and so it is inferred that initial sulphide formation was triggered by the resultant increases in SiO_2 rather than by the addition of significant quantities of crustal S. *Hawkesworth et al.* [1995] noted that Sr isotopic ratios in the Noril'sk lavas increased both with increasing La/Sm and with no change in La/Sm (Figure 12), suggesting that the latter might reflect the high-level addition of evaporite-derived material. The wide ranges of Sr in the Nadezhdinsky (35–518 ppm), in which Sr is decoupled from the other minor and trace elements, may also reflect the addition of crustal Sr from the evaporites (P. C. Lightfoot et al., ms. in preparation). However, such changes appear to have been secondary to the crustal contamination processes associated with the marked depletion in the metal contents of the Nadezhdinsky lavas.

Models for the CWGT infer that, like the Nadezhdinsky, the CWGT were generated by crustal contamination of nearly picritic magmas with 14–16 wt% MgO [*Pedersen*, 1985b; *Lightfoot et al.*, in press]. However, the picrites of West Greenland have much clearer affinities with oceanic basalts (e.g., high Ta/La and high ε_{Nd}) and there is compelling evidence for at least two crustal contamination trends (Figure 10). Contamination appears to have predated significant fractionation of feldspar and mafic minerals, because olivine in the CWGT on Qeqertarssuaq was in equilibrium with a high-Mg parental magma, and yet it shows strong depletion in Ni [*Pedersen, 1985b*], consistent with equilibration of the magma with sulphides prior to olivine crystallisation [e.g., *Lightfoot et al.,* in press]. However, the picrites sampled in West Greenland show much less evidence from both major and trace elements for interaction with lithospheric material than the Tuklonsky picrites sampled at Noril'sk (Figures 9–11, 13). The field relations show that the Qeqertarssuaq lavas were erupted onto Archean gneisses and over a thin onlapping sequence of epicontinental shales. Some of the CWGT contain fragments of bituminous shale [*Pedersen, 1979*], and on Nuussuaq there are almost 5 km of shales which contain significant hydrocarbons [*Pulvertaft, 1989*].

All the CWGT are characterised by lower Ta/La, Ti/Zr, and P/Zr, and ε_{Nd} and higher $^{87}Sr/^{86}Sr$ than the West Greenland picrites (e.g., Figure 11). In detail, two trends fan out from the fields of the picrites: trend 1 is to elevated Th/La (and Th/Nb or Th/Ta) and relatively low ε_{Nd} with increasing La/Sm, and trend 2 is to less elevated Th/La, higher ε_{Nd} values and higher La/Yb and Zr/Y at higher La/Sm than trend 1 (Figure 10). The crustal components responsible for these two trends are the subject of continuing debate, but the high Zr/Y and La/Sm at moderate Th/La may be more consistent with an igneous Proterozoic metabasic contaminant for trend 2. In contrast, Th/La ratios are not readily fractionated by igneous processes, particularly in the upper mantle, and so the high Th/La ratios of trend 1 are strongly indicative of a sedimentary contaminant [e.g., *Hawkesworth et al.,* in press]. The different ε_{Nd} values for trends 1 and 2 are reflected in different model Nd ages of 2.6 Ga and 1.1 Ga for the most contaminated samples on trends 1 and 2, respectively. Pb isotopic compositions indicate that both crustal contaminants had unradiogenic Pb isotopic ratios, but that the time-integrated Th/U of the contaminant responsible for trend 1 was higher than that responsible for trend 2 [*Lightfoot et al.,* in press]. Spatially, trend 1 is dominated by data for the Asûk-Kûgánguaq and the Malîgat rocks from Qeqertarssuaq, whereas data for Nuussuaq scatter across to trend 2, together with just three samples from Qeqertarssuaq (Figure 10).

4.3. Models for Mineralization

A number of observations bear on models of crustal contamination and the interaction between the contaminated magmas and sulphides in Noril'sk and West Greenland.

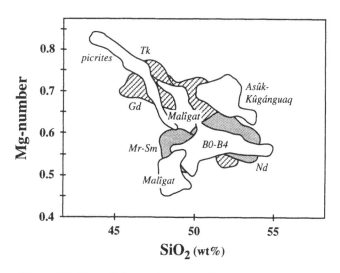

Figure 13. Plot of Mg-number versus SiO_2 content for the lavas of Noril'sk and West Greenland. Data from *Lightfoot et al.* [1990, 1993, 1994, in press, ms. in preparation]; *Hawkesworth et al.* [1995]; *Wooden et al.* [1993]. Shaded fields - Noril'sk; blank fields - West Greenland; abbreviations as for Figure 9. In the CWGT the lowest Cu/Zr ratios tend to be developed in the rocks that have relatively high SiO_2 at high Mg-number. Mg number = $Mg^{2+}/Mg^{2+} + Fe^{2+}$) assuming that 85% of the total Fe is Fe^{2+}.

(i) SiO_2 more markedly with decreasing MgO in the picrites of Noril'sk than in those from West Greenland. This difference can be seen in the shallower trends, and at higher SiO_2, for the Noril'sk picrites on a diagram of Mg-number vs. SiO_2 (Figure 13).

(ii) The lavas that exhibit the strongest evidence for having equilibrated with sulphides (i.e. low Cu/Zr in the Nadezhdinsky and CWGT, Figure 9) are those which appear to have experienced the largest amounts of crustal contamination. Moreover, at least some of the contaminants inferred from minor- and trace-element ratios are likely to have had high SiO_2 but low S contents; for example, granodiorite. It is inferred that this contamination was early and that the associated sulphur saturation was very largely with magmatic S.

(iii) Cu/Zr ratios are more scattered in the Nadezhdinsky than in the CWGT, suggesting that the interaction with sulphides was more variable in the former.

(iv) In West Greenland the contaminated units are ~100 m thick, they are sandwiched between relatively uncontaminated basalts (Figures 8 and 14a), and there are sharp chemical breaks at the tops and bottoms of these units. In contrast, at Noril'sk the rocks that are highly depleted in Cu are ~200 m thick, and above them there is a gradual increase in Cu/Zr and decrease in La/Sm with stratigraphic height in the upper units of the Nadezhdinsky and the Mo-

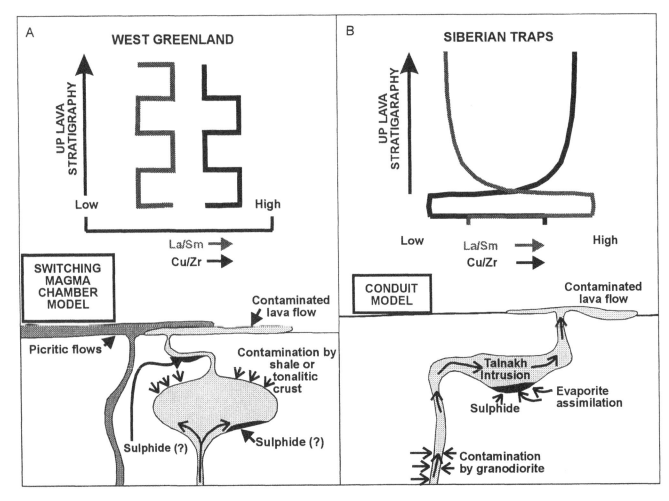

Figure 14. Sketch sections illustrating how the differences in Cu/Zr and La/Sm with height in the lava piles at West Greenland and Noril'sk, may reflect differences in their magma plumbing systems.

rongovsky and Mokulaevsky Formations (~1000 m, Figures 7 and 14b).

The variations in Mg-number with SiO_2 indicate that those rocks with low Cu/Zr (the Nadezhdinsky and the CWGT) have SiO_2 contents that are typically 5% higher than those on, for example, the fractionation trend of the uncontaminated West Greenland picrites (Figure 13). Such increases in SiO_2 in the Asûk and Kûgánguaq rocks of West Greenland are consistent with 20–25% assimilation of a crustal contaminant that had 65–70% SiO_2, which in turn can explain the observed increases in La/Sm and reduction in Ta/La (Figure 11). In the context of this discussion, an increase of 5 wt% SiO_2 appears to have been sufficient to move the magma composition into the field of sulphur saturation (Figure 2 and *Irvine* [1975]), and so we conclude that sulphur saturation can be achieved by crustal contamination without the necessity for the addition of significant amounts of crustal sulphur. The picritic paren-

tal magmas were sulphur-undersaturated, or they too would have fractionated Cu/Zr ratios. If such picrites can contain up to ~2000 ppm dissolved S, as reported for non-picritic basalts from the Wanapum of the Columbia River Basalts [*Thordarson*, 1995], they are only required to reflect more than 10% partial melting of an upper mantle source containing 200 ppm S [see discussion by *Keays*, 1995].

Models for the genesis of the giant Ni, Cu, and PGE sulphide deposits at Noril'sk highlight the importance of equilibration of the sulphide liquid with large volumes of magma. Whereas equilibration has traditionally been hypothesized to occur in deep crustal magma chambers by either settling of the dense immiscible sulphide through the magma column [*Naldrett et al.*, 1992] or batch equilibration in a deep chamber [*Brügmann et al.*, 1993], equilibration may also occur in narrow, horizontal, open-system magma chambers termed chonoliths [*Rad'ko*, 1991;

Naldrett et al., 1995; Torgashin, 1994]. One version of such a model for Noril'sk is illustrated in Figure 15. The early Tuklonsky lavas suffered relatively little crustal contamination, and such magma is inferred to have been parental to the more contaminated magmas. Crustal contamination primarily involved relatively deep-seated granodioritic melts, and the resultant increase in SiO_2 was sufficient to trigger sulphur saturation. Minor and trace element data may then be used to evaluate the likely subsequent contribution from the shallow-level evaporite sequence, in that the evaporites have 2000–2500 ppm Sr and Sr/Sm of ~1500 (P. C. Lightfoot et al., ms. in preparation). Because the Nadezhdinsky lavas have an average Sr/Sm of ~60, they cannot have accommodated more than 0.5% of evaporite; nonetheless, because the evaporites contain ~20 wt% S, the contaminated lavas could have received up to 1000 ppm S from the evaporite sequences.

The sulphide liquid is dense and so in this model it could not be readily expunged from the chonolith [*Naldrett et al., 1995*]. Rather, the chonolith acted as an open-system magma conduit through which successive batches of silicate magma travelled, to be erupted at the surface as the observed CFB sequence. Simplistically, we envisage that the initial magmas equilibrated with the sulphide liquid and were scavenged of Ni, Cu, and PGE to produce the Ni-, Cu-, and PGE-poor lowermost lava flows of the Nadezhdinsky. However, a distinctive feature of the Noril'sk deposits is that the ores are unusually rich; for example, Cu and Ni contents are more than twice those in the ores at Sudbury, Pt more than six times, and Pd more than 24 times the Sudbury ore levels (Figure 1 and *Lightfoot* [1996]). The implication is that the sulphides equilibrated with very large amounts of magma (2500–5000 times their mass [*Brügmann et al., 1993*]) and that subsequent batches of magma further equilibrated with the sulphide liquid as they passed through the chonolith enroute to the surface. As progressively more magma passed through the chonolith, the degree of metal depletion declined and the lavas produced at the surface progressively recovered their siderophile element contents [*Naldrett et al., 1995*] (Figure 14). Apparently, the capacity of the sulphide liquid to scavenge additional metals decreased with time and some parts of the sulphide liquid may have become isolated from the magma. This mechanism is important as it explains why such a large volume of high-grade sulphide mineralization is developed at Noril'sk, and it accounts for the continuous upward recovery in the siderophile-element abundances of the Nadezhdinsky through Morongovsky flows.

In detail, some aspects still have to be reconciled with such models. For example, it is not entirely clear that dense magmatic sulphides at the base of a chonolith can easily react with fresh influxes of silicate magma, and there are step-wise variations in Cu and Ni contents within the Nadezhdinsky lavas [*Fedorenko et al., 1996*; P. C. Lightfoot et al., ms. in preparation] which are not yet explained. A further concern is that in the chonolith model very large volumes of magma (>5000 km^3) are required to have travelled through very small (200 m wide) intrusions. Such factors were highlighted by *Czamanske et al.* [1995], who suggested that crustal contamination took place at greater depths and that the sulphides were subsequently intruded into their present positions.

In West Greenland, the sulphide-laden gabbros of the Hammersdal Complex and the Igdlukunguaq dyke demonstrate that sulphur saturation did occur [*Lightfoot et al., in press*]. Furthermore, many of the gabbroic intrusions are located proximal to faults through which the lavas were erupted and therefore may have had the opportunity to become chonoliths as envisaged for the Noril'sk systems. The recognition of contaminated lavas which have relatively low siderophile-element contents adds further evidence to suggest that the CWGT equilibrated with magmatic sulphides. These are all features that encourage exploration efforts in West Greenland. However, the absence of a progressive recovery in siderophile element abundances within the CWGT (Figure 14) does not correspond with the model presented for Noril'sk, and the lack of recovery may indicate that continued upgrading of any sulphides was not a feature of the West Greenland magma systems. At issue, therefore, is whether such evidence from the lavas is diagnostic of sulphide upgrading, as seen at Noril'sk, and whether it should be an important aspect of any exploration strategy based on basalt chemostratigraphy. At Noril'sk there was opportunity for the magma to assimilate additional sulphur from the evaporitic sediments and large amounts of magma were available to continue to equilibrate with large volumes of sulphide, and these are both conditions for the formation of giant deposits. In the West Greenland CFB, the amount of sulphur available was moderate, although contamination with sediments has been demonstrated (trend 1, Figure 10) and shales locally contain 0.8 wt% S. Some lavas have been depleted strongly in Ni, Cu, and PGE, presumably by interaction with magmatic sulphides, but there is no evidence that the siderophile-element contents of the sulphides ever reached the point at which the siderophile element abundances in the tholeiitic magmas progressively recovered.

4.4. Sulphur Budget

A critical control on the sulphide-forming capacity of a magma is the amount of sulphur available which can then interact with silicate magma and concentrate the Ni, Cu,

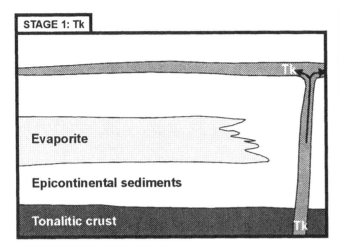
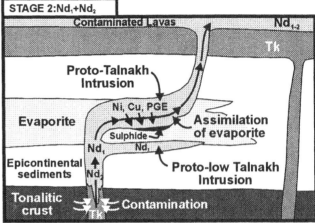
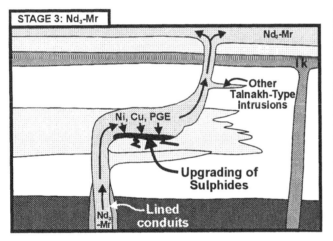
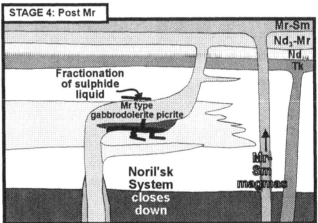

Figure 15. Schematic model for the evolution of the Noril'sk system.

Stage 1: Picritic and tholeiitic lavas of the Tuklonsky Formation were erupted through conduit systems 100 km east of Noril'sk in the region of the Imangda Fault (Figure 2).

Stage 2: Eruption of lavas switched to the Noril'sk Region, and initially focussed along and to the west of the Noril'sk-Kharaelakh Fault [*Naldrett et al.,* 1992]. Contaminated lavas of the Nd_1 were generated by interaction of Tuklonsky magma type with granodiorites, with fractional crystallization. These magmas travelled through the chonolith and underwent a second stage of contamination involving assimilation of evaporitic sediments. The combination of elevated silica contents and abundant sulphur resulted in the fractionation of immiscible sulphides. These sulphides ponded at the base of the chonolith, and reacted with new batches of Nd_1 magma as they passed through the chamber.

Stage 3: Migration of Nd_1 magmas over the sulphides continued, and progressively upgraded the sulphide liquids in Ni, Cu and the PGE. Subsequent Nd_2, Nd_1 and Mr magmas travelled through the conduit and as the sulphides became replete with metals and isolated from the magma, the degree of depletion of the magmas in the metals declined to produce the observed upward increase in Ni, Cu, and PGE abundances with stratigraphic height [e.g., *Naldrett et al.,* 1992; *Brügmann et al.,* 1993].

Stage 4: The Noril'sk conduit shut down and magmatism migrated towards the northeast. The sulphide liquids underwent fractionation to form the Cu-rich ores [*Czamanske et al.,* 1992; *Naldrett et al.,* 1995; *Zientek et al.,* 1994].

and PGE. Assimilation of most crustal rock types is likely to dilute rather than enhance the sulphur contents of mafic magmas, and so models for the formation of giant sulphide deposits typically require the introduction of sedimentary sulphur from compositionally unusual sediments and/or a very efficient segregation and concentration mechanism to ensure the isolation of large volumes of mantle-derived sulphur from the magmas.

The best evidence for the source of the sulphur is arguably from sulphur isotopes, and the ores at Noril'sk typically have $\delta^{34}S$ values of +8 to +14 per mil, compared with 0 per mil in mantle-derived rocks and +20 per mil in the evaporitic sediments [Grinenko, 1985]. Grinenko [1985] further noted that the barren intrusions in the Noril'sk region tend to have sulphur isotopic compositions close to mantle values or displaced to slightly positive values ($\delta^{34}S$ = -2 to +6) and tend to be located in clastic sediments devoid of crustal S. In contrast, the economically mineralized intrusions have more positive $\delta^{34}S$ and tend to be located in crustal rocks that contain abundant sulphates in the form of evaporites. Grinenko [1985] therefore suggested that much of the S was derived in situ from the wall rocks. Using a mantle sulphur isotopic composition of zero and assuming that the evaporites have $\delta^{34}S$ = +20, Grinenko [1985] showed that 20–36% of the S in the noneconomic mineralized intrusions could have been from a crustal source and that the parental mantle-derived magmas had 1200–2300 ppm S. However, using a similar approach for the mineralized intrusions, the calculated initial S contents of the parental magmas were higher than could have been dissolved in the observed volumes of silicate magma. Reversing these calculations suggested that the crustal S in the mineralized intrusions had $\delta^{34}S$ ~+11 to +12, and Grinenko [1985] therefore concluded that contamination of the mineralized intrusions involved a sulphur source at a deeper crustal level. The preferred source was sour gas S, because it was known to be common in the area and has $\delta^{34}S$ ~+10. However, such calculations predate the sulphide upgrading models in which magmatic sulphides interact with large volumes of mantle-derived magma [Naldrett et al., 1995]. In these models, the mass balance considerations no longer require the parental magmas to have had anomalously large dissolved S contents, and hence the crustal S component can have $\delta^{34}S$ values similar to the evaporitic sediments, rather than to the sour gas [cf. Naldrett, 1989].

In West Greenland, mineralized intrusions which are comagmatic with contaminated tholeiites from Nuussuaq (the Igdlukunguaq dyke and the Hammersdal Complex) have negative $\delta^{34}S$ values of -7 to -11, which demands a crustal contribution of S with negative $\delta^{34}S$ if the parental

picritic magmas had a zero per mil signature. Interestingly, one sample of a thick, simple flow from the Asûk unit has $\delta^{34}S$ = +13.6, which contrasts with most of the other CWGT, which have $\delta^{34}S$ = -5 to +5, and is consistent with the addition of sedimentary S from a shale source.

Based on the sulphur isotope arguments, it is apparent that crustal sulphur has contributed to mineralization at both Noril'sk and West Greenland. However, the extent to which the addition of crust-derived S affected the mineral potential of the intrusions still needs to be explored.

The Ni- and Cu-depleted lavas of the Nadezhdinsky Formation at Noril'sk have a volume of ~5,000 km³, based on isopach maps of their surface distribution [Naldrett et al., 1992]. The available data from the Columbia River Basalts [Thordarson, 1995], experimental studies of the dissolved sulphur contents of tholeiitic melts (500–2000 ppm S for melts with 5–25 wt% FeO) [Buchanan and Nolan, 1979; Haughton et al., 1974], and submarine pillow basalts (1050–1800 ppm S in rocks with 9–12.9 wt% FeO) [Mathez, 1976] are consistent with high-Mg mantle-derived melts containing ~1900 ppm S. However, basaltic magmas may typically have S contents of half that value (i.e., 950 ppm), and so the best estimate of the amount of S in the parental Nadezhdinsky magmas prior to saturation is 4.7 km³, which at 1.8 g/cc translates into 8.5×10^9 tonnes of elemental S derived from the upper mantle.

The deposits at Noril'sk contain 555 million tonnes of ore (and perhaps as much as 1500 Mt). Ore grades average 3.9 wt% Cu, and 2.7 wt% Ni, and mineralization is typically 50% gangue. If 50% of the rock is FeS, then the 555 million tonnes would contain about 17% S and the deposits as a whole would contain 1×10^8 tonnes of S. This is likely to be a minimum value, but it indicates that only 1–2% of the magmatic S available is locked up in the deposits at Noril'sk.

The amounts of crustal S assimilated in the magmas are more difficult to constrain, partly because the lavas vary flow by flow and partly because sulphur isotope data appear to be in conflict with the trace element arguments. As discussed above, the trace element contents of the Nadezhdinsky lavas suggest that no more than 0.5% evaporite was assimilated in even the most Sr-rich Nadezhdinsky rocks. Recent work by Czamanske et al. [1995] confirms that the geological evidence at Talnakh favours the replacement of the evaporite-rich sedimentary package by the Talnakh intrusion (<10 km³) rather than by the static rupturing and dilation of the sediments. Assuming that the intrusion replaced a similar volume of sediments, the amount of S added from those sediments can be estimated crudely. Thus, the amount of crust replaced is ~10–20 × 10^9 tonnes, and if it contained an average of

25% S, the amount of crustal S contributed to the system is 4×10^9 tonnes. Such quantities of S are consistent with the S isotope ratios of the magmatic sulphides [*Grinenko*, 1985], and they represent almost half of the S available in the magmas themselves. The total amount of S available in the Noril'sk system is therefore estimated to be ~12.5 $\times$ 10^9 tonnes, and yet <1% appears to have been scavenged by known mineral deposits. Finally, it is important to reiterate that the shallow-level evaporite-rich sediments were not responsible for the trace element and radiogenic isotope variations that are associated with the observed depletions in Cu, Ni, and PGE in the crustally contaminated lavas and intrusions (Figures 7, 9 and 10). Rather, it is inferred that the initial sulphide formation took place at depth in response to crustal contamination with a granodioritic component (Section 4.2, and *Lightfoot et al.* [1993, 1994]; *Hawkesworth et al.* [1995]). The unresolved question concerns the detailed relations between the sulphides inferred to have separated at depth and those presently observed in the shallow-level intrusions within evaporite-rich sediments.

In West Greenland shales from Qeqertarssuaq have an average S content of 0.8% S, and the gneisses have an average of 82 ppm S. The geochemical data discussed in section 4.2 suggest that the Asûk and Kûgánguaq flows were contaminated with up to 30% shale, implying that the contribution of crustal S in these rocks is 0.3 $\times$ 8000 = 2400 ppm S added to the magma. This amount is similar to that inferred for the S contents of the mantle-derived magmas (~1900 ppm S), indicating that roughly equal amounts of mantle and crustally derived S may have been present in the contaminated lavas and associated comagmatic intrusions. Assuming that the Asûk and Kûgánguaq have a volume of ~10^3 km^3, the amount of crustal S added was 2 $\times$ 10^9 tonnes, and there must have been a total S budget of over 4 billion tonnes available for mineralising processes. The observation that the mineralization at Igdlukunguaq and Hammersdal have δ^{34}S values of -7 to -12 provides some support for the suggestion that about half of the S was crustal in origin and that the rest was mantle-derived.

In the model discussed here, one implication is that although very large sulphide deposits may exist in West Greenland, they may not be of the grade found at Noril'sk. A particular challenge now will be to use PGE data for these rocks to ascertain whether these elements, which have much higher sulphide/silicate partition coefficients, record a similar recovery pattern and to look in more detail at the Malîgat CWGT, which appear to be comagmatic with some of the observed mineralized intrusions. Significantly, these patterns of gradual recovery in the Ni, Cu and

PGE contents in the lavas are not unique to the Siberian Traps, and good examples through over 1000 m of basalt stratigraphy have been recorded in the Osler Volcanic Group in the Keweenawan of Ontario, Canada [*Lightfoot et al.*, 1991; *Naldrett and Lightfoot*, 1993]. The potential of intrusions in the Keweenawan region should therefore continue to intrigue explorationists for some time to come.

5. SUMMARY

There is a clear association between sulphide mineralization and some CFB. Two of the most striking examples are the Permo-Triassic Siberian Traps at Noril'sk and the Tertiary CFB of West Greenland, and although they share a number of common features, there are significant differences. The main points of this discussion may be summarised as follows.

1. Present exposures of the Siberian Traps and West Greenland CFB contain up to 3500 m of predominantly picritic and tholeiitic lavas. However, at Noril'sk the picrites represent ~1% of the preserved sequence, whereas in West Greenland they comprise >50%. In both West Greenland and the Siberian Traps, magmatism has been linked to the presence of mantle hot spots. However, the volumes of magma derived from asthenospheric and lithospheric sources appear to have been different in the two areas.

2. Intrusions contemporaneous with volcanism were developed in both the Siberian Traps and in West Greenland. Many of these intrusions can be correlated chemically with units in the associated lava sequences [*Fedorenko*, 1994; *Naldrett et al.*, 1992; *Hawkesworth et al.*, 1995], and some may have acted as open-system conduits (chonoliths) through which the lavas erupted. At Noril'sk the distribution of the intrusion types is linked broadly to the location of the maximum thickness of comagmatic lava [*Naldrett et al.*, 1992]. A general coincidence is that the mineralized intrusions are associated with major NNW-SSE-trending mantle-penetrating faults which appear to have acted as conduits for the magmas and as loci for the intrusions. In Siberia, the intrusions that carry mineralization are associated with the Noril'sk-Kharaelakh Fault, and in West Greenland, numerous diabase and gabbro intrusions are linked to the Boundary Fault.

3. The CFB of the Noril'sk region and West Greenland were erupted onto epicontinental sedimentary sequences. At Noril'sk these include evaporitic sediments and deeper reservoirs of sour gas and oil. In West Greenland, the magmatism occurred over a prograding deltaic assemblage, such that magmas were erupted onto basement, a shallow sequence of hydrocarbon-rich shales, or a thick

sequence (>5 km) of hydrocarbon-bearing shales [*Pulvertaft*, 1989].

4. Mineralization at Noril'sk is associated with relatively small picritic to gabbroic intrusions. The intrusions contain 50–200 times more sulphide mineralization than can have been generated by in situ settling of immiscible sulphide liquid from the silicate portion of the intrusions, and so they must reflect scavenging of S from significantly larger magma volumes. In addition, the sulphides are enormously rich in ore metals; production plus reserves of >555 million tonnes (and possibly 1000 million tonnes) total 3.9 wt% Cu, 2.7 wt% Ni, 3 ppm Pt, and 12 ppm Pd, which is 2–10 times larger than in other giant deposits like Sudbury [*Lightfoot*, 1996]. Mineralization in West Greenland is spatially associated with the Hammersdal Complex and the Igdlukunguaq dyke. These occurrences have small sulphide contents, but the sulphides have moderate to high Ni, Cu, and PGE abundances [*Ulff Møller*, 1991].

5. Within the lava sequences at both Noril'sk and West Greenland, a broad empirical relation exists in which lavas with high SiO_2, LILE, La/Sm, Th/Nb, and $^{87}Sr/^{86}Sr_0$ also have low Cu, Ni and Cu/Zr [e.g., *Lightfoot et al.*, 1990, 1993, in press; *Wooden et al.*, 1993]. These features indicate that the contaminated lavas tend to have abnormally low siderophile element concentrations, and these low concentrations have typically been ascribed to the equilibration of the magma with a dense sulphide liquid which was removed from the system.

6. The minor and trace element variations in the Noril'sk lavas indicate that the contaminant was granodioritic in composition [e.g., *Lightfoot et al.*, 1993, 1994; *Naldrett et al.*, 1992; *Hawkesworth et al.*, 1995]. Such rocks have low S contents, and so we conclude that the initial sulphide formation responsible for the low Ni, Cu and PGE contents of the contaminated lavas was triggered by the associated increase in SiO_2 rather than the introduction of additional (crustal) S. Subsequent interaction with shallow level evaporite-rich sediments contributed additional S, but it was not responsible for the crustal contamination associated with the distinctive metal depletions in the Nadezhdinsky lavas. Mass balance considerations suggest that as little as 1% of the S available in the mantle-derived magmas and the assimilated evaporite-rich sediments at Noril'sk is locked up in the known deposits. In the West Greenland rocks, two trends of contamination with different crustal contaminants are recognised, and they exhibit similar depletions in Ni and Cu. As at Noril'sk, initial sulphide formation would appear to have been primarily controlled by the associated increase in SiO_2 rather than by the S contents of the crustal contaminants.

7. Many of the intrusions at Noril'sk may have acted as chonoliths, and these conduits can be regarded as the sites where sulphides formed and equilibrated with successive batches of silicate liquid passing through the CFB conduit. The high grades of the ores at Noril'sk are consistent with upgrading of the deposits by the equilibration of the sulphide liquids with the successive batches of magma. In this zone-refining process, a gradual recovery in the Ni, Cu, and PGE contents of the later silicate magmas is predicted, and this recovery is an observed feature of the lava sequence [*Lightfoot et al.*, 1990, 1993, 1994; *Brügmann et al.*, 1993; *Naldrett et al.*, 1992, 1995]. The West Greenland lavas contain discrete units of heavily contaminated and Ni- and Cu-depleted lavas, and these too can be ascribed to equilibration of contaminated magma with sulphide liquid. However, no gradual upward recovery in siderophile-element abundances is evident; rather, the top of the package is marked by the sudden adjustment of the system with the expulsion of picritic lavas. This lack of recovery suggests that the dynamics of the conduit system in West Greenland were quite different from those at Noril'sk, and such differences in conduit system dynamics may have implications for mineral exploration. Finally, we note that the features of the Noril'sk region are also observed in the Keweenawan midcontinent rift in a sequence of contaminated lavas (Osler Volcanic Group) which have overall low Ni/MgO and an upward decline in the degree of contamination [*Lightfoot et al.*, 1990; *Naldrett and Lightfoot*, 1993]. The similarities between this sequence and that at Noril'sk are particularly encouraging from the perspective of mineral exploration.

Acknowledgments. The authors are indebted to Falconbridge Exploration who supported the cost of our program on the West Greenland CFB. We particularly appreciate the advice and assistance from Kevin Olshefsky and Tony Green with this research program and discussions with Reid Keays. We are also grateful for the encouragement and support of Tony Naldrett, Valeri Fedorenko, and Nick Gorbachev with the Noril'sk aspect of this study. Much of our work relies on the high-quality data acquired by Will Doherty of the Ontario Geological Survey and the Geological Survey of Canada. The manuscript was improved by the constructive comments of Gerry Czamanske, Bruce Doe, Valeri Fedorenko, and John Mahoney. The manuscript was prepared by Janet Dryden. Diagrams were drafted by Steve Josey.

REFERENCES

Aplonov, S., An aborted Triassic ocean in West Siberia, *Tectonics*, 7, 1103–1122, 1988.

Brügmann, G. E., A. J. Naldrett, M. Asif, P. C. Lightfoot, and N. S. Gorbachev, Siderophile and chalcophile metals as tracers of the evolution of the Siberian Traps in the Noril'sk Region, Russia, *Geochim. Cosmochim. Acta*, 57, 2001–2018, 1993.

Buchanan, D. L., and J. Nolan, Solubility of sulfur and sulfide immiscibility in synthetic tholeiitic melts and their relevance to Bushveld Complex rocks, *Can. Mineral., 17*, 483–494, 1979.

Campbell, I. H., and A. J. Naldrett, The influence of silicate:sulphide ratios on the geochemistry of magmatic sulphides, *Econ. Geol., 74*, 1503–1505, 1979.

Campbell, I. H., G. K. Czamanske, V. A. Fedorenko, R. I. Hill, V. Stepanov, and V. E. Kunilov, Synchronism of the Siberian Traps and the Permian-Triassic boundary, *Science, 258*, 1760–1763, 1992.

Clarke, D. B., and A. K. Pedersen, Tertiary volcanic province of west Greenland, in *Geology of Greenland*, edited by A. Escher and W. S. Watt, pp. 365–385, Geological Survey of Greenland, Copenhagen, 1976.

Clarke, D. B., The Tertiary volcanic province of Baffin Bay, *GAC Spec. Pap., 16*, 445–460, 1991.

Czamanske, G. K., V. E. Kunilov, M. L. Zientek, L. C. Cabri, A. P. Likhachev, L. C. Calk, and R. L. Oscarson, A proton-microprobe study of magmatic sulphide ores from the Noril'sk Talnakh district, Siberia, *Can. Mineral., 30*, 249–287, 1992.

Czamanske, G. K., T. E. Zen'ko, V. A. Fedorenko, L. C. Calk, J. R. Budahn, J. H. Bullock, T. L. Fried, B.-S. W. King, and D. F. Siems, Petrography and geochemical characterization of ore-bearing intrusions of the Noril'sk Type, Siberia: with discussion of their origin, *Res. Geol., 18*, 1–48, 1995.

Dalrymple, B. G., G. K. Czamanske, V. A. Fedorenko, O. N. Simonov, M. A. Lanphere, and A. P. Likhachev, A reconnaissance ^{40}Ar/^{39}Ar geochronological study of ore-bearing and related rocks, Siberian Russia. *Geochim. Cosmochim. Acta, 59*, 2071–2083, 1995.

Devey, C. W., and P. C. Lightfoot, Volcanological and tectonic control of stratigraphy and structure in the western Deccan Trap. *Bull. Volcanol., 48*, 195–207, 1986.

Fedorenko, V. A., The petrochemical series of volcanic rocks of the Noril'sk region. *Geol. Geophys., 6*, 78–88, 1981.

Fedorenko, V. A., Evolution of magmatism as reflected in the volcanic sequence of the Noril'sk Region, in *Proceedings of the Sudbury–Noril'sk Symposium, Spec. Vol. 5*, edited by P.C. Lightfoot and A. J. Naldrett, pp. 171-184, Ontario Geological Survey, Toronto, 1994.

Fedorenko, V. A., P. C. Lightfoot, A. J. Naldrett, G. K. Czamanske, C. J. Hawkesworth, J. L. Wooden, and D. S. Ebel, Petrogenesis of the flood-basalt sequence at Noril'sk, North Central Siberia, *Int. Geol. Rev., 38*, 99–135, 1996.

Gallagher, K., and C. J. Hawkesworth, Mantle plumes, continental magmatism and asymmetry in the South Atlantic, *Earth Planet. Sci. Lett., 123*, 105–117, 1994.

Glazkovsky, A. A., G. I. Gorbunov, and F. A. Sysoev, Deposits of nickel, in *Ore Deposits of the USSR, Vol. II*, edited by V. I. Smirnov, pp. 3–79, English translation by D. A. Brown, 1977.

Grinenko, L. N., Sources of sulfur of the nickeliferous and barren gabbro-dolerite intrusions of the northwest Siberian platform, *Int. Geol. Rev.*, 695–708, 1985.

Hald, N., and A. K. Pedersen, Lithostratigraphy of the Early Tertiary volcanic rocks of central West Greenland, *Rapp. Gronlands geol. Unders., 69*, 17–24, 1975.

Hald, N., Lithostratigraphy of the Malîgat and Hareöen forma-

tions, West Greenland basalt group, on Hareöen and Western Nugssuaq, *Rapp. Gronlands geol. Unders., 79*, 9–16, 1977a.

Hald, N., Normally magnetized lower Tertiary lavas on Nugssuaq, central West Greenland, *Rapp. Gronlands geol. Unders., 79*, 5–7, 1977b.

Hamlyn, P. R., R. R. Keays, W. E. Cameron, A. J. Crawford, and H. M. Waldron, Precious metals in magnesian low-Ti lavas: Implications for metallogenesis and sulfur saturation in primary magmas, *Geochim. Cosmochim. Acta, 49*, 1797–1811, 1985.

Harris, N. B. W., J. A. Pearce, and A. G. Tindle, Geochemical characteristics of collision-zone magmatism, in *Collision Tectonics, Spec. Publ. 19*, edited by M. P. Coward and A. C. Ries, pp. 67–81, Geological Society of London, 1986.

Haughton, D. R., P. L. Roeder, and B. Skinner, Solubility in submarine basalt glass, *J. Geophys. Res., 81*, 4269–4276, 1974.

Hawkesworth, C. J., P. C. Lightfoot, V. A. Fedorenko, S. Blake, A. J. Naldrett, W. Doherty, and N. W. Gorbachev, Magma differentiation and mineralisation in the Siberian continental flood basalts, *Lithos, 34*, 61–88, 1995.

Hawkesworth, C. J., S. P. Turner, D. Peate, F. McDermott and P. van Calsteren, Elemental U and Th variations in island arc rocks: implications for U-series isotopes, *Chem. Geol.*, in press, 1997.

Hergt, J. M., D. W. Peate, and C. J. Hawkesworth, The petrogenesis of Mesozoic Gondwana low Ti flood basalts, *Earth Planet. Sci. Lett., 105*, 134–148, 1991.

Holm, P. M., R. C. O. Gill, A. K. Pedersen, J. G. Larsen, N. Hald, T. F. D. Nielsen and M. F. Thirwall, The tertiary picrites of West Greenland: contributions from Icelandic and other sources, *Earth Planet. Sci. Lett., 115*, 227–244, 1993.

Irvine, T. N., Crystallisation sequence of the Muskox Intrusion and other layered intrusions: II Origin of the chromitite layers and similar deposits of other magmatic ores, *Geochim. Cosmochim. Acta, 39*, 991–1020, 1975.

Keays, R. R., The role of komatiitic and picritic magmatism and S-saturation in the formation of ore deposits, *Lithos, 34*, 1–18, 1995.

Klock, W., H. Palme, and H. J. Tobschall, Trace elements in natural metallic iron from Disko Island, Greenland, *Contrib. Mineral. Petrol., 93*, 273-282, 1986.

Lightfoot, P. C., A. J. Naldrett, N. S. Gorbachev, W. Doherty, and V. A. Fedorenko, Geochemistry of the Siberian Trap of the Noril'sk Area, USSR, with implications for the relative contributions of crust and mantle to flood basalt magmatism, *Contrib. Mineral. Petrol., 104*, 631–644, 1990.

Lightfoot, P. C., G. Chai, A. E. Hodges, and D. Rowell, An update report on the certification of the OGS in-house MRB standard-reference materials, in *Summary of Field Work and Other Activities, Misc. Publ. 157*, pp. 231-236, Ontario Geological Survey, Toronto, 1991.

Lightfoot, P. C., C. J. Hawkesworth, J. Hergt, A. J. Naldrett, N. S. Gorbachev, and V. A. Fedorenko, Remobilisation of the continental lithosphere by a mantle plume: major-, trace-element and Sr-, Nd-, and Pb-isotope evidence from picritic and tholeiitic lavas of the Noril'sk District, Siberian Trap, Russia, *Contrib. Mineral. Petrol.*, *114*, 171–188, 1993.

Lightfoot, P. C., A. J. Naldrett, N. S. Gorbachev, V. A. Fedorenko, C. J. Hawkesworth, J. Hergt, and W. Doherty, Chemostratigraphy of Siberian Trap lavas, Noril'sk District, Russia: implications for the source of flood basalt magmas and their associated Ni-Cu mineralisation, in *Proceedings of the Sudbury–Noril'sk Symposium, Spec. Vol. 5*, edited by P.C. Lightfoot and A. J. Naldrett, pp. 283-312, Ontario Geological Survey, Toronto, 1994.

Lightfoot, P. C., The giant nickel deposits and Sudbury and Noril'sk (abstract), *1996 Prospectors and Developers Assoc. Canada, Nickel in a Nutshell Workshop*, 1–12, 1996.

Lightfoot, P. C., C. J. Hawkesworth, K. Olshefsky, T. Green, W. Doherty and R. R. Keays, Geochemistry of Tertiary tholeiites and picrites from Qeqertarssuaq (Disko Island) and Nuussuaq, West Greenland with implications for the mineral potential of comagmatic intrusions. *Contrib. Mineral. Petrol.*, in press, 1997.

Macdougall, J. D. (Ed.), *Continental Flood Basalts*, 341 pp., Kluwer Academic Publishers, Dordrecht, 1988.

MacLean, W. H., Liquidus phase relations in the FeS-FeO-Fe_3O_4-SiO_2 systems and their applications in geology, *Econ. Geol.*, *64*, 865–884, 1969.

Mahoney, J., J. D. Macdougall, G. W. Lugmair, A. V. Murali, M. Sankar Das, and Gopalan, K., Origin of the Deccan Traps flows at Mahabaleshwar inferred from Nd and Sr isotopic and chemical evidence, *Earth Planet. Sci. Lett.*, *60*, 47–60, 1982.

Mathez, E. A., Sulfur solubility and magmatic sulfides in submarine basalt glass, *J. Geophys. Res.*, 81, 4269–4276, 1976.

Morgan, W. J., Hotspot tracks and the opening of the Atlantic and Indian oceans, in *The Sea*, Vol. 7, edited by C. Emiliani, pp. 443–487, Wiley Interscience, New York, 1981.

Naldrett, A. J., *Magmatic Sulfide Deposits*, 196 pp., Oxford University Press, New York, 1989.

Naldrett, A. J., P. C. Lightfoot, V. A. Fedorenko, W. Doherty, and N. S. Gorbachev, Geology and geochemistry of intrusions and flood basalts of the Noril'sk Region, USSR, with implications for the origin of the Ni-Cu ores, *Econ. Geol.*, *87*, 975–1004, 1992.

Naldrett, A. J., and P. C. Lightfoot, Ni-Cu-PGE ores of the Noril'sk region, Siberia: A model for giant magmatic sulfide deposits associated with flood basalts, in *Giant Ore Deposits, Spec. Publ. 2*, edited by B. H. Whiting, C. V. Hodgsen, and R. Mason, pp. 81–123, Society Economic Geologists, Littleton, CO, 1993.

Naldrett, A. J. V. A. Fedorenko, P. C. Lightfoot, N. S. Gorbachev, W. Doherty, M. Asif, S. Lin, and Z. Johan, A model for the formation of the Ni-Cu-PGE deposits of the Noril'sk, *Trans. Inst. Mining Metal.*, *104*, B18–B36, 1995.

Peach, C. L., E. A. Mathez, and R. R. Keays, Sulphide melt-silicate melt distribution coefficients for noble metals and other chalcophile elements as deduced from MORB: Implications for partial melting, *Geochim. Cosmochim. Acta*, *54*, 3379–3389, 1990.

Pedersen, A. K., New investigations of the native iron bearing volcanic rocks of Disko, central West Greenland, *Rapp. Gronlands geol. Unders.*, *75*, 48–51, 1975.

Pedersen, A. K., Non-stoichiometric magnesian spinels in shale xenoliths from a native iron-bearing andesite at Asûk, Disko, central West Greenland, *Contrib. Mineral. Petrol.*, *67*, 331–340, 1978.

Pedersen, A. K., A shale buchite xenolith with Al-Armalcolite and Native Iron in a lava from Asûk, Disko, Central West Greenland, *Contrib. Mineral. Petrol.*, *69*, 83–94, 1979.

Pedersen, A. K., Lithostratigraphy of the Tertiary Vaigat Formation on Disko, central West Greenland, *Rapp. Gronlands geol. Unders.*, *124*, 30 pp., 1985a.

Pedersen, A. K., Reaction between picrite magma and continental crust: early Tertiary silicic basalts and magnesian andesites from Disko, West Greenland, *Gronlands geol. Unders. Bull.*, *15*, 126 pp, 1985b.

Pedersen, A. K., and Pulvertaft, T. C. R., The nonmarine Cretaceous of the West Greenland Basin, onshore West Greenland, *Cret. Res.*, *13*, 1–10, 1992.

Pulvertaft, T. C. R., Reinvestigation of the Cretaceous boundary fault in Sarqaqdalen, Nugssuaq, central West Greenland, *Rapp. Gronlands geol. Unders.*, *145*, 28–32, 1989.

Rad'ko, V. V., Model of dynamic differentiation of intrusive traps in the northwestern Siberian platform, *Soviet geol. and Geophys.*, *32 (7)*, 70–77, 1991.

Renne, P. R. and A. R. Basu, Rapid eruption of the Siberian Trap Flood Basalts at the Permo-Triassic Boundary, *Science*, *253*, 176–179, 1991.

Richards, M. A., A. R. Duncan and V. E. Courtillot, Flood basalts and hot-spot tracks: plume heads and tails, *Science*, *246*, 103–107, 1989.

Ripley, E. M., Sulphur isotopic abundances of the Dunka Road Cu-Ni deposit, Duluth Complex, Minnesota, *Econ. Geol.*, *76*, 619–620, 1981.

Shaw, D. M., J. Dostal, and R. R. Keays, Additional estimates of continental surface Precambrian shield compositions in Canada, *Geochim. Cosmochim. Acta*, *40*, 73–83, 1976.

Simonov, O. N., V. A. Lul'ko, Yu. N. Amosov, and V. M. Salov, Geological structure of the Noril'sk District, in *Proceedings of the Sudbury–Noril'sk Symposium, Spec. Vol. 5*, edited by P.C. Lightfoot and A. J. Naldrett, pp. 161-170, Ontario Geological Survey, Toronto, 1994..

Smirnov, M. F., The Noril'sk nickeliferous intrusions and their sulfide ores, 58 pp., Nedra Press, Moscow, 1966.

Taylor, S. R., and S. M. McLennan, *The Continental Crust: its Composition and Evolution*, 312 pp., Blackwell Scientific Publications, 1985.

Thordarson, Th., Volatile release and atmospheric effects of basaltic fissure eruptions, Ph.D. thesis, 580 pp., Univ. of Hawaii at Manoa, 1995.

Torgashin, A. S., Geology of the massive and copper ores of the western part of the Oktyabr'sky Deposit, in *Proceedings of the Sudbury–Noril'sk Symposium, Spec. Vol. 5*, edited by P.C. Lightfoot and A. J. Naldrett, pp. 231-242, Ontario Geological Survey, Toronto, 1994.

Turner, S., and Hawkesworth, C. J., The nature of the continental mantle lithosphere: constraints from the major element composition of continental flood basalts, *Chem. Geol.*, *120*, 295–314, 1995.

Turner, S., C. Hawkesworth, K. Gallagher, K. Stewart, D. Peate and M. Mantovani, Mantle plumes, flood basalts, and thermal models for melt generation beneath continents: Assessment of a conductive heating model and application to the Paraná, *J. Geophys. Res.*, *101*, 11503–11518, 1996.

Ulff-Møller, F., Magmatic Pt-Ni mineralisation in the West Greenland Basalt Province: A compilation of the results of prospecting by Greenex A/S in 1985-1988, *Gronlands geologiske Undersogelse, Open File*, *91/1*, 37 pp., 1991.

Wendlandt, R. F., Sulphide saturation of basalts and andesite melts at high pressures and temperatures, *Am. Mineral.*, *67*, 877–885, 1982.

White, R., and D. McKenzie, Magmatism at rift zones: the generation of volcanic continental margins and flood basalts, *J. Geophys. Res.*, *94*, 7685–7729, 1989.

White, R., and D. McKenzie, Mantle plumes and flood basalts, *J. Geophys. Res.*, *100*, 17,543–17,585, 1995.

Wooden, J. L., G. K. Czamanske, V. A. Fedorenko, N. T. Arndt, C. Chauvel, R. M. Bouse, B. W. King, R. J. Knight, and D. F. Siems, Isotopic and trace-element constraints on mantle and crustal contributions to Siberian continental flood basalts, Noril'sk Area, Siberia, *Geochim. Cosmochim. Acta*, *57*, 3677–3704, 1993.

Zientek, M. L., A. P. Likhachev, V. E. Kunilov, S.-J. Barnes, A. L. Meier, R. R. Carlson, P. H. Briggs, T. L. Fries, and B. M. Adrian, Cumulus processes and the composition of magmatic ore deposits: examples from the Talnakh District, Russia, in *Proceedings of the Sudbury–Noril'sk Symposium, Spec. Vol. 5*, edited by P.C. Lightfoot and A. J. Naldrett, pp. 373-392, Ontario Geological Survey, Toronto, 1994.

C.J. Hawkesworth, Department of Earth Sciences, The Open University, Milton Keynes, MK7 6AA, U.K.

P.C. Lightfoot, Department of Earth Sciences, The University of Toronto, 22 Russell St, Toronto, Ontario, M5S 3B1, Canada

Emplacement of Continental Flood Basalt Lava Flows

Stephen Self, Thorvaldur Thordarson[1], and Laszlo Keszthelyi[2]

Department of Geology and Geophysics and Hawaii Center for Volcanology
School of Ocean and Earth Science and Technology, University of Hawaii at Manoa, Honolulu, Hawaii

We propose that continental flood basalt (CFB) lavas were predominantly emplaced as inflated compound pahoehoe flow fields via prolonged, episodic eruptions. Our most detailed observations come from the ~14.7 Ma Roza flow field of the Columbia River Basalt (CRB) Group. The Roza flow field seems to be typical of many flood basalt lavas. Individual flows show a wide range of pahoehoe surface features and a three-part internal structure in vesicularity and other textural parameters. This three-fold division into an upper crust, core, and basal crust appears to be diagnostic of the inflation process and is ubiquitous in basaltic lava flows over a remarkable range of sizes. The pahoehoe surface features and indications of inflation are inconsistent with rapid emplacement of these lava flows. Instead, we interpret the observations to imply that the Roza, and other CFB flows, were emplaced over an extended period of time. From the thickness of the upper crust, which we suggest formed while the flow was actively inflating, and an empirical expression for the rate of crust growth of Hawaiian inflated sheet flows, we estimate that individual Roza flows were emplaced over 5 to 50 months and that the Roza flow field was constructed over a period of 6 to 14 years. However, even with this longer eruption duration, the average lava effusion rate of ~4000 m^3/s is similar to that of the highest-effusion-rate eruption in recorded history (the 1783-4 Laki eruption in Iceland). Our observations of lava characteristics in other CRB flows and in the Deccan Traps suggest that this emplacement style is typical of many, if not most, CFB flows. Initial estimates of the volatile release from the Roza eruption indicate that prodigious amounts of S, Cl, and F were injected into the upper troposphere and lowermost stratosphere; thus this single flood basalt eruption could have had a significant effect on the global atmosphere If other flood basalt eruptions produced similar amounts of volatiles, volatile release might provide a link between flood basalt eruptions and mass extinctions.

---

[1] Now at CSIRO, Exploration and Mining, Private Bag, PO Wembley WA 6014, Australia.
[2] Also at Hawaii Volcano Observatory, United States Geological Survey, P. O. Box 51, Hawaii Volcanoes National Park, HI 96718.

Large Igneous Provinces: Continental, Oceanic, and Planetary Flood Volcanism
Geophysical Monograph 100
Copyright 1997 by the American Geophysical Union

1. INTRODUCTION

The physical volcanology of continental flood basalt (CFB) lava flows has received relatively little attention until recently. The emphasis of most previous research on flood basalt provinces has been directed at defining the gross chemical stratigraphy of the lava piles, identifying different possible mantle sources and crustal or mantle contaminants for these huge volumes of basalt, and investigating how these factors relate to the

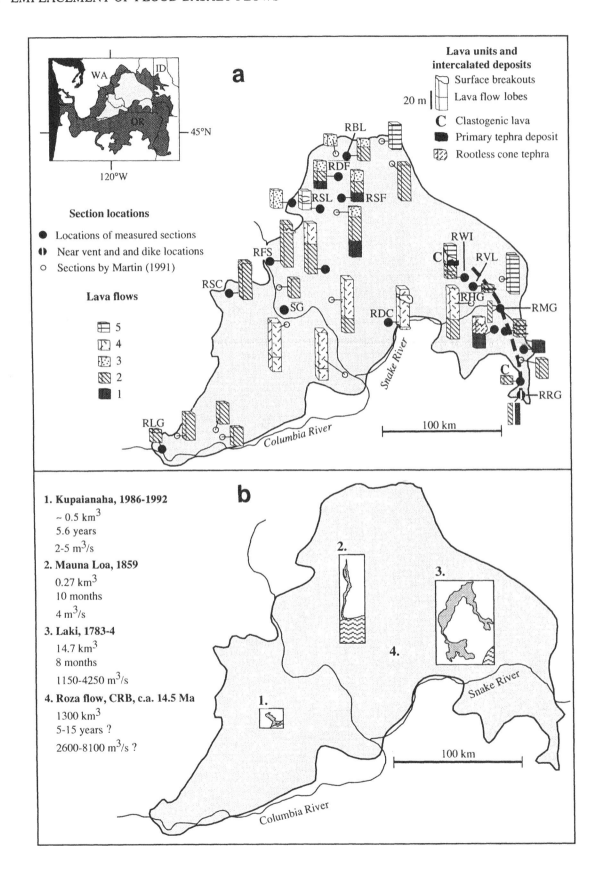

a

Lava units and
intercalated deposits

Surface breakouts
20 m Lava flow lobes
C Clastogenic lava
Primary tephra deposit
Rootless cone tephra

Section locations

● Locations of measured sections
◑ Near vent and and dike locations
○ Sections by Martin (1991)

Lava flows

5
4
3
2
1

RBL
RDF
RSL RSF
RWI
C RVL
RFS
RHG
RSC RMG
SG
RDC
C
RRG
RLG
Columbia River
Snake River
100 km

b

1. **Kupaianaha, 1986-1992**

~ 0.5 km^3
5.6 years
2-5 m^3/s

2. **Mauna Loa, 1859**

0.27 km^3
10 months
4 m^3/s

3. **Laki, 1783-4**

14.7 km^3
8 months
1150-4250 m^3/s

4. **Roza flow, CRB, c.a. 14.5 Ma**

1300 km^3
5-15 years ?
2600-8100 m^3/s ?

2.
3.
4.
1.
Snake River
100 km
Columbia River

picture of global plate tectonic and mantle dynamics [e.g., *Wright et al.*, 1973, *Macdougall*, 1988; *White and McKenzie*, 1989; *Campbell and Griffiths*, 1990; *Kent et al.*, 1992; *Hooper and Hawkesworth*, 1993; *Arndt et al.*, 1993; *Anderson*, 1994; *Peng et al.*, 1994; *Turner and Hawkesworth*, 1995; papers in this monograph]. Furthermore, earlier studies aimed specifically at the physical emplacement of flood basalt lava flows [*Shaw and Swanson*, 1970; *Swanson et al.*, 1975; *Long and Wood*, 1986; *Reidel and Tolan*, 1992] did not have the benefit of recent advances in our understanding of lava flow emplacement mechanics [e.g., *Hon et al.*, 1994].

The emplacement of continental flood basalt lavas also deserves attention in light of the apparent correlation between the ages of flood basalt eruptions and mass extinctions throughout the Phanerozoic [*Courtillot et al.*, 1986, 1988; *Rampino and Stothers*, 1988; *Renne et al.*, 1992, 1995; *Stothers*, 1993; *Courtillot*, 1994]. Although the apparent agreement in ages is highly suggestive of a link, no causal relationships can be established without first understanding the eruptions that form flood basalt provinces and examining their potential for releasing volatiles into the atmosphere.

Here we describe the flow morphology and internal structures of Columbia River Basalt lavas, indicate how these observations are inconsistent with the previously accepted emplacement model and then introduce a new model and describe how the flow features are explained better by it. We also describe what is known about vents for flood basalt lava flows and eruption rates. We end by discussing some of the implications of our work for the environmental impact of flood basalt volcanism. For the purposes herein, we define a flood basalt province as an area greater than 100,000 km^2 covered with at least 1 km thickness of basalt lavas and with individual flows of lengths in excess of 100 km and volumes of more than 100 km^3.

In exploring the physical volcanology of flood basalts, we have concentrated on the Columbia River Basalt (CRB) Group as it is the youngest and best-studied CFB

province [*Hooper*, 1982]. Maps of the areal extent of individual flows and basic parameters such as flow volumes [e.g., *Tolan et al.*, 1989] are not available from any other flood basalt province, making the CRB the only province where one can quantitatively discuss the emplacement of flood basalt lava flows. This fact highlights the immense value of previous studies by many workers in understanding the CRB lava sequence.

The major lava production in the CRB occurred between 16.5 and 14.5 Ma (the Grand Ronde Basalt and Wanapum Formations), when ~ 90% of the total volume of the province was erupted [*Tolan et al.*, 1989]. Even within the CRB, the older units are exposed infrequently as they are covered by thick stacks of overlying lava flows. Our information to date is largely from one of the best-studied voluminous flows, the ~14.8 Ma, 1300 km^3, Roza Member of the Wanapum Formation (Figure 1; Tables 1 and 2) [*Swanson et al.*, 1975, 1979; *Tolan et al.*, 1989; *Martin*, 1989, 1991]. Observations made on other CRB lava flows during this study and reports from older CFB provinces [e.g., *Walker*, 1971; *Keszthelyi et al.*, 1997] suggest that the Roza is representative of the type of flood basalt flow that shows a simple internal structure (Type I of *Long and Wood* [1986]).

2. PHYSICAL DESCRIPTION OF CONTINENTAL FLOOD BASALT LAVA FLOWS

2.1. *Terminology*

Before describing CFB lava flows, we must define our terminology. In choosing terminology we have used two criteria: (1) simplicity and (2) ability to convey the concepts relevant to the emplacement of CFB lava flows. As such, the terms we use here are not necessarily the most useful for describing outcrops or for field mapping in ancient lava sequences. However, the terminology serves well in describing active and young flows in Hawaii and Iceland.

We divide the products of an eruption into three

Figure 1. (a) Map of known distribution of Roza member, Colombia River flood basalt province, showing (inset) extent of Columbia River Basalt Group (dark shading) and Roza lava flow field (light shading). Dashed line delineates Roza eruptive fissure [*Swanson et al.*, 1975]. Dots mark locations of sections measured within the Roza lava; columns show divisions of lava into flow lobes and the five lava flows that make up the flow field [after *Martin*, 1989; *Thordarson*, 1995]. *Martin's* Roza chemical subtypes are shown here as lava flows designated as follows: Subtype IA and B combined, IIA, IIB, III, IV are flows 1, 2, 3, 4, 5, respectively. Letter designations are outcrop location codes referred to in text and Table 2; SG is Sentinel Gap location. Maps of individual Roza lava flows are presented on Figure 14. (b) Comparison of the extent of Roza lava field with those of three historic basaltic pahoehoe flow fields shown at same scale (see insets). For each eruption, volume, duration, and average total volumetric flux of lava is given. (Data: Kilauea [*Mattox et al.*, 1993, and present authors]; Mauna Loa 1859 pahoehoe flow [*Rowland and Walker*, 1990]; Laki [*Thordarson and Self*, 1993]; Roza [*Tolan et al.*, 1989; *Thordarson*, 1995].)

TABLE 1. Widespread Columbia River Basalt Group Lava Units Erupted Between 16.5 and 12 Ma with Occurrence of Physical Lava Features

Formation Member/Unit/Flow	Age[a] (Ma)	Area (km[b])	Volume (km[c])	Pahoehoe Aa[d]	Lobes[e] B/U	Cone/ crust[f]	Lava rise sutures	Tumuli	Elevated tree molds
Saddle Mountain Basalt									
Pomona	12	20,550	760	P		X			
Umatilla		15,110	720	—		—	—	—	—
Wanapum Basalt									
Priest Rapids/									
Rosalia	14.5	57,300	2,800	P^a	B	X	X	X	X
Roza		40,350	1,300	P	B+U	X	X	X	X
Frenchman Springs									
Sentinel Gap		38,760	1,190	P	B+U	X	X	X	
Sand Hollow	15.3	67,110	2,660	P	B+U	X	X	X	X
Silver Falls		28,840	710	P	B			X	
Gingko		37,170	1,570	P	B+U			X	X
Grande Ronde Basalt^2									
N$_2$^{3}		114,460	27,900						
Sentinel Bluffs	15.6			P^a	X	X		X	X
Slack Canyon				P	X	X		X	X
Fields Springs									
Winter Water				P	B	X		X	X
Umtanum				P					
Ortley				—	—	—	—	—	—
Armstrong Canyon				—	—	—	—	—	—
R$_2$		117,730	53,100	P	B				
N$_1$		102,340	31,400						
R$_1$	16.5	96,650	36,200	P^a/A					

[a]Stratigraphy, isotopic age, and volume after *Tolan et al.* [1989] and *Reidel et al.* [1989].

[b]Each Grande Ronde paleomagnetic unit consists of many flows.

[c]Subdivisions of Grande Ronde Basalt Group based on paleomagnetic polarity: N = normal, R = reversed.

[d]Dominant upper and lower surface textural characteristic of flows; P=pahoehoe; A=aa; P^a denotes pahoehoe with rubbly flow top material.

[e]Presence of pahoehoe lobes and toes at base (B) or upper surface (U) of flows.

[f]Clear division of flow into core and upper crust zones based on vesicularity and jointing characteristics (see text).

X Feature noted in flows; — flow not examined.

"levels": flow field, lava flow, and flow lobe. In simplest terms, a flow field is a field of lava flows and each lava flow is made up of a number of lobes.

2.1.1 *Flow fields and lava flows.* A flow field is the aggregate product of a single eruption or vent and is built up of one or more lava flows [e.g., *Pinkerton and Sparks*, 1976; *Kilburn and Lopes*, 1991; *Mattox et al.*, 1993]. For our purposes, a flow field is the product of a single eruption within a flood basalt province and is usually identified on the basis of the chemistry of the constituent flows.

As per the definition in the *Glossary of Geology* [*Bates and Jackson*, 1987] and usage in Hawaii, we use the term "lava flow" to describe the product of a single continuous outpouring of lava. In principle, each flow roughly corresponds to one episode of an eruption. For example, most named flows within the current Kilauea eruption formed after a short pause in the effusion of lava [*Mattox et al.*, 1993]. Cooling and collapses during the pauses can render the previous lava pathways unusable before the eruption resumes, so that the new lava is forced to flow over different areas. If a new lava flow covers an older lava flow while it is still hot, it may be difficult later to distinguish the two flows because the lavas can weld together and cool as a single unit [*Walker*, 1989]. It is also possible for an eruption to simultaneously form two or more separate lava flows. Although defining the exact limits of an ancient lava flow can be difficult to impossible in the field, some term is required to describe the units that build up a large flow field. As such, in our usage, each lava flow in a flood basalt province is regional in scale and formed by a single continuous outpouring of lava. While these definitions seem simple, it is not straightforward to apply

TABLE 2. Total Thickness, Core and Upper Crust Thickness, and Surface and Internal Features of Lava Flows of the Roza Member at Selected Locations on the Columbia River Plateau.

Location Name	Location Code[a]	Lava Flow[b]	Total thickness (m)	Lava core thickness (m)	Upper crust thickness (m)	Ratio core/ total th.[c]	Surface features[d]	Internal features[e]
Asotin Creek	RAC	1	11.5	7.5	4.0	0.65	s	hz
Banks Lake	RBL	2	15.3	8.4	6.9	0.55	s,l,t,it	hz,hv,pv,t,vc
		3	7.2	3.8	3.4	0.53		
Black Butte	RBB	1	17.3	12.3	5.0	0.71		
		2	31.0	20.0	11.0	0.65		
		3*	3.0	1.5	1.5	0.50		
		3*	3.0	1.7	1.3	0.57		
Devils Canyon	RDC	4	28.8	21.3	7.5	0.74	s,l,it	hz
		5	6.5	4.0	2.5	0.62		
Dry Falls	RDF	2	11.0	6.5	5.5	0.59	t,l,s	hz,hs,pv,vc
		3	12.9	5.7	7.2	0.44		
Frenchman Springs	RFS	4	42.5	29.0	13.5	0.68	l,t,s,it,r	is
Horton Grade	RHG	2	25.0	14.3	10.7	0.57	t,l	
		4	32.1	17.1	15.0	0.53		
Moses Coulee	RMC	3	20.6	12.4	8.2	0.60	s,p	hz
Drumheller Channels	RPR	2	52.0	37.7	14.3	0.73	t,s,it	hs,hz,pv
		4	9.0	4.5	4.5	0.50		
Lyle	RLG	2	10.0	5.5	4.5	0.55	l,t,s,it	is,hz,tm
		2*	4.0	2.5	1.5	0.63		
Summer Falls	RSF	1	12.0	6.5	5.5	0.54	s	hz,hs,pv,vc
		2	12.0	5.0	7.0	0.42		
		3	11.0	5.5	6.5	0.50		
Valentine Ridge Rd.	RVL	2	3.5	2.2	1.3	0.63	l,t	
Wanapum Village	RWV	2	20.4	10.0	10.4	0.49	l,t,it	hz,vc
Selah Creek, Yakima Canyon	RSC	2	39.0	29.2	9.8	0.75	l,s,r,t	hs,hz,pv,vc

[a] Location code of *Martin* [1991] used where applicable; Figure 1 shows location for most of these exposures. Table 1 of *Thordarson et al.* [1996a] gives map references for these locations.
[b] Flow designation 1-5 equivalent to *Martin's* [1989,1991] subtypes; thus, IA and B = 1, IIA = 2, IIB=3, III=4, IV=5.
[c] Ratio of lava core thickness to total flow thickness for each flow; range is betwen 0.4 and 0.75.
[d] Morphologic features of flow surfaces noted at each exposure.
[e] Internal flow features noted at each exposure.
it = inflation tumulus
r = pahoehoe ropes
l = pahoehoe lobe (1-100 m in long dimension)
s = pahoehoe sheet lobe (> 100 m in long dimension)
p = pillow lava
t = pahoehoe toe
hs = horizontal vesicular sheets in lava core
pv = pipe vesicles near base of flow
hz = horizontal vesicle zones in upper lava crust
tm = tree molds
is = inflation suture or pit
vc = vesicle cylinders in lava core
* Possibly local surface breakout lobes from flow below.

them to the earlier mapping efforts in the various CFB provinces.

In the CRB, a detailed stratigraphy based on the superposition, chemistry, and paleomagnetic character of the lavas has been built up successfully over the years [e.g., *Mackin*, 1961; *Waters*, 1961; *Bingham and Walters*, 1965; *Schminke*, 1967; *Wright et al.*, 1973, 1989; *Swanson et al.*, 1979; *Reidel*, 1983; *Hooper et al.*, 1984; *Beeson* et al., 1985; *Mangan et al.*, 1985, 1986; *Reidel et al.*, 1989; *Landon and Long*, 1989]. Similar work is ongoing in other CFB provinces. Because the stratigraphy is (necessarily) based largely on the chemical composition of the lavas, there has been some confusion about what constitutes the product of a single eruption. Flow contacts can often be seen within individual chemically defined stratigraphic units. Although a chemical stratigraphy is absolutely vital in sorting out the history of a flood basalt province, even more detail is needed to decipher the physical emplacement processes that formed each lava flow field.

It appears that each stratigraphic member in the Wanapum and Saddle Mountain Formations of the CRB Group (Table 1) is the product of a separate eruption. Each member has a distinct chemical composition and is usually separated from other members by a thin weathering horizon [*Tolan et al.*, 1989]. However, current knowledge does not permit separation of individual lava flow fields within the full extent of the Grande Ronde Formation, which constitutes ~85% of the volume of the CRB province [*Tolan et al.*, 1989]. The Grande Ronde is presently divided by different schemes into (a) paleomagnetically and chemically defined units [e.g., *Swanson et al.*, 1979; *Mangan et al.*, 1986; *Reidel et al.*, 1989] and (b) stratigraphically defined units (flow groups and flows) in the Grande Ronde N_2 magneto-stratigraphic unit [e.g., *Landon and Long*, 1989]. The "flow" subdivision of *Landon and Long* probably represents flow fields or, in some cases, individual lava flows, but correlation of their units cannot be carried widely across the whole outcrop area of the CRB, due in large part to a lack of exposure.

Within the Wanapum members, chemical subunits with subtle differences are often recognized (e.g., the chemical subtypes I-IV in the Roza Member [*Martin*, 1989, 1991] and six "basalts" of the Frenchman Springs Member [*Tolan et al.*, 1989]). These may be the result of slight compositional variations in the lava produced along the length of a fissure system during the life of an eruption. As such, they may generally correspond to our usage of the term "lava flow."

2.1.2. *Flow lobe*. We use this term to describe an individual package of lava that is surrounded by a chilled crust. Flow lobes in the CRB and other CFB provinces can vary in size from tens of centimeters to many kilometers in scale. Small, 10-50 cm thick, 30-100 cm long lobes are usually called toes. Lobes only rarely emanate directly from the vent. They are most often fed from the interior of other, usually larger, lobes.

Many CRB flows are built up of lobes on the scale of hundreds to thousands of meters across that are much wider than they are thick and have relatively flat upper surfaces. Such lobes have been referred to as sheet flows [*Hon et al.*, 1994]. Because these sheets are often regional in scale and are the product of a single continuous outpouring of lava, large sheet-like lobes blur the distinction between lobe and flow. When emphasizing the fact that the entire sheet is composed of a single lobe we use the term "sheet lobe."

A flow composed of a single lobe has been called a "simple lava flow" whereas a flow composed of two or more lobes is called "compound" [*Walker*, 1971]. In common usage, "compound lava flow" has come to imply that the flow is built up of many, overlapping, subequal-sized lobes. We do not make any such implication when using the term "compound"; the lobes can be one or two orders of magnitude different in linear dimension and differ more in terms of volume. The designations "simple" and "compound" are often applied on the basis of only a few outcrops of a lava flow; in our experience, what is referred to as a simple flow in the CRB is usually a large sheet lobe. If followed for a great enough distance (in some cases for tens of kilometers) these sheet lobes terminate against other lobes of the same lava flow. Thus, a lava flow can be compound, although in many localities only a single large lobe is exposed, giving the impression of a simple lava flow. It should also be noted that "simple lava flow" and "sheet lobe" are identical to the Type I lava flow morphology defined in the CRB by *Long and Wood* [1986]. Type II and III flows, which are also common in the CRB, are variants having more complicated, horizontal vesicular zones and/or jointing in their upper crusts and interiors.

2.2. External Features of Continental Flood Basalt Lava Flows

Most continental flood basalt provinces appear to be built up of hundreds of thick (20-100 m) flow fields, each consisting largely of sheet flows. On average, these flows traversed very shallow slopes, (e.g., ~0.1% in the CRB [*Tolan et al.*, 1989]). The flows appear to be hundreds of kilometers long and some have volumes reaching into the thousands of cubic kilometers. Well-documented cases of the extent of individual flow fields, e.g., the Roza Member of the CRB (Figure 1), show the huge size of these lava bodies when compared with historic lava flows. It is uncertain, however, whether a flow field like the Roza covers the entire 40,300 km^2 area within the boundary shown on Figure 1 or whether, as is likely to be the case, there were areas that were not covered by the lava (kipukas).

Upon closer examination, it can be seen that most flood basalt lavas are compound pahoehoe flows. The pahoehoe nature of many major flows in the Columbia River Basalts can be seen in their smooth upper and lower surfaces [e.g., *Mackin*, 1961; *Swanson and Wright*, 1980; *Reidel and Tolan*, 1992]. In many cases, ropes, 30-50 cm scale toes, and other features characteristic of pahoehoe flows are preserved at the flow tops and bottoms (Figure 2; Tables 1 and 2). Pillow lava sequences are often found at the base of CRB flows, occasionally forming almost the whole flow thickness [*Swanson*, 1967; *Schminke*, 1967; *Swanson and Wright*, 1980; *Long and Wood*, 1986]. These pillows are essentially pahoehoe lobes that formed as the lava invaded lacustrine and riverine environments across the Columbia Plateau. Littoral examples of CRB pillow sequences are found along the Oregon coast [*Snavely et al.*, 1973].

Several workers have described aa flows in the CRB, especially in near-vent areas of the Grande Ronde Formation [e.g., *Swanson and Wright*, 1980; *Reidel*, 1983]. In our investigations, we have found only a few cases of true aa in the CRB pile, even in the Grande Ronde source area. To be considered true aa, lava flows should possess (a) spinose rubble (clinker) at the flow top and bottom, (b) elongate, ragged-shaped vesicles, and (c) entrain upper and basal clinker into the flow interior. Many CRB flows have a thin, rubbly flow top [*Reidel*, 1983]. However, they are unlike true aa flows in that the rubble is largely composed of disrupted pahoehoe crust and in that the flows have smooth bases. Such disrupted upper crusts are a common feature of pahoehoe flows in Hawaii, Iceland, Australia, and elsewhere, where they occur in patches on generally undisrupted pahoehoe flow surfaces. These patches are often transitional forms of pahoehoe (e.g., slabby pahoehoe) and are suggestive of either more viscous lava or locally more rapid emplacement. In ancient flows, if the only available outcrops of a lava have surface rubble the flows may have been described as aa.

The compound nature of CRB lava flows can be demonstrated by the fact that most of the chemically defined lava flow packages consist of multiple, physically distinct, lobes. The dominant lobe type is the kilometer-scale sheet lobe that is significantly larger than anything seen in Hawaii or Iceland (Figure 3). Where the contact between two lobes is visible, an overlying lobe usually fills in the gap between them. From a distance, this can give the impression of a single continuous sheet when in reality there are distinct lateral discontinuities.

In most outcrops of the Roza Member and other CRB flow fields, there are two to four distinct sheet lobes [*Martin*, 1989, *Thordarson*, 1995], each probably a part of a separate lava flow. Margins of sheets can rarely be observed because the sheets are usually larger in scale than the outcrop. Smaller toes and lobes are commonly associated with these large sheets [*Finnemore et al.*, 1993; *Thordarson*, 1995] (Figures 2, 4, and 5). In some cases the lobes are clearly associated with the overlying unit, with small toes emplaced in front of a larger sheet that eventually overran them (Figure 2c, d). In other cases they are associated with the underlying unit, with small lobes having oozed out from the larger sheet (Figure 4b).

Other lateral variations are identifiable in the CRB. Most sheet lobes do not have truly flat tops, but instead have hummocks and swales of 1-5 m amplitude and 10-50 m wavelength (Figures 3 and 5). These undulations have been overlooked because (1) they are relatively small compared to the dimensions of the average CRB sheet lobe (typically 1-20% of the total thickness and 0.1-10% of the lobe width) and (2) the contacts between lobes are often very poorly exposed. This is because the upper and lowermost parts of the lobes are vesicular and erode more easily than the dense interiors. The 1-5 m surface topography is very often obscured by talus and vegetation.

When exposures are adequate, axial cracks can be found at the crest of many hummocks which are identical in size and shape to tumuli and other inflation features found on pahoehoe flows across the globe. Tumuli are broad and whaleback-shaped rises, and usually have axial and medial cracks that form as the brittle crust is uplifted [*Walker*, 1991; *Hon et al.*, 1994]. On Hawaiian flows, tumuli are usually on the order of 5-20 m in length, 5-10 m in width, and 2-5 m in height. However, larger tumuli on the order of 50-1000 m in length and width and 10-20 m in height are common on flows in the continental United States [e.g., *Nichols*, 1936; *Theilig*, 1986; *Keszthelyi and Pieri*, 1993; *Chitwood*, 1994]. The Undara flow in Queensland, Australia, has a 40-km-long tumulus (inflation-ridge) [*Atkinson*, 1996]. Tumuli on all these scales are found in the tops of CRB sheet flows (Figure 5) and are often associated with small breakouts. As noted above, some

flows have much broader surface undulations (Figure 3) that appear analogous to inflation ridges and plateaus. Other features in CRB flows (Figure 6) are analogous to suture zones recognized on other pahoehoe flow fields [*Walker*, 1991].

The abundance of compound pahoehoe lava flows in the CRB and other flood basalt provinces is significant because pahoehoe and pillow lavas have only been observed to form at low volumetric fluxes [*Rowland and Walker*, 1990; *Griffiths and Fink*, 1992; *Gregg and Fink*, 1995]. The compound nature of the lava flow fields, with small lobes and toes at the bases of the flows, also indicates that they were emplaced in a series of lobes separated in time and space, arguing for a drawn-out emplacement history. Compound pahoehoe lavas are also common in the Deccan Traps [e.g., *Agashe and Gupte*, 1971; *Phadke and Sukhtankar*, 1971; *Walker*, 1971; *Keszthelyi et al.*, 1997] and other CFB provinces.

2.3. Internal Structure of Continental Flood Basalt Lava Flows

The internal structures within sheet lobes also provide clues to their style of emplacement. There are three key types of internal structures: (a) vesicle patterns, (b) jointing style, and (c) petrographic texture. Of these, jointing is the least informative because fractures form only after the lava has solidified. These internal structures divide each sheet lobe into (1) an upper crust, (2) lava core, and (3) basal zone (Figures 7 and 8) [*Thordarson*, 1995; *Self et al.*, 1996].

This three-part division does not change from the near-vent outcrops to exposures hundreds of kilometers from the source. The divisions can be recognized at every one of the many outcrops of the Roza and other CRB flows that we have examined. The same divisions are also seen in other pahoehoe lava flows of various thicknesses in Hawaii (Figure 7d), Iceland (Figures 7b,c), the Deccan Traps [*Keszthelyi et al.*, 1997], and elsewhere [e.g., *Aubele et al.*, 1988]. Contacts between

the three zones are not knife-sharp, but can usually be located to within 10 cm, even in flows many tens of meters thick.

2.3.1. *Upper crust.* The upper crust is defined by relatively high vesicularity and small prismatic joints at the very top with irregular jointing beneath and hypohyaline (50-90% glass) to hypocrystalline (10-50% glass) textures. The jointing in the upper crust (Figures 7 and 8) can be highly variable, including types previously described as hackly, curvilinear, wine-glass, etc. We generally avoid use of the terms entablature and colonnade because in flood basalts they have become associated with specific genetic models (e.g., water cooling for the entablature [*Long and Wood*, 1986]).

Vesicularity usually decreases while vesicle size increases downward into the flow. Approximately horizontal layers of increased vesicularity can often be found in the upper crust (Figure 9a). The crystallinity grades from a cm-thick glassy rind at top to dominantly microcrystalline downward over several meters (Figure 10a). In most cases in the CRB and elsewhere, the upper crust constitutes 40-50% of the total flow thickness (Table 2).

2.3.2. *Lava core.* The core of a sheet lobe is characterized by very few primary vesicles, regular jointing, and holocrystalline texture (90-100% crystals). The crystals are fine to medium grained and most of the porosity in the core comes from diktytaxitic voids between the crystals (Figure 10b). Jointing tends to be quite regular, and well-developed columnar jointing (colonnade) is sometimes present. The core typically makes up 40-60% of the flow thickness (Table 2).

Megascopic vesicles are usually confined to the late stage residuum formed during in-situ crystallization. Macroscopic bodies of this vesicular late-stage residuum are confined to the lava core and have been reported from many thick basalt flows of all ages [e.g., *Greenhough and Dostal*, 1992; *Puffer and Horter*, 1993]. As the lava crystallizes, volatiles are concentrated into

---

Figure 2. Pahoehoe lava lobes and associated features in CRB flow in Washington state. (a) 60-m-high section through Frenchman Springs member lava (above upper dashed line) and a flow lobe at the top of Grande Ronde Basalt N_2 Sentinel Bluffs unit, probably equivalent to Museum flow group of *Landon and Long* [1989], between and below dashed lines. Near Lower Monument Dam (area of RDC, Figure 1). (b) Boxed area to right of person in (a) exposes a section through a pahoehoe toe at the base of the Museum flow lobe (center; scale is 10 cm across) with centimeter-thick chilled selvage. Note jointed vesicular crust of underlying flow (below level of scale). (c) Basal 70 cm of lower Roza sheet lobe at Dry Falls, showing glassy selvage, s, and thin vesicular crust, c, pipe vesicles, p, and vesicle cylinders, vc. Photograph looks down onto fallen block lying on the ground. Arrow shows original up direction. (d) 40 cm long pahoehoe toe with ropy surface at base of Roza flow in Yakima Canyon (RSC on Figure 1). (e) Upper 5 m of a flow consisting of shelly pahoehoe lobes, seen in section, overlain by massive base of another pahoehoe flow (above dashed line); both Frenchman Springs units at Wallula Gap. Hammer (circled) is 35 cm long.

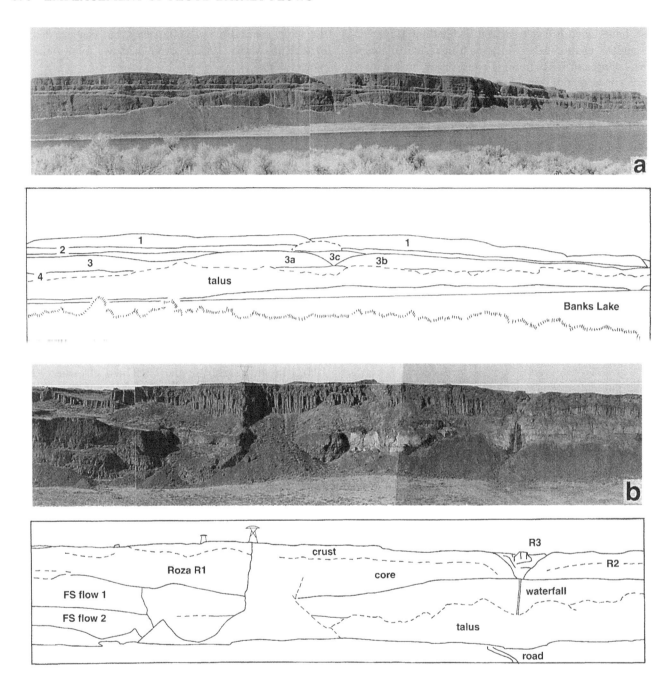

Figure 3. Sheet flow lobes in the CRB. (a) 220-m-high section displays the geometry of large (>1 km scale) sheet lobes in several lavas of Grande Ronde N_2 (probably Sentinel Bluffs) unit, west side of Banks Lake (Grande Coulee), Washington (near RBL on Figure 1). At least four lava flows (1-4) can be seen but it is uncertain how many flow fields are represented. Lobe 3c fills the hollow between two 50 m thick, previously emplaced sheet lobes, 3a and b; lobes of flow 3 cooled together as a single cooling unit because cooling joints pass from one lobe into the adjacent lobes. Lobe 3a is constructed of two tumulus-like bodies and thins in the middle where less thickening by inflation occurred. Section shown is about 3 km long and site is >200 km from the suspected vent area for these flows. (b) Inflated sheet lobes in Roza and Frenchman Springs (FS) members at Frenchman Spring Coulee (RFS on Figure 1). Note that Roza thickens into the hollow between two FS lobes that come together in the center of the photo. On the right, a third Roza lobe (R3) fills in the gap between two earlier sheet lobes (R1, R2) at the waterfall. Upper part of Roza has been scoured by catastrophic glacial floods that excavated the coulee; distinction between crust and core is expressed by different jointing (see text). Cliff is 120 m high and each sheet lobe varies from 20-50 m thick.

the remaining melt, producing a vesicular siliceous sludge that often concentrates into segregation veins. This segregated vesicular material is less dense than the still fluid surrounding lava and rises in cylindrical conduits toward the upper crust, preserved as vesicle cylinders [*Goff*, 1996] (Figure 9b, c). When these conduits reach the solidifying roof of the flow, they are deflected into horizontal sheets. This lateral spreading of the highly vesicular residuum is a result of a mechanical barrier and cannot be attributed to the residuum reaching a neutral buoyancy level. In some cases the buoyant residuum attempts to form diapirs up through the viscoelastic crust. This can result in large, bell-jar shaped, 5-30 cm diameter gas cavities that we call "megavesicles." However, it must be noted that large coalesced megavesicles can also occur within the upper crust.

2.3.3. *Basal zone.* The basal zone is almost always only 0.5-1 m thick, is hypohyaline (50-90% glass) (Figure 10c), slightly vesicular, and sometimes has poorly developed platy jointing (e.g., Figure 2c). In thick flows, the basal zone usually forms much less than 10% of the flow thickness. A 1-3 cm thick, quenched glassy selvage with stretched or round vesicles can often be found at the very base of the flow. Pipe vesicles occur in the base of many CRB lava flows (Figure 2c), consistent with emplacement on very low slopes [*Walker*, 1987].

2.4. *Vents for Continental Flood Basalt Lavas*

Little has been written about the vents from which flood lavas are derived. In part this is simply because, other than feeder dikes, vent structures or edifices appear to be elusive features in CFB provinces. However, the Roza linear vent system of the CRB, identified as coeval with the Roza flow by field and chemical characteristics [*Swanson et al.*, 1975; *Martin*, 1989] is well documented. Identification of this vent system strongly suggests that flood basalt lavas are largely fissure-fed. Features ranging from dikes to small shield-like edifices are exposed within a narrow zone about 5 km wide along a 150-km-long swath of country (Figure 11). An outlier to this zone, the vent-like structures at Pomeroy quarry, Washington (Figure 3 of *Swanson et al.* [1975]), are possibly sections through rootless cones (pseudocraters) in a lower Roza sheet lobe that was buried beneath later Roza lobes (Th. Thordarson, unpublished data).

Deposits of welded spatter and fountain-fed (clastogenic) lava, identified by ghosts of agglutinated clasts, along the trend of the Roza fissure provide most of the evidence for location of the vent system [*Swanson et al.*, 1975]. The central region has several small (50 m long × 5 m high) outcrops of clastogenic lava that give the impression of widespread fallout of material on either side of a fissure and are convincing evidence of proximity to the lava source. At some locations, e.g., Potter White Hill (site 15 of *Swanson et al.* [1975]), the clasts are coarse (30-40 cm), suggesting that a vent was nearby. The best exposed accumulations of Roza spatter and scoria are in two exposures forming an oblique, 1-km-long section across the strike of the fissure at Winona, Washington [*Thordarson and Self*, 1996]. The structures appear to be parts of 5-6 m high scoria ramparts with associated welded spatter, and although no convincing evidence exists of a section across an actual vent, the structures are buried beneath >10 m of fountain-fed lava, suggestive of a nearby source.

The detailed geochemical study of *Martin* [1989] showed that each chemical subtype of the Roza lava is found in a restricted portion of the fissure system, suggesting that lava effusion migrated over time (Figure 11). We speculate, based on the behavior of historic fissure eruptions, that only one or two segments of the Roza fissure system, each several kilometers in length, would have been active at any one time. Furthermore, as in Hawaii and Iceland, each active segment should have contracted to a few point-sources of lava in a matter of days to weeks given the thermal instability inherent in fissure systems [*Wilson and Head*, 1981; *Whitehead and Helfrich*, 1991].

Reidel and Tolan [1992] described another example of a vent structure from the R_1 magneto-stratigraphic unit of the Grande Ronde Formation. A dike-fed, approximately 100-m-wide lava lake of unknown long dimension is exposed in cross section, with associated shelly pahoehoe overspills and scoria and Pele's tear deposits extending to a few hundred meters on either side of the lake. This structure is of average size by Hawaiian standards; e.g., the Kupaianaha lava pond, active from 1986-1991, was ≤100 m in diameter. Yet this Grande Ronde vent was part of a system that produced a lava flow thought to extend for >300 km and have a volume of ~2,000 km³.

Having presented a compilation of the field data available on CRB lavas, we move on to the (differing) interpretations.

3. EARLY MODEL OF FLOOD BASALT EMPLACEMENT

The pioneering, but highly exploratory, work by *Shaw and Swanson* [1970] on the emplacement of the CRB

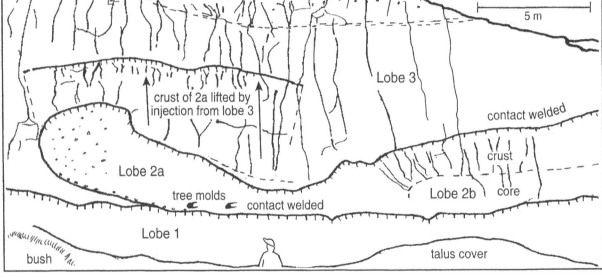

Figure 4. Flow lobes in CRB lavas demonstrating locally compound nature of flows. (a) Sketch of small compound pahoehoe lobes (1-3) in a Wanapum Basalt lava flow 0.5 km north of Soap Lake, Washington (RSL on Figure 1). Note that inflation of lobe 3 disrupted part of the crust of previously emplaced lobe 2a. Individual columnar joints passing through the upper package of lobes and welding of lobe contacts indicate that they cooled together as single unit. (b) A 60-m-long lobe fed by breakout through crust of underlying sheet lobe in a GRB N_2 lava flow on east side of Banks Lake, Washington (near RBL on Figure 1). Circular holes, h, are molds of tree trunks that were lying on the crust of lobe as it inflated. Note that two low-amplitude tumuli-like bodies formed in lobe above to either side of the thickest part of break-out lobe. Pahoehoe rubble is disrupted glassy, vesicular flow-top material. (c) Roza member at Dry Falls (RDF on Figure 1) showing two of the three sheet flow lobes found in this area. Note that the lower lava flow lobe gradually thins away from camera and that the thinning is compensated by thickening of upper lobe. Total thickness of Roza at this exposure is ~23 m.

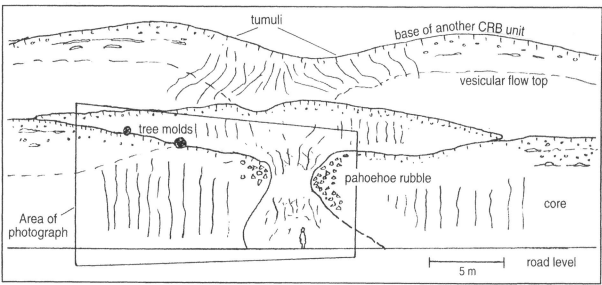

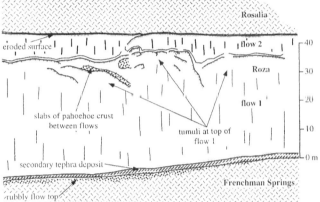

Figure 5. Tumuli in CRB and Kilauea pahoehoe flows. (a) Section through a small tumulus (T) with associated breakout toes and lobes (L) in top of a sheet lobe, overlain by massive base of another sheet lobe (above upper dashed line) that fills in topography; exposed at Frenchman Springs coulee (RFS on Figure 1). (b) Tumulus on active Kalapana flow, Kilauea, in 1990. Note breakout toes and upturned pahoehoe crustal plates. In cross section this would resemble the tumulus shown in (a). (c) Larger elongate tumulus of 5 m amplitude and 40 m length on active Wahaula flow, Kilauea, 1991. (d) Large tumuli of 10 m amplitude and 50 m length in upper surface of a lava flow in the compound Roza members, east wall of Devil's Canyon (RDL on Figure 1). Another lava flow is present above tumuli, with Rosalia flow of Priest Rapids member overlying; a Frenchman Springs unit underlies Roza flow field. Sketch shows interpretation of photo.

lavas has become the entrenched and (until recently) unchallenged dogma that it was never intended to be. In this section, we describe the constraints and assumptions on which *Shaw and Swanson* [1970] based their work, the resulting emplacement model, and how new observations over the past 25 years have forced re-evaluation of some key assumptions. We also show that

field data are inconsistent with some predictions based on this emplacement model.

Shaw and Swanson [1970] made one of the earliest attempts to combine fluid mechanics with field observations such as lava flow thicknesses, topographic slope, and an estimate of lava rheology in order to calculate velocity and style of flow. The key observation that constrained their modeling was that the CRB lavas do not show evidence of measurable crystallization during transport from their source vents. Glassy rinds of flows 300-500 km from the vents are just as glassy as those within a few tens of kilometers of the vents. This suggested to *Shaw and Swanson* that the flows must have traveled very rapidly, attaining their full areal extent before any significant cooling took place.

Shaw and Swanson [1970] also assumed that the glassy nature of the lava selvages indicated that the lavas were erupted well above the liquidus temperature, allowing the lava to cool substantially during transport without crystallization. The high assumed lava temperatures (~1200°C) led to the use of very low viscosities (50 Pa s) in their calculations, which further corroborated their assumption of rapid emplacement. Combined with the great observed flow thicknesses, this resulted in calculated flow velocities of several km/hr, flow in the turbulent regime, and eruption durations on the order of weeks. *Shaw and Swanson* [1970] did suggest longer eruptive durations from extensive fissures as an alternative scenario, but this idea was not generally adopted [e.g., *Hooper*, 1982].

Although a picture of cataclysmic floods of lava charging across the Columbia Plateau is perhaps appealing, good evidence exists that the CRB flood basalt lava flows were not emplaced in this manner. First, the phenocrysts seen in many CRB lava flows, including the Roza flow [e.g., *Swanson and Wright*, 1980; *Martin*, 1989], are now known to be inconsistent with eruption temperatures well above the liquidus. The crystallization of metal oxides in the glassy margins of dikes and lobes suggests eruption temperatures of 1090-1070°C (Table 3) [*Thordarson*, 1995; *Ho and Cashman*, 1995]. The viscosity of the lava can be estimated from the liquid composition and temperature [e.g., *Bottinga and Weill*, 1972; *Shaw*, 1972] and then adjusted for the effect of entrained crystals and bubbles [e.g., *Pinkerton and Stevenson*, 1992]. For the range of glass (and presumably liquid) compositions from the CRB lavas (49-56 wt.% SiO_2 [*Mangan et al.*, 1986]), estimated viscosities at the point of eruption are one to two orders of magnitude higher than the 50 Pa s used by *Shaw and Swanson* [1970]. For the Roza lava, estimated viscosities are 500-700 Pa s (Table 3). These higher viscosities make it unlikely that even 20-m-thick CRB lava flows would have been turbulent (Figure 12). However, lava flows >30 m thick theoretically should have been turbulent.

Turbulent lava flows should erode their bases by both thermal and mechanical erosion, resulting in distinct channels with bell-jar-shaped cross sections [e.g., *Jarvis*, 1995]. CRB flows have no evidence for such downcutting or channelization, being remarkably sheet-like (see Figure 3). Delicate, centimeter-scale surface textures are preserved between thick flows in the CRB, indicating that no erosion took place at the contact (see Figures 2 and 4). No field evidence collected to date supports turbulent emplacement for any of the flows examined.

Emplacement of CRB lava flows cannot be generally explained by rapid laminar flow, either. Rapid flow leads to high strain rates which, in the absence of unusually low viscosities, lead to the production of aa (or aa-like) flow surfaces [*Peterson and Tilling*, 1980]. The calculated viscosities of CRB lavas (Table 3) are in fact even higher than those of typical Hawaiian lavas. Furthermore, aa flows, with disrupted crusts, are inherently thermally inefficient. Adopting the thermal model of *Crisp and Baloga* [1994] for aa lava flows, it can be shown that even if CRB flows were emplaced in less than half a day, they would probably have cooled and crystallized significantly. Only under very special and unlikely conditions can aa lava flows 300-500 km long be produced by laminar flow [*Keszthelyi and Self*, 1996].

The single greatest problem with the *Shaw and Swanson* [1970] rapid emplacement model for flood basalt lava flows was the assumption that the final thickness of the frozen lava flow was approximately the thickness of the flow while it was moving. We now present a new model based on the idea that most CRB lava flows thickened by one to two orders of magnitude during emplacement by the process of inflation.

4. A NEW MODEL FOR THE EMPLACEMENT OF FLOOD BASALT LAVA FLOWS

The dominant flow morphology in the CRB and other flood basalt provinces is the thick, compound, pahoehoe sheet flow. Sheet lobes within the ongoing eruption of Kilauea Volcano, Hawaii, form exclusively by inflation (endogenous growth), i.e., the injection of liquid lava under a solidifying crust [*Hon et al.*, 1994]. Comprehension of the inflation process and the realization that inflation is the fundamental and universal process through which pahoehoe lava flows grow in thickness have led to a revolution in our thinking about the formation of compound pahoehoe flow fields in general and flood basalt sheet lobes in particular.

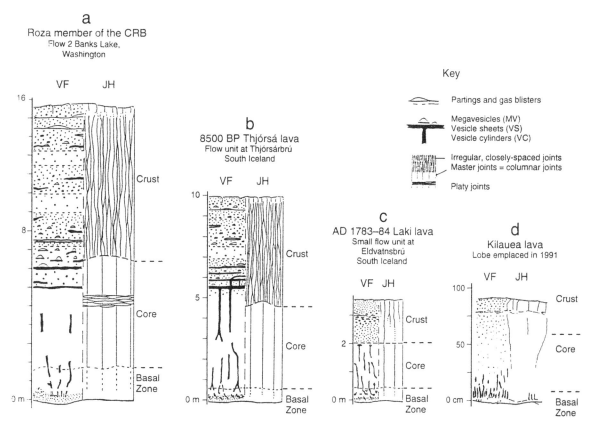

Figure 7. Examples of measured sections through sheet flows of various thicknesses in flow fields of different dimensions. (a) 16-m-thick sheet lobe within Roza member at Banks Lake, Washington (RBL on Figure 1). (b) 10-m-thick flow unit within the 21 km³ Thjorsa lava in south Iceland. (c) 3.5-m-thick flow unit within the 15 km³ Laki lava flow, south Iceland. (d) 80-cm-thick flow unit within the 1991-2 Wahaula flow, Kilauea. Note the overall structural symmetry of these flow units. VF indicates vesiculation features; JH indicates jointing habits. Details of internal features are as in Figure 8b. Note presence of pipe vesicles at the base of each flow. Modified from *Thordarson* [1995].

4.1. *The Inflation Process and Inflation Features*

The inflation process is involved in the formation of all pahoehoe lava bodies from 20-cm-thick toes to 100-m-thick sheets of lava. Pahoehoe toes and lobes initially grow as inflating, liquid-filled balloons. The liquid lava is held in by a partially cooled, viscoelastic skin of lava. Over a period of a few minutes a cold brittle crust will begin to form on top of the viscoelastic skin [*Hon et al.*, 1994; *Keszthelyi and Denlinger*, 1996]. The brittle crust is extensively fractured due to

contraction during cooling and has essentially no tensile strength [*Hon et al.*, 1994]. Thus the cooled, fractured, solid lava does not play an important mechanical role in the emplacement of pahoehoe flows. Instead, it is the resistance of the viscoelastic skin to continued stretching that constrains inflation rates. The motion of the viscoelastic skin is accommodated in the overlying brittle crust by the widening of the cooling-induced fractures [e.g., *Walker*, 1991]. If the rate of inflation is too high, the skin may burst and fluid lava is able to break out from inside the inflating lobe.

Figure 6. Features in CRB lavas thought to be related to thickening by inflation (lava rise). (a) Margins of small inflation suture or pit in a Roza sheet lobe at Jasper Canyon, Blue Lake, Washington. Note curved foliation of stretched vesicles; jagged plates of lava fill the suture. (b) Margins of inflation suture in a Frenchman Springs lava flow at Frenchman Springs Coulee. Note foliation of stretched vesicles and small toes filling upper part of suture. (c) Large lava rise suture or pit in Roza flow (note person in center of feature) showing subhorizontal interleaved plates of lava between two abutting parts of an inflated flow lobe. Subhorizontal cracks formed by extension during inflation extend into massive lava cone on right and left. Both (b) and (c) photographed at Frenchman Spring Coulee (RFS on Figure 1).

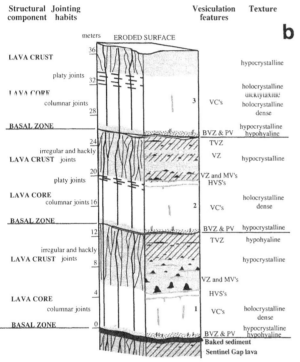

Figure 8. (a) Outcrop of Roza member at Summer Falls (RSF in Figure 1) showing lava flows 1, 2, and 3 (top eroded). Solid line indicates boundary between individual flows and broken line shows boundary between lava core (Co) and lava crust (Cr), most easily discernible in photograph by changes in jointing. Flow below Roza is Sentinel Gap flow of the Frenchman Springs Member. (b) Stratigraphic section measured through the compound Roza member at Summer Falls showing division of each of three lava flows (1, 2, 3) into crust, core, and basal zone. Right side of the stratigraphic column shows vesiculation features and textural properties in each flow unit. Left side shows jointing pattern in the lava. VZ denotes a vesicular zone, where the prefix B = basal and T = top. MV indicates megavesicles; PV and VC stand for pipe vesicles and vesicle cylinders, respectively. HVS denotes horizontal vesicle sheet. Vertical scale is in meters. Modified from *Thordarson* [1995].

Inflation in pahoehoe lavas forms a distinctive set of external morphologic features, the simplest of which are lobes, toes, and (in the presence of surface water) pillows. When inflation is localized, tumuli and other inflation features can form at the surface. Inflation also forms pits (the "lava-rise pits" of *Walker* [1991]) where a small section of the flow does not inflate and the surrounding lava is raised up around it. The pits often contain pahoehoe toes and small lobes where lava has oozed from the marginal cracks. Such pits are typically only 2-10 m across and 2-4 m deep in Hawaii, and the CRB flows have similar size features, but larger inflation pits can be found elsewhere, such as in the Toomba and Kinrara flows, Queensland [*Stephenson et al.*, 1996]. In the CRB, the enigmatic features in Figure 6, previously thought to be spiracles caused by steam-induced fracturing [*Waters*, 1961], can be explained best as filled sutures or pits where two opposing lobes or parts of lobes inflated and thickened together (the lava rise sutures of *Walker* [1991]).

The three-part internal division of the thick pahoehoe sheet lobes in the CRB is also consistent with an origin by inflation. Note that the internal structure of a true aa flow (described in section 2.2) is very different from what we have commonly observed in the Roza and other CRB lava flows. Figure 13 shows how we believe the three-part division comes about. In an inflating lava flow, the upper and lower crusts form by freezing of the lava while the flows are receiving an influx of fresh lava. The core of the flow cools and crystallizes only after the sheet has stagnated. The clearest evidence of this process comes from the vesicle patterns, as previously suggested for the origin of vesicle zonation in thin lava flows [*Aubele et al.*, 1988].

The size distribution and the shape of the vesicles in the upper crust suggest that they are primary, having formed during the eruption process [*Mangan et al.*, 1993] with modification due to bubble rise and coalescence. Horizontal vesicular zones preserved in the crust can be interpreted to form when the sheet is depressurized by a sudden major breakout and bubbles form inside the sheet [cf., *Hon et al.*, 1994]. An alternative possibility is that the vesicular zones are the result of a more bubble-rich batch of lava passing through the sheet lobe. In either case, the bubbles migrate upward (and coalesce) only to be trapped against the downward-growing upper crust of the lava flow. Previous modeling of these horizontal vesicular zones required episodic bubble formation at the base of a stagnant lava flow [e.g., *McMillan et al.*, 1989], a situation difficult to explain physically, or unrealistically high lava viscosities [*Manga*, 1996]. The jointing in the upper crust may be more irregular, in part

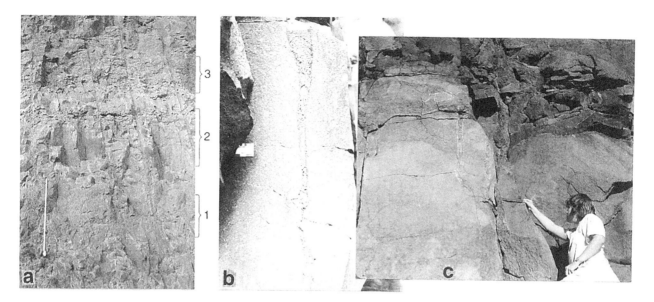

Figure 9. Vesiculation features in CRB lavas. (a) Horizontal vesicular zones (numbered 1-3) in the crust of a sheet lobe in a Grande Ronde Basalt N_2 flow at Armour Draw, Moses Coulee, Washington. Tape is extended to 1 m. (b) Individual vesicle cylinder in core of a Frenchman Springs sheet flow near Lyle, Washington (RLG on Figure 1). (c) Vesicle cylinders (above hand) passing into vesicular sheets (outlines dashed for clarity), upper part of lava core of Levering flow, Sentinel Bluffs unit of Grande Ronde N_2 at Sentinel Gap, Washington (location SG, Figure 1).

due to jostling of the brittle chilled lava that takes place during inflation.

The features in the lava core are best explained by slow, stagnant cooling. The bulk of the core is dense because the bubbles have had time to migrate to the top of the flow. Also, coherent cylinders and sheets of vesicular residuum form only after the flow has stagnated because the flow of lava through the core would disrupt the passive convective patterns that lead to the cylindrical diapirs. Furthermore, columnar jointing is thought to require cooling under stagnant conditions [*Swanson and Wright*, 1980].

The thin basal zone may appear to be inconsistent with a protracted emplacement, but recent measurements of cooling at the base of pahoehoe flows show that the base cools much slower than was expected from any cooling model [*Keszthelyi*, 1995a].

4.2. Eruption Duration and Eruption Rates for the Roza Member of the Columbia River Basalt Group

The model for the formation of the internal divisions within an inflated pahoehoe sheet lobe (Figure 13) provides a means to estimate the duration of the effusive activity that fed a lobe. Assuming that the Roza flow field was emplaced as an inflated pahoehoe flow field, we now discuss its eruption duration and the range of volumetric effusion rates. We argue that many other CRB (and other CFB) lava flows were emplaced in a similar fashion, but we do not at this time have the field data to quantify their eruption durations or effusion rates.

Our model indicates that the boundary between the vesicular upper crust and the dense lava core marks the time when the flux of fresh lava into a lobe ended and the fluid interior of the lobe became stagnant (Figure 13c). By calculating the time required for the upper crust to form, it is possible to estimate the duration over which the lobe was being fed fresh lava. In Hawaii, the growth of thickness of the upper crust conforms to the empirical equation

$$H_c = 0.0779 \, t^{1/2} \qquad (1)$$

where H_c is the thickness of the upper crust in meters, 0.0779 is an empirically determined constant, and t is time in hours [*Hon et al.*, 1994]. We expect some differences in the cooling rate of CRB and Hawaiian lavas because of differences in rainfall and thermal properties (heat capacity, diffusivity, and latent heat of crystallization). We are currently investigating these differences using a modified version of the thermal model of *Keszthelyi and Denlinger* [1996] and preliminary results suggest that the effects are small (<25% errors in estimated duration).

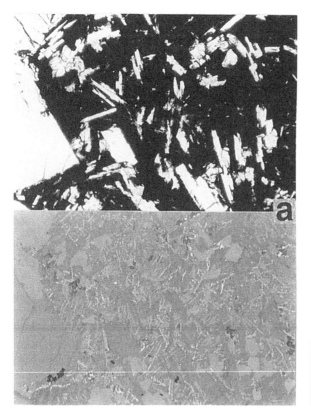

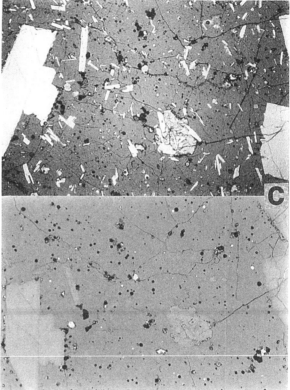

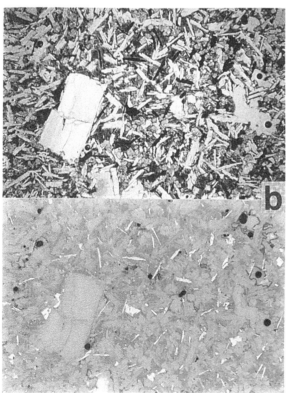

Figure 10. Photomicrographs of thin sections of lava samples from the upper crust, core, and basal zone of the Roza member. (a) Upper crust of lower Roza sheet flow at Summer Falls (RSF on Figure 1). Upper photo plane polarized light (PPL), lower photo reflected light (RL); field of view 1.3 mm across (100 X). Note opaque tachylitic glass between plagioclase pheno- and microcrysts in PPL. In RL the tachylite glass is seen to contain numerous dendritic opaque oxides, a reflection of rapid quenching (but slower than basal selvage). (b) Core of same lobe as in (a). Upper photo PPL; lower RL: field of view 5.2 mm across (25X). Note phenocrysts of plagioclase and coarsely crystalline groundmass, including plagioclase, olivine, pyroxene, and acicular opaque oxide microphenocrysts, especially well seen in RL. (c) Basal glassy selvage (rapidly cooled and quenched sideromelane) of the lower Roza sheet lobe at Frenchman Springs Coulee (RFS on Figure. 1). Field of view 5.2 mm across (25X); view changed slightly between top (PPL) and lower (RL) photos. Note sparse opaque oxide microcrysts. Phenocrysts and microphenocrysts are plagioclase except for a pyroxene/plagioclase glomerocryst at lower right of center.

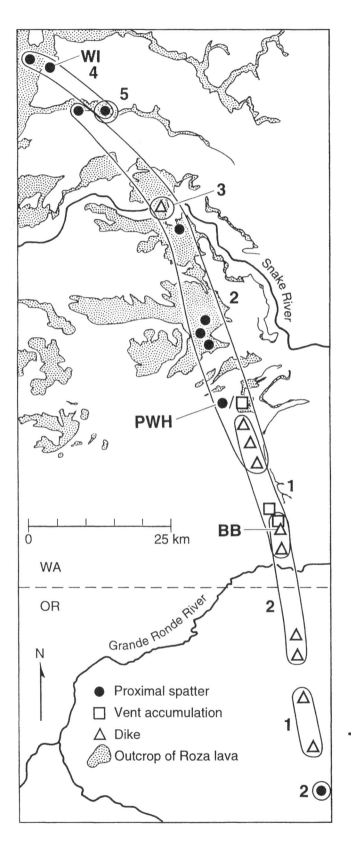

Assuming that equation (1) applies to CRB flows as well as those in Hawaii, the thicknesses of the upper crusts of individual Roza sheet lobes (Table 2) indicate that they were active for months. Most outcrops of the Roza consist of two or more sheet lobes. Because it is not likely that an underlying lobe would actively inflate with an overlying active flow, it is possible to sum the durations calculated for each lobe to arrive at a minimum time required to form the flows at each outcrop. For example, the outcrop at Summer Falls shown in Figure 8 has three sheet lobes that we calculate were active for 6.8, 11, and >9 months, based on the thickness of their respective upper crusts [*Thordarson*, 1995]. Thus this outcrop records over 2 years of effusive activity. Note that this result assumes that the overlying flow arrived the moment the underlying flow stopped inflating, which may be a reasonable approximation. The contacts between Roza sheets are often welded together, suggesting that the overlying unit arrived at least before the underlying unit cooled completely.

While it is relatively straightforward to estimate the duration of activity at each outcrop, it is not as clear how to sum these durations. Unfortunately, it seems that no location records the entire Roza eruption because no single outcrop contains all the chemical subtypes found by *Martin* [1989, 1991]. To estimate the duration of the entire Roza eruption, we sum the longest recorded duration of each of the five lava flows identified within the Roza flow field (see Table 2; Figure 1). This leads to a total eruption duration of about 14 years (Table 4).

There are uncertainties in this calculation. We have already noted that estimated durations may be in error by as much as 25% due to differences in the cooling rates in the CRB versus Hawaii. It is also possible that more than one of the lava flows of the Roza flow field was active at some time in the eruption. Finally, the chemical differences between the flows are subtle, so it is possible that some flows are misidentified at some outcrops. However, because there is an outcrop (Horton Grade) that appears to record 6.4 years of activity, we feel that using the average durations of each lava flow (for a total eruption duration of 5.9 years (Table 4)), is excessively conservative. In any case, we feel confident

Figure 11. Sketch map of features along the Roza fissure and fissure segments thought to be related to each Roza lava flow (1-5) [after *Swanson et al.*, 1975; *Martin*, 1989; and work reported in this study]. PWH = Potter White Hill; WI = Winona; BB = Big Butte; WA = Washington; OR = Oregon.

TABLE 3. Physical and Chemical Properties of the Roza and Gingko Lava Flows, Wanapum Basalts, CRB.

	Physical			
Property	Dike selvage (MgO)	Lava selvage (MgO)	Dike selvage (CaO)	Lava selvage (CaO)
Roza flow				
Temperature[a] (°C)	1095	1084	1105	1100
Viscosity[b] (Pa s): 1% H_2O	200	380	175	190
0.5% H_2O	350	735	310	330
Gingko flow[c]				
Temperature (°C)	1090	1070		
Viscosity (Pa s)	630			

	Chemical									
Oxide	SiO_2	TiO_2	Al_2O_3	Fe_2O_3	MnO	MgO	CaO	Na_2O	K_2O	P_2O_5
Roza[d]	50.23	3.12	13.40	15.25	0.22	4.47	8.61	2.75	1.29	0.68
Gingko[e]	51.55	3.08	14.38	14.19*	0.23	4.16	8.03	2.34	1.23	0.58

[a] Temperature calculated using the empirical geothermometers of *Helz and Thornber* [1987] based on compositions (MgO or CaO) of Roza dike and lava flow glassy selvages [after *Thordarson*, 1995].

[b] Viscosity calculated using the method of *Bottinga and Weill* [1972] for 1.0 and 0.5 wt% H_2O, estimated reasonable values for Roza magma and lava, respectively [*Thordarson*, 1995].

[c] Temperature from *Ho and Cashman* [1995] based on MgO glass geothermometer. Viscosity from *Ho and Cashman* [1996].

[d] Average (n=73) whole-rock major element analysis of Roza chemical subtype IIA of *Martin* [1989], the most voluminous type, equivalent to flow 3 in this paper.

[e] Average (n=38) whole-rock major element analysis of Gingko flow [*Beeson et al.,* 1985].

* Total iron as FeO.

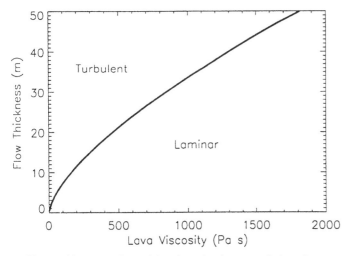

Figure 12. Plot of transition from laminar to turbulent flow as a function of flow thickness and lava viscosity. Curves computed for a slope of 0.1% (0.05°), appropriate for most of the flows in the CRB [*Shaw and Swanson*, 1970] and a reasonable value for other CFB provinces. Estimated viscosities for CRB lavas, for realistic eruption temperatures, a reasonable range of volatile contents, and corrected for entrained crystals and bubbles, range from 120 to 5500 Pa s. This plot indicates that only flows with initial emplacement thicknesses greater than 20-25 m in the viscosity range of the Roza lava could theoretically have flowed turbulently.

in suggesting that the emplacement of the 1300 km^3 Roza flow field took on the order of a decade.

Even with the eruption continuing over a period on the order of 10 years, the estimated volumetric flux of lava during the Roza eruption is very large when compared to historical eruptions. We calculate an average total eruption rate of ~4000 m^3/s for the Roza. This is roughly equal to the peak eruption rate of the 1783-84 Laki eruption in Iceland, the largest historical basaltic eruption [*Thordarson and Self*, 1993]. These high total eruption rates for the Roza are not unreasonable given the potential lengths of the fissures involved. A 4-km-long fissure segment active at one time would give average eruption rates on the order of 1 m^3/s per meter length of fissure (or ~3000 kg/s per meter length of fissure), which is a typical value for many Hawaiian eruptions and only about half the peak eruption rate of the Laki fissure. Fissure segments 4 km long need be active for only an average of 3.2 months in order to migrate across the 150-km-long Roza fissure system in 10 years. For comparison, the Laki fissure system was 27 km long and produced lava for 8 months, though ~60% of the lava erupted from five fissures totaling 13.5 km in length in just 1.5 months [*Thordarson and Self*, 1993].

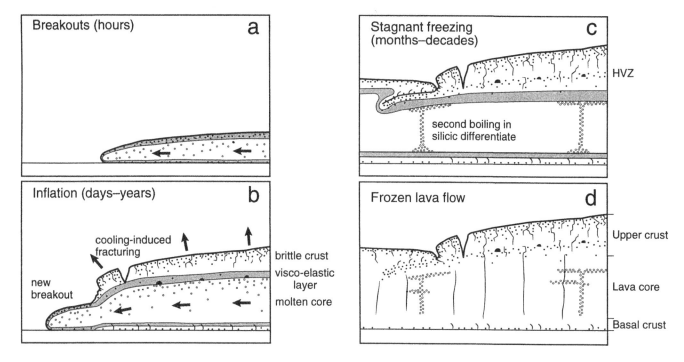

Figure 13. Schematic cross sections of emplacement of a generic inflating pahoehoe sheet flow. Vertical scale varies from 1-5 m for Hawaiian flows to 5-50 m for the CRB flows. (a) Flow arrives as a small, slow-moving, lobe of molten lava held inside a stretchable, chilled viscoelastic skin with brittle crust on top. Bubbles are initially trapped in both the upper and basal crusts. (b) Continued injection of lava into the lobe results in inflation (lifting of the upper crust) and new breakouts. During inflation, bubbles rising from the fluid core become trapped in the viscoelastic mush at the base of the upper crust, forming horizontal vesicular zones. The growth of the lower crust, in which pipe vesicles develop, is much slower. Relatively rapid cooling and motion during inflation results in irregular jointing in the upper crust. (c) After stagnation, diapirs of vesicular residuum form vertical cylinders and horizontal sheets within the crystallizing lava core. Slow cooling of the stationary liquid core forms more regular joints. (d) Emplacement history of flow is preserved in vesicle distribution and jointing pattern of frozen lava.

Data from the vent areas of the Roza also support analogies to Laki and other historical fissure eruptions. Significant amounts of volatiles were degassed at or near the vent [*Thordarson et al.*, 1996], so some mechanism to promote volatile loss, such as high fire-fountains, must have occurred. The dearth of vent edifices supports the notion of high fire-fountaining. Such fire-fountains generally do not produce high cones; instead they produce fields of welded spatter fallout and rootless lava flows, as in the 1986 Izu-Oshima eruption [*Sakaguchi et al.*, 1988].

More generally, the vent structures in the CRB are not extraordinary when compared with historical basaltic eruptive vents elsewhere. Dike widths are similar and spatter deposits are remarkably comparable in scale. Thus, historical fissure eruptions can probably be used as analogs of the Roza and other CRB fissures, even though the total length of the Roza system is larger than that of any known historical eruption. No evidence indicates that the activity producing the immense flows

of the CRB was significantly different than what has been observed historically, except that high total eruption rates were maintained for much longer because of the longer fissure system.

4.3. *Flow Field Evolution and Thermal Efficiency*

Working from the ideas proposed above, we produce a revised picture for the emplacement of the Roza flow field (Figure 14). We envision the advance of each Roza flow as being composed of a slowly advancing (~1-10 cm/s), broad front (many kilometers wide) of small (~20-50 cm tall) pahoehoe lobes and toes. These lobes inflate and coalesce over a distance of perhaps several tens of meters and over a time scale of hours. This part of the flow would be little different from the pahoehoe sheet flows in Hawaii [*Hon et al.*, 1994] or Iceland [*Thordarson and Self*, 1993], except for being significantly wider. The flow lobes would coalesce and continue to inflate, producing broad, flat sheet-like lobes

TABLE 4. Estimated Duration of the Roza Eruption Using Equation (1) Based on the Average and Maximum Crustal Thicknesses Measured in Roza Lava Flows.

Lava flow	Maximum t (years)	Average t (years)	Location of maximum t	Other locations
1	0.57	0.44	RSF	RAC, RBB
2	3.84	1.69	RPR	RBB, RBL, RHG, RDF, RSF, RVL, RWV, RSC
3	1.26	0.58	RMC	RBB, RBL, RDF, RSF, RSL
4	4.22	1.83	RHG	RDC, RFS, RPR
5	4.22	1.33	RC	RDC, RLG
Total years	14.10	5.90		

See Tables 1 and 2 for the location key and measured crust thicknesses.

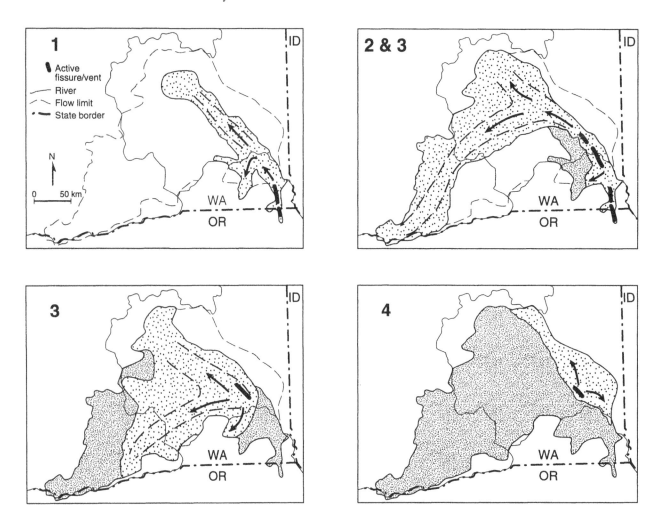

Figure 14. Cartoon map-view of a possible scenario for the development of the Roza compound pahoehoe flow-field based on work of *Martin* [1989; 1991] and *Thordarson* [1995]. Lava travels from the vent under an insulating crust in a preferred pathway (dashed lines). This flux of lava is used to both inflate the sheet flow and to feed new breakouts, which are virtually the only place where incandescent lava is exposed. Active flows are diverted around earlier-formed flows unless the active flows are inflated to a thickness greater than earlier units. 1-5 = the five major flows that form the Roza flow field; active fissure segment at each stage shown by thick bar.

up to several kilometers wide. These sheet lobes would continue to inflate and feed the lava flow front for many months, reaching a final thickness of up to 50 m or more. It is possible that several sheet flows could be fed simultaneously from different fissure segments. As any given lava flow became inactive, as a result of either its section of the fissure shutting down or having the lava diverted to a new flow, it would become a temporary barrier for new flows from either the same or subsequent eruptions.

This picture must still answer the primary constraint on the previous emplacement model of *Shaw and Swanson* [1970]; the lava must not cool significantly during its transport across hundreds of kilometers of surface flowage. *Ho and Cashman* [1995] suggested on the basis of three data points that the Gingko flow of the Frenchman Springs Member of the CRB cooled an average of 0.06-0.11°C/km (Table 3). *Keszthelyi* [1995b] showed that this kind of thermal efficiency would theoretically be feasible for lava tubes in the CRB to achieve. However, we have found no evidence so far for cylindrical, drained lava tubes in the CRB.

We do not find the lack of evidence of lava tubes problematic. On the shallow slopes at the time of CRB flow emplacement, it is unlikely that lava tubes could have drained. Furthermore, inflating sheet flows are expected to form broad preferred internal pathways with elliptical cross sections, not cylindrical conduits. This is commonly the case on the low slopes of the coastal flats on Kilauea and was proposed much earlier for the Buckboard Mesa flow, Nevada, by *Lutton* [1969]. Thermal modeling of such sheet-like preferred pathways suggests that lava can be transported hundreds of kilometers with no detectable cooling [*Keszthelyi and Self*, 1996].

5. POTENTIAL FOR ATMOSPHERIC EFFECTS FROM FLOOD BASALT ERUPTIONS

For flood basalt eruptions to have had widespread climatic impact, the gases emitted by the volcanic activity must have reached high enough into the atmosphere to be transported widely around the Earth. The height to which volcanic gases can be carried from a fissure eruption depends critically on the mass eruption rate per unit length of fissure [*Stothers et al.*, 1986; *Stothers*, 1989], the volatile content of the magma [*Wilson and Head*, 1981], and the moisture content of the atmosphere at the time [*Woods*, 1993]. Because there are so many variables when modeling an ancient eruption like that which formed the Roza flow, at

present we can form only a general idea of the heights to which plumes might reach above the fissures. At eruption rates of $\sim 10^2$ kg/s per meter length of fissure, estimated plume heights would be 3-6 km above the fire-fountains; and at $\sim 3 \times 10^3$ kg/s per meter length of fissure, 8-11 km [*Woods*, 1993]. Mass eruption rates along the fissures may have waxed and waned over two orders of magnitude, as at the 1783 Laki eruption [*Thorarinsson*, 1968; *Thordarson and Self*, 1993], and thus significantly higher plumes would be expected at some periods in the eruption.

Clearly, from the first-order considerations presented above, it is plausible that the Roza, and presumably other flood basalt eruptions, could inject volcanic gases into the uppermost troposphere and even the lower stratosphere (presently ~ 12-13 km altitude at mid-latitudes). Once aloft, the volatiles (the most important of which is sulfur as SO_2 or H_2S) would behave much like the products of any volcanic eruption and form sulfate aerosols. Concentrations of S, F, and Cl were measured in glassy samples and glass inclusions from phenocrysts collected from dikes, near-vent spatter, lava selvages, and lava cores (Figure 15) [*Thordarson and Self*, 1996]. These analyses and a mass balance calculation show that 66% of the total sulfur was lost during the eruption process and that a significant fraction was also degassed during the flow and crystallization of the lava. The degassing during crystallization would not have formed an upper atmospheric aerosol perturbation as it would have been confined to the boundary layer of the lower troposphere, and it should not have been transported globally. However, the local and regional impact of such volcanic smog can be dramatic, as it was during the Laki eruption [*Thorarinsson*, 1981; *Thordarson and Self*, 1997].

The estimated mass of volatiles degassed at the vents during the Roza eruption is staggering; 9,000 Mt of SO_2, 1300 Mt of HF, and 400 Mt of HCl. For comparison, the 1991 Mount Pinatubo eruption released an estimated 20 Mt of SO_2 [e.g., *McCormick et al.*, 1995]. Thus, if the Roza flow field was erupted over a period of 10 years, it would be approximately equivalent to four times the Pinatubo upper atmospheric SO_2 perturbation every month, maintained for a decade. Along with sulfuric acid aerosols, the formation of acid droplets in the lower atmosphere from the F and Cl would lead to extensive acid rain. It should also be noted that considerably more voluminous flows (up to 4000 km³) are known from regional mapping and correlation in the Grande Ronde Basalt Formation [*Reidel et al.*, 1989] and that initial analyses on other Wanapum Formation basalts also

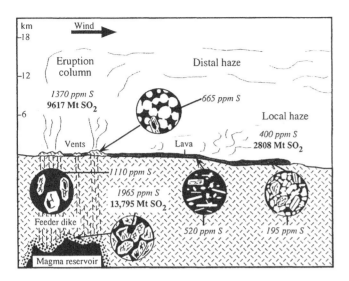

Figure 15. Schematic illustration of Roza eruption (not to scale) and degassing budget, based on example of Laki eruption [*Thordarson et al.*, 1996]. The amount of S retained in samples from various eruption stages and the total amount of SO_2 dissolved in the Roza magma prior to eruption are given, as well as the estimated SO_2 yield at the vents (causing a widespread (distal) haze or aerosol cloud) and from the lava flow (causing a low altitude (local) haze or dry fog).

indicate the potential for large sulfur releases during those eruptions [*Ewart*, 1987].

The climatic and environmental consequences of such emissions maintained over periods of years to decades cannot at present be modeled with any confidence, but must have been severe. Whether the atmospheric and environmental impact of flood basalt volcanism alone could cause mass extinctions is not yet clear. Certainly, individual flood lava events lasting even decades are unlikely to have had such an extreme effect, but, as typified by the CRB, flood basalt provinces are composed of hundreds of such eruptions. During the period of peak output, 50% of the Grande Ronde lavas were emplaced in approximately 300,000 years from about 16.0 to 15.7 Ma [*Baksi*, 1989]. Even in this time of peak activity, individual eruptions would have average recurrence intervals of 5,000-10,000 years, which may have given the environment sufficient time to recover between eruptions. The most reasonable statement, given current knowledge, is that a continental flood basalt eruption probably could not cause mass extinctions, but a series of them during the growth of a CFB province would have been able to stress the environment to such an extent that any other major perturbation would have had a more extreme effect.

6. CONCLUSIONS

We have shown what we believe to be incontrovertible field evidence that the Roza Member of the CRB is a compound pahoehoe flow field. We also find pahoehoe flow fields to be common throughout the CRB and in other flood basalt provinces. This evidence is in the form of surface textures, morphologic features at the flow tops, and the internal structures within the lava flows. The three-part internal structure we have described herein is ubiquitous in inflated pahoehoe lava flows of all scales across the globe and is different from the internal structure of rapidly emplaced aa or aa-like lava flows. We have found no evidence for rapid, turbulent emplacement as envisioned by *Shaw and Swanson* [1970] in the CRB. Instead, the features we have described in the Roza and other CFB lava flows are easily explained by their emplacement as inflating sheet flows. An emplacement model based on pahoehoe flow fields in Hawaii, Iceland, and elsewhere is able to explain the gross geometry, internal structure, and evidence for minimal cooling observed in the Roza and other CFB lava flows.

Having said this, we caution the reader that our simple descriptions and cartoon drawings are not intended to explain, in detail, every feature seen in every outcrop. For example, some flows in the CRB have rubbly upper crusts that may have formed in a manner transitional between pahoehoe and aa, (e.g., slabby pahoehoe). Instead, our model is intended to show that the inflation process is generally applicable to flood basalt lava flows. Detailed studies are needed to decipher lava flow emplacement history at specific locations.

Examination of the physical features of the Roza Member has allowed us to quantitatively speculate on its emplacement. An emplacement model, combined with the cooling model of *Hon et al.* [1994], permits an estimate of the duration of active flow recorded at each outcrop of the Roza flow field. Translating these local durations to the total eruption duration is not straightforward. We expect that many of the lava flows at various outcrops were active simultaneously, and we have evidence that no single outcrop records the entire Roza eruption. However, because the examined outcrops record activity for up to 6.4 years, we suggest that the Roza eruption lasted for about a decade.

Using a 10-year eruption duration, we arrive at an average effusion rate for the Roza of ~4000 m^3/s. This average rate is as high as the peak rate of the largest historical basaltic eruption, the 1783-1784 Laki eruption in Iceland. Because the Roza fissure system is 150 km long, the high total effusion rate can be accommodated by moderate activity along a small fraction of the fissure system. We therefore suggest that the Roza eruption probably consisted of a number of shorter fissure segments. Each segment may have fountained for just a few months before activity migrated to a different part of the fissure system.

The measured release of immense volumes of sulfur, chlorine, and fluorine at flood basalt vents over periods on the order of a decade may have had a strong, detrimental effect on global climate. This provides a plausible, though not yet proven, process to link continental flood basalt eruptions to mass extinctions [*Courtillot*, 1994].

Whereas the causes of flood basalt volcanism and the sources of magmas that form flood basalt provinces appear to be relatively well understood [e.g., *Carlson*, 1991], our knowledge of the physical processes occurring during flood basalt events, such as venting mechanisms, lava flow emplacement, and degassing mechanisms, is much poorer. Our initial studies on CRB lava flows strongly suggest that the previous model of enormous flow rates and catastrophic eruption mechanisms of flood basalt lavas is largely untenable and that CFB lavas in general were emplaced more gradually as pahoehoe sheet flows forming extensive lava flow fields, though still at very high total effusion rates when compared to most basaltic eruptions witnessed by man.

We hope that this work will spur further, more detailed studies on the physical volcanology of flood basalt lava flows.

Acknowledgments. Support for this work was provided by NSF grants EAR-9118755 and 9316881, NASA grants NAG5-1839 and NAGW-3721, by a NASA Global Change Fellowship to ThTh, and by an NSF post-doctoral fellowship to LPK. We thank George Walker, Mark Murphy, Phil Long, Mike Rampino, Steve Reidel, Terry Tolan, and Sara Finnemore for various inputs to this study. Ken Hon and John Wolff are thanked for critical and helpful reviews. This is SOEST contribution no. 4169.

REFERENCES

Agashe, L. V., and R. B. Gupte, Mode of eruption of the Deccan Traps basalts, *Bull. Volcanol., 35*, 591-601, 1971.

Anderson, D. L., Sublithospheric mantle as the source of continental flood basalts - the case against the continental lithosphere and plume head reservoirs, *Earth Planet. Sci. Lett., 123*, 269-280, 1994.

Arndt, N. T., G. K. Czamanske, J. L. Wooden, and V. A. Fedorenko, Mantle and crustal contributions to continental flood volcanism, *Tectonophysics, 223*, 39-52, 1993.

Atkinson, F. A., Some remarkable features of flows from Undara, *AGU Chapman Conference on Long Lava Flows, Conference Abstract Volume*, James Cook University of North Queensland, Townsville, pp. 4-5, 1996.

Aubele, J. C., L. S. Crumpler, and W. E. Elson, Vesicle zonation and vertical structure of basalt flows, *J. Volcanol. Geotherm. Res., 35*, 349-374, 1988.

Baksi, A. K., Reevaluation of the timing and duration of extrusion of the Imnaha, Picture Gorge, and Grande Ronde Basalts, Columbia River Basalt Group, in *Volcanism and Tectonism in the Columbia River Flood-Basalt Province, Spec. Pap. 239*, edited by S. P. Reidel, and P. R. Hooper, pp. 1-20, Geological Society of America, Boulder, CO, 1989.

Bates, R. L., and J. A. Jackson (Eds), *Glossary of Geology*, 3rd edition, 788 pp., American Geological Institute, Alexandria, VA, 1987.

Beeson, M. H., K. R. Fecht, S. P. Reidel, and T. L. Tolan, Regional correlations within the Frenchman Springs Member of the Columbia River Basalt Group: New insights into the middle Miocene tectonics of northwestern Oregon, *Oregon Geol., 47*, 87-96, 1985.

Bingham, J. W., and K. L. Walters, Stratigraphy of the upper part of the Yakima Basalt in Whitman and eastern Franklin Counties, Washington, *U.S. Geol. Surv. Prof. Pap. 525-C*, 87-90, 1965.

Bottinga, Y., and D. F. Weill, The viscosity of magmatic silicate liquids: A model for calculation, *Am. J. Sci., 272*, 438-475, 1972.

Campbell, I. H., and R. W. Griffiths, Implications of mantle plume structure for the evolution of flood basalts, *Earth Planet. Sci. Lett., 99*, 79-93, 1990.

Carlson, R. W., Physical and chemical evidence for the cause and source characteristics of flood basalt volcanism, *Austr. J. Earth Sci., 38, 525-544*, 1991.

Chitwood, L. A., Inflated basaltic lava: examples of processes and landforms from central and southeast Oregon, *Oregon Geol., 56*, 11-21, 1994.

Courtillot, V. E., Mass extinctions in the last 300 million years: one impact and seven flood basalts, *Israeli J. Earth Sci., 43*, 259-266, 1994.

Courtillot, V. E., G. Féraud, H. Maluski, D. Vandamme, M. G. Moreau, and J. Besse, The Deccan flood basalts and the Cretaceous/Tertiary boundary, *Nature, 333*, 843-846, 1988.

Courtillot, V. E., J. Besse, D. Vandamme, R. Montigny, J. J. Jaeger, and H. Capetta, Deccan flood basalts at the Cretaceous-Tertiary boundary, *Earth Planet. Sci. Lett., 80*, 361-374, 1986.

Crisp, J., and S. Baloga, Influence of crystallization and entrainment of cooler material on the emplacement of basaltic aa lava flows, *J. Geophys. Res., 99*, 11,819-11,831, 1994.

Ewart, J. W., Sulfur in the Frenchman Springs member of the Wanapum Basalt in Washington and Oregon, *Geol. Soc. Amer.*

Abstracts with Programs, 19, Cordilleran Section, 376, 1987.

Finnemore, S. L., S. Self, and G. P. L. Walker, Inflation features in lava flows of the Columbia River Basalts (abstract), *Eos Trans. AGU, 74(46)*, Fall Meeting Suppl., 555, 1993.

Goff, F., Vesicle cylinders in vapor-differentiated basalt flows, *J. Volcanol. Geotherm. Res., 71*, 167-185, 1996.

Gregg T. K. P., and J. H. Fink, Quantification of submarine lava-flow morphology through analog experiments, *Geology, 23*, 73-76, 1995.

Greenhough J. D., and J. Dostal, Cooling history and differentiation of a thick North Mountain Basalt flow (Nova Scotia, Canada), *Bull. Volcanol., 55*, 63-73, 1992.

Griffiths, R. W., and J. H. Fink, Solidification and morphology of submarine lavas: A dependence on extrusion rate, *J. Geophys. Res., 97*, 19 729-19 737, 1992.

Helz, R. T., and C. R. Thornber, Geothermometry of Kilauea Iki lava lake, Hawaii, *Bull. Volcanol., 49*, 651-668, 1987.

Ho, A., and K. V. Cashman, Geothermometry of the Gingko Flow, Columbia River Basalt Group (abstract), *Eos Trans AGU, 76(46)*, Fall Meeting Suppl., 679, 1995.

Ho, A., and K. V. Cashman, Temperature constraints on a flow of the Columbia River Basalt Group (abstract), *AGU Chapman Conference on Long Lava Flows, Conference Abstract Volume*, James Cook University of North Queensland, Townsville, p. 22-23, 1996.

Hon, K., J. Kauahikaua, R. Denlinger, and K. Mackay, Emplacement and inflation of pahoehoe sheet flows: Observations and measurements of active lava flows on Kilauea Volcano, Hawaii, *Geol. Soc. Am. Bull., 106*, 351-370, 1994.

Hooper, P. R., The Columbia River Basalts, *Science, 215*, 1463-1468, 1982.

Hooper P. R., and C. J. Hawkesworth, Isotopic and geochemical constraints on the origin and evolution of Columbia River Basalts, *J. Petrol., 34*, 1203-1246, 1993.

Hooper, P. R., W. D. Kleck, C. R. Knowles, S. P. Reidel, and R. L. Thiessen, Imnaha Basalt, Columbia River Basalt Group, *J. Petrol., 25*, 473-500, 1984.

Jarvis, R. A., On the cross-sectional geometry of thermal erosion channels formed by turbulent lava flows, *J. Geophys. Res., 100*, 10,127-10,140, 1995.

Kent, R. W., M. Storey, A. D. and Saunders, Large igneous provinces—sites of plume impact or plume incubation, *Geology, 20*, 891-894, 1992.

Keszthelyi, L., Measurements of the cooling at the base of pahoehoe flows, *Geophys. Res. Lett., 22*, 2195-2198, 1995a.

Keszthelyi, L., A preliminary thermal budget for lava tubes on the Earth and planets, *J. Geophys. Res. 100*, 20 411-20 420, 1995b.

Keszthelyi, L., and R. Denlinger, The initial cooling of pahoehoe lava flows, *Bull. Volcanol., 58*, 5-18, 1996.

Keszthelyi, L. P., and D. C. Pieri, Emplacement of the 75-km-long Carrizozo lava flow field, south-central New Mexico, *J. Volcanol. Geotherm. Res., 59*, 59-75, 1993.

Keszthelyi, L., and S. Self, Some thermal and dynamical considerations for the emplacement of long lava flows, *AGU Chapman Conference on Long Lava Flows, Conference Abstract Volume*, James Cook University of North Queensland, Townsville, pp. 36-38, 1996.

Keszthelyi, L., S. Self, and Th. Thordarson, Application of recent studies on the emplacement of basaltic lava flows to the Deccan Traps, *Mem. Geol. Soc. India*, in press, 1997.

Kilburn, C. R. J., Pahoehoe and aa lavas: A discussion and continuation of the model of Peterson and Tilling, *J. Volcanol. Geotherm. Res., 11*, 373-382, 1981.

Kilburn, C. R. J., and R. M. C. Lopes, General patterns of flow field growth: aa and blocky lavas, *J. Geophys. Res., 96*, 19,721-19,732, 1991.

Landon, R. D., and P. E. Long, Detailed stratigraphy of the N2 Grande Ronde Basalt, Columbia River Basalt Group, in the central Columbia Plateau, in *Volcanism and Tectonism in the Columbia River Flood-Basalt Province, Spec. Pap. 239*, edited by S. P. Reidel, and P. R. Hooper, pp. 55-66, Geological Society of America, 1989.

Long, P. E., and B. J. Wood, Structures, textures, and cooling histories of Columbia River basalt flows, *Geol. Soc. Am. Bull., 97*, 1144-1155, 1986.

Lutton, R. J., Internal structure of the Buckboard Mesa Basalt, *Bull. Volcanol., 33*, 579-593, 1969.

Macdougall, J. D. (Ed.), *Continental Flood Basalts*, 341 pp., Kluwer Academic Publishers, Dordrecht, 1988.

Mackin, J. H., A stratigraphic section in the Yakima Basin and the Ellensburg Formation in south-central Washington, *Wash. Div. Mines Geol., Dept. Conserv., Rept. Invest. 19*, 45 pp., 1961.

Manga, M., Waves of bubbles in basaltic magmas and lavas, *J. Geophys. Res., 101*, 17,457-17,465, 1996.

Mangan, M. T., K. V. Cashman, and S. Newman, Vesiculation of basaltic magma during eruption, *Geology, 21*, 157-160, 1993.

Mangan, M. T., T. L. Wright, D. A. Swanson, and G. R. Byerly, Major oxide, trace element, and glass chemistry pertinent to regional correlation of Grande Ronde Basalt flows, Columbia River Basalt Group, Washington, *U.S. Geol. Surv. Open-File Rep. 85-747*, 74 pp., 1985.

Mangan, M. T., T. L. Wright, D. A. Swanson, and G. R. Byerly, Regional correlation of Grande Ronde Basalt flows, Columbia River Basalt Group, Washington, Oregon, and Idaho, *Geol. Soc. Am. Bull., 97*, 1300-1318, 1986.

Martin, B. S., The Roza Member, Columbia River Basalt Group: Chemical stratigraphy and flow distribution, in *Volcanism and Tectonism in the Columbia River Flood-Basalt Province, Spec. Pap. 239*, edited by S. P. Reidel, and P. R. Hooper, pp. 85-104, Geological Society of America, 1989.

Martin, B. S., Geochemical variations within the Roza member, Wanapum basalt, CRBG: Implications for the magmatic process affecting continental flood basalts, unpublished Ph.D. thesis, University of Massachusetts, 513 pp., 1991.

Mattox, T. N., C. Heliker, J. Kauahikaua, and K. Hon, Development of the 1990 Kalapana flow field, Kilauea Volcano, Hawaii, *Bull. Volcanol., 55*, 407-413, 1993.

McCormick, M. P., L. W. Thompson, C. R. Trepte,

Atmospheric effects of the Mt Pinatubo eruption, *Nature, 373*, 399-404, 1995.

McMillan, K., P. E. Long, and R. W. Cross, Vesiculation in Columbia River basalts, in *Volcanism and Tectonism in the Columbia River Flood-Basalt Province, Spec. Pap. 239,* edited by S. P. Reidel, and P. R. Hooper, pp. 157-167, Geological Society of America, 1989.

Nichols, R. L., Flow-units in basalt, *J. Geol., 44*, 617-630, 1936.

Peng Z. X., J. J. Mahoney, P. Hooper, C. Harris, and J. Beane, A role for lower continental crust in flood basalt genesis? Isotopic and incompatible element study of the lower six units of the western Deccan Traps, *Geochim. Cosmochim. Acta, 58*, 267-288, 1994.

Peterson, D. W., and R. I. Tilling, Transition of basaltic lava from pahoehoe to aa, Kilauea Volcano, Hawaii: Field observations and key factors, *J. Volcanol. Geotherm. Res., 7*, 271-293, 1980.

Phadke, A. V., and Sukhtankar, R. K., Topographic studies of Deccan Trap hills around Poona, India, *Bull. Volcanol., 35*, 709-718, 1971.

Pinkerton, H., and R. S. J. Sparks, The 1975 sub-terminal lavas, Mount Etna: a case history of the formation of a compound lava field, *J. Volcanol. Geotherm. Res., 1*, 167-182, 1976.

Pinkerton, H., and R. Stevenson, Methods of determining the rheological properties of lava from their physico-chemical properties, *J. Volcanol. Geotherm. Res., 53*, 47-66, 1992.

Puffer, J. H., and D. L. Horter, Origin of pegmatitic segregation veins within flood basalts, *Geol. Soc. Am. Bull., 105*, 738-748, 1993.

Rampino, M. R., and R. B. Stothers, Flood basalt volcanism during the past 250 million years, *Science, 241*, 663-668, 1988.

Reidel, S. P., Stratigraphy and petrogenesis of the Grande Ronde Basalt from the deep canyon country of Washington, Oregon, and Idaho, *Geol. Soc. Am. Bull., 94*, 519-542, 1983.

Reidel, S. P., and T. L. Tolan, Eruption and emplacement of flood basalt: An example from the large-volume Teepee Butte Member, Columbia River Basalt Group, *Geol. Soc. Am. Bull., 104*, 1650-1671, 1992.

Reidel, S. P., T. L. Tolan, P. R. Hooper, M. H. Beeson, K. R. Fecht, R. D. Bentley, and J. L. Anderson, The Grande Ronde Basalt, Columbia River Basalt Group: stratigraphic descriptions and correlations in Washington, Oregon, and Idaho, in *Volcanism and Tectonism in the Columbia River Flood-Basalt Province, Spec. Pap. 239,* edited by S. P. Reidel, and P. R. Hooper, pp. 21-53, Geological Society of America, 1989.

Renne, P. R., M. Ernesto, I. G. Pacca, R. S. Coe, J. M. Glen, M. Prévot, and M. Perrin, The age of Parana flood volcanism, rifting of Gondwanaland, and the Jurassic-Triassic boundary, *Science, 258*, 975-981, 1992.

Renne, P. R., Z. Zichao, M. A. Richards, M. T. Black, and, A. R. Basu, Synchrony and Causal Relations between Permian-Triassic Boundary Crises and Siberian Flood Volcanism, *Science, 269*, 1413-1416, 1995.

Rowland, S. K., and G. P. L. Walker, Pahoehoe and aa in Hawaii: Volumetric flow rate controls the lava structure, *Bull. Volcanol., 52*, 615-628, 1990.

Sakaguchi, K., A. Takada, K. Uto, and T. Soya, The 1986 eruption and products of Izu-Oshima Volcano, Japan, *Bull. Volcanol. Soc. Japan, 33*, Special Issue, Prof. K. Nakamura Memorial Volume, 20-32, 1988 (in Japanese).

Schminke, H. U., Stratigraphy and petrography of four Upper Yakima Basalt Flows in south-central Washington, *Geol. Soc. Am. Bull., 78*, 1385-1422, 1967.

Self, S., Th. Thordarson, L. Keszthelyi, G. P. L. Walker, K. Hon, M. T. Murphy, P. Long, and S. Finnemore, A new model for the emplacement of Columbia River Basalts as large inflated pahoehoe lava flow fields, *Geophys. Res. Lett., 23*, 2,689-2,692, 1996.

Shaw, H. R., Visocosities of magmatic silicate liquids: an empirical method of prediction, *Am. J. Sci., 272*, 870-893, 1972.

Shaw, H. R., and D. A. Swanson, Eruption and flow rates of flood basalts, in *Proc. Second Columbia River Basalt Symposium,* edited by E. H. Gilmour and D. Stradling, pp. 271-299, Eastern Washington State College Press, Cheney, 1970.

Snavely, P. D., N. S. MacLeod, and H. C. Wagner, Miocene tholeiitic basalts of coastal Oregon and Washington and their relations to coeval basalts of the Columbia plateau, *Geol. Soc. Am. Bull., 84*, 387-424, 1973.

Stephenson, P. J., Burch-Johnson, D. Stanton, Long lava flows in North Queensland—context, characteristics, emplacement (abstract), *AGU Chapman Conference on Long Lava Flows, Conference Abstract Volume,* James Cook University of North Queensland, Townsville, pp. 86-87, 1996.

Stothers, R. B., Turbulent atmospheric plumes above line sources with an application to volcanic fissure eruptions on the terrestrial planets, *J. Atmos. Sci., 46*, 2662-2670, 1989.

Stothers, R. B., Flood basalts and extinction events, *Geophys. Res. Lett, 20*, 1399-1402, 1993.

Stothers, R. B, J. A. Wolff, S. Self, and M. R. Rampino, Basaltic fissure eruptions, plume heights, and atmospheric aerosols, *Geophys. Res. Lett, 13*, 725-728, 1986.

Swanson, D. A., Yakima Basalt of the Tieton River area, south-central Washington, *Geol. Soc. Am. Bull., 78*, 1077-1110, 1967.

Swanson, D. A., and T. L. Wright, The regional approach to studying the Columbia River Basalt Group, *Memoir Geol. Soc. India, 3*, 58-80, 1980.

Swanson, D. A., T. L. Wright, and R. T. Helz, Linear vent systems and estimated rates of magma production and eruption for the Yakima Basalt on the Columbia Plateau, *Am. J. Sci., 275*, 877-905, 1975.

Swanson D. A., T. L. Wright, P. R. Hooper, and R. D. Bentley, Revisions in stratigraphic nomenclature of the Columbia River Basalt Group, *U.S. Geol. Surv. Bull., 1457-G*, 59 pp., 1979.

Theilig, E., Formation of Pressure Ridges and Emplacement of Compound Basaltic Lava Flows, Ph.D. thesis, Arizona State University, Tempe. 212 pp., 1986.

Thorarinsson, S., On the rate of lava and tephra-production and upward migration of magma in four Icelandic eruptions, *Geol.*

Rundsch., 57, 705-717, 1968.

Thorarinsson, S., Greetings from Iceland: ash-falls and volcanic aerosols in Scandinavia, *Geograf. Ann., 18*, 63,109-63,118, 1981.

Thordarson, Th., Volatile release and atmospheric effects of basaltic fissure eruptions, Ph.D. thesis, University of Hawaii at Manoa, Honolulu, 580 pp., 1995.

Thordarson, Th., and S. Self, The Laki (Skaftár Fires) and Grímsvötn eruptions in 1783-1785, *Bull. Volcanol., 55*, 233-263, 1993.

Thordarson, Th., and S. Self, Sulfur, chlorine and fluorine degassing and atmospheric loading by the Roza eruption, Columbia River Basalt Group, Washington, *J. Volcanol. Geotherm. Res.*, 74, 49-73, 1996.

Thordarson Th., and S. Self, Atmospheric and environmental effects of the 1783-84 Laki eruption, *Global and Planetary Change*, 1997, in revision.

Thordarson Th., S. Self, N. Oskarsson, T. Hulsebosch, Sulfur, chlorine and fluorine degassing and atmospheric loading by the 1783-84 Laki (Skaftar Fires) eruption in Iceland, *Bull. Volcanol.*, 58, 205-225, 1996.

Tolan, T. L., S. P. Reidel, M. H. Beeson, J. L. Anderson, K.R. Fecht, and D.A. Swanson, Revisions to the estimates of the areal extent and volume of the Columbia River Basalt Group, in *Volcanism and Tectonism in the Columbia River Flood-Basalt Province, Spec. Pap. 239*, edited by S. P. Reidel, and P. R. Hooper, pp. 1-20, Geological Society of America, 1989.

Turner, S., and C. Hawkesworth, The nature of the sub-continental mantle: Constraints from the major element composition of Continental Flood Basalts, *Geology, 120*, 295-314, 1995.

Walker, G. P. L., Compound and simple lava flows and flood basalts, *Bull. Volcanol., 35*, 579-590, 1971.

Walker, G. P. L., Pipe vesicles in Hawaiian basaltic lavas: Their origin and potential as a paleoslope indicator, *Geology, 15*, 84-89, 1987.

Walker, G. P. L., Spongy pahoehoe in Hawaii: a study of vesicle distribution patterns in basalt and their significance, *Bull. Volcanol., 51*, 199-209, 1989.

Walker, G. P. L., Structure, and origin by injection under surface crust, of tumuli, "lava rises," "lava-rise pits," and "lava inflation clefts" in Hawaii, *Bull. Volcanol., 53*, 546-558, 1991.

Waters, A. C., Stratigraphic and lithologic variations in the Columbia River Basalt, *Am. J. Sci., 259*, 583-611, 1961.

White, R. S, and D. McKenzie, Magmatism at rift zones: the generation of volcanic continental margins and flood basalts, *J. Geophys. Res., 94*, 7685-7729, 1989.

Whitehead, J. A., and K. R. Helfrich, Instability of flow with temperature-dependent viscosity: A model of magma dynamics, *J. Geophys. Res., 96*, 4145-4155, 1991.

Wilson, L. and J. V. Head, Ascent and eruption of basaltic magma on the Earth and Moon. *J. Geophys. Res., 86*, 2971-3001, 1981.

Woods, A. W., A model of the plumes above basaltic fissure eruptions, *Geophys. Res. Lett., 20*, 1115-1118, 1993.

Wright, T. L., M. J. Grolier, D. A. Swanson, Chemical variation related to the stratigraphy of the Columbia River Basalt, *Geol. Soc. Am. Bull., 84*, 371-386, 1973.

Wright, T. L., M. T. Mangan, and D. A. Swanson, Chemical data for flows and feeder dikes of the Yakima Basalt Subgroup, Columbia River Group, Washington, Oregon, Idaho, and their bearing on a petrogenetic model, *U.S. Geol. Surv. Bull., 1821*, 71 pp., 1989.

Laszlo Keszthelyi, Stephen Self, Thorvaldur Thordarson, Department of Geology and Geophysics, School of Ocean and Earth Science and Technology, University of Hawaii, 2525 Correa Rd., Honolulu, HI 96822 USA.

Large Igneous Provinces: A Planetary Perspective

James W. Head, III

Department of Geological Sciences, Brown University, Providence, Rhode Island

Millard F. Coffin

Institute for Geophysics, The University of Texas at Austin, Austin, Texas

Large igneous provinces (LIPs) are common on the Moon, Mars and Venus, and their presence, characteristics, and geologic and temporal settings offer a potentially important perspective for interpreting LIPs on Earth. On the Moon, shallow magma reservoirs and large shield volcanoes are unknown. The relatively low-density, thick anorthositic crust creates a density trap for rising basaltic magmas which are thought to collect in reservoirs at the base of the crust; reservoir overpressurization causes dikes to propagate to the surface. Dikes sufficiently large to reach the surface are likely to result in large-volume, high-effusion-rate eruptions; single eruptive phases are predicted theoretically and observed in the maria to be several hundred to over 10^3 km^3. On Mars, massive shield volcanoes have formed on the stable lithosphere over hot spots lasting over a billion years; shield heights are up to 25 km above the adjacent plains. Volumes of single edifices are of the order of 1.5 x 10^6 km^3, comparable to the total volumes of many basalt provinces on Earth. The impact cratering record on Venus suggests that Venus underwent rapid and massive planet-wide volcanic resurfacing about 300 m.y. ago, an event possibly related to the overturn of a depleted mantle layer resulting from the vertical accretion of a basaltic crust. This hypothesized event could be the equivalent of a planet-wide LIP and underlines the possibility of episodic and catastrophic LIPs throughout planetary history, resembling mantle overturn events proposed for Earth. The planetary record, in concert with the detailed examination of examples on Earth, can be of use in developing and testing models for the emplacement of LIPs, and may help to distinguish plate tectonic influences from those linked to deeper interior (mantle and core) processes.

INTRODUCTION AND BASIC CHARACTERISTICS OF TERRESTRIAL LARGE IGNEOUS PROVINCES

Recently, attention has been drawn to large igneous provinces on Earth, which are defined as regions of voluminous emplacement of predominantly mafic extrusive and intrusive rock whose origins lie in processes other than "normal" seafloor spreading [e.g., *Coffin and Eldholm*, 1992]. Large igneous provinces are characterized by transient large-scale intrusive and extrusive activity, including continental flood basalt (CFB) provinces (e.g., the Deccan Traps), volcanic passive margins (e.g., the Vøring Margin), oceanic plateaus (e.g., the Ontong Java Plateau), ocean basin flood basalts (e.g., the Caribbean Flood Basalts), and large seamount chains (e.g., Hawaiian-

Large Igneous Provinces: Continental, Oceanic, and Planetary Flood Volcanism
Geophysical Monograph 100

Emperor) [*Coffin and Eldholm*, 1992]. Commonly analyzed separately in the past, recent studies [e.g., *Coffin and Eldholm*, 1992, 1993] have shown that there are important temporal, spatial, and compositional relationships among terrestrial large igneous provinces, informally referred to as LIPs.

These studies, and numerous others that document individual occurrences (see references in this volume and those of *Coffin and Eldholm* [1994]), show that the genesis and evolution of LIPs are closely linked to mantle dynamics, that some LIPs represent major global events (large volumes of lava and associated intrusions are often produced in short episodes, which had potentially major effects on the global environment), and that emplacement of some LIPs may be related to changes in rate and direction of plate motion. Their formation may be episodic, but modification and destruction of older examples, and sedimentation and inaccessibility of others, makes this difficult to determine. Although several models have been proposed for the emplacement of LIPs (primarily associated with mantle plumes) [e.g., *White and McKenzie*, 1989; 1995; *Griffiths and Campbell*, 1991; *Larson*, 1991a, b], these models are not yet well constrained by observations. At present only a limited, but growing (see articles in this volume), amount of quantitative data is available to assess associated mantle and crustal processes; to determine LIP dimensions, durations, rates of emplacement, crustal structure, and relationship to tectonism; and to predict environmental effects of LIP formation. For example, recent workers [*Self et al.*, 1996, and this volume] have presented evidence that Columbia River flood basalt lavas may have been emplaced more gradually as inflating pahoehoe sheet flows forming very extensive flow fields rather than single very high-effusion-rate eruptions.

THE PERSPECTIVE FROM THE PLANETARY GEOLOGICAL RECORD

Large igneous provinces are also common on the terrestrial planetary bodies (Figure 1) other than the Earth [e.g., *Basaltic Volcanism Study Project*, 1981; *Taylor*, 1994], and their presence, characteristics, and geologic and temporal settings offer a potentially important perspective for understanding LIPs on Earth. For example, unlike the Earth, the majority of which is covered by water and thus virtually unknown at high spatial resolution, global imaging coverage exists for the solid surfaces of the Moon and Mars, and the Magellan project imaged over 98% of Venus at ~200 m resolution. In addition, exposure and preservation are excellent due primarily to fewer erosional agents, minimal erosional rates, and relatively stable lithospheres.

Stable lithospheres also mean that longer time intervals are available for study (Figure 1). The age of over one-half of the Earth's surface (the ocean crust) is less than 5% of the age of the planet; the majority of the surface of the Moon and Mars, however, dates to the first half of solar system history. Terrestrial planetary bodies, by virtue of their number, offer multiple examples for study. Thus, LIPs might be studied in different places on one planet and among several planets. Similarly, the terrestrial planets provide an opportunity to assess how different environmental conditions (e.g., different crustal and thermal structure) might influence the formation and effects of LIPs. Furthermore, the segmented, laterally moving, and constantly renewing terrestrial lithosphere both insulates and obscures the view of many mantle convection processes and, indeed, is an active influence on these processes. One-plate planets [*Solomon*, 1977] such as the Moon, Mars, Mercury, and Venus can illustrate the long-term influences of mantle plumes and their variations under different thermal conditions in space and time. The multiple, well-exposed LIPs on the planets can also help to reveal the relation of plumes to tectonic structure. For example, Venus has tens of thousands of kilometers of exposed rift zones [*Senske et al.*, 1992] which display a wide variety of igneous centers [*Senske et al.*, 1992; *Magee Roberts and Head*, 1993], many of which are LIPs.

The planetary record can be instructive in terms of the chronology and episodicity of large igneous events and provinces. The extended historical record available for study (e.g., the first half of solar system history; Figure 1) permits an assessment of changes in the style of LIPs with time (potentially linked to thermal evolution, for example), and the frequency at any given time. Although radiometric dates from the planetary record are sparse, clues from well-exposed deposit morphology can sometimes even be used to estimate single-event duration [*Head and Wilson*, 1980]. Finally, the planetary record can offer a temporally complete perspective on many processes associated with LIPs. For example, lateral plate movement on Earth in the case of the Hawaiian-Emperor seamount chain (and other hotspot-related chains) helps to illustrate many stages in hotspot development by spreading the signature out into a series of volcanic edifices; this same process, however, destroys the signature of the initial plume which presumably has been subducted under Kamchatka. On the planets, particularly Venus, the start-to-finish processes of mantle plumes can be studied (e.g., the relation of thermal uplift, tectonics, and volcanism in a single example and from examples in different stages of formation) [e.g., *Stofan et al.*, 1992; *Keddie and Head*, 1994a] and compared to Earth. In summary, the planetary record, in concert with detailed examination of examples on Earth, should help to

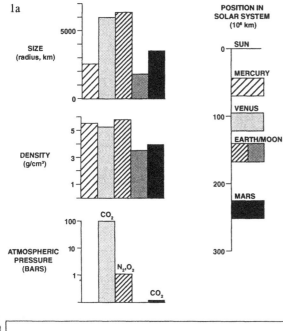

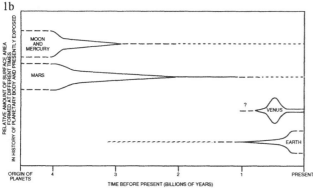

Figure 1. Characteristics and history of terrestrial planetary bodies. (a) Basic characteristics of the terrestrial (silicate-rich, relatively high-density) planetary bodies of the inner solar system. Phases of planetary evolution and many of the present geologic characteristics of these bodies may derive from these factors [e.g., *Head and Solomon*, 1981]. Although there are no direct trends in size or presence and nature of an atmosphere as a function of distance from the Sun, the general density characteristics have been interpreted in terms of temperature/pressure gradients in the collapsing solar nebula. (b) Geologic histories of the terrestrial planetary bodies. Generalized plot of the approximate percentage of presently exposed surface area that formed at different times in the history of the planet. For example, more than one-half of the Earth is seafloor formed in the last 200 million years, whereas the surface area represented by units formed in the first two-thirds of Earth history is very small. On the Moon, Mercury and Mars the majority of the surfaces (e.g., the lunar highlands and maria) formed during the the first third of the history of the solar system. Volcanic activity continued beyond this period on Mars. On Venus, like the Earth, the majority of the surface formed relatively recently, but apparently due to processes different than the plate recycling typical of the Earth.

develop and test models for the emplacement of LIPs, and to distinguish plate tectonic influences from those linked to deeper mantle and core processes.

Terrestrial planetary bodies show a wide variety of characteristics: e.g., size, density, gravity, presence/absence and nature of atmospheres, thermal evolution, and starting conditions (Figure 1). Obviously, all comparisons to the terrestrial record must keep these variations and differences in mind, as well as the positive aspects of comparative planetology described above. The planetary record is not, of course, a panacea. In many cases available information for specific aspects of different planets (e.g., the detailed crustal thickness and structure on Mars and Venus) is limited, resulting in some uncertainties involving correlations, relationships, and causal factors. Nonetheless, the information provided by specific examples and the perspective provided by considering different conditions on different planets should contribute to our understanding of the formation and evolution of LIPs. The purpose of this paper is to present a range of specific examples and to explore the potential application of these examples to current problems in understanding LIPs on Earth.

PLANETARY EXAMPLES

We proceed in order of increasing planetary size (Figure 1), first describing the general crustal, lithospheric, tectonic, and temporal setting of basaltic volcanism on the Moon, Mercury, Mars, and Venus, and then discussing specific examples of large-volume basaltic magmatism on each planetary body and the potential relevance to LIPs on Earth.

The Moon

The Moon's diameter is about one-quarter that of the Earth. The Moon is of lower density, has not retained an atmosphere, is characterized by vertical tectonics of an unsegmented lithosphere (not lateral plate tectonics), and now has a very thick lithosphere. Most of its geological surface activity took place in the first half of solar system history (Figure 1) [e.g., *Head and Solomon*, 1981]. Information about the Moon comes from remote observations and surface exploration, including returned samples and scientific stations [see *Heiken et al.*, 1991].

Basaltic volcanic deposits on the Moon consist largely of lunar maria which cover about 17% of the surface, primarily on the near side (Figure 2a). The total area of the lunar maria (6.3×10^6 km^2 [*Head*, 1975b; 1976]) is considerably larger than typical terrestrial LIPs but only slightly larger than the area of the Ontong Java LIP (Figure 3). The lunar maria were emplaced over about 1.5-2

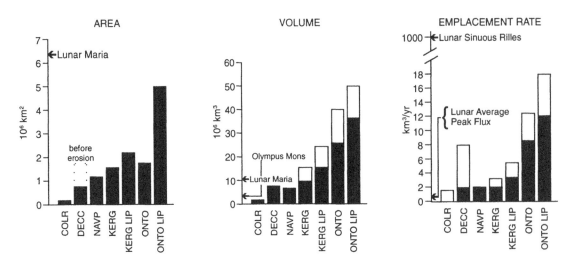

Figure 2. The Moon. (a) Global distribution of the lunar maria. Mare deposits younger than about 3.9 Ga are shown in black; buried mare plains (possible cryptomaria) inferred from remote sensing data and clustering of dark-halo craters are stippled. Selected multi-ringed basins are O = Orientale; H = Humorum; SZ = Schiller-Zucchius; I = Imbrium; N = Nectaris; C = Crisium; B = Balmer; S = Smythii; LF = Lomonosov-Fleming; AK = Al-Khwarizmi-King; T3 – Tsiolkovsky-Stark, M – Milne, M3 – Moscoviense [from *Head and Wilson*, 1992a, modified from *Schultz and Spudis*, 1979, 1983]. (b) Block diagram illustrating the relationship between rising mantle diapirs, anorthositic crust, the neutral buoyancy zone, and surface emplacement of magma by dikes [after *Head and Wilson*, 1992]. Diapirs (1) reaching the base of thick crust (2) on the farside and parts of the nearside stall and propagate dikes into the crust, most of which solidify and do not reach the surface, except occasionally in deeper craters (3). Propagation of dikes from diapirs stalled under thin crust (1) causes outflow of lava and filling of basins. With continued thermal evolution, the lithosphere thickens and a rheological barrier to rising diapirs forms (4); only dikes with extreme overpressure reach the surface (5, 6) forming very high-effusion-rate lavas and sinuous rilles. (c) Rimae Prinz sinuous rilles (27°N, 317°) interpreted to have been formed by very high effusion rates and thermal erosion (Lunar Orbiter LO V M191). Width of image is 65 km. (d) Oblique view from lunar orbit of extensive lava flows in Mare Imbrium. Late in the volcanic filling of the Imbrium impact basin, lava flows erupted from the southwestern edge of the basin (to the lower left, but outside of this image) and flowed for up to 1200. In this image, flows up to about 20 km in width and about 30 m in height can be seen extending from middle left (the direction of the source) to upper right (the middle of the basin). Ridges extending from upper left to lower right are tectonic features formed largely after lava flow emplacement. Mons La Hire, the mountain in the center, is about 20 km in width. Apollo mapping camera frame AS15-1555.

billion years, largely in the first half of solar system history [*Wilhelms*, 1987], but the total volume was relatively small, about 1×10^7 km^3 [*Head*, 1975b]. This value for the total planet is comparable to the volume of the Deccan flood basalt deposits alone, but considerably less than the total present volume of the terrestrial oceanic crust, about 1.7×10^9 km^3 (Figure 3). The average lunar global magma flux was low, about 10^{-2} km^3/a, even at peak periods of mare emplacement (in the Imbrian Period, 3.8–3.2 Ga). This average global flux is comparable to the present local output rates for such individual terrestrial volcanoes as Kilauea or Vesuvius. Output rates for individual eruptions on the Moon were occasionally extremely high; several individual eruptions associated with sinuous rilles may have emplaced more than 10^3 km^3 of lava in about a year [*Hulme*, 1973], a single event that would represent the equivalent of about 70,000 years of the average flux!

Volcanic features manifesting large-volume eruptions include the individual maria themselves, extensive flow fronts, some stretching for distances of over 1200 km [*Schaber*, 1973], volcanic complexes that might signal the location of hotspots, and sinuous rilles, which have been attributed to high-effusion-rate eruptions involving thermal erosion of the substrate. Interestingly, no large shield volcanoes, such as those seen on the Earth (e.g., Hawaii), Mars (e.g., Olympus Mons), or Venus (e.g., Sapas Mons), are observed on the Moon; large caldera-like features are also extremely rare [Head and Wilson, 1991].

The lunar maria are of diverse sizes and shapes [*Head*, 1975a], and individual mare occurrences might be thought of as equivalent to some terrestrial LIPs (Figure 3), particularly those maria that tend to be concentrated within large impact basins of various states of preservation [*Head*, 1975a, b; 1976]. Indeed, *Alt et al.* [1988] proposed that

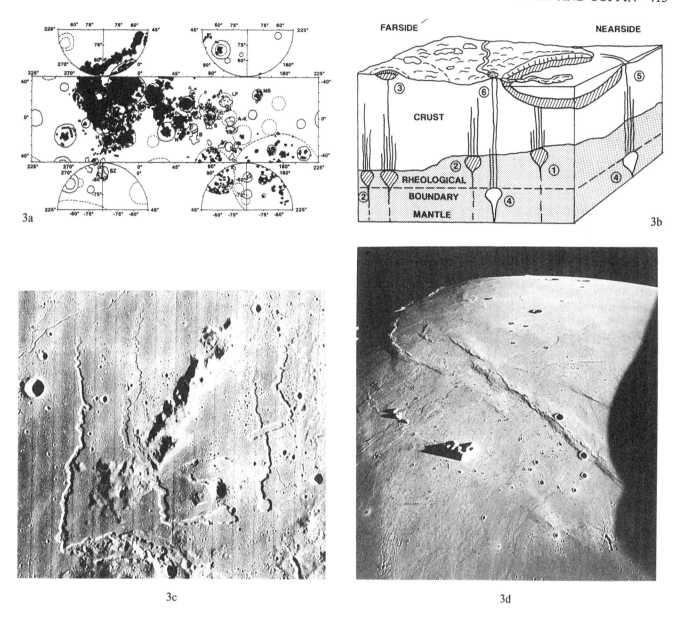

3a

3b

3c

3d

Figure 3. Terrestrial large igneous province areas, volumes, and crustal emplacement rates (averaged over 0.5-4.0 m.y. emplacement duration) relative to some planetary examples. Modified from *Coffin and Eldholm* [1994]. COLR = Columbia River Basalts; DECC = Deccan Traps; NAVP = North Atlantic volcanic province; KERG = Kerguelen Plateau; KERGLIP = Kerguelen Plateau large igneous province; ONTO = Ontong Java Plateau; ONTO LIP = Ontong Java Plateau large igneous province. For volume and emplacement rate, black-only oceanic plateau columns indicate off-ridge emplacement and black plus white columns depict on-ridge emplacement.

flood basalts that form within plates, with no apparent tectonic cause, are the terrestrial equivalents of the lunar maria. In their model, an impact crater on Earth large enough to cause pressure-release melting would be quickly flooded to form a lava lake (equivalent to the lunar maria) and these events, in turn, would initiate hotspots, which would develop into persistent low-pressure cells within the mantle [*Alt et al.*, 1988]. Does the lunar record support this model? Although early theories suggested a causal relationship between lunar impact basin formation and basaltic mare filling, the results of the Apollo and Luna exploration programs and models of basin formation and evolution [*Solomon et al.*, 1982; *Bratt et al.*, 1985] showed that generation of basalts via impacts was unlikely and that impact basin formation and filling by mare basalt are separated in time. In the case of the 900-km-diameter Orientale impact basin, vast quantities of substrate were impact-melted by the basin-forming event to produce a sheet of high-albedo plains lining the basin interior and floor, and are estimated to have a volume of ~200,000 km³ [*Head*, 1974]. This unit has a compositional affinity to the non-mare target rocks [*Head et al.*, 1993] and is distinctly different in composition and age from the adjacent basaltic maria deposits, which span an interval of several hundred million years [*Greeley et al.*, 1993]. In most other mare basins, the vast majority of the exposed volcanic plains were emplaced over several hundred million years following the impact event [*Basaltic Volcanism Study Project*, 1981]. There is no evidence for the production of basin-sized lunar basaltic "lava lakes which crystallized from the surface down" [*Alt et al.*, 1988]. The stratigraphy of lunar maria infilling documents both the long and sequential development of extrusive events, and the difference in age between the basin-forming event and its basaltic lava filling. Localization of the maria in the basins apparently was due to passive variations in crustal thickness and ponding in topographic lows [*Head and Wilson*, 1992a], processes discussed further below.

Although the equivalence of an impact origin of a basin and its fill on the Moon and LIPs on Earth proposed by *Alt et al.* [1988] is not supported by evidence from the Moon, impacts on Earth could potentially initiate volcanism. The small size of the Moon (and correspondingly very different pressure gradient), its thicker crust, and its variable lithospheric thickness could all inhibit melting relative to a comparable event on Earth. Convincing arguments have been put forth to indicate that impact-initiated volcanism was not a factor in the large (~200 km diameter) Sudbury basin formed in continental crust on Earth [*Grieve et al.*, 1991a]. Similar-size impacts into thin crust and lithosphere typical of a young oceanic floor setting could conceivably produce pressure-release melting and associated volcanism [*Rogers*, 1982]. Craters typically formed during the time of emplacement of most well-documented LIPs (e.g., the last 250 m.y. [*Coffin and Eldholm*, 1994]) are characterized by relatively small size, shallow depths of excavation, and lack of significant lava fill [*Grieve et al.*, 1991b; but see also *Oberbeck et al.*, 1993]. Large-scale rifting and deep-source plume volcanism are more likely candidates for LIP formation and evolution during this time period. In early Earth history, however, very large impacts into ocean crust and thin lithosphere may have been sites of extensive volcanism caused by mantle uplift and decompressional melting [e.g., *Grieve*, 1980; *Frey*, 1980; *Grieve and Parmentier*, 1984].

Other large volcanic accumulations on the Moon include the extensive lava flow fronts of Mare Imbrium which were emplaced at least a billion years after the formation of the impact basin. These occur in three phases which extend 1200, 600, and 400 km from the southwestern edge of the basin into its interior. The three flow units have a total volume of >4 × 10⁴ km³, and very high effusion rates are implied by their lengths and volumes; effusion rates and flow volumes are comparable to some of those reported for the Columbia River flood basalts [*Schaber*, 1973; *Tolan et al.*, 1989], although fractal analyses raise the possibility that the Imbrium flows could have been emplaced as numerous thin pahoehoe flows [*Bruno et al.*, 1992]. The fact that these units are some of the youngest on the Moon suggests that other more degraded flows filling the earlier lunar maria may also have been emplaced similarly. Examination of isolated mare basalt ponds in the highlands fringing the continuous maria has shown that typical volumes range from 100 to 1200 km³, values similar to those of terrestrial flood basalt eruption units [*Yingst and Head*, 1994; 1995; *Tolan et al.*, 1989]. Thus, many of the individual eruptions that make up the maria may be equivalent to units within flood basalts and LIPs on Earth, but the eruption frequency seems to have been much less; the lunar maria were emplaced over many hundreds of millions of years, rather than a few million years as was apparently the case in most terrestrial examples.

Another unusual characteristic of lunar maria relative to LIPs on Earth is sinuous rilles (Figure 2c), which are meandering channels preferentially located along the edges of the maria [*Schubert et al.*, 1970]. They range up to about 3 km wide and from a few kilometers to more than 300 km long. Sinuous rilles are generally an order of magnitude larger and often much more sinuous than terrestrial lava channels. Many characteristics of lunar sinuous rilles unexplained by simple lava channel, tube or other models [e.g., *Oberbeck et al.*, 1969, *Greeley*, 1971; *Spudis et al.*, 1987] can be accounted for by thermal erosion [*Hulme*, 1973, 1982; *Carr*, 1974]. The length, width, and depth of large sinuous rilles and the nature of their source regions provide important information on eruption conditions. For a 50-km-long rille in the Marius Hills, *Hulme* [1973] calculated an effusion rate of 4 × 10⁴ m³/s, an eruption duration of about one year, and a total magma volume of about 1200 km³. The sizes of source depressions of sinuous rilles provide independent evidence for extremely high-effusion-rate eruptions of long duration [*Wilson and Head*, 1980; *Head and Wilson*, 1980]. On the basis of these studies, key factors in the formation of sinuous rilles by thermal erosion are (1) turbulent flow, requiring high effusion rates and aided by low yield strength and (2) sustained flow (implying very long-duration eruptions and thus very high eruption volumes) to cause the continued downcutting of the rille to the observed depths. Thus, eruptions that caused many of the large sinuous rilles on the Moon were apparently characterized by rapid effusion of low-yield-strength lavas for prolonged periods, producing flows of extremely high volumes (in the range 300–1200 km^3), comparable to those in terrestrial flood basalt provinces (e.g., the ~1375 km^3 Roza Member of the Columbia River Basalt [*Martin*, 1989]). In contrast, typical eruption volumes for shield-related flows on Earth are much less than a cubic kilometer [*Peterson and Moore*, 1987], with the largest historic lava flow (Laki) being about 15 km^3 [*Jonsson*, 1983; *Thordarson and Self*, 1993].

Several mare-related areas show unusual concentrations of volcanic features on the Moon [*Guest*, 1971; *Whitford-Stark and Head*, 1977]. Two of the most significant of these (Figure 3) are the Marius Hills area (35,000 km^2), which displays 20 sinuous rilles and over 100 domes and cones, and the Aristarchus Plateau/Rima Prinz region (40,000 km^2) which is dominated by 36 sinuous rilles (Figure 2c). The high concentration of sinuous rilles suggests that these complexes are the sites of multiple high-effusion-rate, high-volume eruptions and that these centers may be the surface manifestation of hotspots [*Head and Wilson*, 1992a] and thus possible analogs to terrestrial LIPs. The thick crust (about 60–80 km) and lithosphere (in excess of the thickness of the crust) characteristic of the Moon (and thus the greater depths of magma sources) may make these candidate hotspots less recognizable and more analogous to continental volcanic provinces. In addition, lava flow deposits on the Moon are much more widely dispersed from their sources.

In summary, the lunar maria are comparable in scale to some terrestrial LIPs (Figure 3) but on the basis of available data appear to have been emplaced over much longer periods of time (e.g., 10^8 to 10^9 years rather than 10^6 to 10^7 years). Many individual eruptions, however, appear to be similar in volume and eruption rates to those in flood basalt provinces [*Tolan et al.*, 1989]. Little evidence exists for shallow magma reservoirs and repeated small-volume eruptions that would build up large shield volcanoes. The observed characteristics seem to call for large batches of magma erupted over short periods of time from relatively deep sources but separated in time by significant intervals. How can these characteristics be accounted for in terms of the nature of the source regions and the modes of emplacement?

One model [*Head and Wilson*, 1992a] begins with the observations that the basaltic maria are superposed on the ancient, globally continuous, and thick low-density anor-

thositic highland crust, the latter derived primarily from global-scale melting associated with planetary accretion. The low-density highland crust provided a density barrier [*Solomon*, 1975] to ascending mantle plumes and basaltic melts. In this view, rising diapirs and magma bodies tended to collect at the base of the 60–80 km thick crust (Figure 2b). Following sufficient overpressurization of source regions by partial melting or arrival of additional material into the reservoir, individual dikes propagated toward the surface. Thus, the thick highland crust created a deep zone of neutral buoyancy for rising magma that could only be overcome by overpressurization events which caused dikes to propagate to the surface.

In this model, whether intrusion or eruption occurred was determined by variations in overpressurization and crustal thickness. Low levels of overpressurization resulted in intrusion into the lower crust, forming dikes which cooled and solidified. Dikes characterized by sufficient overpressurization to approach the surface could have several fates. Overpressurization events large enough to propagate dikes to the surface to cause eruptions are predicted to involve very large volumes of magma [*Head and Wilson*, 1992a], comparable to those associated with many observed lava flows, such as the flows extending hundreds of kilometers into Mare Imbrium [*Schaber*, 1973] and those associated with sinuous rilles. Intrusion close enough to the surface to produce a distinctive near-surface stress field often resulted in the production of linear graben-like features along the strike of the dike and small associated effusions and eruptions. In the case of the linear graben Rima Parry V, small spatter cones are aligned along the central part of the graben [*Head and Wilson*, 1994a]. Dikes propagating to slightly deeper levels may not create near-surface stress fields sufficient to form graben, but subsequent degassing may form chains of pit craters over the site of the dike.

The model predicts that the relationship between the size of the magma source and highland crustal thickness was such that dikes propagated to the near-surface and surface relatively infrequently (Figure 2b). Thus, most dikes had sufficient time to cool before the next dike was emplaced. Frequent emplacement of dikes to create a shallow magma reservoir was very difficult on the Moon. The lack of Hawaii-like shield volcanoes and the paucity of caldera-like features are thus attributed to the difficulty in producing shallow magma reservoirs which result in emplacement of many individual flows, edifice-forming flows, and associated calderas [*Head and Wilson*, 1991]. In addition, the same lack of multiple, continuous dike emplacement events of sufficient magnitude to reach the surface over short periods of time meant that the lunar maria tended to be produced from relatively large eruption events spaced over very long intervals, in contrast to terrestrial LIPs.

The lunar situation described in this model is analogous in many ways to basaltic magma bodies interacting with terrestrial continental crust. On Earth, zones of neutral buoyancy [e.g., *Glazner and Ussler*, 1988] stall buoyantly rising basaltic magma bodies within the crust. Overpressurization events can cause the same features seen on the Moon, as exemplified by many of the basaltic volcanic fields in the western United States [*e.g., Crumpler et al.*, 1994], and indeed large-scale flood basalts can be emplaced that are comparable in size to the large lunar flows [*Tolan et al.*, 1989]. The low melting temperature of the continental crust relative to that of the more refractory lunar anorthositic crust means that stalled basaltic magma bodies in continental crust may cause associated and large-scale crustal melting, resulting in a geochemical and petrologic complexity unknown on the Moon. The continental crust and the lunar highlands illustrate the role of large-scale density barriers impeding the creation of significant shallow basaltic reservoirs, such as those observed at seafloor spreading centers and in large edifices such as Hawaii. Complex shallow reservoirs do exist in continental crust, however, where local conditions of melt generation and, unlike on the Moon, sustained supply rates exist (as in continent margin subduction zones and hotspot traces or rifting environments). In these cases, composite volcanoes are common. No known analog of these features exists on the Moon and Venus, but several examples may be present on Mars (e.g., Hecates Tholus [*Mouginis-Mark et al.*, 1982; *Wilson and Head*, 1994; *Hodges and Moore*, 1994]).

Mercury

Mercury remains one of the most enigmatic and promising planets in the inner solar system in terms of understanding the relationship of its unusual interior to its volcanic and magmatic history [*Chapman*, 1988]. Information about Mercury comes from the Mariner 10 mission and Earth-based observations [see *Vilas et al.*, 1988]. Mercury is about one-third the diameter but approximately the same density as the Earth, has not retained an atmosphere, and is characterized by vertical and some lateral tectonics of a largely unsegmented lithosphere (not lateral plate tectonics). Most geological surface activity took place in the first third of solar system history (Figure 1) [*Head and Solomon*, 1981; *Vilas et al.*, 1988]. The very high density of Mercury relative to its size has been attributed both to initial temperature-pressure

conditions in the inner part of the condensing solar nebula, which favored retention of refractory components [e.g., *Goettel*, 1988], and to the effects of a giant impact event stripping off a low density crust and upper mantle after core formation [*Cameron et al.*, 1988].

Mercury is poorly explored in terms of photographic coverage and remote sensing data [*Chapman*, 1988]. Knowledge of internal structure is meager, although a high-density core comprising well over one-half Mercury's diameter (about the size of the Earth's Moon) is likely. In addition, prominent albedo variations such as those that distinguish the lunar maria from the heavily cratered highlands are not apparent on Mercury. Smooth plains are present, but a possible volcanic origin cannot readily be distinguished from plains produced by ponding of impact ejecta, a process known to occur in the light plains surrounding impact basins on the Moon [*Oberbeck*, 1975; *Oberbeck et al.*, 1975; *Wilhelms*, 1976]. The stratigraphy and geologic history of Mercury suggest that major volcanic provinces were emplaced in the first third of solar system history [*Spudis and Guest*, 1988], but the details are insufficient to provide a basic characterization of such provinces or an understanding of their mode of emplacement. If these plains are indeed of volcanic origin, their general lack of associated volcanic features [*Strom et al.*, 1975; *Trask and Strom*, 1976] suggests possible flood basalt emplacement.

Mars

Information about Mars [e.g., *Kieffer et al.*, 1992a] comes from Earth-based observations, extensive spacecraft exploration (including orbiters and landers [e.g., *Kieffer et al.*, 1992b; *Snyder and Moroz*, 1992]), and meteorites believed to be ejected from Mars by impacts and transported to Earth [e.g., *Longhi et al.*, 1992]. Mars is about one-half the diameter and of much lower density than the Earth, has a thin CO_2 atmosphere, and is characterized by vertical tectonics of an unsegmented lithosphere (not lateral plate tectonics); most of its major geological surface activity took place in the first half of solar system history, with some volcanism and significant eolian activity continuing well into the last half of solar system history (Figure 1, 4a) [e.g., *Head and Solomon*, 1981; *Kieffer et al.*, 1992b]. The total area of Mars covered by volcanic material has been estimated to be about 58% of the surface ($\sim 0.84 \times 10^8$ km^2) [*Tanaka et al.*, 1988], and the total volume of surface extrusion to be 2×10^8 km^3 [*Greeley*, 1987] (Figure 3). The corresponding intrusive volume is not known but is likely to be larger by at least a factor of 10, the ratio typical of the continental

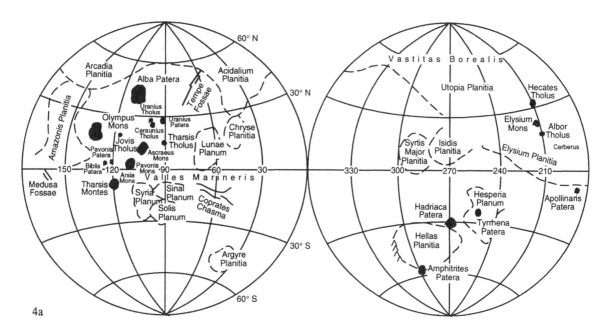

Figure 4. Mars. (a) Generalized topography and geography. The dashed line extending from upper left (near Arcadia Planitia) to lower right (near Apollinaris Patera) across both hemispheres separates the northern lowlands from the southern highlands; other closed dashed lines are impact basin depressions (e.g., Argyre Plainita, Hellas Planitia) or volcanic provinces (e.g., Hesperia Planum; Lunae Planum). Black spots are shield volcanoes of various sizes; the concentration of shield volcanoes in the left hemisphere is the Tharsis region (see Figure 4b for enlargement) and in the right hemisphere is the Elysium region. (b) Geologic sketch map of the Tharsis region. Topography is indicated by contour lines at 5 km intervals with tick marks pointing downslope. Stars mark summits of the major shield volcanoes (Figure 4a), Olympus Mons and the Tharsis Montes (Arsia, Pavonis, and Ascraeus), which commonly reach elevations in excess of 20 km; upper contours are omitted. Width of diagram is about 4000 km at the equator and the units are discussed in the text. Map is from *Head and Solomon* [1981] from data of *Wise et al.* [1979] and *Scott and Carr* [1978]. (c) Oblique view of Olympus Mons volcano, one of the large lava shields in the Tharsis region (Figure 4a, b); summit is about 25 km above the base of the volcano and is characterized by a complex caldera and two nearby impact craters. Flows emanating from near the summit extend down the flanks and often cascade over the several-kilometer-high scarp at the base of the volcano. Viking Orbiter photograph VO 641A52. (d) Stratigraphic sequence showing context and main events in the evolution of Tharsis in relation to global processes [from *Banerdt et al.*, 1992]. Locations of regions are shown in Figure 4a, b. Absolute ages are from the time-scale models of Hartmann-Tanaka (H/T) [*Hartmann*, 1978] and Neukum-Wise (N/W) [*Neukum and Wise*, 1976], as summarized by *Tanaka et al.* [1992].

regions on Earth [e.g., *Crisp*, 1984; see *Wilson and Head*, 1994]. Volcanism has decreased over geologic time from broad regional resurfacing to local activity; areal resurfacing rates have steadily decreased from ~1 km^2 a^{-1} to ~10^{-2} km^2 a^{-1} [*Tanaka et al.*, 1992].

On Mars, in contrast to the Moon, large shield volcanoes have been emplaced that resemble those on Venus and the Earth in morphology. They exhibit a wide range of rift zone development, internal deformation related to lithospheric loading and flexure, flank and slope failure, and summit caldera development [*Carr*, 1973, 1981; *Hodges and Moore*, 1994; *Wilson and Head*, 1994; *Crumpler et al., 1996*]. Their scales are quite different,

however (Figure 4b, c). Martian shields possess breadths of many hundreds of kilometers, and their heights are commonly a factor of three greater than Hawaii (up to 25 km!). Volumes of individual shields are gigantic (Figure 3). Olympus Mons (Figure 4c) has a volume of about 2 × 10^6 km^3 (above its base), compared to 1 × 10^5 km^3 (above its base) for the island of Hawaii (which is composed of several different shields) and 1.1 × 10^6 km^3 for the whole Hawaiian-Emperor seamount chain [*Barger and Jackson*, 1974]. Volumes of other single edifices are of the order of 1.5 × 10^6 km^3, comparable to extrusive volumes estimated for the Karoo, Paraná, Deccan and North Atlantic basalt provinces on Earth (Figure 3). Martian caldera structures

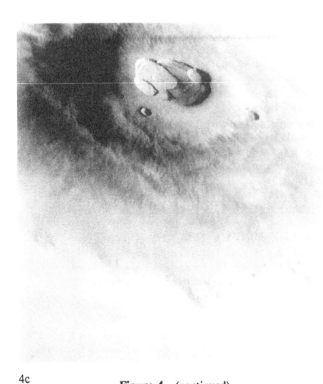

4b

4c

Figure 4. (continued)

in longer cooling-limited flows and wider dikes characterized by higher effusion rates [*Wilson and Head*, 1994]. Because the lithosphere has been stable and has not moved laterally over the majority of martian history, regions of melting in the mantle (e.g., mantle plumes) concentrate their effusive products in a single area, rather than having them spread out in conveyor-belt-like fashion, as in the case of the Hawaiian-Emperor seamount chain on the Pacific Ocean floor. Thus, melt products accrete vertically into huge accumulations [*Carr*, 1973], loading the lithosphere and causing flexure, deformation, and edifice flank failure.

The extreme height of martian volcanoes also appears to be related to lithospheric structure. *Comer et al.* [1985] examined deformational structures surrounding several Tharsis-region volcanoes (shown as large black spots in Figure 4a) to assess lithospheric flexure caused by volcano loading and to estimate the thickness of the elastic lithosphere. They found that elastic lithosphere thicknesses are in the range of 20–50 km for regions surrounding the majority of the Tharsis shields. The lithosphere appears to be at least 150 km thick in the region of Olympus Mons. Thus, one factor contributing to the large height of the martian volcanoes is the relatively thick elastic lithosphere during their formation; the volcanic load and underlying lithosphere did not subside at a rate that would limit their heights. In addition, variations in lithospheric thickness in

4d

EPOCH	GLOBAL PROCESSES		REGIONAL ACTIVITY				ABSOLUTE AGES (Ga)	
	Planetary differentiation	Impact structures	Tharsis	Olympus Mons	Valles Marineris	Elysium	H/T	N/W
Late Amazonian	:	:		:	:	:	:	0 — 0
Middle Amazonian	:	:		I	:	I		0.70 — 0.25
Early Amazonian	:	:		I	:	I	I	2.50 — 0.70
Late Hesperian	I			I			I	3.55 — 1.80
Early Hesperian		:		I		I		3.70 — 3.10
Late Noachian		:		I		I		3.80 — 3.50
Middle Noachian		I		I		I		4.30 — 3.85
Early Noachian	I	I		I		I		4.50 — 3.92
								4.60 — 4.60

Figure 4. (continued)

are also much larger than those typical of Earth [*Head and Wilson*, 1994b; *Crumpler et al., 1996*; *McGuire et al.*, 1996]. How do we account for these differences?

On Mars, low gravity and low atmospheric pressure at the surface result in a crustal bulk density profile different from other planets [*Wilson and Head*, 1994], which means that magma reservoirs are predicted to be deeper than on Earth by a factor of about four. The lower gravity results space and time can be very important in the construction and subsequent modification of volcanic edifices. For the volcanoes forming the Galápagos Archipelago, *Feighner and Richards* [1994] showed that lithospheric thickness is related to volcano size and structure across the archipelago between areas of effective elastic lithospheric thickness of 6 and 12 km. *McNutt et al.* [1989] demonstrated that the thermal and mechanical state of the lithosphere apparently controls the expression of weak plumes such as the Marquesas. In a study relevant to terrestrial shield volcanoes and to the history of lava emplacement in LIPs, *McGovern and Solomon* [1993] modeled lithospheric flexure and time-dependent stress and faulting on the Tharsis volcanoes and demonstrated sufficiently large flexural stresses in several examples to cause failure by faulting. Such stresses in turn could have influenced the subsequent path of magma ascent and emplacement.

One of the most impressive global-scale features on Mars is the Tharsis region, a LIP comprising ~20% of the surface area of Mars that dwarfs those presently known on Earth in size, associated features, and duration (Figure 3). Tharsis, which forms a broad dome or rise about 4000 km in diameter rising as much as 10 km above surrounding terrain, dominates the western hemisphere of Mars (Figure 4b). Its area of $>6.5 \times 10^6$ km^2 is larger than the largest known terrestrial LIP and totals over one-half the total area of the lunar maria (Figure 3). The Tharsis rise is composed of areally extensive volcanic plains spanning a wide range of ages; massive superposed shield volcanoes (e.g., >500 km wide and up to 25 km high) are associated with tectonic features that include radial fractures and graben extending beyond the rise and perhaps associated with uplift, and concentric wrinkle ridges indicating crustal shortening. The volcanic deposits clearly associated with this province, the western volcanic assemblage [*Tanaka et al.*, 1992], cover an area of 1.4×10^7 km^2 (Figure 3); more-degraded deposits may also be volcanic.

On the basis of geologic mapping at a variety of scales [*Scott and Tanaka*, 1986, and summarized by *Tanaka et al.*, 1992] a general stratigraphy and chronology for Tharsis has begun to emerge (Figure 4d). In contrast to many terrestrial LIPs which formed over 10^5–10^6 years, these data point to volcanic and tectonic activity in the Tharsis region spanning 10^8–10^9 years. Ancient cratered terrain bounds Tharsis to the south and is exposed at high elevations within Tharsis, suggesting extensive uplift. Plains units interpreted as volcanic and major shield volcanoes dominate the rest of Tharsis. Undivided plains (pu in Figure 4b) make up the vast majority of surface units and extend north of Tharsis. Ridged plains (pr) are characterized by many mare-ridge type features indicating

shortening. Major volcanic edifices and structures dot the surface of Tharsis and four of these (indicated by stars on their summits in Figure 4b) exceed 25 km in elevation above the surrounding terrain (Figure 4c). Younger volcanic plains units (pt) surround the major central shield volcanoes, their most likely source. Tectonic features are abundant and the most prominent of these, the Valles Marineris rift system, extends several thousand kilometers from central Tharsis toward the east. Its floor is indicated by unit cf, canyon floor materials, in Figure 4b.

Tharsis rise development involved complex episodic tectonism and intimately associated volcanism on both local and regional scales. Early fractured plains (Noachian and Hesperian Epochs, Figure 4d) made up mostly of volcanic rocks erupted during the early stages of Tharsis activity are cut by the most intense deformation in Tharsis, represented by fault systems that are radial and concentric to volcanic centers such as Tharsis Montes and Alba Patera. These faults formed during the Noachian and Hesperian epochs; concentric ridge systems representing local shortening were formed mainly at distances greater than 2000 km from the center of the Tharsis rise, primarily during the Late Noachian and Early Hesperian Epochs. The latest faulting occurred in the Amazonian Epoch (Figure 4b, d) primarily in association with the active volcanic centers mentioned above [*Tanaka et al.*, 1992]. The Tharsis Montes (from south to north, Arsia, Pavonis and Ascraeus Montes; Figure 4a, b), composed of three massive shield volcanoes aligned in a row along the crest of the Tharsis rise, are the primary sources for the volcanic Tharsis Montes Formation (largely unit pt in Figure 4b), which covers an area of almost 7×10^6 km^2 and is composed of lobate sheet flows, some of which extend almost 1500 km from the source shields [*Schaber et al.*, 1978; *Plescia and Saunders*, 1979]. Olympus Mons, a similar shield volcano to the west of Tharsis Montes, is the source area for Upper Amazonian lava flows, some of the youngest on Mars (Figure 4b, c).

Theories to account for the Tharsis rise abound [see discussions by *Schubert et al.*, 1992; *Banerdt et al.*, 1992; and *Tanaka et al.*, 1992]. Initial ideas centered on an area of convective upwelling producing a very large mantle plume which generated uplift and volcanism [e.g., *Carr*, 1974], an idea supported by calculations of mantle convection under martian conditions in which a limited number of convection cells are favored [e.g., *Schubert et al.*, 1990]. The interpreted topographic uplift, however, could not be explained by these dynamic processes alone. Isostatic uplift caused by lateral migration and intrusion of material thermally eroded from the base of the crust of the northern lowlands was favored by *Wise et al.* [1979], whereas *Schultz et al.* [1982] suggested preferential concentration of volcanism along early impact-basin ring structures. Debate also centers on the relative role of volcanism and uplift, with some workers preferring major uplift and relatively minor volcanism [e.g., *Plescia and Saunders*, 1980] and others suggesting that Tharsis resulted from an extended period of regional volcanism in an area of thin lithosphere [e.g., *Solomon and Head*, 1982]. The inability of stress models to account simply for the extensive radial graben systems has led workers to accept the idea that more than one mechanism of lithospheric deformation is required; simple isostatic or flexural loading models do not satisfy all observations. The present conundrum is that stress models seem to require two different events, but the geologic evidence suggests that the radial graben formed essentially simultaneously [e.g., *Banerdt et al.*, 1992].

Finnerty et al. [1988] constructed a quantitative petrologic model for Tharsis which was extended to a more general model for the evolution of Tharsis [*Phillips et al.*, 1990]. Using melt partitioning data and models for likely martian mantle compositions, they showed that extraction of basalt melt from the mantle and subsequent crustal intrusion and extrusion could have resulted in a net volume increase in the crust-mantle column, producing a prominent topographic rise with no net increase in mass. Much of the support for the uplift would come from the source-region residuum, and most of the magma produced by the required melting must end up as intrusions in the crust and upper mantle. Although consistent with many of the major characteristics of Tharsis, these models do not easily satisfy the gravity data.

What currently supports the Tharsis region, some several billion years after its initial activity? Gravity data show an extremely large free-air anomaly [e.g., *Esposito et al.*, 1992]; simple isostatic compensation is essentially ruled out and dynamic support by active mantle flow is very unlikely because of the difficulty of maintaining such large-scale and consistent mantle flow for several billion years. Many models have been proposed, and the most likely have the Tharsis rise partially supported by the elastic strength of the lithosphere, with additional support from the buoyancy of a crustal root at depths of about 50–100 km [e.g., *Banerdt et al.*, 1992]. The early history of the Tharsis rise might have involved a transient mode of support (e.g., a convective plume, or an upper mantle density-deficit induced by thermal or chemical factors) and a regional crustal thickness about 25–30 km in excess of that estimated in global-scale models. Subsequent reduction and removal of the transient support were accompanied by the general cooling of the planet, leaving a superisostatic load on a cooling, thickening lithosphere.

A second large domal area on Mars, the Elysium rise, is about 2000 km across (Figure 4a). Although smaller than Tharsis, it also has high concentrations of volcanism and tectonism [*Greeley and Guest*, 1987; *Mouginis-Mark et al.*, 1984], including several volcanic formations, and three major shield volcanoes (Albor Tholus, Hecates Tholus, and Elysium Mons). *Hall et al.* [1986] argued on the basis of thermal and mechanical arguments that flexural uplift preceded or was contemporaneous with the emplacement of the majority of the volcanic deposits.

Why does Mars have two prominent, long-lasting, extremely large igneous provinces? Convective planforms in the martian mantle were modeled using numerical simulations of fully three-dimensional convection in a spherical shell [e.g., *Schubert et al.*, 1990, 1992]. These models suggest that cylindrical plumes are the most probable form of upwelling in the mantle and that downwelling occurs in an interconnected network of planar sheets; the number of upwelling plumes is a function of the geometry of heating. Increase in bottom heating causes a decrease in the number of upwellings and an increase in their intensity, with very substantial bottom heating producing only six plumes. Gradual cooling of the planet, and the core in particular, means that the planform and style of convection likely changed with time, with fewer, more vigorous plumes earlier and more, smaller plumes later. In addition, the temperature-dependence of mantle viscosity will have an influence on plume structure and abundance. Although the trend in early history might have been toward a small number of vigorous plumes, a variable lithospheric thickness and a thickening lithosphere with time [e.g., *Comer et al.*, 1985; *Solomon and Head*, 1990] might hide the surface effects of all but the most prominent plumes.

What are some possible lessons for those who study LIPs on Earth? First, it is clear that LIPs can achieve massive proportions and form over long periods. Tharsis covers 20% of the surface area of the planet Mars and was active for several billions of years. In addition, the martian LIPs confirm that scale and total duration of igneous emplacement can change as a function of time and thermal evolution (large-scale planetary cooling). Early plumes might have been less numerous, larger, and more vigorous due to a larger role of bottom heating. Planetary thermal evolution will also influence lithospheric thickness and the surface manifestation of plume impingement; thus we should anticipate considerable variability in LIPs through time. The abundance of large shield volcanoes within Tharsis, each of which would qualify as a LIP on Earth (Figure 3), also suggests that individual plumes are likely within a larger diffuse upwelling such as may have formed Tharsis as a whole. Given the incompleteness of the terrestrial record, the martian record suggests that some single LIPs on Earth might be only one "tree" in a larger "forest" of a megaplume. Finally, the petrogenetic effects of shallow melting and the resulting residuum might leave depleted mantle signatures that could persist for hundreds of millions to billions of years, even on a planet as dynamic as Earth.

Venus

Venus is approximately the same diameter and density as the Earth and is Earth's closest planetary neighbor (Figure 1). These similarities have led to frequent comparison of Venus with the Earth and the idea that Venus might be a "sibling" or possibly even a "twin." Venus offers an important test of major ideas about planetary evolution in terms of the role of planetary size and initial position in the solar system [Head and Solomon, 1981], crustal formation and evolution [Head, 1990b], and mechanisms of lithospheric heat transfer [Solomon and Head, 1982]. Although Venus has many similarities with the Earth, it also has important differences. It has a thick, dense CO_2 atmosphere, rotates very slowly, and has essentially no magnetic field. The high surface pressure is approximately comparable to that on the ocean floor below about a kilometer depth, and average surface temperatures are around 475°C, precluding the presence of liquid water and resulting in the preservation of landforms in their near-pristine state. Information on the nature of Venus has come from Earth-based observations and a variety of planetary probes, including flybys, orbiters, balloons, atmospheric probes and landers [e.g., Hunten et al., 1983; Cruikshank, 1983]. Data from the recent Magellan mission provided high resolution radar image coverage for almost the whole planet and altimetry and gravity data [Saunders et al., 1992]; this information, together with data from previous US and Soviet missions, has resulted in a more comprehensive view of the geology and geophysics of Venus (Figure 5a).

Approximately 80% of the surface of Venus is made up of volcanic plains and a wide variety of volcanic landforms (Figure 5b-e), probably largely basaltic in composition [Head et al., 1992]. Unlike on the Earth, volcanic landforms are not distributed along elongated plate boundaries and hotspot traces; rather, they are broadly distributed over the whole planet and also clustered in a large region (Beta-Atla-Themis) making up about 20% of the surface [Crumpler et al., 1993]. Although Venus has folded mountain belts [Crumpler et al., 1986], global rift zones [Senske et al., 1992], and features that resemble Earth's convergent plate boundaries [Head, 1990a; McKenzie et al., 1992], Magellan revealed no evidence for the extensive global plate-tectonic boundaries and crustal structures that

would indicate ongoing crustal spreading and recycling [Solomon and Head, 1982; 1991; Solomon et al., 1992].

The ~80% of the surface area of Venus estimated to be covered by volcanic plains (~3.68×10^8 km^2 [Head et al., 1992]) can be combined with an estimate of the average plains thickness of about 2.5 km based on stratigraphic relationships [Head et al., 1996a] to predict the total volume of surface extrusion of about 9.2×10^8 km^3 (Figure 3). On the basis of impact crater counts on volcanic units, volcanism has apparently decreased over geologic time from a period of global resurfacing to much less voluminous local activity, with average effusion rates changing from about 5 km^3/a to <1 km^3/a [Head et al., 1992].

On the Earth, typical basaltic melts are positively buoyant, but can stall at a neutral buoyancy zone (NBZ) representing a near-surface, low-density horizon related to weathering and gas-exsolution porosity [e.g., Ryan, 1988]. The results of theoretical modeling indicate that very high atmospheric pressure on Venus reduces volatile exsolution and magma fragmentation, serving to inhibit the formation of NBZs and shallow magma reservoirs [Head and Wilson, 1986]. For a range of common terrestrial magma volatile contents (<0.5 wt% H_2O, <0.35 wt% CO_2), magma ascending and erupting near or below the mean planetary radius (MPR) on Venus should not stall to produce shallow magma reservoirs. In this case, magma should ascend directly to the surface; such eruptions should be characterized by relatively high total volumes and effusion rates, comparable to those observed in terrestrial flood-basalt provinces (Figure 3) [Head and Wilson, 1992b].

Because atmospheric pressure changes considerably with elevation on Venus, the same range of volatile contents results in the production of NBZs and magma reservoirs at elevations well above MPR. For the same range of volatile contents at higher elevations (about 2 km above MPR), about half of the cases treated by Head and Wilson [1992b] result in direct ascent of magma to the surface and half in the production of NBZs. In general, NBZs and shallow magma reservoirs on Venus are predicted to appear as gas content increases and, because of the high atmospheric pressure, to be nominally shallower on Venus than on Earth. The shallowest depths for NBZs are about 1 km and depths increase slowly with increasing CO_2 content and rapidly with increasing H_2O content. For a fixed volatile content, NBZs become deeper with increasing elevation. Over the range of elevations (-1 to +4.5 km) treated by Head and Wilson [1992b], depths differ by a factor of 2–4, which is about the same factor as that induced by variations in CO_2. NBZ reservoirs can become deeper than reservoirs on Earth produced with similar volatile contents if common terrestrial volatile contents are exceeded. To a

5a

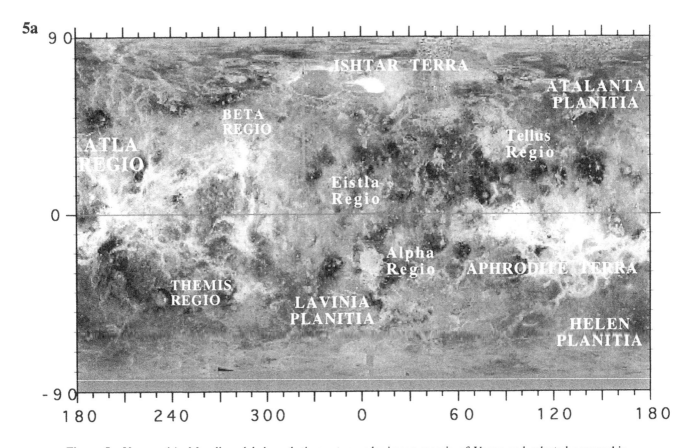

Figure 5. Venus. (a) Magellan global synthetic-aperture radar-image mosaic of Venus and selected geographic features. The bright areas have generally high radar backscatter (rough at centimeter scales; rift zones, deformed highlands) and the dark areas have low backscatter and relatively smooth (volcanic plains and lowlands). Cylindrical projection. One degree of longitude is about equivalent to 100 km at the equator. (b) A portion of Baltis Vallis, an example of a sinuous channel on Venus (width of image is about 150 km). Baltis Vallis, about 2-3 km wide, starts at about 45°N, 186° and winds for about 6800 km to the vicinity of 12°N, 168° (Figure 5a). Dark areas are volcanic plains and brighter linear parts are scarps and surface roughness at centimeter-meter scale. The network-like pattern of ridges is caused by post-emplacement subsidence and deformation of lavas. Portion of C1-MIDR 45N159. North is at the top. (c) A large lava flow field on the flanks of Sif Mons, located just to the south of this image in Eistla Regio (Figure 5a). Flows, appearing bright in the radar image, emanate from the summit and stream down the flanks for hundreds of kilometers, often being trapped in narrow linear graben about 1-2 km wide. Width of image is about 125 km. Portion of C1-MIDR 30N351. North is at the top. (d) Sapas Mons, a typical shield volcano on Venus [see *Keddie and Head*, 1994] is located in western Atla Regio (Figure 5a). The 2.4-km-high edifice is comprises radar-bright (upper flanks of the volcano) and radar-dark (lower flanks) radial flows. The two dark summit features are steep-sided scalloped-margined domes. Image is about 650 km wide. Portion of F-MIDRP 10N188. North is at the top. (e) Linear graben related to radial fracturing around a central reservoir and interpreted to be the surface manifestation of wide dikes at shallow depth; note large lava flows emanating from several of these. Located southeast of Atla Regio at about 15°S, 215°. Width of black boxes is 20 km. Portion of C1-MIDR 15S215. North is at the top.

first order, the characteristics and global distribution of volcanic landforms that are largely extrusive [*Keddie and Head*, 1994a, b] and structures that reflect intrusive activity [*Grosfils and Head*, 1995] support the idea that neutral buoyancy contributes to major aspects of volcano growth and development on Venus.

How do these different conditions influence the formation of LIPs? Populating the >80% of the surface of Venus comprised of volcanic plains are more than 1500 edifices or volcanic sources in excess of 20 km diameter [*Head et al.*, 1992]. Over 150 of these are major shield volcanoes (Figure 5d) in excess of 100 km in diameter. The lack of a

5b

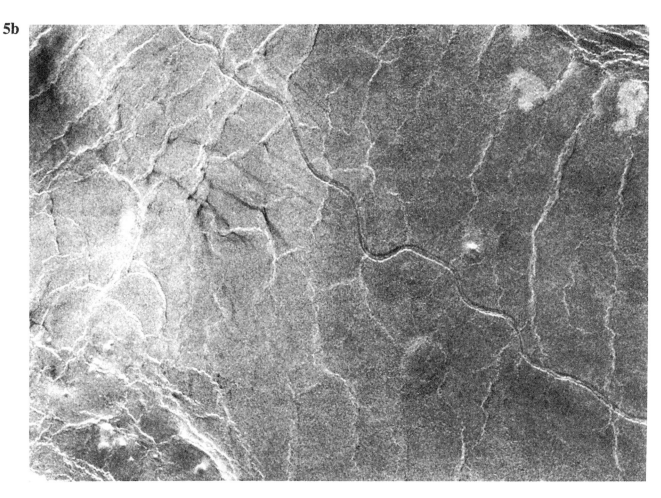

Figure 5. (continued)

hydrosphere and insignificant erosion on Venus mean that the early record of the volcano may be exposed and that different phases in its evolution can sometimes be more readily outlined than is commonly the case on Earth, particularly where flow lengths have decreased with time, leaving exposed sequential phases of volcano evolution [e.g., *Keddie and Head*, 1994a]. The heights of these shield volcanoes are considerably less than those on the Earth and Mars, typically less than about 2 km above the surrounding plains [*Keddie and Head*, 1994b]. Several factors help to account for these differences and illustrate how LIPs may be produced on Venus. One, the environment on Venus (surface temperature and pressure) favors larger primary magma reservoirs which will cause the wide dispersal of conduits that build edifices, resulting in broader, flatter structures. Two, models of shallow NBZ reservoir locations during edifice growth show that, for Earth, the center of the magma chamber remains at a constant depth below the growing summit of the edifice,

thus keeping pace with the increasing elevation [*Head and Wilson*, 1992b]. In contrast, on Venus, because of the major gradient in atmospheric pressure with altitude, the chamber's center becomes deeper relative to the summit of the growing edifice. Although the chamber's elevation does rise with time, the rise rate is low. Therefore, magma reservoirs on Venus will remain in the pre-volcano substrate longer, and in many cases may not emerge into the edifice at all. In addition, the lower rate of vertical migration implies that, for a given magma supply rate, magma reservoirs would tend to stabilize, undergo greater lateral growth, and become larger on Venus than Earth. Thus, the proportion of the available magma going into production of the edifice relative to that intruded into the substrate is smaller on Venus than Earth. The resulting large reservoirs would encourage multiple and more widely dispersed source vents and large volumes for individual eruptions. All of these factors result in volcanic edifices that are low and broad, with reservoirs predomi-

5c

Figure 5. (continued)

5d

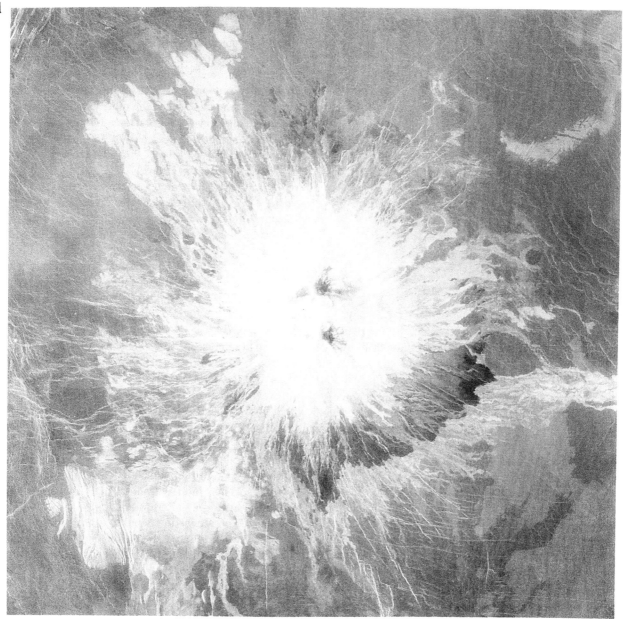

Figure 5.   (continued)

nantly in the substrate, rather than the edifice. Because of the inhibition of volatile exsolution in the terrestrial submarine environment and its influence on volcanic landforms [e.g., *Head et al.*, 1996b], these factors could also be important in the formation of LIPs on the seafloor and in the initial stages of Hawaiian-type edifice formation.

On Venus, over 150 large radiating lineament systems with a radius in excess of 100 km have been mapped

[*Grosfils and Head*, 1994]. These structures are interpreted to be the surface manifestations of dike swarms radiating away from a central source, analogous to the giant radiating dike swarms on the Earth such as the Mackenzie dike swarm in Canada. On Venus, however, lack of erosion permits the surface equivalent of the deeply eroded Mackenzie-type swarms to be studied [e.g., *Ernst et al.*, 1995]. Theoretical analysis and predictions of the characteristics of dikes emplaced in unbuffered (declining

5e

Figure 5. (continued)

driving pressure) and buffered (constant driving pressure) environments [*Parfitt and Head*, 1993] show that buffered driving pressure can readily account for the very wide (exceeding 100 m) and very long (in excess of 2000 km) dikes observed on the Earth and Venus in some mafic dike swarms. In addition, some dikes emplaced in buffered conditions will grow vertically and laterally until they reach the surface at some distance from the magma reservoir. The high driving pressure and large dike widths typical of these conditions mean that eruptions produced would be characterized by large volumes of lavas emplaced at very high rates, potentially producing flood basalts [*Parfitt and Head*, 1993] (Figure 5e).

An interesting class of features with associated large-scale effusive activity was discovered in Earth-based and Venera mission radar images. Named coronae, these features range in size from 60 to over 2000 km in diameter and are characterized by circular to elongate outlines and one or more discontinuous annuli comprising concentric compressional and/or extensional troughs and ridges [*Pronin and Stofan*, 1990]. Magellan provided data for a global census of coronae [*Stofan et al.*, 1992; *Head et al.*, 1992] that revealed over 300 of these features widely distributed over the surface [*Magee Roberts and Head*, 1993] and showed the details of their structure [*Stofan et al.*, 1992; *Squyres et al.*, 1992]. The prevailing interpretation is that these features represent the surface manifestation of hotspots or plumes [e.g., *Stofan et al.*, 1991, 1992; *Squyres et al.*, 1992; *Janes et al.*, 1992] and that their diversity in structure and associated volcanism reflects differences in plume size and intensity and the

thermal structure of the overlying lithosphere [e.g., *Erickson and Arkani-Hamed*, 1992]. Ascending mantle diapirs elevate and deform the lithosphere as they approach the base of the lithosphere. There, they flatten, spread laterally, and cool; on the surface the raised plateau subsides to produce a central depression and a deformed annulus and moat. One of the most impressive aspects of the documentation of these features is that they potentially represent hundreds of examples of hotspots whose history can be mapped because of the lack of erosion and incomplete coverage by later deposits.

A step in this direction was taken by *Magee Roberts and Head* [1993], who studied the temporal and spatial relationships of areally extensive volcanic flow fields associated with coronae. They showed that large-scale flow fields (average of 1.1×10^5 km^2, but up to 1.5×10^6 km^2; Figure 3) formed a significant stage in the evolution of at least 41% of all coronae and that the timing and scale of many coronae flow fields are consistent with the arrival and pressure-release melting of material in the head of a mantle plume. They also showed that those coronae with associated large-scale volcanic activity were preferentially located in areas of rifting (Figure 5c) and interpreted this to mean that the intersection of mantle upwellings with zones of extension is the most significant factor in the volume of melt produced and erupted from a corona.

Among the features observed on the volcanic plains of Venus are sinuous channels [*Baker et al.*, 1992], many of which resemble sinuous rilles on the Moon. Lunar sinuous rilles, which range up to about 300 km in length, are thought to have formed from thermal erosion associated with high-effusion-rate eruptions [e.g., *Hulme*, 1982]. In some cases on Venus the deposits from the lavas that are proposed to have eroded the sinuous channels can be identified and distinguished from the surrounding plains. Many of the channels are much longer than 300 km and, indeed, one of them, Baltis Vallis (Figure 5b), is over 6800 km in length! These extreme lengths imply very high effusion rates and large-volume eruptions and may indicate that the lavas associated with these features were of an unusual composition, temperature, and/or viscosity (e.g., possibly komatiites [see *Head et al.*, 1994] or carbonatites [*Kargel et al.*, 1994]). Clearly, the eruption products forming these channels would qualify as LIPs.

Large lava flow fields approaching and exceeding the dimensions of many terrestrial LIPs are observed in numerous other settings on Venus. Mylitta Fluctus is one such field that originates along a rift zone and covers about 3×10^5 km^2, extending down into Lavinia Planitia, a lowland basin. Maximum flow lengths range from 400 to 1000 km, flow widths from 30 to 100 km, and the total volume of the flow field is of the order of 2×10^4 km^3 [*Magee Roberts et al.*, 1992]. *Magee and Head* [1995] documented the morphology, morphometry, stratigraphy and distribution of the global population of large flow fields on Venus in excess of 5×10^4 km^2. The largest of the 208 such flow fields is 1.6×10^6 km^2 and the average area is 2.2×10^5 km^2 (Figure 3); collectively, the flow fields cover an area of 4.0×10^7 km^2, about 11% of the plains regions of Venus. The most common source vents for the large flow fields are coronae, large volcanic shields, and fissures and fractures within rifts and fracture belts. Most flow fields are associated with zones of extension, such as major rift zones and fracture belts, and the emplacement of the flow fields tended to postdate the onset of extension. In reference to terrestrial flood basalts and the discussion about the relative importance of large-scale mantle upwelling (e.g., plume heads) versus lithospheric extension causing enhanced decompressional melting, the Venus data support the idea that lithospheric extension and thinning accompany the formation of the majority of flood-basalt lavas there. In addition, examination of a 6800-km-long rift zone interpreted to have originated from passive rifting in response to stresses linked to adjacent downwelling shows that extension occurred generally prior to the eruption of large-scale volcanic flow fields, comparable to some terrestrial flood basalts [*Magee Roberts et al.*, 1992; *Magee and Head*, 1995]. This is in contrast to the Columbia River and Deccan Basalt Groups, where evidence has been presented that eruption of the main tholeiitic phase preceded significant extension and crustal thinning [*Hooper*, 1990].

Many rift zones on Venus are associated with broad rises resembling the Tharsis and Elysium regions on Mars. The Beta, Atla, and Western Eistla regions are each up to 2000-3000 km in diameter and rise up to several kilometers above the surrounding plains (Figure 5a). They are characterized by rift systems which cross (Western Eistla) or radiate away from (Beta and Atla) the central high, and large shield volcanoes are located on the summit and flanks of the rise. Positive gravity anomalies are consistent with mantle upwelling [*Senske et al.*, 1992]. These rises appear to represent a scale of mantle upwelling much larger than that related to individual volcanoes and coronae [e.g., *Head et al.*, 1992], although several coronae reach tremendous dimensions (e.g., Heng-O, 1060 km, and Artemis, ~2500 km). The global distribution of volcanic landforms revealed by the Magellan mission showed a concentration of volcanic edifices and sources in an area comprising about 9.2×10^7 km^2, or about 20% of the surface of Venus [*Head et al.*, 1992; *Crumpler et al.*, 1993, 1996b]. This area of regionally abundant volcanic sources

also corresponds to the location of three major rifted rises, Beta Regio, Atla Regio, and Themis Regio (thus the term BAT region). This concentration suggests several scales of upwelling and instabilities (relatively small for the individual volcanic source regions, a thousand kilometers for the broad rises, and perhaps many thousands of kilometers for the BAT region). The BAT region covers a comparable planetary surface area percentage (20%) to that of the Tharsis rise on Mars. The ages of its individual components (rifting, rises, volcanoes) largely postdate the earliest plains emplacement [*Crumpler et al.*, 1994; *Basilevsky and Head*, 1995a, b], and thus it may represent mantle convection patterns linked to the aftermath of the collapse of a negatively buoyant, depleted mantle layer remaining from the extraction of the basaltic crust [e.g., *Head*, 1995].

Preliminary analysis of the global stratigraphy of Venus suggests that the dominant geologic processes and styles of volcanism have changed over time [*Basilevsky and Head*, 1995a,b; 1996]. The oldest terrain exposed is known as tessera (tile in Greek, for the similarity of the terrain texture to parquet floor tiles). This terrain is high-standing and very complexly deformed, somewhat continent-like, and comprises about 8% of the surface of Venus [*Ivanov and Head*, 1996]. Tessera is embayed by two major plains units and is thus probably much more widespread in the subsurface than its present outcrop would suggest. The oldest of the two major plains units (ridged plains) covers most of the surface of Venus. Most ridged plains are homogeneous and flow-unit boundaries generally are not traceable; however, ridged plains are characterized by numerous sinuous channels, suggesting large-volume, high-effusion-rate eruptions. The stratigraphically younger regional plains unit is characterized by smoother lobate flows (such as those at Mylitta Fluctus) usually emanating from discrete sources. Although individual flows are volumetrically significant and often akin to flood basalts, they do not have the distinctive sinuous channels of the ridged plains and thus appear to have a different mode of emplacement. Large volcanic edifices representing individual volcanic sources, and the emplacement of hundreds of flows from subjacent localized magma reservoirs are the most recent features and these are superposed on most other types of plains units [*Crumpler et al.*, 1997]. Thus, following tessera formation, two units that could be interpreted as LIP-related were formed: the ridged plains, where large-scale, sinuous channel-type emplacement occurred, and the smooth/lobate plains, where flood-basalt-like provinces were produced, often in conjunction with rift zones. This major change over time (together with substantial changes observed in the thermal evolution of Mars and the Moon) suggests that the style, number, and size of LIPs may vary over the long-term geologic record of the Earth.

What causes these changes on Venus? The size-frequency distribution of exposed impact craters shows that the average surface age (crater retention age) is about 300–500 Ma, much more similar to that of the Earth than the smaller terrestrial planets (Figure 1). Even more surprisingly, the areal distribution of craters cannot be distinguished from a completely spatially random population [Schaber et al., 1992; Phillips et al., 1992; Strom et al., 1994] and nearly all of the craters have not been modified by post-emplacement volcanism. On the basis of these results, it was hypothesized that Venus underwent a global tectonic and volcanic resurfacing event about 300–500 m.y. ago that eradicated the previous cratering record. Subsequent to that event (thought to have lasted about 10^7–10^8 years), volcanism was relatively minor in volume and areal distribution (on the basis of the small number of craters modified by volcanic activity) [Schaber et al., 1992; Strom et al., 1994]. Many mechanisms have been proposed to explain this hypothesized event [e.g., see review of Solomon, 1993], including episodic plate tectonics [e.g., Turcotte, 1993]. Among these hypotheses [e.g., Schaber et al., 1992; Parmentier and Hess, 1992; Head et al., 1994] are mechanisms that call for near-global volcanic resurfacing, in effect a planet-wide LIP! In the scenario proposed by Parmentier and Hess [1992], vertical crustal accretion leads to formation of a thick, melt-depleted mantle layer that evolves chemically and thermally over geologic time; the depleted mantle layer ultimately becomes negatively buoyant and founders. This event is predicted to occur over a geologically short time [Parmentier and Hess, 1992] and the foundering, downwelling depleted mantle layer is hypothesized to have deformed much of the crust into tessera terrain, while the complementary upwelling fertile mantle underwent massive pressure-release melting to produce voluminous sinuous-channel-related flood basalts over a relatively short time ($\leq 10^8$ yrs) [Head et al., 1994]. Subsequently, volcanism waned but was locally significant in rifts (where local flood basalt units were emplaced, such as Mylitta Fluctus) and near hotspots, where magma reservoirs evolved to produce volcanic edifices (e.g., Sapas Mons).

Three important observations can be made relative to the study of LIPs on Earth. One, the hypothesized depleted-mantle-layer overturn event on Venus could be the equivalent of a planet-wide LIP. Approximately 80% of the planet (3.68×10^8 km^2; Figure 3) may have been resurfaced over a very short time during the emplacement of the ridged plains [Basilevsky and Head, 1995a, b].

Although volumes are not well known, they may be of the order of 9.2×10^8 km^3 [Head et al., 1992]. Effusion rates are likely to have been very high for many plains units emplaced through sinuous channels, but estimates are difficult to make. Integrated fluxes were very high, however. If the ridged plains were emplaced in about 10^8 years, the average flux would be 5–7 km^3/a (Figure 3), which is about three orders of magnitude more than the peak lunar mare flux, approximately comparable to the typical extrusive component of the Earth at present (intraplate and plate boundary) and more than five times greater than the flux typical of the last tens of millions of years for Venus. Two, the great contrast in magma flux between the resurfacing event and subsequent activity shows that major changes can take place in the geologic history of a planet; different types of LIPs can occur in relatively rapid succession, and periods can occur when virtually none are emplaced. Three, if either the hypothesis concerning the buildup and collapse of the depleted mantle layer or the episodic-plate-tectonic hypothesis is correct, this means that large-scale planetary heat loss can be cataclysmic and episodic, a phenomenon not considered in monotonic thermal evolution models.

One implication of the tectonic and volcanic record of Venus is that the crust forms and evolves [Head, 1990b] primarily in a vertical sense rather than in a lateral sense, as is the case in terrestrial oceanic plate spreading, although hypotheses for episodic plate tectonics on Venus have been proposed [e.g., Turcotte, 1993]. This concept of vertical crustal accretion has important implications for the production of LIPs on Venus and the general evolution of secondary crust over geologic time [Head et al., 1994], as discussed below.

Other Planetary Bodies

Outside the orbits of the terrestrial planets lie the asteroid belt and the outer gas giant planets and their satellites. Some meteorites and asteroids show evidence for differentiation and basaltic volcanism [e.g., Taylor et al., 1993], phases of which may have been volumetrically significant [e.g., Wilson and Keil, 1996]. Outer planet satellites are predominantly low-density bodies composed primarily of water and related ices [e.g., Burns and Matthews, 1986]. One exception is the innermost of the Galilean satellites of Jupiter, Io, which is approximately the same size and density as the Earth's moon. In one of the most spectacular predictions [Peale et al., 1979] and discoveries [Morabito et al., 1979] of planetary exploration, images returned by Voyager showed numerous active volcanic eruptions on Io [Smith et al., 1979].

Through a combination of pyroclastic eruptions and lava flows, Io appears to be resurfaced at the phenomenally high rate of 10^{-4} to 1 cm/yr [*Nash et al.*, 1986]. Further exploration by the Galileo mission will provide evidence for the nature of changes on Io in the last 17 years and the relation of these resurfacing rates and styles to LIPs.

SUMMARY, RELEVANCE TO TERRESTRIAL LIPS, AND OUTSTANDING QUESTIONS

Environments, Associations, Settings of Formation and Style of Emplacement

The planetary record provides a perspective on the three main categories of terrestrial LIPs: oceanic plateaus, continental flood basalts, and volcanic passive margins. Multiple analogs to continental flood basalts (e.g., Moon) and to oceanic plateaus (e.g., Mars, Venus) exist on the terrestrial planets; the rift-related LIPs on Venus provide important information about sequence and timing of emplacement in relation to volcanic passive margins. The role of large-volume, long-distance, lateral dike emplacement of flood basalts is illustrated by coronae on Venus; these provide probable analogs to the now-eroded giant radiating dike swarms of Earth and associated, but largely eroded, flood basalts. The great range of scales of upwellings on Venus and Mars, the influence of crustal thickness and composition, and lithospheric thickness variations in space and time also are significant for studies of terrestrial environments. Venus shows the potential significance of vertical crustal accretion and the influence of the complementary depleted mantle layer on further petrogenetic evolution, as well as the possibility of episodic, cataclysmic planetary-scale resurfacing. Theoretical analysis of environments on Venus also illustrates that differences in thermal structure can cause fundamental differences in the volume of melt produced in buoyant upwellings in both rise [e.g., *Sotin et al.*, 1989] and plume [e.g., *Erickson and Arkani-Hamed*, 1992] environments. It has been proposed that many terrestrial LIPs are analogous to the lunar maria and resulted from impact-related pressure-release melting [e.g., *Alt et al.*, 1988]. The lunar geologic record shows that this is unlikely at the present time because of the generally small size of impact projectiles, but that it may have been more significant in past Earth history when larger impacts occurred, especially in oceanic settings.

Influence of Surface Environment

The example of Venus suggests that the external environment into which magma is extruded can have a major influence on the occurrence, depth, and size of magma reservoirs and thus on the possibility of flood basalts [*Head and Wilson*, 1992b]. This consideration implies that intraplate submarine reservoirs and extrusions (more Venus-like) may be different from those in subaerial environments on Earth, as indeed might many LIPs formed earlier in Earth's history, when atmospheric pressure may have been higher.

Controls on Mode and Location

The lunar record shows how a low-density crustal layer analogous to continental crust on Earth can act as a filter to plumes and associated volcanic activity, obscuring their surface manifestation and even precluding the construction of shallow magma reservoirs and large shield volcanoes. Variations in lithospheric thickness on the terrestrial planets illustrate how thermal structure can influence the occurrence and mode of large-volume extrusions and how changes in lithospheric thickness with time can alter the style and abundance of flood basalts.

Relation to Internal Structure

The wide range of plume-like features with associated large extrusive components suggests that plume sources could possibly extend from the upper mantle to the core-mantle boundary. The extremely large and long-lasting provinces on Mars (Tharsis) and Venus (BAT) strongly suggest that even larger instabilities occur than those commonly associated with individual terrestrial plumes. Better knowledge of variations in the volcanic flux associated with plume-like features on Venus will help us to understand terrestrial plume and mantle structure [e.g., *Bercovici and Mahoney*, 1994]. Another potentially important perspective comes from the vertical crustal accretion hypothesis for Venus; the formation and accumulation of a complementary melt-depleted mantle layer can significantly alter the nature of further melt production and can also influence crustal buoyancy and stability. The lunar highland crust density barrier also illustrates that in some cases basalts may localize at the base of the crust, as is believed to occur for terrestrial flood basalt magmas. Venus demonstrates that large-scale melting in the mantle (the hypothesized large-scale mantle overturn) may have had very long-lasting effects on mantle convection patterns and volcanism [e.g., *Crumpler et al.*, 1993].

Implications for Plume Structure

Although the compositional, thermal and mechanical structure of the crust and lithosphere of the planets is not the same as on the Earth today, the planetary record can provide a frame of reference for questions such as plume structure (plume heads and tails), plume incubation versus plume impact [e.g., *Kent et al.*, 1992], internal mantle structure [e.g., *Bercovici and Mahoney*, 1994], and plume duration (e.g., on Venus and Mars).

Large Igneous Province Substructure

Eroded, giant radiating dike swarms on Earth (e.g., the Mackenzie dike swarm) and their uneroded counterparts on Venus [e.g., *Ernst et al.*, 1995] show that flood basalts need not occur only above a plume head; dike thicknesses, lengths, and flow rates are such that flood basalts can occur several thousand kilometers away from a plume, through lateral transport of magma in dikes and its eruption due to buffered conditions in the magma reservoir [*Parfitt and Head*, 1993].

Areas and Volumes

Planetary LIPs have a wide range of areas and volumes (Figure 3), showing that terrestrial LIPs are not unique in this respect. In addition, volumes range up to that of Tharsis on Mars, and scales on Venus exceed the present surface area of the oceanic crust. These examples suggest the possibility of larger terrestrial LIPs than presently recognized (e.g., terrestrial superplumes [*Larson*, 1991a, b]). The volumes of the larger planetary LIPs and the stationary lithosphere of most terrestrial planets suggest that voluminous partial melting of mantle has occurred (e.g., see figure 10 of *Coffin and Eldholm* [1994]) and that residual, depleted mantle layers must play an important role in the continued evolution of planetary upper mantles.

Duration and Rates of Emplacement

Planetary LIPs are seen in which volumes were very high and eruption durations were both short (the large out-flows and sinuous rilles on the Moon) and long (the shield volcanoes on Mars). In addition, the Tharsis rise on Mars shows that mantle melting anomalies can last billions of years, producing prodigious LIPs, and the Venus global resurfacing model suggests that large-scale mantle overturn may provide short-term ($\leq 10^8$ yr) pulses of global-scale igneous provinces. On the basis of the planetary perspective, high eruption rates may be due to a variety of conditions, including trapping of melt at density barriers and subsequent overpressurization of reservoirs [*Head and Wilson*, 1992a], buffered conditions in magma reservoirs [*Parfitt and Head*, 1993], higher temperatures

and degrees of melting in the earlier history of planets, periods of possible cataclysmic resurfacing (e.g., on Venus [*Head et al.*, 1994]), and periods of anomalous rates of mantle convection. The cataclysmic resurfacing hypothesis for Venus also illustrates the possibility that thermal evolution may not be steady-state and monotonic but rather episodic [see also *Condie*, 1995].

Petrogenetic Evolution

Do extraterrestrial LIPs contain differentiates of basalt, and if so, where do these occur in the sequence? In situ geochemical analyses on Venus suggest that tholeiites and possibly more alkaline basalts [*Surkov et al.*, 1987] form the vast volcanic plains. The extensive sinuous channels have been interpreted as evidence for possible komatiites [*Head et al.*, 1994] and carbonatites [*Kargel et al.*, 1994]. Steep-sided domes [*Pavri et al.*, 1992] and large deposits of viscous-appearing deposits [*Moore et al.*, 1992] have been observed. Unfortunately, widespread and detailed chemical analyses have not yet been made on the planets.

Influence on the Atmosphere and Environment

Voluminous and prolonged volcanic outpourings can make important contributions to the atmosphere of planets throughout their evolution, as on Mars [e.g., *Greeley*, 1987]. If large volumes of flood basalts are extruded over very short periods, outgassed volatiles, heat flux, and voluminous particulate matter can influence short-term chemistry and circulation of the atmosphere and long-term climate evolution. Potentially the most dramatic example of this is the widespread volcanic resurfacing hypothesized for Venus. For example, *Bullock and Grinspoon* [1996] showed that an increased flux of volcanism such as that interpreted to be associated with the proposed global resurfacing would precipitate a climatic catastrophe leading to much higher temperatures and pressures.

Relation to Geologic History

A most important perspective from the planets is that the characteristics and rates of geologic processes as a function of time and thermal evolution have experienced large-scale changes. The geological processes dominating the geologic record over the last several hundred million years on the planets (i.e., the temporal equivalent of the Phanerozoic on the Earth) are not the same as those operating in the earlier history of Mars, Venus, Mercury, and the Moon. Thus, we should anticipate potentially major changes in the style of volcanic extrusion as a function of geologic time on Earth. It is clear from the planetary record that coincident with the general thermal evolution of the planet, changes can occur in the mantle convection planform, the development and scale of mantle instabilities, and the conditions of melting in the Earth's crust. Much of the evidence from the planets and from thermal evolution models suggests a more important role for LIPs in the earlier history of the Earth.

Origin of LIPs: A Planetary Perspective

The planetary perspective provides many examples of LIPs in a diverse range of geological environments. This underlines the fact that there is no single origin for LIPs, but that, taken together, they can help to understand the nature and significance of large-scale melting in the shallow interiors of planets [e.g., *Coffin and Eldholm*, 1994]. For example, the broad rifted rises of Venus and Mars and their associated LIPs serve as potential analogs for early continental breakup and the early stages of crustal spreading on Earth. Continued analysis of data from the planetary record will help to provide perspective on terrestrial LIP dimensions, durations, and rates of emplacement, as well as mantle and crustal structure and processes, relationship to tectonism, environmental effects and petrological and geochemical characteristics and evolution.

Acknowledgments. Research for this paper was supported by a grant from the National Aeronautics and Space Administration (NAGW-2185) to JWH. MFC acknowledges the support of the Industrial Liaisons Program of the University of Oslo, Norway. Thanks are extended to Mary Ellen Murphy, Anne C. Côté, Anne McKay and Peter Neivert for help in preparation of the manuscript, and to G. J. Taylor and Laszlo Kesthelyi for very helpful reviews. Special thanks are extended to John Mahoney for his excellent scientific and editorial suggestions. University of Texas Institute for Geophysics contribution number 1269.

REFERENCES

Alt, D., J. M. Sears, and D. W. Hyndman, Terrestrial maria: The origins of large basaltic plateaus, hotspot tracks and speading ridges, *J. Geol., 96,* 647-662, 1988.

Baker, V. R., G. Komatsu, T. J. Parker, V. C. Gulick, J. S. Kargel, and J. S. Lewis, Channels and valleys on Venus: Preliminary analysis of Magellan data, *J. Geophys. Res., 97,* 13,421-13,444, 1992.

Banerdt, W. B., M. P. Golombek, and K. L. Tanaka, Stress and tectonics on Mars, in *Mars,* edited by H. H. Kieffer, B. M. Jakosky, C. W. Snyder, and M. S. Matthews, pp. 249-297, The University of Arizona Press, Tucson, 1992.

Barger, K. E., and E. D. Jackson, Calculated volumes of individual shield volcanoes along the Hawaiian-Emperor Chain, *J. Res., U.S. Geol. Surv., 2,* 545-550, 1974.

Basaltic Volcanism Study Project, *Basaltic Volcanism on the Terrestrial Planets,* 1286 pp., Pergamon Press, New York, 1981.

Basilevsky, A. T., and J. W. Head, Global stratigraphy of Venus: Analysis of a random sample of thirty-six test areas, *Earth, Moon and Planets, 66,* 285-336, 1995a.

Basilevsky, A. T., and J. W. Head, Regional and global stratigraphy of Venus: A preliminary assessment and implications for the geologic history of Venus, *Planet. Space Sci., 43,* 1523-1553, 1995b.

Basilevsky, A. T., and J. W. Head, Evidence for rapid and widespread emplacement of volcanic plains on Venus: Stratigraphic studies in the Baltis Vallis region, *Geophys. Res. Lett.,* 1497-1500, 1996.

Bercovici, D., and J. Mahoney, Double flood basalts and plume head separation at the 660-kilometer discontinuity, *Science, 266,* 1367-1369, 1994.

Bratt, S. R., S. C. Solomon, and J. W. Head, The evolution of impact basins: Cooling, subsidence, and thermal stress, *90,* 12,415-12,433, 1985.

Bruno, B. C., G. J. Taylor, S. K. Rowland, P. G. Lucey, and S. Self, Lava flows are fractals, *Geophys. Res. Lett., 19,* 305-308, 1992.

Bullock, M. A., and D. H. Grinspoon, The stability of climate on Venus, *J. Geophys. Res., 101,* 7521-7529, 1996.

Burns, J. A., and M. S. Matthews, editors, *Satellites,* The University of Arizona Press, Tucson, 1021 pp, 1986.

Cameron, A. G. W., B. Fegley Jr., W. Benz, and W. L. Slattery, The strange density of Mercury: Theoretical considerations, in *Mercury,* edited by F. Vilas, C. R. Chapman, M. S. Matthews, pp. 692-708, The University of Arizona Press, Tucson, 1988.

Carr, M. H., Volcanism on Mars, *J. Geophys. Res., 78,* 4049-4062, 1973.

Carr, M. H., The role of lava erosion in the formation of lunar rilles and martian channels, *Icarus, 22,* 1-23, 1974.

Carr, M. H., *The Surface of Mars,* Yale University Press, New Haven and London, p. 232, 1981.

Chapman, C. R., Mercury: Introduction to an end-member planet, in *Mercury,* edited by F. Vilas, C. R. Chapman, and M. S. Matthews, pp. 1-23, The University of Arizona Press, Tucson, 1988.

Coffin, M. F., and O. Eldholm, Volcanism and continental breakup: A global compilation of large igneous provinces, in *Magmatism and the Causes of Continental Break-Up, Spec. Publ. 68,* edited by B. C. Storey, T. Alabaster, and R. J. Pankhurst, pp. 21-34, The Geological Society, London, 1992.

Coffin, M. F., and O. Eldholm, Scratching the surface: Estimating dimensions of large igneous provinces, *Geology, 21,* 515-518, 1993.

Coffin, M. F., and O. Eldholm, Large igneous provinces: Crustal structure, dimensions, and external consequences, *Rev. Geophysics, 32,* 1-36, 1994.

Comer, R., S. Solomon, and J. W. Head, Mars: Thickness of the lithosphere from the tectonic response to volcanic loads, *J. Geophys. Res., 23*, 61-92, 1985.

Condie, K. C., Episodic ages of greenstones: A key to mantle dynamics?, *Geophys. Res. Lett., 22*, 2215-2218, 1995.

Crisp, J. A., Rates of magma emplacement and volcanic output, *J. Volcanol. Geotherm. Res., 20*, 177-211, 1984.

Cruikshank, D. P., The Development of Studies of Venus, in *Venus*, edited by D. M. Hunten, L. Colin, T. M. Donahue, and V. I. Moroz, pp. 1-9, The University of Arizona Press, Tucson, 1983.

Crumpler, L. S., J. W. Head, and D. B. Campbell, Orogenic belts on Venus, *Geology, 14*, 1031-1034, 1986.

Crumpler, L. S., J. W. Head, and J. C. Aubele, Relation to major volcanic center concentration on Venus to global tectonic patterns, *Science, 261*, 591-595, 1993.

Crumpler, L. S., J. C. Aubele, and C. D. Condit, Volcanoes and neotectonic characteristics of the Springerville volcanic field, Arizona, New Mexico Geological Society Guidebook, 45th Field Conference, Mogollon Slope, West-Central New Mexico and East-Central New Mexico, 147-164, 1994.

Crumpler, L. S., J. W. Head, and J. C. Aubele, Calderas on Mars: Characteristics, structural evolution, and associated flank structures, in *Volcano Instability on the Earth and Other Planets, Spec. Publ. 110*, edited by W. J. McGuire, A. P. Jones, and J. Neuberg, eds., pp. 307-348, The Geological Society, London, 1996.

Crumpler, L. S., J. C. Aubele, D. A. Senske, S. T. Keddie, K. P. Magee, and J. W. Head, Volcanoes and centers of volcanism on Venus, *Venus II, University of Arizona Press* (in press), 1997.

Erickson, S. G., and J. Arkani-Hamed, Impingement of mantle plumes on the lithosphere: Contrast between Earth and Venus, *Geophys. Res. Lett., 19*, 885-888, 1992.

Ernst, R. E., J. W. Head, E. Parfitt, E. Grofils, and L. Wilson, Giant radiating dyke swarms on Earth and Venus, *Earth-Sci. Rev., 39*, 1-58, 1995.

Esposito, P. B., W. B. Banerdt, G. F. Lindal, W. L. Sjogren, M. A. Slade, B. G. Bills, D. E. Smith, and G. Balmino, Gravity and topography, in *Mars*, edited by H. H. Kieffer, B. M. Jakosky, C.W. Snyder, and M. S. Matthews, pp. 209-248, The University of Arizona Press, Tucson, 1992.

Feighner, M., and M. Richards, Lithospheric structure and compensation mechanisms of the Galapagos Archipelago, *J. Geophys. Res., 99*, 6711-6729, 1994.

Finnerty, A. A., R. J. Phillips, and W. B. Banerdt, Igneous processes and the closed system evolution of the Tharsis region of Mars, *J. Geophys. Res., 93*, 10,225-10,235, 1988.

Frey, H., Crustal evolution of the early Earth: The role of major impacts, *Precamb. Res., 10*, 195-216, 1980.

Glazner, A., and W. Ussler, Trapping of magma at midcrustal density discontinuities, *Geophys. Res. Lett., 15*, 673-675, 1988.

Goettel, K. A., Present bounds on the bulk composition of Mercury: Implications for the planetary formation process, in *Mercury*, edited by F. Vilas, C. R. Chapman, M. S. Matthews, pp. 613-621, The University of Arizona Press, Tucson, 1988.

Greeley, R., Lunar Hadley Rille: Considerations of its origin, *Science, 172*, 722-725, 1971.

Greeley, R., Release of juvenile water on Mars: Estimated amounts and timing associated with volcanism, *Science, 236*, 1653-1654, 1987.

Greeley, R., and J. E. Guest, Geologic map of the eastern equatorial region of Mars, *U.S. Geol. Surv. Misc. Inv. Series Map I-1802-B*, 1987.

Greeley, R., M. J. S. Belton, L. R. Gaddis, J. W. Head, S. D. Kadel, A. S. McEwen, S. L. Murchie, G. Neukum, C. M. Pieters, J. M. Sunshine, and D. A. Williams, Galileo imaging observations of lunar maria and related deposits, *J. Geophys. Res., 98*, 17,183-17,205, 1993.

Grieve, R. A. F., Impact bombardment and its role in proto-continental growth on the early Earth, *Precamb. Res., 10*, 217-248, 1980.

Grieve, R. A. F., and E. M. Parmentier, Impact phenomena as factors in the evolution of the Earth, *Proc. 27th Int. Geol. Cong, 19*, VNU Science Press, 99-114, 1984.

Grieve, R. A. F., P. B. Robertson, and M. R. Dence, Constraints on the formation of ring impact structures, based on terrestrial data, in *Multi-Ring Basins*, edited by P. H. Schultz and R. B. Merrill, pp. 37-57, Pergamon, New York, 1991a.

Grieve, R. A. F., D. Stoffler, and A. Deutsch, The Sudbury Structure: Controversial or misunderstood?, *J. Geophys. Res., 96*, 22,753-22,764, 1991b.

Griffiths, R. W., and I. H. Campbell, Interaction of mantle plume heads with the Earth's surface and onset of small-scale convection, *J. Geophys. Res., 96*, 18,295-18,310, 1991.

Grosfils, E. B., and J. W. Head, The global distribution of giant radiating dike swarms on Venus: Implications for the global stress state, *Geophys. Res. Lett., 21*, 701-704, 1994.

Grosfils, E. B., and J. W. Head, Radiating dike swarms on Venus: Evidence for emplacement at zones of neutral buoyancy, *Planet. Space Sci., 43*, 1555-1560, 1995.

Guest, J. E., Centers of igneous activity in the maria, in *Geology and Geophysics of the Moon*, edited by G. Fielder, pp. 41-53, Elsevier, Amsterdam, 1971.

Hall, J. L., S. C. Solomon, and J. W. Head, Elysium Region, Mars: Tests of lithospheric loading models for the formation of tectonic features, *J. Geophys. Res., 91*, 11,377-11,392, 1986.

Hartmann, W. K., Martian cratering V: Toward an empirical Martian chronology, and its implications, *Geophys. Res. Lett., 5*, 450-452, 1978.

Head, J. W., Orientale multi-ringed basin interior and implications for the petrogenesis of lunar highland samples, *The Moon, 11*, 327-356, 1974.

Head, J. W., Mode of occurrence and style of emplacement of lunar mare deposits, in *Origin of Mare Basalts, Proc. Lunar Sci. Institute Conf., 234*, 61-65, 1975a.

Head, J. W., Lunar mare deposits: Areas, volumes, sequence, and implication for melting in source areas, in *Origin of Mare Basalts, Proc. Lunar Sci. Institute Conf., 234*, 66-69, 1975b.

Head, J. W., Lunar volcanism in space and time, *Rev. Geophys. Space Sci., 14*, 265-300, 1976.

Head, J. W., The formation of mountain belts on Venus: Evidence for large-scale convergence, underthrusting and crustal imbrication in Freyja Montes, Ishtar Terra, *Geology, 18,* 99-102, 1990a.

Head, J. W., Processes of crustal formation and evolution on Venus: An analysis of topography and crustal thickness variations, *Earth, Moon and Planets, 50/51,* 25-55, 1990b.

Head, J. W., Processes of crustal and depleted mantle layer loss on Venus: Evidence from basins in tesserae, uplands, and plains, *Lunar Planet. Sci. Conf., 26,* 577-578, 1995.

Head, J. W., and S. C. Solomon, Tectonic evolution of the terrestrial planets, *Science, 213,* 62-76, 1981.

Head, J. W., and L. Wilson, The formation of eroded depressions around the sources of lunar sinuous rilles: Observations, *Lunar and Planet. Sci. Conf., 11,* 426-428, 1980.

Head, J. W., and L. Wilson, Volcanic processes and landforms on Venus: Theory, predictions, and observations, *J. Geophys. Res., 91,* 9407-9446, 1986.

Head, J. W., and L. Wilson, Absence of large shield volcanoes and calderas on the Moon: Consequence of magma transport phenomena?, *Geophys. Res. Lett., 18,* 2121-2124, 1991.

Head, J. W., and L. Wilson, Lunar mare volcanism: Stratigraphy, eruption conditions, and the evolution of secondary crusts, *Geochim. Cosmochim. Acta, 55,* 2155-2175, 1992a.

Head, J. W., and L. Wilson, Magma reservoirs and neutral buoyancy zones on Venus: Implications for the formation and evolution of volcanic landforms, *J. Geophys. Res., 97,* 3877-3903, 1992b.

Head, J. W., and L. Wilson, Lunar graben formation due to near-surface deformation accompanying dike emplacement, *Planet. Space Sci., 41,* 719-727, 1994a.

Head, J. W., and L. Wilson, Mars: Formation and evolution of magma reservoirs, *Lunar Planet. Sci. Conf., 25,* 527-528, 1994b.

Head, J. W., L. S. Crumpler, J. C. Aubele, J. E. Guest, and R. S. Saunders, Venus volcanism: Classification of volcanic features and structures, associations, and global distribution from Magellan data, *J. Geophys. Res., 97,* 13,153-13,197, 1992.

Head, J. W., S. Murchie, J. F. Mustard, C. M. Pieters, J. Neukum, A. McEwen, R. Greeley, E. Nagel, and M. J. S. Belton, Lunar impact basins: New data for the western limb and far side (Orientale and South Pole-Aitken Basins) from the first Galileo flyby, *J. Geophys. Res., 98,* 17,149-17,181, 1993.

Head, J. W., E. M. Parmentier, and P. C. Hess, Venus: Vertical accretion of crust and depleted mantle and implications for geological history and processes, *Planet. Space Sci., 42,* 803-811, 1994.

Head, J. W., A. T. Basilevsky, L. Wilson, and P. C. Hess, Evolution of volcanic styles on Venus: Change but not Noachian?, *Lunar Planet. Sci. Conf., 27,* 525-526, 1996a.

Head, J. W., L. Wilson, and D. K. Smith, Mid-ocean ridge eruptive vent morphology and structure: Evidence for dike widths, eruption rates, and evolution of eruptions and axial volcanic ridges, *J. Geophys. Res., 101,* 28,265-28,280, 1996b.

Heiken, G. H., D. T. Vaniman, and B. M. French (Eds.), *Lunar Sourcebook: A User's Guide to the Moon,* 736 pp., Cambridge University Press, New York, 1991.

Hodges C. A., and H. J. Moore, *Atlas of Volcanic Landforms on Mars,* U. S. Geol. Surv. Prof. Pap. 1534, p. 194, 1994.

Hooper, P. R., The timing of crustal extension and the eruption of continental flood basalts, *Nature, 345,* 246-249, 1990.

Hulme, G., Turbulent lava flow and the formation of lunar sinuous rilles, *Mod. Geol., 4,* 107-117, 1973.

Hulme, G., A review of lava flow processes related to the formation of lunar sinuous rilles, *Geophys. Surveys, 5,* 245-279, 1982.

Hunten, D. M., L. Colin, T. M. Donahue, and V. I. Moroz (Eds.), *Venus,* 1143 pp., The University of Arizona Press, Tucson, 1983.

Ivanov, M. A., and J. W. Head, Tessera terrain on Venus: A survey of the global distribution, characteristics, and relation to surrounding units from Magellan data, *J. Geophys. Res., 101,* 14,861-14,908, 1996.

Janes, D. M., S. W. Squyres, D. L. Bindschadler, G. Baer, G. Schubert, V. L. Sharpton, and E. R. Stofan, Geophysical models for the formation and evolution of coronae on Venus, *J. Geophys. Res., 97,* 16,055-16,067, 1992.

Jónsson, J., Vocanic eruption in historical time on the Reykjanes Peninsula, southwest Iceland (in Icelandic with English summary), *Náttúrufraedingurlun, 52,* 127-139, 1983.

Kargel, J. S., R. L. Kirk, B. Fegley Jr., and A. H. Treiman, Carbonate-sulfate volcanism on Venus?, *Icarus, 112,* 219-252, 1994.

Keddie, S. T., and J. W. Head, Sapas Mons, Venus: Evolution of a large shield volcano, *Earth, Moon and Planets, 65,* 129-190, 1994a.

Keddie, S. T., and J. W. Head, Height and altitude distribution of large volcanoes on Venus, *Planet. Space Sci., 42,* 455-462, 1994b.

Kent, R. W., M. Storey, and A. D. Saunders, Large igneous provinces: Site of plume impact or plume incubation?, *Geology, 20,* 891-894, 1992.

Kieffer, H. H., B. M. Jakosky, C. W. Snyder, and M. S. Matthews (Eds.), *Mars,* 1498 pp., The University of Arizona Press, Tucson, 1992a.

Kieffer, H. H., B. M. Jakosky, and C. W. Snyder, The planet Mars: From antiquity to the present, in *Mars,* edited by H. H. Kieffer, B. M. Jakosky, C. W. Snyder, and M. S. Matthews, pp. 1-33, The University of Arizona Press, Tucson, 1992b.

Larson, R. L., Latest pulse of Earth: Evidence for a mid-Cretaceous superplume, *Geology, 19,* 547-550, 1991a.

Larson, R. L., Geological consequences of superplumes, *Geology, 19,* 963-966, 1991b.

Longhi, J., E. Knittle, J. R. Holloway, and H. Wänke, The Bulk Composition, Mineralogy and Internal Structure of Mars, in *Mars,* edited by H. H. Kieffer, B. M. Jakosky, C. W. Snyder, and M. S. Matthews, pp. 184-208, The University of Arizona Press, Tucson, 1992.

Magee Roberts, K., and J. W. Head, Large-scale volcanism associated with coronae on Venus: Implications for formation and evolution, *Geophys. Res. Lett., 20,* 1111-1114, 1993.

Magee, K., and J. W. Head, The role of rifting in the generation of melt: Implications for the origin and evolution of the Lada Terra-Lavinia Planitia region of Venus, *J. Geophys. Res., 100,* 1527-1552, 1995.

Magee Roberts, K., J. E. Guest, J. W. Head, and M. G. Lancaster, Mylitta Fluctus, Venus: Rift-related, centralized volcanism and the emplacement of large-volume flow units, *J. Geophys. Res., 97,* 15,991-16,015, 1992.

Martin, B. S., The Roza Member, Columbia River Basalt Group; Chemical stratigraphy and flow distribution, in *Volcanism and Tectonism in the Columbia River Flood-basalt Province, Spec. Pap. 239,* edited by S. Reidel and P. Hooper, Geological Society of America, , pp. 85-104, 1989.

McGovern, P., and S. Solomon, State of stress, faulting, and eruption characteristics of large volcanoes on Mars, *J. Geophys. Res., 98,* 23,553-23,579, 1993.

McGuire, W. J., A. P. Jones, and J. Neuberg (Eds.), *Volcano Instability on the Earth and Other Planets, Spec. Publ. 110,* 388 pp., The Geological Society, London, 1996.

McKenzie, D., P. G. Ford, C. Johnson, B. Parsons, D. Sandwell, S. Saunders, and S. C. Solomon, Features on Venus generated by plate boundary processes, *J. Geophys. Res., 97,* 13,533-13,544, 1992.

McNutt, M., K. Fischer, S. Kruse, and J. Natland, The origin of the Marquesas fracture zone ridge and its implications for the nature of hot spots, *Earth Planet. Sci. Lett., 91,* 381-393, 1989.

Moore, H. J., J. J. Plaut, P. M. Schenk, and J. W. Head, An unusual volcano on Venus, *J. Geophys. Res., 97,* 13,479-13,493, 1992.

Morabito, L. A., S. P. Synnott, P. Kupferman and S. A. Collins, Discovery of currently active extraterrestrial volcanism, *Science, 204,* 972, 1979.

Mouginis-Mark, P. J., L. Wilson, and J. W. Head, Explosive volcanism at Hecates Tholus, Mars: Investigation of eruption conditions, *J. Geophys. Res., 87,* 9890-9904, 1982.

Mouginis-Mark, P. J., L. Wilson, J. W. Head, S. H. Brown, J. L. Hall, and K. Sullivan, Elysium Planitia, Mars: Regional geology, volcanology and evidence for volcano/ground ice interactions, *Earth, Moon, and Planets, 30,* 149-173, 1984.

Nash, D. B., M. H. Carr, J. Gradie, D. M. Hunten and C. F. Yoder, Io, in *Satellites,* J. A. Burns and M. S. Matthews, editors, pp. 629-688, The University of Arizona Press, Tucson, 1986.

Neukum, G., and D. U. Wise, Mars: A standard crater curve and possible new time scale, *Science, 194,* 1381-1387, 1976.

Oberbeck, V. R., The role of ballistic erosion and sedimentation in lunar stratigraphy, *Rev. Geophys., 13,* 337-362, 1975.

Oberbeck, V. R., W. L. Quaide and R. Greeley, On the origin of lunar sinuous rilles, *Mod. Geol., 1,* 75-80, 1969.

Oberbeck, V. R., F. Hörz, R. H. Morrison, W. L. Quaide, and D. E. Gault, On the origin of the lunar smooth-plains, *The Moon, 12,* 19-54, 1975.

Oberbeck, V. R., J. R. Marshall, and H. Aggarwal, Impacts, tillites, and the breakup of Gondwanaland, *J. Geology, 101,* 1-19, 1993.

Parfitt, E. A., and J. W. Head, Buffered and unbuffered dike emplacement on Earth and Venus: Implications for magma reservoir size, depth, and rate of magma replenishment, *Earth, Moon, and Planets, 61,* 249-281, 1993.

Parmentier, E. M., and P. C. Hess, Chemical differentiation of a convecting planetary interior: Consequences for a one plate

planet such as Venus, *Geophys. Res. Lett., 19,* 2015-2018, 1992.

Pavri, B., J. W. Head, K. B. Klose, and L. Wilson, Steep-sided domes on Venus: Characteristics, geologic setting, and eruption conditions from Magellan data, *J. Geophys. Res., 97,* 13,445-13,478, 1992.

Peale, S. J., P. Cassen and R. T. Reynolds, Melting of Io by tidal dissipation, *Science, 203,* 892-894, 1979.

Peterson, D. W., and R. B. Moore, Geologic history and evolution of geologic concepts, Island of Hawaii, in *Volcanism in Hawaii, Prof. Pap, 1350,* edited by R. W. Decker, T. L. Wright, and P. H. Stauffer, pp. 149-189, U.S. Geological Survey, Washington, D.C., 1987.

Phillips, R. J., N. H. Sleep, and W. B. Banerdt, Permanent uplift in magmatic systems with application to the Tharsis region of Mars, *J. Geophys. Res., 95,* 5089-5100, 1990.

Phillips, R. J., R. F. Raubertas, R. E. Arvidson, I. C. Sarkar, R. R. Herrick, N. Izenberg, and R. E. Grimm, Impact craters and Venus resurfacing history, *J. Geophys. Res., 97,* 15,923-15,984, 1992.

Plescia, J. B., and R. S. Saunders, The chronology of the Martian volcanoes, *Proc. Lunar Planet. Sci. Conf., 10,* 2841-2859, 1979.

Plescia, J. B., and R. S. Saunders, Estimation of the thickness of the Tharsis lava flows and implications for the nature of the topography of the Tharsis plateau, *Proc. Lunar Planet. Sci. Conf., 11,* 2423-2426, 1980.

Pronin, A. A., and E. R. Stofan, Coronae on Venus: Morphology, classification and distribution, *Icarus, 87,* 452-474, 1990.

Rogers, G. C., Oceanic plateaus as meteorite impact signatures, *Nature, 299,* 341-342, 1982.

Ryan, M. P., The mechanics and three-dimensional internal structure of active magmatic systems: Kilauea volcano, Hawaii, *J. Geophys. Res., 93,* 4213-4248, 1988.

Saunders, R. S., and 26 others, Magellan Mission Summary, *J. Geophys. Res., 97,* 13,067-13,091, 1992.

Schaber, G. G., Lava flows in Mare Imbrium: Geologic evaluation from Apollo orbital photography, *Proc. Lunar Planet. Sci. Conf. 4,* 73-92, 1973.

Schaber, G. G., K. C. Horstman, and A.L. Dial Jr., Lava flow materials in the Tharsis regions of Mars, *Proc. Lunar Planet. Sci. Conf. 9,* 3433-3458, 1978.

Schaber, G. G., R. G. Strom, H. J. Moore, L. A. Soderblom, R. L. Kirk, D. J. Chadwick, D. D. Dawson, L. R. Gaddis, J. M. Boyce, and J. Russell, Geology and distribution of impact craters on Venus: What are they telling us?, *J. Geophys. Res., 97,* 13,257-13,301, 1992.

Schubert, G., R. E. Lingenfelter, and S. J. Peale, The morphology, distrbution, and origin of lunar sinuous rilles, *Rev. Geophys. Space Phys., 8,* 199-224, 1970.

Schubert, G., D. Bercovici, and G. A. Glatzmaier, Mantle dynamics in Mars and Venus: Influence of an immobile lithosphere on three-dimensional mantle convection, *J. Geophys. Res., 95,* 14,105-14,130, 1990.

Schubert, G. G., R. G. Strom, H. J. Moore, L. A. Soderblum, R. L. Kirk, D. J. Chadwick, D. D. Dawson, L. R. Gaddis, J. M.

Boyce, and J. Russell, Geology and distribution of impact craters on Venus: What are they telling us?, *J. Geophys. Res., 97,* 13,257-13,302, 1992.

Schultz, P. H., and P. D. Spudis, Evidence for ancient mare volcanism, *Proc. Lunar Planet. Sci. Conf., 10,* 2899-2918, 1979.

Schultz, P. H., and P. D. Spudis, The beginning and end of lunar mare volcanism, *Nature, 302,* 233-236, 1983.

Schultz, P. H., R. A. Schultz, and J. R. Rogers, The structure and evolution of ancient impact basins on Mars, *J. Geophys. Res., 78,* 9803-9820, 1982.

Scott, D. H., and M. H. Carr, Geologic Map of Mars, scale 1:25,000,000, Map I-1083, *U. S. Geol. Surv. Misc. Invest. Ser.,* 1978.

Scott, D. H., and K. L. Tanaka, Geologic map of the western equatorial region of Mars, scale 1:15,000,000, Map I-1802A, *U. S. Geol. Surv. Misc. Invest. Ser.,* 1986.

Self, S., T. Thordarson, L. Keszthelyi, G. P. L. Walker, K. Hon, M. T. Murphy, P. Long, and S. Finnemore, A new model for the emplacement of the Columbia River Basalt as large, inflated pahoehoe sheet lava flow fields, *Geophys. Res. Lett., 23,* 2689-2692, 1996.

Senske, D., G. G. Schaber, and E. R. Stofan, Regional topographic rises on Venus: Geology of western Eistla region and comparison to Beta Regio and Alta Regio, *J. Geophys. Res., 97,* 13,395-13,420, 1992.

Smith, B. A., and 21 others, The Jupiter system through the eyes of Voyager, *Science, 204,* 951-972, 1979.

Snyder, C. W., and V. I. Moroz, Telescopic Observations: Visual, Photographic, Polarimetric, in *Mars,* edited by H. H. Kieffer, B. M. Jakosky, C. W. Snyder, and M. S. Matthews, pp. 71-119, The University of Arizona Press, Tucson, 1992.

Solomon, S. C., Mare volcanism and lunar crustal structure, *Proc. Lunar Sci. Conf., 6,* 1021-1042, 1975.

Solomon, S. C., The relationship between crustal tectonics and internal evolution on the Moon and Mercury, *Phys. Earth Planet. Inter., 15,* 135-145, 1977.

Solomon, S. C., The geophysics of Venus, *Physics Today,* 48-55, 1993.

Solomon, S. C., and J. W. Head, Mechanisms for lithospheric heat transport on Venus: Implications for tectonic style and volcanism, *J. Geophys. Res., 87,* 9236-9246, 1982.

Solomon, S. C., and J. W. Head, Heterogeneities in the thickness of the elastic lithosphere of Mars: Constraints on heat flow and internal dynamics, *J. Geophys. Res., 95,* 11,073-11,083, 1990.

Solomon, S. C., and J. W. Head, Fundamental issues in the geology and geophysics of Venus, *Science, 252,* 252-260, 1991.

Solomon, S. C., R. P. Comer, and J. W. Head, The evolution of impact basins: Viscous relaxation of topographic relief, *J. Geophys. Res., 87,* 3975-3992, 1982.

Solomon, S. C., S. E. Smrekar, D. L. Bindschadler, R. E. Grimm, W. M. Kaula, G. E. McGill, R. J. Phillips, R. S. Saunders, G. Schubert, S. W. Squyres, and E. R. Stofan, Venus tectonics: An overview of Magellan observations, *J. Geophys. Res., 97,* 13,199-13,255, 1992.

Sotin, C., D. A. Senske, J. W. Head, and E. M. Parmentier, Terrestrial spreading centers under Venus conditions: Evaluation of a crustal spreading model for western Aphrodite Terra, *Earth Planet. Sci. Lett., 95,* 321-333, 1989.

Spudis, P. D., and J. E. Guest, in *Mercury,* edited by F. Vilas, C. R. Chapman, M. S. Matthews, pp. 118-164, The University of Arizona Press, Tucson, 1988.

Spudis, P. D., G. A. Swann and R. Greeley, The formation of Hadley Rille and implications for the geology of the Apollo 15 region, *Proc. Lunar Planet. Sci. Conf, 18,* 243-254, 1987.

Squyres, S. W., D. M James, G. Baer, D. L. Bindschadler, G. Schubert, V. L. Sharpton, and E. R. Stofan, The morphology and evolution of coronae on Venus, *J. Geophys. Res., 97,* 13,611-13,634, 1992.

Stofan E. R., D. L. Bindschadler, J. W. Head, and E. M. Parmentier, Corona structures on Venus: Models of origin, *J. Geophys. Res., 96,* 20,933-20,946, 1991.

Stofan E. R., V. L. Sharpton, G. Schubert, G. Baer, D. L. Bindschadler, D. M. Janes, and S. W. Squyres, Global distribution and characteristics of coronae and related features on Venus: Implications for origin and relation to mantle processes, *J. Geophys. Res., 97,* 13,347-13,378, 1992.

Strom, R. G., N. J. Trask, and J. E. Guest, Tectonism and volcanism on Mercury, *J. Geophys. Res., 80,* 2478-2507, 1975.

Strom, R. G., G. G. Schaber, and D. D. Dawson, The global resurfacing of Venus, *J. Geophys. Res., 99,* 12,899-12,926, 1994.

Surkov, Yu. A., et al., Uranium, thorium, and potassium in Venusian rocks at the landing site of Vega 1 and 2, *J. Geophys Res., 92, suppl.,* E537-E540, 1987.

Tanaka, K. L., N. K. Isbell, D. H. Scott, R. Greeley, and J. E. Guest, The resurfacing history of Mars: A synthesis of digitized, Viking-based geology. *Proc. Lunar Planet. Sci. Conf., 18,* 665-678, 1988.

Tanaka, K. L., D. H. Scott, and R. Greeley, Global stratigraphy, in *Mars,* edited by H. H. Kieffer, B. M. Jakosky, C. W. Snyder, and M. S. Matthews, pp. 354-382, The University of Arizona Press, Tucson, 1992.

Taylor, G. J., K. Keil, T. McCoy, H. Haack, and E. R. D. Scott, Asteroid differentiation: Pyroclastic volcanism to magma oceans, *Meteoritics, 28,* 34-52, 1993.

Taylor, S. R., Large-scale basaltic volcanism on the Moon and Venus, in *Volcanism,* edited by K. V. Subbarao, pp. 1-20, 1994.

Thordarson, Th., and S. Self, The Laki (Skaftár Fires) and Grímsvötn eruptions in 1783-1785, *Bull. Volcanol., 55,* 233-263, 1993.

Tolan, T., S. Reidel, M. Beeson, J. Anderson, K. Fecht, and D. Swanson, Revisions to the estimates of the areal extent and volume of the Columbia River Basalt Group, in *Volcanism and Tectonism in the Columbia River Flood-basalt Province, Spec. Pap. 239,* edited by S. Reidel and P. Hooper, Geological Society of America, , pp. 1-20, 1989.

Trask, N. J., and R. G. Strom, Additional evidence for mercurial volcanism, *Icarus, 28,* 559-563, 1976.

Turcotte, D. L., An episodic hypothesis for venusian tectonics, *J. Geophys. Res., 98,* 17,061-17,068, 1993.

Vilas, F., C. R. Chapman, M. S. Matthews (Eds.), *Mercury,* The University of Arizona Press, Tucson, pp. 1-752, 1988.

White, R. S., and D. McKenzie, Magmatism at rift zones: The generation of volcanic continental margins and flood basalts, *J. Geophys. Res., 94,* 7685-7729, 1989.

White, R. S., and D. McKenzie, Mantle plumes and flood basalts, *J. Geophys. Res., 100,* 17,543-17,585, 1995.

Whitford-Stark, J. L., and J. W. Head, The Procellarum volcanic complexes: Contrasting styles of volcanism, *Proc. Lunar Sci. Conf., 8,* 2705-2724, 1977.

Wilhelms, D. E., Mercurian volcanism questioned, *Icarus, 28,* 551-558, 1976.

Wilhelms, D. E., The geologic history of the Moon, *U.S. Geol. Surv. Prof. Paper 1348,* 302 pp., 1987.

Wilson, L., and J. W. Head, The formation of eroded depressions around the sources of lunar sinuous rilles: Theory, *Lunar Planet. Sci. Conf., 11,* 1260-1262, 1980.

Wilson, L., and J. W. Head, Mars: Review and analysis of volcanic eruption theory and relationships to observed landforms, *Rev. Geophys., 32,* 221-263, 1994.

Wilson, L., and K. Keil, Volcanic eruptions and intrusions on the asteroid 4 Vesta, *J. Geophys. Res., 101,* 18,927-18,940, 1996.

Wise, D. U., M. P. Golombek, G. E. McGill, Tharsis province of Mars: Geologic sequence, geometry, and a deformation mechanism, *Icarus, 38,* 456-472, 1979.

Yingst, R. A., and J. W. Head, Lunar mare deposit volumes, composition, age, and location: Implications for source areas and modes of emplacement, *Lunar Planet. Sci. Conf., 25,* 1531-1532, 1994.

Yingst, R. A., and J. W. Head, Spatial and areal distribution of lunar mare deposits in Mare Orientale and South Pole/Aitken Basin: Implications for crustal thickness relationships, *Lunar Planet. Sci. Conf., 26,* 1539-1540, 1995.

M. F. Coffin, Institute for Geophysics, The University of Texas at Austin, Austin, TX 78759-8397 USA. (mikec@coffin.ig.utexas.edu)

J.W. Head, III, Department of Geological Sciences, Brown University, Box 1846, Providence, RI 02912 USA. (James_Head_III@Brown.edu)